POPULATION GENETICS
AND
MICROEVOLUTIONARY
THEORY

POPULATION GENETICS AND MICROEVOLUTIONARY THEORY

Alan R. Templeton

Department of Biology
Washington University
St. Louis, Missouri

A John Wiley & Sons., Inc., Publication

Copyright © 2006 by John Wiley & Sons, Inc. All rights reserved.

Published by John Wiley & Sons, Inc., Hoboken, New Jersey
Published simultaneously in Canada

No part of this publication may be reproduced, stored in a retrieval system or transmitted in any form or by any means, electronic, mechanical, photocopying, recording, scanning or otherwise, except as permitted under Sections 107 or 108 of the 1976 United States Copyright Act, without either the prior written permission of the Publisher, or authorization through payment of the appropriate per-copy fee to the Copyright Clearance Center, 222 Rosewood Drive, Danvers, MA 01923, (978) 750-8400, fax (978) 646-8600. Requests to the Publisher for permission should be addressed to the Permissions Department, John Wiley & Sons, Inc., 111 River Street, Hoboken, NJ 07030, (201) 748-6011, fax (201) 748-6008.

To order books or for customer service please, call 1(800)-CALL-WILEY (225-5945).

Limit of Liability/Disclaimer of Warranty: While the publisher and author have used their best efforts in preparing this book, they make no representations or warranties with respect to the accuracy or completeness of the contents of this book and specifically disclaim any implied warranties of merchantability or fitness for a particular purpose. No warranty may be created or extended by sales representatives or written sales materials. The advice and strategies contained herein may not be suitable for your situation. You should consult with a professional where appropriate. Neither the publisher nor author shall be liable for any loss of profit or any other commercial damages, including but not limited to special, incidental, consequential, or other damages.

For general information on our other products and services or for technical support, please contact our Customer Care Department within the United States at (800) 762-2974, outside the United States at (317) 572-3993 or fax (317) 572-4002.

Wiley also publishes its books in a variety of electronic formats. Some content that appears in print may not be available in electronic format. For more information about Wiley products, visit our web site at www.wiley.com.

Library of Congress Cataloging-in-Publication Data:

Templeton, Alan Robert.
    Population genetics and microevolutionary theory/ Alan R. Templeton
       p. cm.
    Includes bibliographical references and index.
    ISBN-13: 978-0-471-40951-9 (cloth)
    ISBN-10: 0-471-40951-0 (cloth)
    1. Population genetics.   2. Evolution (Biology)   I. Title.

    QH455.T46 2006
    576.5′8–dc22   2006042030

Printed in the United States of America

10 9 8 7 6 5

To Bonnie
and to the Memory of Hampton Carson

CONTENTS

PREFACE ix

1. SCOPE AND BASIC PREMISES OF POPULATION GENETICS 1

PART I. POPULATION STRUCTURE AND HISTORY 19

2. MODELING EVOLUTION AND THE HARDY–WEINBERG LAW 21

3. SYSTEMS OF MATING 48

4. GENETIC DRIFT 82

5. GENETIC DRIFT IN LARGE POPULATIONS AND COALESCENCE 118

6. GENE FLOW AND POPULATION SUBDIVISION 168

7. GENE FLOW AND POPULATION HISTORY 204

PART II. GENOTYPE AND PHENOTYPE 247

8. BASIC QUANTITATIVE GENETIC DEFINITIONS AND THEORY 249

9. QUANTITATIVE GENETICS: UNMEASURED GENOTYPES 274

10. QUANTITATIVE GENETICS: MEASURED GENOTYPES 297

PART III. NATURAL SELECTION AND ADAPTATION   341

11. NATURAL SELECTION   343

12. INTERACTIONS OF NATURAL SELECTION WITH OTHER EVOLUTIONARY FORCES   372

13. UNITS AND TARGETS OF SELECTION   407

14. SELECTION IN HETEROGENEOUS ENVIRONMENTS   453

15. SELECTION IN AGE-STRUCTURED POPULATIONS   497

APPENDIX 1.   GENETIC SURVEY TECHNIQUES   540

APPENDIX 2.   PROBABILITY AND STATISTICS   555

REFERENCES   582

PROBLEMS AND ANSWERS   612

INDEX   681

PREFACE

I have been teaching population genetics for 30 years, and during that time the importance and centrality of this field to modern biology have increased dramatically. Population genetics has always played a central role in evolutionary biology as it deals with the mechanisms by which evolution occurs within populations and species, the ultimate basis of all evolutionary change. However, as molecular genetics matured into genomics, population genetics was transformed from a discipline receiving new techniques from molecular genetics into a discipline providing the basic analytical methods for many aspects of genomics. Moreover, an increasing number of students are interested in the problems of species extinction and of environmental degradation and change. Population genetics offers many basic tools for conservation biology as well. As a result, the audience for population genetics has increased substantially, and I have witnessed a sixfold increase in the enrollment in my population genetics course over the past several years. This book is written with this expanded audience in mind. Many examples are given from conservation biology, human genetics, and genetic epidemiology, yet the focus of this book remains on the basic microevolutionary mechanisms and how they interact to create evolutionary change. This book is intended to provide a solid basis in population genetics both for those students primarily interested in evolutionary biology and genetics as well as for those students primarily interested in applying the tools of population genetics, particularly in the areas of conservation biology, human genetics, and genomics. Without a solid foundation in population genetics, the analytical tools emerging from population genetics will frequently be misapplied and incorrect interpretations can be made. This book is designed to provide that foundation both for future population and evolutionary geneticists and for those who will be applying population genetic concepts and techniques to other areas.

One theme throughout this book is that many important biological phenomena emerge from the interactions of two or more factors. As a consequence, evolution must be viewed with a multidimensional perspective, and it is insufficient to examine each evolutionary force one by one. Two highly influential mentors strengthened this theme in my work: Charles Sing and Hampton Carson. Charlie was my Ph.D. advisor and continues to be a mentor, collaborator, and friend. Charlie always stressed the importance of interactions in biology and genetics, and he was and is concerned with the "big picture" questions. I cannot thank Charlie enough for his continuing intellectual challenges and for his friendship.

Hamp Carson also stressed the importance of interacting forces in evolution and genetics. Hamp was both my undergraduate research mentor and my postdoctoral advisor, as well as a long-time collaborator and friend. Hamp died at the age of 91 as this book was nearing completion. He lived a full and highly productive life, and I dedicate this book in his memory to honor his life and accomplishments.

Many of my graduate students, both current and former, contributed significantly to this book. Indeed, the impetus for writing this book came largely from two former graduate students, Delbert Hutchison and Keri Shingleton. When Delbert and Keri were at Washington University as graduate students, they also served as teaching assistants in my population genetics course. My lectures did not follow any of the existing textbooks, so first Delbert, and then Keri, wrote out detailed lecture notes to help the students. These notes also formed the backbone of this book, and both Delbert and Keri strongly urged me to take their notes and transform them into a book. This is the book that resulted from that transformation.

Many of my graduate students read drafts of the chapters and offered many suggestions that were incorporated into the book. I thank the following graduate students for their valuable input: Corey Anderson, Jennifer Brisson, Nicholas Griffin, Jon Hess, Keoni Kauwe, Rosemarie Koch, Melissa Kramer, Taylor Maxwell, Jennifer Neuwald, James Robertson, and Jared Strasburg. In addition, many of my former graduate students and colleagues read drafts of this book and often used these drafts in teaching their own courses in population genetics. They also provided me with excellent feedback, both from themselves and from their students, so I wish to thank Reinaldo Alves de Brito, Keith Crandall, Delbert Hutchison, J. Spencer Johnston, and Eric Routman. I also want to thank three anonymous reviewers for their comments and suggestions on the first six chapters of this book. Finally, I used drafts of this book as my text in my population genetics class at Washington University. Many of the students in this class, both graduate and undergraduate, provided me with valuable feedback, and I thank them for their help.

1

SCOPE AND BASIC PREMISES OF POPULATION GENETICS

Population genetics is concerned with the origin, amount, and distribution of genetic variation present in populations of organisms and the fate of this variation through space and time. The kinds of populations that will be the primary focus of this book are populations of sexually reproducing diploid organisms, and the fate of genetic variation in such populations will be examined at or below the species level. Variation in genes through space and time constitute the fundamental basis of evolutionary change; indeed, in its most basic sense, **evolution** is the genetic transformation of reproducing populations over space and time. Population genetics is therefore at the very heart of evolutionary biology and can be thought of as the science of the mechanisms responsible for **microevolution,** evolution within species. Many of these mechanisms have a great impact on the origin of new species and on evolution above the species level (macroevolution), but these topics will not be dealt with in this book.

BASIC PREMISES OF POPULATION GENETICS

Microevolutionary mechanisms work upon genetic variability, so it is not surprising that the fundamental premises that underlie population genetic theory and practice all deal with various properties of deoxyribonucleic acid (DNA), the molecule that encodes genetic information in most organisms. [A few organisms use ribonucleic acid (RNA) as their genetic material, and the same properties apply to RNA in those cases.] Indeed, the theory of microevolutionary change stems from just three premises:

1. DNA can replicate.
2. DNA can mutate and recombine.
3. Phenotypes emerge from the interaction of DNA and environment.

The implications of each of these premises will now be examined.

Population Genetics and Microevolutionary Theory, By Alan R. Templeton
Copyright © 2006 John Wiley & Sons, Inc.

DNA Can Replicate

Because DNA can replicate, a particular kind of gene (specific set of nucleotides) can be passed on from one generation to the next and can also come to exist as multiple copies in different individuals. Genes therefore have an existence in time and space that transcends the individuals that temporarily bear them. The biological existence of genes over space and time is the *physical basis of evolution*.

The physical manifestation of a gene's continuity over time and through space is a reproducing population of individuals. Individuals have no continuity over space or time; individuals are unique events that live and then die and cannot evolve. But the genes that an individual bears are potentially immortal through DNA replication. For this potential to be realized, the individuals must reproduce. Therefore, to observe evolution it is essential to study a population of reproducing individuals. A reproducing population does have continuity over time as one generation of individuals is replaced by the next. A reproducing population generally consists of many individuals, and these individuals collectively have a distribution over space. Hence, a reproducing population has continuity over time and space and constitutes the physical reality of a gene's continuity over time and space. Evolution is therefore possible only at the level of a reproducing population and not at the level of the individuals contained within the population.

The focus of population genetics must be upon reproducing populations to study microevolution. However, the exact meaning of what is meant by a population is not fixed but rather can vary depending upon the questions being addressed. The population could be a local breeding group of individuals found in close geographic proximity or it could be a collection of local breeding groups distributed over a landscape such that most individuals only have contact with other members of their local group but that on occasion there is some reproductive interchange among local groups. Alternatively, a population could be a group of individuals continuously distributed over a broad geographical area such that individuals at the extremes of the range are unlikely to ever come into contact, or any other grouping of individuals up to and including the entire species. Within this hierarchy of populations found within species, much of population genetics focuses upon the **local population, or deme**, a collection of interbreeding individuals of the same species that live in sufficient proximity that they share a system of mating. Systems of mating will be discussed in more detail in subsequent chapters, but for now the **system of mating** refers to the rules by which individuals pair for sexual reproduction. The individuals within a deme share a common system of mating. Because a deme is a breeding population, individuals are continually turning over as births and deaths occur, but the local population is a dynamic entity that can persist through time far longer than the individuals that temporarily comprise it. The local population therefore has the attributes that allow the physical manifestation of the genetic continuity over space and time that follows from the premise that DNA can replicate.

Because our primary interest is on genetic continuity, we will make a useful abstraction from the deme. Associated with every local population of individuals is a corresponding local population of genes called the **gene pool**, the set of genes collectively shared by the individuals of the deme. An alternative and often more useful way of defining the gene pool is that the gene pool is the population of potential gametes produced by all the individuals of the deme. Gametes are the bridges between the generations, so defining a gene pool as a population of potential gametes emphasizes the genetic continuity over time that provides the physical basis for evolution. For empirical studies, the first definition is primarily used; for theory, the second definition is preferred.

BASIC PREMISES OF POPULATION GENETICS

The gene pool associated with a local population is described by measuring the numbers and frequencies of the various types of genes or gene combinations in the pool. At this lowest meaningful biological level of a deme, **evolution** is defined as a change through time of the frequencies of various types of genes or gene combinations in the gene pool. This definition is not intended to be an all-encompassing definition of evolution. Rather, it is a narrow and focused definition of evolution that is useful in much of population genetics precisely because of its narrowness. This will therefore be our primary definition of evolution in this book. Since only a local population at the minimum can have a gene pool, only populations can evolve under this definition of evolution, not individuals. Therefore, evolution is an emergent property of reproducing populations of individuals that is not manifested in the individuals themselves. However, there can be higher order assemblages of local populations that can evolve. In many cases, we will consider collections of several local populations that are interconnected by dispersal and reproduction, up to and including the entire species. However, an entire species in some cases could be just a single deme or it could be a collection of many demes with limited reproductive interchange. A species is therefore not a convenient unit of study in population genetics because species status itself does not define the reproductive status that is so critical in population genetic theory. We will always need to specify the type and level of reproducing population that is relevant for the questions being addressed.

DNA Can Mutate and Recombine

Evolution requires change, and change can only occur when alternatives exist. If DNA replication were always 100% accurate, there could be no evolution. A necessary prerequisite for evolution is genetic diversity. The ultimate source of this genetic diversity is mutation. There are many forms of mutation, such as single-nucleotide substitutions, insertions, deletions, transpositions, duplications, and so on. For now, our only concern is that these mutational processes create diversity in the population of genes present in a gene pool. Because of mutation, alternative copies of the same homologous region of DNA in a gene pool will show different states.

Mutation occurs at the molecular level. Although many environmental agents can influence the rate and type of mutation, one of the central tenets of Darwinian evolution is that mutations are random with respect to the needs of the organism in coping with its environment. There have been many experiments addressing this tenet, but one of the more elegant and convincing is replica plating, first used by Joshua and Esther Lederberg (1952) (Figure 1.1). Replica plating and other experiments provide empirical proof that mutation, occurring on DNA at the molecular level, is not being directed to produce a particular phenotypic consequence at the level of an individual interacting with its environment. Therefore, we will regard mutations as being random with respect to the organism's needs in coping with its environment.

Mutation creates allelic diversity. **Alleles** are simply alternative forms of a gene. In some cases genetic surveys focus on a region of DNA that may not be a gene in a classical sense; it may be a DNA region much larger or smaller than a gene or a noncoding region. We will use the term **haplotype** to refer to an alternative form (specific nucleotide sequence) among the homologous copies of a defined DNA region, whether a gene or not. The allelic or haplotypic diversity created by mutation can be greatly amplified by the genetic mechanisms of recombination and diploidy. In much of genetics, recombination refers to meiotic crossing

4 SCOPE AND BASIC PREMISES OF POPULATION GENETICS

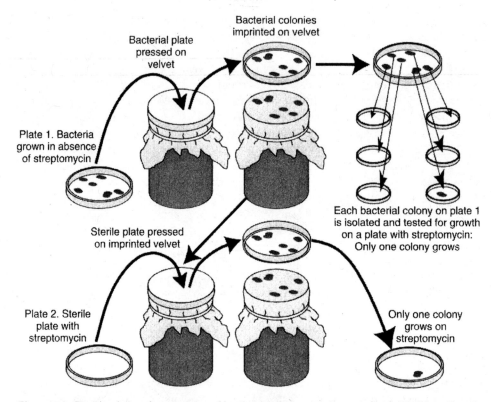

Figure 1.1. Replica plating. A suspension of bacterial cells is spread upon a Petri dish (plate 1) such that each individual bacterium should be well separated from all others. Each bacterium then grows into a colony of genetically identical individuals. Next, a circular block covered with velvet is pressed onto the surface of plate 1. Some bacteria from each colony stick to the velvet, so a duplicate of the original plate is made when the velvet is pressed onto the surface of a second Petri dish (plate 2), called the replica plate. The medium on the replica plate contains streptomycin, an antibiotic that kills most bacteria from the original strain. In the example illustrated, only one bacterial colony on the replica plate can grow on streptomycin, and its position on plate 2 identifies it as the descendant of a particular colony on plate 1. Each bacterial colony on plate 1 is then tested for growth on a plate with the antibiotic streptomycin. If mutations were random and streptomycin simply selected preexisting mutations rather than inducing them, then the colonies on plate 1 that occupied the positions associated with resistant colonies on plate 2 should also show resistance, even though these colonies had not yet been exposed to streptomycin. As shown, this was indeed the case.

over, but we use the term *recombination* in a broader sense as any genetic mechanism that can create new combinations of alleles or haplotypes. This definition of recombination encompasses the meiotic events of both independent assortment and crossing over and also includes gene conversion and any nonmeiotic events that create new gene combinations that can be passed on through a gamete to the next generation. Sexual reproduction and diploidy can also be thought of as mechanisms that create new combinations of genes.

As an illustration of the genetic diversity that can be generated by the joint effects of mutation and recombination, consider the MHC complex (major histocompatibility complex, also known in humans as HLA, human leukocyte antigen) of about 100 genes on the same chromosome. Table 1.1 shows the number of alleles found at 20 of these loci as of 1997 in human populations (Bodmer and Bodmer 1999). As can be seen, mutational changes at these

Table 1.1. Numbers of Alleles Known in 1997 at 20 Loci within Human *MHC* (*HLA*) Region

Locus	Number of Alleles
MHC-1	83
MHC-B	186
MHC-C	42
MHC-E	5
MHC-G	7
MHC-DRA	2
MHC-DRB1	184
MHC-DRB3	11
MHC-DRB4	9
MHC-DRB5	12
MHC-DQA1	18
MHC-DQB1	31
MHC-DOB	1
MHC-DMA	4
MHC-DMB	5
MHC-DNA	1
MHC-DPA1	10
MHC-DPB1	77
TAP1	5
TAP2	4
Total	698

loci have generated from 1 to 186 alleles per locus with a total of 698 alleles over all 20 loci. However, these loci can and do recombine. Hence, recombination has the potential of combining these 698 alleles into 1.71×10^{21} distinct gamete types (obtained by multiplying the allele numbers at each locus). Sexual reproduction has the potential of bringing together all pairs of these gamete types in a diploid individual, resulting in over 10^{42} genotypes and over 10^{33} distinct possible antigenic phenotypes (Bodmer and Bodmer 1999). And this is only from 20 loci in one small region of one chromosome of the human genome! Given that there are only about 6×10^9 humans in the world, everyone on the world (with the exception of identical twins) will have a unique *MHC* genotype when these 20 loci are considered simultaneously. But of course, humans differ at many more loci than just these 20. As of 2004, about 6 million polymorphic nucleotides were known in the human genome. Assuming that most of these are biallelic, each polymorphic nucleotide defines three genotypes, so collectively the number of possible genotypes defined by these known polymorphic sites is $3^{6,000,000} = 10^{2,862,728}$ genotypes. To put this number into perspective, the mass of our entire galaxy in grams is 1.9×10^{44} (Weinberg 1977), a number far smaller than the number of potential genotypes that are possible in humanity just with the known genetic variation. Hence mutation and recombination can generate truly astronomical levels of genetic variation.

The distinction between mutation and recombination is often blurred because recombination can occur within a gene and thereby create new alleles or haplotypes. For example, 71 individuals from three human populations were sequenced for a 9.7-kb region within the *lipoprotein lipase* locus (*LPL*) (Nickerson et al. 1998). This represents just about a third of this one locus. Eighty-eight variable sites were discovered, and 69 of these sites were used to define 88 distinct haplotypes or alleles. These 88 haplotypes arose from at least 69 mutational events (a minimum of one mutation for each of the 69 variable nucleotide sites) coupled with about 30 recombination per gene conversion events (Templeton et al. 2000a). Thus, intragenic recombination and mutation have together generated 88 haplotypes as inferred using only a subset of the known variable sites in just a third of a single gene in a sample of 142 chromosomes. These 88 haplotypes in turn define 3916 possible genotypes—a number considerably larger than the sample size of 71 people!

Studies such as those mentioned above make it clear that mutation and recombination can generate large amounts of genetic diversity at particular loci or chromosomal regions, but they do not address the question of how much genetic variation is present within species in general. How much genetic variation is present in natural populations was one of the defining questions of population genetics up until the mid-1960s. Before then, most of the techniques used to define genes required genetic variation to exist. For example, many of the early important discoveries in Mendelian genetics were made in the laboratory of Thomas Hunt Morgan during the first few decades of the twentieth century. This laboratory used morphological variation in the fruitfly *Drosophila melanogaster* as its source of material to study. Among the genes identified in this laboratory was the locus that codes for an enzyme in eye pigment biosynthesis known as *vermillion* in *Drosophila*. Morgan and his students could only identify *vermillion* as a genetic locus because they found a mutant that coded for a defective enzyme, thereby producing a fly with bright red eyes. If a gene existed with no allelic diversity at all, it could not even be identified as a locus with the techniques used in Morgan's laboratory. Hence, *all* observable loci had at least two alleles in these studies (the "wildtype" and "mutant" alleles in Morgan's terminology). As a result, even the simple question of how many loci have more than one allele could not be answered directly. This situation changed dramatically in the mid-1960s with the first applications of molecular genetic surveys (first on proteins, later on the DNA directly; see Appendix 1, which gives a brief survey of the molecular techniques used to measure genetic variation). These new molecular techniques allowed genes to be defined biochemically and irrespective of whether or not they had allelic variation. The initial studies (Harris 1966; Johnson et al. 1966; Lewontin and Hubby 1966), using techniques that could only detect mutations causing amino acid changes in protein-coding loci (and only a subset of all amino acid changes at that), revealed that about a third of all protein-coding loci were polymorphic (i.e., a locus with two or more alleles such that the most common allele has a frequency of less than 0.95 in the gene pool) in a variety of species. As our genetic survey techniques acquired greater resolution (Appendix 1), this figure has only gone up.

These genetic surveys have made it clear that many species, including our own, have literally astronomically large amounts of genetic variation. The chapters in Part I of this book will examine how premises 1 and 2 combine to explain great complexity at the population level in terms of the amount of genetic variation and its distribution in individuals, within demes among demes, and over space and time. Because it is now clear that many species have vast amounts of genetic variation, the field of population genetics has become less concerned with the amount of genetic variation and more concerned with its phenotypic and evolutionary significance. This shift in emphasis leads directly into our third and final premise.

Phenotypes Emerge from Interaction of DNA and Environment

A **phenotype** is a measurable trait of an individual (or as we will see later, it can be generalized to other units of biological organization). In Morgan's day, genes could only be identified through their phenotypic effects. The gene was often named for its phenotypic

BASIC PREMISES OF POPULATION GENETICS   7

effect in a highly inbred laboratory strain maintained under controlled environmental conditions. This method of identifying genes led to a simple-minded equation of genes with phenotypes that still plagues us today. Almost daily, one reads about "the gene for coronary artery disease," "the gene for thrill seeking," and so on. Equating genes with phenotypes is reinforced by metaphors appearing in many textbooks and science museums to the effect that DNA is the "blueprint" of life. However, DNA is not a blueprint for anything; that is not how genetic information is encoded or processed. For example, the human brain contains about 10^{11} neurons and 10^{15} neuronal connections (Coveney and Highfield 1995). Does the DNA provide a blueprint for these 10^{15} connections? The answer is an obvious "no." There are only about three billion base pairs in the human genome. Even if every base pair coded for a bit of information, there is insufficient information storage capacity in the human genome by several orders of magnitude to provide a blueprint for the neuronal connections of the human brain. DNA does not provide phenotypic blueprints; instead the information encoded in DNA controls dynamic processes (such as axonal growth patterns and signal responses) that always occur in an environmental context. There is no doubt that environmental influences have an impact on the number and pattern of neuronal connections that develop in mammalian brains in general. It is this interaction of genetic information with environmental variables through developmental processes that yield phenotypes (such as the precise pattern of neuronal connections of a person's brain). Genes should never be equated to phenotypes. Phenotypes emerge from genetically influenced dynamic processes whose outcome depends upon environmental context.

In this book, phenotypes are always regarded as arising from an interaction of genotype with environment. The marine worm *Bonellia* (Figure 1.2) provides an example of this interaction (Gilbert 2000). The free-swimming larval forms of these worms are sexually

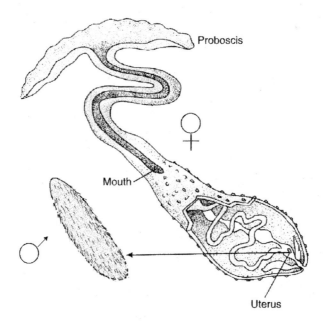

Figure 1.2. Sexes in *Bonellia*. The female has a walnut-sized body that is usually buried in the mud with a protruding proboscis. The male is a ciliated microorganism that lives inside the female. Adapted from Fig. 3.18 from *Genetics*, 3rd Edition, by Peter J. Russell. Copyright © 1992 by Peter J. Russell. Reprinted by permission of Pearson Education, Inc.

undifferentiated. If a larva settles alone on the normal mud substrate, it becomes a female with a long (about 15-cm) tube connecting a proboscis to a more rounded part of the body that contains the uterus. On the other hand, the larva is attracted to females, and if it can find a female, it differentiates into a male that exists as a ciliated microparasite inside the female. The body forms are so different they were initially thought to be totally different creatures. Hence, the same genotype, depending upon environmental context, can yield two drastically different body types. The interaction between genotype and environment in producing phenotype is critical for understanding the evolutionary significance of genetic variability, so the chapters in Part II will be devoted to an exploration of the premise that phenotypes emerge from a genotype-by-environment interaction.

As a prelude to why the interaction of genotype and environment is so critical to evolution, consider the following phenotypes that an organism can display:

- Being alive versus being dead: the phenotype of **viability** (the ability of the individual to survive in the environment)
- Given being alive, having mated versus not having mated; the phenotype of **mating success** (the ability of a living individual to find a mate in the environment)
- Given being alive and mated, the number of offspring produced; the phenotype of **fertility** or **fecundity** (the number of offspring the mated, living individual can produce in the environment)

The three phenotypes given above play an important role in microevolutionary theory because collectively these phenotypes determine the chances of an individual passing on its DNA in the context of the environment. The collective phenotype produced by combining these three components required for passing on DNA is called **reproductive fitness**. Fitness will be discussed in detail in Part III. Reproductive fitness turns premise 1 (DNA can replicate) into reality. DNA is not truly self-replicating. DNA can only replicate in the context of an individual surviving in an environment, mating in that environment, and producing offspring in that environment. Hence, the phenotype of reproductive fitness unites premise 3 (phenotypes are gene-by-environment interactions) with premise 1. This unification of premises implies that the probability of DNA replication is determined by how the genotype interacts with the environment. In a population of genetically diverse individuals (arising from premise 2 that DNA can mutate and recombine), it is possible that some genotypes will interact with the environment to produce more or fewer acts of DNA replication than other genotypes. Hence, the environment influences the relative chances for various genotypes of replicating their DNA. As we will see in Part III, this influence of the environment (premise 3) upon DNA replication (premise 1) in genetically variable populations (premise 2) is the basis for natural selection and one of the major emergent features of microevolution: **adaptation** to the environment, which refers to attributes and traits displayed by organisms that aid them in living and reproducing in specific environments. Adaptation is one of the more dramatic features of evolution, and indeed it was the main focus of the theories of Darwin and Wallace. Adaptation can only be understood in terms of a three-way interaction among all of the central premises of population genetics.

This book uses these three premises in a progressive fashion: Part I utilizes premises 1 and 2, which are molecular in focus, to explain the amount and pattern of genetic variation under the assumption that the variation has no phenotypic significance. Part II focuses upon premise 3 and considers what happens when genetic variation does influence phenotype. Finally, Part III considers the emergent evolutionary properties that arise from the interactions of all three premises and specifically focuses upon adaptation through natural selection. In this manner, we hope to achieve a thorough and integrated theory of microevolutionary processes.

METHODOLOGICAL APPROACHES IN POPULATION GENETICS

Evolutionary processes have produced an immense array of biological diversity on this planet, with species displaying complex and intricate adaptations to their environments. Understanding this diversity and complexity, its origins, and its implications ranging from the molecular through ecological levels is a daunting challenge. To meet this challenge, the study of population genetics requires an appreciation of a broad range of scientific approaches. We will make use of four approaches in this book:

- Reductionism
- Holism
- Comparative analysis
- Monitoring of natural populations

Reductionism

At one end of the above range of methodologies is the reductionist approach. Reductionism seeks to break down phenomena from a complex whole into simpler, more workable parts to find underlying rules, laws, and explanations. The reductionist approach is based upon the assumption that many complex features of a system can be explained in terms of a few components or rules contained within the system itself; that is, the explanation for the observed complexity lies within the *content* of the system. In this manner, simplicity (the parts contained within the system) generates complexity (the attributes of the whole system). Reductionism seeks necessary and sufficient explanations for the phenomenon under study. Such content-oriented explanations based upon reductionism are said to be proximate causes for the phenomenon of interest.

For example, why do people die? A reductionist approach would look at each instance of death and attempt to describe why that particular person died at that particular time in terms of the status of that individual's body at the time of death. One would get different answers for different individuals, and one would not need to look beyond the health status of a particular individual to obtain the proximate answer. Death is explained exclusively in terms of the content of the individual's body and nothing external to the body is considered. Taking such a reductionist approach, the three leading proximate causes of death in the year 2000 in the United States are (1) heart disease (29.6% of all deaths that year), (2) cancer (23%), and (3) cerebrovascular disease (7%) (Mokdad et al. 2004).

Much of population genetic theory and practice are reductionist in approach. One of the primary tools for implementing the reductionist approach is the controlled experiment in which all potential variables save one are ideally fixed, thereby allowing strong inference about how the single remaining variable factor causes effects of interest in the system under study. The controlled experiment fixes the context to allow inference about the content of a system varying with respect to a single factor. The experimental approach has been widely applied in population genetics and has proven to be a powerful tool in elucidating causal factors in microevolution. Note, however, that the strong inferences made possible by this approach are limited by the fixed contexts of the experiment, so generalizations outside of that context need to be made with great caution. Moreover, potential interactions with variables that have been experimentally fixed lie outside the domain of inference of the experimental approach. Indeed, in the ideal controlled experiment in which only a single factor is varying, all interaction effects are eliminated from the domain of inference, so some potentially important biological phenomena are not amenable to inference in a controlled experiment.

The reductionist approach is used in both experimental and theoretical population genetics. In modeling microevolution, the complexity of an evolving population is often simplified by reducing the number of variables and ignoring many biological details. With such simplification, laws and complex evolutionary patterns can be elucidated from a few components or factors that are contained within the population itself. Part I uses a reductionist approach to explain the fates and patterns of genetic diversity observed in populations in terms of simple attributes of the population itself. This reductionist approach yields an explanation of many important microevolutionary phenomena, often confirmed by appropriate controlled experiments. However, reductionism alone is insufficient to understand all of microevolution.

Holism

As a complement to the reductionist approach that simplicity generates complexity, the holistic approach is based upon the assumption that simple patterns exist in nature that emerge when underlying complex systems are placed into a particular *context* (simplicity emerges from complexity). The explanation of these emergent patterns often depends not upon knowing the detailed content of the component systems but rather upon the context in which these components are placed in a higher level interacting whole. These context-dependent explanations that do not depend upon detailed content reveal what is commonly called ultimate causation.

For example, why do people die? A holistic approach would look at multiple variables that define the health context of a population of individuals. One would not be trying to explain why a particular individual died at a particular time, but rather one would be trying to access the importance of context variables as predictors of death at the level of the whole population. Taking such a holistic approach, the three leading ultimate causes of death in the year 2000 in the United States are (1) tobacco consumption (18.1%), (2) being overweight (poor diet and physical inactivity, 16.6%), and (3) alcohol consumption (3.5%) (Mokdad et al. 2004). The ultimate explanation of causes of death does not depend upon the cause of death of any particular individual. The ultimate answers as to why people die also depend *not* upon the state of their bodies at the time of death (content) but rather upon the environmental context (tobacco, diet, physical activity, alcohol) into which their bodies have been placed.

It is critical to note that reductionist and holistic approaches are complementary, not antagonistic. Both approaches provide answers that are meaningful, albeit at different biological levels. A practicing physician would be most concerned with the particular health status of his or her patients. Such a physician would be prescribing specific treatments for specific individuals based on studies and knowledge of proximate causation. However, a public health official would focus more on ultimate causation and would try to augment the health of the U.S. population by encouraging less tobacco use and reducing the number of overweight people. Both answers to why people die are valid and both answers can be used in making health-related decisions. The reductionist and holistic answers each lead to insights and details that are not addressed by the other.

Moreover, reductionist and holistic approaches can converge. A controlled experiment can allow two or more factors to vary, not just one, and can be designed to look at the interactions of the variables. This allows one to study the effect of one variable in the context of another variable. Similarly, a holistic study can be designed that controls (fixes) some variables, resulting in ultimate answers that focus on the content defined by the remaining variables. For example, one can do "case–control" studies by assembling two groups of people, say one group of smokers and one group of nonsmokers, who are matched on several other variables (age, gender, etc.). Such studies have revealed that smoking increases the risk of individuals developing heart disease, cancer, and cerebrovascular disease, thereby forging a link between the studies on ultimate and proximate causations of death in the U.S. population. In this manner, the gap between reductionism and holism and between proximate and ultimate causation can often be narrowed.

All too often, reductionism and holism are presented as alternative, antagonistic approaches in biology. This legacy is particularly true for studies on the inheritance of traits, which has often been phrased as a debate between nature (content) and nurture (context). As discussed earlier in this chapter, this is a false dichotomy. Premise 3 tells us that traits emerge from the *interaction* of genotypes with environments, and modern studies on trait variation often seek to examine both content (the genes affecting trait variation) and context (the environments in which the genes are expressed). As soon as we deal with the phenotypic significance of genetic variation in Part II, an exclusively molecular, reductionist focus is no longer appropriate. Rather we must take an organismal, holistic focus in the context of an environment.

Of the traits that emerge from the interaction of genotypes with environment are those traits related to the ability of an individual to reproduce and pass on genes to the next generation. As already discussed in this chapter and in detail in Part III, the evolutionary mechanism of natural selection emerges from the interaction of genotypes with environments. Many explanations of ultimate causation in evolutionary biology depend upon natural selection. Again and again, the traits expressed by particular individuals or in particular populations or species are explained in the ultimate sense in terms of arguments of how particular environmental contexts result in natural selection favoring the trait. Population genetics deals in part with the mechanism of natural selection (Part III), and hence population genetics is an essential component of any explanation of ultimate causation based upon evolutionary change induced by natural selection. However, the population genetic approach to mechanisms such as natural selection explicitly uses both reductionism and holism simultaneously. For example, in population genetics natural selection is discussed in terms of the specific genes *contained* within the organisms being selected and the mapping of these genes to phenotype in the context of an environment, with the evolutionary response modulated by the other evolutionary forces contained within the population as discussed in Part I. Such an integrated reductionist/holistic approach will be the emphasis in Parts II and III.

Comparative Analysis

An evolutionary process occurs over time; therefore evolving populations (and the genes contained within those populations) have a history. The comparative approach to biological science makes active use of this history. This is a scientific method used extensively in biology, mostly at the species level and above. Traditionally, an evolutionary tree is constructed for a group of species. Then other data about these organisms (anatomy, developmental pathways, behavior, etc.) are overlaid upon the evolutionary tree. In this manner, it is possible to infer how many evolutionary transitions occurred in characters of interest, the locations of transitions within the evolutionary tree, and patterns of evolutionary associations among characters. Contrasts between those organisms on either side of a transitional branch are those that are most informative about the character of interest because the sharing of evolutionary history for all other traits is maximized by this contrast. A comparative contrast bears some similarity to a controlled experiment in reductionist empirical science because the contrast is chosen to minimize confounding factors.

For example, Darwin's finches comprise a group of 14 species of songbirds living on the Galápagos Islands and Cocos Island off the coast of Equador that were collected by Charles Darwin and other members of the *Beagle* expedition in 1835. These 14 species have drawn the attention of many evolutionary biologists because of the remarkable diversity in the shape and size of their beaks, which range from sharp and pointed to broad and deep (Figure 1.3). Why do these 14 species show such remarkable diversity in beak shape and size? Both the proximate and ultimate answers to this question have been studied using the comparative method. Petren et al.(1999) estimated an evolutionary tree of these finches from molecular genetic differences, with the resulting tree shown in Figure 1.3. Abzhanov et al. (2004) compared beak development in the six species of the genus *Geospiza* from this evolutionary tree and also compared the expression patterns of a variety of growth factors that are known to influence avian craniofacial development. By overlaying these data upon the molecular genetic tree, they produced evolutionary contrasts that separated out the effects of beak size and beak shape. Most of the growth factors they examined showed no significant pattern of change on this evolutionary tree. The expression patterns of bone morphogenetic proteins 2 and 7, coded for by the *Bmp2* and *Bmp7* genes, respectively, correlated with beak size but not with beak shape. The expression patterns of bone morphogenetic factor 4, coded for by the *Bmp4* gene, strongly correlated with beak shape changes on this evolutionary tree. Because the comparative study implicated *Bmp4* expression as being an important proximate cause of beak shape diversity, Abzhanov et al. (2004) next performed controlled experiments to test this hypothesis within a reductionist framework. They attached the chicken *Bmp4* gene to a viral vector and infected developing cells with this virus to alter the expression of the *Bmp4* gene. In this manner, they were able to alter the beak shape of chick embryos in a manner that mimicked the types of changes observed in the evolution of Darwin's finches.

This work on *Bmp4* expression does not, however, provide the ultimate answer as to why Darwin's finches show much diversity in beak size and shape. The comparative approach can also be used to address the ultimate question of what environmental factors, if any, caused this beak diversity and underlying patterns of *Bmp4* expression to have evolved through natural selection. Perhaps this beak diversity evolved on the South American mainland, and the Galápagos Islands were simply colonized by finches with preexisting beak diversity. In this case, the ultimate answer would lie in evolution in the mainland and have little to do with the context of being on the Galápagos Islands. Alternatively, if all 14 species evolved on the Galápagos Islands, then the ultimate answer would lie specifically in the context of the Galápagos Islands. The evolutionary tree in Figure 1.3 shows that all 14 species evolved on the Galápagos Islands, so the ultimate answer should lie in the environments found on these islands. This shows that just having an evolutionary tree allows some hypothesis about ultimate causation to be tested directly. The comparative analysis clearly indicates that the

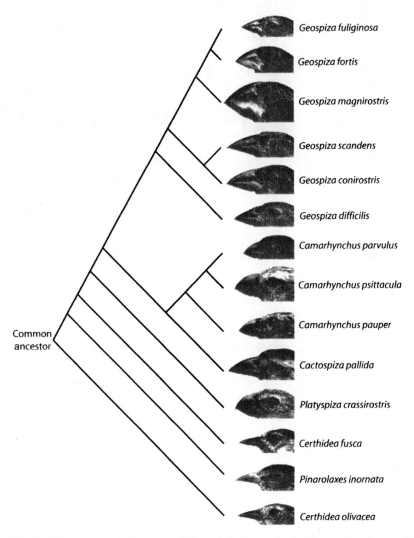

Figure 1.3. Evolutionary tree of 14 species of Darwin's finches estimated from molecular genetic data. Modified from Fig. 3 in Petren et al. (1999). Copyright ©1999 by the Royal Society of London.

ultimate explanation lies in the environments found on the Galápagos Islands and not on the mainland.

Because beaks are used to procure and process food, diet is a logical environmental factor for studies on how natural selection may have shaped beak diversity in these finches. Fieldwork has revealed much about the ecology of Darwin's finches (Grant 1986), including their diets. Different species eat items of different sizes, an example of which is shown in Figure 1.4. This dietary data can also be overlaid upon the evolutionary tree of the finches, and it reveals a strong correlation in shifts of diet with transitions in beak size and shape. Note that this comparative analysis reveals a strong association between content (the beak size and shape of individual species) and context (the dietary environment). Such a

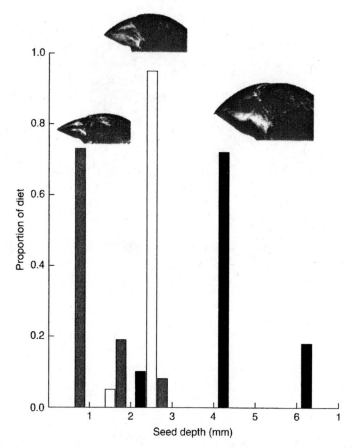

Figure 1.4. Proportions of various seed sizes in diet of three of Darwin's finches: *Geospiza magnirostris* (solid bars), *Geospiza fortis* (open bars), and *Geospiza fuliginosa* (gray bars). Redrawn with permission from Fig. 35 in P. R. Grant, *Ecology and Evolution of Darwin's Finches* (1986). Copyright ©1986 by Princeton University Press.

content–context association in evolutionary history suggests the hypothesis that the beak diversity is being shaped by natural selection as adaptations for different diets.

These studies on Darwin's finches reveal that the comparative approach can be used to test and formulate hypotheses of both proximate and ultimate causation. One of the more exciting developments in population genetics during the last part of the twentieth century was the development of molecular techniques that have allowed the application of comparative approaches *within* species. As illustrated above, it is now possible to trace the evolutionary history of species through molecular genetic studies. However, this evolutionary history can often be inferred for the genetic variation found within a species as well. In this manner, population genetic studies on genetic variation within a species can now include the evolutionary history of that genetic variation. This opens the door to comparative approaches within species. Such intraspecific comparative approaches are used throughout this book, and they represent a particularly powerful way of uniting reductionism and holism within population genetics.

Monitoring Natural Populations

Many hypotheses in population genetics can be tested by monitoring natural populations. One of the simplest types of monitoring is a one-time sample of individuals of unknown relationship coupled with some sort of genetic survey (using one or more of the techniques described in Appendix 1). Such simple genetic surveys allow one to estimate and test most of the evolutionary forces described in Part I. Just as genetic surveys of present-day species can allow an evolutionary tree of those species to be estimated (e.g., Figure 1.3), so can a genetic survey of present-day genes and/or populations allow an evolutionary history of those genes and/or populations to be estimated. Moreover, the genetic survey data can be overlaid with phenotypic data to test hypothesis about how genetic variation influences phenotypic variation, as will be shown in Part II. Finally, Part III shows that many tests for the presence or past operation of natural selection are possible from such genetic survey data.

The monitoring of natural populations can be extended beyond a simple one-time survey of genetic variation of individuals of unknown relationship. For example, one can sample families (parents and offspring) instead of individuals or follow a population longitudinally through time to obtain multigeneration data. Such designs allow more hypotheses to be tested. For example, Boag (1983) sampled parents and offspring of the Darwin finch *Geospiza fortis* and plotted the beak depth of the offspring against the average beak depth of their two parents (Figure 1.5). As will be shown in Part II, such data can be used to

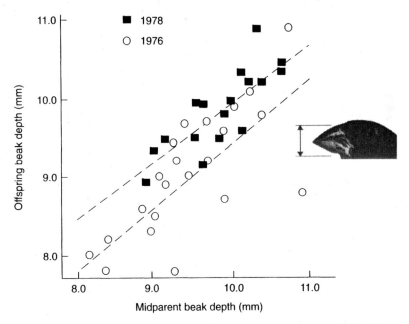

Figure 1.5. Relationship between beak depth of offspring and average beak depth of their parents (midparent beak depth) in medium ground finch, *G. fortis*, as measured in two years, 1976 and 1978. The lines show a fitted least-squares regression to these data (Appendix 2). As will be explained in Chapter 9, the nearly identical, positive slopes of these lines indicate that genetic variation in these populations contribute in a major way to variation in beak depth. From Fig. 1 in Boag (1983). Copyright ©1983 by The Society for the Study of Evolution.

16 SCOPE AND BASIC PREMISES OF POPULATION GENETICS

estimate the contribution of genetic variation to variation in the trait of beak depth even in the absence of a molecular genetic survey. In this case, the plots shown in Figure 1.5 reveal that the intraspecific variation observed in beak depth in *G. fortis* is strongly influenced by genetic variation within this population.

Population genetics is concerned with the fate of genes over space and time within a species, and this fate can be observed or estimated by monitoring populations over space and time. Such monitoring over space and time also allows population geneticists to make use of natural experiments. For example, natural selection arises out of how individuals interact with their environment, but environments themselves often change over space and time. Although not a controlled experiment in the strict reductionist sense, spatial and temporal environmental contrasts can sometimes provide a similar inference structure. To see how, consider again Darwin's finches. The comparative method implied that the variation in beak size and shape reflected adaptations to dietary differences. However, this answer of ultimate causation raises yet other questions about ultimate causation: Why did some or all of the current species evolve a different diet from that of the common ancestral finch and why do the current species display such a variety of diets? These questions of ultimate causation can be addressed through the use of natural experiments involving environmental contrasts in time and space. For example, in 1977 the Galápagos Islands suffered a severe drought. By monitoring both the finch populations and the environment in which they live, it was discovered that this drought had a major impact on both the abundance of the seeds eaten by these finches and the characteristics of the seeds. For example, there was a dramatic shift from small and soft seeds to large and hard seeds during the drought for the seeds eaten by the medium ground finch, *G. fortis* (Figure 1.6). The inference from the comparative method that beak size and shape are adaptive to diet leads to the prediction that this drought-induced shift in diet would result in natural selection on the beaks in *G. fortis*. This prediction is testable by monitoring the population before and after the drought. There

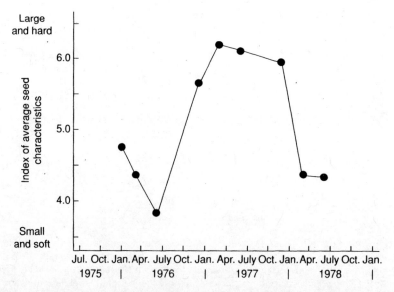

Figure 1.6. Characteristics of average seed available as food to medium ground finches (*G. fortis*) before, during, and after 1977 drought. Reprinted from Fig. 1 in P. T. Boag and P. R. Grant, *Science* 214: 82–85(1981). Copyright ©1981 by the AAAS.

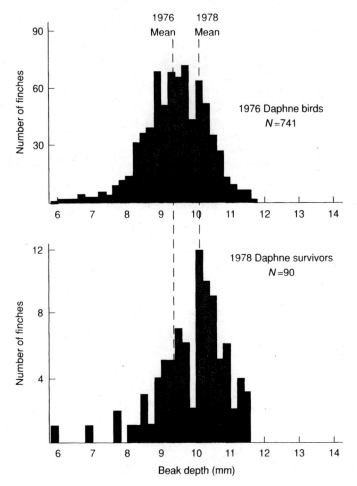

Figure 1.7. Frequency distributions of beak depth in *G. fortis* on island of Daphne Major before (1976) and after (1978) a drought. Dashed lines indicate the mean beak depths in 1976 and 1978. Redrawn with permission from Fig. 59 in P. R. Grant, *Ecology and Evolution of Darwin's Finches* (1986). Copyright © 1986 by Princeton University Press.

was a significant shift upward in beak depth in the survivors of the drought relative to the predrought population (Figure 1.7), a shift consistent with the hypothesis that increased beak depth is an adaptation to the larger and harder seeds that were available during the drought. Given that variation in beak depth is strongly influenced by genetic variation in this population (Figure 1.5), another prediction is that natural selection operated on this population to cause evolution in this population in response to the drought. This prediction can also be tested by looking at the beak depths of the finches hatched in the years before and after the drought, and indeed the predicted genetic shift is observed (Figure 1.8).

Subsequent environmental changes confirmed that changes in seed availability induce selection on beak shape and size (Grant and Grant 1993, 2002). Even though the subsequent environmental shifts were different from those induced by a drought, this environmental heterogeneity over time did replicate the testable prediction that beak shape and size are subject to natural selection due to interactions with the available seed environment. These

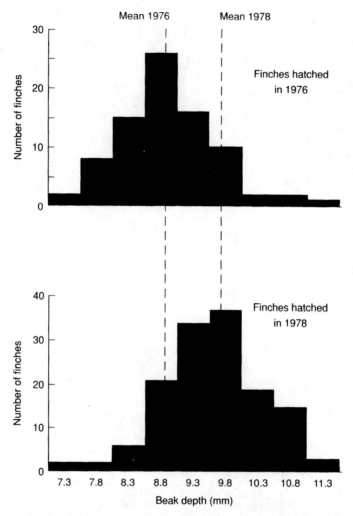

Figure 1.8. Beak depth in *G. fortis* hatched year before drought (1976) versus year after drought (1978). Dashed lines indicate the mean beak depth for the finches born before and after the drought.

natural experiments from monitoring populations reinforce the inference made from the comparative method that beak size and shape are adaptations to diet. Moreover, these temporal natural experiments suggest that beak size and shape would not remain static once an ancestral finch colonized these islands but rather would evolve because the seed environment is subject to change over time. Moreover, the seed environment varies from island to island, so this selective hypothesis could also explain some of the diversity of beak size and shape between finch species that primarily live on different islands.

These studies on Darwin's finches illustrate that the monitoring of natural populations can be a powerful method of inference in population genetics. Note that studies on Darwin's finches have utilized reductionist controlled experiments, reductionist comparative studies, holistic comparative studies, and monitoring of natural populations. The best studies in population genetics tend to integrate multiple methods of inference that are complementary and reinforcing to one another.

POPULATION STRUCTURE AND HISTORY

2

MODELING EVOLUTION AND THE HARDY–WEINBERG LAW

Throughout this book we will construct models of reproducing populations to investigate how various factors can cause evolutionary changes. <u>In this chapter, we will construct some simple models of an isolated local population</u>. These models use a reductionistic approach to eliminate many possible features in order to focus our inference upon one or a few potential microevolutionary factors. The models will also provide insights that have been historically important to the acceptance of the neo-Darwinian theory of evolution at the beginning of the twentieth century and are of increasing importance to the application of genetics to human health and other contemporary problems at the beginning of the twenty-first century.

HOW TO MODEL MICROEVOLUTION

Given our definition that evolution is a change over time in the frequency of alleles or allele combinations in the gene pool, any model of evolution must include at the minimum the passing of genetic material from one generation to the next. Hence, our fundamental time unit will be the transition between two consecutive generations at comparable stages. We can then examine the frequencies of alleles or allelic combinations in the parental versus offspring generation to infer whether or not evolution has occurred. All such transgenerational models of microevolution have to make assumptions about three major mechanisms:

- Mechanisms of producing gametes
- Mechanisms of uniting gametes
- Mechanisms of developing phenotypes.

In order to specify how gametes are produced, we have to specify the genetic architecture. **Genetic architecture** refers to the number of loci and their genomic positions, the number

---

Population Genetics and Microevolutionary Theory, By Alan R. Templeton
Copyright © 2006 John Wiley & Sons, Inc.

of alleles per locus, the mutation rates, and the mode and rules of inheritance of the genetic elements. For example, the first model we will develop assumes a genetic architecture of a single autosomal locus with two alleles with no mutation. The genetic architecture provides the information needed to specify how gametes are produced. For a single-locus, two-allele autosomal model with no mutation, we need only to use Mendel's first law of inheritance (the law of equal segregation of the two alleles in an individual heterozygous at an autosomal locus) to specify how genotypes produce gametes. Other single-locus genetic architectures can display different modes of inheritance, including X-linked loci (with a haplo–diploid, sex-linked mode of inheritance), Y chromosomal loci (with a haploid, unisexual paternal mode of inheritance in humans), or mitochondrial DNA (with a haploid, maternal mode of inheritance in humans). We can also examine genetic architectures that depend upon more than one locus, in which case mixed modes of inheritance are possible and in which Mendel's second law (independent assortment) and/or recombination frequencies of linked loci may enter into the rules by which gametes are produced. We can even have deviations from the standard rules of inheritance. For example, we may specify that a locus is subject to deviations from Mendel's first law of 50–50 segregation in the production of gametes from heterozygotes. In a multilocus model we may specify that unequal crossing over can occur, thereby producing variation in the number of genes transmitted to the gametes. The assumptions about genetic architecture that we make obviously limit the types of evolutionary processes that we can model. Hence, the specification of genetic architecture is a critical first step in any model of microevolution.

Because our focus is upon sexually reproducing diploid organisms, the transition from one generation to the next involves not only the production of gametes but also the pairing of gametes to form new diploid zygotes. Hence, we need to specify the mechanisms or rules by which gametes are paired together in the reproducing population. These mechanisms of uniting gametes are called **population structure**. **Population structure** includes the following:

- System of mating of the population
- Size of the population
- Presence, amount, and pattern of genetic exchange with other populations
- Age structure of the individuals within the population

All of these factors can have an impact on which gametes are likely to be paired and transmitted to the next generation through newly formed zygotes. As with genetic architecture, we can make assumptions about population structure that vary from the simple to the complex, depending upon the types of phenomena we wish to examine for evolutionary impact. The system of mating can be simply a random pairing of individuals or can be influenced by degrees of biological relatedness or other factors. We can choose to ignore the impact of population size by assuming size to be infinite or we can examine small populations in which the population size has a major impact on the probability of two gametes being united in a zygote. We can model a single deme in which all uniting gametes come from that deme or we can allow gametes from outside the deme to enter at some specified rate or probability, which in turn could be a function of geographical distance, ecological barriers, and so on. We can assume discrete generations in which all individuals are born at the same time and then reproduce at the same time followed by complete reproductive senescence or death. Alternatively, we can assume that individuals can reproduce at many times throughout their life and can mate with individuals of different ages and offspring can coexist with their parents. Until we specify these parameters of population structure, we cannot model microevolution because the uniting of gametes is a necessary step in the transmission of genes from one generation to the next in sexually reproducing organisms.

In most species, the zygote that results from uniting gametes is not capable of immediate reproduction but rather must grow, develop, survive, and mature reproductively. All of this takes place in an environment or suite of environments. From premise 3 in Chapter 1 (phenotypes are gene-by-environment interactions), we know that actual DNA replication depends upon the phenotypes of the individuals bearing the DNA. Hence, we also need to specify **phenotypic development**, that is, the mechanisms that describe how zygotes acquire phenotypes in the context of the environment. Assumptions can range from the simple (the genetic architecture has no impact on phenotype under any of the environments encountered by individuals in the population) to the complex (phenotypes are dynamic entities constantly changing as the external environment changes and/or as the individual ages with changing patterns of epistasis and pleiotropy throughout).

All models of microevolution must make assumptions about the mechanisms of producing gametes, uniting gametes, and developing phenotypes. Without such assumptions, it is impossible to specify the genetic transition from one generation to the next. Quite often, models are presented that do not explicitly state the assumptions being made about all three mechanisms. This does not mean that assumptions are not being made; rather, they are being made in an implicit fashion. Throughout this book an effort is made to state explicitly the assumptions being made about all three of these critical components of transferring DNA from one generation to the next in a reproducing population. We will do this now for our first and simplest model of evolution, commonly called the Hardy–Weinberg model.

HARDY–WEINBERG MODEL

One of the simplest models of population genetics is the Hardy–Weinberg model, named after two individuals who independently developed this model in 1908 (Hardy 1908; Weinberg 1908). Although this model makes several simplifying assumptions that are unrealistic, it has still proven to be useful in describing many population genetic attributes and will serve as a useful base model in the development of more realistic models of microevolution. Hardy was an English mathematician, and his development of the model is mathematically simpler but yields less biological insight than the more detailed model of Weinberg, a German physician. Both derivations will be presented here because each has advantages over the other for particular problems that will be addressed later in this book.

Both derivations start with a common set of assumptions, as summarized in Table 2.1. We now discuss each of the assumptions given in that table. Concerning the *mechanisms of producing gametes*, both men assumed a single autosomal locus with two alleles and with no mutation. Meiosis was assumed to be completely normal and regular, so that Mendel's first law of equal segregation could predict the gametes produced by any genotype. There are also no maternal or paternal effects of any sort, so it makes no difference which parent contributes a gamete bearing a specific allele.

Concerning the *mechanisms of uniting gametes*, both men assumed a single population that has no genetic contact with any other populations; that is, an isolated population. Within this closed population, Hardy assumed the individuals are monoecious (each individual is both a male and a female) and self-compatible; Weinberg allowed the sexes to be separate but assumed that the sex of the individual has no impact on any aspect of inheritance or genetic architecture. The system of mating in both derivations is known as **random mating**

Table 2.1. Assumptions of Hardy–Weinberg Model

Mechanisms of producing gametes (genetic architecture)	One autosomal locus, two alleles, no mutation, Mendel's first law
Mechanisms of uniting gametes (population structure)	
System of mating	Random
Size of population	Infinite
Genetic exchange	None (one isolated population)
Age structure	None (discrete generations)
Mechanisms of developing phenotypes	All genotypes have identical phenotypes with respect to their ability for replicating their DNA

and means that the probability of two genotypes being mates is simply the product of the frequencies of the two genotypes in the population. Note that random mating is defined solely in terms of the genotypes at the locus of interest; there is no implication in this assumption that mating is random for any other locus or set of loci or for any phenotypes not associated with the locus of interest. For example, humans do not mate at random for a number of phenotypes (gender, skin color, height, birthplace, etc.), but as long as the genetic variation at the locus of interest has no impact on any of these phenotypes, the assumption of random mating can still hold. Hence, random mating is an assumption that is specific to the genetic architecture of interest and that does not necessarily generalize to other genetic systems found in the same organisms.

Concerning the other aspects of population structure, both derivations make the assumption that the population is of infinite size, thereby eliminating any possible effects of finite population size upon the probability of uniting gametes. Both men ignored the effects of age structure by assuming discrete, nonoverlapping generations. Finally, concerning the mechanisms of developing phenotypes, nothing was explicitly assumed, but implicitly both derivations require that under the range of environments in which the individuals of the population are living and reproducing there is no phenotypic variation for viability, mating success, and fertility. In terms of their ability to replicate DNA, all genotypes have identical phenotypes. This means that all genotypes have the same reproductive fitness, so there is no natural selection in this model.

To examine the population genetic implications of these assumptions upon a reproducing population, we need to go through a complete generation transition. In both derivations, we will start with a population of reproductively mature adults. The essence of this model (and many others in population genetics) is to follow the fate of genes from this population of adults through producing gametes, mating to unite gametes (zygote production), and then zygotic development to the adults of the next generation. We will then examine the gene pools associated with these two generations of adults to see if any evolution has occurred.

Because we are dealing with a single autosomal locus with two alleles (say A and a) and no additional mutation, adult individuals are of three possible genotypes: AA, Aa, and aa. We will characterize the adult population by their genotypes and the frequencies of these genotypes in the total population (see Figure 2.1). Let these three genotype frequencies be G_{AA}, G_{Aa}, and G_{aa}, where the subscript indicates the genotype associated with each

HARDY–WEINBERG MODEL   25

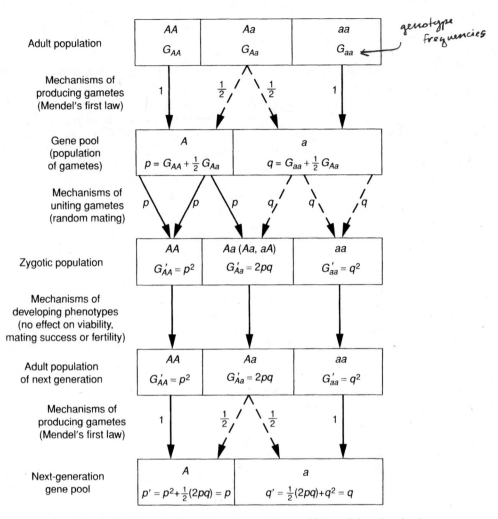

Figure 2.1. Derivation of Hardy–Weinberg law for single autosomal locus with two alleles, *A* and *a*. In going from adults to gametes, solid lines represent Mendelian transition probabilities for homozygotes, and dashed lines represent Mendelian transition probabilities for heterozygotes. In going from gametes to zygotes, solid lines represent gametes bearing the *A* allele, and dashed lines represent gametes bearing the *a* allele.

frequency. Because these three genotypes represent a mutually exclusive and exhaustive set of possible genotypes, these three genotype frequencies define a probability distribution over the genotypes found in the adult population (see Appendix 2 for a discussion of probability distributions). This means that $G_{AA} + G_{Aa} + G_{aa} = 1$. This probability distribution of *genotype frequencies* represents our fundamental description of the *adult population*.

At this point, the derivations of Hardy and of Weinberg diverge. We will first follow Hardy's and then return to Weinberg's. The population of adult individuals can produce gametes. As discussed in Chapter 1, the population of potential gametes produced by these individuals defines the gene pool (Figure 2.1). Because of our assumptions about genetic architecture and no mutation, all we need is Mendel's first law to predict the frequencies of the various haploid genotypes (gametes) found in the gene pool from the frequencies of the diploid adult genotypes. Two and only two haploid gametic types are possible: A and a. The frequencies of these gametes (which for a one-locus model are called **allele frequencies**) also define a probability distribution over the gamete types found in the gene pool. This probability distribution of *gamete frequencies* represents our fundamental description of the *gene pool*. We will let p be the frequency of gametes bearing the A allele in the gene pool and q the frequency of gametes bearing the a allele in the gene pool. Because p and q define a probability distribution over the gene pool, $p + q = 1$, or $q = 1 - p$. Hence, we need only one number, say p, to completely characterize the gene pool in this model. A critical question is: Can we predict the allele (gamete) frequencies from the genotype frequencies? Under our assumptions of the mechanisms for producing gametes, the answer is "yes" and all we need to use is Mendel's first law of equal segregation. Under Mendel's law, the probability of an AA genotype producing an A gamete is 1 and the probability of an AA genotype producing an a gamete is 0. Similarly, the probability of an aa genotype producing an A gamete is 0 and the probability of an aa genotype producing an a gamete is 1 under standard Mendelian inheritance. Finally, Mendel's first law predicts that the probability of an Aa genotype producing an A gamete is $\frac{1}{2}$ and the probability of an Aa genotype producing an a gamete is $\frac{1}{2}$. These Mendelian probabilities are transition probabilities that describe how one goes from adult genotypes to gamete types. Hence, the transition from the adult population to the gene pool is determined completely by these transmission probabilities (our mathematical descriptor of the mechanisms of producing gametes). As can be seen from Figure 2.1, these transition probabilities from diploidy to haploidy allow us to predict the gene pool state completely from the adult population genotype state. In particular, all we have to do is multiply each transmission probability by the frequency of the genotype with which it is associated and then sum over all genotypes for each gamete type. Thus, $1 \times G_{AA}$ is the frequency of A gametes coming from AA individuals, $\frac{1}{2} \times G_{Aa}$ is the frequency of A gametes coming from Aa individuals, and $0 \times G_{aa} = 0$ is the frequency of A gametes coming from aa individuals. Hence, the total frequency of the A allele in the gene pool is $1 \times G_{AA} + \frac{1}{2} \times G_{Aa} + 0 \times G_{aa} = G_{AA} + \frac{1}{2} G_{Aa} = p$. Similarly, the frequency of the a allele in the gene pool is $0 \times G_{AA} + \frac{1}{2} \times G_{Aa} + 1 \times G_{aa} = G_{aa} + \frac{1}{2} G_{Aa} = q = 1 - (G_{AA} + \frac{1}{2} G_{Aa}) = 1 - p$ (see Figure 2.1). Note that the Mendelian transmission probabilities (the 0's, 1's, and $\frac{1}{2}$'s used above) and the genotype frequencies (the G's) completely determine the allele frequencies in the gene pool. In general, gamete frequencies can always be calculated from genotype frequencies given a knowledge of the mechanisms of producing gametes. Letting g_j be the frequency of gamete type j in the gene pool (either an allele for a single-locus genetic architecture or a multiallelic gamete for a multilocus genetic architecture), the general formula for calculating a gamete frequency is

$$g_j = \sum_{\text{genotypes}} \text{probability (genotype } k \text{ producing gamete } j) \times \text{(frequency of genotype } k)$$

(2.1)

where "genotype k" is simply a specific genotype possible under the assumed genetic architecture. The equations previously used to calculate p and q are special cases of equation 2.1 for a single autosomal locus with two alleles. This equation makes it clear that two types of information are needed to calculate gamete frequencies:

- Information about the mechanisms of producing gametes which determine the probability of a specific genotype producing a specific gamete type
- Genotype frequencies of the population of interest

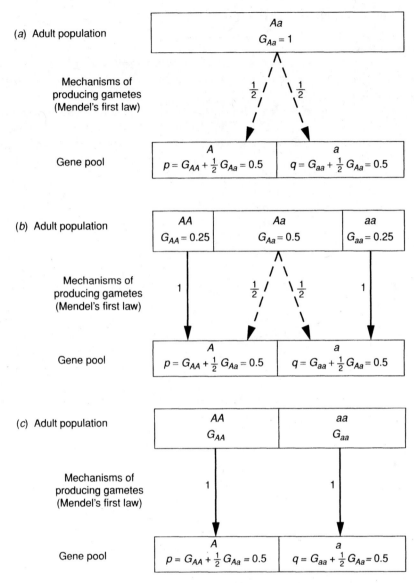

Figure 2.2. Different adult populations sharing a common gene pool.

It is always possible to calculate the gamete frequencies from the genotype frequencies given a knowledge of the mechanisms of producing gametes. Is it also possible to calculate the genotype frequencies from the gamete frequencies given a knowledge of the mechanisms of producing gametes? The answer is "no." To see this, consider a population of adults consisting only of Aa individuals (Figure 2.2a). In this population, $G_{AA} = 0$, $G_{Aa} = 1$, and $G_{aa} = 0$. Hence, $p = G_{AA} + \frac{1}{2} \times G_{Aa} = 0 + \frac{1}{2} \times 1 = 0.5$. Now, consider a population with $G_{AA} = 0.25$, $G_{Aa} = 0.5$, and $G_{aa} = 0.25$ (Figure 2.2b). For this population, $p = G_{AA} + \frac{1}{2} \times G_{Aa} = 0.25 + \frac{1}{2} \times 0.5 = 0.5$. Now consider the population shown in Figure 2.2c, in which $G_{AA} = 0.5$, $G_{Aa} = 0$, and $G_{aa} = 0.5$. In this population, $p = G_{AA} + \frac{1}{2} \times G_{Aa} = 0.5 + \frac{1}{2} \times 0 = 0.5$. Hence, three very different populations of adults all give rise to identical gene pools! This shows that there is no one-to-one mapping between genotype frequencies and gamete frequencies. Although gamete frequencies can always be calculated from genotype frequencies given a knowledge of the rules of inheritance, genotype frequencies are *not* uniquely determined by gamete frequencies and the rules of inheritance. Obviously, we need additional information to predict genotype frequencies from gamete frequencies. This is where population structure comes in.

Hardy and Weinberg made assumptions about population structure that remove as potential evolutionary factors mutation, genetic contact with other populations, population size, and age structure. All that is left in their simplified model is system of mating. Under Hardy's formulation, random mating means that two gametes are randomly and independently drawn from the gene pool and united to form a zygote. By a random draw, Hardy meant that the probability of a gamete being drawn is the same as its frequency in the gene pool. Hence, if the proportion of the gametes bearing the A allele is p, then the probability of choosing a gamete with an A allele is p. Similarly, the probability of drawing an a gamete is q. Individuals are monoecious in Hardy's model, and every individual contributes equally to both male and female gametes. Hence, although the second gamete drawn from the gene pool must be from the opposite sex of the first, all individuals are still equally likely to be the source of the second gamete. Moreover, Hardy regarded the number of gametes that could be produced by an individual as effectively infinite, so that drawing the first gamete from the gene pool has no effect upon drawing the second. The assumption of random mating also stipulates that this second gamete is drawn independently from the gene pool, which means that the probabilities are identical on the second draw and that the joint probability of both gametes is simply the product of their respective allele frequencies. Table 2.2 shows how these gamete frequencies are multiplied to yield zygotic genotype frequencies. Note, in calculating the frequency of the Aa genotype, there are two ways of creating a heterozygous zygote; the A allele could come from the paternal parent and a from the maternal or vice versa. The Hardy–Weinberg assumptions imply that parental origin of an allele has no

Table 2.2. Multiplication of Allele Frequencies to Yield Zygotic Genotypic Frequencies under Hardy–Weinberg Model of Random Mating

			Male gametes	
		Allele	A	a
		Frequency	p	q
	Allele	Frequency	AA $p \times p = p^2$	Aa $p \times q = pq$
Female gametes	A	p		
	a	q	aA $q \times p = qp$	aa $q \times q = q^2$

Summed frequencies in zygotes:
AA: $G'_{AA} = p^2$
Aa: $G'_{Aa} = pq + qp = 2pq$
aa: $G'_{aa} = q^2$

Note: The zygotic genotype frequencies are indicated by G'_k

effect. Hence, the two types of heterozygotes, each with frequency pq, are pooled together into a single Aa class with frequency $2pq$.

As the zygotes develop and mature into adults capable of contributing genes to the next generation, there is no change in their relative frequencies because of the implicit assumption of no phenotypic variation in viability, mating success, or fertility. Hence, Hardy showed that the genotype frequencies of the next generation could be predicted from allele frequencies given knowledge of the system of mating. From Figure 2.1 or Table 2.2, these predicted genotype frequencies are

$$G'_{AA} = p^2 \quad G'_{Aa} = 2pq \quad G'_{aa} = q^2$$

This array of genotype frequencies is known as the **Hardy–Weinberg law**.

We did not make any assumptions in this derivation about the initial genotypic frequencies, for example, G_{AA}. The initial adult population does not have to have Hardy–Weinberg genotype frequencies for the zygotes to have Hardy–Weinberg frequencies; all that is required is random mating of the adults regardless of their genotype frequencies. Hence, it takes *only one generation of random mating* to achieve Hardy–Weinberg genotype frequencies regardless of the starting genotype frequencies.

Weinberg's derivation differed from Hardy's at the point of modeling uniting gametes. To Weinberg, random mating meant that the probability of two genotypes being involved in a mating event was simply the product of their respective genotype frequencies. Given a mating, offspring genotypes would be produced according to standard Mendelian probabilities. Hence, in Weinberg's derivation, the mechanisms of producing gametes and the mechanisms of gametic union are utilized in an integrated fashion, as shown in Table 2.3. Note that this table makes an additional assumption not needed under the monoecious version of Hardy, namely, that the genotype frequencies are identical in both sexes. With

Table 2.3. Weinberg's Derivation of Hardy–Weinberg Genotype Frequencies

Mating Pair	Frequency of Mating Pair	Mendelian Probabilities of Offspring (Zygotes)		
		AA	Aa	aa
$AA \times AA$	$G_{AA} \times G_{AA} = G_{AA}^2$	1	0	0
$AA \times Aa$	$G_{AA} \times G_{Aa} = G_{AA}G_{Aa}$	$\frac{1}{2}$	$\frac{1}{2}$	0
$Aa \times AA$	$G_{Aa} \times G_{AA} = G_{AA}G_{Aa}$	$\frac{1}{2}$	$\frac{1}{2}$	0
$AA \times aa$	$G_{AA} \times G_{aa} = G_{AA}G_{aa}$	0	1	0
$aa \times AA$	$G_{aa} \times G_{AA} = G_{AA}G_{aa}$	0	1	0
$Aa \times Aa$	$G_{Aa} \times G_{Aa} = G_{Aa}^2$	$\frac{1}{4}$	$\frac{1}{2}$	$\frac{1}{4}$
$Aa \times aa$	$G_{Aa} \times G_{aa} = G_{Aa}G_{aa}$	0	$\frac{1}{2}$	$\frac{1}{2}$
$aa \times Aa$	$G_{aa} \times G_{Aa} = G_{Aa}G_{aa}$	0	$\frac{1}{2}$	$\frac{1}{2}$
$aa \times aa$	$G_{aa} \times G_{aa} = G_{aa}^2$	0	0	1
Total offspring		G'_{AA}	G'_{Aa}	G'_{aa}

Summing zygotes over all mating types:

$G'_{AA} = G_{AA}^2 + \frac{1}{2}[2G_{AA}G_{Aa}] + \frac{1}{4}G_{Aa}^2 = [G_{AA} + \frac{1}{2}G_{Aa}]^2 = p^2$

$G'_{Aa} = \frac{1}{2}[2G_{AA}G_{Aa}] + 2G_{AA}G_{aa} + \frac{1}{2}G_{Aa}^2 + \frac{1}{2}[G_{Aa}G_{aa}] = 2[G_{AA} + \frac{1}{2}G_{Aa}][G_{aa} + \frac{1}{2}G_{Aa}] = 2pq$

$G'_{aa} = \frac{1}{4}G_{Aa}^2 + \frac{1}{2}[2G_{Aa}G_{aa}] + G_{aa}^2 = [G_{aa} + \frac{1}{2}G_{Aa}]^2 = q^2$

this additional assumption, the end result of Weinberg's derivation is the same as Hardy's: The zygotic genotype frequencies (and hence the adult genotype frequencies of the next generation under the assumptions made here) are again given by $G'_{AA} = p^2$, $G'_{Aa} = 2pq$, and $G'_{aa} = q^2$.

We now address the important question of whether or not microevolution has occurred in this model; that is, are the allele frequencies in the offspring generation different or the same as the allele frequencies of the parent generation. Given that the adults of the offspring generation have the genotype frequencies $G'_{AA} = p^2$, $G'_{Aa} = 2pq$, and $G'_{aa} = q^2$, the allele frequencies in the pool of gametes they produce (say p' for A and q' for a) are calculated from equation 2.1 as

$$p' = p^2 + \frac{1}{2}(2pq) = p^2 + pq = p(p+q) = p \qquad (2.2)$$

and $q' = q$ (also shown in Figure 2.1). The allele frequencies p and p' make a contrast at comparable stages in two successive generations (here at the stage of producing gametes), and this contrast allows us to see if evolution has occurred. Because $p = p'$, by definition there has been no evolution. Hence, the Hardy–Weinberg model predicts that allele frequencies are stable over time and that no evolution is occurring under this set of assumptions. Because of this stability over time, Hardy–Weinberg genotype frequencies are often called the Hardy–Weinberg *equilibrium*. As noted earlier, it takes only one generation of random mating to achieve Hardy–Weinberg frequencies, and once achieved the population will remain in this state until one or more assumptions of the Hardy–Weinberg model are violated.

EXAMPLE OF HARDY–WEINBERG LAW

As an illustration of the application of this model, consider a human population of Pueblo Indians scored for genetic variation at the autosomal blood group locus *MN* (Figure 2.3). This locus has two common alleles in most human populations, the *M* allele and the *N* allele. Genetic variation at this locus determines your MN blood group type, with a very simple genotype-to-phenotype mapping: *MM* genotypes have blood group M, *MN* genotypes have blood group MN, and *NN* genotypes have blood group N. Hence, it is easy to characterize the genotypes of all individuals in a population by determining their MN blood group type. Figure 2.3 shows the number of individuals with each of the possible genotypes at this locus in a sample of 140 Pueblo Indians (Boyd 1950). The first step in analyzing a population is to convert the genotype numbers into genotype frequencies by dividing the number of individuals of a given genotype by the total sample size. For example, 83 Pueblo Indians had the *MM* genotype out of the total sample of 140, so the frequency of the *MM* genotype in that sample is $83/140 = 0.593$. Figure 2.3 then shows how the allele frequencies are calculated in the pool of potential gametes, yielding p (the frequency of M in this case) $= 0.757$ and $q = 0.243$. We can also apply the other definition of gene pool to this sample: The gene pool is the population of genes collectively shared by all the individuals. Since this is a diploid locus, the 140 Pueblo Indians collectively share 280 copies of genes at the *MN* locus. The 166 copies found in the 83 *MM* homozygotes are all M, and half of the 92 copies found in the 46 *MN* heterozygotes are M. Hence, the total number of M alleles in this sample of 280 genes is $166 + \frac{1}{2} \times 92 = 212$. The frequency of the M allele is therefore $212/280 = 0.757$. As this shows, either way of conceptualizing the gene pool leads to the same answer.

Continuing with Figure 2.3, we can see that the zygotic frequencies should be 0.573 for *MM*, 0.368 for *MN*, and 0.059 for *NN* if this population were randomly mating. Recall that

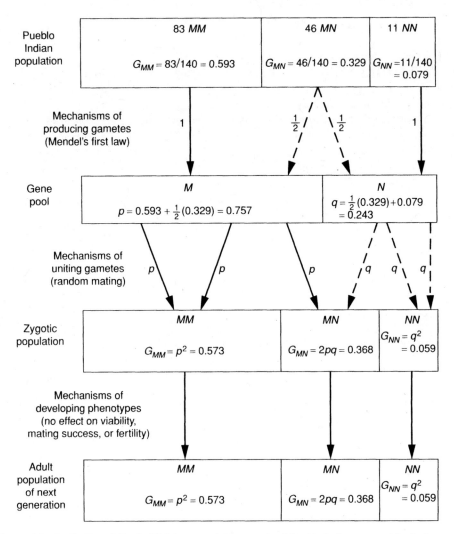

Figure 2.3. Application of Hardy–Weinberg model to sample of Pueblo Indians scored for their genotypes at autosomal *MN* blood group locus.

random mating in this case simply means that the individuals are choosing mates at random with respect to their MN blood group types; it does *not* mean that mating is random for every trait! For example, this population is evenly split between males and females, so the frequency of the female genotype *XX* (where *X* designates the human X chromosome) is 0.5 and the frequency of the male genotype *XY* is 0.5 (where *Y* designates the human Y chromosome). Because sex is determined by the X and Y chromosomes as wholes and these chromosomes do not normally recombine, we effectively can treat gender as determined by a single locus with two alleles, *X* and *Y*. The frequency of *X* gametes in the Pueblo Indian gene pool is $0.5 + \frac{1}{2}(0.5) = 0.75$ and the frequency of *Y* gametes is $\frac{1}{2}(0.5) = 0.25$. Therefore, we would expect the Hardy–Weinberg genotype frequencies of

$$G_{XX} = (0.75)^2 = 0.5625 \qquad G_{XY} = 2(0.75)(0.25) = 0.375$$
$$G_{YY} = (0.25)^2 = 0.0625$$

Obviously, this population is not at Hardy–Weinberg equilibrium for the X and Y chromosomes, and the reason is straightforward: Mating is not random for these genetic elements. Instead, the only cross that can yield offspring is $XX \times XY$, a gross deviation from the Hardy–Weinberg model portrayed in Table 2.3. Because of this highly nonrandom system of mating, the X and Y chromosomes can never achieve Hardy–Weinberg frequencies. Hence, systems of mating can be *locus specific* and Hardy–Weinberg frequencies are only for loci that have a *random system of mating*. Other genetic systems found in the same individuals in the same population may deviate from Hardy–Weinberg because mating is not random for that genetic system.

Recall that when the assumptions of Hardy–Weinberg are met, the population goes to Hardy–Weinberg genotype frequencies in a single generation and remains at those frequencies. Hence, if the Pueblo Indian population had been randomly mating for the MN blood groups in the past and if the other assumptions of Hardy–Weinberg are at least approximately true, we would expect the adult genotype frequencies of the next generation shown in Figure 2.3 to hold for the current adult population as well. This observation provides a basis for testing the hypothesis that this, or any population, has Hardy–Weinberg frequencies. The statistical details and a worked example of such a test are provided in Box 2.1.

IMPORTANCE OF HARDY–WEINBERG LAW

At first, the Hardy–Weinberg law may seem a relatively minor, even trivial, accomplishment. Nevertheless, this simple model played an important role in the development of both genetics and evolution in the early part of the twentieth century. Mendelian genetics had been rediscovered at the start of the twentieth century, but many did not accept it. One of the early proponents of Mendelian genetics was R. C. Punnett (of "Punnett square" fame). Punnett made a presentation at a scientific meeting in which he argued that the trait of brachydactyly (short fingers) was inherited as a Mendelian dominant trait in humans. Udny Yule, a member of the audience, raised the objection that one would expect a 3 : 1 ratio of people with brachydactyly to those without if the Mendelian model were true, and this clearly was not the case. Punnett suspected that there was an error in this argument, but he could not come up with a response at the meeting. Later Punnett explained the problem to his mathematician friend, G. H. Hardy, who immediately proceeded to derive his version of the Hardy–Weinberg law. Hardy's derivation made it clear that Yule had confused the family Mendelian ratio of 3 : 1 (which was for the offspring of a specific mating between two heterozygotes for the dominant trait) with the frequency in a population. Suppose in our earlier derivation that the A allele is dominant over a for some trait. Then the Hardy–Weinberg law predicts that the ratio of frequencies of those with the dominant trait to those with the recessive trait in a random-mating population should be $p^2 + 2pq : q^2$. There is no constraint upon this ratio to be 3 : 1 or any of the other family ratios expected under Mendelian inheritance. Rather, this population ratio can vary continuously as p varies from 0 to 1.

The predicted ratio of individuals with dominant to recessive traits also provided a method for predicting the frequency of carriers for genetic disease. Many genetic diseases in humans are recessive, so now let a be a recessive disease allele. Only two phenotypic categories could be observed in these early Mendelian studies: the dominant phenotype, associated with the genotypes AA and Aa, and the recessive, associated with the genotype aa. Thus, there was seemingly no way to predict how many people were carriers (Aa) as

BOX 2.1 TESTING TO SEE IF A POPULATION IS IN SINGLE-LOCUS HARDY–WEINBERG

We first estimate the allele frequencies using either equation 2.1 or the gene-counting method and then calculate the expected Hardy–Weinberg genotype frequencies. These steps have already been done for the Pueblo Indians, as shown in Figure 2.3. Next, we convert the expected Hardy–Weinberg genotype frequencies into expected genotype numbers by multiplying each frequency by the total sample size, which is 140 in this case. For example, the expected number of *MM* homozygotes under Hardy–Weinberg for the Pueblo Indian sample is $0.573 \times 140 = 80.22$. Similarly, the expected numbers of *MN* and *NN* genotypes are 51.52 and 8.26, respectively. Now we can calculate a standard chi-square statistic (see Appendix 2):

$$\sum_{\text{genotypes}} \frac{[\text{Obs}(i) - \text{Exp}(i)]^2}{\text{Exp}(i)} = \frac{(83 - 80.22)^2}{80.22} + \frac{(46 - 51.52)^2}{51.52} + \frac{(11 - 8.26)^2}{8.26} = 1.59$$

(2.3)

where Obs(i) is the observed number of individuals with genotype i and Exp(i) is the expected number of individuals with genotype i under Hardy–Weinberg (in this case i can be *MM*, *MN*, or *NN*). If the null hypothesis of Hardy–Weinberg is true, we expect the statistic calculated in equation 2.3 to have a value such that there is a high probability of the statistic having that or a higher value when in fact the population is at Hardy–Weinberg. To calculate this probability, we need the degrees of freedom associated with the chi-square statistic. In general, the degree of freedom is the number of categories being tested (three genotype categories in this case) minus 1 minus the number of independent parameters that had to be estimated from the data being tested to generate the expected numbers. In order to generate the Hardy–Weinberg expected values, we first had to estimate the allele frequencies of *M* and *N* from the data being tested. However, recall that $q = 1 - p$, so that once we know p, we automatically know q. This means that the data are used to estimate only one independent parameter (the parameter p). Therefore, the degree of freedom is $3 - 1 - 1 = 1$. We can now look up the value of 1.59 with one degree of freedom in a chi-square table or statistical calculator and find that the probability of getting a value of 1.59 or larger if the null hypothesis of Hardy–Weinberg were true is 0.21. Generally, such probabilities have to be less than 0.05 before the null hypothesis is rejected. Hence, we fail to reject the null hypothesis of Hardy–Weinberg for this sample of Pueblo Indians scored for the *MN* locus. It would have been simpler to say that the Pueblo Indian population is in Hardy–Weinberg, but we have not actually demonstrated this. Our sample is relatively small, and perhaps with more extensive sampling we would reject Hardy–Weinberg. Hence, all that we have really demonstrated is that we fail to reject Hardy–Weinberg for our current sample. Statistical tests never prove that a null hypothesis is true; the test either rejects or fails to reject the null hypothesis.

they could not be distinguished phenotypically from the *AA* homozygotes. However, if we assume Hardy–Weinberg is true, then the frequency of individuals affected with the genetic disease (which is observable) is q^2. Hence, we can estimate q is this case as

$$\hat{q} = \sqrt{G_{aa}}$$

(2.4)

Given $\hat{q}$, the frequency of carriers of the genetic disease can be estimated as $2(1 - \hat{q})\hat{q}$. Note, in this case, we cannot actually test the population for Hardy–Weinberg because we only have two observable categories and we have estimated one parameter from the data to be tested (equation 2.3). Therefore, the degrees of freedom are $2 - 1 - 1 = 0$. Zero degrees of freedom means we have insufficient information in the data to test the model (Appendix 2). Equation 2.4 should never be used when all genotypic classes are observable because it is valid only in the special case of Hardy–Weinberg genotype frequencies. In contrast, equation 2.1 makes no assumptions about Hardy–Weinberg and is true for any set of genotype frequencies. Therefore, when all genotypic classes are observable, equation 2.1 should be used instead of equation 2.4 because equation 2.1 will always give you the right answer whereas equation 2.4 will only give the right answer in a specific special case. Nevertheless, equation 2.4 played an important role throughout much of the twentieth century in genetic counseling in predicting heterozygous carrier frequencies for autosomal recessive genetic diseases when all genotypic classes were not observable.

The Hardy–Weinberg law also predicts no evolution; that is, the allele frequencies remain constant over time. At first this may also seem to be a rather uninteresting result, but this observation was critical for the acceptance of the Darwin–Wallace concept of natural selection. The publication of Darwin's book *The Origin of Species* in 1859 strongly established the concept of descent with modification within biology. However, Darwin's (and Wallace's) explanation for the origin of adaptations via natural selection was less universally accepted. Darwin felt that the Scottish engineer Fleeming Jenkin raised one of the most serious objections to the theory of natural selection in 1867. At this time, the dominant idea of inheritance was that of "blending inheritance" in which the traits of the father and mother are blended together, much as mixing two different colors of paint together results in a new color that represents equal amounts of the original colors. Jenkin pointed out that half of the heritable variation would be lost every generation under blending inheritance; hence, a population should quickly become homogeneous. Recall from Chapter 1 that heritable variation is a necessary prerequisite for all evolution, so evolution itself would grind to a halt unless mutation replenished this loss at the same rate. Darwin and Wallace had based their theories of natural selection upon the tenet that mutation creates new variation at random with respect to the needs of the organism in coping with its environment. It seemed implausible that half of the genetic material could mutate at random every generation and the organisms still survive. Hence, Jenkin's argument seemed to imply that either genetic variation would quickly vanish and all evolution halt or that natural selection required levels of mutation that would result in extinction. This problem even led Darwin in his 1868 book *The Variation of Animals and Plants under Domestication* to speculate that mutation might be directed by the environment. By the beginning of the twentieth century, many neo-Lamarkian ideas based upon directed mutations were popular alternatives to natural selection of random mutations.

Jenkin's argument was finally put to rest by the Hardy–Weinberg law. The Hardy–Weinberg model, by ignoring many potential evolutionary forces (Table 2.1), focuses our attention upon the potential evolutionary impact of Mendelian inheritance alone. By demonstrating that Mendelian inheritance results in a population with a constant allele frequency, it was evident that Mendelian genetic variation is not rapidly lost from a population. Indeed, under the strict assumptions of Hardy–Weinberg, genetic variation persists indefinitely. Thus, even though the Hardy–Weinberg model is one of no evolution, this model was critical for the acceptance of natural selection as a plausible mechanism of evolutionary change under Mendelian inheritance.

In general, this book is concerned about evolutionary change. In modeling evolution, Hardy–Weinberg is a useful null model of evolutionary stasis. Indeed, much of the rest of this book is devoted to relaxing one or more of the assumptions of the original Hardy–Weinberg model and seeing whether or not evolution can result. In this sense, Hardy–Weinberg serves as a valuable springboard for the investigation of many forces of evolutionary change. In the remainder of this chapter we consider just one slight deviation from the original Hardy–Weinberg model, and we will investigate the evolutionary implications of this slight change.

HARDY–WEINBERG FOR TWO LOCI

The original Hardy–Weinberg model assumed a genetic architecture of one autosomal locus with two alleles. We will now consider a slightly more complicated genetic architecture of two autosomal loci, each with two alleles (say A and a at locus 1 and B and b at locus 2). Otherwise, we will retain all other assumptions of the original Hardy–Weinberg model. However, there is one new assumption. Recall from Chapter 1 that our second premise is that DNA can mutate and recombine. We will retain the Hardy–Weinberg assumption of no mutation, but we will allow recombination (either independent assortment if the two loci are on different autosomes or crossing over if they are on the same autosome).

Because our main interest is on whether or not evolutionary change occurs, we will start with the gene pool and go to the next generation's gene pool (Figure 2.4), rather than going from adult population to adult population as in Figures 2.1 and 2.3. Given two loci with two alleles each and the possibility of recombination between them, a total of four gamete types are possible (AB, Ab, aB, and ab). The gene pool is characterized by four gamete frequencies (Figure 2.4), symbolized by g_{xy}, where x indicates the allele at locus 1 and y indicates the allele at locus 2. Just as p and q sum to 1, these four gamete frequencies also sum to 1 because they define a probability distribution over the gene pool. The transition from this gene pool to the zygotes is governed by the same population structure (rules of uniting gametes) as given in the single-locus Hardy–Weinberg. In particular, the assumption of random mating means that gametes are drawn independently from the gene pool, with the probability of any given gamete type being equal to its frequency. The probability of any particular genotype is simply the product of its gamete frequencies, just as in the single-locus Hardy–Weinberg model. In Figure 2.4 we are not keeping track of the paternal or maternal origins of any gamete, so both types of heterozygotes are always pooled and therefore the product of the gamete frequencies for heterozygous genotypes is multiplied by 2. For example, the frequency of the genotype AB/Ab is $2g_{AB}g_{Ab}$. Note that there are two types of double heterozygotes, AB/ab (the cis double heterozygote with a random-mating frequency of $2g_{AB}g_{ab}$) and Ab/aB (the trans double-heterozygote with a random-mating frequency of $2g_{Ab}g_{aB}$). Although the cis and trans double heterozygotes share the double-heterozygous genotype, completely different gamete types produce the cis and trans double-heterozygosity. As we will soon see, the cis and trans double-heterozygous genotypes contribute to the gene pool in different ways. Hence, we will keep the cis and trans double-heterozygote classes separate.

The rules for uniting gametes in the two-locus model are the same as for the single-locus model, the only difference being that there are now 10 genotypic combinations. As with the single-locus model, if we know the gamete frequencies and know that the mating is at random (along with the other population structure assumptions of Hardy–Weinberg),

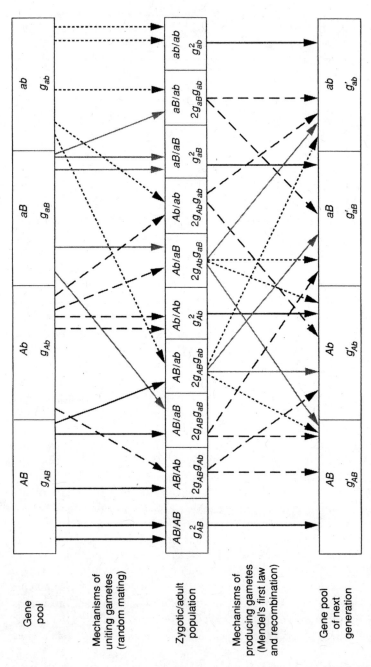

Figure 2.4. Derivation of Hardy–Weinberg law for two autosomal loci with two alleles each: A and a at locus 1 and B and b at locus 2. In going from gametes to zygotes, solid lines represent gametes bearing the AB alleles and are assigned the weight g_{AB}, dashed lines represent gametes bearing the Ab allele and are assigned the weight g_{Ab}, grey lines represent gametes bearing the aB gametes and are assigned the weight g_{aB}, and dotted lines represent gametes bearing the ab alleles and are assigned the weight g_{ab}. In going from adults to gametes, solid lines represent Mendelian transition probabilities of 1 for homozygotes, dashed lines represent Mendelian transition probabilities of $\frac{1}{2}$ for single heterozygotes, dotted lines represent nonrecombinant Mendelian transition probabilities of $\frac{1}{2}(1-r)$ (where r is the recombination frequency between loci 1 and 2) for double heterozygotes, and grey lines represent recombinant Mendelian transition probabilities of $\frac{1}{2}r$ for double heterozygotes.

we can predict the zygotic genotype frequencies. If we further assume that there are no phenotypic differences that affect viability, mating success, or fertility, we can also predict the next generation's adult genotype frequencies from the gamete frequencies.

The similarities to the single-locus model end when we advance to the transition from the next generation's adult population to the gene pool of the next generation (Figure 2.4). At this point, some new rules are encountered in producing gametes that did not exist at all in the single-locus model (Figure 2.1). As before, homozygous genotypes can only produce gametes bearing the alleles for which they are homozygous (this comes from the assumptions of normal meiosis and no mutations). As before, genotypes heterozygous for just one locus produce two gamete types, with equal frequency as stipulated by Mendel's first law. However, genotypes that are heterozygous for both loci can produce all four gamete types, and the probabilities are determined by a combination of Mendel's first law and recombination (Mendel's second law of independent assortment if the loci are on different chromosomes or the recombination frequency if on the same chromosome). Hence, the transition from genotypes to gametes requires a new parameter, the recombination frequency r, which is $\frac{1}{2}$ if the loci are on different chromosomes and $0 \leq r \leq \frac{1}{2}$ if the loci are on the same chromosome.

The addition of recombination produces some qualitative differences with the single-locus model. First, in the single-locus model, an individual could only pass on gametes of the same types that the individual inherited from its parents. But note from Figure 2.4 that the cis double heterozygote AB/ab, which inherited the cis AB and ab gamete types from its parents, can produce not only the cis gamete types, each with probability $\frac{1}{2}(1-r)$, but also the trans gamete types Ab and aB, each with probability $\frac{1}{2}r$. Similarly, the trans double heterozygote can produce both cis and trans gamete types (Figure 2.4). Thus, recombination allows the double heterozygotes to produce gamete types that they themselves did not inherit from their parents. This effect of recombination is found only in the double-heterozygote class, but this does not mean that recombination only occurs in double heterozygotes. Consider, for example, the single heterozygote AB/Ab. If no recombination occurs in meiosis, this genotype will produce the gamete types AB and Ab with equal frequency. Hence, the total probability of gamete type AB with no recombination is $\frac{1}{2}(1-r)$, and similarly it is $\frac{1}{2}(1-r)$ for Ab. Now consider a meiotic event in which recombination did occur. Such a recombinant meiosis also produces the gamete types AB and Ab with equal frequency, that is, with probability $\frac{1}{2}r$ for each. However, in the recombinant AB gamete the A allele that is combined with the B allele originally came from the Ab gamete that the AB/Ab individual inherited from one of its parents. Hence, recombination has occurred, but because we do not distinguish among copies of the A alleles, we see no observable genetic impact. Hence, the total probability of an AB gamete, regardless of the source of the A allele, is $\frac{1}{2}(1-r) + \frac{1}{2}r = \frac{1}{2}$, and the total probability of an Ab gamete, regardless of the source of the A allele, is $\frac{1}{2}(1-r) + \frac{1}{2}r = \frac{1}{2}$. Thus, recombination is occurring in all genotypes but is observable only in double heterozygotes.

The qualitative difference from the single-locus model that causes some genotypes to produce gamete types that they themselves did not inherit leads to yet another qualitative difference: The two definitions of gene pool given in Chapter 1 are no longer equivalent. If we define the gene pool as the shared genes of all the adult individuals, we obtain the gamete frequencies from the pool of gametes produced by their parents (the g_{xy}'s in Figure 2.4). On the other hand, if we define the gene pool as the population of potential gametes produced by all the adult individuals, the effects of recombination enter and we obtain the g'_{xy}'s in Figure 2.4. To avoid any further confusion on this point, the term "gene pool" in this book will always refer to the population of potential gametes unless otherwise stated. The general population genetic literature often does not make this distinction because in the standard single-locus Hardy–Weinberg model it is not important. Quite frequently there is a time difference of one generation among the models of various authors depending upon which definition of gene pool they use (usually implicitly). Therefore, readers have to be careful in interpreting what various authors mean by gene pool when dealing with multilocus models or other models in which these two definitions may diverge.

The most important qualitative difference from the single-locus model involves the potential for evolution. As seen before, the single-locus Hardy–Weinberg model goes to equilibrium in a single generation of random mating and then stays at equilibrium, resulting in no evolution. To see if this is the case for the two-locus model, we now use equation 2.1 to calculate the gamete frequency of the AB gamete using the weights implied by the arrows in Figure 2.4 going from adults to gametes:

$$\begin{aligned}
g'_{AB} &= 1 \cdot g^2_{AB} + \frac{1}{2}(2g_{AB}g_{Ab}) + \frac{1}{2}(2g_{AB}g_{aB}) + \frac{1}{2}(1-r)(2g_{AB}g_{ab}) + \frac{1}{2}r(2g_{Ab}g_{aB}) \\
&= g_{AB}[g_{AB} + g_{Ab} + g_{aB} + (1-r)g_{ab}] + rg_{Ab}g_{aB} \\
&= g_{AB}[g_{AB} + g_{Ab} + g_{aB} + g_{ab}] + rg_{Ab}g_{aB} - rg_{AB}g_{ab} \\
&= g_{AB} + r(g_{Ab}g_{aB} - g_{AB}g_{ab}) = g_{AB} - rD
\end{aligned} \tag{2.5}$$

where $D = (g_{AB}g_{ab} - g_{Ab}g_{aB})$. The parameter D is commonly known as **linkage disequilibrium**. However, because it can exist for pairs of loci on different chromosomes that are not linked at all, a more accurate but more cumbersome term is **gametic-phase imbalance**. Because the term linkage disequilibrium dominates the literature, we will use it throughout the book, but with the caveat that it can be applied to unlinked loci.

Similarly, the other three gamete types can be obtained from equation 2.1 as

$$\begin{aligned}
g'_{Ab} &= 1 \cdot g^2_{Ab} + \frac{1}{2}(2g_{AB}g_{Ab}) + \frac{1}{2}(2g_{Ab}g_{ab}) + \frac{1}{2}(1-r)(2g_{Ab}g_{aB}) + \frac{1}{2}r(2g_{AB}g_{ab}) \\
&= g_{Ab} + rD \\
g'_{aB} &= 1 \cdot g^2_{aB} + \frac{1}{2}(2g_{AB}g_{aB}) + \frac{1}{2}(2g_{aB}g_{ab}) + \frac{1}{2}(1-r)(2g_{Ab}g_{aB}) + \frac{1}{2}r(2g_{AB}g_{ab}) \\
&= g_{aB} + rD \\
g'_{ab} &= 1 \cdot g^2_{ab} + \frac{1}{2}(2g_{Ab}g_{ab}) + \frac{1}{2}(2g_{aB}g_{ab}) + \frac{1}{2}(1-r)(2g_{AB}g_{ab}) + \frac{1}{2}r(2g_{Ab}g_{aB}) \\
&= g_{ab} - rD
\end{aligned} \tag{2.6}$$

At this point, we can now address our primary question: Is evolution occurring? Recall our definition from Chapter 1 of evolution as a change in the frequencies of various types of genes or gene combinations in the gene pool. As is evident from equations 2.5 and 2.6, as long as $r > 0$ (that is, some recombination is occurring) and $D \neq 0$ (there is some linkage disequilibrium), $g_{xy} \neq g'_{xy}$: *Evolution is occurring*! Thus, a seemingly minor change from one to two loci results in a major qualitative change of population-level attributes.

No evolution occurs in this model if $r = 0$. In that case, the two-locus model is equivalent to a single-locus model with four possible alleles. Thus, some multilocus systems can be treated as if they were a single locus as long as there is no recombination. On the other hand, recombination can sometimes occur *within* a single gene. As mentioned in

Chapter 1, the genetic variation within a 9.7-kb segment of the *lipoprotein lipase (LPL)* gene in humans was shaped in part by about 30 recombination events (Templeton et al. 2000a). Thus, in some cases the evolutionary potential created by recombination must be considered even at the single-locus level. In the case of *LPL*, we are looking at two or more different polymorphic nucleotide sites within the same gene and not, technically speaking, at different loci. However, the qualitative evolutionary potential is still the same as long as the polymorphic sites under examination can recombine, regardless of whether those sites are single nucleotides within a gene or traditional loci.

No evolution also occurs in this model if $D = 0$. Here, D will equal zero when the two-locus gamete frequencies are the product of their respective single-locus allele frequencies. To see this, let p_A be the frequency of the A allele at locus 1 and p_B the frequency of the B allele at locus 2. These single-locus allele frequencies are related to the two-locus gamete frequencies by

$$p_A = g_{AB} + g_{Ab} \qquad p_B = g_{AB} + g_{aB} \tag{2.7}$$

Now consider the product of the A and B allele frequencies:

$$\begin{aligned} p_A p_B &= (g_{AB} + g_{Ab})(g_{AB} + g_{aB}) \\ &= g_{AB}^2 + g_{AB}g_{aB} + g_{AB}g_{Ab} + g_{Ab}g_{aB} \\ &= g_{AB}(g_{AB} + g_{aB} + g_{Ab}) + g_{Ab}g_{aB} \\ &= g_{AB}(1 - g_{ab}) + g_{Ab}g_{aB} \\ &= g_{AB} - g_{AB}g_{ab} + g_{Ab}g_{aB} \\ &= g_{AB} - D. \end{aligned} \tag{2.8}$$

Solving equation 2.8 for D yields

$$D = g_{AB} - p_A p_B \tag{2.9}$$

and similar equations can be derived in terms of the other three gamete frequencies. <u>Equation 2.9 suggests another biological interpretation of D; *it is the deviation of the two-locus gamete frequencies from the product of the respective single-locus allele frequencies*.</u> Equation 2.9 also makes it clear that D will be zero when the two-locus gamete frequency is given by the product of the respective single-locus allele frequencies. This can also be seen by evaluating the original formula for linkage disequilibrium under the assumption that the two-locus gamete frequencies are given the product of their respective allele frequencies: $D = g_{AB}g_{ab} - g_{Ab}g_{aB} = (p_A p_B)(p_a p_b) - (p_A p_b)(p_a p_B) = p_A p_B p_a p_b - p_A p_b p_a p_B = 0$.

The two-locus gamete frequencies will be products of the single-locus allele frequencies when knowing what allele is present at one locus in a gamete does not alter the probabilities of the alleles at the second locus; that is, the probabilities of the alleles at the second locus are simply their respective allele frequencies regardless of what allele occurs at the first locus. When $D \neq 0$, knowing which allele a gamete bears at one locus does influence the probabilities of the alleles at the second locus. In statistical terms, $D = 0$ means that there is no association in the population between variation at locus 1 with variation at locus 2. When $D = 0$, equations 2.5 and 2.6 show that the gamete frequencies (and hence the genotype frequencies) are constant, just as they were in the single-locus Hardy–Weinberg model. Thus, when $D = 0$ the population is at a nonevolving equilibrium, given the other standard Hardy–Weinberg assumptions. We can now understand why D is called *disequilibrium*. When D is not zero and there is recombination, the population is evolving and is not at a two-locus Hardy–Weinberg equilibrium. The larger D is in magnitude, the greater this deviation from two-locus equilibrium.

Evolution occurs when $r > 0$ and $D \neq 0$, and we now examine the evolutionary process induced by linkage disequilibrium in more detail. From Figure 2.4 or equations 2.5 and 2.6, we see that linkage disequilibrium in the original gene pool ($g_{AB}g_{ab} - g_{Ab}g_{aB}$) influences the next generation's gene pool. Similarly, the linkage disequilibrium in the next generation's gene pool will influence the subsequent generation's gene pool. The linkage disequilibrium in the next generation's gene pool in Figure 2.4 is

$$\begin{aligned} D_1 &= [g'_{AB}g'_{ab} - g'_{aB}g'_{Ab}] \\ &= [(g_{AB} - rD)(g_{ab} - rD) - (g_{aB} + rD)(g_{Ab} + rD)] \\ &= D(1 - r) \end{aligned} \quad (2.10)$$

Using equation 2.10 recursively, we can see that D_2 (the linkage disequilibrium in the gene pool two generations removed from the original gene pool) is $D(1 - r)^2$. In general, if we start with some initial linkage disequilibrium, say D_0, then D_t, the linkage disequilibrium after t generations of random mating, is

$$D_t = D_0(1 - r)^t. \quad (2.11)$$

Equation 2.11 reveals that the evolution induced by linkage disequilibrium is both gradual and directional, as illustrated in Figure 2.5. Because $r \leq \frac{1}{2}$, the quantity $(1 - r)^t$ goes to

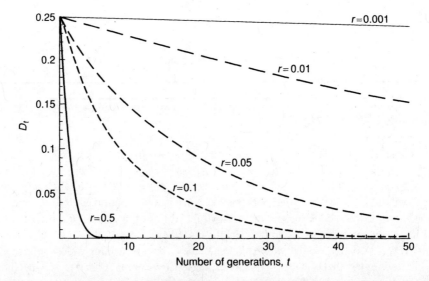

Figure 2.5. Decay of linkage disequilibrium with time in generations as function of different recombination rates r starting with initial value of $D_0 = 0.25$.

Figure ... ...rium in autosomal *ApoAI, ApoCIII, ApoAIV* gene region of human genome between insertion/deletion polymorphism (indel) and five restriction sites. Data from Haviland et al. (1991).

and the *Pvu*II site. Hence, linkage disequilibrium in this DNA region is not a reliable guide to physical proximity.

The example shown in Figure 2.7 illustrates that evolutionary history can obscure information; in this case, we cannot infer relative positional information from the magnitude of linkage disequilibrium. This has important implications in medical genetics. Many studies will look at a few markers within a gene region and then look for disease associations. The rationale for such studies is not that the markers being scored are actually causing the disease; rather, the hope is that one or more of the markers will be in linkage disequilibrium with the mutation that actually causes the disease or contributes to disease risk (this topic will be discussed in more detail in Chapter 10). This approach has found many associations between markers and diseases in humans and illustrates just one practical application of linkage disequilibrium. However, the causative mutation is not necessarily physically closest to the marker showing the strongest disease association. As a hypothetical example, suppose the three rather evenly spaced restriction site markers *Xmn*I, *Sst*I, and *Pvu*II were used to look for disease associations between the *ApoAI, CIII, AIV* region and coronary artery disease (such associations do indeed exist). Suppose further that the insertion/deletion polymorphism was a causative mutation by affecting, say, the expression of *ApoAI*. However, using these three markers, the effect of the insertion/deletion polymorphism would be detected as an association through linkage disequilibrium with the *Pvu*II site—the site that is actually the most distant site from the hypothetical causative mutation among all the markers surveyed! Hence, an investigator might be tempted to conclude that the causative site is close to the *ApoAIV* locus and is not in the *ApoAI* locus. Ignoring evolutionary history can lead to many false conclusions in medical genetics. Our understanding of the present must be predicated upon a knowledge of how the current genetic variation arose during its evolutionary past. When dealing with a DNA region in which there is a poor or no correlation between linkage disequilibrium and physical position, we must always be cautious in interpreting marker association data in genetic disease studies.

The linkage disequilibrium found in small DNA regions can, if used properly, actually help us in our search for disease associations. For example, loci coding for apoproteins have been associated not only with coronary artery disease but also with Alzheimer's dis-

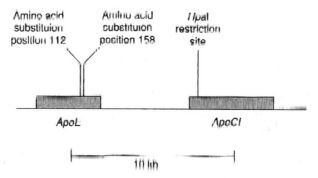

Figure 2.8. Genetically variable sites in autosomal ApoE, ApoCI region of human genome that have been associated with risk for Alzheimer's disease.

ease, a mental dementia that afflicts many people as they age. Figure 2.8 shows another autosomal apoprotein coding region in the human genome, this segment coding for apoprotein E and apoprotein CI. As shown, there are two polymorphic sites in the coding region of the *ApoE* locus, each associated with an amino acid change. There is extensive linkage disequilibrium between these two sites such that only three of the four possible gamete types defined by these two polymorphisms exist and $|D'| = 1$ (indeed, there is no evidence that recombination has ever occurred between these polymorphic nucleotide sites). These three gamete types define three distinct alleles at the *ApoE* locus named $\varepsilon 2$, $\varepsilon 3$, and $\varepsilon 4$. Many studies have revealed an association between the $\varepsilon 4$ allele and a high risk for Alzheimer's disease (such an association does *not* mean that $\varepsilon 4$ causes Alzheimer's disease). Nearby, a *Hpa*I restriction site polymorphism was found in the *ApoCI* locus that also has a significant association with risk for Alzheimer's disease (Chartier-Harlin et al. 1994), with people bearing the allele in which the enzyme *Hpa*I cuts this site having increased risk. Note that both of these single-locus studies subdivide people into two risk categories (high and low) for Alzheimer's disease. However, the *Hpa*I site is also in linkage disequilibrium with the *ApoE* sites, so does this *ApoCI* site actually provide new risk information or is it redundant with the information previously documented in *ApoE*? When the *ApoE* alleles and the *ApoCI* restriction site are considered simultaneously, three, not two, risk categories for Alzheimer's disease are revealed (Table 2.4) (Templeton 1995). So when combined, these two apoprotein loci do indeed provide more refined information about risk for Alzheimer's disease than either locus separately. Moreover, note from Table 2.4 that the *Hpa*I restriction site is associated with both the highest and the lowest risk categories, depending upon which *ApoE* allele it is combined with. On the basis of the associations of the *Hpa*I restriction site alone, we would have placed into the "high" *Hpa*I risk category both those people

Table 2.4. Risk for Alzheimer's Disease as Associated with $\varepsilon 2$, $\varepsilon 3$, and $\varepsilon 4$ Alleles at *ApoE* Locus and *Hpa*I Restriction Site Polymorphism in *ApoCI* Locus

Risk Category	ApoE	ApoCI
High	$\varepsilon 3$ or $\varepsilon 4$	*Hpa*I cuts
Medium	$\varepsilon 3$	*Hpa*I does not cut
Low	$\varepsilon 2$	*Hpa*I cuts

Note: See Figure 2.8.

with *highest* and *lowest* risk to Alzheimer's disease. Thus, ignoring evolutionary history as manifested in linkage disequilibrium could have led to erroneous medical advice in this case.

As shown in this chapter, the Hardy–Weinberg law, a seemingly simple model, nevertheless leads to many important insights about the evolutionary process. This model played an important role in the establishment of Mendelian genetics and natural selection during the first half of the twentieth century. The two-locus version of this law is currently playing a critical role in medical genetics in the twenty-first century. The difference in the potential for evolutionary change between the one-locus and two-locus versions shows that we must be cautious in generalizing inferences from our reductionist models. It is therefore critical to examine what happens when some of the other assumptions of the original Hardy–Weinberg model are altered or relaxed. In the next chapter, we focus upon one of these critical assumptions: system of mating.

3

SYSTEMS OF MATING

As defined in Chapter 1, a deme is a collection of interbreeding individuals of the same species that live in sufficient proximity that they share a common system of mating—the rules by which pairs of gametes are chosen from the local gene pool to be united in a zygote. Sufficient proximity depends upon the geographical range of the group of individuals and their ability to disperse and interbreed across this range. These geographical factors will be dealt with in Chapter 4. Here we simply note that, depending upon the geographical scale involved and the individuals' dispersal and mating abilities, a deme may correspond to the population of the entire species or to a subpopulation restricted to a small local region within the species' range. A deme is not defined by geography but rather by a shared system of mating. The Hardy–Weinberg model assumes one particular system of mating—random mating—but many other systems of mating exist. Moreover, as shown in Chapter 2, it is possible for different loci or complexes of loci within the same deme to have different systems of mating. It is therefore more accurate to say that a deme shares a common system of mating for a particular genetic system or locus. The purpose of this chapter is to investigate some alternatives to random mating and their evolutionary consequences.

INBREEDING

In its most basic sense, **inbreeding** is mating between biological relatives. Two individuals are related if among the ancestors of the first individual are one or more ancestors of the second individual. Because of shared common ancestors, the two individuals could share genes at a locus that are identical copies of a single ancestral gene (via premise 1—DNA can replicate). Such identical copies due to shared ancestry are said to be **identical by descent**. In contrast, the same allele can arise more than once due to recurrent mutation. Identical

---

Population Genetics and Microevolutionary Theory, By Alan R. Templeton
Copyright © 2006 John Wiley & Sons, Inc.

copies of a gene due to recurrent mutation from different ancestral genes are said to be **identical by state**.

Virtually all individuals within most species are related to all other individuals if you go far enough back in time. For example, computer simulations using reasonable assumptions about humanity's demographic history indicate that all humans living today share at least one common ancestor who lived sometime between 55 CE (Common Era) and 1415 BCE (Before the Common Era) (Rohde et al. 2004). Thus, all humans are biological relatives if we could trace our ancestry back a few thousand years. In practice, we often know pedigree relationships only for a few generations into the past. Given our ignorance about long-term pedigrees, how do we decide who is a relative and who is not? The solution to this practical problem is to regard some particular generation or set of individuals as the reference population whose members are regarded as unrelated. We assume that we can ascertain the biological relatedness of any two individuals in the current population by going back to but not beyond the individuals in that reference population. By assumption, all the genes in this reference population are regarded as *not* being identical by descent. If two identical genes today are traced back to different genes in the reference population, this pair of genes is regarded as being identical by state and not by descent.

There are several alternative ways of measuring inbreeding within this basic concept of mating between known relatives. Many of these alternatives are incompatible with one another because they focus on measuring different biological phenomena that are associated with matings between relatives. Unfortunately, all of these alternative ways of measuring "inbreeding" are typically called "inbreeding coefficients" in the population genetic literature. This lack of verbal distinctions between different biological concepts has resulted in confusion and misunderstanding. Jacquard (1975) tried to clarify this confusion in an excellent article entitled "Inbreeding: One Word, Several Meanings," but the many meanings of the word *inbreeding coefficient* are still rarely specified in much of the population genetic literature. The responsibility for making the distinctions among the several distinct and mutually incompatible inbreeding coefficients therefore often falls upon the reader. Consequently, it is important to be knowledgeable of the more common concepts of inbreeding, which we will examine in this chapter.

Definitions of Inbreeding

Pedigree Inbreeding. When two biological relatives mate, the resulting offspring could be homozygous for an allele through identity by descent. In other words, the gene at a particular autosomal locus being passed on by the father could be identical to the homologous gene being passed on by the mother because both genes are identical copies of a single piece of DNA found in a common ancestor. The amount of inbreeding in this case is measured by F (the first of many inbreeding coefficients), defined as the probability that the offspring is homozygous due to identity by descent at a randomly chosen autosomal locus. Offspring for whom $F > 0$ (that is, offspring with a finite chance of being homozygous at a locus through identity by descent) are said to be **inbred**. Because F is a probability, it can range in value from 0 (no chance for any identity by descent) to 1 (all autosomal loci are identical by descent with certainty). The probability F can be calculated for an individual by applying Mendel's first law of 50–50 segregation to the pedigree of that individual.

As an example, consider the pedigree in Figure 3.1, which shows an offspring produced by a mating between two half sibs. For pedigree data, the reference population is simply the set of individuals for which no further pedigree information exists. In Figure 3.1, the

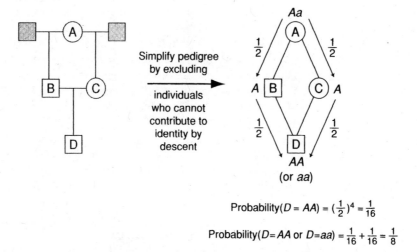

Figure 3.1. Mating between two half sibs (individuals B and C) who share a common mother (individual A, who is heterozygous *Aa*) to produce an inbred offspring (individual D). The left side of the figure portrays the pedigree in the standard format of human genetics, where squares denote males, circles denote females, horizontal lines connecting a male and female denote a mating, and vertical lines coming off from the horizontal mating lines indicate the offspring. The right side of the figure shows how this pedigree is simplified for the purposes of calculating the inbreeding coefficient F by deleting all individuals from the pedigree who are not common ancestors of the offspring of interest (individual D in this case). Shading in the pedigree on the left indicates the deleted individuals. The Mendelian probabilities associated with transmitting the *A* allele are indicated in the simplified pedigree.

reference population consists of individual A and the two males with whom she mated. These three individuals are assumed to be unrelated, and any alleles they carry, even if identical, are not considered to be identical by descent but rather to be identical by state. In Figure 3.1, there is only one shared ancestor (A) common to both the mother (C) and the father (B). Assuming that the common ancestor herself has no inbreeding in the pedigree sense, her two alleles at an autosomal locus cannot be identical by descent and are indicated by *A* and *a* (they may be identical by state). The probability that the common ancestor (A) passes on the *A* allele to her son (B) is $\frac{1}{2}$ from Mendel's first law, and likewise the probability that she passes on the *A* allele to her daughter (C) is $\frac{1}{2}$. Both the son (B) and daughter (C) also received an allele at this locus from their fathers, who are not common ancestors and cannot contribute to identity by descent. Therefore, the only way for the offspring (D) to be identical by descent for this locus is for both the father (B) and the mother (C) to pass on the allele they inherited from their common ancestor (A), and each of these gamete transmissions also has a probability of $\frac{1}{2}$ under Mendel's first law (Figure 3.1). Because the four segregation probabilities shown in Figure 3.1 are all independent, the probability that all four occurred as shown is $(\frac{1}{2})^4 = \frac{1}{16}$ = probability that individual D is homozygous by descent for allele *A*. The common ancestor (A) also had a second allele *a*, and the probability that individual D is homozygous by descent for allele *a* is likewise $\frac{1}{16}$. Hence, the total probability of individual D being identical by descent at this locus is $\frac{1}{16} + \frac{1}{16} = \frac{1}{8}$ since the event of D being *AA* is mutually exclusive from the event of D being *aa*. By definition, the pedigree inbreeding coefficient for individual D is therefore $F = \frac{1}{8}$.

The calculation of F can become much more difficult when there are many common ancestors and ways of being identical by descent and when the common ancestors themselves

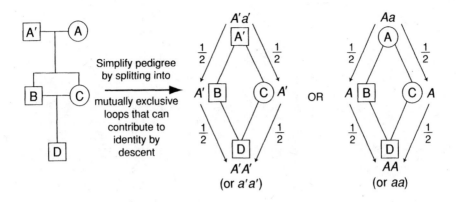

Figure 3.2. Inbreeding associated with mating of two full sibs.

are inbred in the pedigree sense. However, the basic principles are the same: Nothing more than Mendel's first law is applied to the pedigree to calculate the pedigree inbreeding coefficient F. For example, consider the case of two full sibs mating to produce an inbred offspring (Figure 3.2). In this case, the inbred offspring can be homozygous by descent for an allele from its grandmother or grandfather (Figure 3.2). Since an individual can be homozygous by descent for an allele from one and only one of the common maternal/paternal ancestors, identity by descent for an allele from the grandmother is mutually exclusive from identity by descent for an allele from the grandfather. Hence, the total probability of identity by descent, regardless of which common ancestor provided the allele, is the sum of the identity probabilities associated with the grandmother and grandfather, each of which is $\frac{1}{8}$ (Figure 3.2). Hence, $F = \frac{1}{8} + \frac{1}{8} = \frac{1}{4}$ for the offspring of two full sibs.

Of course, some pedigrees have many more common ancestors and pathways of potential identity by descent, making the calculation of F more difficult than the simple examples shown in Figures 3.1 and 3.2. The algorithms used to make these calculations for more complicated pedigrees were worked out many centuries ago by the Roman Catholic Church. Dispensations for incestuous marriages were needed to be granted before the Church could recognize such marriages. Therefore, priests needed to work out the degree of inbreeding that would occur in the offspring from such a marriage in order to distinguish degrees of consanguinity that are dispensable from those that are not (Cavalli-Sforza and Bodmer 1971). Today, many computer programs use these same algorithms to calculate F.

It is critical to note that the pedigree inbreeding coefficient F is applied to a particular *individual* coming from a specified union with a specified pedigree. Therefore F is an individual concept and not a population concept at all. Indeed, a single population often consists of individuals showing great variation in their F's. For example, a captive herd of Speke's gazelle (*Gazella spekei*) was established at the St. Louis Zoo between 1969 and 1972 from one male and three females imported from Africa (Templeton and Read 1994). Assuming that these four imported animals are unrelated (that is, the four founding animals constitute the reference population), their initial offspring would all have $F = 0$. However, because there was only one male in the original herd, the most distant relationship among captive-bred animals is that of a half sib (all the initial captive-bred offspring must share the same father). As a consequence, once the initial founders had died or were too old to breed, the *least inbred* mating possible among the captive-born animals would be between half

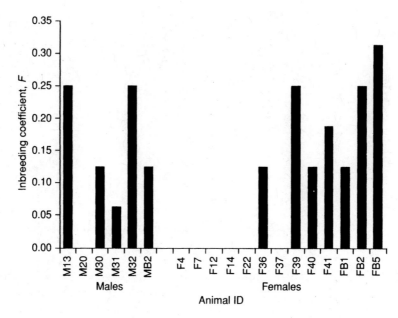

Figure 3.3. Pedigree inbreeding coefficients for all individuals from captive herd of Speke's gazelle.

sibs, with $F = \frac{1}{8} = 0.125$ (Figure 3.1). Moreover, in the initial decade of captive breeding, some father–daughter matings and other highly consanguineous matings occurred as well, resulting in a herd by 1979 (now split between zoos in St. Louis and Texas) that consisted of 19 individuals with a broad spread of individual F's ranging from 0 to 0.3125 (Figure 3.3).

Recall from Chapter 2 that the system of mating used in the Hardy–Weinberg law is a population concept applied to the level of a deme and to a particular locus. Random mating as a concept is meaningless for specific individuals within a deme. Figure 3.3 illustrates that F refers to *individuals*, not the *deme*. Hence, pedigree inbreeding (the one most people think of when they encounter the word "inbreeding") does not—indeed, cannot—measure the system of mating of a deme. This means that F cannot be used to look for deviations from the Hardy–Weinberg assumption. However, this does not mean that pedigree inbreeding has no population genetic or evolutionary implications.

One of the most important evolutionary implications of pedigree inbreeding (F) is that it displays strong interactions with rare, recessive alleles and epistatic gene complexes. Consider first a model in which a recessive allele is lethal when homozygous. Let B be the sum over all loci of the probability that a gamete drawn from the gene pool bears a recessive lethal allele at a particular locus. Because B is a sum of probabilities of non–mutually exclusive events, B can be greater than 1. Indeed, the simplest biological interpretation of B is that it is the average *number* of lethal alleles over all loci borne by a gamete in the gene pool. When pedigree inbreeding occurs, then BF is the rate of occurrence of both gametes bearing lethal alleles that are identical by descent, thereby resulting in the death of the inbred individual. Of course, an individual can die from many causes, not just due to identity by descent for a lethal allele. The only way for an individual to live is (1) not to be identical by descent for a lethal allele and (2) not to die from something else, either genetic or environmental.

Under the assumption that B is a small number, the number of times an inbred individual will be identical by descent for a lethal allele will follow a distribution known as the Poisson distribution (Appendix 2). The only way for the individual not to die of identity by descent for a lethal gene is to have exactly zero lethal genes that are identical by descent and therefore homozygous. This probability equals e^{-BF} under the Poisson distribution. Let $-A$ be the natural logarithm of the probability of not dying from any cause other than being homozygous for a lethal recessive allele that is identical by descent. Then, the probability of not dying from something else is e^{-A}. To be alive, both events must be true, so the probability of being alive is $e^{-BF}e^{-A} = e^{-A-BF}$. Therefore, we have the expected mathematical relationship

$$\ln \text{(probability of an inbred individual with } F \text{ being alive)} = -A - BF \qquad (3.1)$$

Note that equation 3.1 predicts that viability (the probability of being alive at a given age) should decrease with increasing inbreeding (as measured by F). This is an example of **inbreeding depression**, the reduction of a beneficial trait (such as viability or birth weight) with increasing levels of pedigree inbreeding. Inbreeding depression does not always occur with pedigree inbreeding, nor is it necessarily associated with any of the other definitions of inbreeding. However, inbreeding depression is a common phenomenon in mammals (Ralls et al. 1988), including humans, so we need to examine the application of equation 3.1.

One complication of applying equation 3.1 is that any one individual is either dead or alive, so the realized probability for any one individual is either 0 or 1 regardless of F. However, equation 3.1 predicts a linear relationship between the natural logarithm of the probability of being alive with F, so in the model this probability can take on intermediate values between 0 and 1. Although F is defined for an individual, equation 3.1 cannot be meaningfully applied to an individual. We must therefore extend the concept of pedigree inbreeding up to the level of a deme before we can make use of equation 3.1.

To illustrate how to do this, consider the 1979 population of Speke's gazelle whose individual F's are portrayed in Figure 3.3. As can be seen, several animals have identical levels of pedigree inbreeding: Seven animals share an $F = 0$, five share an $F = 0.125$, and four share an $F = 0.25$. Although any one animal is either dead or alive at a given age, the proportion of animals alive at a given age in a cohort that shares a common level of pedigree inbreeding varies between 0 (everyone in the cohort is dead) and 1 (everyone in the cohort is alive). Hence, the probability of an inbred individual with a specific F being alive at a given age is estimated by the proportion of the cohort that all share that same specific F that are alive at the given age. Complications can arise due to small sample sizes within certain cohorts, but small sample size corrections can be used to deal with these difficulties (Templeton and Read 1998). Equation 3.1 is now implemented by doing a regression (Appendix 2) of the natural logarithm of the cohort viability at a given age against the various F's associated with different cohorts to estimate A and B. For example, for the Speke's gazelle herd up to 1982, a regression of the natural logarithm of survivorship up to 30 days after birth upon F yields $A = 0.23$ and $B = 2.62$, and survivorship up to one year (the approximate age of sexual maturity in this species) yields $A = 0.42$ and $B = 3.75$ (Figure 3.4). This means that the average gamete from this population behaved as if it bore 3.75 alleles that would kill before one year of age any animal homozygous for such an allele.

In general, lethality can arise from several other genetic causes under inbreeding besides homozygosity for a recessive, lethal allele. For example, homozygosity for an allele may lower viability but may not necessarily be absolutely lethal. Nevertheless, homozygosity for such deleterious alleles could reduce the average survivorship for a cohort of animals

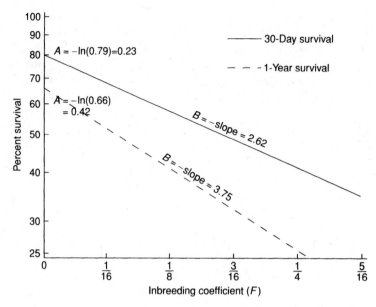

Figure 3.4. Inbreeding depression in captive herd of Speke's gazelle. From Templeton and Read (1984).

sharing a common F. Alternatively, some homozygous combinations of alleles at different loci may interact to reduce viability through epistasis. For example, knockout (complete loss of function) mutations were induced for virtually all of the 6200 genes in the yeast (*Saccharomyces cerevisiae*) genome. Yeast can exist in a haploid phase that genetically mimics the state of $F = 1$ for every locus, so the effects of these knockout mutants could be studied in the equivalent of a homozygous state. Given the compact nature of the yeast genome, it was anticipated that most of these knockouts would have lethal consequences; that is they would behave as recessive, lethal alleles. Surprisingly, more than 80% of these knockout mutations were not lethal and seemed "nonessential" (Tong et al. 2001). However, when yeast strains were constructed that bore pairs of mutants from this nonessential class, extensive lethality emerged from their interactions (Tong et al., 2001, 2004). Similarly, a detailed analysis of the genetic causes of the inbreeding depression found in the captive population of Speke's gazelle revealed that epistasis between loci was a significant contributor to the observed B (Templeton and Read, 1984; Templeton, 2002a). The yeast experiments and the results obtained with the Speke's gazelle make it clear that B should be regarded as the number of "lethal equivalents" rather than the number of actual lethal alleles. The term *lethal equivalents* emphasizes that we really do not know the genetic architecture underlying inbreeding depression from these regression analyses, but lethal equivalents do allow us to measure the severity of inbreeding depression in a variety of populations using the standard reference model of equation 3.1.

Because each diploid animal results from the union of two gametes and, by definition, the only animals that survive are those not homozygous for any lethal equivalent, a living animal is expected to bear about $2B$ lethal equivalents in heterozygous condition. In the original non-inbred population of Speke's gazelles, the average number of lethal equivalents for one-year survivorship borne by the founding animals of this herd is therefore 7.5 lethal equivalents per animal. Studies on inbreeding in humans from the United States and Europe yield values of $2B$ between 5 and 8 (Stine 1977). These numbers mean that most humans, just like most Speke's gazelles, are bearers of multiple potentially lethal genetic diseases or gene combinations. Consequently, there is a large potential for inbreeding depression and other deleterious genetic effects in most human populations when pedigree inbreeding does occur. For example, cousin matings represent only 0.05% of matings in the United States (Neel et al. 1949), but 18–24% of albinos and 27–53% of Tay-Sachs cases (a lethal genetic disease) in the United States come from cousin matings (both of these are autosomal recessive traits, with the recessive allele being rare). This same pattern is true for many other recessive genetic diseases. Hence, even small amounts of pedigree inbreeding in a population that is either randomly mating or even avoiding system-of-mating inbreeding (to be discussed next) can increase the incidence of some types of genetic disease by orders of magnitude in the pedigree-inbred subset of the population.

Because of inbreeding depression and the tendency for increased incidence of genetic disease in consanguineous matings in humans, "inbreeding"—regardless of the exact definition being used—is often viewed as something deleterious for a population. The idea that inbreeding is deleterious has raised many concerns for endangered species, as such species often are reduced to small sizes, which as we have seen leads to pedigree inbreeding. Studies on pedigree inbreeding depression, such as those performed for Speke's gazelle, demonstrate that these concerns are real, and much of applied conservation genetics focuses on dealing with inbreeding in its various senses and consequences. However, is inbreeding always deleterious? The answer appears to be no. For example, many higher plants have extensive self-mating, the most extreme form of inbreeding, and this inbreeding can be adaptive under many conditions (Holsinger 1991). To understand the ultimate cause for why inbreeding is not always deleterious, we must turn our attention from inbreeding at the level of an individual to inbreeding at the level of a deme's system of mating.

Inbreeding as Deviation from Random-Mating Expectations. To obtain a system of mating measure of inbreeding at the deme level, we must examine deviations from Hardy–Weinberg genotype frequencies that are due to nonrandom mating. First, recall the random-mating model for the simple one-locus, two-allele (A and a) model shown in Table 2.2. Note that in Table 2.2 the genotype frequencies are obtained by multiplying the allele frequencies associated with the male and female gametes. Now, suppose that gametes are put together in such a way that there is a deviation from the product rule of Hardy–Weinberg in producing genotype frequencies but that the marginal allele frequencies remain the same. Let λ be this deviation parameter from the simple product of the gamete frequencies, as shown in Table 3.1. Note from Table 3.1 that λ only affects the genotype frequencies and not the gamete frequencies. This is because λ is designed to measure how gametes come together to form genotypes for a given set of gamete frequencies. Also, note that λ is applied to a deme and measures deviations from Hardy–Weinberg genotype frequencies in that deme. In contrast, F is defined for individuals, not demes, and measures the probability of identity by descent for that individual and not the system of mating of the deme as a whole. Biologically, λ is quite different from F.

Mathematically, λ is also quite different from F. Recall that F is a probability and like all probabilities is defined only between 0 and 1 inclusively. In contrast, as can be seen in Box 3.1, λ is the covariance (see Appendix 2) between uniting gametes. A covariance is proportional to the correlation coefficient (Appendix 2) and can take on both positive and negative values and is mathematically noncomparable to a probability such as F. If $\lambda > 0$, there is a positive correlation between uniting gametes in excess of random-mating expectations. This means that the alleles borne by the uniting gametes are more likely to share the same allelic state than expected under random mating. If $\lambda < 0$, there is a negative correlation between uniting gametes, and the alleles borne by the uniting gametes are less

Table 3.1. Multiplication of Allele Frequencies Coupled with Deviation from Resulting Products as Measured by λ to Yield Zygotic Genotypic Frequencies under System of Mating That Allows Deviation from Random Mating.

			Male gametes		
		Allele	A	a	
		Frequency	p	q	
	Allele Frequency				Marginal allele frequencies in deme
Female gametes	A	p	AA $p^2 + \lambda$	Aa $pq - \lambda$	$(p^2 + \lambda) + (pq - \lambda)$ $= p^2 + pq = p(p + q)$ $= p$
	a	q	aA $qp - \lambda$	aa $q^2 + \lambda$	$(qp - \lambda) + (q^2 - \lambda)$ $= qp + q^2 = q(p + q)$ $= q$
Marginal allele frequencies in deme			$(p^2 + \lambda) + (qp - \lambda)$ $= p^2 + qp = p(p + q)$ $= p$	$(pq - \lambda) + (q^2 + \lambda)$ $= pq + q^2 = q(p + q)$ $= q$	

Summed frequencies in zygotes:

AA: $G'_{AA} = p^2 + \lambda$
Aa: $G'_{Aa} = pq - \lambda + qp - \lambda = 2pq - 2\lambda$
aa: $G'_{aa} = q^2 + \lambda$

Note: The zygotic genotype frequencies are indicated by G'_k.

likely to share the same allelic state than expected under random mating. Random mating occurs when there is no correlation between uniting gametes ($\lambda = 0$). The actual correlation between uniting gametes is $\lambda/(pq)$ (see Box 3.1). The correlation coefficient (Appendix 2) has a standardized range of -1 to $+1$ inclusively, in contrast to the covariance that has no standardized range. Hence, it is more convenient to measure deviations from Hardy–Weinberg at the deme level in terms of the correlation of uniting gametes as opposed to the covariance of uniting gametes. Accordingly, we define the inbreeding coefficient to be $f \equiv \lambda/(pq)$, defined as the correlation of uniting gametes within the deme. From Table 3.1, we can now see that the genotype frequencies that emerge from this system of mating can be expressed as

$$G'_{AA} = p^2 + \lambda = p^2 + pq\left(\frac{\lambda}{pq}\right) = p^2 + pqf$$
$$G'_{Aa} = 2pq - 2\lambda = 2pq - 2pq\left(\frac{\lambda}{pq}\right) = 2pq - 2pqf = 2pq(1 - f) \quad (3.2)$$
$$G'_{aa} = q^2 + \lambda = q^2 + pq\left(\frac{\lambda}{pq}\right) = q^2 + pqf$$

Because f is a correlation coefficient, it can take on both positive and negative values (as well as zero, the random-mating case). Generally, when f is positive, the system of mating of the deme is described as one of inbreeding, and when f is negative, the system of mating of the deme is described as one of avoidance of inbreeding. However, regardless of whether or not f is positive or negative, f is called the inbreeding coefficient.

BOX 3.1 THE CORRELATION OF UNITING GAMETES

In order to show that λ is the covariance among uniting gametes, we must first define a random variable to assign to the gametes. In our simple genetic model, the gametes bear only one of two possible alleles, A and a. Let x be a random variable that indicates the allele borne by a male gamete such that $x = 1$ if the male gamete bears an A allele and $x = 0$ if the male gamete bears an a allele. Similarly, let y be a random variable that indicates the allele borne by a female gamete such that $y = 1$ if the female gamete bears an A allele and $y = 0$ if the female gamete bears an a allele. Let p be the frequency of A-bearing gametes in the gene pool. Because we are dealing with an autosomal locus, p is the frequency of A for both male and female gametes.

Using these definitions and the standard formula for means, variances, and covariances (Appendix 2), we have

$$\text{Mean}(x) = \mu_x = 1 \times p + 0 \times q = p$$
$$\text{Mean}(y) = \mu_y = 1 \times p + 0 \times q = p$$
$$\text{Variance}(x) = \sigma_x^2 = (1 - \mu_x)^2 \times p + (0 - \mu_x)^2 \times q$$
$$= (1 - p)^2 p + (-p)^2 q = pq$$
$$\text{Variance}(y) = \sigma_y^2 = pq$$
$$\text{Covariance}(x, y) = (1 - \mu_x)(1 - \mu_y)(p^2 + \lambda) + (1 - \mu_x)(0 - \mu_y)(2pq - 2\lambda)$$
$$+ (0 - \mu_x)(0 - \mu_y)(q^2 + \lambda)$$
$$= q^2(p^2 + \lambda) - pq(2pq - 2\lambda) + p^2(q^2 + \lambda)$$
$$= \lambda(q^2 + 2pq + p^2)$$
$$= \lambda$$

Hence, λ is the covariance between uniting gametes under a system of mating that produces the genotype frequencies given in Table 3.1. Because covariances do not have a standardized range whereas correlations do, it is usually more convenient to measure the nonrandom associations between uniting gametes through their correlation coefficient rather than their covariance. The correlation coefficient is (Appendix 2)

$$\rho_{x,y} = \frac{Cov(x, y)}{\sqrt{\sigma_x^2 \sigma_y^2}} = \frac{\lambda}{pq}$$

Although inbreeding as measured by f alters the genotype frequencies from Hardy–Weinberg (equations 3.2), it does not cause any change in allele frequency. The frequency of the A allele in the final generation in Table 3.1 is

$$p' = 1 \times (p^2 + pqf) + {}^1/_2[2pq(1 - f)] = p^2 + pqf + pq(1 - f)$$
$$= p^2 + pqf + pq - pqf = p^2 + pq = p(p + q) = p$$

Because the allele frequencies are not changing over time in Table 3.1, inbreeding as measured by f is not an evolutionary force by itself at the single-locus level (that is, system-of-mating inbreeding alone does not change the frequencies of alleles in the gene pool).

Another interpretation of f is suggested by equations 3.2: In addition to f being a correlation coefficient, f is also a direct measure of the deviation of heterozygote genotype frequencies from Hardy–Weinberg expectations. Note that the frequency of heterozygotes in equations 3.2 is $2pq(1-f)$, and recall from Chapter 2 that the expected frequency of heterozygotes under Hardy–Weinberg is $2pq$. Hence, an alternative mathematical definition of f is

$$f = 1 - \frac{\text{observed frequency of heterozygotes in deme}}{\text{expected frequency of heterozygotes under Hardy–Weinberg}} \quad (3.3)$$

From equation 3.3, we can see that a positive correlation between uniting gametes leads to a heterozygote deficiency in the deme (typically called an inbreeding system of mating), no correlation yields Hardy–Weinberg frequencies (random mating), and a negative correlation (typically called avoidance of inbreeding) yields an excess of heterozygotes in the deme.

In most of the population genetic literature, both f and F are called inbreeding coefficients and are often assigned the same mathematical symbol. That will not be the case in this book. The symbol F, which will be called pedigree inbreeding, refers to a specific individual, measures that individual's probability of identity by descent for a randomly chosen autosomal locus, and ranges from 0 to 1. In contrast, f will be called system-of-mating inbreeding, refers to a deme, measures deviations from Hardy–Weinberg genotype frequencies, and ranges from -1 to $+1$ (Table 3.2). Because f and F are both called inbreeding coefficients and frequently assigned the same symbol in much of the literature, it is not surprising that these two extremely different definitions of inbreeding have often been confused. We will illustrate the difference between these two inbreeding coefficients by returning to the example of the captive herd of Speke's gazelle.

Recall that the captive herd of the Speke's gazelle was founded at the St. Louis Zoo with one male and three females between 1969 and 1972. Because there was only one male, all animals born in this herd were biological relatives. Under the assumption that the four founding animals were unrelated (our reference population), all of these original founders and the offspring between them have $F = 0$; that is, these individuals were not "inbred." By 1982, these older animals had all died off and all animals in the herd had $F > 0$. Given

Table 3.2. Contrast between Pedigree Inbreeding Coefficient F and System-of-Mating Inbreeding Coefficient f

Property	F	f
Data used to calculate	Pedigree data for specific individuals	Genotype frequency data for specific locus and deme
Type of mathematical measure	Probability	Correlation coefficient
Range of values	$0 \leq F \leq 1$	$-1 \leq f \leq 1$
Biological level of applicability	Individual	Deme
Biological meaning	Expected chance of identity by descent at randomly chosen autosomal locus for specific individual caused by biological relatedness of individual's parents	System of mating of deme measured as deviations from random-mating genotype frequency expectations

that all animals bred in captivity had to be at least half sibs of one another (there was only one founding male), this inbred state of the descendants of the original founders and their offspring was inevitable *regardless* of system of mating. The average F in 1982 was 0.149 relative to the founder reference population, making this captive herd one of the most highly inbred populations of large mammals known. An isozyme survey (Appendix 1) was also performed on these same animals in 1982. For example, at the polymorphic *general protein* locus (GP), the observed heterozygosity was 0.500, but the expected heterozygosity under random mating was 0.375. Hence, for this locus, $f = -0.333$. Several other polymorphic isozyme loci were scored, all yielding $f < 0$, with the average f over all loci being -0.291. This highly negative f indicates a strong avoidance of system-of-mating inbreeding.

We now have what appears to be a contradiction, at least for those who confuse f and F. This herd of gazelles is simultaneously one of the most highly inbred (pedigree sense F) populations of large mammals known and is also strongly avoiding inbreeding (system-of-mating sense f). There is no paradox here except verbally; the two types of inbreeding and inbreeding coefficients are measuring completely different biological attributes. The negative f indicates that the breeders of this managed herd were avoiding inbreeding in a system-of-mating sense within the severe constraints of this herd of close biological relatives.

If inbreeding were being avoided at the level of system of mating, then why did every individual in the herd have an $F > 0$? Keep in mind that "random mating" means that females and males are paired together "at random" regardless of their biological relationship. In any finite population, there is always a finite probability of two related individuals being paired as mates under random mating. The smaller the population, the more likely it is to have biological relatives mate at random. Hence, random mating ($f = 0$) implies some matings among biological relatives that will yield $F > 0$ in any finite population. Indeed, even avoidance of inbreeding ($f < 0$) can still result in matings among biological relatives in a finite population. For example, many human cultures (but not all) have incest taboos that often extend up to first cousins. Assuming a stable sized population of N adults with an average and variance of two offspring per family (the number of offspring being Poisson), then $f = -1/(N - 10)$ when relatives up to and including first cousins are excluded as mates but mating is otherwise random (Jacquard 1974). Note that as N increases, f approaches 0. This means that although incest taboos are common in human societies, the Hardy–Weinberg law fits very well for most loci within most large human demes. However, some human demes are small. Suppose $N = 50$ (a small local human population, but still found in some hunter/gathering societies); then $f = -0.025$ under this nonrandom system of mating. Nevertheless, such small human populations typically contain many inbred ($F > 0$) individuals despite their incest taboos in choosing mates ($f < 0$).

Consider, for example, a set of religious colonies in the upper great plains of North America that are descendants of a small group of anabaptist Protestants who originally immigrated from the Tyrolean Alps (Steinberg et al. 1966). There has been very little immigration into these religious colonies from other human populations, so they represent a genetic isolate. Internally, their system of mating is one of strong avoidance of mating between close relatives as incest is considered a sin. Despite this strong avoidance of pedigree inbreeding, the average F for one isolated subsect was 0.0255. This makes this population one of the more highly inbred human populations known despite a system of mating that avoids inbreeding. The reason for this seeming contradiction is that these colonies were founded by relatively few individuals, so virtually everyone in the colony today is related to everyone else. Hence, the pedigree inbreeding is due to the small population size at the time the colonies were founded and not due to the system of mating (such "founder effects" will be discussed in more detail in the next chapter). Indeed, if these individuals truly mated at random, then the average F under random mating would be 0.0311, a value considerably larger than the observed average F of 0.0255. As this population reveals, avoidance of inbreeding in the system-of-mating sense does *not* necessarily result in no pedigree inbreeding ($F = 0$) but rather results in lower levels of pedigree inbreeding than would have occurred under random mating. The strong avoidance of inbreeding in this human population also results in large deviations from Hardy–Weinberg expectations. For example, a sample from this population scored for the MN blood group had 1083 individuals with genotype *MM*, 1220 with *MN*, and 260 with *NN*. Using the test given in Chapter 2, the resulting chi square is 9.68 with one degree of freedom, which is significant at the 0.002 level. Hence, unlike most other human populations (see Chapter 2 for two examples), this religious colony does not have Hardy–Weinberg genotype frequencies for the *MN* locus. Instead, there is a significant excess of heterozygotes (only 1149 are expected under random mating, versus the 1220 that were observed). Using equation 3.2, this results in $f = -0.0615$. Thus, this religious colony started from a small number of founders is highly inbred in the pedigree sense ($F = 0.0255$), even though the population is strongly avoiding inbreeding in the system-of-mating sense ($f = -0.0615$). The two inbreeding coefficients F and f are most definitely not the same either mathematically or biologically in this human population.

Another human example is given in Figure 3.5 (Roberts 1967) that illustrates how small founding population size can result in pedigree inbreeding despite strong avoidance of system-of-mating inbreeding. Twenty people colonized the remote Atlantic island of Tristan da Cunha in the early 1800s, with a few more migrants coming later (more details will be given in the next chapter). Despite a strong incest taboo among these Christian colonists and a system of mating characterized by $f < 0$, individuals with pedigree inbreeding began to

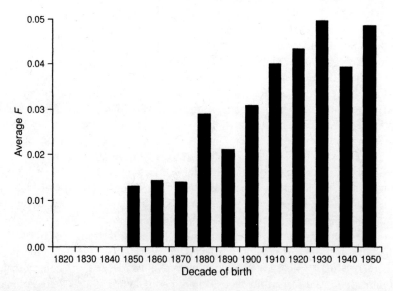

Figure 3.5. Average pedigree inbreeding coefficient for human population on Tristan da Cunha as function of decade of birth.

Table 3.3. **First Eight Marriages between Biological Relatives on Tristan da Cunha Showing Date of Marriage, Number of Available Women of Marriageable Age,[a] and Number of Available Women Not Related to Groom**

Marriage between Relatives	Date of marriage	Number of available women	Number of non relatives
1	1854	7	3
2	1856	9	2
3	1871	1	0
4	1876	1	0
5	1884	7	1
6	1888	8	0
7	1893	3	0
8	1898	1	0

[a] Sixteen years and over, single, and not a sister of the groom.

be born by the 1850s (under the assumption that all colonists and migrants were unrelated), with more and more extreme pedigree inbreeding occurring as time passed by (Figure 3.5). How could this human population, so strongly avoiding system-of-mating inbreeding under their taboo against incest, become a population with one of the highest levels of average pedigree inbreeding known in humanity? Table 3.3 gives the answer. The first marriage between biological relatives that resulted in an inbred offspring occurred in 1854. At the time of this marriage, there were only seven women of marriageable age (16 years or older) who were available (single and not a sister of the groom), and only three of them were not related to the man involved in this union. The number of women of marriageable age that were not relatives of the grooms in subsequent years quickly went down to zero (Table 3.3). Hence, for those wishing to remain on the island and marry, there was no choice but to marry a relative, although more distant relatives were chosen compared to random-mating expectations (hence, $f < 0$).

The potential discrepancy between f and F gets more extreme as the population size gets smaller. Because the founder population of the Speke's gazelle herd was just four individuals, the effect of finite size on f versus F was much larger than on the human populations living in the North American religious colonies or on Tristan da Cunha. Indeed, despite extreme avoidance of system-of-mating inbreeding, pairing the *least* related gazelles still meant that most matings were between half sibs ($F = 0.125$), the most distantly related animals that existed in the population once the original founders and their offspring had died. Thus, even strong avoidance of inbreeding in terms of system of mating can result in many inbred individuals in the pedigree sense. The breeding program for the Speke's gazelle has sometimes mistakenly been called a program of "deliberate" inbreeding. The system of mating was under the control of the breeders, so the only deliberate choice, as shown by f, is a strong avoidance of inbreeding. The accumulation of high levels of f in the individuals that constitute this population is not deliberate but rather is the inevitable consequence of the small founding population size of this herd. Because F is a probability, F has to be greater than or equal to zero. It is therefore mathematically impossible to measure avoidance of inbreeding with F, so the fact that $F > 0$ for every animal in the herd tells one nothing about the system of mating. Indeed, the average F of a population is often due more to its finite size than to its system of mating, as illustrated both by the Speke's gazelle herd and the Tristan da Cunha human population. Consequently, the average value of F for a deme, $\bar{F}$, is used as a measure not of system of mating but of another evolutionary force called genetic drift. We will discuss this biological application of $\bar{F}$ in more detail in Chapter 4. (Note, $\bar{F}$ in most of the population genetic literature is also simply called the inbreeding coefficient and is usually given the same symbol as F and f—readers beware!)

Biologically and mathematically F and f are quite distinct (Table 3.2), but neither is an evolutionary force by itself. We already saw that $f \neq 0$ does not cause any change in allele frequency (Table 3.1), and F cannot be an evolutionary force because it is defined for a specific individual and not a deme or population. Recall from Chapter 1 that only populations evolve—not individuals—so F cannot by definition be used to describe evolutionary change ($\bar{F}$, being a population average, is a population-level parameter and therefore can be used to describe evolutionary change, as we will see in Chapter 4). Nevertheless, both F and f have important population consequences that can affect the course of evolution when coupled with other evolutionary agents. We already discussed one important evolutionary consequence associated with F in the previous section, inbreeding depression in a population. We now examine some of the population and evolutionary consequences of system-of-mating inbreeding, f.

Even small deviations from Hardy–Weinberg as measured by f can have major impacts on the genotype frequencies found in local populations. For example, consider the impact of system-of-mating inbreeding ($f > 0$) on the incidence of a rare, recessive autosomal trait, that is, a trait expressed only in individuals who are homozygous for a particular allele. This genetic category is of considerable interest because many genetic diseases in humans and other species are associated with a recessive allele that is rare in the deme. We can immediately see the impact of f on the frequency of such a recessive trait by looking at Table 3.1, now regarding a as the recessive allele. The frequency of individuals with the recessive phenotype is, from Table 3.1 or from equations 3.2, $q^2 + pqf$. Suppose first that $q = 0.001$ (most genetic disease alleles are rare) and $f = 0$ (random mating). Then, the frequency of affected individuals is $(0.001)^2 = 0.000001$, or 1 in a million. Now let $f = 0.01$, a seemingly minor deviation from random mating. Then $q^2 + pqf = 0.000001 + (0.999)(0.001)(0.01) = 0.000011$. Thus, a 1% inbreeding level causes an 11-fold increase in the incidence of the recessive trait in the population. Even small deviations from random mating cause large changes in genotype frequencies when rare alleles are involved. As we shall see in Chapter 12, this change in genotype frequencies means that rare, recessive, deleterious alleles are subject to stronger natural selection in an inbreeding population ($f > 0$) than in a random-mating population ($f = 0$). When an inbreeding system of mating persists for many generations, this greater exposure to selection means that recessive, deleterious alleles can be reduced to lower frequencies by natural selection than they would have been in a random-mating population. Similar considerations lead to the prediction of reduced numbers of lethal equivalents when system-of-mating inbreeding persists for many generations.

For example, certain human populations, such as the Tamils in India, have favored first-cousin marriages for centuries, thereby resulting in $f > 0$ for the populations and the production of most individuals with $F > 0$ within the populations. Despite the high levels of pedigree inbreeding among the individuals from these populations, their B values are not significantly different from zero (Rao and Inbaraj 1980). Hence, unlike the general population in the United States with B values between 2.5 and 4, human populations that have practiced inbreeding as a system of mating for centuries show no detectable pedigree inbreeding depression! As will be shown in detail in Chapter 12, there is a strong evolutionary interaction between f and F that is mediated by natural selection. For now, it is important to keep in mind that inbreeding in either of the senses used is deleterious in some contexts but not in others.

ASSORTATIVE MATING

System-of-mating inbreeding (f) represents a deviation from random mating in which the biological relatedness among individuals affects their probability of becoming mates. Individuals can also have their probability of mating influenced by the traits or phenotypes displayed by potential mates. One such deviation from random mating based on individual phenotype is called **assortative mating**. Under assortative mating, individuals with similar phenotypes are more likely to mate than expected under random pairing in the population. This results in a positive correlation between the trait values for mating pairs in the population. Although assortative mating can arise from an individual's preference to mate with another individual with a similar phenotype, assortative mating can arise from many factors other than mate choice. Consider treehoppers (a type of insect) in the genus *Enchenopa* (Wood and Keese 1990). These treehoppers can feed on a number of different host plants, and the treehoppers that fed and developed upon a particular type of host plant preferentially mate with other treehoppers that fed and developed upon the same kind of host plant. Does this mean that treehoppers *prefer* as mates other individuals that fed on the same type of host plant? Not at all. Rather, it was found that the various types of host plants affect the rate at which individuals become sexually mature, so that males and females that came from the same type of host plant tended to become sexually mature at the same time. Because female treehoppers are receptive for only a short period of time after achieving maturity and because males die rapidly after maturity, most of the individuals available as potential mates at any given time are those that feed on the same type of host plant. As a consequence, the treehoppers may have no individual preference at all for other treehoppers that used the same host plant, yet there would still be assortative mating for the trait of host plant use because of the influence of the host plant upon the timing of sexual maturity. To see if the assortative mating was due to the impact of the host plant on developmental time rather than individual mating preferences, individuals from different host plants were experimentally manipulated to become mature at the same time. No assortative mating by host plant occurred under those conditions. Hence, the assortative mating in this case arose from the effects of host plants on developmental times and did not involve any mate choice or preference on the part of individual treehoppers.

The treehopper example illustrates the fact that assortative mating is simply a positive phenotypic correlation among mating individuals at the level of the deme: It may arise from individual-level mating preferences, but such correlations can also arise from many other factors that have nothing to do with individual-level mating preferences. Regardless of its cause, we need to consider the evolutionary consequences of assortative mating for a phenotype. To do so, we need to develop a deme-level model of assortative mating.

Simple Model of Assortative Mating

To investigate the evolutionary consequences of assortative mating for a genetic system, we first consider a simple model of an autosomal locus with two alleles (A and a) such that there is a 1 : 1 genotype–phenotype relationship with each genotype having a distinct phenotype. If

64 SYSTEMS OF MATING

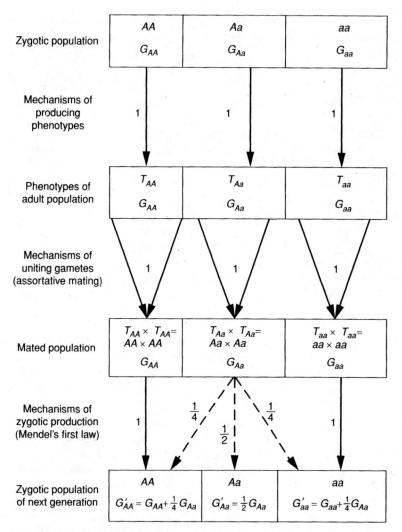

Figure 3.6. Model of 100% assortative mating for phenotype determined by single autosomal locus with two alleles and with all three possible genotypes having distinct phenotypes. The G's refer to genotype frequencies and T's to phenotype frequencies.

the mating is 100% assortative for these phenotypes (that is, individuals only mate with other individuals with identical phenotypes), we obtain the generation-to-generation transition shown in Figure 3.6.

Note that the genotype frequencies in the initial population in Figure 3.6 are not necessarily in Hardy–Weinberg equilibrium. However, regardless of whether or not we start in Hardy–Weinberg, the genotype frequencies will change each generation as long as there are some heterozygotes. In particular, the frequency of heterozygotes is halved each generation. Thus, this deviation from random mating causes a major qualitative difference with the original Hardy–Weinberg model given in Chapter 2: Genotype frequencies do not immediately go to equilibrium. Rather, because the heterozygote genotype frequency is halved every generation, the deme gradually approaches a genotypic equilibrium in which the population is entirely homozygous. At this equilibrium, the initial heterozygote frequency is evenly split between the two homozygous genotypes (from Mendel's first law), yielding an equilibrium array of genotype frequencies of

$$G_{AA}(\text{equilibrium}) = G_{AA} + \frac{1}{2} G_{Aa}$$
$$G_{Aa}(\text{equilibrium}) = 0 \qquad (3.4)$$
$$G_{aa}(\text{equilibrium}) = G_{aa} + \frac{1}{2} G_{Aa}$$

The inbreeding coefficient f measures deviations from Hardy–Weinberg genotype frequencies, so f can also measure the impact of assortative mating on the population. Suppose the initial zygotic genotype frequencies shown in Figure 3.6 were in Hardy–Weinberg equilibrium. Then $G_{Aa} = 2pq$ and $f = 0$. After one generation of assortative mating, we have that $G'_{Aa} = \frac{1}{2}(2pq)$ from Figure 3.6. Hence, $1 - f = \frac{1}{2}(2pq)/(2pq) = \frac{1}{2}$, so $f = \frac{1}{2}$. As this population continues to mate assortatively, the heterozygote deficiency relative to Hardy–Weinberg becomes more extreme until, at equilibrium, there are no heterozygotes. At this equilibrium $f = 1$. Note that if one were only examining this locus at a single generation, it would be indistinguishable from inbreeding because of an inbreeding-like deficiency of heterozygotes relative to Hardy–Weinberg expectations.

This example shows that calling f an inbreeding coefficient can sometimes be misleading. A nonzero value of f can arise in situations where there is no inbreeding (system-of-mating sense) at all. Jacquard (1975) recommended that f be called the "coefficient of deviation from random mating" because f measures deviations from random-mating genotype frequencies, regardless of the biological cause of that deviation. This accurate, but somewhat cumbersome, name for f has not become generally adopted in the population genetic literature, so once again readers and students of this literature must always keep in mind that the inbreeding coefficient f may have nothing to do with system-of-mating inbreeding in particular cases.

As shown above, assortative mating causes deviations from Hardy–Weinberg genotype frequencies. Does assortative mating also cause evolution; that is, does assortative mating alter gamete frequencies over time? Recall, the general definition of the frequency of the A allele is $p = 1 \times G_{AA} + \frac{1}{2} \times G_{Aa}$. After one generation of assortative mating, we see from Figure 3.6 that the frequency of A will now be $1 \times (1 \times G_{AA} + \frac{1}{4} \times G_{Aa}) + \frac{1}{2} \times (\frac{1}{2} \times G_{Aa}) = 1 \times G_{AA} + \frac{1}{2} \times G_{Aa} = p$. From equations 3.4, we see at equilibrium that the frequency of A will be $1 \times (1 \times G_{AA} + \frac{1}{2} \times G_{Aa}) + \frac{1}{2} \times (0) = 1 \times G_{AA} + \frac{1}{2} \times G_{Aa} = p$. Hence, just like inbreeding, assortative mating by itself does not alter allele frequencies and does not cause evolutionary change *at the single-locus level*. Once again, assortative mating mimics the effects of system-of-mating inbreeding.

Assortative mating also mimics the effects of pedigree inbreeding with respect to increasing the phenotypic frequencies of recessive traits. As an example, consider the phenotype of profound, early-onset deafness in humans. Such deafness can be caused by disease, accidents, and genes. Genetic causes explain about 50% of these cases, with homozygosity for recessive alleles at any one of over 35 different loci leading to deafness (Cavalli-Sforza and Bodmer 1971; Marazita et al. 1993). However, the allele frequencies associated with deafness at most of these loci are extremely rare. The one exception is the *GJB2* locus that encodes the gap-junction protein connexin 26. This locus has an allele with $q \approx 0.01$ in U.S. and European populations (Green et al. 1999; Storm et al. 1999). Homozygosity for this allele accounts for about half of the genetic cases of profound early-onset deafness.

Under random mating, we expect $q^2 = 0.0001$, so about 1 in 10,000 births should yield a deaf child due to homozygosity for this allele. The actual incidence of deafness due to this recessive allele is 3–5 in 10,000 births. The reason is that there is strong assortative mating for deafness, with the phenotypic assortment rate being over 80% in the United States, 92% in England, and 94% in Northern Ireland (Aoki and Feldman 1994). Early-onset deaf children are often sent to special schools or classes and hence socialize mostly with one another. Also, they communicate best with one another. As a result, there is a strong tendency to marry, which in turn results in increased homozygosity for alleles yielding deafness beyond random expectations.

The impact of assortative mating for deafness upon the frequency of homozygotes at the *GJB2* locus is reduced by the fact deafness can be caused by many other loci and nongenetic factors as well. In general, the impact of assortative mating in yielding an f is proportional to both the phenotypic correlation between mates (which is high for deaf people) *and* the correlation between genotype and phenotype. This latter correlation is reduced when the phenotype has many distinct genetic and nongenetic causes, as is the case for deafness. As a consequence, many of the deaf people who marry are deaf for different reasons and not necessarily because they are homozygous for the same alleles associated with deafness. Hence, identical phenotypes of a married deaf couple does not imply identical genotypes, which reduces the impact of assortative mating upon the frequency of homozygotes at any specific locus that contributes to deafness. For the *GJB2* locus in particular, only about one-fourth of the people with early-onset deafness are homozygous for the recessive allele at this locus. Therefore, of the couples in which both individuals are deaf, we would expect only $(\frac{1}{4})^2 = \frac{1}{16}$ to have the same homozygous genotypes at the *GJB2* locus. The actual figure is closer to 1 in 6 (Koehn et al. 1990). The reason for this discrepancy is that there is another attribute of assortative mating that leads to increased homozygosity for the deafness allele at the *GJB2* locus even when one or both parents are deaf for a reason other than homozygosity for the deafness allele at *GJB2*. How can this be? To answer this question, we must look at the impact of assortative mating upon a multilocus genetic architecture.

Creation of Linkage Disequilibrium by Assortative Mating

Within a single deme, assortative mating can only have an effect on genotype frequencies when a genotype–phenotype correlation exists. However, as we saw above with the single-locus model, assortative mating does not alter allele frequencies and is therefore not an evolutionary force at the single-locus level. This conclusion is drastically altered as soon as we go to models of assortative mating for phenotypes influenced by two or more loci.

Table 3.4 presents a 100% assortative mating model for a two-locus genetic architecture. In this model, two autosomal loci exist, each with two alleles (*A* and *a* at one locus and *B* and *b* at the second), with r being the recombination frequency between them. We further assume that each capital-letter allele contributes $+1$ to the phenotype and each small-letter allele contributes 0. To obtain the total phenotype of an individual genotype, we add up the phenotypic contributions of each allele over this pair of loci. Hence, the phenotype of the genotype *AB/AB* is $+4$, and so on, as shown in Table 3.4. Mating is 100% assortative in this model in the sense that individuals mate only with other individuals that have identical phenotypes. There are 10 two-locus genotypes (Table 3.4), which can be paired together in 55 distinct ways under random mating. However, only 18 of these 55 mating types are allowed under 100% assortative mating.

Table 3.4. One Hundred Percent Assortative Mating for Two-Loci, Two-Allele Genetic Architecture with Additive Phenotypic Effects

Mating Type	Mate Phenotype	$\frac{AB}{AB}$	$\frac{AB}{Ab}$	$\frac{AB}{aB}$	$\frac{AB}{ab}$	$\frac{Ab}{aB}$	$\frac{Ab}{Ab}$	$\frac{aB}{aB}$	$\frac{aB}{ab}$	$\frac{ab}{ab}$
$\frac{AB}{AB} \times \frac{AB}{AB}$	4	1	—	—	—	—	—	—	—	—
$\frac{AB}{AB} \times \frac{AB}{Ab}$	3	$\tfrac{1}{4}$	$\tfrac{1}{2}$	—	—	—	$\tfrac{1}{4}$	—	—	—
$\frac{AB}{Ab} \times \frac{AB}{aB}$	3	$\tfrac{1}{4}$	$\tfrac{1}{4}$	$\tfrac{1}{4}$	—	$\tfrac{1}{4}$	—	—	—	—
$\frac{AB}{aB} \times \frac{AB}{aB}$	3	$\tfrac{1}{4}$	—	$\tfrac{1}{2}$	—	—	—	$\tfrac{1}{4}$	—	—
$\frac{AB}{ab} \times \frac{AB}{ab}$	2	$\tfrac{1}{4}(1-r)^2$	$\tfrac{1}{2}(1-r)r$	$\tfrac{1}{2}(1-r)r$	$\tfrac{1}{2}(1-r)^2$	$\tfrac{1}{2}r^2$	$\tfrac{1}{4}r^2$	$\tfrac{1}{4}r^2$	$\tfrac{1}{2}(1-r)r$	$\tfrac{1}{4}(1-r)^2$
$\frac{AB}{ab} \times \frac{Ab}{aB}$	2	$\tfrac{1}{4}(1-r)r$	$\tfrac{1}{4}(1-r)^2 + \tfrac{1}{4}r^2$	$\tfrac{1}{4}(1-r)^2 + \tfrac{1}{4}r^2$	$\tfrac{1}{2}(1-r)r$	$\tfrac{1}{2}(1-r)r$	$\tfrac{1}{4}(1-r)r$	$\tfrac{1}{4}(1-r)r$	$\tfrac{1}{4}(1-r)^2 + \tfrac{1}{4}r^2$	$\tfrac{1}{4}(1-r)r$
$\frac{AB}{ab} \times \frac{Ab}{Ab}$	2	—	$\tfrac{1}{2}(1-r)$	—	—	$\tfrac{1}{2}r$	$\tfrac{1}{2}(1-r)$	—	$\tfrac{1}{2}r$	—
$\frac{AB}{ab} \times \frac{aB}{aB}$	2	—	—	$\tfrac{1}{2}(1-r)$	$\tfrac{1}{2}r$	$\tfrac{1}{2}r$	—	$\tfrac{1}{2}(1-r)$	—	—
$\frac{Ab}{aB} \times \frac{Ab}{aB}$	2	$\tfrac{1}{4}r^2$	$\tfrac{1}{2}(1-r)r$	$\tfrac{1}{2}(1-r)r$	$\tfrac{1}{2}r^2$	$\tfrac{1}{2}(1-r)^2$	$\tfrac{1}{4}(1-r)^2$	$\tfrac{1}{4}(1-r)^2$	$\tfrac{1}{2}(1-r)r$	$\tfrac{1}{4}r^2$
$\frac{Ab}{aB} \times \frac{Ab}{Ab}$	2	—	—	—	—	$\tfrac{1}{2}(1-r)$	$\tfrac{1}{2}(1-r)$	—	$\tfrac{1}{2}(1-r)$	—
$\frac{Ab}{aB} \times \frac{aB}{aB}$	2	—	$\tfrac{1}{2}r$	—	—	$\tfrac{1}{2}r$	—	$\tfrac{1}{2}(1-r)$	$\tfrac{1}{2}(1-r)$	—
$\frac{Ab}{Ab} \times \frac{Ab}{Ab}$	2	—	—	—	—	—	1	—	—	—
$\frac{Ab}{Ab} \times \frac{aB}{aB}$	2	—	—	—	—	1	—	—	—	—
$\frac{aB}{aB} \times \frac{aB}{aB}$	2	—	—	—	—	—	—	1	—	—
$\frac{Ab}{ab} \times \frac{Ab}{ab}$	1	—	—	—	—	—	$\tfrac{1}{4}$	—	—	$\tfrac{1}{4}$
$\frac{Ab}{ab} \times \frac{aB}{ab}$	1	—	—	—	—	$\tfrac{1}{4}$	—	—	$\tfrac{1}{4}$	$\tfrac{1}{4}$
$\frac{aB}{ab} \times \frac{aB}{ab}$	1	—	—	—	—	—	—	$\tfrac{1}{4}$	$\tfrac{1}{2}$	$\tfrac{1}{4}$
$\frac{ab}{ab} \times \frac{ab}{ab}$	0	—	—	—	—	—	—	—	—	1

Offspring Genotypes

Table 3.4 presents what is known as a *transition matrix* in mathematics. The rows in such a transition matrix are probabilities that describe all the possible offspring outcomes from each mating type. We have already encountered such a transition matrix in Chapter 2 in the derivation of Weinberg's version of the Hardy–Weinberg law (Table 2.2). Such transition matrices can be used to predict the changes in genotype frequencies over many generations. Sometimes the mathematics needed to make such predictions can appear quite complicated, but all we need to do here is note a few features of the transition matrix given in Table 3.4 to make some important evolutionary insights into assortative mating. Note that the probabilities in each row sum to 1, so each row represents a probability distribution over a mutually exclusive and exhaustive set (see Appendix 2). However, note that 5 of the 18 rows have only one entry, a probability of 1 for a particular offspring genotypic class. The first row in Table 3.4 is one such example. This row describes the predicted offspring frequencies from the mating *AB/AB* × *AB/AB*. Under our model of inheritance (which assumes no mutations), all offspring from this mating type are *AB/AB*, and that is reflected in Table 3.4 by the probability of 1 appearing in the *AB/AB* column. Note that the offspring genotype in this case is the same as the genotypes of both parents. Hence, once a zygote is conceived with the genotype *AB/AB*, the resulting adult and *all of its descendants* can only mate exclusively with other *AB/AB* individuals and will only produce *AB/AB* offspring. In mathematical jargon, having the genotype of *AB/AB* is called an "absorbing state" because once a zygote enters that genotypic state, all of its descendants will also be in that state. Similarly, the very last row of Table 3.4 also defines an absorbing state, consisting of the mating type *ab/ab* × *ab/ab*.

An examination of Table 3.4 reveals three other rows with a single entry of 1. At first glance, these may seem to be the same situation as described in the preceding paragraph. However, one of these rows with a 1 in the +2 phenotypic category is associated with the mating type *Ab/Ab* × *aB/aB* (Table 3.4) producing with probability 1 the offspring genotype of *Ab/aB*. Note that the offspring from this mating type has a genotype that is different from both parents. As a consequence, the offspring of this mating type do not remain in the parental mating-type class, but rather they *have* to engage in matings associated with other rows. By looking at the rows in Table 3.4 that involve the offspring genotype *Ab/aB*, four are found and none of these rows are associated with an entry of a single 1. Therefore, the mating type *Ab/Ab* × *aB/aB* is not an absorbing state. The descendants of the offspring in this class are expected to go to many other mating-type classes in future generations.

The remaining two rows with a single entry of 1 are associated with matings of identical homozygous genotypes that produce offspring with the same genotype, just like the two absorbing states we identified above. One of these remaining two rows is the mating type *Ab/Ab* × *Ab/Ab*, which produces offspring of genotype *Ab/Ab*. Note from Table 3.4 that individuals with the genotype *Ab/Ab* have a +2 phenotype and can therefore mate with any other individuals with a +2 phenotype (those individuals with genotypes *Ab/Ab*, *AB/ab*, *Ab/aB*, or *aB/aB*). Hence, not all offspring of *Ab/Ab* × *Ab/Ab* parents will necessarily remain within this mating-type class the next generation. If they mate with any other +2 genotype, they can produce a wide variety of offspring types. Hence, the mating type *Ab/Ab* × *Ab/Ab* is not necessarily an absorbing state (and similarly the mating type *aB/aB* × *aB/aB*). Nevertheless, these mating types have the potential for being absorbing if, somehow, they were the only genotype in the +2 phenotypic category. To see if this potential could ever be realized involves mathematics beyond the scope of this book, but interested readers should see Ghai (1973). A verbal description of Ghai's mathematical results is given in the next paragraph.

Note that two of the genotypes with the +2 phenotype are double heterozygotes, which become increasingly rare as assortative mating proceeds (recall that assortative mating

Table 3.5. Equilibrium Populations under Two-Locus Model of 100% Assortative Mating

Genotypes	Initial Gene Pool		
	$p_A = p_B$	$p_A < p_B$	$p_A > p_B$
AB/AB	p_A	p_A	p_B
Ab/Ab	0	0	$p_A - p_B$
aB/aB	0	$p_B - p_A$	0
ab/ab	p_b	p_b	p_a

decreases heterozygote genotype frequencies in general). As time goes by, the +2 phenotypic class increasingly consists of just the homozygous Ab/Ab and aB/aB genotypes. Whenever matings occur between the Ab/Ab and aB/aB genotypes, the offspring are removed from the genotypic classes of the parents, as noted above. Consequently, it is impossible to have a stable equilibrium with both the Ab/Ab and aB/aB genotypes persisting in the population. However, if one of the Ab/Ab and aB/aB genotypes is rarer than the other, a greater proportion of the individuals of the rarer genotype will mate with the other +2 genotype. The result of these complex dynamics is that at most only one of the Ab/Ab and aB/aB genotypes has the potential for becoming an absorbing state as the population moves toward an equilibrium, and the winner is the one bearing the alleles that were more frequent in the initial population.

The other mating types in Table 3.4 define "transient states." For all the rows that do not have a single entry of 1, there is always a nonzero value for one or more of the two to three absorbing states (assuming $r > 0$; if $r = 0$, then this model becomes effectively a single-locus model). Thus, every generation some of the transient mating types will produce progeny that enter an absorbing state, and once there they are stuck in that state forever under the assumptions of the model. The mathematical consequence of this is straightforward: The genotype frequencies of AB/AB and ab/ab can only increase with time while the genotype frequencies of all other genotypes can only decrease, with the possible exception of either Ab/Ab or aB/aB, but not both. This process will continue until all genotypes are in absorbing states. This results in three possible equilibrium populations. The initial frequencies of the A and B alleles determine to which of these three possible states the population evolves, as shown in Table 3.5.

Suppose we started with an initial population with a gene pool of g_{AB}, g_{Ab}, g_{aB}, and g_{ab} and therefore initial allele frequencies of $p_A = g_{AB} + g_{Ab}$ and $p_B = g_{AB} + g_{aB}$. In general, we can see from Table 3.5 that the population will evolve to a gene pool with a different set of gamete frequencies than the initial gene pool. Hence, *assortative mating by itself causes evolutionary change by creating linkage disequilibrium*. Note that linkage disequilibrium D exists (D equals $p_A p_b$ or $p_a p_B$) in the equilibrium assortative mating population because there are only two or three gamete types at equilibrium, depending upon the initial conditions. Even if we had started with all four possible gamete types with no linkage disequilibrium, assortative mating in this case would have generated linkage disequilibrium.

Table 3.4 represents an extreme version of assortative mating under a simplistic genotype–phenotype model. However, it does capture some general consequences of assortative mating under less extreme conditions:

1. As we saw with the single-locus models, *assortative mating increases the frequency of homozygotes at the expense of heterozygotes*.

2. Multiple equilibria exist and the evolution of the population is determined by the state of its initial gene pool. This also is a general feature of multilocus assortative mating models and illustrates that *historical factors are a determinant of the course of evolution*. In evolution, the present is constrained by the past. As we will see throughout this book, models that are more complex than the original Hardy–Weinberg model often have multiple possible evolutionary outcomes in which historical factors play a dominant role.

3. *Assortative mating can create and maintain linkage disequilibrium*. Note that in Table 3.5 two gamete types usually dominate the predicted equilibrium gene pool: *AB* and *ab*. Both *A* and *B* contribute +1 to the phenotype, and both *a* and *b* contribute 0 to the phenotype. Hence, the disequilibrium created by assortative mating in this case places alleles together in gametes that have similar phenotypic effects. This is also a general feature of the evolutionary impact of assortative mating: *Assortative mating causes gametes to have alleles at different loci that cause similar phenotypic effects*.

As the above features reveal, our earlier conclusion of assortative mating not being an evolutionary force is abandoned as soon as we leave the realm of one-locus models. Indeed, *at the multilocus level, assortative mating is an extremely powerful microevolutionary force.* Also note that assortative mating in the extreme model given in Table 3.4 splits the original population into genetic subsets (*AB/AB* and *ab/ab* and sometimes *Ab/Ab* or *aB/aB* but never both) that are reproductively isolated from one another. Because of its ability to split a population into genetically differentiated and isolated subsets, assortative mating is also considered to be a powerful force not only in microevolution but also in the origin of species.

The assortative mating for deafness is less extreme than that given in Table 3.4, and moreover the genotype–phenotype relationship is quite different. Nevertheless, assortative mating for deafness will also create linkage disequilibrium in which bearers of a deaf allele at one locus tend to also be bearers of deaf alleles at other loci (Aoki and Feldman 1994). This disequilibrium also augments the incidence of genetic deafness in human populations. For example, let us return to the *GJB2* locus. As shown earlier, we expect only about $\frac{1}{16}$ of the deaf couples to be homozygous for this locus when this locus is considered by itself. Given the extreme rarity of alleles for deafness at all the other loci, we would normally dismiss the possibility that the remaining $\frac{15}{16}$ of the deaf couples would have any substantial risk of having deaf children. Thus, we would normally expect only about 1 in 16 children of deaf couples to be deaf. But as noted earlier, the actual figure is closer to 1 in 6 (Koehn et al. 1990). What is going on here?

The answer is linkage disequilibrium. Let *A* and *a* be alleles at the *GJB2* locus such that *a* is a recessive allele for deafness. Let *B* and *b* be alleles at an independently acting autosomal deafness locus such that individuals who are homozygous for the *b* allele are deaf. Suppose a mating occurs between two deaf individuals such that one parent is deaf because of homozygosity for the *a* allele (*aa*) and the other parent is deaf because of homozygosity for the *b* allele (*bb*). Given that both the *a* and *b* alleles are rare in the general population, we would normally expect them not to be found together in the same individual. In that case, we expect both parents to be homozygous for the dominant allele at the other deafness locus (that is, *aaBB* and *AAbb*), and their offspring therefore would have the genotype *AaBb* and not be deaf. But note that this offspring, when he or she reproduces, can produce *ab* gametes (the frequency depending upon *r*). Hence, the assortative mating for deafness makes it likely that two very rare alleles in the general population will be placed together upon the same gamete; that is, assortative mating has changed the human gene pool by creating linkage disequilibrium among loci contributing to deafness. This increases the chances that marriages among deaf people in subsequent generations will yield deaf children. Indeed, such matings may have already occurred in the ancestry of the deaf parents. In particular, given that a is the most common of the alleles causing deafness, it is likely under assortative mating for the parent who is bb to have had an ancestor who was aa. This past assortative mating in turn increases the probability of such crosses as $Aabb \times aaBB$. In this cross, half of the children of the deaf couple would be expected to be deaf, even though the two parents are deaf because of homozygosity for alleles at independently acting loci! The linkage disequilibrium that is induced by assortative mating, along with the increased homozygosity also induced by assortative mating, explains why autosomal recessive deafness occurs far more frequently in humans than expected under random mating.

Assortative Mating Versus Inbreeding

As seen above, both assortative mating and inbreeding increase homozygosity and both can be measured by f. Despite these similarities between assortative mating and inbreeding, there are important differences. One of the critical differences between these two systems of nonrandom mating is that assortative mating by itself is a powerful force for evolutionary change at the multilocus level, creating linkage disequilibrium among alleles having similar phenotypic effects. As we noted earlier, inbreeding is not an evolutionary force by itself at the single-locus level, but how about at the multilocus level? Table 3.6 presents the most extreme form of inbreeding possible in a two-locus model—complete selfing. Under 100% selfing, an individual can only mate with itself (many plant species and a few animals have this system of mating), so for a two-locus model with 10 possible genotypes there are only 10 possible mating types (Table 3.6). In examining Table 3.6 for absorbing versus transient states, you should discover that this model of 100% selfing has four universal absorbing states, in contrast to the two observed with the model of 100% assortative mating given in Table 3.4. The four absorbing states correspond to selfing of the four homozygous genotypes: AB/AB, Ab/Ab, aB/aB, and ab/ab. At equilibrium, these are the only genotypes present in the population, and obviously the frequencies of the gametes AB, Ab, aB, and ab correspond to the respective homozygous genotype frequencies. Hence, we already see a big difference between 100% selfing and 100% assortative mating; all four gamete types can exist under selfing whereas only two or three can exist at equilibrium under the 100% assortative mating model given in Table 3.4. Also recall that if the initial population had no disequilibrium ($D = 0$), 100% assortative mating would create linkage disequilibrium (Table 3.5). In contrast, detailed mathematical analysis of the transition matrix defined by Table 3.6 reveals that no disequilibrium is created by 100% selfing (Karlin 1969); the gamete frequencies remain unchanged when there is no initial disequilibrium, so even this extreme form of inbreeding is not an evolutionary force by itself at the two-locus level in this situation. This is a general difference between inbreeding and assortative mating: *Assortative mating actively generates linkage disequilibrium; inbreeding does not.*

However, this does not mean that inbreeding has no effect on multilocus evolution. Consider what happens when the initial population starts with some linkage disequilibrium. We saw in Chapter 2 that under random mating the disequilibrium dissipates according to the equation $D_t = D_0(1 - r)^t$, where D_0 is the initial disequilibrium, r is the recombination frequency, and D_t is the amount of disequilibrium remaining after t generations of random

Table 3.6. One Hundred Percent Selfing for Two Loci, Two-Allele Genetic Architecture

Mating Type	Offspring Genotypes									
	$\frac{AB}{AB}$	$\frac{AB}{Ab}$	$\frac{AB}{aB}$	$\frac{AB}{ab}$	$\frac{Ab}{aB}$	$\frac{Ab}{Ab}$	$\frac{aB}{aB}$	$\frac{Ab}{ab}$	$\frac{aB}{ab}$	$\frac{ab}{ab}$
$\frac{AB}{AB} \times \frac{AB}{AB}$	1	—	—	—	—	—	—	—	—	—
$\frac{AB}{Ab} \times \frac{AB}{Ab}$	$\frac{1}{4}$	$\frac{1}{2}$	—	—	—	$\frac{1}{4}$	—	—	—	—
$\frac{AB}{aB} \times \frac{AB}{aB}$	$\frac{1}{4}$	—	$\frac{1}{2}$	—	—	—	$\frac{1}{4}$	—	—	—
$\frac{AB}{ab} \times \frac{AB}{ab}$	$\frac{1}{4}(1-r)^2$	$\frac{1}{2}(1-r)r$	$\frac{1}{2}(1-r)r$	$\frac{1}{2}(1-r)^2$	$\frac{1}{2}r^2$	$\frac{1}{4}r^2$	$\frac{1}{4}r^2$	—	—	$\frac{1}{4}(1-r)^2$
$\frac{Ab}{aB} \times \frac{Ab}{aB}$	$\frac{1}{4}r^2$	$\frac{1}{2}(1-r)r$	$\frac{1}{2}(1-r)r$	$\frac{1}{2}r^2$	$\frac{1}{2}(1-r)^2$	$\frac{1}{4}(1-r)^2$	$\frac{1}{4}(1-r)^2$	—	—	$\frac{1}{4}r^2$
$\frac{Ab}{Ab} \times \frac{Ab}{Ab}$	—	—	—	—	—	1	—	—	—	—
$\frac{aB}{aB} \times \frac{aB}{aB}$	—	—	—	—	—	—	1	—	—	—
$\frac{Ab}{ab} \times \frac{Ab}{ab}$	—	—	—	—	—	$\frac{1}{4}$	—	$\frac{1}{2}$	—	$\frac{1}{4}$
$\frac{aB}{ab} \times \frac{aB}{ab}$	—	—	—	—	—	—	$\frac{1}{4}$	—	$\frac{1}{2}$	$\frac{1}{4}$
$\frac{ab}{ab} \times \frac{ab}{ab}$	—	—	—	—	—	—	—	—	—	1

mating. In deriving this equation, we had noted that disequilibrium is dissipated only in the offspring of double heterozygotes because recombination is effective in creating new recombinant gamete types only in double heterozygotes. Because inbreeding reduces the frequency of heterozygotes in general, it also reduces the frequency of double heterozygotes in particular. Therefore, the opportunity for recombination to dissipate disequilibrium is reduced under inbreeding simply because there are fewer double heterozygotes. In the extreme case of 100% selfing (Table 3.6), the equilibrium population consists only of homozygotes, so no dissipation of linkage disequilibrium is possible as the population nears genotypic equilibrium. Therefore, when starting with an initial population with some linkage disequilibrium, there is a dynamic race between recombination dissipating linkage disequilibrium through the increasingly rare double heterozygotes and the approach to the equilibrium population in which no further dissipation is possible. Under 100% selfing, the approach to homozygosity is so rapid that not all of the initial linkage disequilibrium is dissipated, meaning that some linkage disequilibrium can persist indefinitely under 100% selfing (Karlin, 1969). Under models of less extreme inbreeding, linkage disequilibrium is eventually reduced to zero, just as it is under random mating, but at a reduced rate relative to its decay under random mating. Consequently, by itself, inbreeding, in all but its most extreme form of 100% selfing, is not ultimately an evolutionary force at either the single-or two-locus level. This is in great contrast to assortative mating, which is a powerful multilocus evolutionary force.

There is another major difference between assortative mating and inbreeding at the multilocus level. Assortative mating affects the genotype frequencies only of those loci that contribute to the phenotype that affects mating and other loci that are in linkage disequilibrium with them. In contrast, inbreeding (in the system-of-mating sense) is based on choosing mates by pedigree relationship; hence, all loci are affected by inbreeding. Recall that inbreeding in the U.S. population increases the incidence of all genetic diseases associated with rare, autosomal recessive alleles. In contrast, children from marriages between deaf people have no increased risk for genetic diseases other than deafness and diseases associated with deafness through pleiotropy. Hence, *assortative mating is locus specific whereas system-of-mating inbreeding alters the genotype frequencies at all loci.*

The fact that different genetic elements in the same populations can display different systems of mating is illustrated by studies on the fly *Sciara ocellaris* (Perondini et al. 1983). This species shows much chromosomal variability involving the structural modification of single bands of the polytene chromosomes. These band variants define what behaves as a single-locus, two-allele genetic architecture. By examining the genotypes obtained from mated pairs, five of these polymorphic band systems fit well to a random-mating model, seven showed assortative mating, and three deviated from random mating in other ways. Note that the deviations from random mating cannot be attributed to inbreeding in this case because all loci do not show a positive f as expected under inbreeding. What, then, is the system of mating for these flies? This question is unanswerable because we must first define the genetic system of interest. These flies are randomly mating for some polymorphisms and assortatively mating for others. There is no such thing as *the* system of mating for a population; rather, there is only the system of mating for a specific genetic architecture within the population.

Assortative Mating, Linkage Disequilibrium, and Admixture

As pointed out above, assortative mating directly affects the genotype and gamete frequencies of the loci that contribute to the phenotype for which assortative mating is occurring and of any loci in linkage disequilibrium with them. Therefore, if there is extensive disequilibrium among loci throughout the genome, assortative mating can potentially affect the genotype frequencies at many loci even though those loci do not directly affect the assorting phenotype. Indeed, this circumstance is not a particularly rare one. To see why, we need to consider another source of linkage disequilibrium. In Chapter 2, we saw that mutation will create linkage disequilibrium. We also saw that this disequilibrium can have important evolutionary and applied implications when dealing with DNA regions that show little recombination with the mutational site. However, now we will consider another factor that can create linkage disequilibrium, but unlike mutation, this factor can create massive amounts of disequilibrium between loci scattered all over the genome—even loci on different chromosomes. This evolutionary factor is admixture.

Admixture occurs when two or more genetically distinct subpopulations are mixed together and begin interbreeding. As will be detailed in subsequent chapters, many species have local populations that have little to no genetic interchange with one another for many generations, causing those subpopulations to become genetically differentiated from one another (that is, they acquire different allele frequencies at many loci). Such reproductive isolation is often temporary, and events can occur that bring such differentiated subpopulations back together in the same area, allowing them to interbreed. This mixing together of previously differentiated subpopulations induces linkage disequilibrium in the admixed population even if no disequilibrium existed in either of the original subpopulations. For example, suppose a species consists of two subpopulations that have different allele frequencies at two loci, say the A locus (with alleles A and a) and the B locus (with alleles B and b). Let p_1 be the frequency of A in subpopulation 1 and p_2 be its frequency in subpopulation 2. Because we defined these populations to be genetically differentiated at the A locus, we know that $p_1 \neq p_2$. Similarly, let k_1 and k_2 be the frequencies of the B allele in the two subpopulations, once again with $k_1 \neq k_2$. We also assume that there is no linkage disequilibrium between the A and the B loci within either subpopulation. Finally, suppose these two subpopulations are brought together to form a new, admixed population such that a fraction m of the genes are derived from subpopulation 1 and $1 - m$ are derived from subpopulation 2 in this new admixed population. Then, in the first generation in which these two subpopulations are mixed together, the gamete frequencies in the newly created hybrid population are

$$\begin{aligned} g_{AB} &= mp_1k_1 + (1-m)p_2k_2 \\ g_{Ab} &= mp_1(1-k_1) + (1-m)p_2(1-k_2) \\ g_{aB} &= m(1-p_1)k_1 + (1-m)(1-p_2)k_2 \\ g_{ab} &= m(1-p_1)(1-k_1) + (1-m)(1-p_2)(1-k_2) \end{aligned} \quad (3.5)$$

From these gamete frequencies, we can calculate the linkage disequilibrium (see equation 2.4) in this newly admixed population to be (after some algebra)

$$D_{\text{admixture}} = m(1-m)(p_1 - p_2)(k_1 - k_2) \quad (3.6)$$

Note that as long as the two original subpopulations are genetically differentiated at the A and B loci ($p_1 \neq p_2$ and $k_1 \neq k_2$), $D_{\text{admixture}} \neq 0$. An example of this is shown in Figure 3.7. Thus, hybridization or admixture between two subpopulations creates linkage disequilibrium between all pairs of loci that had different allele frequencies in the original subpopulations.

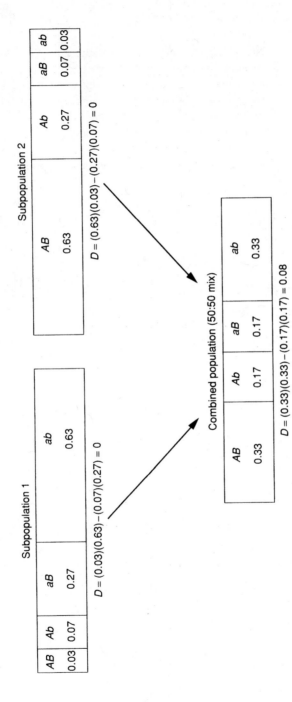

Figure 3.7. Creation of linkage disequilibrium (D) by admixture between two populations with differentiated gene pools.

Now suppose that locus A influences a trait for which there is assortative mating. Assortative mating would alter the genotype frequencies at the A locus within both subpopulations, but assortative mating would have no impact on the B locus genotype frequencies within either subpopulation because there is no linkage disequilibrium within either subpopulation. However, once admixture occurs, there is now disequilibrium between the A and B loci. Now, assortative mating on the A trait would also have an impact on the B locus genotype frequencies in the admixed population. One aspect of this impact is to increase homozygosity above random-mating expectations in the admixed population. Recall from Chapter 2 that a single generation of random mating is sufficient to establish one-locus Hardy–Weinberg genotype frequencies. However, when two or more previously differentiated subpopulations are mixed together with assortative mating on a trait that differs in initial frequency among the subpopulations, Hardy–Weinberg genotype frequencies are established only slowly even at loci not affecting the trait causing assortment. This gradual approach to Hardy–Weinberg genotype frequencies occurs as the linkage disequilibrium caused by admixture between these loci and the loci influencing the assorting trait gradually breaks down. A second impact stems directly from the first; decreased frequencies of heterozygotes reduce the dissipation of linkage disequilibrium. Hence, multilocus Hardy–Weinberg equilibrium is also delayed relative to random-mating expectations, and this prolongs the time it takes to achieve single-locus Hardy–Weinberg genotype frequencies. A third impact is that the mixed subpopulations do not fuse immediately, but rather the total population is stratified into genetically differentiated subcomponents that reflect the original historical subpopulations.

As an example, consider the colonization of North America following the voyages of Columbus. This colonization brought together European and sub-Saharan African human subpopulations that had allele frequency differences at many loci. Once brought together in the New World, admixture began. However, there was assortative mating for skin color. The skin color differences between Europeans and sub-Saharan Africans are due to about five loci (Cavalli-Sforza and Bodmer 1971). There is no evidence that humans tend to mate nonrandomly for most blood group loci, but Europeans and Africans show many allele frequency differences for such loci. As a result, there is strong disequilibrium between the skin color loci and blood group loci in the U.S. population as a whole. Assortative mating for skin color has therefore led to the persistence of allele frequency differences at blood group loci in modern-day European Americans and African Americans (two of the current subcomponents of the U.S. population that trace, imperfectly, to the original European and African subpopulations that came to North America) despite nearly 500 years of hybridization. Hence, the U.S. population in toto is not a Hardy–Weinberg population. When regarded as a single entity, the U.S. population has too many homozygotes for almost all loci and extensive linkage disequilibrium throughout the genome.

What is even more remarkable about admixed populations is that even a nongenetic trait that leads to assortative mating can maintain allele frequency differences for many generations. All that is needed is for a phenotype causing assortative mating to be associated with the different historical subpopulations. For example, for reasons to be discussed in the next chapter, people of the Amish religion in the United States have different allele frequencies at many loci from the surrounding non-Amish populations due to the manner in which their initial populations were founded a couple of centuries ago. There is assortative mating by religion in the United States, and this has led to the persistence of these initial genetic differences between the Amish and their non-Amish neighbors over these centuries.

The interaction of assortative mating with population stratification based upon historical subpopulations has practical implications in DNA forensics. A common use of DNA screening in forensics is to calculate the probability that a sample taken from a crime scene is from a particular suspect. For example, suppose the suspect is homozygous for a marker, say A, that has a frequency p in the total population. Under Hardy–Weinberg, the probability that a sample from a randomly chosen individual would match that of the suspect is p^2. The smaller p^2, the more likely it is that our suspect is indeed the perpetrator. Hence, rare alleles provide the greatest discrimination. However, under assortative mating in a population stratified by historical subpopulations (the reality in the United States), the probability of a random match is $p^2 + 2pqf$. As we noted with inbreeding, it is the genotype frequencies associated with the rare alleles that are most sensitive to even small values of f, so it is possible to obtain very different probabilities depending upon which population genetic model is used: random mating or assortative mating in an admixed population. The situation gets more complicated when we screen for several genetic markers simultaneously. Under random mating, unlinked marker genotype frequencies can usually be multiplied to obtain the overall probability of a match because there is little to no disequilibrium among unlinked markers in a randomly mating population. Assortative mating can cause the persistence of linkage disequilibrium among many loci throughout the genome, both linked and unlinked. This greatly complicates the multilocus calculations. One solution is to define different reference populations that represent the population stratification, but this has the difficulty that a particular individual (in this case, the suspect) has his or her own unique genealogy that often reflects a degree of admixture among the reference populations. The best solution seems to look at so many loci that, regardless of which population model is used, the probability of a random match is extremely small.

All of these examples reveal that assortative mating, particularly at the multilocus level, is a powerful evolutionary force. We now turn our attention to the opposite of assortative mating—preferential mating of individuals with dissimilar phenotypes—to see if this system of mating is likewise a powerful evolutionary force.

DISASSORTATIVE MATING

Disassortative mating is the preferential mating of individuals with dissimilar phenotypes. This means that there is a negative correlation between the phenotypes of mating individuals. For example, the major histocompatibility complex (MHC), mentioned in Chapter 1, is found not only in humans but in mice as well. In mice, genetic variation in MHC induces odor differences. There is disassortative mating at this gene complex in mice that is due to olfactory discrimination of potential mates (Potts and Wakeland 1993). Recent evidence indicates that this may be occurring in humans as well (Wedekind et al. 1995; Wedekind and Füri 1997). Several college males were asked to wear the same t-shirt for two days. A group of college females were then asked to smell the t-shirts and rank them (no pun intended). The t-shirts with the most pleasant (or least obnoxious) smells turned out to be shirts worn by males that shared fewer *MHC* alleles with the women than expected by chance alone. Moreover, the women indicated that the preferred smells were similar to those of their boyfriends or husbands. A subsequent experiment revealed that both men and women prefer odors from others most dissimilar to themselves for *MHC*. More direct evidence that *MHC* affects mating preferences in humans comes from studies in a population of Hutterites, a small, isolated religious sect that has maintained genealogical records since about 400 members originally migrated from Europe to North America in the 1870s. Hutterite couples were less likely to share *MHC* alleles than by chance, even after corrected for the nonrandom mating among colony lineages and close inbreeding taboos (Ober et al. 1997). Note that the baseline system of mating for detecting disassortative mating in this case is not random mating but rather avoidance of system-of-mating inbreeding.

To see the impact of disassortative mating, consider the simple one-locus, two-allele model with a one-to-one genotype–phenotype mapping and 100% disassortative mating given in Figure 3.8. As with assortative mating, the genotype frequencies in general will change over the generations in this model of disassortative mating. For example, if we start out at Hardy–Weinberg frequencies with $p = 0.25$, then the initial random-mating heterozygote frequency of 0.375 is altered by a single generation of disassortative mating to 0.565. Note that disassortative mating induces a heterozygote excess with respect to Hardy–Weinberg expectations ($f < 0$)—exactly the opposite of assortative mating. Moreover, unlike assortative mating, the allele frequencies in general are changing. In the example given above, the allele frequency is changed from 0.25 to 0.326 in a single generation.

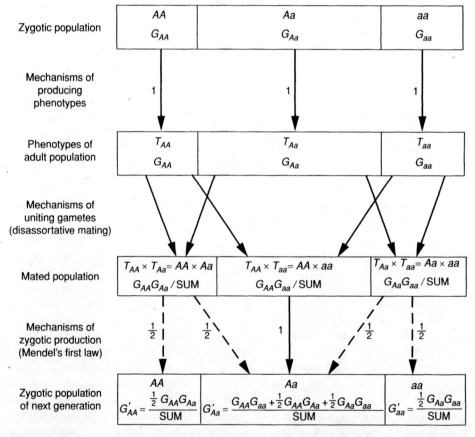

Figure 3.8. Model of 100% disassortative mating for phenotype determined by single autosomal locus with two alleles and with all three possible genotypes having distinct phenotypes. The G's refer to genotype frequencies and T's to phenotype frequencies. SUM $= G_{AA} \times G_{Aa} + G_{AA} \times G_{aa} + G_{Aa} \times G_{aa}$ and is needed to standardize the mating probabilities to sum to 1.

Therefore, disassortative mating can be a powerful evolutionary force even at the single-locus level. Moreover, disassortative mating is a powerful factor in maintaining genetic polymorphisms. In the example given above, we saw that if we started at $p = 0.25$, then p would increase to 0.326 in a single generation of disassortative mating. Similarly, using the equations shown in Figure 3.8, we can see that if we started with $p = 0.75$, then p would decrease to 0.674 in a single generation of disassortative mating. Thus, disassortative mating is pushing the allele frequencies to an intermediate value in this case. Indeed, we can keep iterating the disassortative mating model given in Figure 3.8 to find that it quickly stabilizes at $G_{AA} = 0.4175$, $G_{Aa} = 0.5361$, and $G_{aa} = 0.0464$, yielding an equilibrium allele frequency of A of 0.6856 and an equilibrium f of $1 - G_{Aa}/(2pq) = -0.2435$ (note that once again f is measuring not the avoidance of system-of-mating inbreeding in this case but rather the impact of disassortative mating). This is a true equilibrium, but different initial conditions will go to different equilibria, although all are polymorphic. In general, *disassortative mating results in stable equilibrium populations with intermediate allele frequencies and $f < 0$.*

The reason for this intermediate stability is that any individual with a rare phenotype has an inherent mating advantage under a disassortative mating system: The rarer you are, the more other individuals in the deme have a dissimilar phenotype to you and hence "prefer" you as a mate; "prefer" is in quotation marks because, just as with assortative mating, disassortative mating does not always result from an active choice of mates but can arise from other causes. For example, many plants show disassortative mating for gametophytic self-sterility loci (S loci) (Vekemans and Slatkin 1994). These loci have alleles such that a grain of pollen bearing a particular allele at such an S locus can only successfully fertilize an ovum from a plant not having that same allele. This results in strong disassortative mating at S loci.

Because rare phenotypes have such a large mating advantage in disassortative mating systems, new mutations associated with novel phenotypes have a tendency to increase in frequency. As a result, loci with disassortative mating systems tend to be polymorphic, not just for two alleles, but for many alleles. We already saw in Chapter 1 that MHC in humans has enormous levels of genetic variation, and the same is true for the homologous MHC in other species, such as mice. The disassortative mating at these complexes is thought to be one of the contributors (but not the sole contributor) to these high levels of polymorphism. Similarly, 45 alleles at a self-sterility locus were found in a rare plant species (*Oenothera organensis*) with a total population size of only 500–1000 individuals (Wright 1969)! Moreover, molecular analyses of S alleles indicate that they have persisted as polymorphisms for millions of years, even remaining polymorphic through speciation events (Ioerger et al. 1991; Vekemans and Slatkin 1994). The same is true for MHC. For example, some of the alleles at human *MHC-DRB* loci mentioned in Chapter 1 have persisted as distinct allelic lineages since the time of the split of the Old World monkeys from apes (over 35 million years ago) (Zhu et al. 1991). This means that you can be heterozygous for two alleles at an *MHC* locus such that the allele you inherited from your father is evolutionarily more closely related to an allele from a pigtail macaque (*Macaca nemestrina*—an Old World monkey) than it is to the allele you inherited from your mother! Hence, disassortative mating is a strong evolutionary force even at the single-locus level that tends to maintain stable polymorphisms for long periods of time.

As noted above, disassortative mating also causes deviations from Hardy–Weinberg genotype frequencies and in particular causes an excess of heterozygosity. In this sense, disassortative mating "mimics" avoidance of system-of-mating inbreeding ($f < 0$), but as with assortative mating, disassortative mating is locus specific and does not have a general impact over all loci as does avoidance of inbreeding. For example, the Speke's gazelle population surveyed in 1982 had a significant negative f for every locus examined and the f's were statistically indistinguishable across loci (Templeton 1994). This indicates that the deviation from Hardy–Weinberg in this population was due to avoidance of inbreeding. In contrast, disassortative mating, just as we saw previously with assortative mating, is expected to affect only the loci contributing to the phenotype leading to disassortative mating and any other loci that are in linkage disequilibrium with the phenotypically relevant loci. We certainly see this for humans. For example, many human populations have Hardy–Weinberg frequencies for blood group loci (recall the examples given in Chapter 2), but humans generally show assortative mating for stature and disassortative mating for *MHC*. As we saw earlier for assortative mating, there is no such thing as the system of mating for a deme in general; some systems of mating are specific to particular genetic systems and can be overlaid upon systems of mating that have more generalized effects, such as inbreeding or avoidance of inbreeding. Hence, disassortative mating has very different evolutionary impacts than avoidance of inbreeding, so these two systems of mating should not be equated. Moreover, as we saw with the Hutterite population, disassortative mating can cause locus-specific deviations beyond those caused by avoidance of inbreeding.

We saw earlier how assortative mating can split a deme into genetically isolated subpopulations by creating linkage disequilibrium. Although disassortative mating also tends to place together on the same gamete alleles with opposite phenotypic effects, it is not nearly as effective as assortative mating in generating and maintaining linkage disequilibrium. We saw earlier that the rate at which disequilibrium breaks down depends upon the frequency of double heterozygotes. Because disassortative mating results in heterozygote excesses, it accentuates the effectiveness of recombination in breaking down linkage disequilibrium. Moreover, because dissimilar individuals preferentially mate, there is no possibility of subdividing a deme into genetic/phenotypic subdivisions that are reproductively isolated. Hence, in contrast to assortative mating, disassortative mating prevents demes from genetically subdividing.

We also saw earlier that assortative mating can cause the persistence of genetic differences between historically differentiated demes long after contact has been established between the demes. In contrast, disassortative mating rapidly destroys genetic differences between historical subpopulations that come together and mate disassortatively—even for nongenetic traits. For example, the Makiritare and Yanomama Indians lived contiguously in South America prior to 1875 but apparently did not interbreed much (Chagnon et al. 1970). As a consequence, most villages of these two tribes have significant genetic differentiation at many loci. Indeed, at several loci, the Makiritare have alleles that are not even present in the Yanomama gene pool. This situation began to change in one area when the Makiritare made contact with Europeans and acquired steel tools. The Yanomama, being more in the interior, did not have contact with non-Indians until the 1950s, and even in the 1970s most Yanomama still had no outside contact. Hence, the Yanomama depended upon the Makiritare for steel tools. The Makiritare demanded sexual access to Yanomama women in exchange for the tools, siring many children who were raised as Yanomama. This also caused much hatred. One group of Yanomama eventually moved away to an area called Borabuk, but before they left they ambushed the Makiritare and abducted many Makiritare women, who once captive had an average of 7.3 children as compared to 3.8 children per Yanomama woman. Because of this history, there were effectively two generations in which most offspring in the Borabuk Yanomama were actually Yanomama–Makiritare hybrids.

This extreme disassortative mating between Yanomama and Makiritare, although not based upon any genetic traits, has led to the current Borabuk Yanomama being virtually identical genetically to the Makiritare, although culturally they are still Yanomama (Chagnon et al. 1970).

As we have seen in this chapter, systems of mating can be potent evolutionary forces, both by themselves and in interactions with other evolutionary factors. In subsequent chapters we will examine additional interactions between system of mating and other evolutionary forces.

4
GENETIC DRIFT

In deriving the Hardy–Weinberg law in Chapter 2, we assumed that the population size was infinite. In some derivations of the Hardy–Weinberg law, this assumption is not stated explicitly, but it enters implicitly by the act of equating allele and genotype *probabilities* to allele and genotype *frequencies*. The allele frequency is simply the number of alleles of a given type divided by twice the total population size for an autosomal locus in a diploid species. Likewise, the genotype frequency is simply the number of individuals with a specified genotype divided by the total population size. But in deriving the Hardy–Weinberg law in Chapter 2, we actually calculated allele and genotype probabilities. For example, we calculated the probability of drawing an A allele from the gene pool as p and then stated that this is also the frequency of the A allele in the next generation. This stems from the common definition that the probability of an event is the frequency of the event in an infinite number of trials. But what happens when the population is finite in size? For example, suppose a population has a gene pool with two alleles, say H and T, such that the probability of drawing either allele is 0.5 (i.e., $p = q = \frac{1}{2}$). Now suppose that $2N$ (a finite number) gametes are drawn from this gene pool to form the next generation of N diploid adults. Will the frequency of H and T be 0.5 in this finite population?

You can simulate this situation. For example, let $N = 5$ (corresponding to a sample of 10 gametes), and place 10 coins in a box, shake the box, and count the number of heads (i.e., allele H). Suppose that after doing this experiment one time, six heads were observed. Hence, the frequency of the H allele in this simulation was $6/10 = 0.6$, which is not the same as 0.5, the probability of H. Figure 4.1a shows the results of doing this coin flip simulation 20 times, and you are strongly encouraged to do this experiment yourself. As you can see from Figure 4.1a or from your own simulations, when population size is finite, the frequency of an allele in the next generation is often not the same as the probability of drawing that allele from the gene pool. Given our definition of evolution as a change in allele frequency, we note that this random sampling error can induce evolution. Random changes

---

Population Genetics and Microevolutionary Theory, By Alan R. Templeton
Copyright © 2006 John Wiley & Sons, Inc.

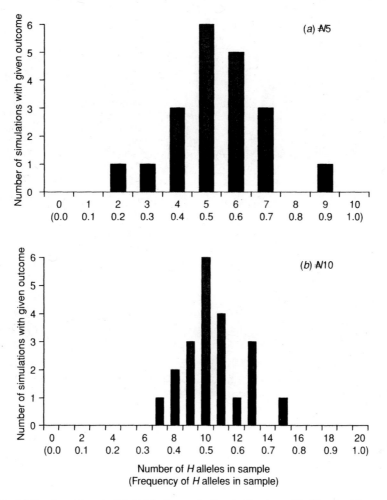

Figure 4.1. (a) Number of heads (H allele) observed after shaking box with 10 coins. The experiment was repeated 20 times. (b) Number of heads (H allele) observed after shaking box with 20 coins. The experiment was repeated 20 times.

in allele frequency due to sampling error in finite populations are known as **genetic drift**. Genetic drift is an evolutionary force that can alter the genetic makeup of a population's gene pool through time and shows that the Hardy–Weinberg "equilibrium" and its predicted stability of allele frequencies do not hold exactly for any finite population. The purpose of this chapter is to investigate the evolutionary properties and significance of genetic drift.

BASIC EVOLUTIONARY PROPERTIES OF GENETIC DRIFT

The coin flip simulation shown in Figure 4.1a illustrates that finite population size can induce random changes in allele frequencies due to sampling error. But what exactly is the relationship between genetic drift and finite population size? To answer this question, repeat the coin flip experiment but now use 20 coins for the simulation of a diploid population of $N = 10$ and $2N = 20$ gametes (double our previous population size). Figure 4.1b shows the results of such a simulation repeated 20 times. As can be seen from Figure 4.1b, there are random deviations from 0.5, so genetic drift is still operating in this larger but still finite population. By comparing Figures 4.1a and 4.1b, an important property of genetic drift is revealed: The simulated frequencies are more tightly clustered around 0.5 when $2N = 20$ (Figure 4.1b) than when $2N = 10$ (Figure 4.1b). This means that on the average the observed allele frequencies deviate less from the expected allele frequency when the population size is larger. Thus, the amount of evolutionary change associated with random sampling error is *inversely* related to population size. The larger the population, the less the allele frequency will change on the average. Hence, genetic drift is most powerful as an evolutionary force when N is small.

In the coin box experiments, the outcome was about equally likely to deviate above and below 0.5 (Figures 4.1a and b). Hence, for a large number of identical populations, the overall allele frequency remains 0.5, although in any individual population, it is quite likely that the allele frequency will change from 0.5. The fact that deviations are equally likely above and below 0.5 simply means that there is *no direction* to genetic drift. Although we can see that finite population size is likely to alter allele frequencies due to sampling error, we cannot predict the precise outcome or even the direction of the change in any specific population.

The coin box simulations given in Figure 4.1 only simulate one generation of genetic drift starting with an initial allele frequency of 0.5. The coin box simulations do not simulate the impact of drift over multiple generations because the probability of a coin flip producing an H allele remains unchanged at 0.5. However, suppose drift caused the allele frequency to change from 0.5 to 0.6 in one particular population. How about the next generation? Is it equally likely to be above or below 0.5, as it was in the first generation and will always be in our coin flip simulations? The answer is no, drift at one generation is always centered around the allele frequency of the previous generation, and allele frequencies in more ancient generations are irrelevant. Thus, after the allele frequency drifts to 0.6 from 0.5, the probability of drawing an H allele is now 0.6 and sampling error in the second generation is centered around 0.6 and not 0.5. This in turn means that after two generations of drift and given that the first generation experienced a deviation above 0.5, it is no longer true that deviations will be equally likely above and below 0.5. Once the population drifted to a frequency of 0.6, the next generation's allele frequency is more likely to stay above 0.5. Under genetic drift, there is *no tendency to return to ancestral allele frequencies*. With each passing generation, it becomes more and more likely to deviate from the initial conditions.

The action of drift over several generations can be simulated using a computer in which each generation drifts around the allele frequency of the previous generation. Figure 4.2 shows the results of 20 replicates of simulated drift in diploid populations of size 10 ($2N = 20$) over multiple generations, and Figure 4.3 shows the results in populations of size 25 ($2N = 50$). In both cases, the initial allele frequency starts at 0.5, but with increasing generation number, more and more of the populations deviate from 0.5 and by larger amounts. As can be seen by contrasting Figure 4.2 ($N = 10$) with Figure 4.3 ($N = 25$), the smaller population size tends to have the more radical changes in allele frequency in a given amount of time, as was shown in Figure 4.1 for one generation. However, Figure 4.3 shows that even with the larger population size of 25, substantial changes have occurred by generation 10. In general, we expect to obtain larger and larger deviations from the initial conditions with increasing generation time. Figures 4.2 and 4.3 show that N determines the rate of change caused by drift and that even large populations can be affected by drift if given enough time. The evolutionary changes in allele frequencies caused by genetic drift *accumulate with time*.

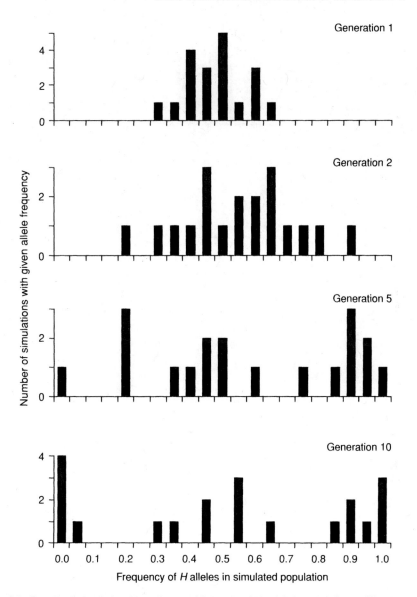

Figure 4.2. Results of simulating 20 replicates of finite population of size 10 ($2N = 20$) for 10 generations starting from initial gene pool of $p = \frac{1}{2}$. The distribution of allele frequencies is shown after 1, 2, 5, and 10 generations of genetic drift.

Also note in these simulations (particularly for $N = 10$) that eventually populations tend to go to allele frequencies of 0 (loss of the allele) or 1 (fixation of the allele). Genetic drift, like any other evolutionary force, can only operate as an evolutionary force when there is genetic variability. Hence, as long as p is not equal to 0 or 1, drift will cause changes in allele frequency. However, once an allele is lost or fixed, genetic drift can no longer cause allele frequency changes (all evolution requires genetic variation). Once lost or fixed, the allele stays lost or fixed, barring new mutations or the reintroduction of allelic variation

86 GENETIC DRIFT

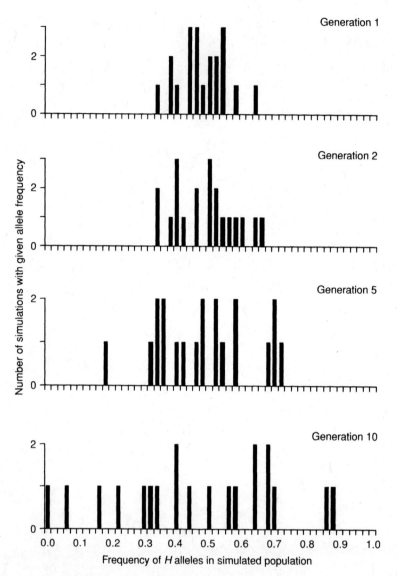

Figure 4.3. Results of simulating 20 replicates of finite population of size 25 ($2N = 50$) for 10 generations starting from initial gene pool of $p = \frac{1}{2}$. The distribution of allele frequencies is shown after 1, 2, 5, and 10 generations of genetic drift.

by genetic interchange with an outside population. Genetic drift is like a room with flypaper on all the walls. The walls represent loss and fixation, and sooner or later (depending upon population size, which in this analogy is directly related to the size of the room), the fly (allele frequency) will hit a wall and be "stuck." Genetic drift causes a *loss of genetic variation* within a finite population.

In Figures 4.1, 4.2, and 4.3, we simulated several replicates of the initial population. Now suppose that several subpopulations are established from a common ancestral population such that they are all genetically isolated from one another (that is, no gametes are exchanged between the subpopulations). Population subdivision into isolated demes is called **fragmentation**. Figures 4.1, 4.2, and 4.3 can therefore also be regarded as simulations of population fragmentation of a common ancestral population such that the fragmented subpopulations are all of equal size. Note that the ancestral gene pool is the same ($p = 0.5$) in all the populations simulated. Therefore, these same figures allow us to examine the role of genetic drift upon fragmented populations. Now we shift our focus from the evolution within each fragmented deme to the evolution of changes between subpopulations. Because of the genetic isolation under fragmentation, drift will operate independently in each subpopulation. Because of the randomness of the evolutionary direction of drift, it is unlikely that all the independent subpopulations will evolve in the same direction. This is shown in Figures 4.1, 4.2, and 4.3 by regarding the replicate elements of the histograms as isolated subpopulations. The spread of these histograms around the initial allele frequency shows that different subpopulations evolve away from 0.5 in different directions and magnitudes. Thus, although each subpopulation began with the same allele frequency, they now have many different allele frequencies. Genetic drift causes an *increase of allele frequency differences* among finite subpopulations.

All of these properties of genetic drift have been demonstrated empirically by Buri (1956), as shown in Figure 4.4. He initiated 107 populations of eight males and eight females of the fruit fly *Drosophila melanogaster*, all with two eye color alleles (*bw* and *bw^{75}*) at equal frequency. He then followed the evolutionary fate of these replicate populations for 19 generations. Note the following from his experimental results shown in Figure 4.4:

- When allele frequencies are averaged over all 107 populations, there is almost no change from the initial allele frequencies of 0.5. *Drift has no direction.*
- The chances of any particular population deviating from 0.5 and the magnitude of that deviation increase with each generation. *Evolutionary change via drift accumulates with time.*
- With increasing time, more and more populations become fixed for one allele or the other. By generation 19, over half of the populations had lost their genetic variation at this locus. Ultimately, all populations are expected to become fixed. *Drift causes the loss of genetic variability within a population.*
- As alleles are lost by drift, it is obvious that many copies of the remaining allele have to be identical by descent. For example, the original gene pool had 16 *bw* alleles in it, but those populations that are fixed for the *bw* allele have 32 copies of that allele. Moreover, some of these copies of the *bw* allele found in a fixed subpopulation are descended via DNA replication from just one of the original *bw* copies found in the initial population. When two or more copies of a gene are of the same allelic state and descended via DNA replication from a single common gene in some initial reference population, the genes are said to be *identical by descent*, as noted in Chapter 3. In the subpopulations fixed for the *bw* allele in Buri's experiments (and similarly for those fixed for *bw^{75}*), many individuals will be homozygous for alleles that are identical by descent. This means that the copy of the gene the individual received from its mother is identical by descent to the copy it received from its father. As fixation proceeds, homozygosity from identity by descent tends to increase with each succeeding generation subject to drift. *Drift causes the average probability of identity by descent to increase within a population.*

88 GENETIC DRIFT

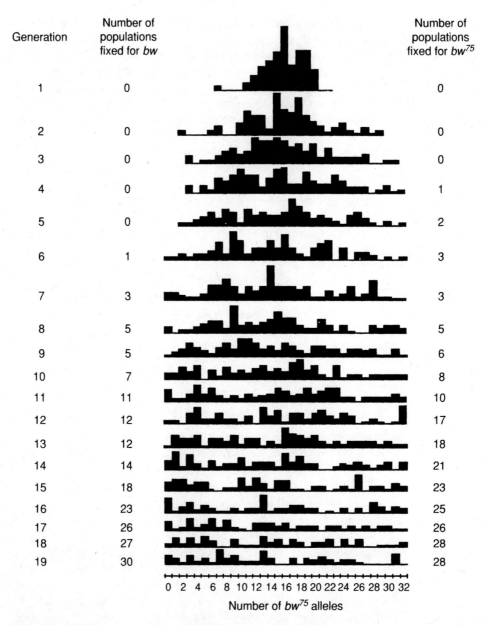

Figure 4.4. Allele frequency distributions in 107 replicate populations of *D. melanogaster*, each of size 16 and with discrete generations. From Fig. 6 in Buri (1956). Copyright ©1956 by The Society for the Study of Evolution.

- All populations started out with identical gene pools, but with time, the populations deviate not only from the ancestral condition but also from each other. At generation 19, for example, 30 populations are fixed for *bw* and 28 for *bw*^{75}. These populations no longer share any alleles at this locus, even though they were derived from genetically identical ancestral populations. *Drift causes an increase of genetic differences between populations.*

FOUNDER AND BOTTLENECK EFFECTS

As shown in the previous section, genetic drift causes its most dramatic and rapid changes in small populations. However, even a population that is large most of the time but has an occasional generation of very small size can experience pronounced evolutionary changes due to drift in the generation of small size. If the population size grows rapidly after a generation of small size, the increased population size tends to decrease the force of subsequent drift, thereby freezing in the drift effects that occurred when the population was small. These features are illustrated via computer simulation in Figure 4.5. Figure 4.5a shows four replicate simulations of genetic drift in populations of size 1000, over 100 generations, with an initial allele frequency of 0.5. Figure 4.5b shows parallel simulations, but with just one difference: At generation 20 the population size was reduced to 4 individuals and then immediately restored to 1000 at generation 21. In contrasting Figure 4.5a with 4.5b, the striking difference is the radical change in allele frequency that occurs in each population during the transition from generation 20 to 21, reflecting drift during the generation of small size. However, there is relatively little subsequent change from the allele frequencies that existed at generation 21. Thus, the pronounced evolutionary changes induced by the single generation of small population size are "frozen in" by subsequent population growth and have a profound and continuing impact on the gene pool long after the population has grown large. These computer simulations show that genetic drift can cause major evolutionary change in a population that normally has a large population size as long as either

- the population was derived from a small number of founding individuals drawn from a large ancestral population (**founder effect**) or
- the population went through one or more generations of very small size followed by subsequent population growth (**bottleneck effect**).

We will now consider some examples of founder and bottleneck effects.

There are many biological contexts in which a founder event can arise. For example, there is much evidence that Hawaiian *Drosophila* are on rare occasions blown to a new island on which the species was previously absent (Carson and Templeton 1984). Because this is such a rare event, it would usually involve only a single female. Most *Drosophila* females typically have had multiple matings and can store sperm for long periods of time. A single female being blown from one island to another would often therefore carry over the genetic material from two or three males. Hence, a founder size of 4 or less is realistic in such cases. (Single males could also be blown to a new island, but no population could be established in such circumstances.) If the inseminated female found herself on an island for which the ecological niche for which she was adapted was unoccupied, the population size could easily rebound by one or two orders of magnitude in a single generation, resulting in a situation not unlike that shown in Figure 4.5b.

Founder events are also common in humans. The village of Salinas, located in a remote mountainous area in the Dominican Republic, had a population of about 4300 people in 1974 (Imperato-McGinley et al. 1974). Seven generations prior to that time, the village was much smaller. One of the founders at that time was a man by the name of Altagracia Carrasco who had several children by at least four different women. Because the population size was small at that time, the alleles carried by Altagracia constituted a sizable portion of the total gene pool. Indeed, this is the general impact of a founder event: The alleles carried

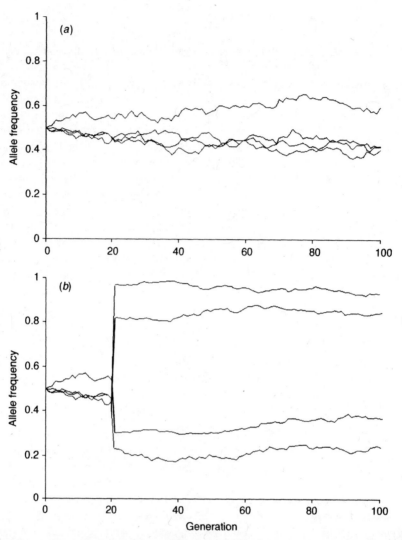

Figure 4.5. Computer simulation of genetic drift in four replicate populations starting with initial allele frequency of 0.5 over period of 100 generations. (*a*) Population size is kept constant at 1000 individuals every generation. (*b*) Same replicates are repeated until generation 20, at which point the population size is reduced to 4 individuals. The population size then rebounds to 1000 individuals at generation 21 and remains at 1000 for the remainder of the simulation to simulate a bottleneck effect.

by the founding *individuals* automatically become frequent in the founding *population* and subsequent generations. In the case of the Salinas human population, even an allele carried exclusively by this one man would by necessity come to be in relatively high frequency due to the founder effect. It turns out that Altagracia did indeed have a unique allele; he was a heterozygote for a single base substitution of thymine for cytosine in exon 5 of the autosomal *5-alpha-reductase-2* gene, causing a tryptophan (TGG) replacement of arginine (CGG) at amino acid 246 of the enzyme (Cai et al. 1996). This amino acid change in turn results in an enzyme with low catalytic activity, such that individuals homozygous for this allele have a severe deficiency in the activity of the enzyme 5-alpha-reductase-2. Although autosomal, being deficient for this enzyme has its most dramatic effects in XY individuals because this enzyme normally catalyzes the conversion of the male hormone testosterone to dihydrotestosterone (DHT). During development of an XY individual, a testosterone signal causes the developing gonads to differentiate into testes rather than ovaries. However, DHT is the hormone required for full masculinization of the external genitalia. As a consequence, XY individuals who are homozygous for this allele develop testes, but most of their remaining sexual differentiation occurs along the female pathway, the default route of development in mammals. As a consequence of the high frequency of this allele in the village of Salinas, many XY babies are born that have testes internally but externally appear to be female and are subsequently raised as girls. However, these homozygotes only have an enzyme deficient in activity, not a complete absence. When these "girls" reach puberty, their testes start producing such high levels of testosterone that the low amount of active 5-alpha-reductase-2 that they have can at long last produce sufficient DHT to trigger the development of the external male genitalia. Hence, shortly after these girls enter puberty, they are transformed into males. Although this is an unusual situation in most human populations, it is so common in the village of Salinas because of the founder effect that the townspeople even have a word for these girls who turn into men: *guevedoces* ("penis at 12" years of age).

Our second example is one of a founder effect followed by bottleneck effects (Roberts 1967, 1968). Tristan da Cunha is an isolated island in the Atlantic Ocean. With the exile of Napoleon on the remote island of St. Helena, the British decided to establish a military garrison in 1816 on the neighboring though still distant island of Tristan da Cunha. In 1817 the British Admiralty decided that Tristan was of no importance to Napoleon's security, so the garrison was withdrawn. A Scots corporal, William Glass, asked and received permission to remain on the island with his wife, infant son, and newborn daughter. A few others decided to remain and were joined later by additional men and women, some by choice and some due to shipwrecks. Altogether, there were 20 initial founders. The population size grew to 270 by 1961, mostly due to reproduction but with a few additional immigrants. The growth of this population from 1816 to 1960 is shown in Figure 4.6.

Because there is complete pedigree information over the entire colony history, the gene pool can be reconstructed at any time as the percentage of genes in the total population derived from a particular founding individual (Figure 4.7). This method of portraying the gene pool can be related to our standard method of characterizing the gene pool through allele frequencies by regarding each founder as homozygous for a unique allele at a hypothetical locus. Then, the proportion of the genes derived from a particular founder represents the allele frequency at the hypothetical locus of that founder's unique allele in the total gene pool.

The top histogram in Figure 4.7 shows the gene pool composition in 1855 and 1857. Note from the population size graph in Figure 4.6 that a large drop in population size occurred between these years. This was caused by the death in 1853 of William Glass, the original founder. Following his death, 25 of his descendants left for America in 1856. This bottleneck was also accentuated by the arrival of a missionary minister in 1851. This minister soon disliked the island, preaching that its only fit inhabitants were "the wild birds of the ocean." Under his influence, 45 other islanders left with him, thereby reducing the population size from 103 at the end of 1855 to 33 in March 1857. Note that in going from 1855 to 1857 the gene pool composition changes substantially; the relative contributions of some individuals show sharp decreases (founders 1 and 2) whereas others show sharp increases (founders 3, 4, 9, 10, 11, and 17). Moreover, the genetic

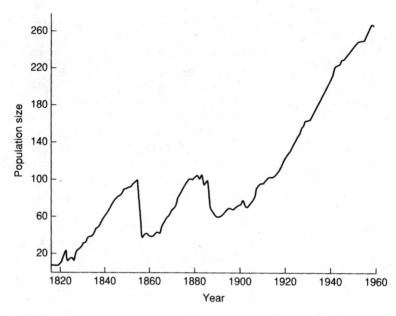

Figure 4.6. Population size of Tristan da Cunha on December 31 of each year from 1816 to 1960. Adapted from M. D. F. Roberts *Nature* 220: 1084–1088 (1968). Copyright ©1968, by Macmillan Publishers.

contributions of many individuals are completely lost during this bottleneck (founders 6, 7, 12–16, 19, and 20). Thus, the gene pool is quite different and less diverse after the first bottleneck.

Figure 4.6 reveals that the population grew steadily between 1857 and 1884. With the exception of a few new immigrant individuals (founders 21–26), the basic shape of the gene pool histograms changes very little in those 27 years (the second histogram from the top in Figure 4.7). In particular, note that there is much less change in these 27 years than in the 2 years between 1855 and 1857. Hence, the changes induced by the first bottleneck were "frozen in" by subsequent population growth.

Figure 4.6 shows that a second, less drastic bottleneck occurred between 1884 and 1891. The island has no natural harbor, so the islanders had to row out in small boats to trade with passing vessels. In 1884, a boat manned by 15 adult males sank beneath the waves with the resulting death of everyone on board, making Tristan the "Island of Widows." Only 4 adult men were left on the island, 2 very aged, leading many of the widows and their offspring to leave the island. This reduced the population size from 106 in 1884 to 59 in 1891. The third histogram in Figure 4.7 shows the impact of this second bottleneck on the island's gene pool. As with the first bottleneck, some individual contributions went up substantially (founders 3, 4, and 22), others went down (founders 9 and 10), and many were lost altogether (founders 21 and 24–26).

After this second bottleneck there was another phase of steady population growth (Figure 4.6). The shapes of the gene pool histograms change little from 1891 to 1961 during this phase of increased population growth (the bottom histogram in Figure 4.7, which excludes the impact of a few additional immigrants). Once again, this shows how subsequent population growth freezes in the changes induced by drift during the bottleneck.

FOUNDER AND BOTTLENECK EFFECTS   93

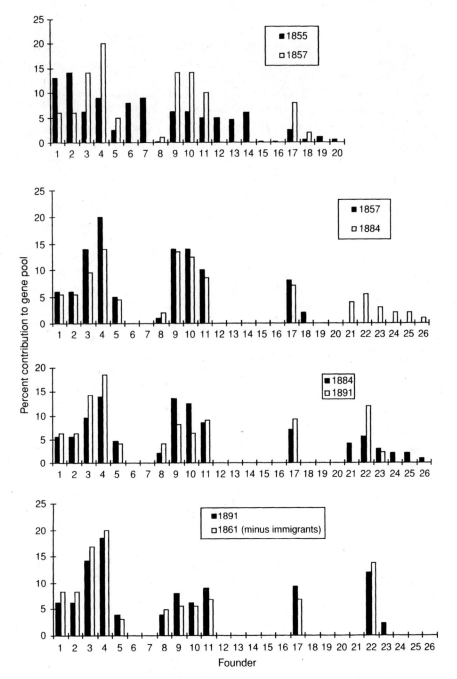

Figure 4.7. Gene pool changes over time in Tristan da Cunha population. The gene pool is estimated from pedigree data as the proportion of the total gene pool that is derived from a particular founder (indicated by numbers on the *x* axis). Each histogram contrasts the gene pool at two times, as indicated by the legends.

As discussed in Chapter 3, the founder and bottleneck effects on Tristan da Cunha also led to pedigree inbreeding, despite a system of mating of avoidance of inbreeding (see Figure 3.4). This is yet another effect of genetic drift: Finite population size leads to an increase in the mean inbreeding coefficient (the average probability of uniting gametes bearing alleles identical by descent) with time. Each bottleneck accentuates this accumulation of $\bar{F}$ because the number of founders contributing to the gene pool goes down after each bottleneck event, making it more likely that the surviving individuals must share a common ancestor. Thus, founder and bottleneck effects usually *increase* pedigree inbreeding.

GENETIC DRIFT AND DISEQUILIBRIUM

Just as drift causes changes in allele frequencies, it also changes multilocus gamete frequencies. Genetic drift tends to create linkage disequilibrium and associations between loci by chance alone. As we consider more and more loci simultaneously, we subdivide any finite gene pool into more and more gametic categories, thereby tending to make any one particular gamete type more rare. Sampling error is a strong force of evolutionary change for any gamete type that is rare in a gene pool, so in general genetic drift is a more powerful force for altering gamete frequencies at the multilocus level than at the single-locus level.

The increased sensitivity of linkage disequilibrium to drift compared to allele frequencies is illustrated by a study of 34 X-linked microsatellite loci in the United Kingdom and in 10 regions in Scotland (Vitart et al. 2005)—the urban region of Edinburgh and 9 rural regions. The rural regions had smaller populations sizes (but still in the tens of thousands or above) and less gene flow. There was little overall differentiation among these subpopulations in allele frequencies, but these areas showed large differences in the amount of linkage disequilibrium. Because several X-linked loci were available, one convenient measure of overall disequilibrium is the map distance in centimorgans at which the linkage disequilibrium on the X chromosome is half the difference between its maximum and minimum values, called the LD half distance. These are plotted in Figure 4.8. The large populations in the United Kingdom and in Edinburgh had no overall linkage disequilibrium by this measure, but several of the rural regions did have significant overall disequilibrium. Thus, differentiation could be observed at the level of linkage disequilibrium even though it was absent in terms of single-locus allele frequencies.

Founder and bottleneck effects are particularly effective in creating linkage disequilibrium and chance associations. If the loci are closely linked, the specific associations created by a founder or bottleneck episode can persist for many generations. For example, linkage disequilibrium occurs in a human population living in southern Italy west of the Apennine mountain range (Filosa et al. 1993). A 3-Mb telomeric region of the human X chromosome contains the genes for the enzyme glucose-6-phosphate dehydrogenase (G6PD) and red/green color vision. Nearly 400 distinct mutants are known at the *G6PD* locus that result in a deficiency of G6PD activity, which in turn can cause hemolytic anemia in individuals hemizygous or homozygous for a deficient allele. This population west of the Apennines has a unique deficiency allele (*Med1*), indicating both a founder effect and the relative genetic isolation of this area. Most remarkably, all *Med1* G6PD-deficient males also had red/green color blindness (which is controlled by a small complex of tightly linked genes). Interestingly, on the nearby island of Sardinia, there is also remarkable homogeneity for *G6PD*-deficient alleles (Frigerio et al. 1994), consistent with a founder effect most likely due to Phoenician contact with the island in the fifth century BCE (Filippi et al. 1977). But

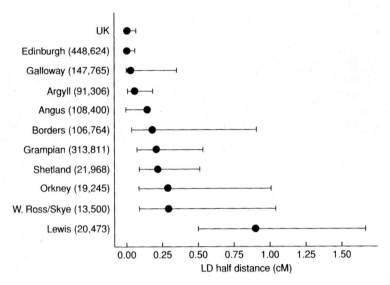

Figure 4.8. Linkage disequilibrium (LD) half distances in centimorgans in 10 Scottish regions and unrelated U.K. subjects on basis of 34 X-linked microsatellite markers. The estimates of the LD half distances are indicated by blackened circles, and the lines indicate the 90% confidence intervals. The number in parentheses is the 2001 census size for the 10 Scottish regions. Modified from Fig. 4 in V. Vitart et al., *American Journal of Human Genetics* 76: 763–772 (2005). Copyright © 2005 by The American Society of Human Genetics.

in contrast to the southern Italian population, the Sardinian population has a nearly complete absence of color blindness in *G6PD*-deficient males. Hence, both of these populations influenced by founder effects display significant linkage disequilibrium between *G6PD* and red/green color blindness, but in completely opposite directions! Such is the randomness in associations created by genetic drift.

In the previous chapter, we discovered that there were strong interactions between assortative and disassortative mating systems when populations with distinct allele frequencies are brought together and begin to interbreed. We saw earlier in this chapter that isolated subpopulations will always diverge from one another in terms of allele frequencies due to genetic drift. Divergence at single-locus allele frequencies was sufficient to generate linkage disequilibrium under admixture, as shown in Chapter 3. Now we see that isolated subpopulations will diverge not only in terms of allele frequencies but also in terms of multilocus associations within each of their gene pools (as illustrated by the opposite associations of G6PD-deficient alleles and color blindness in southern Italy versus Sardinia). The linkage disequilibrium induced by genetic drift accentuates the evolutionary impact of admixture because it makes the initial gene pools of the subpopulations even more divergent from one another. Hence, if two or more subpopulations have become genetically differentiated from one another due to drift and then are brought back together, assortative mating on any trait (whether genetic or not) that is correlated with the historic subpopulations will have a major impact on preserving both their allele frequency differences and their internal patterns of linkage disequilibrium long after genetic contact between the subpopulations has been reestablished.

The evolutionary interaction between assortative mating and drift-induced disequilibrium has major implications for studies on associations between genetic markers and disease in countries such as the United States in which several genetically differentiated populations have been placed together with patterns of assortative mating that are correlated with the historical origins of the populations. Pooling all the populations together for studying clinical associations is a poor strategy because the extensive linkage disequilibrium induced by admixture (Chapter 3) can create spurious associations between markers scattered throughout the entire genome with the disease, making disequilibrium useless for mapping causative loci. For example, hypertension (high blood pressure) is more common in African Americans than in Eruopean Americans (Dressler 1996). These two American subpopulations also differ in allele frequencies at several loci that determine blood group antigens (Workman 1973). For example, the R^o allele at the Rh blood group locus has a frequency between 0.4 and 0.6 in various African American populations, whereas this same allele has a frequency of less than 0.03 in Americans of English ancestry (Workman 1973). Therefore, if one pooled African and European Americans together in a study on hypertension, one would find a strong association of hypertension with the R^o allele, even though there is no evidence that the Rh locus is directly related to hypertension.

Pooling can also obscure genetic associations by mixing together disequilibria that by chance are in opposite directions. For example, we just saw that linkage disequilibrium exists within southern Italians and within Sardinians for $G6PD$ and color blindness. Therefore, association studies *within* either of these populations that used color blindness as a marker would correctly map the genetic region containing the genetic disease G6PD deficiency, a form of hemolytic anemia associated with deficient alleles at the $G6PD$ locus. However, suppose we wanted to see if X-linked color blindness was associated with X-linked G6PD deficiency in the country of Italy (which includes Sardinia). If we pooled all Italians and Sardinians together, we would not get much of an association, because many Italian populations show no disequilibrium between these two X-linked markers, and those that do show disequilibrium in opposite directions, as previously noted. Hence, pooling would mix informative and noninformative subpopulations and would mix opposite patterns of disequilibria that would cancel each other out. The true signal of genome location due to disequilibrium obtainable *within* either subpopulation would be diluted or even nullified in a *pooled* population, making the accurate mapping of G6PD deficiency difficult or impossible by use of the color blindness locus as a marker.

Instead of pooling, one should stratify analyses of associations between marker loci and diseases. However, one must keep in mind that genetic drift causes disequilibrium to be random in direction, and therefore the association of a disease with a particular marker allele has no universal validity. Such association studies should be used to determine genome location only of the disease-influencing genes, and the marker alleles should never be used to make disease predictions in other human populations that may not have the same pattern of linkage disequilibrium. For example, by performing separate analyses of association in the southern Italian and Sardinian populations (as was actually done), strong associations would be found between color blindness and G6PD deficiency, thereby implying that the marker locus of color blindness is closely linked to a gene influencing the disease G6PD deficiency (as indeed they are). However, suppose we had only studied the southern Italian population west of the Apennines. In that population, we would have found that color blind males were more likely to have G6PD deficiency because of that population's disequilibrium. If we then generalized from this study and incorrectly claimed that color blindness is a genetic marker indicating a predisposition to G6PD deficiency, what would happen when we applied this "knowledge" to the people living in Sardinia? There, we would be telling the people with low risk for G6PD deficiency (color blind males) that they had high risk!

Because the disequilibrium induced by drift is random in direction and admixture in humans is incomplete due to assortative mating for all sorts of traits, disease–marker association studies in humans should not be generalized beyond the actual populations studied. This is an important caution in this age in which "genes for disease xxx" are announced almost daily, yet almost all of these "genes" are actually identified through association studies with no direct proof of cause and effect.

GENETIC DRIFT, DISEQUILIBRIUM, AND SYSTEM OF MATING

The patterns of linkage disequilibrium found within a population can be strongly influenced by the interaction between genetic drift and assortative mating. Recall the two-locus model of assortative mating given in Table 3.4. This simple model and multilocus assortative mating models in general have multiple evolutionary equilibria with different patterns of linkage disequilibrium, as can be calculated from the equilibrium gamete frequencies shown in Table 3.5. In the deterministic model given in Chapter 3, the equilibrium to which a population evolves is determined only by the initial conditions of the gene pool. For the particular model shown in Table 3.4, the evolutionary outcome depended upon the relative magnitudes of the allele frequencies of the A and B alleles (Table 3.5). When the populations are regarded as being of finite size, genetic drift can also strongly influence the evolutionary outcome. This is particularly true when the initial population starts near a boundary condition that separates different potential evolutionary outcomes. For example, suppose we had a finite population whose system of mating and genetic architecture was described by Table 3.4 and that had a gene pool with $p_A \approx p_B$. Because genetic drift alters allele frequencies, it would be virtually impossible to predict which of the three evolutionary outcomes shown in Table 3.5 would actually evolve. Chance events in the initial few generations that made one allele more common than another would be amplified by the evolutionary force of assortative mating in the later generations. Only when the gene pool has been brought sufficiently close to one of the equilibrium states can assortative mating exert strong evolutionary pressures to make it extremely unlikely that drift would take the population into the domain of a different equilibrium. In general, *when multiple equilibria exist in a population genetic model, genetic drift can play a major role in determining which evolutionary trajectory is realized.*

Disassortative mating can also strongly interact with drift-induced linkage disequilibrium, particularly after founder or bottleneck effects. For example, the fruit fly *D. melanogaster* has a pheromone system leading to strong disassortative mating that is genetically controlled by just a handful of loci scattered over the genome (Averhoff and Richardson 1974, 1975). As seen in Chapter 3, disassortative mating is expected to maintain high heterozygosity at these pheromone loci and at loci in disequilibrium with them. However, as we also saw in Chapter 3, disassortative mating rapidly destroys linkage disequilibrium. Therefore, in a large population, disassortative mating at these pheromone loci is not expected to have much affect at other loci. *Drosophila melanogaster* has only a few chromosomes (an X chromosome, two major autosomes, and a very small autosome), no recombination in males, and low levels of recombination within these chromosomes in females. As a consequence of these genome and recombinational features, virtually every locus in the entire genome will be induced to have linkage disequilibrium with at least one of the pheromone loci when a severe bottleneck or founder event occurs in *D. melanogaster*. Hence, disassortative mating at the pheromone loci will also effectively cause disassortative mating at all loci for a few generations after the bottleneck or founder event. Although this disequilibrium will rapidly decrease, if the small population size does not persist for many generations, this combination of disassortative mating and temporary disequilibrium ensures that very little overall genetic variation will be lost due to drift as a consequence of the bottleneck or founder event. As a result, *D. melanogaster* is buffered against severe losses of allelic variation under temporary population bottlenecks. In contrast, other species of *Drosophila* (e.g., *D. pseudoobscura*; Powell and Morton 1979) do not have this pheromone mating system, and most species of *Drosophila* have much more recombination per unit of DNA in their genomes than *D. melanogaster*. As a result, the evolutionary impact of founder or bottleneck effects can vary tremendously from one species to the next, even within the same genus. This heterogeneity in evolutionary response to founder effects has indeed been empirically demonstrated in *Drosophila* (Templeton 1996a, 1999b) and serves to remind us that evolutionary outcomes are not determined by a single evolutionary force such as genetic drift but rather arise from an *interaction* of multiple evolutionary factors (in this case, an interaction between drift, recombination, and system of mating). Hence, there is really no such thing as *the* founder effect; rather, there are *many* types of founder effects with diverse evolutionary impacts depending upon their context.

EFFECTIVE POPULATION SIZE

As seen above, finite population size has many important evolutionary consequences: increasing the average amount of identity by descent, increasing the variance of allele frequencies through time and across populations, and causing the loss or fixation of alleles. As also shown by the above examples and simulations, the rate at which these effects occur is roughly inversely proportional to population size. In an idealized population, we can derive a precise quantitative relationship between these evolutionary effects of genetic drift and population size. As with Hardy–Weinberg, the idealized model for studying genetic drift contains many assumptions that are biologically unrealistic but that make the mathematics more tractable. In particular, our idealized case assumes the following:

- A diploid population of hermaphroditic, self-compatible organisms
- A finite population size of N breeding adults, with no fluctuation in population size from generation to generation
- Random mating
- Complete genetic isolation (no contact with any other population)
- Discrete generations with no age structure
- All individuals contributing the same number of gametes on the average to the next generation (no natural selection)
- The sampling variation in the number of gametes contributed to the next generation given by a Poisson probability distribution

The last assumption is often implicit rather than explicit and at first may seem hard to understand, but it is critical for modeling drift. Because the population size is constant (N) and diploid, each individual must on the average pass on two gametes to the next generation. However, by chance alone, some individuals may successfully pass on 0, 1, 2, 3, or more gametes to the next generation of N adults. Obviously, to the extent that some individuals pass on fewer than average gametes (particularly zero gametes), the effects of sampling error (genetic drift) are going to be accentuated. Hence, it is necessary to make a specific assumption about how uneven successful reproduction is going to be among the adults, and the Poisson model is a mathematically convenient choice.

Under these idealized conditions, it is possible to define accurately how finite population size influences the rate of change in $\bar{F}$ (the average probability of identity by descent in the population), variance of allele frequencies, or other genetic feature of the population. Of course, real populations deviate from one or more of the assumptions we have made in our idealized case. **Effective population size** allows us to measure the strength of genetic drift as an evolutionary force in these nonideal situations. Suppose, for example, we are examining how the average probability of identity by descent is accumulating in a finite population such as the Tristan da Cunha population in Figure 3.4. This human population violates every one of the assumptions that we made for our idealized population. But using the initial founders as our reference point (the $\bar{F}$'s in Figure 3.4 were calculated under the assumption of no identity by descent in the original founders), we can calculate the average generation times and then observe what $\bar{F}$ is at a particular generation. We then ask the question, what value of N do we need to use in our idealized model to yield the same value of $\bar{F}$ after the same number of generations starting from the same initial condition of $\bar{F} = 0$? Whatever value of N that is needed to accomplish this feat is the effective size. The advantage of an effective size is that it allows us to directly measure the strength of genetic drift as an evolutionary force across all real populations using a common reference. This is important because different real populations could violate different assumptions to different degrees. Without a common reference, it would be difficult to make assessments about the role of drift in these populations. For example, suppose there is a founder event of 20 individuals in a population of dioecious animals (that is, any individual is either male or female but not both as in our idealized reference population) such that there are 10 males and 10 females, but otherwise this population is exactly like our idealized population. Now suppose there is another founder population of 25 self-compatible plants with a 50–50 mix of random mating and selfing, but otherwise this plant population is exactly like our idealized population. In which population does genetic drift have the stronger impact on increasing $\bar{F}$? By determining the effective population size with respect to $\bar{F}$, we can answer this question. Hence, *effective population sizes* are measures of the strength of genetic drift in influencing some population genetic feature of interest.

What cannot be emphasized enough is that there is no such thing as *the* effective size of a population. Just as inbreeding is one word with several meanings in population genetics, so is effective population size. As indicated above, effective sizes are calculated with respect to some initial reference population and over a specific time period. Alternatively, the initial population can be regarded as being so distant in the past, that the initial state is no longer relevant and effective size is calculated with respect to a long-term equilibrium state. Regardless, the same population today can yield very different effective sizes depending upon what initial reference time is used and the number of generations involved. Moreover, when real populations deviate from the idealized reference population, the various genetic features used to monitor the impact of drift ($\bar{F}$, allele frequency variance, rate of loss or fixation, etc.) can be affected quite differently. This variation in response to deviations from the idealized population means that each genetic feature being monitored requires its *own* effective size. Therefore, if several genetic features are being monitored and/or different initial conditions and time frames are being considered, a single real population will require many different effective sizes to describe the effects of drift. It is therefore critical to describe the initial

100 GENETIC DRIFT

population, the time frame, and the genetic feature of interest before determining an effective size. Without these descriptors, the concept of an effective size is biologically meaningless.

In this book, we will consider two commonly used effective population sizes:

- The inbreeding effective size N_{ef}, used to describe the average accumulation of identity by descent ($\bar{F}$) in a population via genetic drift
- The variance effective size N_{ev}, used to describe the sampling error as measured by variance across generations that is induced in allele frequency via genetic drift or alternatively the variance in allele frequency across replicate subpopulations induced by genetic drift

Inbreeding Effective Size

When the genetic feature of interest is $\bar{F}$ (the average probability of identity by descent in uniting gametes for an autosomal locus in the population), the strength of genetic drift is measured by the inbreeding effective size. First consider our idealized reference population of constant size N. We start with an initial generation in which we regard all N individuals as being totally unrelated and non-inbred [i.e., $F(0) = 0$, where $F(i)$ denotes the average probability of identity by descent at generation i]. Suppose these N individuals produce N offspring for the next generation (generation 1). This requires that $2N$ gametes be drawn from the gene pool. Because the individuals in generation 0 are regarded as unrelated and non-inbred, any two uniting gametes that come from different individuals at generation 0 cannot be identical by descent. However, because the individuals are by assumption hermaphroditic and self-compatible, it is possible for both of the gametes involved in a fertilization event to come from the same individual in generation 0. Suppose one gamete has already been drawn from the gene pool. What is the chance that the second gamete to be united with it is drawn from the same parental individual? Because we have assumed random mating, any particular individual is equally likely to mate with any of the N individuals in the population (including itself). Hence, the chance of drawing two gametes from the same individual is simply $1/N$. Given that a self-mating has occurred, the probability that both gametes bear copies of the same gene found in the parental selfer is $\frac{1}{2}$ due to Mendelian segregation. Hence, the probability of identity by descent in the population is the probability of a self-mating times the probability that both gametes bear copies of the same gene; that is,

$$\text{Probability of identity by descent in generation } 1 = \frac{1}{N} \times \frac{1}{2} = \frac{1}{2N} = \bar{F}(1) \qquad (4.1)$$

We can immediately measure the force of genetic drift upon $\bar{F}(1)$ in this idealized population: $\bar{F}$ increases proportional to the inverse of $2N$, the number of gametes sampled from the gene pool. This means that the average probability of identity by descent increases by $1/(2N)$ due to genetic drift in this generation.

Now consider the second generation. First, an individual at generation 2 could have been produced by a selfing event among the parents from generation 1. The contribution to the average identity by descent at generation 2 from selfing at generation 1 is once again simply $1/(2N)$. This also means that the probability that a pair of uniting gametes at generation 2 is not identical by descent due to selfing at generation 1 is $1 - 1/(2N)$. In addition, even if gametes are not identical by descent due to selfing of generation 1 parents, the gametes can still be identical by descent due to inbreeding in the previous generation. The probability that two gametes chosen at random from generation 1 are identical by descent due to previous inbreeding is $\bar{F}(1)$. Note that this is the probability that two gametes bear identical copies of the gene derived from a single individual in the initial generation. Hence, the total probability of identity by descent at generation 2 is

$$\bar{F}(2) = \frac{1}{2N} + \left(1 - \frac{1}{2N}\right)\bar{F}(1) \qquad (4.2)$$

Equation 4.2 states that the probability of identity by descent at generation 2 is due in part to new inbreeding due to selfing induced by drift among the parents at generation 1 [the $1/(2N)$ term on the right side of equation 4.2] plus identity due to uniting gametes being derived from the same grandparent at the initial generation [the $\bar{F}(1)$ term] weighted by the probability that these gametes are not already identical due to selfing [the $1 - 1/(2N)$ term].

There was nothing about the derivation of equation 4.2 that limits it just to generation 2, so in general we can write

$$\bar{F}(t) = \frac{1}{2N} + \left(1 - \frac{1}{2N}\right)\bar{F}(t-1) \qquad (4.3)$$

At this point, the mathematics becomes simpler if we focus upon "heterozygosity" rather that $\bar{F}$. Here, heterozygosity is not the observed heterozygosity nor the expected heterozygosity under random mating; rather, it is the average over all individuals of the probability that an individual receives two genes from its parents at an autosomal locus that are not identical by descent relative to the initial reference generation (the individuals at generation 0 in this model). Thus, individuals who are homozygous for genes that are identical by state relative to generation 0 but are not identical by descent due to inbreeding subsequent to the initial generation are regarded as heterozygotes. For example, consider the experiments of Buri shown in Figure 4.4. The initial generation consists of a gene pool with 16 bw alleles and 16 bw^{75} alleles. By specifying this initial generation as the reference generation for calculating $\bar{F}$ in subsequent generations, we regard all 16 copies of the bw allele as different "alleles" even though they are identical in state, and likewise all 16 copies of the bw^{75} allele are regarded as distinct alleles. Because of this convention, if a fly in a later generation is homozygous for bw but the two bw alleles that the fly is bearing trace to different copies of the 16 bw alleles present in the initial generation, this fly is regarded as being heterozygous. On the other hand, if the two bw alleles trace back to the same bw allele present in the initial generation, then the fly is regarded as being "homozygous." Hence, in measuring the impact of drift through an inbreeding effective size, population geneticists make a strict distinction between identity by descent relative to the reference generation versus identity by state. This sometimes creates confusion, because this definition of heterozygosity is typically not distinguished verbally from other concepts of heterozygosity (as you can see, many words in population genetics have several meanings).

Mathematically, heterozygosity in this context is simply the average probability that two uniting gametes in the population are *not* identical by descent relative to the initial reference generation; that is,

$$H(t) = 1 - \bar{F}(t) \qquad (4.4)$$

Substituting equation 4.3 into the $\bar{F}$ term of equation 4.4, we have

$$H(t) = 1 - \frac{1}{2N} - \left(1 - \frac{1}{2N}\right)\bar{F}(t-1)$$

$$= \left(1 - \frac{1}{2N}\right)[1 - \bar{F}(t-1)]$$

$$= \left(1 - \frac{1}{2N}\right)H(t-1) \qquad (4.5)$$

Because our initial reference generation has, by definition, $\bar{F}(0) = 0$, then $H(0) = 1$. Hence, by using equation 4.5 recursively, we have

$$H(1) = \left(1 - \frac{1}{2N}\right), H(2) = \left(1 - \frac{1}{2N}\right)^2, \ldots, H(t) = \left(1 - \frac{1}{2N}\right)^t \qquad (4.6)$$

We now substitute equation 4.4 into 4.6 to express everything in terms of $\bar{F}$:

$$1 - \bar{F}(t) = \left(1 - \frac{1}{2N}\right)^t \qquad (4.7)$$

Solving equation 4.7 for $\bar{F}$ yields

$$\bar{F}(t) = 1 - \left(1 - \frac{1}{2N}\right)^t \qquad (4.8)$$

Equation 4.8 is a simple mathematical function of N and generation time t. However, what happens when the actual population deviates from any or all of the idealized assumptions? No matter what these deviations may be, after t generations there will be some realized level of inbreeding, $\bar{F}(t)$, in the actual population. Then, the inbreeding effective size (N_{ef}) of the actual population is defined as that number which makes the following equation true:

$$\bar{F}(t) = 1 - \left(1 - \frac{1}{2N_{ef}}\right)^t \qquad (4.9)$$

Solving for N_{ef} in equation 4.9, the inbreeding effective size is defined as

$$N_{ef} = \frac{1}{2\left\{1 - [1 - \bar{F}(t)]^{1/t}\right\}} \qquad (4.10)$$

Equations 4.9 and 4.10 make it clear that the **inbreeding effective size** is the number needed to make the rate of accumulation of identity by descent due to drift equal to that of the rate found in an idealized population of size N_{ef}. Note that N_{ef} is determined exclusively from $\bar{F}(t)$ and t; it has no direct dependence upon the actual population size.

With complete pedigree data on a population, it is possible to calculate F for every individual and therefore $\bar{F}(t)$ as well. Under these circumstances it is possible to determine the inbreeding effective size directly from equation 4.10 without the use of secondary equations or approximations. An example of such a population was given in Chapter 3, the captive population of Speke's gazelle founded by one male and three females between 1969 and 1972. The average probability of identity by descent in the Speke's gazelle breeding herd in 1979, regarding the four founders as unrelated and non-inbred (and hence the reference generation), was 0.1283. The average number of generations of these animals from the founders was 1.7 generations. Substituting $t = 1.7$ and $\bar{F}(1.7) = 0.1283$ into equation 4.10, the inbreeding effective size of the 1979 herd is 6.4, a number which is considerably less than the 1979 census size of 19 breeding animals. This low size is attributable to the founder effect: Although there were 19 breeding animals available in 1979, their genes were derived from only four founders and therefore they accumulated inbreeding at a very fast rate, as indicated by the low inbreeding effective size. In 1979, a new management program was instituted that included avoidance of breeding between close biological relatives (Templeton and Read 1983). The first generation bred from the 19 animals available in 1979 consisted of 15 offspring with an average probability of identity by descent of 0.1490. Using this value for $\bar{F}(t)$ and augmenting t by 1 to yield $t = 2.7$, equation 4.10 yields an inbreeding effective size of 8.6. Note that the inbreeding effective size is still smaller than the census size of 15, once again showing the persistent effects of the initial founding event. However, note that the inbreeding effective size *increased* (from 6.4 to 8.6) even though the census size *decreased* (from 19 to 15) when our inference is confined to the animals bred under the new program. This increase in inbreeding effective size reflects the impact of the avoidance of inbreeding (in a system-of-mating sense), another deviation from our idealized assumption of random mating.

In order to focus more directly upon the genetic impact of the Templeton and Read (1983) breeding program that was initiated in 1979, we can now regard the 19 animals available in 1979 as the reference generation. However, there is one complication here. Because these gazelles have separate sexes, selfing is impossible. Selfing was the fundamental source of new inbreeding in the derivations of equations 4.1 and 4.2, but with separate sexes the fundamental source of new inbreeding is the sharing of a grandparent rather than a single parent (selfing). This delays the effects of inbreeding by one generation. Hence, if we regarded the 19 animals as completely unrelated and non-inbred, their offspring would also be regarded as non-inbred under any system of mating with separate sexes. Therefore, we will make the reference generation the parents of the 19 animals available in 1979 and not just the four original founders from 1969–1972. This allows the possibility of inbreeding in the offspring of these 19 animals but reduces the impact of the original founder event and allows us to see more clearly the impact of the breeding program that was initiated in 1979. Using the parents of the 19 animals available in 1979 as the reference generation, the average probability of identity by descent of the first generation born under the new breeding program is 0.0207. Since the grandparents of the 15 animals born under the new breeding program are now the reference generation, $t = 2$, so equation 4.10 yields an inbreeding effective size of the first generation of Speke's gazelle born under the new breeding program to be 48.1. Note that the inbreeding effective size is much *greater* than the census size of 15. This observation reveals yet another widespread fallacy—that effective population sizes have to be smaller than census sizes. This statement, widespread in much of the conservation biology literature, is not true. Always remember that inbreeding effective sizes are determined solely by $\bar{F}$ and t with no direct dependency upon census size. Since the idealized reference population has random mating ($f = 0$), a breeding program such as that designed for the Speke's gazelle that has strong avoidance of inbreeding (recall from Chapter 3 that $f = -0.291$ under the system of mating established under the new breeding program) will greatly reduce the rate at which average pedigree inbreeding ($\bar{F}$) accumulates, thereby greatly augmenting the inbreeding effective size. Hence, effective sizes can be much larger or much smaller than the census size, depending upon the type of deviations that are occurring from the idealized reference population.

Note also that we now have two very different inbreeding effective sizes (8.6 and 48.1) for the same 15 animals. The first size tells us how rapidly inbreeding has accumulated since the initial founder event; the latter size focuses on the impact of the new breeding program. Hence, both numbers give valuable information. Unfortunately, it is commonplace in the population genetic literature to refer to "the" effective population size as if there were only one effective size for a population. As the Speke's gazelle example shows, there can be many different effective sizes for the same population as a function of different (and meaningful) reference generations. As we will now see, there are also different genetic types of effective size as well.

Variance Effective Size

Genetic drift causes random deviations from the allele frequency of the previous generation. Drift also causes variation in allele frequency across replicate subpopulations (e.g., Figure 4.4). These random deviations in allele frequency either across generations or across replicate subpopulations are commonly measured by the variance in allele frequency. To see how the variance of allele frequency can be used to measure the strength of genetic drift, we will consider evolution induced by genetic drift at a single autosomal locus with two alleles, A and a, in a population of size N that satisfies all of our idealized assumptions. Exactly $2N$ gametes are sampled out of the gene pool each generation in the idealized population. The idealized set of assumptions means that the sample of A and a alleles due to drift follows the binomial distribution (Appendix 2), the same type of distribution simulated in Figures 4.1–4.3. Let p be the original frequency of A, and let x be the number (not frequency) of A's in the finite sample of $2N$ gametes. Then, with an idealized population of size N, the binomial probability that $x = X$ (where X is a specific realized value to the number of A alleles in the sampled gametes from the gene pool) is

$$\text{Probability}(x = X) = \binom{2N}{X} p^X q^{2N-X} \qquad (4.11)$$

where $q = 1 - p$. The mean or expected value of x in the binomial is $2Np$ and the variance of x is $2Npq$ (Appendix 2). The allele frequency of A in the next generation is $x/(2N)$, so

$$\text{Expected allele frequency} = E\left(\frac{x}{2N}\right) = \frac{E(x)}{2N} = \frac{2Np}{2N} = p \qquad (4.12)$$

The fact that the expected allele frequency does not change from one generation to the next reflects the fact that drift has no direction. Because there is no direction to drift, it is impossible to predict whether deviations caused by drift will be above or below the previous generation's allele frequency. Although no change is expected *on the average*, we know from the previous computer simulations (Figures 4.1–4.3) and examples (Figure 4.4) that any particular population is likely to experience an altered allele frequency. The expected squared deviation from the original allele frequency (that is, the variance in allele frequency) measures this tendency of drift to alter the allele frequency in any particular population. This variance is

$$\text{Variance in allele frequency} = \text{Var}\left(\frac{x}{2N}\right) = \frac{\text{Var}(x)}{(2N)^2} = \frac{2Npq}{(2N)^2} = \frac{pq}{2N} \quad (4.13)$$

When populations deviate from one or more of our idealized set of assumptions, the **variance effective size** of one generation relative to the previous generation is defined to be the number, N_{ev}, that makes the following equation true:

$$\text{Variance in allele frequency} = \frac{pq}{2N_{ev}} \quad (4.14)$$

The variance in equation 4.14 can be interpreted either as the expected square deviation of the allele frequency in a particular population from its initial value of p or as the variance in allele frequencies across identical replicate populations all starting with an initial allele frequency of p. As can be seen from Figure 4.4, the allele frequencies in replicate populations become increasingly spread out with increasing time. Mathematically this means that the variance in allele frequency increases with time under drift, in terms of either the variance across replicate populations or the expected variance within a single population. When the reference generation is more than one generation in the past, we must first determine how the variance of allele frequency changes over multiple generations. This determination is done in Box 4.1. From that box, we see that

$$\text{Variance in allele frequency after } t \text{ generations} = \text{Var}(p_t) = pq\left[1 - \left(1 - \frac{1}{2N}\right)^t\right] \quad (4.15)$$

where p is the allele frequency at generation 0, $q = 1 - p$, t is the number of generations from the initial reference population, and N is the number of individuals in the idealized population undergoing drift (or the size of each subpopulation when dealing with isolated replicates). As t increases in equation 4.15, the variance in allele frequency also increases. This confirms our earlier observation that drift accumulates with time and that deviations from the initial conditions become more and more likely. Also note that as t goes to infinity, the variance in allele frequency goes to pq. In terms of an experimental setup like Buri's (Figure 4.4), the variance of pq is obtained when all populations have become fixed for either the A or a allele, such that a proportion p of the populations are fixed for the A allele (and therefore have an allele frequency of 1 for A) and q are fixed for the a allele (and therefore have an allele frequency of 0 for A). Under these conditions, the average allele

BOX 4.1 THE VARIANCE IN ALLELE FREQUENCY OVER TIME

Let x_i be the number of A alleles in an idealized finite population of size N at generation i. We will first consider the impact of two successive generations of drift upon the variance of the allele frequency at generation i, $p_i = x_i/(2N)$. Because genetic drift has no direction, we know that the expected value of p_i is p for every generation i, where p is the frequency of the A allele in the initial reference population (generation 0). By definition, the variance of p_i, σ_i^2, is

$$\sigma_i^2 = E(p_i - p)^2$$

This variance can be expressed in terms of the allele frequencies of two successive generations as

$$\begin{aligned}\sigma_i^2 &= E(p_i - p_{i-1} + p_{i-1} - p)^2 \\ &= E(p_i - p_{i-1})^2 + 2E(p_i - p_{i-1})(p_{i-1} - p) + E(p_{i-1} - p)^2 \\ &= E(p_i - p_{i-1})^2 + 2E(p_i - p_{i-1})(p_{i-1} - p) + \sigma_{i-1}^2\end{aligned}$$

The mathematical assumptions made about our idealized population give the drift process the statistical property of being a Markov process. This means that the probabilities in going from generation $i-1$ to generation i are conditionally independent of all previous generations once the outcome at generation $i-1$ is given. This in turn means that we can separate the expectation operator into two components: $E = E_{i-1} E_{i|i-1}$, where E_{i-1} is the expectation at generation $i-1$ and $E_{i|i-1}$ is the expectation at generation i given the outcome at generation $i-1$. Using this Markovian property, the variance at generation i can be expressed as

$$\begin{aligned}\sigma_i^2 &= E_{i-1} E_{i|i-1}(p_i - p_{i-1})^2 + 2E_{i-1}\left[(p_i - p_{i-1})E_{i|i-1}(p_{i-1} - p)\right] + \sigma_{i-1}^2 \\ &= E_{i-1}\frac{p_{i-1}(1 - p_{i-1})}{2N} + 2E_{i-1}\left[(p_i - p_{i-1})E_{i|i-1}(p_{i-1} - p)\right] + \sigma_{i-1}^2 \\ &= \frac{1}{2N} E_{i-1}\left[p_{i-1} - p^2 - (p_{i-1} - p)^2\right] + 0 + \sigma_{i-1}^2 \\ &= \frac{1}{2N}\left[p - p^2 - E_{i-1}(p_{i-1} - p)^2\right] + \sigma_{i-1}^2 \\ &= \frac{pq}{2N} + \sigma_{i-1}^2\left(1 - \frac{1}{2N}\right)\end{aligned}$$

We already saw that $\sigma_i^2 = pq/(2N)$, so plugging this initial condition into the above equation, we have that

$$\begin{aligned}\sigma_i^2 &= \frac{pq}{2N} + \frac{pq}{2N}\left(1 - \frac{1}{2N}\right) \\ &= pq\left[1 - \left(1 - \frac{1}{2N}\right)^2\right]\end{aligned}$$

By recursion, we have

$$\sigma_i^2 = pq\left[1 - \left(1 - \frac{1}{2N}\right)^i\right]$$

frequency of A over all replicate populations is $p \cdot 1 + q \cdot 0 = p$ and the variance of the allele frequency is

$$\text{Var(allele frequency)} = E\left(\frac{x}{2N} - p\right)^2 = p(1-p)^2 + q(0-p)^2 = pq(q+p) = pq \quad (4.16)$$

Hence, equation 4.15 also tells us that eventually drift causes all initial genetic variation to become lost or fixed, the only way to achieve the maximum theoretical variance of pq. Equation 4.15 also provides us with our primary definition of the variance effective size; namely, if the actual variance in allele frequency after t generations is σ_t^2, then the *variance effective size* of generation t relative to generation 0 is defined to be the number, N_{ev}, that makes the following equation true:

$$\sigma_t^2 = pq\left[1 - \left(1 - \frac{1}{2N_{ev}}\right)^t\right] \quad (4.17)$$

that is,

$$N_{ev} = \frac{1}{2\left\{1 - \left[1 - \sigma_t^2/(pq)\right]^{1/t}\right\}} \quad (4.18)$$

The variance effective size measures how rapidly allele frequencies are likely to change and/or how rapidly isolated subpopulations diverge from one another under genetic drift. As with inbreeding effective size, variance effective size is defined solely in terms of the genetic feature of interest (in this case the variance of allele frequency), the reference population, and time and *not* in terms of census size. The variance effective size also measures how rapidly a population loses genetic variation under drift as measured relative to the maximum variance in allele frequency (equation 4.16). With regard to this last biological interpretation, there is another effective size called the **eigenvalue effective size** that directly measures the rate at which alleles become fixed or lost. Because of its mathematical complexity, we will not deal with the eigenvalue effective size in any detail in this book. We simply note that the eigenvalue and variance effective sizes both tend to have similar values because they are measuring the same biological phenomenon (loss of variation due to drift) although through different measures. The existence of the eigenvalue effective size also serves as a reminder that many types of effective size exist. Once again, there is no such thing as "the" effective size; there are many different effective sizes each with its own unique biological meaning. Effective size is defined with respect to a genetic feature of interest and a reference generation. As the genetic feature of interest changes and as the reference generation changes, the effective size changes. Hence, a population can be characterized by several different effective sizes simultaneously for a variety of genetic parameters and reference generations.

To show this, consider once again the captive herd of Speke's gazelle. There are no replicate populations of this herd, but because we know the entire pedigree, we can simulate drift at a hypothetical autosomal locus keeping the pedigree structure constant in order to create the analogue of multiple replicate populations (MacCleur et al. 1986). Using the original four founding animals as the reference generation, the variance across 10,000 simulated replications of the actual pedigree structure resulted in $\sigma^2/pq = 0.135$ for the census population of 15 animals born under the Templeton and Read (1984) breeding program. Recall that these animals are 2.7 generations from the original founding animals, so $t = 2.7$. Using these values in equation 4.18, the variance effective size of these 15 animals relative to the original founders is 9.6. Recall that the inbreeding effective size for these same 15 animals relative to the original founders was 8.6. Thus, the variance effective size is 12% larger than the inbreeding effective size for these particular animals relative to the reference founding generation. There is no a priori reason for these two effective sizes to be equal, and the difference between inbreeding and variance effective sizes in this case means that these animals accumulated pedigree inbreeding faster than they lost genetic variation, primarily due to the system of mating that existed prior to the initiation of the Templeton and Read (1984) program.

We can eliminate the impact of the system of mating that existed prior to the Templeton and Read (1984) program by using as the reference population the parents of the 15 animals from the first generation of the breeding program. Using the exact pedigree structure and computer-simulated replicates, the increase in variance in that single generation implies a variance effective size of 20.1. As with the inbreeding effective size calculated earlier, note that the variance effective size in this case is also larger than the census size of 15. Thus, the breeding program is preserving genetic variation in this population at a rate above that of the census size. Once again, the common wisdom that effective sizes are always smaller than census sizes is violated in this case for both variance and inbreeding effective sizes. However, although the variance effective size is only modestly larger than the census size, recall that the inbreeding effective size for these same 15 animals was 48.1; that is, the inbreeding effective size is now 239% larger than the variance effective size for the same population of gazelles. When the four founding animals were the reference generation, the variance effective size was *larger* than inbreeding effective size (9.6 vs. 8.6), but now it is much *smaller* (20.1 vs. 48.1). This illustrates the critical importance of the reference generation for both inbreeding and variance effective sizes. Moreover, the large discrepancy between inbreeding and variance effective sizes under the Templeton and Read (1984) breeding program serves as a caution that these two types of effective sizes are biologically extremely distinct and that the idea of "the" effective size is meaningless and misleading.

The difference between the inbreeding and variance effective numbers for the Speke's gazelle relates to a widespread misconception that inbreeding *causes* a loss of allelic diversity. This misconception stems from observations like that made with the Speke's' gazelle herd; this herd is becoming inbred in an identity-by-descent sense and is losing allelic variation. The effective size calculations show, however, that these two genetic processes are occurring at different rates, particularly after the initiation of the Templeton and Read (1984) breeding program. Inbreeding, in the system-of-mating sense (f), has no direct impact on gamete frequencies and hence does not promote or retard the loss of genetic variation by itself (Chapter 3). How about average pedigree inbreeding ($\bar{F}$)? Consider an infinitely large selfing population. Such a population will have a high value of $\bar{F}$ but no loss of allelic variation due to drift. Thus, inbreeding, as measured by either f or $\bar{F}$, has no direct impact on losing allelic variation. What is really going on is a correlation between inbreeding and loss of allelic variation; genetic drift causes a reduction in genetic variability, as measured by N_{ev} and genetic drift increases inbreeding in the sense of average probability of identity by descent, as measured by N_{ef}, leading to a negative correlation between genetic variation and inbreeding in many finite populations. Hence, *genetic drift, not inbreeding, is the actual cause of loss of allelic variation in a finite population.* This distinction between cause and correlation is critical because correlated relationships can be violated in particular instances. For example, population subdivision (to be discussed in Chapter 6) can under some circumstances decrease the rate of loss of genetic variation but increase the rate of accumulation of identity by descent; that is, subdivided populations in toto can be more inbred but have higher levels of overall allelic diversity than a single panmictic population of size equal to the sum of all of the subpopulations. Hence, in some realistic biological situations, factors that increase pedigree inbreeding result in a greater retention of allelic diversity. Hence, *inbreeding does not cause a loss of allelic diversity in finite populations; rather, the loss of allelic diversity is due to finite size itself.*

In general, inbreeding effective sizes are primarily sensitive to the number of *parents* and their reproductive characteristics (or other generations even more remote in the past, depending upon the reference generation), whereas variance effective size is primarily sensitive to the number of *offspring* and their attributes. As an example, the simulations of the bottleneck effect shown in Figure 4.5b dealt with populations that were ideal except for fluctuating population size. In particular, $N = 1000$ up to and including generation 20, but population size dropped to 4 in generation 21 and then went back up to 1000 in generation 22. Suppose we start with generation 20 as our reference generation. What are the inbreeding and variance effective sizes of generation 21? First, we calculate inbreeding effective size. Since gametes are randomly sampled from all 1000 parents to produce the 4 individuals of generation 21, the probability that two uniting gametes are derived from the same parent is simply 1/1000. Because there is no inbreeding in the parents (by definition, since we made it the reference generation), the probability that two such gametes involved in a selfing event are identical by descent is $\frac{1}{2}$. Hence, $\bar{F}(21) = 1/2000$, which means that the inbreeding effective size of generation 21 is 1000 even though the census size is 4. Now consider the variance effective size. Since only eight gametes are sampled, the variance of p in generation 21 is $pq/8$, which implies a variance effective size of 4 at generation 21 using generation 20 as the reference generation. Hence, the same four individuals at generation 21 have two very different effective sizes (1000 versus 4) for the genetic parameters of identity by descent versus variance of allele frequencies relative to generation 20.

In nonideal populations (that is, real populations), inbreeding and variance effective sizes are generally different. Hence, the term "effective size of the population" is meaningless unless the genetic feature of interest and the reference generation are both specified. We will now examine the differences between inbreeding and variance effective sizes in more detail.

Some Contrasts between Inbreeding and Variance Effective Sizes

Population geneticists have derived several equations that relate inbreeding and variance effective sizes to actual population size under a variety of deviations from the idealized assumptions. For example, suppose the population is not constant but fluctuates from generation to generation but with all other idealized assumptions holding true. In analogy to equation 4.6, the heterozygosities at various generations are given by

$$H(1) = \left(1 - \frac{1}{2N(0)}\right)$$

$$H(2) = \left(1 - \frac{1}{2N(0)}\right) \times \left(1 - \frac{1}{2N(1)}\right)$$

$$\vdots$$

$$H(t) = 1 - \bar{F}(t) = \prod_{i=1}^{t}\left(1 - \frac{1}{2N(i-1)}\right) \tag{4.19}$$

rather than $[1 - (1/2N)]^t$ for $H(t)$ as in equation 4.6. Using equation 4.7, we have

$$1 - \bar{F}(t) = \prod_{i=1}^{t}\left(1 - \frac{1}{2N(i-1)}\right) \equiv \left(1 - \frac{1}{2N_{ef}}\right)^t \tag{4.20}$$

so

$$t \times \ln\left(1 - \frac{1}{2N_{ef}}\right) = \sum_{i=1}^{t} \ln\left(1 - \frac{1}{2N(i-1)}\right) \tag{4.21}$$

where ln is the natural logarithm. If all the N's are large numbers, then using a Taylor's series expansion from calculus, $\ln[1 - 1/(2N)] \approx 1/(2N)$. Under the assumption that all N's are large, an approximation to equation 4.21 is

$$\frac{1}{2N_{ef}} = \sum_{i=1}^{t} \frac{1}{2N(i-1)} \tag{4.22}$$

Solving equation 4.22 for N_{ef}, an *approximation* to the inbreeding effective size is (Crow and Kimura 1970)

$$N_{ef} = \frac{t}{\frac{1}{N(0)} + \frac{1}{N(1)} + \cdots + \frac{1}{N(t-1)}} \tag{4.23}$$

that is, the inbreeding effective size is approximately the harmonic mean of the population sizes of the previous generations going back to the initial reference generation. Harmonic means are very sensitive to low values. As a consequence, a single generation of low population size can allow drift to have a major impact on the accumulation of identity by descent, as we have already seen for founder and bottleneck effects (e.g., Figure 3.4).

Now consider the impact of fluctuating population size in an otherwise idealized population upon variance effective size. The equation analogous to equation 4.17 now becomes

$$\sigma_t^2 = pq\left\{1 - \left[\prod_{i=1}^{t}\left(1 - \frac{1}{2N(i)}\right)\right]\right\} \tag{4.24}$$

In a proof similar to that given for inbreeding effective size, the variance effective size is approximately given by

$$N_{ev} = \frac{t}{\frac{1}{N(1)} + \frac{1}{N(2)} + \cdots + \frac{1}{N(t)}} \qquad (4.25)$$

As with the inbreeding effective size given earlier, the variance effective size is approximated by the harmonic mean of the population sizes. However, there is one critical difference between equations 4.23 and 4.25: In equation 4.23 the inbreeding effective size is a function of the population sizes from generation 0 to generation $t - 1$; in equation 4.25 the variance effective size is a function of the population sizes from generation 1 to generation t. This shift in the generational indices between 4.23 and 4.25 reflects the greater dependence of inbreeding effective size upon the parents (generations 0 through $t - 1$) versus the greater dependence of variance effective size upon the offspring (generations 1 through t). Whether or not this shift makes much of a difference depends upon the sizes at generations 0 and t. For example, in the Speke's gazelle captive population, a natural reference generation would be the four founding animals because we had no detailed information about any previous generation. As we already saw, the founder effect of this generation of four animals had a pronounced effect on the inbreeding effective sizes of subsequent generations. However, the reference founder generation of 4 ($t = 0$) would not directly affect the variance effective size of this population, resulting in variance effective sizes that are larger than inbreeding effective sizes for this herd when the four founders are the reference generation, as we have previously seen.

Consider now another single deviation from our idealized situation—a deviation from random mating. As we saw earlier with the Speke's gazelle, the inbreeding effective size of the first generation of 15 animals bred under the Templeton and Read (1984) breeding program relative to their grandparents was 48.1 whereas the variance effective size was 20.1. This great excess of inbreeding effective size over census size and variance effective size was primarily due to the system of mating characterized by extreme avoidance of inbreeding (recall from Chapter 3 that $f = -0.291$ under the Templeton and Read program). Avoidance of inbreeding had a strong and direct impact upon the rate of accumulation of identity by descent but had only a modest impact upon variance effective size. Likewise, we would expect a system of mating that promoted inbreeding as a deviation from Hardy–Weinberg to decrease inbreeding effective size relative to variance effective size. To see this analytically, we will use one of the mathematical interpretations of f found in the population genetic literature (e.g., Li 1955).

Suppose a population has a system of mating that promotes inbreeding such that a fraction f of individuals are completely inbred (that is, they are homozygous through identity by descent) and the remainder of the population $(1 - f)$ reproduce through random mating. This mixture results in the same genotype frequencies as other interpretations of f (equations 3.2) but is easier to work with analytically in investigating interactions with drift (although this interpretation of f is defined only for $f \geq 0$). First, consider the impact of f upon the rate of accumulation of F in an otherwise ideal population of constant size N. The analogue of equation 4.3 is now

$$\bar{F}(t) = f + (1 - f)\left[\frac{1}{2N} + \left(1 - \frac{1}{2N}\right)\bar{F}(t - 1)\right] \qquad (4.26)$$

The first term on the right-hand side of equation 4.26 represents the direct impact of the inbreeding system of mating upon identity by descent, and the part in the brackets is identical to equation 4.3 and represents the impact of genetic drift in the random-mating fraction $(1 - f)$. Just as we did for equation 4.3, we can use equation 4.26 to predict the average identity by descent at generation t in terms of an initial reference generation for which we set $\bar{F}(0) = 0$:

$$\bar{F}(t) = 1 - \left[(1-f)\left(1 - \frac{1}{2N}\right)\right]^t \tag{4.27}$$

Substituting equation 4.27 into equation 4.10, we have that inbreeding effective size under this inbreeding system of mating characterized by f is

$$N_{ef} = \frac{N}{1 + f(2N - 1)} \tag{4.28}$$

Now consider the impact of f upon the variance of allele frequency given an initial frequency of p. The random-mating fraction of the population has the standard variance of $pq/(2N)$. However, the inbred fraction is totally homozygous through identity by descent, so individuals from this portion can only pass on one type of gamete per individual. Hence, the variance in the inbred fraction is pq/N. Therefore, the total variance in allele frequency induced by drift is

$$\text{Variance in allele frequency} = (1-f)\frac{pq}{2N} + f\frac{pq}{N} = \frac{pq(1+f)}{2N} \tag{4.29}$$

The variance effective size under this deviation from random mating is therefore [see Li (1955) for more details]

$$N_{ev} = \frac{N}{1 + f} \tag{4.30}$$

Note that both equations 4.28 and 4.30 yield effective sizes equal to N when $f = 0$. This makes sense because we are assuming the populations are ideal with the possible exception of system of mating. When $f = 0$, the populations are completely ideal and hence there is no distinction between an effective size of any type and the census size when all idealized conditions are satisfied.

When $f > 0$, however, these two types of effective size are not generally the same. To illustrate the difference between equations 4.28 and 4.30, consider the special case when $f = 0.1$ and N varies from 2 to 200, as plotted in Figure 4.9. As can be seen from this figure, there is a uniform reduction in the variance effective size relative to the census size for all values of N such that N_{ev} is a linear function of N. In great contrast, N_{ef} quickly levels off close to a value of 5. Why this difference that becomes more extreme with increasing N? The reason is that the system of mating, regardless of population size, is augmenting $\bar{F}$ by 0.1 per generation in this case. An ideal random-mating population of size 5 would also create 0.1 new identity by descent every generation [$1/(2N) = 0.1$ when $N = 5$]. Hence, as the census size of this nonrandomly mating population becomes increasingly large, the impact of genetic drift becomes increasingly small such that the total rate of accumulation of average identity by descent converges to the rate due exclusively to nonrandom mating.

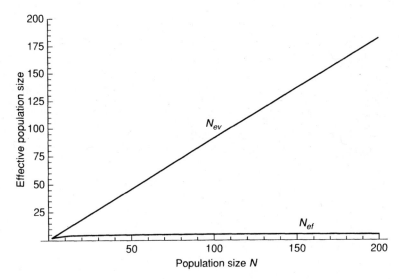

Figure 4.9. Inbreeding effective size (thick line) and variance effective size (thin line) as function of population size N. The population is assumed to satisfy all the idealized assumptions except for random mating. The population is assumed to have an inbreeding system of mating such that $f = 0.1$.

This results in an asymptotic inbreeding effective size of 5. Equations 4.28 and 4.30 and Figure 4.9 reinforce our earlier conclusion that inbreeding and variance effective sizes can be influenced in extremely different ways by biologically plausible and common deviations from our set of idealized assumptions.

There are many other equations in the population genetic literature that are either special cases or approximations to equations 4.10 and 4.18 that focus upon the impact of specific deviations or sets of deviations from our idealized set of assumptions. For example, Crow and Kimura (1970) give equations for several special cases of inbreeding and variance effective sizes. We will only consider one: when the offspring distribution no longer has a Poisson distribution with a mean and variance of two offspring per individual (the number required for a stable population size in a sexually reproducing diploid species) but rather has a mean of k offspring per individual with a variance of v. Let a population of N individuals fulfill all of the other idealized assumptions. Then, the inbreeding and variance effective sizes of this population of N individuals relative to the previous generation of N_0 individuals (note that $N = N_0 k/2$) is (Crow and Kimura 1970)

$$N_{ef} = \frac{2N - 1}{k - 1 + (v/k)[1 - k/(2N)]} \qquad N_{ev} = \frac{2N}{1 - f_0 + (v/k)(1 + f_0)} \qquad (4.31)$$

where f_0 is the deviation from Hardy–Weinberg genotype frequencies in the parental generation. Given that our idealized assumptions include random mating, one may wonder why f_0 is appearing. The reason is that finite population size alone induces deviations from Hardy–Weinberg expected frequencies (we already saw this in testing for Hardy–Weinberg genotype frequencies with a chi-square statistic with finite samples; even with random mating, a finite sample is rarely exactly at the Hardy–Weinberg expectations). For an idealized monoecious population such as we have here, the expected value of f_0 induced by genetic drift is $-1/(2N_0 - 1) = -1/(4N/k - 1)$ (Crow and Kimura 1970). We will use this expected value of f_0 in all evaluations of equation 4.31.

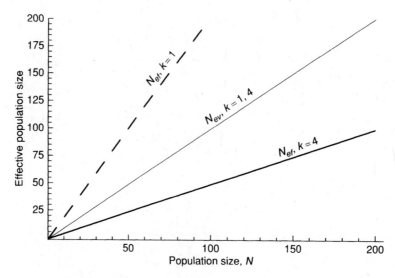

Figure 4.10. Inbreeding effective size (thick lines) and variance effective size (thin lines) as function of population size N. The population is assumed to satisfy all the idealized assumptions except that it is either increasing with $k = 4$ (solid lines) or decreasing with $k = 1$ (dashed lines). The variance effective sizes for $k = 4$ and $k = 1$ are virtually identical and appear as a single thin solid line.

One simple way of deviating from the idealized set of assumptions is to have k be a number other than 2. If $k < 2$, the population size is declining with time; if $k > 2$, the population size is increasing. To focus upon the impact of population decline or growth upon the effective sizes, we retain the assumption of a Poisson offspring distribution ($v = k$) and plot equations 4.31 for the special cases of $k = 4$ and $k = 1$ over the range of N from 2 to 200 (Figure 4.10). As can be seen, the impact of population increases or declines have very little impact on the variance effective size relative to the previous generation, with $N_{ev} \approx N$ in both cases. In contrast, the inbreeding effective size is very sensitive to population growth or decline. When the population is declining, the inbreeding effective size is larger than N, and when the population is increasing, it is smaller than N. This difference in the behavior of these two effective sizes reflects the observation made earlier that inbreeding effective size is primarily sensitive to the parental attributes and numbers (which are less than N when the population is growing and more than N when the population is declining) whereas the variance effective size is primarily sensitive to the offspring attributes and numbers (which is N by definition in both cases).

We now focus on the impact of deviating from the Poisson assumption by plotting the effective sizes as a function of v, the variance in number of offspring, for populations with $k = 4$ (Figure 4.11) and $k = 1$ (Figure 4.12) when $N = 100$. Figures 4.11 and 4.12 reveal that both types of effective size decrease as v increases; that is, the larger the variance in offspring number, the lower both types of effective population size. A contrast of Figures 4.11 and 4.12 reveals an interaction between the demographic parameters of k and v with N_{ef} and N_{ev}. When the population size is increasing, N_{ev} is more sensitive to changes in v than is N_{ef} (Figure 4.11), but the opposite is true when population size is declining (Figure 4.12). One consequence of this interaction is that as v gets larger, the difference between these two types of effective size becomes smaller (for example, contrast the results at $v = 1$ versus $v = 10$ in Figures 4.11 and 4.12). These two figures reinforce our conclusion that inbreeding

EFFECTIVE POPULATION SIZE   115

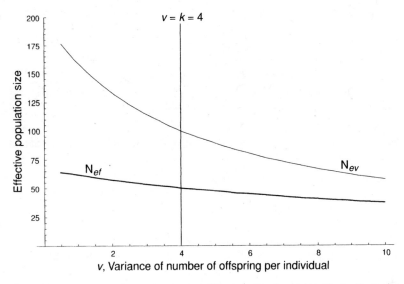

Figure 4.11. Relationship between inbreeding (thick line) and variance (thin line) effective sizes with variance in number of offspring per individual in otherwise idealized population of size 100 relative to previous generation and with average number of four offspring per parental individual. The intersections of the effective size curves with the straight line marked $v = k = 4$ show the effective sizes obtained under the ideal assumption of a Poisson offspring distribution.

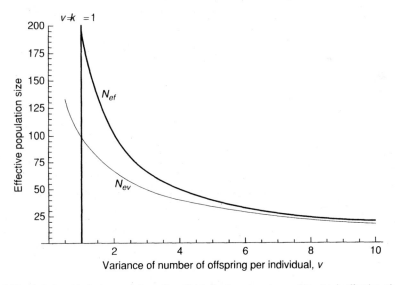

Figure 4.12. Relationship between inbreeding (thick line) and variance (thin line) effective sizes with variance in number of offspring per individual in otherwise idealized population of size 100 relative to previous generation and with average number of one offspring per parental individual. The intersections of the effective size curves with the straight line marked $v = k = 1$ show the effective sizes obtained under the ideal assumption of a Poisson offspring distribution.

and variance effective sizes are distinct biological measures that should never be equated and each is affected in complex but distinct fashions when deviations from our idealized set of assumptions occur.

The population literature contains many other equations for effective size that examine the impact of one or more deviations from the idealized set of assumptions. These equations can then be used to estimate various effective population sizes by the measurement of specific ecological and demographic variables (such as k or v or the harmonic means of census sizes). Although this patchwork approach breaks the problem of estimating effective population size into more easily managed chunks, one should never forget that equations such as 4.23, 4.25, 4.28, 4.30, or 4.31 are only special cases or approximations to an effective size that include the impact of some ecological and demographic factors but totally ignore others. Given the strong interactions that exist among the factors influencing effective sizes, an estimation procedure that ignores some factors is biologically unreliable. For example, suppose we measured a population's census size to be 100 and estimated the demographic parameter k to be 1. All unmeasured factors are generally assumed to correspond to the ideal case, so these two measurements would cause us to estimate the inbreeding effective size to be 200 and the variance effective size to be 100 (the intersection with the $v = k = 1$ line in Figure 4.12). However, suppose that the unmeasured parameter v was in reality 10. In that case, both effective sizes would be about 25 (Figure 4.12). Hence, by ignoring the variance in offspring number in this case, we would have seriously overestimated both types of effective sizes and would have erroneously concluded that the inbreeding effective size was much larger than the variance effective size. Because of the strong interactions among factors, the failure to estimate even a single one can seriously affect the entire estimation procedure.

One way of avoiding the problem of a patchwork estimation procedure of included and missing factors is to return to the original definitions of effective size in terms of a genetic feature (equations 4.10 and 4.18) and then measure how these genetic features change over time (Husband and Barrett 1992, 1995; Nunney 1995). For example, equation 4.17 can be expressed as

$$\frac{\sigma_t^2}{pq} = V_s = \left[1 - \left(1 - \frac{1}{2N_{ev}}\right)^t\right] \approx \frac{t}{2N_{ev}} \quad (4.32)$$

when N_{ev} is large and where V_s is the standardized variance of allele frequency at time t. Krimbas and Tsakas (1971) showed that one could sample a population at two time periods, say times 0 and t (where time is measured in generations), and estimate the frequency of allele i at an autosomal locus in these two samples as x_i at time 0 and y_i at time t. Krimbas and Tsakas (1971) then showed that one could estimate V_s by

$$\hat{V}_s = \frac{1}{A} \sum_{i=1}^{A} \frac{(x_i - y_i)^2}{x_i(1 - x_i)} \quad (4.33)$$

where A is the number of alleles at the locus being surveyed. Combining equation 4.33 with the approximation given in equation 4.32, an approximate estimator of the variance effective size is

$$\hat{N}_{ev} = \frac{t}{2\hat{V}_s} \quad (4.34)$$

The estimators in equations 4.33 and 4.34 have some poor statistical properties. For example, if an allele found at time t was not present in the initial sample due to incomplete sampling, $x_i = 0$ and equation 4.33 is undefined. Also, equation 4.34 is not corrected for several potential sampling biases. Accordingly, Nei and Tajima (1981) suggested the following alternatives

$$\hat{V}_{sa} = \frac{1}{A} \sum_{i=1}^{A} \frac{(x_i - y_i)^2}{\frac{1}{2} x_i(1 - x_i) - x_i y_i} \qquad \hat{N}_{eva} = \frac{t}{2\left[\hat{V}_{sa} - 1/(2n_0) - 1/(2n_t)\right]} \qquad (4.35)$$

Subsequent to the work of Nei and Tajima (1981), there have been many statistical refinements and additions (e.g., multiple loci, confidence intervals for N_{ev}) to this approach (Jehle et al. 2001; Jorde and Ryman 1995; Kitada et al. 2000; Luikart et al. 1999; Waples 1989; Williamson and Slatkin 1999).

Just as one can combine temporal sampling of a population with equation 4.17 to obtain an estimate of variance effective size, one can also in theory combine temporal sampling with equation 4.6 to obtain an estimator of inbreeding effective size through temporal changes in the estimated heterozygosity (for example, see Cornuet and Luikart 1996). However, estimators of inbreeding effective size through temporal changes in sampled heterozygosity do not seem to be as successful as those for variance effective size based on temporal changes in sampled allele frequencies (Luikart et al. 1998). Part of this lack of success stems from the fact that the statistical estimators of heterozygosities tend to have more sampling error than allele frequencies for a given sample size. However, most of the difficulty arises from the fact that the heterozygosity in equation 4.6 is defined in terms of non–identity by descent, but what is observable in a genetic survey of a sample is non–identity by state. Accordingly, one must make additional assumptions about the mutational process (Cornuet and Luikart 1996) that are often not testable from the data being analyzed.

Although the direct genetic estimators of effective size avoid the problem of missing important ecological or demographic variables by going directly to the original definitions of effective sizes, the genetic estimators still make biological assumptions that could be violated (for example, that no loci are being influenced by natural selection and that the population is genetically isolated). Moreover, the estimators of the genetic variables being monitored often have poor statistical properties. Finally, these procedures inherently are limited to estimating effective sizes over short time intervals since they must use the first sample as the reference generation. There are alternative methods for using genetic data to estimate effective sizes over long periods of evolutionary time, as will be shown in the next chapter.

There has been much concern with estimating various types of effective population size because genetic drift is an important evolutionary force in its own right and because the impact of other evolutionary forces in natural populations (such as natural selection) is always overlaid upon genetic drift because all real populations are finite. Moreover, genetic drift has direct implications in conservation biology for the preservation of the most fundamental level of biodiversity: genetic diversity within a species. Genetic diversity is the ultimate basis of all evolutionary change, so biodiversity at this level has an upwardly cascading effect on biodiversity at all other levels: species, communities, and ecosystems. Genetic drift and its quantification through effective sizes are therefore important topics within population genetics.

5

GENETIC DRIFT IN LARGE POPULATIONS AND COALESCENCE

As shown in the previous chapter, the impact of drift as an evolutionary force is proportional to $1/(2N)$ for a diploid system in an idealized finite population of size N. From this, one might be tempted to think that drift is only important in small populations or populations that have experienced bottleneck/founder events, thereby making their effective sizes small. But drift can be important in large populations as well. In this chapter, we will consider two circumstances in which drift can play a major evolutionary role in populations of any size, including extremely large populations. These two circumstances are newly arisen mutations and **neutral mutations** that have no impact on the bearer's ability to replicate and pass on DNA to the next generation. We will also examine genetic drift of neutral alleles backward in time; that is, we will start with the present and look at the drift-induced evolutionary process going from the present to the past. This time-reversed approach is called coalescence, and it offers many new insights into the evolutionary impact of genetic drift.

NEWLY ARISEN MUTATIONS

The first circumstance in which genetic drift can play an important role in both small and large populations involves the evolutionary fate of newly arisen mutations. Whenever a mutation first occurs, it is normally found in only one copy in one individual. Recall that every individual in our idealized population produces an average of two offspring (to maintain constant population size) under a Poisson probability distribution and assuming that all mutations are neutral. Under this probability distribution, the probability that this single mutant will leave no copies the next generation is the probability of having i offspring $(e^{-2}2^i/i!$ under a Poisson, see Appendix 2) times the probability that none of the offspring

---

Population Genetics and Microevolutionary Theory, By Alan R. Templeton
Copyright © 2006 John Wiley & Sons, Inc.

received the mutant allele ($[\frac{1}{2}]^i$ under Mendel's first law); that is,

$$\text{Prob}(0 \text{ copies}) = \sum_{i=0}^{\infty} \frac{e^{-2}2^i}{i!} \left(\frac{1}{2}\right)^i = e^{-2} \sum_{i=0}^{\infty} \frac{1}{i!} = e^{-1} = 0.37 \qquad (5.1)$$

Thus, *irrespective of population size*, over a third of the mutations are lost in the first generation due to drift! Hence, the genetic variants that become available for subsequent evolutionary processes are strongly influenced by two random factors in all populations regardless of size: mutation and genetic drift.

Although total population size does not play much of a role in the survival of a new mutant in the first few generations after it occurred, population growth rates can play a major role. This can be modeled by letting the average number of offspring be k. If $k > 2$, the population is expanding in size; if $k < 2$, the population is declining in size. Then,

$$\text{Prob}(0 \text{ copies}) = \sum_{i=0}^{\infty} \frac{e^{-k}k^i}{i!} \left(\frac{1}{2}\right)^i = e^{-k/2} \sum_{i=0}^{\infty} \frac{e^{-k/2}(k/2)^i}{i!} = e^{-k/2} \qquad (5.2)$$

Hence, as k increases, the chance of a mutant allele surviving the first generation goes up. For example, if $k = 4$, then the probability of zero copies is reduced from the stable size value of 0.37 to 0.14, whereas if the population is declining with $k = 1$, then the probability of losing a new mutant during the first generation is 0.61.

This impact of population growth rate upon the evolutionary fate of new mutations has interesting implications when a founder or bottleneck event is followed by a period of rapid population growth (a population flush). Such founder-flush models (Carson 1968; Carson and Templeton 1984) produce a strange combination; the founder event causes a low variance effective size relative to the pre–founder event generation. As we saw in Chapter 4, this means that there is some loss of the old alleles that were present in the pre–founder event ancestral population. However, the flush that follows the founder event results in $k > 2$, so new mutations that arise during the flush phase are more likely to persist in the population. Hence, genetic variation is initially reduced, but it is then replenished at a rapid rate. In contrast, if a founder event is followed by little or no population growth (k close to 2), there will be no rapid replenishment of lost variation and continued small population size over many generations will cause additional losses of ancestral alleles. Therefore, a founder event followed by slow or no population growth results in reduced levels of genetic variation, whereas a founder-flush event results in a rapid turnover of variation in a population with less serious reductions of overall levels of genetic variation. The rapid change in the available allelic forms under the founder-flush model may trigger large evolutionary changes in the population when combined with natural selection (Carson 1968; Slatkin 1996; Templeton 1980a), as will be discussed later in this book. This also reinforces our earlier conclusion that not all founder or bottleneck events have the same population genetic consequences. Earlier, we saw that system of mating and amount of recombination have a major effect on modulating the evolutionary consequences of a founder or bottleneck event; now we see that the population growth rate that exists shortly after a founder or bottleneck event is also a critical modulator of the evolutionary consequences. It is incorrect to talk about *the* founder effect in evolution; there are many types of founder effects in terms of their population genetic consequences depending upon the genetic and demographic context in which they occur (Templeton 1980a).

NEUTRAL ALLELES

The other major circumstance under which drift has a major evolutionary impact irrespective of population size is when the mutant alleles are neutral. A **neutral allele** is functionally equivalent to its ancestral allele in terms of its chances of being replicated and passed on to the next generation. Sewall Wright first elaborated the evolutionary importance of genetic drift for neutral alleles in *small* populations in the 1920s. Wright assumed neutrality only for mathematical convenience because he was primarily interested in how genetic drift interacted with nonneutral alleles subject to natural selection in populations with small variance effective sizes. He did not pursue the evolutionary role that drift could play irrespective of population size upon neutral alleles. The importance of genetic drift upon neutral alleles in all populations irrespective of size only became generally recognized in the 1960s. Two important breakthroughs in molecular biology occurred during that decade that contributed significantly to our understanding of genetic variation and the importance of drift upon neutral alleles:

- Amino acid sequencing
- Protein electrophoresis

Now we consider the role that each of these developments played for focusing attention upon neutral alleles.

Amino acid sequencing allowed comparisons between homologous genes in different species, revealing the fate of genes over long periods of evolutionary time. **Homology** refers to traits (including amino acid or DNA sequences) found in two or more individuals (or chromosomes for DNA sequences) that have been derived, with or without modification, from a common ancestral form. Genes within a species' gene pool are homologous when all the different copies of a gene occupying a specific locus are derived, ultimately, from a common ancestral gene through DNA replication events (premise 1 from Chapter 1).

Just as genes can be homologous within a species, genes can also be homologous across species. The genes at a locus found in one species could be derived from the same ancestral gene in the distant past as the genes occupying a locus in a second species. If so, the genes are said to be homologous regardless of the species from which they were sampled. To make use of the interspecific homology of genes, a method was needed to score genetic variability across species. During the first half of the twentieth century, the primary techniques for scoring genetic variation were based upon some sort of breeding experiment (e.g., the classical "test-cross" of Mendelian genetics), and such breeding designs were primarily limited to assaying intraspecific variation. The ability to purify certain proteins and perform amino acid sequencing of these proteins advanced considerably in the 1950s and 1960s. As a result, starting in the 1960s, data sets became available on the amino acid sequences of homologous proteins (and hence homologous genes) across many species. Table 5.1 shows an example of the type of data being generated at this time (Dayhoff 1969). This table gives the pairwise amino acid sequence differences between the α chains of hemoglobin (an essential portion of the protein complex that transports oxygen in our blood) in six species.

Note that the pairwise differences in Table 5.1 for humans versus sharks and carps versus sharks are all approximately the same. Yet, humans overall are phenotypically very much *unlike* sharks whereas carps are more similar to sharks (at least from the human perspective). The differences observed at the molecular level did not seem to correspond to the differences

Table 5.1. Pairwise Amino Acid Sequence Differences in α Chain of Hemoglobin in Five Vertebrate Species

	Mouse	Chicken	Newt	Carp	Shark
Human	16	35	62	68	79
Mouse		39	63	68	79
Chicken			63	72	83
Newt				74	84
Carp					85

Note: Each entry represents the number of amino acid positions at which the two species differ in this protein. The columns are ordered by evolutionary relatedness such that the species at the top of any column is equally distant in evolutionary time from all species listed to its left.

observed in overall phenotypes. Instead, these pairwise molecular contrasts reflected the historical relationships of these species, which are shown in Figure 5.1 as inferred from the fossil record. The total lengths of the branches between any pair of current species and their common ancestral node in Figure 5.1 are proportional to the time since the two species diverged from their common ancestor. The columns in Table 5.1 are arranged such that the elements of each column represent an evolutionary divergence of the same length of time. Note that humans and carps shared a common ancestor after the lineage leading to the shark had already split off. Therefore, in terms of *time*, humans and carps are equally removed from sharks. Hence, the near constancy of the pairwise amino acid differences in a column seems to indicate that amino acid differences accumulate proportional to time, regardless of

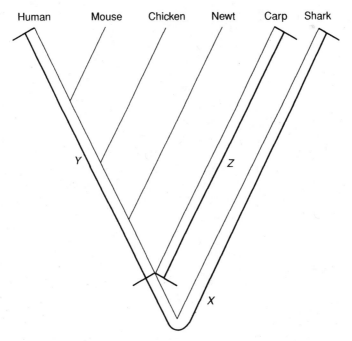

Figure 5.1. Evolutionary tree of six species whose α-chain hemoglobin amino acid sequence differences are given in Table 5.1.

the amount of overall phenotypic evolution that occurred between species. The observation of uniformity in the rate of amino acid replacement (first noted by Zuckerkandl and Pauling 1962) leads to the proposal of the **molecular clock**, that molecular changes in genes and their protein products accumulate more or less constantly over time.

The idea of the molecular clock was strongly supported in an influential paper by King and Jukes (1969). This paper was quite controversial at the time because many evolutionary biologists thought that changes in species (phenotypic and molecular) should reflect the influence of changes in environment or niche space. Yet here was evidence, at the molecular level, that only time since divergence seemed important. Molecular evolution appeared to proceed at a constant rate, in great contrast to morphological evolution. This seemed to undercut the importance of natural selection at the molecular level and indicated that some other evolutionary force may predominate at the molecular level. As we shall shortly see, King and Jukes argued that their molecular clock data could be best explained by a new variant of genetic drift theory articulated by Kimura (1968a).

We now know that much of the molecular data are not as clocklike as they initially appeared to be. For example, suppose the number of pairwise differences from the shark to the common ancestor of humans and carps is X (which is not directly observable from the data since the protein of the ancestor cannot be directly determined) and the pairwise differences from that common ancestor to the human and to the carp are Y and Z, respectively, as shown in Figure 5.1. Assume now that the total distance between any two points in the evolutionary tree is approximately the sum of the intermediate branch lengths. Then the number of pairwise differences between humans and sharks is $X + Y$ and the number of differences between carps and sharks is $X + Z$ (which are observable entries in the first column of Table 5.1). Note that these two numbers are always going to appear to have similar values because they share X in common. Today, statistical procedures exist to correct for this artifact, and the molecular clock has been found to be frequently violated or subject to much larger errors than suspected in the 1960s. Nevertheless, the idea of a molecular clock was of tremendous importance to the formulation of scientific thought at the time. Although not always strictly clocklike in its behavior, the accumulation of molecular divergence with time is now well established and provides a useful albeit rough measure of the timing of evolutionary events (as will be seen in Chapter 7).

The molecular clock challenged the preconceptions of many biologists about how *interspecific* genetic variation accumulated during the evolutionary process. Also in the 1960s, a second advance in molecular biology was challenging preconceptions about the amount of *intraspecific* genetic variability: the technique of protein electrophoresis (details of this technique are given in Appendix 1). Previously, most genetic survey techniques required variation at a locus to even identify the locus at all, thereby ensuring that *all* detectable loci had two or more alleles. Thus, the simple question of how many genes within a species were variable and how many were not was unanswerable by direct means. This situation changed with the advent of protein electrophoresis (Appendix 1). Protein electrophoresis allowed one to observe homologous gene products regardless of whether or not there was underlying allelic variation in amino acid sequence within species. To see why this was important, we need a brief history lesson.

Two schools of thought originated quite early in the development of population genetics and both traced their roots to Thomas Hunt Morgan's famed *Drosophila* laboratory that made so many of the fundamental advances in genetics in the first half of the twentieth century. One school, the "classical school," was associated with Morgan himself and some of his students, such as H. J. Muller. Morgan and Muller worked primarily with highly inbred strains of the fruit fly *Drosophila melanogaster* and scored genetic variation that affected morphological traits (e.g., eye color, wing vein patterns). They observed that most individuals were alike for these morphological traits and hence developed the idea that there is little genetic variation in natural populations, with most individuals being homozygous for a "wildtype" allele at each locus. Occasionally, mutants occur, but these are generally deleterious and are rapidly eliminated from the gene pool by natural selection. Even more rarely, a mutant is advantageous, in which case it rapidly increases in frequency in the gene pool and becomes the new wildtype allele. In either event, most loci in most individuals are homozygous for the wildtype allele at most times; that is, natural populations have very little genetic variation under this model.

Alfred Sturtevant was another student, and later an associate, of Morgan. A Russian student by the name of Theodosius Dobzhansky came to Morgan's laboratory and soon began working with Sturtevant. In the 1930s Dobzhansky and Sturtevant began to look at genetic variation in natural populations, as opposed to laboratory stocks, They scored genetic variation in *Drosophila* using the then new cytogenetic technique of staining the giant polytene chromosomes of the larval salivary glands (Figure 5.2). Such stained chromosomes can reveal up to 5000 bands over the entire genome, providing a degree of genetic resolution unparalleled at the time. One form of variation that was readily observable with this technique was large structural rearrangements of the chromosomes caused by inverted segments. They soon established that these inversions were inherited as Mendelian units, just like the genes being studied by Morgan and Muller. In contrast to the morphological traits studied by Morgan and Muller in laboratory strains, they found extensive genetic diversity in these inversions in natural populations with no obvious wildtype. Subsequent studies revealed selection operating upon these alternative chromosome inversion types, leading to the "balanced school" which postulated that natural populations had high levels of genetic diversity and that this diversity is maintained by selection (so-called balancing selection).

The debate over the amount of genetic variation in natural populations continued until the mid-1960s because until then the techniques used to define genes usually required genetic variation to exist. This situation changed dramatically in the mid-1960s with the first applications of protein electrophoresis. The initial electrophoresis studies (Harris 1966; Johnson et al. 1966; Lewontin and Hubby 1966) indicated that about a third of all protein-coding loci were polymorphic (i.e., a locus with two or more alleles such that the most common allele has a frequency of less than 0.95 in the gene pool) in a variety of species. This amount of variation was much higher than many had expected, and it had to be an underestimate because protein electrophoresis can only detect a subset of the mutations causing amino acid changes in protein-coding loci (Appendix 1).

These protein electrophoresis surveys made it clear that most species have large amounts of genetic variation. Interestingly, these surveys did not settle the debate between the classical and balanced schools. At first glance, it would seem that the observation of much polymorphism in natural populations should support the balanced school. What happened instead was that the observation of high levels of genetic variation transformed the debate from the *amount* of genetic variation to the *significance* of genetic variation. Recall that the balanced school argued not only that high levels of genetic variation existed in natural populations but also that the genetic diversity was maintained by balancing selection and was therefore not neutral. The classical school now became the school of neutral evolution, arguing that most of the genetic diversity being observed by the molecular techniques did not have any phenotypic impact at all. The wildtype allele of Morgan became a set of

Figure 5.2. Polytene X chromosome from *D. melanogaster*.

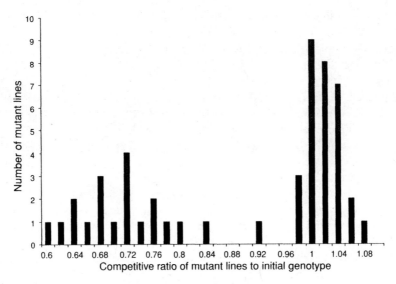

Figure 5.3. Distribution of competitive ratios of 50 mutation accumulation lines to founding nonmutant genotype in yeast *S. cerevisiae*. Modified from Fig. 1.A in C. Zeyl and J. DeVisser, *Genetics* 157:53–61 (2001). Copyright © 2001 by The Genetics Society of America.

functionally equivalent alleles under the neutral theory. In this manner, the proponents of the classical school acknowledged the presence of variation but not that it was maintained by natural selection.

An empirical demonstration of neutral alleles is provided by the work of Zeyl and DeVisser (2001). They founded 50 replicate populations from a single founding genotype of the yeast *Saccharomyces cerevisiae*. They then transferred these replicate populations to new media for several generations, going through single genetic individuals at each transfer. By going through single genetic individuals, mutations that occurred in the lineage leading to these individuals would accumulate and would not be eliminated by selection. However, any mutation that resulted in complete or nearly complete lethality or sterility would be eliminated under their experimental protocol, so their results cannot be used to estimate the incidence of extremely deleterious mutations. They kept replicates of the founding genotype in a dormant state that would not accumulate mutations, and then they tested their 50 replicate lines against the original, nonmutated genotype for competitive growth in their experimental media. A ratio of competitive growth of a mutant line to the nonmutant founder of 1 corresponds to neutrality. Ratios greater than 1 imply that a line carries beneficial mutations, and ratios less than 1 imply deleterious mutations under the laboratory environment. Figure 5.3 summarizes their results. As can be seen, the most common class of mutant lines (18% of the total) had a ratio of 1, implying neutrality. Indeed, the distribution of selective effects is bimodal, with one mode being centered around neutrality and the other distributed over a broad range of deleterious effects. Thus, the concept of a set of functionally equivalent neutral alleles has empirical validity in this case and in many others.

The leading proponent and developer of the neutral theory was Motoo Kimura. In 1968, Kimura published two papers that put forth a model of evolution of neutral alleles via genetic drift that explained *both* the observation of high levels of genetic variation and the molecular clock (Kimura 1968a,b). Kimura showed that many aspects of the evolution of truly neutral genes through genetic drift did not depend upon population size at all. How can this be, given that in Chapter 4 we saw that the strength of genetic drift as an evolutionary force is inversely proportional to population size (in an ideal population)?

The answer to this question lies in the fact that neutral theory deals not with drift alone but also with the interactions between genetic drift and mutation. Kimura considered an ideal population of size N in which all $2N$ genes at an autosomal locus are neutral. These $2N$ genes are not the same as $2N$ alleles. Alleles refer to different *types* of genes at a locus. Genes, in this context, refer to the individual copies at a locus. Therefore, in 10 individuals there are necessarily 20 "genes" that in theory could be grouped into 1–20 different allelic classes. As shown in the previous chapter, drift in the absence of mutation will eventually fix one gene (and therefore one allelic type) and lose all the others. Under neutrality, all genes are equally likely to be fixed by definition, meaning that each of the original $2N$ genes has a probability of fixation of $1/(2N)$.

Kimura now introduced mutation into his model. Let μ be the mutation rate of neutral alleles in the population. He also assumed the **infinite-alleles model**, in which all mutations yield a new allele. Then the rate of production of new neutral alleles is the number of genes at the locus in the population times the mutation rate, $2N\mu$.

Kimura next considered the balance of drift versus mutation. He pointed out that large populations have a much smaller probability of fixation ($\frac{1}{2N}$) but a greater rate of mutant production ($2N\mu$) than do small populations. These two effects balance one another out so that the overall rate of molecular evolution (the rate at which new alleles are produced times the probability of one going to fixation) is

$$\text{Rate of neutral evolution} = \frac{1}{2N} \times 2N\mu = \mu \tag{5.3}$$

Note that in equation 5.3 there is no effect of the population size on the rate of neutral evolution by drift! Hence, *drift is an important evolutionary force for neutral alleles in all populations, not just small ones.*

Kimura used equation 5.3 to explain the molecular clock, and his explanation was taken up by King and Jukes (1969). If the neutral mutation rate is relatively constant (an *internal* property of the genetic system as opposed to an *external* property arising from changing environments), then the rate of molecular evolution is independent of what is going on in the environment. This explains why we see an apparent molecular clock as long as the alleles are neutral (functionally equivalent). One problem with this explanation was that mutation rates are generally measured per generation but the clock was measured in absolute time (years). Kimura overcame this difficulty by claiming that mutation rates are constant in absolute time and not in generation time.

Kimura also used neutrality to explain why there is so much polymorphism in natural populations. Kimura noted that it takes time (many generations) for a neutral allele to go from being a new mutation to fixation, and during this time the population is polymorphic, albeit in a transient sense. But quite often, one or more copies of the allele going to fixation will actually mutate to new neutral alleles before the original version actually goes to fixation. Hence, if the population is large enough, large amounts of neutral variation will be present at any given time, even though the specific alleles are in constant turnover due to mutation and drift.

The impact of drift and mutation upon polymorphism can be quantified by returning to equation 4.3, the equation that describes how the average probability of identity by descent in an idealized population changes from one generation to the next due to the force of genetic drift. There is no mutation in equation 4.3, so we need to add a model of mutation in order to investigate how the balance of mutation and drift can explain intraspecific polymorphisms. We will assume an infinite-alleles, neutral mutation model. Under the infinite-alleles model, the only way to be "identical" by descent (which is synonymous to identical by state under this model) is to have no mutational events occurring between the generations in addition to being inbred in the pedigree sense.

Given that neutral mutations occur with a probability μ, then the probability that no mutation occurs is $1 - \mu$. Since two gametes are needed to produce an individual, identity by descent requires that no mutation occurred in the production of either gamete. Under the assumption that all meiotic events are independent, the probability that both the male and female gametes experience no mutation is $(1 - \mu)^2$. We can now incorporate these new probabilities into equation 4.3 to obtain

Average probability of identity by descent at generation t

= probability of identity by descent due to genetic drift

× probability of no mutation in both gametes

$$\bar{F}(t) = \left[\frac{1}{2N} + \left(1 - \frac{1}{2N}\right)\bar{F}(t-1)\right](1-\mu)^2 \tag{5.4}$$

As N decreases, $\bar{F}$ increases, but as μ increases, $\bar{F}$ decreases. Hence, equation 5.4 tells us that drift increases the average probability of identity by descent, whereas mutation decreases it. Over time, the average probability of identity by descent may reach an equilibrium $\bar{F}_{eq}$, reflecting a balance between the two antagonistic forces of drift and mutation. This equilibrium can be determined from equation 5.4 by setting $\bar{F}(t) = \bar{F}(t-1)$ to yield

$$\bar{F}_{eq} = \frac{1}{2N\left[1/(1-\mu)^2 - 1\right] + 1} \tag{5.5}$$

Using a Taylor's series expansion, $(1-\mu)^{-2} \approx 1 + 2\mu + 3\mu^2 + 4\mu^3 + \cdots$. If μ is very small, terms of μ^2 and higher can be ignored, so a good approximation to equation 5.5 is

$$\bar{F}_{eq} = \frac{1}{4N\mu + 1} \tag{5.6}$$

To deal with populations that deviate from our idealized set of assumptions, we need to substitute N_{ef} (Chapter 4) for N in equation 5.6. Because $\bar{F}_{eq}$ is the equilibrium level of average identity by descent in the population and because identity by state equals identity by descent under the infinite-alleles model, $\bar{F}_{eq}$ also has the interpretation of being the expected homozygosity under random mating. Therefore $1 - \bar{F}_{eq} = H_{eq}$ is the expected heterozygosity. Let $\theta = 4N_{ef}\mu$, where θ measures the proportional strength of mutation (μ) to genetic drift ($1/N_{ef}$). Then equation 5.6 can be recast as

$$1 - \bar{F}_{eq} = H_{eq} = 1 - \frac{1}{\theta + 1} = \frac{\theta}{\theta + 1} \tag{5.7}$$

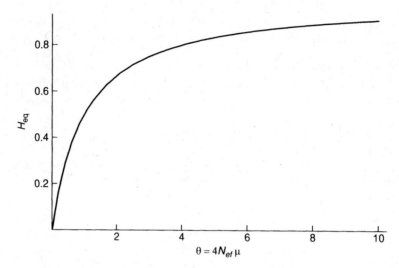

Figure 5.4. Relationship between equilibrium expected heterozygosity (H_{eq}) under infinite-alleles model and ratio of strength of neutral mutation rate (μ) to genetic drift ($1/N_{ef}$) as measured by $\theta = 4N_{ef}\mu$.

Equation 5.7 gives the expected heterozygosity for neutral alleles as a function of the ratio of the strength of mutation versus drift (θ). Figure 5.4 shows a plot of expected heterozygosity versus θ. As can be seen, H_{eq} can take on any value between 0 and 1. Thus, depending upon the exact value of θ, the neutral theory can explain any degree of genetic variability found in a population.

CRITIQUES OF NEUTRAL THEORY

The neutral theory showed that genetic drift was an important evolutionary force in all populations, regardless of size. Under neutrality, drift and mutation could together explain much of the observed patterns in genetic variation both between and within species. However, a drawback of applying the neutral theory was that it explained the genetic observations in terms of parameters that were unknown or difficult to estimate. For example, the predictions of the neutral theory with respect to the molecular clock depend upon μ, the *neutral* mutation rate. The neutral theory always acknowledged that some mutations are not neutral, and in particular many new mutations are deleterious, just as shown in Figure 5.3. Hence, μ in the neutral theory is *not* the mutation rate, but rather is the rate of mutation to functionally equivalent alleles only. Directly estimating mutation rates is difficult enough, but no one has devised a method of directly measuring the mutation rate to just neutral alleles. The situation is even worse for explaining the amount of intraspecific variation under the neutral theory; as shown by equation 5.7, the neutral explanation depends upon the values of both μ and of N_{ef}, the inbreeding effective size. As shown in the last chapter, estimating the inbreeding effective size is also a difficult task in most situations. In the absence of direct estimates, one could always set the values of μ and of N_{ef} to explain any particular rate of interspecific evolution or intraspecific heterozygosity *under the assumption of neutrality*. Thus, the neutral theory had great explanatory power but was difficult to test.

This difficulty became apparent in the 1970s when many population geneticists attempted to test the neutral theory. It proved extremely difficult to make observations that unambiguously discriminated neutrality from selection. For example, the molecular clock is explained under neutrality by assuming that μ is a constant over time for the gene under study. However, given that μ is the neutral mutation rate, μ is not expected to be constant across genes that are subject to different degrees of functional constraint even if the mutation rate *per nucleotide* is constant in all genes (that is, the mutation rate that includes both neutral and nonneutral mutations). For example, the α chain of the hemoglobin molecule carries oxygen to the cells and interacts with a variety of other globin chains during the course of development. Hence, it seems reasonable that most amino acid substitutions in this molecule would have deleterious consequences for the functioning of the molecule, leading to a low neutral amino acid replacement mutation rate. In contrast, the protein fibrinopeptide is involved in blood clotting, but only a small portion of the molecule is actually involved in this process; the bulk of the molecule seems to have little function other than to be cleaved off and discarded when the clotting process is initiated. Consequently, fibrinopeptide should have a higher neutral mutation rate than α-hemoglobin and therefore a faster molecular clock under the neutral theory. Figure 5.5 shows the observed amino acid sequence evolution in

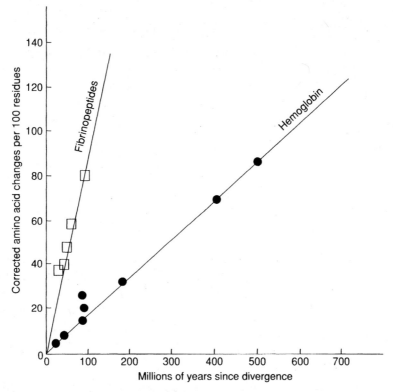

Figure 5.5. Rates of amino acid substitution in fibrinopeptides and α-hemoglobin. The approximate time of divergence between any two species being compared is given on the *x* axis as estimated from paleontological data. The *y* axis gives the number of amino acid changes per 100 residues for a comparison of the proteins of two species, after correction for the possibility of multiple substitutions at a single amino acid site. From Fig. 3 in R. E. Dickerson, *Journal of Molecular Evolution* 1:26–45 (1971), Copyright © 1971 by Springer. Modified with permission of Springer Science and Business Media.

these two molecules, along with the estimated rate of molecular divergence, which should be the neutral mutation rate under the neutral theory. As can be seen, fibrinopeptide evolves much more rapidly than the more functionally constrained molecule of α-hemoglobin, as predicted by the neutral theory.

This line of argument for neutrality has been greatly extended with data sets based upon DNA sequencing. For example, we now know that third codon positions usually evolve more rapidly than the first and second positions. This fits the functional constraint argument for neutrality because a third-position mutation is more likely to be synonymous than mutations at the first and second positions due to the degeneracy of the genetic code; hence, a third-position mutation should be more likely to be neutral. Similarly, sometimes a functional gene gets duplicated, but the duplicate is nonfunctional and is therefore called a pseudogene. In general, pseudogenes evolve more rapidly than their functional ancestral gene, a pattern also consistent with the greater probability of a neutral mutation in the pseudogene versus the functional gene.

This pattern of faster evolution in molecules with less functional constraint is a test supporting neutrality only if this same pattern is inconsistent with the hypothesis that molecular evolution is driven by natural selection. Is it? Back in the 1930s, the English population geneticist and statistician R. A. Fisher argued that the smaller the effect of a mutation upon function, the more likely it would be that the mutation would have an advantageous effect on the phenotype of fitness (for now, we define **fitness** as the quantitative ability of an individual to replicate and pass on DNA). We will discuss Fisher's reasoning in more detail in Chapter 10, but for now we simply point out that Fisher's theory, formulated long before the neutral theory, also results in the prediction that molecules with low levels of functional constraint should evolve rapidly due to natural selection because more mutations will be of small phenotypic effect and thereby are more likely to be selectively advantageous. Hence, interspecific clock patterns such as that shown in Figure 5.5 are compatible both with the neutral theory and Fisher's theory of selectively driven advantageous mutations of small phenotypic effect.

Now consider testing the neutral theory with observations on intraspecific levels of genetic variation. Equation 5.7 allows any level of variation to be explained as a function of μ and N_{ef}. Thus, at first glance it would appear that the neutral theory is virtually unassailable by observations on the amounts of intraspecific variation and thereby not testable with such data. But is it? If μ is regarded as a constant for a given gene or set of genes (to explain the molecular clock), then any variation in expected heterozygosity levels across species for the same gene or set of genes would have to be due to differences in N_{ef} under neutrality. Figure 5.6 shows a plot of expected heterozygosities (determined by protein electrophoresis) in several species versus the logarithm of the estimated actual population size of those species. Note that all the observed heterozygosities are less than 0.25. Under the neutral theory, the predicted relationship shown in Figure 5.4 maps the heterozygosities found in Figure 5.6 onto a very narrow range of θ values and hence of N_{ef} values. Indeed, the bulk of the observed expected heterozygosities shown in Figure 5.6 are between 0.01 and 0.10, implying only a 10-fold range of variation in N_{ef} across species whose actual population sizes span 14 orders of magnitude! Although N_{ef} is not the same as census size, a discrepancy of 13 orders of magnitude seems difficult to explain under neutrality.

One way out of this apparent dilemma of too little variation in expected heterozygosities across species is to turn attention to μ. Recall that μ is not the mutation rate but the neutral mutation rate. But what exactly is a neutral allele? As originally used by Kimura, a neutral

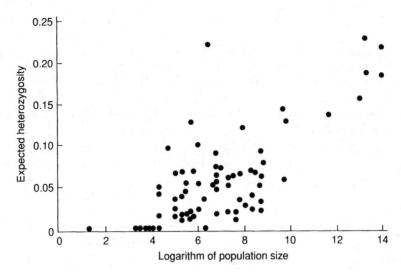

Figure 5.6. Relationship between population size (measured on logarithmic scale) and expected heterozygosity from protein electrophoretic data, as observed in a large number of species. From Fig. 1 in M. Nei and D. Graur, *Evolutionary Biology* 17:73–118 (1984). Copyright © 1984 by Springer. Modified with permission of Springer Science and Business Media.

allele is one that has absolutely no impact on the phenotype of fitness under any conditions. However, one of Kimura's colleagues, Tomoko Ohta, focused on mutations that influenced fitness but only slightly (Ohta 1976). In terms of Figure 5.3, Ohta was concerned with all the mutations in the upper mode centered around neutrality and not just the absolutely neutral ones. She argued that these mutations of small selective impact would be effectively neutral in very small populations because genetic drift would dominate the weak effects of selection. Hence, the effective neutral mutation rate would be high in small populations because it would include not only the absolutely neutral mutations but also many of the slightly selected mutations. As population size increased and the force of drift decreased, more and more of these mutations of slight effect would come under the influence of selection. Hence, as N_{ef} increases, the effectively neutral mutation rate μ would decrease. Because of these opposite effects, the product $N_{ef}\mu$ would show a much-dampened range of variation. With Ohta's inclusion of mutations with slight fitness effects, a wide range of N_{ef} values should give similar H values under this modified neutral theory, as is observed (Figure 5.6).

The trouble with Ohta's explanation for the proponents of neutrality is that it undermines the prediction of the molecular clock. The μ in equation 5.3 is now a function of N_{ef}, so discrepancies in the clock across species and across time are now expected as no one expects N_{ef} to remain constant across species and over time. As a consequence, although Kimura (1979) briefly embraced Ohta's idea of slightly selected mutations, he quickly abandoned it and went back to a model of absolute neutrality in order to preserve the molecular clock, but at the price of not having a satisfactory explanation for the patterns of intraspecific variation.

Because of difficulties such as those mentioned above, Kimura's neutral theory is still controversial. Nevertheless, most population geneticists now accept the idea that genetic drift is a major player in the evolution of genes at the molecular level. The ideas of a strict molecular clock and of absolute neutrality do not appear to explain the observed data well. As illustrated by the split between Kimura and Ohta over slightly selected mutations, many of the difficulties in the neutral theory arise from trying to explain simultaneously patterns of interspecific and intraspecific evolution. Initially, this was a particularly troublesome area for testing the predictions of the neutral theory because the primary technique of observing interspecific evolution (amino acid sequencing) was noncomparable to the primary technique of observing intraspecific variation (protein electrophoresis, which does not detect all amino acid substitutions and does not yield direct data on the number of amino acid substitutions). Moreover, for technical reasons, the proteins most amenable to amino acid sequencing were often not the same as the proteins most amenable to protein electrophoresis screening. As we will see in the next section of this chapter, the advent of first restriction mapping and then DNA sequencing has greatly altered this situation, allowing joint studies of the inter- and intraspecific evolution of many genes. As a consequence, there are now many additional ways of testing the neutral theory. The focus of much current research is now not testing the neutral theory per se but rather using the neutral theory as a convenient null hypothesis. The interesting evolutionary biology usually emerges when that null hypothesis is rejected, and the focus of the research shifts to why. We will examine these new tests of the neutral theory in Chapter 12 after natural selection has been discussed.

THE COALESCENT

Up to now we have been taking a forward-looking approach to see what will happen in the next generation given the current generation. Frequently, however, we make observations about present patterns and want to know how evolution occurred in the past to produce the present. Therefore, we also need to look at evolution backward in time from the present. This backward approach is in many ways the more practical one when dealing with natural populations rather than experimental ones. With natural populations, we do not know the future, but we can survey genetic variation in current populations and then use the present-day observations to make inferences about the evolutionary past. Also, we often do not have the luxury or opportunity of conducting forward-looking experiments, particularly when we want to study evolution over long time scales. Hence, we need a theory that deals with time backward from the present in order to test hypotheses about evolution over long time scales. One class of backward-looking models in population genetics is called *coalescent theory* (Kingman 1982a,b). We will introduce coalescent theory by modeling genetic drift in this time-backward sense.

Basic Coalescent Process

Recall from Chapter 1 our first population genetic premise: DNA replicates. In a forward sense, this means that one molecule of DNA can become two or more molecules in the future. In the backward sense, this means that two or more molecules of DNA observed today can coalesce into a single copy of DNA in the past (Figure 5.7). We say a **coalescent event** occurs when two lineages of DNA molecules merge back into a single DNA molecule at some time in the past. Hence, a coalescent event is simply the time inverse of a DNA replication event.

To see how genetic drift can be recast in terms of coalescent events, consider a hypothetical drift experiment consisting of just six haploid individuals (Figure 5.8). This population

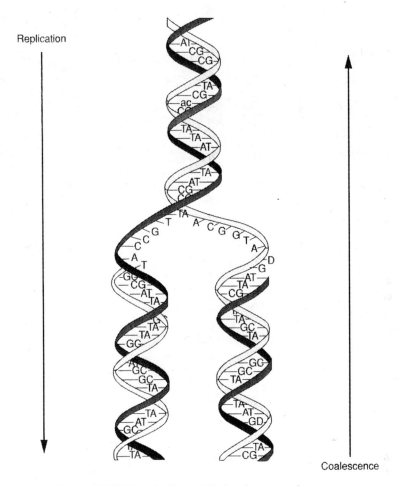

Figure 5.7. DNA replication and its time-inverse coalescence.

has six copies of any homologous DNA segment at any given time. As this finite population reproduces, DNA replicates. By chance alone, some molecules get more copies into the next generation than others. Eventually, fixation occurs (at generation 10 in Figure 5.8) such that all copies are now descended from a single DNA molecule.

Now consider the drift process as observed backward in time. Suppose generation 10 is the current generation and we survey all six genes. We now look at the genealogical history of those six genes. Note first that we cannot observe any part of the drift process that led to the extinction of a gene lineage prior to generation 10 (the parts in grey in Figure 5.8). Because we are starting only with the genes present in the current population (generation 10), we have no way of knowing about the gene lineages that no longer exist. (In some cases, we can get information about past genes from fossils, as will be discussed in Chapter 7, but such cases are exceptional.) Moreover, once all our current gene lineages have coalesced back to a common ancestor (generation 3 in Figure 5.8), we no longer have genetic variation in the coalescent process. Therefore, we have no information about evolutionary events that occurred prior to the appearance of the common ancestor because all observable aspects of evolution involve genetic variation. We do not know exactly how many other genes

134 GENETIC DRIFT IN LARGE POPULATIONS AND COALESCENCE

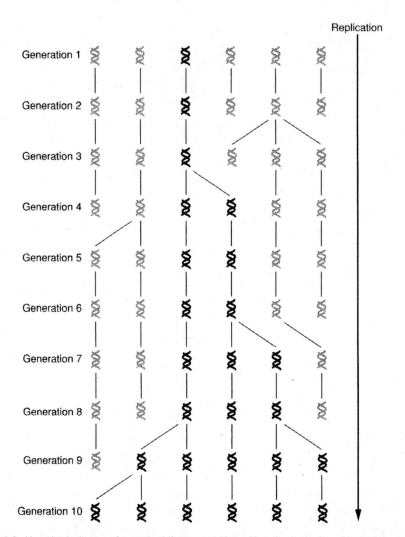

Figure 5.8. Hypothetical case of genetic drift in population with only six copies of homologous gene. Each vertical or diagonal line indicates a DNA replication event, going from the top to the bottom. A DNA molecule with no lines coming from its bottom did not pass on any descendants. By generation 10, all six copies are descended from a single DNA molecule. This common ancestral molecule and all its descendants are shown in bold compared to all the other DNA lineages that went extinct by generation 10.

existed at that generation of ultimate coalescence. We also do not know how many copies of the "winning" gene (the one that became the ancestor of all current genes) existed in the coalescence generation or in previous generations. Finally, we do not know how long the winning gene existed in the population prior to it being the ancestor of all current variation. Hence, in the backward experiment, we cannot observe as much as in the forward experiment. The most we can observe about the coalescent process is shown in Figure 5.9.

In general, consider taking a sample of n genes from a population. The word "genes" in coalescent models refers to the different copies of a homologous stretch of DNA. Because drift inevitably causes fixation in the future sense, this means in the backward sense that

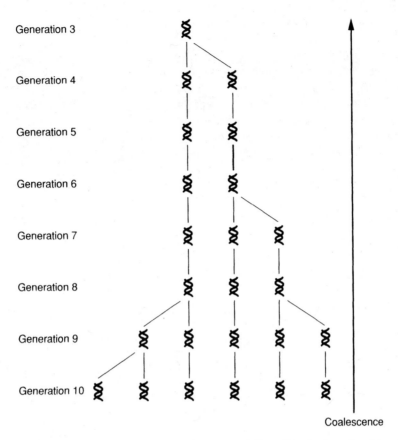

Figure 5.9. Hypothetical case of coalescent process in population with only six copies of homologous gene. Each combination of a vertical with a diagonal line indicates a coalescent event, going from the bottom to the top. This figure shows the same population illustrated in Figure 5.8 but only that part of the genealogical structure associated with the variation present at generation 10 back to the generation of coalescence of all six copies present at generation 10.

eventually all of the genes can be traced back in time to a common ancestral gene from which all current copies are descended. Thus, what we see as we look backward in time is a series of coalescent events, with each event reducing the number of DNA lineages (as illustrated in Figure 5.9). This will proceed until all sampled copies in the present coalesce to a common ancestral DNA molecule that existed at some time in the past.

For example, all the copies of mitochondrial DNA (mtDNA) found in living humans must eventually coalesce into a single ancestral mtDNA. Because mtDNA is inherited as a maternal haploid in humans, this ancestral mtDNA must have been present in a female. Some scientists and much of the popular media have dubbed this bearer of our ancestral mtDNA as "mitochondrial Eve" and have treated this as a startling discovery about human evolution, as will be detailed in Chapter 7. However, the existence of a mitochondrial Eve is trivial under coalescence. Finite population size (and all real populations are finite) ensures that *all copies of any homologous piece of DNA present in any species have been derived from a single common ancestral DNA molecule*—indeed, this is the very definition of genic homology (descent from a common ancestor). To say that humans have a mitochondrial

Eve is to say only that all human mtDNA is homologous. When "Eve" is called the ancestor of us all, it only means that our mtDNA is descended from her mtDNA and not necessarily any other piece of the human genome.

All genes that are homologous have a common ancestor at some point in the past. Assuming that all the genes are neutral, what can we say about how finite population size (genetic drift) influences the coalescent process? Consider first the simple case of a random sample of just two genes. The probability that these two genes coalesce in the previous generation is the same as being identical by descent due to pedigree inbreeding from the previous generation under random mating. As shown earlier, this probability is $1/(2N_{ef})$ for a diploid gene. In general, the probability of coalescence in the previous generation is $1/(xN_{ef})$, where x is the ploidy level. Therefore, the probability that the two genes did not coalesce in the previous generation is $1 - 1/(xN_{ef})$. Hence, the probability of coalescence exactly t generations ago is the probability of no coalescence for the first $t - 1$ generations in the past followed by a coalescent event at generation t:

$$\text{Prob(coalescence at } t) = \left(1 - \frac{1}{xN_{ef}}\right)^{t-1} \left(\frac{1}{xN_{ef}}\right) \tag{5.8}$$

Then

$$\text{Average time to coalescence} = \sum_{t=1}^{\infty} t \left(1 - \frac{1}{xN_{ef}}\right)^{t-1} \left(\frac{1}{xN_{ef}}\right) = xN_{ef} \tag{5.9}$$

As shown in Box 5.1 for a sample of n genes from a large population, the average coalescent time to the common ancestor of all n genes is $2xN_{ef}(1 - 1/n)$. These equations for expected coalescence time reveal another property of the evolutionary impact of drift: Drift determines how rapidly genes coalesce to a common ancestral molecule. The smaller the inbreeding effective size, the more rapidly coalescence is expected to occur.

Also note from Box 5.1 that the expected time for ultimate coalescence approaches $2xN_{ef}$ as the sample size (n) increases. Hence, the expected time to coalescence to a single molecule for any sample of genes is bounded between xN_{ef} and $2xN_{ef}$. Therefore, if you are interested in "old" events, you do not need large samples of genes to include some of the oldest coalescent events in your sample. For example, with just a sample of 10 genes, the expected coalescence time of your sample is 90% of the expected coalescence time for the entire population. Indeed, just a sample of two genes yields 50% of the expected coalescence time for the entire population! On the other hand, the expected time to the first coalescent event in a sample of n genes (that is, the coalescent event that reduces the number of gene lineages from n to $n - 1$) is shown in Box 5.1 to be $2xN_{ef}/[n(n-1)]$, which is very sensitive to the sample size n. For example, suppose we take a sample of 10 genes at a diploid locus ($x = 2$); then the expected range between the first and last coalescent events in this sample is $0.0444N_{ef}$ to $3.6N_{ef}$. Now, if we increase our sample to 100 genes, then the coalescent events in our sample are expected to be between $0.0004N_{ef}$ and $3.96N_{ef}$. Note that by increasing our sample by an order of magnitude we had only a minor impact on the expected time to the oldest coalescent event (going from $3.6N_{ef}$ to $3.96N_{ef}$) but changed the expected time to the first coalescent event by two orders of magnitude ($0.0444N_{ef}$ to $0.0004N_{ef}$). Hence, if you are interested only in the old events in a gene's history, then small samples are sufficient, but if you are interested in recent events as well, then large sample sizes are critical.

BOX 5.1 THE n COALESCENT

Suppose a sample of n homologous autosomal genes is taken from a large population of size N such that n is much smaller than N. Under these conditions, it is unlikely that more than a single pair of genes will coalesce at any given generation. Recall that the probability that a pair of genes coalesced in the previous generation in an idealized population is $1/(xN)$, where x is the ploidy level. The number of pairs of genes in a sample of n is given by the binomial coefficient:

$$\text{Number of pairs of genes} = \binom{n}{2} = \frac{n!}{(n-2)!2!} = \frac{n(n-1)}{2}$$

We do not care which pair coalesced first, simply that any pair coalesced. Hence, the probability that one pair coalesced the previous generation is simply the product of the probability that a pair coalesced times the number of pairs:

$$\text{Prob(coalescence in previous generation)} = \binom{n}{2}\frac{1}{xN} = \frac{n(n-1)}{2xN}$$

Hence,

$$\text{Prob(no coalescence in previous generation)} = 1 - \frac{n(n-1)}{2xN}$$

If no coalescence occurred in the previous generation, there are still n gene lineages and the above probabilities still apply to the next generation in the past, and so on until the first pair coalesced. Therefore,

$$\text{Prob(first coalescence in } t \text{ generations)} = \left(1 - \frac{n(n-1)}{2xN}\right)^{t-1} \frac{n(n-1)}{2xN}$$

and the expected time to the first coalescence is

$$E(\text{time to first coalescence}) = \sum_{t=1}^{\infty} t \left(1 - \frac{n(n-1)}{2xN}\right)^{t-1} \frac{n(n-1)}{2xN} = \frac{2xN}{n(n-1)}$$

The variance of the time to first coalescence, σ_1^2, is given by

$$\sigma_1^2 = \sum_{t=1}^{\infty} \left(t - \frac{n(n-1)}{4N}\right)^2 \left(1 - \frac{n(n-1)}{2xN}\right)^{t-1} \frac{n(n-1)}{2xN}$$

$$= \frac{2xN}{n(n-1)}\left(\frac{2xN}{n(n-1)} - 1\right)$$

Once the first coalescent event has occurred, we now have $n - 1$ gene lineages, and therefore the expected time to the second coalescent event starting from the time at which the first coalescent event took place is

$$E(\text{time to second coalescence} \mid \text{first}) = \sum_{t=1}^{\infty} t\left(1 - \frac{(n-1)(n-2)}{2xN}\right)^{t-1} \frac{(n-1)(n-2)}{2xN}$$

$$= \frac{2xN}{(n-1)(n-2)}$$

In general, the expected time and variance between the $k-1$ coalescent event and the kth event is

$$E(\text{time between } k-1 \text{ and } k \text{ coalescent events}) = \frac{2xN}{(n-k+1)(n-k)}$$

$$\sigma_k^2 = \frac{2xN}{(n-k+1)(n-k)}\left(\frac{2xN}{(n-k+1)(n-k)} - 1\right)$$

As the coalescent process proceeds, we finally get down to the $n-2$ coalescent event that leaves only a single pair of genes. From the equation immediately above, we can see, by letting $k = n-1$, the next and final coalescent event, that the expected time between the second to last and the last coalescent event is xN, the same result obtained for a random pair of genes.

Notice that the time between coalescent events becomes progressively longer as we look farther and father back into the past. Indeed, as we will now see, the time between the next to last coalescent event and the final one that unites all the original DNA lineages into a single ancestral molecule takes nearly as much time as all the $n-1$ coalescent events that preceded it added together. To show this, we note that the total time it takes n genes to coalesce to a single ancestral molecule is the sum of all the times from the first to the last coalescent event and that these times are independent because of the Markovian property (Box 4.1). Hence, we have that

$$E(\text{time to coalescence of all } n \text{ genes}) = \sum_{k=1}^{n-1} \frac{2xN}{(n-k+1)(n-k)} = 2xN\left(1 - \frac{1}{n}\right)$$

$$\text{Var}(\text{time to coalescence of all } n \text{ genes}) = \sum_{k=1}^{n-1} \sigma_k^2 = \sum_{k=1}^{n-1} \frac{2xN}{(n-k+1)(n-k)}$$

$$\times \left(\frac{2xN}{(n-k+1)(n-k)} - 1\right)$$

$$\approx 4x^2N^2 \sum_{k=1}^{n-1} \frac{1}{(n-k+1)^2(n-k)^2}$$

$$= 4x^2N^2 \sum_{i=2}^{n} \frac{1}{(i)^2(i-1)^2}$$

As the above equations reveal, the time to coalescence of a large sample of genes is about $2xN$, so the first $n-2$ coalescent events take about xN generations and the last coalescent event of the remaining pair of DNA lineages also takes xN generations on the average.

The equation for the expected coalescent time for a large sample of genes, $2xN_{ef}$, also indicates that ploidy level has a major impact on the time to ultimate coalescence. For an autosomal region, $x = 2$, so the expected time in a large sample to the most recent common ancestral DNA molecule for an autosomal region is $4N_{ef}$. X-linked DNA is haploid in males and diploid in females, so in a population with a 50–50 sex ratio, $x = 1.5$ and the expected ultimate coalescent time is $3N_{ef}$. The mtDNA is inherited as a haploid element in many animals, so $x = 1$. Moreover, mtDNA is maternally inherited, so only females pass on their mtDNA. Thus, the inbreeding effective size for the total population of males and females, the N_{ef} that is applicable to autosomal and X-linked DNA, is not applicable to mtDNA. Instead, the expected time to ultimate coalescence of mtDNA is influenced only by the inbreeding effective size of females, say $N_{ef\female}$. Thus, with $x = 1$, the expected coalescence time of mtDNA is $2N_{ef\female}$. Similarly, Y-chromosomal DNA is inherited as a paternal haploid, so its expected coalescent time is $2N_{ef\male}$, the inbreeding effective size for males. Because the sex ratio is close to 50–50 in humans, it is commonplace to approximate the sex-specific inbreeding sizes by $\frac{1}{2}N_{ef}$. Thus, a 1 : 1 : 3 : 4 ratio is expected for the relative coalescence times of Y-DNA, mtDNA, X-linked DNA, and autosomal DNA, respectively. However, inbreeding effective sizes are affected by many factors, including the variance of reproductive success (Chapter 4). In humans, and indeed most mammals, the variance of reproductive success is generally larger in males than in females. The higher this variance, the lower the effective size, so in general $N_{ef\female} > N_{ef\male}$. Therefore, Y-DNA is expected to coalesce the most rapidly of all the genetic elements in the human genome.

Templeton (2005) estimated the times to ultimate coalescence of human Y-DNA, mtDNA, 11 X-linked loci, and 12 autosomal loci by using the molecular clock method of Takahata et al. (2001). With this method, one first needs to compare the human genes to their homologue in chimpanzees. Let D_{CH} be the average number of mutations that have accumulated between the chimp and human genes. Given that the fossil record indicates that humans and chimps split about 6 million years ago (MYA) (Haile-Selassie 2001; Pickford and Senut 2001), the time available for this number of mutations to accumulate is 12 million years (6 on the human lineage and 6 on the chimp lineage). Let D_H be the average number of mutations that have accumulated within the human sample of genes between lineages that diverged from the common human ancestral gene. Then, assuming a constant rate of mutational accumulation (the molecular clock, equation 5.3), the estimated time to coalescence of the human gene sample is $12D_H/D_{CH}$. These estimated coalescence times are presented in Figure 5.10. As expected, Y-DNA has the smallest coalescence time, with mtDNA not being much longer. The X-linked loci all have longer coalescence times than mtDNA, and the autosomal loci have the longest times to coalescence on the average. Hence, these data confirm the theoretical prediction that the average time to coalescence should be affected by the level of ploidy and the pattern of inheritance (unisexual versus bisexual).

Note in Figure 5.10 that there is much variation in coalescence times within the X-linked loci and within the autosomal loci. This result indicates that there is a large variance in coalescent times. Equation 5.8 can also be used to calculate the variance in ultimate coalescence time. For example, the variance of time to coalescence of two genes (σ_{ct}^2) is the average or expectation of $(t - 2N_{ef})^2$:

$$\sigma_{ct}^2 = \sum_{t=1}^{\infty} (t - xN_{ef})^2 \left(1 - \frac{1}{xN_{ef}}\right)^{t-1} \left(\frac{1}{xN_{ef}}\right) = xN_{ef}(xN_{ef} - 1)$$
$$= x^2 N_{ef}^2 - xN_{ef} \tag{5.10}$$

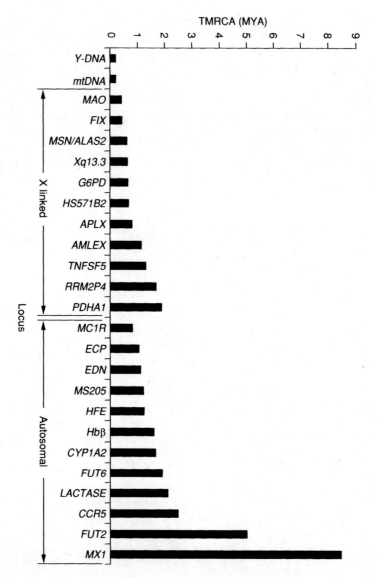

Figure 5.10. Estimated coalescent times (time to the most recent common ancestor, or TMRCA) for 25 human DNA regions. Details and references for the DNA regions studied are given in Templeton (2005).

Note that the variance (equation 5.10) is proportional to N_{ef}^2 and is therefore much larger than the mean (equation 5.9). Hence, the time to coalescence has much variation. Similarly, as shown in Box 5.1, the variance in ultimate coalescence time for a sample of n genes is also proportional to N_{ef}^2.

To see this variance experimentally, consider the times to fixation (the forward analogue to times to coalescence) in Buri's experiment on genetic drift in *Drosophila* (Figure 4.4). These fixation times are plotted in Figure 5.11. Note how "spread out" the observations are in this set of identical replicates: And most of the fixation is yet to occur! In general, coalescence times have a large variance and are skewed toward older times.

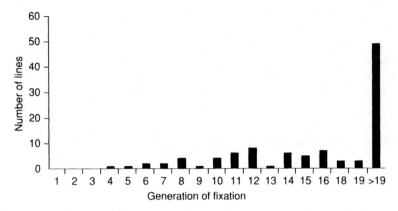

Figure 5.11. Observed times to fixation in the 107 replicate lines of 16 *D. melanogaster* in Buri's experiment on genetic drift. The experiment was terminated after generation 19, and over half the lines were still polymorphic at that point. These lines are indicated in the last column and would obviously take longer than 19 generations to go to fixation.

In Buri's experiment we can directly visualize the enormous variance in coalescence times associated with genetic drift under identical demographic conditions because there were 107 replicates. For most natural populations, you have only one realization of the drift process that is subject to this large variance in coalescent time. This variance in coalescence time is sometimes called *evolutionary stochasticity* to emphasize that it is an inherent property of the evolutionary realization of a single coalescent-drift process and is not related to the sample size or other sources of error. Unfortunately, many attempts to estimate coalescent times use equation 5.3 (the neutral molecular clock equation) and ignore the evolutionary stochasticity of the coalescent process. Kimura and others derived equation 5.3 for interspecific comparisons spanning millions of years (e.g., Table 5.1). On that time scale, the fixation of new mutations (the time forward inverse of coalescence to a single molecule) within a species was regarded as instantaneous and therefore not a contributor to variance. One simply has to wait for a neutral mutation to occur and go to (instantaneous) fixation, which is regarded as a relatively rare, random event. These assumptions yield a Poisson probability distribution (Appendix 2) as the descriptor of the number of mutations accumulated under the neutral molecular clock, so the neutral clock is also subject to evolutionary stochasticity. For the Poisson clock, the mean equals the variance (Appendix 2). However, when dealing with the smaller time scales of interest to a population geneticist, regarding fixation as instantaneous is a poor assumption. As shown by equation 5.10 and Box 5.1, the time to the most recent common ancestral molecule has much greater variance than a Poisson process and therefore must always be regarded as subject to extensive evolutionary error stemming from genetic drift. For example, we mentioned earlier that there is a human mtDNA common ancestral molecule. How long ago in the past must we go to coalesce back to this ancestral molecule? Using the standard molecular clock (equation 5.3) and an estimator of μ of 10^{-8} per year, the time to coalescence of all mtDNA to a common ancestral molecule has been estimated to be 290,000 years ago (Stoneking et al. 1986). (Actually, the mutation rate for mtDNA, much less the neutral mutation rate, is not accurately known, introducing further error into this estimation process, but that error will be ignored for now and μ will be treated as a known constant.) This figure of 290,000 is however subject to much error because of evolutionary stochasticity (the inherent variance associated with the clock and with the coalescent process). When evolutionary stochasticity is taken into account (ignoring sampling error, measurement error, and the considerable ambiguity in μ), the 95% confidence interval around 290,000 is 152,000–473,000 years (Templeton 1993)—a span of over 300,000 years! Hence, genetic drift induces much variance into coalescence times that can never be eliminated by increased sampling because natural populations represent a sample size of 1 of the coalescent process that we can observe for any single-locus or DNA region, such as mtDNA. Thus, coalescent theory teaches us to be humble in our dating of events through intraspecific studies of molecular variation.

Coalescence with Mutation

The basic coalescent model given above ignores mutation. Now we want to include both genetic drift and mutation as potential evolutionary forces. Consider first a sample of just two genes. By adding mutation, the two gene lineages could coalesce, mutate, or do neither in any given generation. As before, the probability that these two genes coalesce in the previous generation is $1/(xN_{ef})$, and the probability that they do not coalesce in the previous generation is $1 - 1/(xN_{ef})$. Assuming an infinite-alleles model of mutation, the probability that two genes are identical by descent is the probability that the two gene lineages coalesce *before* a mutation occurred in either lineage. If the two genes coalesced t generations ago, this means that there were $2t$ DNA replication events at risk for mutation (two gene lineages each undergoing t replication events). Hence, the probability that neither gene lineage experienced any mutation over this entire time period is $(1 - \mu)^{2t}$. Putting this probability together with equation 5.8 that describes the impact of drift, we have

$$\text{Prob(coalescence before mutation)} = \text{Prob(identity by descent)}$$

$$= \left(1 - \frac{1}{xN_{ef}}\right)^{t-1} \left(\frac{1}{xN_{ef}}\right)(1 - \mu)^{2t}$$

$$= \text{Prob(no coalescence for } t - 1 \text{ generations)}$$

$$\times \text{Prob(coalescence at generation } t\text{)}$$

$$\times \text{Prob(no mutation in } 2t \text{ DNA replications)}$$

(5.11)

Now consider the probability that we observe a mutation before coalescence. Suppose the mutation occurred t generations into the past. This means that in the $2t$ DNA replication events being considered, only one experienced a mutation and the other $(2t - 1)$ replication events did not. Because we have two gene lineages, either one of them could have mutated at generation t or we do not care which one it is. Hence, there are two ways for the mutation to have occurred at generation t. The total probability of having a single mutation at generation t in $2t$ DNA replication events is therefore $2\mu(1 - \mu)^{2t-1}$. The probability of no coalescence in these t generations is $[1 - 1/(xN_{ef})]^t$. Putting these two probabilities together, we have

$$\text{Prob(mutation before coalescence)} = \left(1 - \frac{1}{xN_{ef}}\right)^t 2\mu(1 - \mu)^{2t-1} \qquad (5.12)$$

If μ is very small and N_{ef} is very large, then the occurrence of both coalescence and mutation during the same generation can be ignored. Therefore, if we look backward in time until these two gene lineages either coalesce or have a mutation, the conditional probability (see Appendix 2) of a mutation before coalescence given that either mutation or coalescence has occurred is

$$\text{Prob(mutation before coalescence | mutation or coalescence)}$$
$$= \frac{2\mu(1-\mu)^{2t-1}\left(1-1/xN_{ef}\right)^t}{2\mu(1-\mu)^{2t-1}\left(1-1/xN_{ef}\right)^t + [1/(xN_{ef})](1-\mu)^{2t}\left(1-1/xN_{ef}\right)^{t-1}}$$
$$= \frac{2xN_{ef}\mu - 2\mu}{2xN_{ef}\mu - 3\mu + 1} \quad (5.13)$$

If $\mu \ll N_{ef}\mu$ (i.e., a large inbreeding effective size) and defining $\theta = 2xN_{ef}$, equation 5.13 simplifies to

$$\text{Prob(mutation before coalescence | mutation or coalescence)}$$
$$= \frac{2xN_{ef}\mu - 2\mu}{2xN_{ef}\mu - 3\mu + 1} \approx \frac{2xN_{ef}\mu}{2xN_{ef}\mu + 1} = \frac{\theta}{\theta + 1} \quad (5.14)$$

Note that the probability of mutation before coalescence given that one has occurred is the same as the expected heterozygosity under random mating for an autosomal locus ($x = 2$) that we derived in equation 5.7. This makes sense because if we had mutation before coalescence given that mutation or coalescence has occurred, then the two gene lineages being compared must be different alleles (recall our mutational model is the infinite-alleles model). Since we drew the two genes from the population at random, this is equivalent to the expected heterozygosity under random mating. Hence, whether we look at the joint impact of drift and mutation in a time-forward sense (equation 5.7) or a time-backward sense (equation 5.14), we get the same result for the impact of the balance of mutation to genetic drift upon the level of genetic variation present in the population.

Gene Trees Versus Allele or Haplotype trees

With mutation in the model, a distinction can now be made between gene trees and haplotype trees. **Gene trees** are genealogies of genes. They describe how different copies at a homologous DNA region are "related" by ordering coalescent events. Figure 5.12a shows a simple gene tree for a sample of six copies of an homologous DNA region. This figure is a repeat of the gene tree shown in Figure 5.9, but now we allow some of the DNA replication events to have experienced mutation. Looking back through time we can see that the A and $B1$ copies of the gene in Figure 5.12a coalesce prior to their coalescence with the $B2$ copy. However, note that the $B1$ and $B2$ genes share the same haplotype state, whereas the A gene differs from them by a single mutation. Thus, the A and $B1$ genes experienced mutation prior to coalescence, whereas the $B1$ and $B2$ genes coalesced prior to mutation. The gene tree shows precise information here about the gene genealogy, including cases in which two genes are closer genealogically even though they are not identical (e.g., A and $B1$) compared to two genes that are more distant genealogically yet identical in sequence (the $B1$ and $B2$ genes).

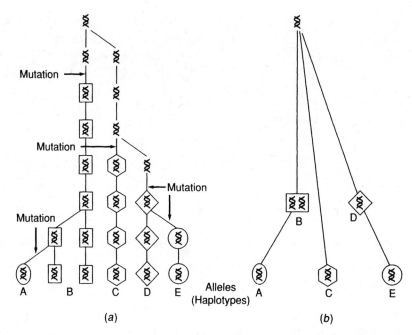

Figure 5.12. Gene trees and allele/haplotype trees. (*a*) The same gene tree portrayed in Figure 5.9, but now with some mutational events added on. Each mutation creates a new, distinguishable allele, as indicated by a change in shape of the box containing the DNA molecule. The extinct, ancestral haplotype is shown only by an unboxed DNA motif. (*b*) Haplotype tree associated with gene tree in (*a*). The only observable coalescent events are those associated with a mutational change, so each line in this tree represents a single mutational change. Letters correspond to different allelic categories that exist in the current population.

Unless pedigree information is available (such as that for the captive Speke's gazelle population or the human population on Tristan da Cunha), such precise information about gene genealogy is usually not known. For example, from sequence data alone, the *B1* and *B2* genes in Figure 5.12*a* are identical and therefore indistinguishable. We would have no way of knowing from sequence data that the *B1* gene is actually genealogically closer to the *A* gene than to its indistinguishable *B2* copy. Generally, our resolution of genealogical relationships is limited to those copies of the DNA that are also distinguishable in sequence state. Such distinguishable copies are created by mutation (under the infinite-alleles model). The gene copies that differ from one another by one or more mutational events are called alleles if the DNA region we are examining corresponds to a functional gene locus, or more generally they are called haplotypes if the DNA region corresponds to something other than a traditional locus—such as a portion of a locus, noncoding DNA, a DNA region that spans several loci, or even an entire genome such as mtDNA. Hence, **haplotypes** refer to alternative states of a homologous DNA region regardless of whether or not that region corresponds to a gene locus; **alleles** are alternative states of a homologous DNA region that corresponds to a gene locus.

Haplotypes or alleles are important to us because they are the only copies of DNA that we can actually distinguish in the absence of pedigree data. The only branches in the gene tree that we can observe from sequence data are those marked by a mutation. We cannot observe the branches in the gene tree that are caused by DNA replication without mutation.

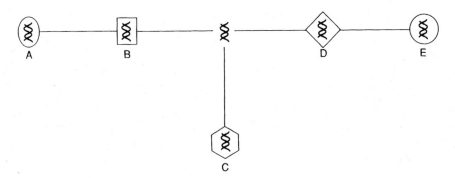

Figure 5.13. Unrooted network corresponding to rooted allele/haplotype tree shown in Figure 5.12b. Each line in this network represents a single mutational change. Letters correspond to different allelic categories that exist in the current population.

Therefore, the tree observable from sequence data retains only those branches in the gene tree associated with a mutational change. This lower resolution tree is called an allele or haplotype tree. The **allele or haplotype tree** is the gene tree in which all branches not marked by a mutational event are collapsed together. The haplotype tree corresponding to Figure 5.12a is shown in Figure 5.12b.

As can be seen by contrasting Figures 5.12a and b, we cannot see all the coalescent events in the haplotype tree. We therefore group genes into their allelic or haplotype classes. The **haplotype tree** is therefore a tree of genetic variation showing how that variation arose and is interrelated. We also know the allele frequencies in our current sample (for example, allele B in Figure 5.12 has two copies in our sample, B1 and B2, all other alleles are present in only one copy). Hence, the coalescent process that we can see only deals with the evolutionary history of alleles or haplotypes and their current frequencies in our sample.

In some cases, we do not know even as much information about the allele tree as that shown in Figure 5.12b. In that figure, the "black" ancestral DNA is portrayed as known, so it constitutes a root for the tree and gives us time polarity. For some allele or haplotype trees, we do not know the root of the tree. This results in an unrooted allele or haplotype tree called a **haplotype network**. Such unrooted networks show the mutational relationships that occurred in evolution to transform one haplotype (allele) into another but do not indicate the temporal orientation of mutational events. Figure 5.13 shows the unrooted haplotype network corresponding to Figure 5.12b.

Some Lessons from the Past: Inversion Trees and Networks

The concept of a haplotype tree is not a new one in population genetics but actually goes back to the very origins of the balanced school. Recall that this school arose from the work of Sturtevant and Dobzhansky on cytogenetic variation observable in the polytene chromosomes of *Drosophila*. As mentioned previously, polytene chromosomes in *Drosophila* provide many "bands" or landmarks in the *Drosophila* genome (see Figure 5.2), so these polytene chromosomes provided the first high-resolution genetic maps available to geneticists. Surveys of natural populations revealed that one type of cytogenetic variant was particularly common in *Drosophila* populations: paracentric inversions. Such inversions arise when two breakpoints occur on the same chromosome arm and the portion of the chromosome bracketed by the breaks attaches to the broken ends in an inverted configuration.

Given that a good cytogenetic preparation of an average chromosome arm in *Drosophila* can reveal close to 1000 bands, it seemed unlikely that any two independently arising inversions would involve exactly the same pair of breakpoints, and much experimental work confirms this as a general rule (Powell 1997) with some exceptions (Ruiz et al. 1997). The infinite-alleles model is therefore generally justified for this genomic-level method of screening for variation. Moreover, once an inversion occurs, it is inherited as a single simple Mendelian unit and recombination is suppressed between the overlapping inverted states in *Drosophila*. Hence, whole chromosome arms behave like a "locus" with each unique inversion acting as an "allele."

Sturtevant and Dobzhansky realized that the evolutionary uniqueness and simple inheritance of inversions made them powerful markers of the evolutionary process. If the same mutated form can arise repeatedly, it can obscure evolutionary history because identity by state no longer necessarily means identity by descent. This phenomenon of independent mutations yielding the same genetic state is now called **homoplasy**. Homoplasy represents a major difficulty when trying to reconstruct evolutionary trees, whether they are haplotype trees or the more traditional species trees of evolutionary biology. We will discuss how to deal with homoplasy later, but for now we note that homoplasy is usually not a problem for intraspecific paracentric inversions in *Drosophila*. Sturtevant and Dobzhansky also reasoned that although paracentric inversions are the most common type of cytogenetic variant found in *Drosophila* populations, the creation of a new inversion is a relatively rare event as indicated by the generally small number of different inversions that are polymorphic. Therefore, they estimated the evolutionary networks of inversions by finding the network that had the fewest cytogenetic mutations (and ideally no homoplasy). Today this principle of constructing evolutionary trees or networks by minimizing the number of mutational changes and homoplasy to explain derived states is known as **maximum parsimony**, and many computer programs are available to estimate evolutionary trees and networks under this criterion.

Figure 5.14 shows one of the first inversion trees estimated by Sturtevant and Dobzhansky (1936), with some later inversion data added, through maximum parsimony (although not called so at the time). This is a tree of inversions on the third chromosome arm found in two "races" (A and B) of the species *Drosophila pseudoobscura*. One interesting feature of the tree shown in Figure 5.14 is that not all of the inversions in this tree are present in the current population. At the center of this tree is a hypothetical A inversion state that maximum parsimony reconstructed as a necessary intermediate between the inversion types Santa Cruz and standard (standard is usually the most common or first encountered polytene banding state, and the alternative banding states are named after the geographical locations at which they were first discovered). Because of this inferred but unobserved intermediate inversion state, the two present-day standard and Santa Cruz banding states are separated by two inversion changes, not one. As this example shows, maximum parsimony estimates not only the evolutionary tree or network but also the unobserved genetic states that are necessary intermediates between existing alleles, haplotypes, or banding patterns.

Each of the arrows in Figure 5.14 represents a single inversion mutation, but note that most of the arrows are single headed and only two are double headed, reflecting how they were originally drawn in 1936. The single-headed arrows indicate time polarity, with the arrow pointing from the older to the newer inversion type. This tree is rooted by noting that the hypothetical A polytene banding state seemed to differ by the fewest number of inversions from the polytene banding state found in the closely related species, *D. miranda*. However, the cytological resolution does not exclude the possibilities that the Santa Cruz or

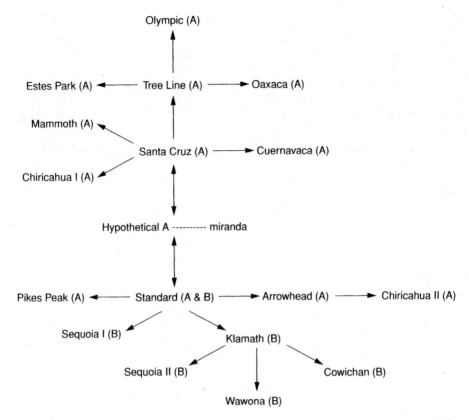

Figure 5.14. Evolutionary tree of inversion types found on third chromosome of individuals sampled from race A and race B (now considered species, as explained in the text) of *D. pseudoobscura* as estimated through maximum parsimony by Sturtevant and Dobzhansky (1936) with subsequent addition inversion data. Each line in the tree represents a single inversion mutation, with the exception of the dashed line connecting hypothetical A to the banding patterns found in the closely related species, *D. miranda*.

standard banding states could have been the evolutionary connection to *D. miranda*. This is why double-headed arrows interconnect these three inversion types, thereby indicating no time polarity among these three.

Using a related species to root the intraspecific inversion network within *D. pseudoobscura* is based on the assumption that the polytene banding states found in the common ancestor of *D. miranda* and *D. pseudoobscura* are older than the polytene banding states found within *D. pseudoobscura*. If this assumption is true, then the root of the total inversion tree that includes *D. pseudoobscura* and *D. miranda* must lie somewhere on the branch connecting the polytene banding states found in *D. miranda* to those found in *D. pseudoobscura*. Therefore, although the root of the combined *D. miranda/D. pseudoobscura* tree is not unambiguously identified, the root of the *D. pseudoobscura* portion of the tree is; namely, the *D. pseudoobscura* root is the polytene banding state (either currently existing or a reconstructed hypothetical) within the *D. pseudoobscura* portion of the network that represents the connection point to the branch leading to *D. miranda*. In Figure 5.14, this root corresponds to hypothetical A (although, as discussed above, the Santa Cruz and standard banding states were also possible connecting points). This method of rooting is called outgroup rooting. In general, **outgroup rooting** defines the root for the evolutionary tree of the entities (species, haplotypes, alleles, inversions, etc.) of primary interest (called the "ingroup") as that specific entity or node in the network that connects the ingroup to another entity or set of entities (called the "outgroup") that are thought to be more evolutionarily distant from all the ingroup entities than any ingroup entity is to another ingroup entity.

Outgroup rooting indicates that hypothetical A is the likely ancestral state for many if not all of the current inversion states found in *D. pseudoobscura*. The inference that certain ancestral forms are no longer present is not surprising to most evolutionary biologists who deal with evolutionary trees of species rather than inversions or haplotypes. Indeed, in evolutionary trees of species, the common ancestor of modern species is almost always assumed to be extinct. What is surprising in Figure 5.14 to many evolutionary biologists is that many of the inversions found in the present-day population are also ancestral forms for other inversions. For example, consider the Santa Cruz inversion type. In the tree of Sturtevant and Dobzhansky, seven other inversions are clearly descendants of Santa Cruz (namely, Chiricahua I, Mammoth, Estes Park, Tree Line, Olympic, Oaxaca, and Cuernavaca). Thus, ancestral and descendant forms coexist in the current population. This is the norm, not the exception, for intraspecific inversion and haplotype trees. All internal nodes in Figure 5.14 represent inversion types that have one or more descendant forms, and only one of the six internal nodes does not currently exist in the population (hypothetical A).

The reason for the persistence of ancestral forms in inversion or haplotype trees is that in general there were multiple identical copies of each inversion or haplotype in the ancestral population (from premise 1 in Chapter 1, DNA can replicate). Hence, when one copy of the ancestral form mutates to create a descendant form, the other identical copies of the ancestral form in general do not simultaneously mutate. This leads to the coexistence in the population of the newly mutated descendant form with many identical copies of its ancestral form.

Such temporal coexistence within polymorphic populations can sometimes be extreme. For example, subsequent research revealed that the two races (A and B) were actually two different species, with race A now called *D. pseudoobscura* and race B called *Drosophila persimilis*. The inversion tree is therefore part intraspecific and part interspecific. In general, allele or haplotype trees can spread *across* species boundaries; such trees therefore have the potential of being in part a tree of genetic variation *within* species and in part a tree of genetic differences *between* species. Moreover, note that the inversion labeled standard is found in *both* species. **Transpecific polymorphisms** occur when some of the haplotypes (or alleles or inversions) found in one species are genealogically more closely related to haplotypic lineages found in a second species than to other haplotypes found in their own species (Figures 5.15a and b). The standard inversion in Figure 5.14 is an example of such a transpecific polymorphism. As stated earlier, all the genetic variation found in a species at a homologous DNA region has to coalesce back to a single DNA molecule if we go back far enough in time, but the phenomenon of transpecific polymorphism makes it clear that the time needed to go back to the most recent common ancestral haplotype may actually be longer than the species has existed! Indeed, Figure 5.10 shows example of this phenomenon. Note that the coalescence time for the *MX1* autosomal locus is about 8 million years, yet the calibration used for the split between humans and chimpanzees was 6 million years. Hence, all current human haplotype variation at this locus coalesces back to an ancestral DNA molecule, but that molecule existed in the gene pool of the common ancestral species of both humans and chimpanzees. We also noted another example of transpecific polymorphism in Chapter 3, the self-sterility (*S*) alleles found in certain plants (Ioerger et al. 1991). A more dramatic example of haplotype trees being far older than the species containing the haplotypes is found in the *MHC* region introduced in Chapter 1. The

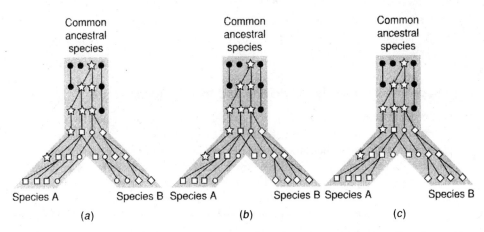

Figure 5.15. Contrasting patterns of species tree versus gene trees: (a) transpecific polymorphism (both species polymorphic); (b) transpecific polymorphism (one species polymorphic); (c) intraspecific monophyly. The inverted Y's represent the splitting of one species into two, with time going from top to bottom (the present). The gene pool for a species at any given time is represented by four genes. The genealogy of the genes is shown by the small lines contained within the Y's. In all panels, black circles indicate past gene lineages that did not survive to the present, open squares indicate a gene lineage that is monophyletic in species A, open diamonds a gene lineage that is monophyletic in species B, open circles a gene lineage showing transpecific polymorphism, and open stars the portion of the gene tree that unites all the gene lineages found in the current species back to a common ancestral molecule.

closest haplotype tree relative of some human haplotypes are haplotypes found in Old World monkeys rather than another human haplotype, indicating transpecific polymorphisms that have existed for at least 35 million years (Zhu et al. 1991). In other words, of the two haplotypes you received from your parents, one of your parental haplotype lineages may coalesce with a haplotype lineage found in a Rhesus monkey before it coalesces with the haplotype lineage you inherited from your other parent!

Finally, the inversion tree shown in Figure 5.14 reveals yet another critical attribute of allele or haplotype trees: Allele/haplotype trees are not necessarily evolutionary trees of species or populations! In this case, because of transpecific polymorphism, there is no clean separation of branches in the inversion tree between the two species *D. pseudoobscura* and *D. persimilis*. Obviously, there are strong associations between the inversion types and the species, but the inversion types found within any given species are *not* descended from just one common ancestral type found exclusively within that species. In modern terminology, the inversions within any given species are not monophyletic within species. A **monophyletic group** consists of all the descendants (inversions, haplotypes, species, etc.) from a single, common ancestral form. For an inversion or haplotype tree to be monophyletic within a species, all haplotypes or inversions found within the species must be a monophyletic group found exclusively within that species (Figure 5.15c). This is not the case in Figure 5.14. Figure 5.14 is properly only an evolutionary tree of the inversion variability found in the third chromosome of the sampled individuals. Such inversion trees may also correspond to trees of species or populations in some circumstances, but this is not always the case. The same is true for haplotype trees (Figures 5.15a and b). For example, the haplotype tree of all human *MHC* haplotypes is far older than the human species and does not define a monophyletic group within humans. Instead, when comparing humans to many other higher primate species, we sometimes obtain a situation like that shown in

Figure 5.15a for the *MHC* locus. You should never equate a haplotype tree to a tree of species or populations with no population genetic analysis. Such analyses will be described in Chapter 7.

Haplotype Trees from Restriction Site and DNA Sequence Data

Impact of Mutation. As the methods of restriction site mapping and then DNA sequencing became more and more available, the ability to estimate haplotype trees was no longer limited to *Drosophila* polytene chromosomes. Indeed, haplotype trees are now a tool available for virtually any species or population. There are some important differences, however, in estimating haplotype trees from modern molecular data as compared to the inversion trees of the drosophilists. The first difference deals with the nature of mutation at the nucleotide level versus the inversion level. When there is a substitution at a nucleotide, there are only three possible states for the new mutant nucleotide site (for example, a site with nucleotide A can only mutate to G, C, and T). Hence, at the nucleotide level, we cannot justify an infinite-alleles model, which was defensible for inversion evolution. However, in general, the DNA regions being surveyed in modern molecular studies include many thousands of base pairs. If mutation is randomly distributed across these thousands of nucleotides and is relatively rare, then the chance of any single nucleotide mutating more than once should be low. Hence, the infinite-alleles model is frequently replaced by the **infinite-sites model**, in which each mutation occurs at a new nucleotide site. Under this model, there is no homoplasy because no nucleotide site can ever mutate more than once.

Unlike the experimental evidence for inversions that frequently justifies the infinite-alleles model (Powell 1997), evidence is strong that the infinite-sites model can sometimes be seriously violated for DNA sequence data. For example, several studies indicate that most mutations in human nuclear DNA come from a small number of highly mutable sites. About a third of all mutations in human nuclear DNA are transitions from 5-methylcytosine to thymine (Cooper 1999; Jones et al. 1992). Methylated cytosines occur exclusively at CG dinucleotides, which are markedly underrepresented in human DNA relative to their expected frequencies under the hypothesis of independence of each nucleotide state (Jones et al. 1992). Hence, in germline methylated regions, a large proportion of the mutations occur in this small subset of CG dinucleotides. Mutagenic sites have also been reported for mononucleotide-repeat regions, DNA polymerase α arrest sites, and other rarely occurring motifs in human DNA (Cooper 1999; Krawczak and Cooper 1991; Nakagawa et al. 1998; Todorova and Danieli 1997; Tvrdik et al. 1998).

A highly nonrandom distribution of polymorphic sites was found in the 9.7-kb region within the *lipoprotein lipase* locus (*LPL*) that was sequenced in 71 individuals as discussed in Chapter 1. Eighty-seven variable sites were discovered in this sample, and 69 of these variable sites were ordered into haplotypes, defining a total of 88 distinct haplotypes. Table 5.2 shows how the 69 ordered polymorphic sites were preferentially located within three known mutagenic motifs in the human genome compared to all remaining nucleotide sites (Templeton et al. 2000a). When haplotype trees were estimated from the *LPL* data, these highly polymorphic sites with known mutagenic motifs were also preferentially associated with high rates of homoplasy (Templeton et al. 2000c). As *LPL* illustrates, the existence of mutagenic motifs undermines the assumption of the infinite-sites model and makes homoplasy a problem that cannot be ignored at the DNA sequence level.

Mutational homoplasy can make the estimation of a haplotype network or tree difficult because it can generate ambiguous or conflicting signals about evolutionary history. For

Table 5.2. Distribution of Polymorphic Nucleotide Sites in 9.7-kb Region of Human LPL Gene over Nucleotides Associated with Three Known Mutagenic Motifs and All Remaining Nucleotide Positions

Motif	Number of Nucleotides in Motif	Number of Polymorphic Nucleotides	Percent Polymorphic
CG	198	19	9.6
Polymerase α arrest sites with motif TG(A/G)(A/G)GA	264	8	3.0
Mononucleotide runs ≥ 5 nucleotides	456	15	3.3
All other sites	8,777	46	0.5

Source: From Templeton et al. (2000a).

example, Fullerton et al. (2000) sequenced 5.5 kb of the human *Apoprotein E* (*ApoE*) region in 96 individuals, revealing 23 *s*ingle-*n*ucleotide *p*olymorphisms (called SNPs in the current jargon of genomics and pronounced "snips"). Figure 5.16 shows the haplotype network estimated for this region under maximum parsimony. Note that there are several loops in this network. Since time does not go in circles, there should be no loops in a true evolutionary history. Rather, each of the loops reflects ambiguity about the evolutionary history of these haplotypes due to homoplasy. Figure 5.17*a* focuses on one of these loops; the one near the top, left-hand corner of Figure 5.16 involving the haplotypes labeled 1, 4, 21, and 26. As shown in Figure 5.17*a*, this loop appears in the *ApoE* haplotype network due to homoplasy at either site 560 or site 624 (site numbers refer to the nucleotide position in the reference sequence used by Fullerton et al. 2000). This homoplasy creates ambiguity in the evolutionary relationship among these haplotypes. Maximum parsimony is only concerned with minimizing the total number of mutations in the network, and there are four ways of breaking this loop (that is, by deleting one of the dashed arrows), each resulting in a total of four mutational steps. Each way of breaking this loop results in a different estimated evolutionary history for these haplotypes, as shown in Figure 5.17*a*. Hence, there are four alternative evolutionary histories for these four haplotypes that are indistinguishable under maximum parsimony. The total haplotype network shown in Figure 5.16 reveals another loop of four mutational changes and a double loop. As shown in Figure 5.17*a*, each of the single loops can be broken in 4 different ways, and the double loop can be broken in 15 different ways (Templeton 1997a). Thus, because of homoplasy, there are $4 \times 4 \times 15 = 240$ different possible and equally parsimonious evolutionary histories for the *ApoE* haplotypes. For the *LPL* haplotypes mentioned above, there is even more homoplasy that results in many thousands of equally parsimonious networks. Thus, homoplasy can create serious difficulties in estimating haplotype networks or trees under maximum parsimony.

Fortunately, coalescent theory indicates that there are other sources of information about the evolutionary history of intraspecific haplotypes besides just the total number of mutations in the tree. For example, the coalescent equation 5.12 indicates that nucleotide divergence becomes more probable with increasing time. This equation can be coupled with a **finite-sites model**, in which a finite set of nucleotides are subject to mutation and multiple mutational events at each site are allowed. The resulting mathematics is beyond the scope of this book, but the basic result is that the more divergent two haplotypes are at the sequence level, in general the longer the time since they shared a common ancestral

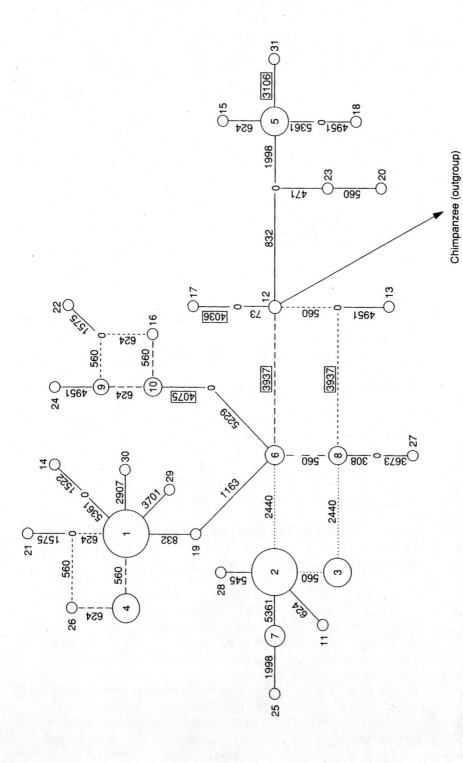

Figure 5.16. Haplotype network of 5.5-kb segment of *ApoE* gene. Circles designate the haplotypes, each identified by a number (1–31) either inside or beside the circle. The relative sizes of the circles indicate the relative frequencies of the haplotypes in the sample. A 0 indicates an inferred intermediate haplotype that was not found in the sample. Each line represents a single mutational change, with the number indicating the mutated nucleotide position. A solid line is unambiguous under maximum parsimony, whereas all dashed lines are ambiguous alternatives under maximum parsimony. Thin lines with long dashes are ambiguous under both maximum and statistical parsimony. Modified from Fig. 2 in S. M. Fullerton et al., *American Journal of Human Genetics* 67:881–900 (2000). Copyright © 2005 by The American Society of Human Genetics.

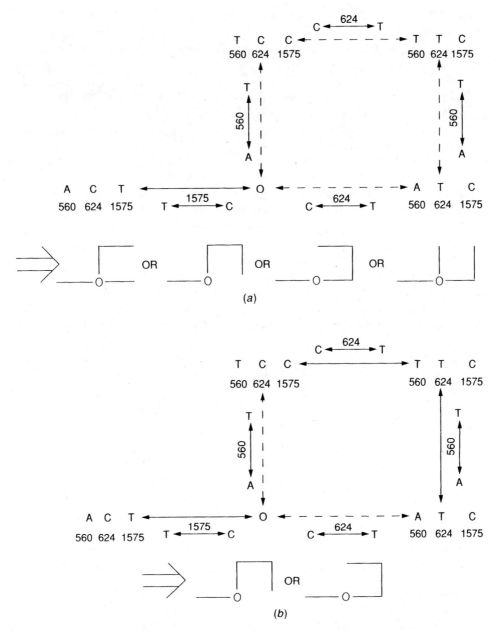

Figure 5.17. Difference between (*a*) maximum parsimony and (*b*) statistical parsimony for four haplotypes found at human *ApoE* locus. Solid lines indicate mutational changes that are fully resolved under the relevant parsimony criterion, dashed lines indicate mutational changes that may or may not have occurred, depending upon the true evolutionary history. The small double-headed arrows indicate the actual nucleotide substitution along with its position number in the reference sequence associated with each potential mutation. The possible evolutionary histories consistent with the criteria are shown underneath the loops.

molecule and the greater the chance of two or more hits at any one nucleotide site (Templeton et al. 1992). This observation implies that there is information about evolutionary history not only in the total number of mutational events but also in how homoplasy is allocated among the inferred branches. For example, we saw that there are four equally parsimonious ways of breaking the loop shown in Figure 5.17a. However, consider breaking the loop between the two haplotypes at the top with the sequences TCC and TTC. These two haplotypes differ by only a single nucleotide out of the 23 variable sites and out of about 5500 total sites. Breaking the loop between these two haplotypes implies an evolutionary history in which three mutations rather than just the one observable difference separate the haplotypes TCC and TTC. Similarly, a break between the haplotypes TTC and ATC implies that these two haplotypes that differ by only one observable nucleotide actually had three mutational events between them before they coalesced. On the other hand, suppose we broke the loop by deleting one of the dashed arrows that connect haplotype ACT to either TCC or ATC. In either case, haplotype ACT differs from both of these alternative connecting haplotypes by two observable mutational changes, and deleting either of these dashed arrows would allocate an extra pair of mutations between one of these more divergent haplotype pairs. Placing an extra pair of mutations between two haplotypes that differ at two nucleotide positions is far more likely than placing an extra pair of mutations between haplotypes that differ at only one nucleotide position under the finite-sites model. A computer program for calculating this probability under a neutral finite-sites model is available at http://darwin.uvigo.es/.

These coalescent probabilities lead to a refinement of maximum parsimony called **statistical parsimony**, in which branches that allocate homoplasies between less divergent haplotypes versus alternatives between more divergent haplotypes are eliminated when the probability of homoplasy is less than 0.05 as a function of the level of observable divergence. (A program for estimating statistically parsimonious networks is available at the same website given above.) For example, under statistical parsimony, there are only two ways of breaking the loop (Figure 5.17b) rather than the four under maximum parsimony (Figure 5.17a). Although this may seem to be only a modest gain in resolving the haplotype tree, many haplotype trees contain loops under maximum parsimony that can be completely resolved by statistical parsimony. Moreover, the ambiguities caused by disjoint loops are multiplicative. For example, using all of the *ApoE* data, we already saw that there are $4 \times 4 \times 15 = 240$ different maximum-parsimony trees for the *ApoE* haplotypes. Statistical parsimony reduces the possibilities to two alternatives for the two single loops and for one of the loops in the double loop (Figure 5.16), yielding a total of $2 \times 2 \times 2 \times 4 = 32$ trees under statistical parsimony. Thus, the number of evolutionary possibilities has been reduced by an order of magnitude in this case by the application of coalescent theory through statistical parsimony.

Coalescent theory provides even more sources of information about evolutionary history. As pointed out with regard to Figure 5.12, the potentially observable aspects of the coalescent process are the haplotype tree and the current haplotype frequencies. Castelloe and Templeton (1994) showed that there are nonrandom associations between the frequency of a haplotype and its topological position in the haplotype tree under a neutral coalescent model. For example, a haplotype that has many copies of itself in the gene pool is much more likely to give rise to a mutational descendant than a rare haplotype simply because there are more copies at risk for mutation. When dealing with haplotypes near the tips of the tree (that is, relatively recent events), the current haplotype frequencies are expected to be close to the frequencies in the recent past. Therefore, if a tip haplotype (one that is connected to only one other haplotype or node in the tree) has an ambiguous connection in the haplotype tree, the relative probabilities of the alternative connections are proportional to the relative frequencies of the potential ancestral haplotypes under neutral coalescence (Crandall and Templeton 1993). For example, haplotype 21 in Figure 5.16 is a tip haplotype, and under statistical parsimony it can be connected to the remainder of the tree either through haplotype 1 or through haplotype 26. As can be seen from the sizes of the circles, haplotype 1 is much more common than haplotype 26 (the actual haplotype frequencies are 0.234 and 0.005, respectively). The haplotype frequencies of 1 and 26 are significantly different, so the frequency information can be used to infer topological probabilities. In particular, under neutral coalescence the probability of haplotype 21 being connected to haplotype 1 is $0.234/(0.234 + 0.005) = 0.98$ and the probability of it being connected to haplotype 26 is $0.005/(0.234 + 0.005) = 0.02$. Hence, we can with great confidence discriminate between these two alternatives.

The genetic properties of the sequenced region provide yet another source of evolutionary information. As indicated above, mutations are not uniformly distributed in many DNA regions, and therefore homoplasy is nonrandomly distributed across the variable sites, as was shown for the *LPL* region. Fullerton et al. (2000) showed that homoplasy was also highly concentrated into two sites in the *ApoE* region, sites 560 and 624, that are both located within an *Alu* element. **Alu** is a family of short interspersed repeated elements of about 300 bp in length found in the genomes of all primates. About 900,000 copies exist in the human genome, accounting for about 5% of the total DNA in our genome (Cooper 1999). *Alu* sequences are thought to promote localized gene conversion (Cooper 1999), so the high rate of homoplasy at these two sites may be due to local gene conversion rather than traditional single-nucleotide mutation (Fullerton et al. 2000). Regardless of the molecular mechanism, the data indicate that homoplasy is concentrated into these two *Alu* sites. This information can be used to resolve the double loop in Figure 15.16. As can be seen from that figure, the double loop is caused either by homoplasy at site 560, one of the *Alu* sites, or sites 2440 and 3937, which otherwise show no homoplasy at all. Moreover, site 3937 is one of the amino acid replacement sites, which are the rarest class of mutations in this region. Therefore, the overall pattern of mutation in this DNA region argues that the double loop is best resolved through homoplasy at site 560. When this resolution is coupled with the one based on haplotype frequencies in the preceding paragraph and the resolutions achieved through statistical parsimony, only one ambiguity remains in the entire haplotype network—the two alternatives of connecting haplotype 22. The two alternative haplotypes to which haplotype 22 may be connected are both rare and do not have significantly different allele frequencies. Moreover, both alternative pathways involve homoplasy at one of the *Alu* sites (560 and 624). Although this one ambiguity cannot be resolved, we have gone from 240 maximum-parsimony solutions of this haplotype tree to only 2 by using coalescent theory and our knowledge of fundamental mutational properties.

We now know from studies in *Drosophila* and humans that different regions of the nuclear genome can vary tremendously in the amount and type of mutation. For example, as mentioned earlier, the human *LPL* region shows extensive mutational homoplasy due to specific types of mutagenic sites, particularly CG dinucleotides (Table 5.2). In contrast, the *ApoE* displays no evidence for any excessive homoplasy at CG dinucleotides (Fullerton et al. 2000), but it does at the *Alu* sites. There is no *Alu* element in the portion of the *LPL* gene that was sequenced, so obviously *Alu*-associated gene conversion could not be a source of homoplasy there. But both *LPL* and *ApoE* had CG dinucleotides, yet the behavior of this motif with regard to homoplasy was different in the two regions. This difference is actually not surprising. The CG dinucleotides are only hypermutable when they are methylated in the germline (El-Maarri et al. 1998), and different parts of the genome vary considerably in their amount of methylation (Cooper 1999). Moreover, the degree of methylated CG mutability seems to vary as a function of the flanking sequences (Cooper 1999). Thus, what is a hypermutable motif in one part of the genome may be unlikely to mutate in another region. It is therefore best to approach the analysis of any new DNA region cautiously, keeping in mind that the infinite-sites model may be violated, but for different reasons in different regions.

Impact of Recombination. The second major difference in the construction of haplotype trees from DNA sequence data versus inversion trees is recombination, particularly when one is dealing with nuclear DNA. As noted earlier, inversions in *Drosophila* act as cross-over suppressors, but many regions of the nuclear genomes of most species are subject to crossing over. Recombination creates far more serious problems for the estimation of haplotype trees than mutational homoplasy. As we saw above, mutational homoplasy makes it difficult to estimate a well-resolved haplotype tree, but recombination can undercut the very idea of an evolutionary tree itself. When recombination occurs, a single haplotype can come to bear different DNA segments that had experienced different patterns of mutation and coalescence in the past. Thus, there is no single evolutionary history for such recombinant haplotypes. When recombination is common and uniform in a DNA region, the very idea of a haplotype tree becomes biologically meaningless. Therefore, the first step in an analysis of a nuclear DNA region should be screening for the presence of recombination. A variety of algorithms exist for such screening (Crandall and Templeton 1999; Posada 2002; Posada and Crandall 2002; Posada et al. 2002; Worobey 2001).

Screening for recombination should be done before estimating the haplotype tree because most computer programs that estimate evolutionary trees *assume* that recombination does not occur. Recombination places the same mutations upon various haplotype backgrounds. This creates the appearance of homoplasy when an attempt is made to estimate an evolutionary tree under the assumption of no recombination, even if no actual multiple mutations occurred at the same nucleotide. For example, Figure 5.18 presents a statistical parsimony network of the *LPL* data under the assumption of no recombination. This network is filled with homoplasy. Some of this homoplasy is undoubtedly due to multiple mutational hits (recall Table 5.2), but some of this homoplasy is highly correlated with the physical position of the sites in the DNA. For example, three sites near the 5' end of the sequenced region (polymorphic sites 7, 8, and 13 in Figure 5.18) show much homoplasy, but the same three mutations appear to occur repeatedly together in the statistical parsimony network. A screen for recombination reveals that this pattern is not due to some mysterious mutational mechanism that causes these three sites to mutate again and again in concert on the same branches, but rather it is due to a **recombination hotspot** (a region in the genome in which many recombination events occur and surrounded by flanking regions with little or no recombination) in the sixth intron of this gene. As a result, the motif associated with sites 7, 8, and 13, located just 5' to the recombination hotspot, has been placed upon several different sequence states defined by the region just 3' to the hotspot (Templeton et al. 2000a). Overall, some 29 recombination events were detected in the *LPL* region, making recombination the major source of homoplasy in the data (Templeton et al. 2000a). Thus, the "tree" portrayed in Figure 5.18 does not really represent the evolutionary history of this DNA region since many of the "mutations" shown in that figure are not true mutations at all but rather are artifacts created by how the tree estimation algorithm fails to deal with

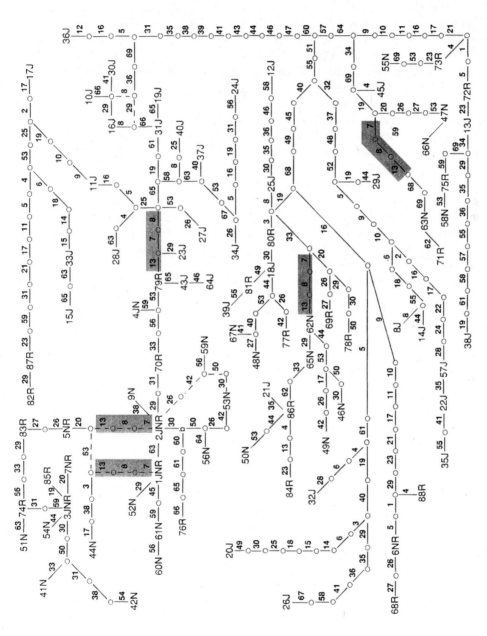

Figure 5.18. Statistical parsimony haplotype network based on 69 variable sites (numbered 1–69, going from 5′ to 3′) for 9.7 kb in the human *LPL* gene. Each line represents a single mutational change, with the number by the line indicating which of the 69 variable sites mutated at that step. Haplotypes present in the sample are indicated by an alphanumeric designation. Intermediate nodes inferred to have existed but not present in the sample are indicated by small circles. Mutational lines that are fully resolved are solid; mutational lines that may or may not have occurred indicating alternative statistically parsimonious solutions are indicated by dashed lines. The appearance of the same site number more than once indicates homoplasy in this case. The three 5′ sites 7, 8, and 13 seem to have mutated to the same state repeatedly but together, as indicated by grey shading.

157

recombination. Because of recombination, the haplotypes in many DNA regions cannot be ordered into an evolutionary tree. Haplotype trees are therefore only a possibility, not a certainty, in DNA regions subject to recombination.

Recombination also undercuts the coalescent theory developed earlier that assumed the infinite-alleles model. For example, equations 5.11 and 5.12 assume through the infinite-alleles model that the only way to create a new haplotype (allele) is through mutation. However, when recombination occurs, a new haplotype can be created by rearranging the phase of previously existing polymorphic sites in the absence of mutation. Hence, haplotype diversity is created by a mixture of recombination and mutation, as is the case for the *LPL* haplotypes. This undercuts the biological interpretation of equation 5.14 that explains the expected haplotype heterozygosity only in terms of genetic drift and mutation. As a consequence, the central coalescent parameter of θ is no longer only a function of drift (inbreeding effective size) and mutation but is also affected by recombination. The coalescent theory results that were derived from the infinite-alleles model are biologically applicable *only* in those DNA regions that have experienced no recombination.

This difficulty with the infinite-alleles model created by recombination was one of the primary motivating factors for going to the infinite-sites model. By defining the DNA region of interest as a single nucleotide, recombination is no longer a source of new genetic diversity. However, to derive the coalescent results described earlier at the single-nucleotide level, it is still required to assume that identify by state is the same as identity by descent. That is the purpose of the infinite-sites model at the single-nucleotide level. Because the infinite-alleles model and the infinite-sites model play similar roles in developing coalescent theory at the levels of a multisite DNA region and a single-nucleotide DNA region, respectively, some authors have mistakenly equated the two (e.g., Innan et al. 2005). In reality, these two models of mutation have extremely different properties, both in strengths and in weaknesses. As just stated, the infinite-alleles model when applied to multisite DNA regions can be undermined by recombination, whereas the infinite-sites model when applied to single nucleotides is not affected by recombination. However, as also noted above, the mutational process at the nucleotide level is characterized by many mutagenic motifs, which in turn undercut, often seriously, the infinite-sites model. In contrast, the infinite-alleles model, in the absence of recombination, is much more robust to the problem of mutational homoplasy. If the infinite-sites model is true, then all mutations must result in new haplotypes, so the infinite-alleles model is also true. However, if a site mutates more than once during the coalescent process, it will only violate the infinite-alleles assumption if the exact same mutation occurs at the same site *and* on exactly the same ancestral haplotype background. This additional restriction means that the infinite-alleles model is more robust to multiple mutational hits at the same nucleotide site than the infinite-sites model.

The enhanced robustness of the infinite-alleles model to mutational homoplasy can be seen in the *ApoE* haplotype tree described above. This 5.5-kb region was screened for possible recombination, but none was found (Fullerton et al. 2000). Hence, the haplotype variation in this DNA region does have a treelike evolutionary history, and this tree is given in Figure 5.19, which presents the same *ApoE* haplotype tree shown in Figure 5.16 but now incorporating all the resolutions discussed earlier that were achieved by statistical parsimony, coalescent frequency testing, and mutational motif analysis. Although recombination plays no role in this DNA region, mutational homoplasy does. The sites on all branches affected by mutational homoplasy are circled in Figure 5.19; that is, all the circled mutations occurred independently two or more times in the evolutionary history of this DNA region. As is evident, *most* branches on this tree are defined by mutations at nucleotide sites that

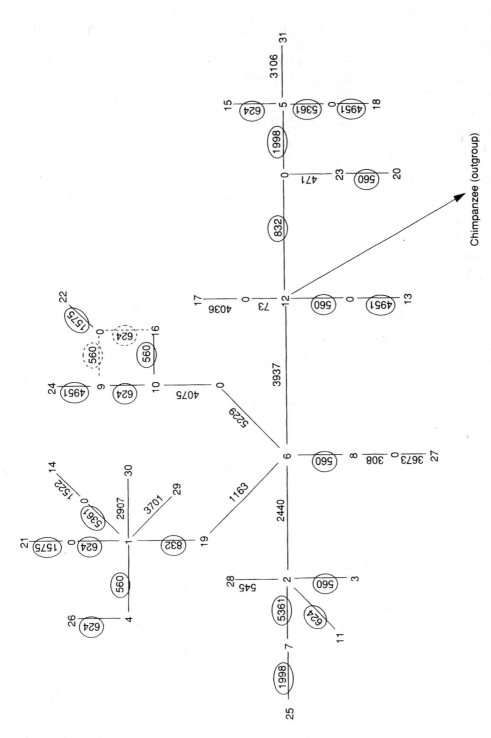

Figure 5.19. Statistical parsimony haplotype tree for *ApoE* gene incorporating additional resolutions based on haplotype frequencies and knowledge of mutational motifs. Only one ambiguity exists in this tree, the connection of haplotype 22 to the remainder of the tree has two possible alternatives, indicated by dashed lines. The nucleotide sites that have experienced multiple, independent mutational events to the same state are circled. The two alternative connections of haplotype 22 both result in homoplasy, but only one of these could have occurred, so they are circled in a dashed line.

experienced multiple, identical mutational events. Each of these circled sites represents a violation of the infinite-sites model that *did not violate* the infinite-alleles model because the independent mutations at a particular site occurred on different ancestral backgrounds. In general, every case of homoplasy detected in any haplotype tree based on DNA sequence data in regions with no recombination represents a violation of the infinite-sites model that did not violate the infinite-alleles model. Certainly, it is possible for the same mutation to occur independently upon different copies of the same ancestral haplotype, thereby violating the infinite-alleles model as well. Such violations are totally invisible to us with DNA sequence data, so some degree of error is unobservable in any haplotype tree analysis. Nevertheless, as shown in Figure 5.19, many violations of the infinite-sites model do not violate the infinite-alleles model and thereby yield detectable homoplasy. Consequently, in the absence of recombination but in the presence of mutational motifs, it is much better to use haplotypes and not nucleotides as the basis for population genetic analyses.

The contrasting strengths and weaknesses of the infinite-alleles versus infinite-sites model leaves us in somewhat of a conundrum: The infinite-sites model applied to nucleotides deals well with the problems caused by recombination but not with mutational homoplasy, whereas the infinite-alleles model applied to longer stretches of DNA deals better with the problems caused by mutational homoplasy but not with recombination. What are we to do? The fact that recombination was concentrated into a hotspot in *LPL* suggests a possible practical solution. It is becoming apparent that much of the nuclear genome in some species is characterized by recombination hotspots that separate DNA regions of low to no recombination (Comeron and Kreitman 2000; Crawford et al. 2004; Eisenbarth et al. 2000; Fullerton et al. 2001; Liu et al. 2004; McVean et al. 2004; Payseur and Nachman 2000; Reich et al. 2002; Rinaldo et al. 2005). It is not clear at this point how many species have their nuclear genome recombination concentrated into hotspots, but at least for humans this pattern exists and allows us to identify many regions of the human genome in which the infinite-alleles model is applicable and to partition the roles of mutation and recombination as contributors to haplotype diversity (Templeton et al. 2000c).

We return to *LPL* as an example of such a partitioning. Although there is no biologically meaningful evolutionary tree for the haplotypes defined by all 69 ordered variable sites found in the entire 9.7-kb *LPL* region, the concentration of recombination within the sixth intron of the *LPL* gene implies that meaningful evolutionary histories should exist for the regions just 5' and 3' of this recombination hotspot. This conclusion is also indicated by the pattern of linkage disequilibrium in the *LPL* region (Figure 5.20). Recombination reduces the magnitude of linkage disequilibrium (Chapter 2), so regions of high recombination generally tend to have low amounts of linkage disequilibrium. Figure 5.20 shows that the region in which almost all of the 29 inferred recombination events occurred has much less linkage disequilibrium both among sites within this region and between these sites and outside sites than is found in either the 5' and 3' flanking sets of variable sites (Templeton et al. 2000a). Accordingly, separate haplotype trees could be estimated for the 5' and 3' regions through statistical parsimony (Figure 5.21).

Extensive homoplasy was observed within each of these flanking regions of low recombination, indicating serious deviations from the infinite-sites model. Some of this homoplasy resulted in "loops" in both estimated flanking region haplotype networks that could not be explained by recombination (Figure 5.21). An analysis of the sites involved in homoplasy and these loops of ambiguity revealed that they were preferentially the same mutagenic sites identified in Table 5.2 (Templeton et al. 2000c). Even with these ambiguities, most of the evolutionary relationships among haplotypes were resolved in the flanking regions,

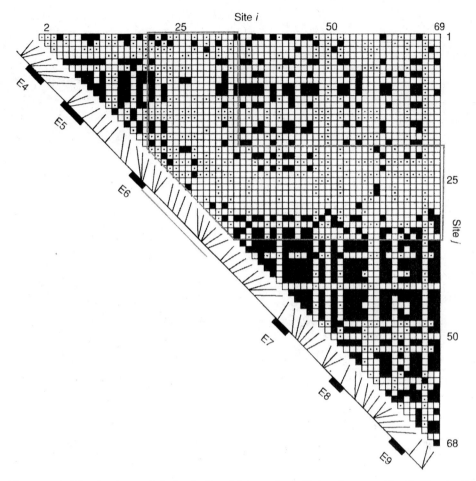

Figure 5.20. Plot showing statistically significant linkage disequilibrium (measured by D', Chapter 2), indicated by blackened square, for pairs of 69 polymorphic and ordered sites in *LPL* gene region. In comparisons of site pairs in which both sites have rare nucleotides, there can be complete disequilibrium (one of the four possible gametes having a count of zero), yet no possibility of statistical significance. Site pairs that lack the power to yield a significant association are indicated by a dot in the center of the square. The diagonal line, with exons 4–9 labeled, indicates the location of each varying site along the portion of the gene that was sequenced. The thick grey lines outline a DNA region that has significantly less pairwise disequilibrium than the remainder of the sequenced region.

so we do have a good, albeit incomplete, picture of haplotype evolution in the 5' and 3' *LPL* regions. Altogether, 29 of the 88 haplotypes in the entire *LPL* region were produced by mutations alone, with no recombination in their evolutionary history. All the remaining haplotypes either were produced by one or more recombination events or were mutational derivatives of ancestral recombinant haplotypes.

Estimating Haplotype Trees with Molecule Genetic Distances.
Screening for recombinants and identifying blocks of low to no recombination can help avoid or overcome many of the problems in using parsimony to estimate haplotype networks or trees, but one major problem remains: finding the maximum or statistical parsimony tree(s) in the first

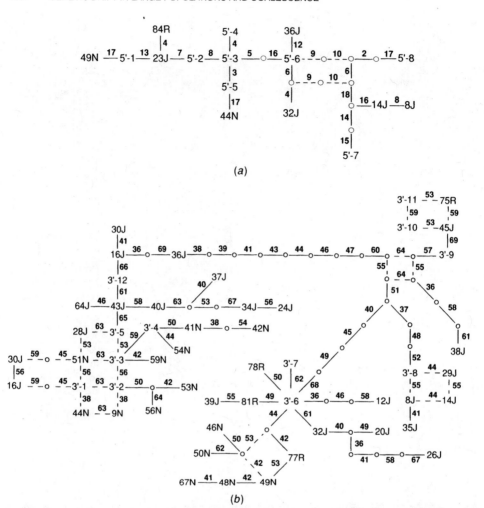

Figure 5.21. Haplotype trees for (a) 5' and (b) 3' regions on either side of sixth-intron recombination hotspot in human *LPL* gene. Each line represents a single mutational change, with the number by the line indicating which of the 69 variable sites (numbered 1–69, going from 5' to 3') mutated at that step. Haplotypes present in the sample are indicated by an alphanumeric designation. Intermediate nodes inferred to have existed but not present in the sample are indicated by small circles. Mutational lines that are fully resolved are solid; mutational lines that may or may not have occurred depending upon the true evolutionary history are indicated by dashed lines.

place. For n distinct haplotypes, there are a maximum of $(2n - 3)!/[2^{n-2}(n - 2)!]$ distinct rooted trees. Just 10 haplotypes have 34,459,425 trees. Once beyond 20–30 haplotypes, most computer programs that implement parsimony cannot examine all the possibilities and instead use certain heuristic algorithms that are not guaranteed to find the parsimonious solutions. Moreover, implementing even these heuristic algorithms with data sets with many haplotypes can be time consuming with even the fastest computers. This is a common problem because at the DNA sequence level it is not difficult to find 30 or more haplotypes even in a relatively small sample.

There exist other tree-estimating algorithms that are much less computationally intensive. Almost all of these are based upon some form of genetic distance, so we first need to examine what a genetic distance means. There are two types of genetic distance in the evolutionary literature: genetic distances between populations and genetic distances between molecules. For now, we are only interested in **molecule genetic distances**, which ideally measure the number of mutational events that occurred in the two molecular lineages being compared back to their coalescence to a common ancestral molecule. Population genetic distances will be discussed in the next two chapters, but we note for now that most population genetic distances can reach even their highest values in the complete absence of mutation. Therefore, these two types of genetic distance should never be equated or confused. Unfortunately, both are typically called "genetic distance" in the evolutionary literature, so readers have to infer which type of distance is being used by context. In this text, we will always use the terms "molecule genetic distance" and "population genetic distance" to make this distinction explicit.

Many mathematically distinct distance measures exist within both classes, and we will not attempt an exhaustive survey. Rather, we will give only the simplest types of molecule genetic distance designed for DNA sequence data. The simplest molecule genetic distance measure for DNA comparisons is just the observed number of nucleotide differences between the molecules being compared. The problem with this simple measure is homoplasy. When the same nucleotide mutates more than once, we still see still this as only a single observable difference or see no difference at all if the second mutation reverts back to the ancestral state within one DNA lineage or if parallel mutational events occurred in both DNA lineages. Because any nucleotide can take on only four distinct states, any time a single nucleotide site undergoes more than one mutation, it is likely that a reversal or parallelism occurred. This causes the observed number of differences to underestimate the actual number of mutational events separating two DNA molecules, which is our ideal standard for a molecule genetic distance.

One of the simplest molecule genetic distances that corrects for multiple mutational hits is the Jukes–Cantor distance (Jukes and Cantor 1969). The derivation of this molecule genetic distance is given in Box 5.2, where it is shown that under the assumptions of neutrality all sites mutate at the same homogeneous rate μ and all mutations are equally likely to go to any of the three alternative nucleotide states:

$$D_{JC} = -\frac{3}{4} \ln\left(1 - \frac{4}{3}\pi\right) \tag{5.15}$$

where D_{JC} is the Jukes–Cantor molecule genetic distance and π is the observed number of nucleotides that are different divided by the total number of nucleotides being compared.

Figure 5.22 shows a plot of the Jukes–Cantor molecule genetic distance as a function of π, the observed proportion of nucleotides that differ between the DNA molecules being compared, as well as a plot of π against itself. As can be seen, when D_{JC} is small, there is little difference between D_{JC} and π. Thus, unobserved mutations become common only when the observed divergence level (π) becomes large. That is, when two DNA molecules show few observed nucleotide differences, it is unlikely that unseen homoplasies have occurred in their evolutionary history. This is exactly the same property used to justify statistical parsimony.

The Jukes–Cantor model assumes neutrality, a constant mutation rate that applies uniformly to all nucleotide sites, and an equal probability of mutating to all three alternative nucleotide states. There are many ways of deviating from this idealized set of assumptions, and there are therefore many other molecule genetic distance measures designed to deal

BOX 5.2 JUKES–CANTOR MOLECULE GENETIC DISTANCE

Consider a single nucleotide site that has a probability μ of mutating per unit time (only neutral mutations are allowed). This model assumes that when a nucleotide site mutates it is equally likely to mutate to any of the three other nucleotide states. Suppose further that mutation is such a rare occurrence that in any time unit it is only likely for at most one DNA lineage to mutate and not both. Finally, let p_t be the probability that the nucleotide site is in the same state in the two DNA molecules being compared given they coalesced t time units ago. Note that p_t refers to identity by state and is observable from the current sequences. Then, with the assumptions made above, the probability that the two homologous sites are identical at time $t+1$ is the probability that the sites were identical at time t and that no mutation occurred times the probability that they were not identical at time t but that one molecule mutated to the state of the other:

$$p_{t+1} = p_t(1-\mu)^2 + \frac{1}{3}(1-p_t)2\mu \approx (1-2\mu)p_t + \frac{1}{3}2\mu(1-p_t)$$

with the approximation requiring μ to be small. The above equation can be rearranged as

$$\Delta p = p_{t+1} - p_t = -2\mu p_t + \frac{1}{3}2\mu(1-p_t) = -\frac{8}{3}\mu p_t + \frac{2}{3}\mu$$

Approximating the above by a differential equation yields

$$\frac{dp_t}{dt} = -\frac{8}{3}\mu p_t + \frac{2}{3}\mu$$

and the solution to this differential equation is

$$p_t = \frac{1}{4}\left(1 + 3e^{-8\mu t/3}\right)$$

A molecule genetic distance ideally measures the total number of mutations that occurred between the two DNA molecules being compared, not their observed number of differences. Jukes and Cantor used the neutral model throughout, so the expected total number of mutations between the two molecules under neutrality is $2\mu t$. Therefore, we want to extract $2\mu t$ from the equation given above to obtain the expected total number of mutations, both observed and unobserved. The extraction proceeds as follows

$$p_t = \frac{1}{4} + \frac{3}{4}e^{-8\mu t/3}$$

$$\frac{3}{4}e^{-8\mu t/3} = p_t - \frac{1}{4}$$

$$-\frac{8}{3}\mu t = \ln\left(\frac{4}{3}p_t - \frac{1}{3}\right)$$

$$2\mu t = -\frac{3}{4}\ln\left(\frac{4}{3}p_t - \frac{1}{3}\right)$$

where ln is the natural logarithm operator. The above equation refers to only a single nucleotide, so p_t is either 0 and 1. Hence, this equation will not yield biologically meaningful results when applied to just a single nucleotide. Therefore, Jukes and Cantor (1969) assumed that the same set of assumptions is valid for all the nucleotides in the sequenced portion of the two molecules being compared. Defining π as the observed number of nucleotides that are different divided by the total number of nucleotides being compared, Jukes and Cantor noted that p_t is estimated by $1 - \pi$. Hence, substituting $1 - \pi$ for p_t yields

$$2\mu t = -\frac{3}{4} \ln\left(1 - \frac{4}{3}\pi\right) \equiv D_{JC}$$

where D_{JC} is the Jukes–Cantor molecule genetic distance.

with more complicated models of mutation. As already shown by the contrast of *LPL* and *ApoE*, the underlying model of mutation that is most appropriate can vary considerably from one DNA region to the next. If all of your Jukes–Cantor distances are less than or equal to 0.05, it does not make much difference which mutation model you use. However, as the molecule genetic distances get larger, they are more sensitive to the underlying mutation model, and the model used can sometimes have a major impact on all subsequent inferences. Therefore, when some of the Jukes–Cantor distances are large, it is important to examine each DNA region carefully and then choose the molecule genetic distance that is appropriate for that region. This involves looking at overall base composition, evidence for transition/transversion biases in the mutational process, mutagenic sites, other sources of mutational rate heterogeneity across nucleotides, and many other potential factors. A computer program to aid in such a search is available at http://darwin.uvigo.es/.

Once the problem of multiple mutational events has been dealt with through an appropriate genetic distance, a haplotype tree can be estimated through one of several algorithms.

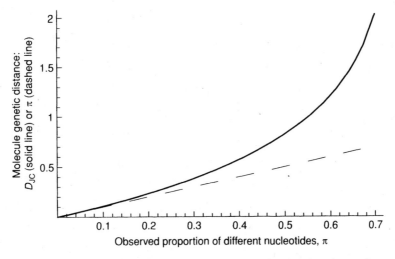

Figure 5.22. Plot of Jukes–Cantor molecule genetic distance (solid line) and π, observed proportion of nucleotides that differ between DNA molecules being compared (dashed line).

> **BOX 5.3 NEIGHBOR-JOINING METHOD OF TREE ESTIMATION**
>
> Step 1 in neighbor joining is the calculation of the net molecule genetic distance of each haplotype from all other haplotypes in the sample. Letting d_{ik} be the molecule genetic distance between haplotype i and haplotype k, the net molecule genetic distance for haplotype i is
>
> $$r_i = \sum_{k=1}^{n} d_{ik}$$
>
> where n is the number of haplotypes in the sample. The net distances are used to evaluate violations of the molecular clock model. For example, suppose one haplotype lineage experienced a much higher rate of accumulation of mutations than the remaining haplotypes. Then the haplotypes from that fast-evolving lineage would all have high r's.
>
> The second step is to use the net distances to create a new set of rate-corrected molecule genetic distances as
>
> $$\delta_{ij} = d_{ij} - \frac{r_i + r_j}{n - 2}$$
>
> The third step is the actual neighbor joining. The two haplotypes with the smallest δ_{ij} are placed on a common branch in the estimated evolutionary tree. That means that haplotypes i and j are now connected to a common node, say u, and u in turn is connected to the remaining haplotypes. Haplotypes i and j are now removed from the estimation procedure and replaced by their common node, u.
>
> The fourth step is to calculate the molecule genetic distances from each of the remaining haplotypes to node u as $d_{ku} = (d_{ik} + d_{jk} - d_{ij})/2$.
>
> The fifth step is to decrease n by 1, reflecting the fact that two haplotypes (i and j) have been replaced by a single node (u). Steps 1–5 are then repeated until only one branch is left, thereby producing the neighbor-joining tree.

One of the more popular ones is the algorithm of **neighbor joining** (Saitou and Nei 1987), which estimates an evolutionary tree by grouping together the entities that are close together with respect to a molecule genetic distance measure. The details for estimating trees under neighbor joining are given in Box 5.3.

Many computer programs exist that can determine the neighbor-joining tree rapidly even for large data sets containing many haplotypes. This is a great advantage over parsimony approaches. Unlike parsimony, neighbor joining always produces only a single tree. This is a great disadvantage. The loops that appear in a statistical parsimony network or the multiple solutions under maximum parsimony are excellent reminders that we do not really know the true evolutionary history of the haplotypes; rather, we are only estimating this history, often with some error or ambiguity. Indeed, it is often best to estimate a haplotype tree by more than one method as another way of assessing this ambiguity. For example, Templeton et al. (2000a) estimated the *LPL* haplotype trees using both statistical parsimony and neighbor joining. The resulting trees had some differences, but the differences were not statistically significant using the tree comparison test of Templeton (1983a, 1987a) that is part of many standard tree estimation packages such as PAUP (Phylogenetic Analysis

Using Parsimony; Swofford 1997). Such congruence indicates that these two very different algorithms for tree estimation are detecting a common evolutionary signal.

Another potential disadvantage of neighbor joining is that it reduces all of the mutational changes at the nucleotide sites into a single number. As we saw with both the *LPL* and *ApoE* examples, the details of the inferred mutational changes at each individual site can teach us much about the roles of recombination and mutation in shaping haplotype diversity and allow tests of specific hypothesis (e.g., mutagenic motifs in the genome and the fit of the infinite-sites model). Such valuable knowledge should not be ignored, so the individual site changes should be mapped onto the neighbor-joining tree and examined thoroughly even though these changes are not explicitly used in estimating the neighbor-joining tree. Many computer programs, such as PAUP, allow this to be done.

The above discussion and examples reveal that estimating haplotype trees can sometimes be a difficult and laborious process, so why bother? We will see in later chapters that both haplotype networks and haplotype trees, even if they contain many ambiguities and some recombination has occurred, can be used as powerful tools in population genetics for investigating a variety of evolutionary processes and patterns. Hence, the effort is often very much worthwhile.

6

GENE FLOW AND POPULATION SUBDIVISION

In deriving the Hardy–Weinberg law in Chapter 2, we assumed that the population was completely isolated. Isolation means that all individuals that contribute to the next generation come from the same population with no input at all from individuals from other populations. However, most species consist of not just one deme but rather many **local populations or subpopulations** consisting of the individuals inhabiting a geographic area from which most mating pairs are drawn that is generally small relative to the species' total geographic distribution. Although most matings may occur within a local population, in many species there is at least some interbreeding between individuals born into different local populations. Genetic interchange between local populations is called **gene flow**. In Chapter 1 we noted that DNA replication implies that genes have an existence in space and time that transcends the individuals that temporally bear them. Up to now, we have been primarily focused upon a gene's temporal existence, but with gene flow we begin to study a gene's spatial existence. In this chapter, we will study the evolutionary implications of gene flow and investigate how a species can become subdivided into genetically distinct local populations when gene flow is restricted. Restricted gene flow leads to variation in the frequency of a gene over space.

GENE FLOW

Gene Flow Between Two Local Populations

We start with a simple model in which two infinitely large local populations experience gene flow by symmetrically exchanging a portion m of their gametes each generation. We will monitor the evolution of these two populations at a single autosomal locus with two neutral alleles (A and a). The basic model is illustrated in Figure 6.1. In this simple model there is no mutation, selection, or genetic drift. For any given local population, we assume that a portion $1 - m$ of the gametes are sampled at random from the same local area and

---

Population Genetics and Microevolutionary Theory, By Alan R. Templeton
Copyright © 2006 John Wiley & Sons, Inc.

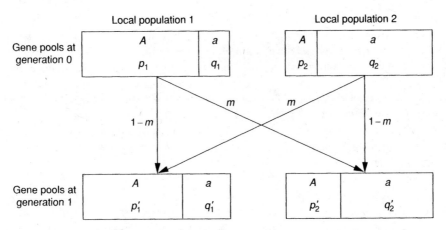

Figure 6.1. Model of symmetrical gene flow between two populations. The boxes represent the gene pools at an autosomal locus with two alleles, A and a, for the two populations over two successive generations, with m of the genes being interchanged between the two localities and $1 - m$ staying within the same locality.

that a portion m of the gametes are sampled at random from the other local population's gene pool (that is, gene flow). Letting p_1 be the initial frequency of the A allele in local population 1 and p_2 be the initial frequency of A in local population 2, the allele frequencies in the next generation in the two local populations are

$$p'_1 = (1-m)p_1 + mp_2 \qquad p'_2 = (1-m)p_2 + mp_1 \qquad (6.1)$$

We can now see if evolution occurred by examining whether or not the allele frequencies in either local population change across the generations:

$$\Delta p_1 = p'_1 - p_1 = (1-m)p_1 + mp_2 - p_1 = -m(p_1 - p_2)$$
$$\Delta p_2 = -m(p_2 - p_1) \qquad (6.2)$$

Equations 6.2 show that gene flow acts as an evolutionary force (that is, gene flow alters allele frequencies) if the following two conditions are satisfied:

- $m > 0$ (the local populations have some genetic exchange and are not completely reproductively isolated) and
- $p_1 \neq p_2$ (the local populations have genetically distinct gene pools).

In other words, gene flow is an evolutionary force when it occurs between populations with distinct gene pools.

It is important to keep in mind that m is defined in terms of the gene pools, and therefore m represents the amount of exchange of *gametes* between the local populations and not necessarily individuals. In some species, gametes are exchanged directly without the individuals moving at all. For example, most trees are wind pollinated, and the pollen (regarding these haploid gametophytes as essentially being gametes) can be blown for hundreds of miles by the wind. Hence, tree populations that are quite distant can still experience gene flow, yet no diploid trees are walking back and forth! For many other species, m requires that individuals move from their local population of birth to a different local population, followed by reproduction in their new location. Because gene flow requires both movement and reproduction, m is not just the amount of dispersal of individuals between local populations, but instead m represents a *complex interaction* between the pattern of dispersal and the system of mating. For example, system-of-mating inbreeding can greatly reduce gene flow, even if the individuals are in physical proximity. As mentioned in Chapter 3, the Tamils of India preferentially marry cousins. As a result of this inbreeding system of mating, the Tamils have little gene flow with other peoples with whom they physically intermingle.

Assortative mating can also greatly reduce the amount of gene flow. The European corn borer, an insect pest, has two pheromone races that apparently had once been geographically separated but are now broadly overlapping (Harrison and Vawter 1977). There is strong assortative mating for pheromone phenotype in these insects with greater than 95% of the matings occurring within the pheromone types (Malausa et al. 2005). Moreover, these races have allele frequency differences at many isozyme loci (Appendix 1) because of their historical isolation. Recall from Chapter 3 that when two previously isolated, genetically differentiated populations make genetic contact with one another, extensive linkage disequilibrium is created in the mixed population:

$$D_{admixture} = m(1 - m)(p_1 - p_2)(k_1 - k_2) \tag{6.3}$$

where the p's refer to the allele frequencies in the two populations at one locus and the k's are the respective frequencies at a second locus. Thus, in the areas of overlap of the pheromone races, there is linkage disequilibrium between the pheromone loci and all other loci having allele frequency differences between the historical races. Because assortative mating reduces the chances that individuals from the different pheromone races will mate with one another, it also reduces the effective gene flow m for all loci that had different allele frequencies in the historical races. As a result, assortative mating for pheromone type greatly reduces gene flow as an evolutionary force for all differentiated loci. Despite close physical proximity of individual corn borers, the effective m is very small and the races have maintained their differentiation even at isozyme loci that have no direct impact on the pheromone phenotype.

In contrast, disassortative mating enhances m for all loci. As mentioned in Chapter 3, disassortative mating systems give a reproductive advantage to individuals who are phenotypically dissimilar to the majority of individuals in the population. Often, dispersing individuals tend to deviate more on the average from the phenotypic means of the population into which they have dispersed. This gives dispersing individuals a reproductive advantage in their new population, thereby enhancing gene flow for all the genes borne by the dispersing individuals. For example, *Drosophila melanogaster* has a strong disassortative mating pheromone system (Averhoff and Richardson 1974, 1975), just the opposite of the European corn borer. Across the globe *D. melanogaster* is predominately a single, cosmopolitan species showing only modest geographical differentiation (except for some selected loci) even on a continental basis (Singh and Rhomberg 1987).

It is also important to note that assortative or disassortative mating on a *nongenetic* phenotype can influence m for many loci as long as the phenotypic differences influencing system of mating are correlated with the historical gene pool differences. This was already noted for the Amish human populations (Chapter 3), who have assortative mating based on religion and who, as a consequence, have little gene flow with surrounding populations and maintain genetic distinctiveness from their neighbors.

An example involving assortative mating on both genetic and nongenetic phenotypes is provided by gene flow between human populations of European origin and of African origin in the United States and in northeastern Brazil. In North America, European settlers brought in African slaves mainly from 1700 to 1808, with 98% of Africans coming from West and West-Central Africa. Once in North America, gene flow occurred between peoples of European and African origin, even though there was a tendency for assortative mating on the basis of skin color (Chapter 3). The people resulting from matings between individuals of European and African origin have been *socially* classified as "blacks" in North America. Genetically, and phenotypically for skin color, such people of mixed ancestry are intermediate and are no more black than they are "white." The social recognition of just two primary skin color categories is therefore a cultural decision, not a biological one. Nevertheless, this cultural classification has a direct and strong biological impact because it is coupled with assortative mating. The factors of assortative mating by the cultural "skin color" category, the cultural decision to classify people of mixed ancestry as blacks, and the numerical predominance of whites all combine to create an asymmetrical gene flow pattern. Effectively, much more gene flow occurred from whites to blacks than in the other direction in North America. A simplified version of this asymmetrical gene flow is given in Figure 6.2. In that figure, M is the effective amount of gene flow over the entire relevant period of North American history (in contrast to m, which is a per-generation gene flow parameter).

As can be seen from Figure 6.2, this pattern of gene flow (simplified relative to the actual pattern) results in an allele frequency of current African Americans (p_A) of $Mp_E + (1 - M)p_W$, where p_E and p_W are the allele frequencies in the ancestral European and West African populations, respectively. Solving for M in Figure 6.2, we can estimate M from the allele frequencies as

$$M = \frac{p_A - p_W}{p_E - p_W} = \frac{\text{change in allele frequency in African Americans from West Africans}}{\text{initial difference in allele frequency between Europeans and West Africans}}$$

(6.4)

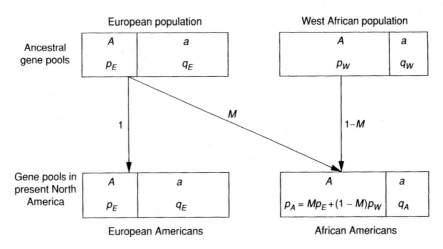

Figure 6.2. Model of asymmetrical gene flow between two populations representing simplified version of gene flow between Europeans and West Africans in North America to produce current African American population. In this model, M represents the cumulative impact of gene flow over several centuries.

For example, at the autosomal Rh blood group locus, the frequency of the $Rh+$ allele is 0.4381 in African Americans (p_A), 0.5512 in West Africans, and 0.0279 in European Americans. Assuming the current West African allele frequencies have not changed much over the last few centuries, $p_W = 0.5512$. Assuming that the current European American allele frequencies are close to the ancestral European immigrant allele frequencies because European Americans have not been so strongly influenced by gene flow, then $p_E = 0.0279$. These Rh allele frequencies yield $M = 0.216$ from equation 6.4. This value is typical of African Americans in North America (Reed 1969). What this number tells us is that the African American gene pool has been affected by gene flow such that it was about 20% European in origin and 80% West African at the time of these studies. However, there is much variation in the degree of admixture among different local populations of African Americans. For example, African Americans living in Columbia, South Carolina, have $M = 0.18$. In contrast, the Gullah-speaking Sea Island African Americans that live in nearby coastal South Carolina have $M = 0.035$. This low amount of admixture is consistent with the history of the Gullah-speaking African Americans, who have been relatively isolated throughout their entire history, have lived in an area that has always had an African American majority, and have retained many aspects of African culture, including their language (Parra et al. 2001).

European and West African populations were also brought into physical contact in northeastern Brazil at about the same time as North America. However, the social definitions of "race," particularly for individuals of mixed ancestry, were and are different from those used in North America. In northeastern Brazil, a number of alternative categories are available for individuals of mixed ancestry, and many individuals who would be socially classified as blacks in North America would not be considered blacks in Brazil. In a study of Brazilians (Franco et al. 1982), the Brazilian authors used the term white in the context of Brazilian culture. The gene pool of these whites was estimated to be 67% of European origin, 20% of West African origin, and 13% of Amerindian origin, using an equation similar to 6.4. In contrast, the "nonwhites" were 58% European, 25% African, and 17% Amerindian. Using just skin color and not social classification, the Brazilian subjects were also characterized from "most Caucasoid" to "most Negroid." In northeast Brazil, the most Caucasoid group is 71% European, in contrast to the nearly 100% found in North American whites. The most Negroid Brazilian group is 28% European—an amount of admixture greater than that of the average African American from North America. No matter how one categorizes the Brazilians in this study, it is obvious that there has been much more gene flow between the European and African gene pools in northeastern Brazil as compared to North America. Hence, the *cultural* systems of mating in the two countries have had a major genetic impact on the composition of their present-day populations despite similar initial founding populations and proportions. This example shows that m or its multigenerational cumulative analogue M is determined not just by physical movement of individuals but also by system of mating as influenced by genetic and nongenetic factors.

Genetic Impact of Gene Flow

We have already seen that allele frequencies are altered when gene flow occurs between genetically distinct populations. Gene flow therefore can be an evolutionary force. In this section, we will see that gene flow causes evolution in a nonrandom, predictable fashion. To show this, we will return to our simple model of symmetrical gene flow given in Figure 6.1. Starting with the initial populations prior to gene flow, their genetic distinctiveness is measured by the difference in their allele frequencies; that is, $d_0 = p_1 - p_2$. After one generation of gene flow, Figure 6.1 shows that

$$p_1' = (1 - m)p_1 + mp_2 = p_1 - m(p_1 - p_2) = p_1 - md_0 \quad (6.5)$$

and similarly,

$$p_2' = p_2 + md_0 \quad (6.6)$$

Hence, the difference in gene pools between the two local populations after a single generation of gene flow is

$$d_1 = p_1' - p_2' = p_1 - md_0 - p_2 - md_0 = d_0(1 - 2m) \quad (6.7)$$

Note that equation 6.7 implies that $|d_1| < |d_0|$ for all $m > 0$ and $d_0 \neq 0$. By using the above equations recursively, the difference in allele frequencies between the two local populations after t generations of gene flow is

$$d_t = d_0(1 - 2m)^t \to 0 \quad \text{as } t \to \infty \quad (6.8)$$

Therefore, *gene flow decreases the allele frequency differences between local populations.*

Now consider a special case of Figure 6.1 in which $p_1 = 0$ and $p_2 = 1$. In this case, the frequency of the A allele in the population 1 gene pool will go from being completely absent to being present with a frequency of m. This evolutionary change caused by gene flow mimics that of mutation. If the mutation rate from a to A were μ, then the evolutionary change caused by mutation in a population initially lacking the A allele would be to introduce that allele with a frequency of μ. Hence, *gene flow can introduce new alleles into a population*, with m being the analog of the mutation rate. One major difference between gene flow and mutation as sources of new genetic variation for a local deme is that in general μ is constrained to take on only very small values, whereas m can be either small or large. A second major difference is that gene flow can introduce variation at many loci simultaneously, whereas mutation generally affects only one locus or nucleotide site at a time. A third major difference is that many new mutations are deleterious (Figure 5.3) and initially occur as single copies, thereby ensuring that many are rapidly lost from the population. In contrast, gene flow introduces genetic variation that has usually been around for more than one generation and can introduce multiple copies of new variants. Hence, there is the potential for a massive influx of new genetic variability through gene flow that can drastically alter a local gene pool, even in a single generation.

The effects of gene flow on genetic variation between and within local populations described above can be summarized as *gene flow decreases genetic variability between local populations and increases genetic variability within a local population*. Recall from Chapter 4 that genetic drift causes an increase in genetic variability between populations (their allele frequencies diverge) and decreases genetic variability within a population (loss and fixation of alleles). Hence, the effects of gene flow on within- and between-population genetic variability are the *opposite* of those of genetic drift.

In Chapter 2, we introduced the idea of population structure as the mechanisms or rules by which gametes are paired together in the reproducing population. We now include in those rules the exchange of gametes among local populations (gene flow). Parallel to this process-oriented definition of population structure, there is also a pattern-oriented definition: **Population structure** is the amount of genetic variability and its distribution within and among local populations and individuals within a species. This definition emphasizes the spatial *patterns* of genetic variation that emerge from the rules of gametic exchange. The pattern of genotypic variability (heterozygosity versus homozygosity) among individuals within a local population is highly dependent upon the system of mating, as we saw in Chapter 3. As mentioned above, the distribution of allelic variation within and among local demes is influenced by both gene flow and genetic drift. Therefore, genetic population structure has three major components:

- System of mating
- Genetic drift
- Gene flow.

Because of the opposite effects of gene flow and genetic drift, the *balance* between drift and gene flow is a *primary determinant* of the genetic population structure of a species.

The concept of genetic population structure (hereafter called population structure) is critical for the remainder of this book. Genotypic variability provides the raw material for all evolutionary change, including that caused by natural selection. Population structure therefore determines the pattern and amount of genetic variability that is available for evolution within a species. As will be seen later, natural selection and other evolutionary forces operate within the constraints imposed by the population structure. Hence, virtually all evolutionary predictions, particularly those related to adaptive evolution, must always be placed in the context of population structure.

Given the central importance of population structure to microevolutionary processes, we need additional tools to measure and quantify it. The tools for measuring system of mating have already been discussed in Chapter 3 and those for drift in Chapters 4 and 5, so now we need to develop measures for the balance between gene flow and drift.

BALANCE OF GENE FLOW AND DRIFT

Recall from Chapter 4 that to measure the impact of genetic drift upon identity by descent, we started with equation 4.3:

$$\bar{F}(t) = \frac{1}{2N} + \left(1 - \frac{1}{2N}\right)\bar{F}(t-1)$$

where N is replaced by the inbreeding effective size for nonideal populations. To examine the balance between drift and mutation, we modified the above equation to yield equation 5.4:

$$\bar{F}(t) = \left[\frac{1}{2N} + \left(1 - \frac{1}{2N}\right)\bar{F}(t-1)\right](1-\mu)^2$$

Because gene flow and mutation behave in an analogous manner with respect to genetic variation within a local deme, a similar modification of equation 4.3 can be used to address the following question: Suppose a local deme of inbreeding effective size N_{ef} is experiencing gene flow at a rate of m per generation from some outside source. What is the probability that two randomly drawn genes from this local deme are identical by descent AND came from parents from the local population? That is, if one of the genes came from a migrant, we no longer regard it as "identical." Effectively, this means that the outside-source population or populations are assumed to share no identity by descent ($F = 0$) with the local deme of interest. The equation for the average probability of identity by descent within the local deme is then

$$\bar{F}(t) = \left[\frac{1}{2N_{ef}} + \left(1 - \frac{1}{2N_{ef}}\right) \bar{F}(t-1) \right] (1-m)^2 \quad (6.9)$$

Equation 6.9 is the probability of identity by descent as a function of genetic drift in the local deme (equation 4.3) times the probability that both of the randomly chosen gametes came from the local deme. At equilibrium, equation 6.9 yields (analogous to equation 5.6)

$$\bar{F}_{eq} = \frac{1}{4N_{ef}m + 1} \quad (6.10)$$

if m is small such that m is much greater than m^2 and m is on the order of magnitude of $1/N_{ef}$ or smaller.

Recall from equation 5.7 that the balance of mutation to genetic drift was measured by $\theta = 4N_{ef}\mu$. The balance of gene flow to genetic drift is measured in equation 6.10 by a similar parameter: $4N_{ef}m$. The similarity between gene flow and mutation can also be framed in terms of a coalescent process. For example, we can determine the conditional probability that two genes randomly drawn from the same subpopulation coalesce back to a common ancestor before either lineage experienced a gene flow event given than either coalescence or gene flow has occurred. In analogy to equation 5.13,

$$\text{Prob(gene flow before coalesence} \mid \text{gene flow or coalescence)} \approx \frac{4N_{ef}m}{4N_{ef}m + 1} \quad (6.11)$$

Since the probability of identity in this model is the probability of coalescence before gene flow given that either gene flow or coalescence has occurred, the equilibrium probability of identity in the gene flow coalescent model is simply 1 minus equation 6.11, which yields equation 6.10. Whether we look backward or forward in time, we obtain the same equilibrium balance of gene flow (proportional to m) to drift (proportional to $1/N_{ef}$) as measured by their ratio of relative strength: $m/(1/N_{ef}) = N_{ef}m$.

Note that our concept of "identity" has been altered once again from what it was in Chapter 4. Wright (1931), who first derived equation 6.10, defined the $\bar{F}$ in equation 6.10 as F_{st}, where the st designates this as identity by descent in the subpopulation relative to the total population. Therefore, we have yet another inbreeding coefficient in the population genetic literature that is distinct mathematically and biologically from the inbreeding coefficients previously used in this book. In particular, F_{st} does not measure identity by descent in the pedigree inbreeding sense (F), or system-of-mating inbreeding (f), or the impact of genetic drift within a single deme upon average identity-by-descent inbreeding ($\bar{F}$); rather, the inbreeding coefficient F_{st} measures the ratio of drift to gene flow to drift and how this ratio influences population structure in a process-oriented sense. In terms of patterns, a high value of F_{st} indicates that there is little genetic variation in a local population relative to the total population, whereas a small value indicates much local variation relative to the total. Hence, in terms of the pattern definition of population structure, F_{st} measures the proportion of genetic variation among individuals drawn from all demes that is due to genetic differences between demes and is one of the most commonly used measures of population structure in the evolutionary genetic literature.

Equation 6.10 shows us how processes can generate patterns of population structure. As m increases (gene flow becomes more powerful), F_{st} decreases (more variation within local demes and less genetic differences between). As $1/N_{ef}$ increases (drift becomes more powerful), F_{st} increases (less variation within local demes and more genetic differences between). These properties are exactly as expected from the impact of genetic drift and gene flow on variation within and between local demes when considered separately. What is surprising from equation 6.10 is that even a small amount gene flow can cause two populations to behave effectively as a single evolutionary lineage. For example, let $N_{ef}m = 1$, that is, one "effective" migrant per generation. Here, $N_{ef}m$ is not the actual number of migrating individuals per generation but rather is an *effective number of migrants* because it depends upon the local inbreeding effective size N_{ef} and not the census number of individuals. Then, $F_{st} = \frac{1}{5} = 0.20$, as shown in Figure 6.3. This means from equation 6.11 that 80% of the gene pairs drawn from the same subpopulation will show gene flow before coalescence. Hence, the genealogical histories of the local demes are extensively intertwined when $N_{ef}m \geq 1$. Note also that $N_{ef}m = 1$ defines a transition point an inflection point in the plot of equation 6.10 against the effective number of migrants (Figure 6.3). With increasing effective number of migrants F_{st} declines only very slowly when $N_{ef}m \geq 1$, but with decreasing effective number of migrants F_{st} rises very rapidly when $N_{ef}m \leq 1$. Because $N_{ef}m = 1$ is at the inflection point of equation 6.10, an effective number of migrants of 1 marks a biologically significant transition in the relative evolutionary importance of gene flow to drift. It is impressive that very few effective migrants are needed (only one or more per generation on average) to cause gene flow to dominate over genetic drift, leading to subpopulations that display great genetic homogeneity with one another.

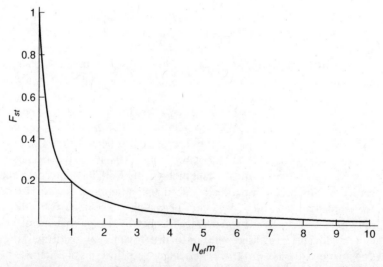

Figure 6.3. Plot of F_{st} in equation 6.10 versus effective number of migrants per generation, $N_{ef}m$. The plot shows the inflection point at $N_{ef}m = 1$, corresponding to $F_{st} = 0.2$.

It is also surprising that the extent of this genealogical mixing depends only upon the effective *number* of migrants ($N_{ef}m$) and not upon the *rate* of gene flow (m). For example, two subpopulations of a billion each would share 80% of their genes by exchanging only one effective individual per generation from equation 6.10, but so would two subpopulations of size 100. However, the rates of gene flow would be greatly different in these two cases: $m = 0.000000001$ for the first case and $m = 0.01$ for the second. Thus, very different rates of gene flow can have similar impacts upon population structure. Alternatively, identical rates of gene flow can have very different impacts on population structure. Suppose, for example, that $m = 0.01$ in both the cases considered above. For the two subpopulations with inbreeding effective sizes of 1 billion, the resulting F_{st} is effectively zero (2.5×10^{-8}) from equation 6.10, whereas for the local populations of inbreeding effective size 100, $F_{st} = 0.20$. The reason why the same number of effective migrants is needed to yield a specific value of F_{st} and not the same rate of gene flow is that F_{st} represents a *balance* between the rate at which genetic drift causes subpopulations to diverge versus the rate at which gene flow makes them more similar. In large populations, divergence is slow, so small amounts of gene flow are effective in counterbalancing drift-induced divergence; as populations become smaller, larger and larger rates of gene flow are needed to counterbalance the increasing rate of drift-induced divergence.

Similarly, it is the product $N_{ef}m$ (the ratio of the strength of gene flow to drift) and not m alone that determines the relative coalescence times of genes within and among local populations. If there is restricted gene flow among demes, it makes sense that the average time to coalescence for two genes sampled within a deme will be less than that for two genes sampled at random for the entire species. In particular, Slatkin (1991) has shown that these relative times are determined by $N_{ef}m$. The exact relationship of coalescence times within and among local populations depends upon the pattern of gene flow. Consider the simple case of a species subdivided into a large number of local demes each of size N_{ef} and each receiving a fraction m of its genes per generation from the species at large. This "island model" of gene flow among multiple local demes of identical inbreeding effective size (Figure 6.4) also leads to equation 6.10 and hence is a straightforward multideme extension of the two-deme model illustrated in Figure 6.1. Slatkin (1991) has shown that

$$N_{ef}m = \frac{\bar{t}_0}{4(\bar{t} - \bar{t}_0)} \tag{6.12}$$

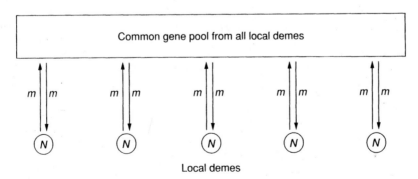

Figure 6.4. Island model of gene flow among multiple local demes each of idealized size N, each contributing fraction m of its gametes to common gene pool that is then distributed at random over all local demes in same proportion.

where $\bar{t}_0$ is the average time to coalescence of two genes sampled from the same subpopulation and $\bar{t}$ is the average time to coalescence of two genes sampled from the entire species. Hence, the ratio of within-deme coalescence time to entire-species coalescence time is

$$\frac{\bar{t}_0}{\bar{t}} = \frac{4N_{ef}m}{1+4N_{ef}m} \tag{6.13}$$

Therefore, F_{st} now has a simple interpretation in terms of coalescence times:

$$F_{st} = \frac{\bar{t} - \bar{t}_0}{\bar{t}} \quad \text{or} \quad \frac{\bar{t}_0}{\bar{t}} = 1 - F_{st} \tag{6.14}$$

For example, $F_{st} = 0.156$ when averaged over 109 loci for local populations of humans scattered throughout the world (Barbujani et al. 1997). From equation 6.10, this F_{st} value yields 1.4 effective migrants per generation. Thus, although there is seemingly little gene flow among human populations at a global scale in an absolute sense (a little over one effective migrant on average per generation among human subpopulations), this is sufficient to place the human species in the domain where gene flow dominates over drift, resulting in human populations at the global level showing little genetic differentiation by the standard criterion used in population genetics (Figure 6.3). An F_{st} value of 0.156 also means that $(1 - F_{st}) \times 100 = 84.4\%$ of the time the gene you inherited from your mother and the gene you inherited from your father trace back to different human subpopulations even when your parents are both from the same human subpopulation (from equation 6.11). Finally, this means from equation 6.13 or 6.14 that the time it takes two genes sampled from the same local human population to coalesce is 84.4% of the time it would take two genes sampled at random over the entire human species to coalesce. In humans, there is little difference between local coalescence times and global coalescence times because of extensive gene flow. Therefore, human subpopulations even at the global level are extensively intertwined genetically. However, this interpretation is made under the assumption that the human F_{st} value is actually due to the equilibrium balance of gene flow and drift and not some other factors. This assumption will be examined in detail in Chapter 7.

Equation 6.9 can be extended to include mutation. If we assume that "identity" can be destroyed by both mutation and gene flow, then the appropriate analogue to equation 6.9 is

$$\bar{F}(t) = \left[\frac{1}{2N_{ef}} + \left(1 - \frac{1}{2N_{ef}}\right)\bar{F}(t-1)\right][(1-\mu)(1-m)]^2 \tag{6.15}$$

If both μ and m are small, then using a Taylor's series we have

$$\bar{F}_{eq} = F_{st} = \frac{1}{4N_{ef}(\mu + m) + 1} \tag{6.16}$$

Equation 6.16 shows that the joint impact of mutation and gene flow is described by the sum of μ and m, once again emphasizing the similar role that the disparate forces of mutation and gene flow have upon genetic variation and identity by descent. Equation 6.16 also shows that when m is much larger than μ (frequently a realistic assumption), gene flow dominates over mutation in interacting with genetic drift to determine population structure.

In all of the above equations, F_{st} was defined in terms of identity by descent, but in many cases all we can really observe is identity by state. To get around the problem of identity by state not always being the same as identity by descent, F_{st} is often measured by the proportional *increase* in identity by state that occurs when sampling within versus between subpopulations (Cockerham and Weir 1987). Let F_s be the probability of identity by state of two genes randomly sampled within a deme, and let F_t be the probability of identity by state of two genes randomly sampled from the total species. If all subpopulations had identical gene pools, then $F_s = F_t$. But with population subdivision, we expect an increase in identity by state for genes sampled within the same subpopulation beyond the random background value of F_t. Note that $1 - F_t$ is the probability that two genes are *not* identical by state with random sampling of the total population. If F_{st} is now regarded as the additional probability of identity by state that occurs beyond random background sampling when we sample two genes from the same local deme, we have

$$F_s = F_t + (1 - F_t)F_{st} \quad F_{st} = \frac{F_s - F_t}{1 - F_t} \tag{6.17}$$

Equation 6.17 provides another way of estimating F_{st} through identity by state by randomly sampling pairs of genes drawn from the total population (F_t) and from within the subpopulations (F_s) (Davis et al. 1990).

Equation 6.17 is also used to extend the models so far discussed to multiple hierarchies. All our models of gene flow so far have assumed a total population subdivided into a series of local demes or subpopulations, which are then all treated equally with respect to gene flow. However, this is often an unrealistic model. For example, most gene flow among human subpopulations occurs between subpopulations living on the same continent (Cavalli-Sforza et al. 1994). As a result, instead of subdividing the global human population simply into local demes, it is more biologically realistic to first subdivide the global human population into continental subpopulations and then subdivide each continental subpopulation into local intracontinental populations. To deal with this hierarchy of three levels, we need three levels of sampling. Therefore, let F_t be the probability of identity by state of two genes randomly sampled from the total human species, F_c be the probability of identity by state of two genes randomly sampled from humans living on the same continent, and F_s be the probability of identity by state of two genes randomly sampled from the same intracontinental subpopulation. Then, in analogy to equation 6.17, we have

$$F_{ct} = \frac{F_c - F_t}{1 - F_t} \quad F_{sc} = \frac{F_s - F_c}{1 - F_c} \tag{6.18}$$

where F_{ct} measures the increase in identity by descent due to sampling within continental subpopulations relative to the total human species and F_{sc} measures the increase in identity by descent due to sampling within intracontinental local populations relative to the continental subpopulations. In this three-hierarchy model F_{st} can be recovered from (Wright 1969)

$$(1 - F_{sc})(1 - F_{ct}) = \left(\frac{1 - F_s}{1 - F_c}\right)\left(\frac{1 - F_c}{1 - F_t}\right) = \left(\frac{1 - F_s}{1 - F_t}\right) = 1 - F_{st} \tag{6.19}$$

Equation 6.19 indicates the total F_{st} can be partitioned into two components: F_{ct}, which measures genetic differentiation among continental subpopulations of humans, and F_{sc},

180   GENE FLOW AND POPULATION SUBDIVISION

which measures genetic differentiation among local populations living in the same continent. Both contribute to F_{st}, as shown by equation 6.19. In particular, for humans, the F_{st} of 0.156 can be partitioned into a component of 0.047 that measures the relative proportion of genetic variation among local populations within continental groups and 0.108 among continental groups (Barbujani et al. 1997). Also note that equation 6.19 defines a chain rule that allows F statistics for measuring population structure to be extended to an arbitrary number of levels (Wright 1969).

WAHLUND EFFECT AND *F* STATISTICS

We have just seen how yet another inbreeding coefficient based upon the concept of identity by descent enters into the population genetic literature, but this time as a measure of how the balance of drift and gene flow influences identity by descent and coalescent times within and between demes in a subdivided species. We also saw in Chapter 4 that genetic drift influences many genetic parameters besides identity by descent, including the variance of allele frequencies across isolated replicate demes. This aspect of drift motivates an alternative definition of F_{st} in terms of variances of allele frequencies across the local demes (Cockerham and Weir 1987).

Consider first a model in which a species is subdivided into n discrete demes where N_i is the size of the ith deme. Suppose further that the species is polymorphic at a single autosomal locus with two alleles (A and a) and that each deme has a potentially different allele frequency (due to past drift in this neutral model). Let p_i be the frequency of allele A in deme i. Let N be the total population size ($N = \sum N_i$) and w_i the proportion of the total population that is in deme i ($w_i = N_i/N$). For now, we assume random mating within each deme. Hence, the genotype frequencies in deme i are

$$\begin{array}{cccc} \text{Genotype} & AA & Aa & aa \\ \text{Frequency} & p_i^2 & 2p_iq_i & q_i^2 \end{array}$$

The frequency of A in the total population is $\bar{p} = \sum w_i p_i$. If there were no genetic subdivision (that is, all demes had identical gene pools), then with random mating the expected genotype frequencies in the total population would be

$$\begin{array}{cccc} \text{Genotype:} & AA & Aa & aa \\ \text{Frequency:} & \bar{p}^2 & 2\bar{p}\bar{q} & \bar{q}^2 \end{array} \quad (6.20)$$

However, in the general case where the demes can have different allele frequencies, the actual genotype frequencies in the total population are

$$\text{Freq}(AA) = \sum_{i=1}^{n} w_i p_i^2$$

$$\text{Freq}(Aa) = 2\sum_{i=1}^{n} w_i p_i q_i \quad (6.21)$$

$$\text{Freq}(aa) = \sum_{i=1}^{n} w_i q_i^2$$

By definition, the variance in allele frequency across demes is

$$\text{Var}(p) = \sigma_p^2 = \sum_{i=1}^{n} w_i (p_i - \bar{p})^2 = \sum_{i=1}^{n} w_i p_i^2 - \bar{p}^2 = \sum_{i=1}^{n} w_i q_i^2 - \bar{q}^2 \qquad (6.22)$$

Substituting equation 6.22 into 6.21, the genotype frequencies in the total population can be expressed as

$$\text{Freq}(AA) = \sum_{i=1}^{n} w_i p_i^2 - \bar{p}^2 + \bar{p}^2 = \bar{p}^2 + \sigma_p^2$$

$$\text{Freq}(aa) = \sum_{i=1}^{n} w_i q_i^2 - \bar{q}^2 + \bar{q}^2 = \bar{q}^2 + \sigma_p^2 \qquad (6.23)$$

$$\text{Freq}(Aa) = 1 - \text{Freq}(AA) - \text{Freq}(aa) = 2\bar{p}\bar{q} - 2\sigma_p^2$$

By factoring out the term $2\bar{p}\bar{q}$ from the heterozygote frequency in equation 6.23, the observed frequency of heterozygotes in the total population can be expressed as

$$\text{Freq}(Aa) = 2\bar{p}\bar{q}\left(1 - \frac{\sigma_p^2}{\bar{p}\bar{q}}\right) = 2\bar{p}\bar{q}(1 - f_{st}) \qquad (6.24)$$

where $f_{st} = \sigma_p^2/(\bar{p}\bar{q})$. Hence, the genotype frequencies can now be expressed as

$$\text{Freq}(AA) = \bar{p}^2 + \bar{p}\bar{q} f_{st}$$

$$\text{Freq}(Aa) = 2\bar{p}\bar{q}(1 - f_{st}) \qquad (6.25)$$

$$\text{Freq}(aa) = \bar{q}^2 + \bar{p}\bar{q} f_{st}$$

Note the resemblance between equation 6.25 and equation 3.1 from Chapter 3. Equation 3.1 describes the deviations from Hardy–Weinberg genotype frequencies induced by system-of-mating inbreeding (f). Because a variance can only be positive, $f_{st} \geq 0$, which implies that the subdivision of the population into genetically distinct demes causes deviations from Hardy–Weinberg that are identical in form to those caused by system-of-mating inbreeding within demes ($f < 0$). This inbreeding coefficient is called f_{st} because it refers to the deviation from Hardy–Weinberg at the total population level caused by allele frequency deviations in the subdivided demes from the total-population allele frequency. This deviation from Hardy–Weinberg genotype frequencies in the species as a whole that is caused by population subdivision is called the **Wahlund effect**, after the man who first identified this phenomenon.

The parameter f_{st} is a standardized variance of allele frequencies across demes. In the extreme case where there is no gene flow at all ($m = 0$), we know from Chapter 4 that drift will eventually cause all populations to either lose or fix the A allele. Since drift has no direction, a portion $\bar{p}$ of the populations will be fixed for A, a portion $\bar{q}$ will be fixed for a, and the variance (equation 6.22) becomes $\bar{p}(1 - \bar{p})^2 + \bar{q}(0 - \bar{p})^2 = \bar{p}\bar{q}$. Therefore, f_{st} is the ratio of the actual variance in allele frequencies across demes to the theoretical maximum when there is no gene flow at all.

From equation 6.24, an alternative expression for f_{st} is given by

$$f_{st} = 1 - \frac{\text{Freq}(Aa)}{2\bar{p}\bar{q}} = \frac{2\bar{p}\bar{q} - \text{Freq}(Aa)}{2\bar{p}\bar{q}} = \frac{H_t - H_s}{H_t} \quad (6.26)$$

where $H_t = 2\bar{p}\bar{q}$ is the expected heterozygosity if the total population were mating at random and H_s is the observed frequency of heterozygotes in the total population, which is the same as the average heterozygosity in the subpopulations (recall that random mating is assumed within each subpopulation). The definition of f_{st} given by equation 6.26 is useful in extending the concept of f_{st} to the case with multiple alleles, as expected and observed heterozygosities are easily calculated or measured regardless of the number of alleles per locus. Equation 6.26 is used more commonly in the literature to measure population structure than equation 6.17, but readers need to be wary as many papers do not explicitly state which of the two definitions of F_{st}/f_{st} is being used. This is unfortunate, because the distinction can sometimes be important.

Equation 6.10 shows that the F_{st} defined in terms of probability of identity by descent can be related to the amount of gene flow, m, under the island model of gene flow in an equilibrium population. In the island model, a species is subdivided into a large number of local demes of equal size and with each local deme receiving a fraction m of its genes per generation from the species at large (Figure 6.4). Under this model, the variance in allele frequency across demes for an autosomal locus with two alleles reaches an equilibrium between drift and gene flow of (Li 1955)

$$\sigma^2 = \frac{\bar{p}\bar{q}}{2N_{ev} - (2N_{ev} - 1)(1 - m)^2} \quad (6.27)$$

Because $f_{st} \equiv \sigma_p^2/(\bar{p}\bar{q})$ for a two-allele system, we have

$$f_{st} = \frac{1}{2N_{ev} - (2N_{ev} - 1)(1 - m)^2} \approx \frac{1}{4N_{ev}m + 1} \quad (6.28)$$

when m is small.

Note that equation 6.28 is *almost* identical to equation 6.10. The only difference is that in equation 6.10 for F_{st} we have N_{ef} and in equation 6.28 for f_{st} we have N_{ev}. This seemingly minor difference between equations 6.10 and 6.28 points out that there are two qualitatively different ways of defining F_{st}:

- F_{st} is measured through identity by descent or identity by state (equations 6.10, 6.11, and 6.17, and the related coalescent equation 6.14) and
- f_{st} is measured through variances of allele frequencies or heterozygosities (equations 6.25, 6.26, and 6.28).

Both f_{st} and F_{st} ideally measure the balance of genetic drift versus gene flow and in many cases are similar in value. However, as we saw in Chapter 4, N_{ef} and N_{ev} can sometimes differ by orders of magnitude under biologically realistic conditions. Accordingly, the two alternative definitions of F_{st} and f_{st} can also differ substantially because they focus upon different impacts of genetic drift. In particular, the two definitions can differ substantially when recent events have occurred that disturb any equilibrium. The identity-by-descent definitions depend upon long-term coalescent properties (e.g., equation 6.13) and are not as sensitive to recent disturbances as are the allele frequency variance definitions, which respond rapidly to altered conditions.

For example, using an estimation procedure based upon equation 6.17 for F_{st}, Georgiadis et al. (1994) obtained an F_{st} value not significantly different from zero for mtDNA for African elephants sampled either within eastern Africa or within southern Africa. Using the same data with an estimation procedure based upon equation 6.26, Siegismund and Arctander (1995) found significant f_{st} values of 0.16 and 0.30 for eastern and southern African elephant populations, respectively. This apparent discrepancy is expected given that human use of the habitat in these regions has fragmented and reduced the elephant populations over the last century in these two regions of Africa. Recall from Chapter 4 that N_{ef} is generally larger than N_{ev} when population size is decreasing. Hence, we expect $4N_{ef}m > 4N_{ev}m$ for these elephant populations, which means we expect $F_{st} < f_{st}$, as observed. Note that F_{st} and f_{st} are *not* two alternative ways of estimating the same population structure parameter. As the elephant example shows, F_{st} and f_{st} are biologically different parameters that measure different aspects of population structure in nonideal (*in sensu* effective size) populations.

Both F_{st} and f_{st} measure the relative balance of drift to gene flow, and in both cases this balance appears as a product of an effective size and m, the amount of gene flow. Because the balance in either equation 6.10 or 6.28 depends only upon the product $N_e m$, you cannot predict the population structure of a species just by knowing either the effective sizes alone or the rate of gene flow alone. Thus, a species with large local population sizes could still show extreme genetic subdivision and a large Wahlund effect if it had low rates of gene flow, and similarly a species subdivided into small local populations could show little subdivision if it had high rates of gene flow. For example, the snail *Rumina decollata* is particularly abundant in the parks and gardens along the Boulevard des Arceaux in the city of Montpellier, France. However, like most snails, it has limited dispersal capabilities, even over this area of about 2.5 acres. Hence, despite thousands of snails living in this small area, the f_{st} among 24 colonies along the Boulevard des Arceaux is 0.294 (Selander and Hudson 1976), nearly twice as large as the f_{st} value of around 0.15 for humans on a global scale. In contrast, the mouse, *Mus musculus*, tends to live in small populations within barns in Texas, but it has much better dispersal capabilities than the snail *Rumina*. The f_{st} among barn subpopulations within the same farm (the "total" population in this case) is 0.021 (Selander et al. 1969), a value an order-of-magnitude smaller than the snail's f_{st} even though the geographical scales are comparable and the total population size of mice is much smaller than that of the snails. These examples show that the amount of genetic subdivision in a species depends upon the *balance* of drift and gene flow and *not* either evolutionary factor considered by itself.

POPULATION SUBDIVISION AND SYSTEM OF MATING

The balance between drift and gene flow is the primary determinant of what fraction of a species' genetic variability is available in local gene pools, but the local system of mating then takes the gene pool variation available at the gametic level and transforms it into genotypic variation at the individual level, as we saw in Chapter 3. Therefore, a full consideration of how genetic variation is distributed between demes, among individuals within a deme, and within individuals (heterozygosity versus homozygosity) requires a model that

184 GENE FLOW AND POPULATION SUBDIVISION

integrates the effects of system of mating, genetic drift, and gene flow—the three major components of population structure. So far we have only considered random mating within the local population in this chapter. Now we consider nonrandom mating within demes by letting f_{is} be the system-of-mating inbreeding coefficient for individuals within a subpopulation. This is the same as the f introduced in Chapter 3, but now we add the subscripts to emphasize that we are no longer considering just one deme in isolation from all other demes. Because f_{is} functions in a manner identical to f in local demes, we have from equations 3.1 that within each deme the genotype frequencies are

$$\text{Freq}(AA \text{ in deme } j) = p_j^2 + p_j q_j f_{is}$$
$$\text{Freq}(Aa \text{ in deme } j) = 2 p_j q_j (1 - f_{is}) \quad (6.29)$$
$$\text{Freq}(aa \text{ in deme } j) = q_j^2 + p_j q_j f_{is}$$

With respect to the total population, the AA genotype frequency is now

$$\begin{aligned}
\text{Freq}(AA) &= \sum_{j=1}^{n} w_j \left(p_j^2 + p_j q_j f_{is} \right) \\
&= \sum_{j=1}^{n} w_j p_j^2 + \sum_{j=1}^{n} w_j \left(p_j - p_j^2 \right) f_{is} \quad (6.30) \\
&= \bar{p}^2 + \sigma_p^2 + f_{is} \left(\bar{p} - \bar{p}^2 - \sigma_p^2 \right) \\
&= \bar{p}^2 + \bar{p}\bar{q} \left[f_{st} + f_{is}(1 - f_{st}) \right]
\end{aligned}$$

Letting $f_{it} = f_{st} + f_{is}(1 - f_{st})$ and performing similar derivations to equation 6.30 for the other genotype frequencies, we have

$$\text{Freq}(AA) = \bar{p}^2 + \bar{p}\bar{q} f_{it}$$
$$\text{Freq}(Aa) = 2\bar{p}\bar{q}(1 - f_{it}) \quad (6.31)$$
$$\text{Freq}(aa) = \bar{q}^2 + \bar{p}\bar{q} f_{it}$$

Equations 6.31 superficially resemble equations 6.25, but equations 6.31 are a function of f_{it} and not f_{st}. It also follows from $f_{it} = f_{st} + f_{is}(1 - f_{st})$ that $1 - f_{it} = 1 - f_{st} - f_{is}(1 - f_s) = (1 - f_{st})(1 - f_{is})$, so the deviation of heterozygote genotype frequency from Hardy-Weinberg at the total population level $(1 - f_{it})$ is partitioned into a component due to the local system of mating $(1 - f_{is})$ and a component due to differences in allele frequencies across local demes $(1 - f_{st})$.

For example, the Yanomama Indians of South America, like many human populations, have incest taboos. Because they live in rather small villages, an incest taboo can have a measurable impact as a deviation from Hardy–Weinberg genotype frequencies within villages, and in this case $f_{is} = -0.01$, indicating an avoidance of system-of-mating inbreeding and a slight excess of observed heterozygosity within villages (Ward and Neel 1976). The small variance effective sizes within villages also result in a substantial amount of differentiation among villages, with $f_{st} = 0.073$ (almost half of the global human f_{st}). Hence, the overall deviation from Hardy–Weinberg genotype frequencies in the total Yanomama population is $f_{it} = f_{st} + f_{is}(1-f_{st}) = 0.073 - 0.01(0.927) = 0.064$. Even though the Yanomama have a local system of mating characterized by an avoidance of inbreeding within villages ($f_{is} = -0.01 < 0$), the Wahlund effect induced by the balance of drift and intervillage gene flow creates an overall heterozygosity deficiency ($f_{it} = 0.064 > 0$) at the tribal level. If one ignored or was unaware of the genetic subdivision of Yanomama into villages, the deviations from expected Hardy–Weinberg genotype frequencies at the tribal level would imply that the Yanomama's system of mating is characterized by inbreeding when in fact it is characterized by avoidance of inbreeding ($f_{is} < 0$). Because of the Wahlund effect, caution is required in interpreting deviations from Hardy–Weinberg expectations: A deficiency of heterozygotes in a sample *could* mean that the population has system-of-mating inbreeding or that the sample included individuals from different local populations with different allele frequencies. Hence, f_{it} confounds local systems of mating with population subdivision.

Of course, a positive f_{it} could be due to both local inbreeding and population subdivision. Consider again the snail *Rumina* with $f_{st} = 0.294$. This snail is also capable of self-fertilization, and a detailed examination of its local system of mating reveals that 85% of the time the snails self-mate (Selander and Hudson 1976). As a result of much selfing within local demes, $f_{it} = 0.775$, yielding $f_{is} = 0.681$. In this snail both local system-of-mating inbreeding (selfing in this case) and population subdivision make large contributions to an overall extreme deficiency of heterozygotes relative to Hardy–Weinberg expectations.

OTHER RELATIONSHIPS OF f_{st} TO N (DRIFT) AND m (GENE FLOW)

Equations 6.10 and 6.28 relate the balance of drift and gene flow (as measured by F_{st} or f_{st}) to underlying quantitative measures of genetic drift (N_{ef} or N_{ev}) and gene flow (m). However, we derived equations 6.10 and 6.28 only for specific models of gene flow: either symmetrical gene flow between two demes or the island model of many demes. The two-deme model is of limited generality, and the island model depends upon several specific and biologically implausible assumptions, such as a portion m of the gametes being extracted from each deme and distributed at random over all other demes regardless of their locations relative to one another. The island model was chosen primarily for its mathematical convenience rather than its biological realism. Changing the underlying assumptions of the model can change the balance between drift and gene flow. Consequently, equations 6.10 and 6.28 do *not* represent the general quantitative relationship between drift and gene flow as forces causing genetic subdivision. Instead, these equations represent only special and highly unrealistic cases. We now consider some alternate models.

One aspect of the island model of gene flow and drift that is unrealistic for many species is the assumption that all dispersing individuals (or gametes) are equally likely to end up in any local deme. In most real species, some pairs of local demes experience much more genetic interchange than others. One common type of deviation from the island model is **isolation by distance**, in which local demes living nearby to one another interchange gametes more frequently than do geographically distant demes. We need look no further than our own species to see isolation by distance. For example, of 2022 marriages recorded in the upper Ina Valley of Japan (Sekiguchi and Sekiguchi 1951), nearly half occurred between people from the same buraku (hamlet), over two-thirds from the same village (including the burakus within the village), and less than a third from more remote locations (Table 6.1). As

Table 6.1. Isolation by Distance in Ina Valley, Japan, as Measured by Location of Spouse's Birthplace for 2022 Marriages

Spouse's birthplace	Percentage of marriages
Within Buraku (hamlet)	49.6
Within village but outside Buraku	19.5
Neighboring villages	19.1
Within Gun (county)	6.4
Within Prefecture (state) but outside Gun	2.9
Outside Prefecture	2.5

Suorce: Sekiguchi and Sekiguchi (1951).

geographical distance between spouses increases, the percentage of marriages decreases, exactly the pattern expected under isolation by distance.

There are many models of isolation by distance in the population genetic literature, and we will only consider a few simple ones. We start with a one-dimensional stepping-stone model in which a species is subdivided into discrete local demes, as with the island model. These local demes are arrayed along a one-dimensional habitat such as a river, valley, or shoreline. One version of this model allows two types of gene flow (Figure 6.5). First, a fraction m_∞ of the gametes leave each deme and disperse at random over the entire species, just as in the island model. Second, a fraction m_1 of the gametes from each deme disperse only to the adjacent demes. Because this is a one-dimensional model, each deme has just two neighbors (Figure 6.5, ignoring the demes at the two ends of the habitat), and we assume symmetrical gene flow at this local geographic level; that is $m_1/2$ go to one of the neighboring demes and the other $m_1/2$ go to the other (Figure 6.5). Then (Weiss and Kimura, 1965)

$$f_{st} = \frac{1}{2N_{ev}\left[1 - \left(1 - \frac{1}{2N_{ev}}\right)\left(1 - \frac{2R_1R_2}{R_1+R_2}\right)\right]} \quad (6.32)$$

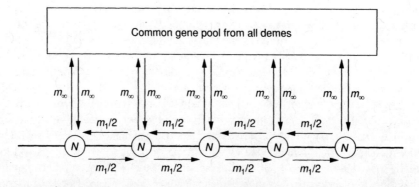

Figure 6.5. One-dimensional stepping-stone model of gene flow between discrete demes. Each deme is of idealized size N and is represented as a circle arrayed on a line. A portion m_1 of the gametes from any one population are exchanged with the two neighboring populations, half going to each neighbor. Moreover, each population contributes a fraction m_∞ of its gametes to a common gene pool that is then distributed at random over all demes in the same proportion.

where the correlations between allele frequencies of demes one (R_1) and two (R_2) steps apart are

$$R_1 = \sqrt{[1 + (1 - m_1)(1 - m_\infty)]^2 - [m_1(1 - m_\infty)]^2}$$

$$R_2 = \sqrt{[1 - (1 - m_1)(1 - m_\infty)]^2 - [m_1(1 - m_\infty)]^2} \qquad (6.33)$$

When $m_1 = 0$, all dispersal is at random over the entire species' geographical distribution (the island model) and equation 6.32 simplifies to $f_{st} = 1/(1 + 4N_{ev}m_\infty)$. Hence, equation 6.28 is a special case of 6.32 when there is no additional gene flow between adjacent demes.

In most real populations (for example, Table 6.1) there is much more dispersal between neighboring demes than between distant demes. In terms of our stepping-stone model, this means that m_∞ is very small relative to m_1, and under these conditions equation 6.32 is approximately

$$f_{st} = \frac{1}{1 + 4N_{ev}\sqrt{2m_1 m_\infty}} \qquad (6.34)$$

Note from equation 6.34 that even when m_∞ is very small relative to m_1, the long-distance dispersal parameter m_∞ still has a major impact on genetic subdivision because the impact of gene flow is given by the product of m_1 and m_∞. Consequently, equation 6.34 is extremely sensitive to long-distance dispersal even when such dispersal is very rare compared to short-distance dispersal. For example, let $N_{ev} = 100$ and $m_1 = 0.1$. Then $f_{st} = 0.053$ if $m_\infty = 0.01$ and $f_{st} = 0.276$ if $m_\infty = 0.001$. Note in this example that large differences in f_{st} are invoked by changes in long-distance dispersal even though long-distance dispersal is 10–100 times less common than short-distance dispersal. At first this sensitivity to rare, long-distance dispersal may seem counterintuitive, but the reason for it can be found in equations 6.2. From those equations, we saw that the evolutionary impact of gene flow upon allele frequencies depends upon two factors: (1) how much interchange is actually occurring between two demes and (2) how genetically distinct the demes are in their allele frequencies. When much dispersal occurs between neighboring demes, the allele frequencies in those neighboring demes are going to be very similar. Hence, even though there is a large amount of exchange between neighbors relative to long-distance dispersal, exchange between neighboring demes has only a minor evolutionary impact upon allele frequencies. On the other hand, when long-distance dispersal occurs, it generally brings in gametes to the local deme that come from a distant deme with very different allele frequencies. As a consequence, long-distance dispersal has a large evolutionary impact when it occurs. This trade-off between frequency of genetic interchange and magnitude of the evolutionary impact given genetic interchange explains why both m_1 and m_∞ contribute in a symmetrical fashion to f_{st}.

The importance of long-distance dispersal (m_∞) upon overall gene flow even when it is rare relative to short-distance dispersal means that gene flow is difficult to measure accurately from dispersal data. Many methods of monitoring dispersal directly are limited to short geographical scales, and it is usually impossible to quantify the amount of long-distance dispersal. Yet these rare, long-distance dispersal events can have a major impact on a species' genetic population structure.

Models of isolation by distance have also been developed for the case in which a species is continuously distributed over a habitat and not subdivided into discrete local demes (Malécot 1950). In models of a species with continuously distributed individuals, the density of the individuals, δ (actually a variance effective density), replaces the population size of the local demes (N in the discrete model shown in Figure 6.5) and σ, the standard deviation of the geographical distance between birthplace of parent and offspring, replaces $\sqrt{m_1}$ for short-distance dispersal. The parameter m_∞ is retained as the long-distance dispersal parameter that measures random dispersal over the entire species range. Then, at equilibrium

$$f_{st} \approx \frac{1}{1 + 4\delta\sigma\sqrt{2m_\infty}} \qquad (6.35)$$

which is the continuous analogue of equation 6.34.

These models have also been extended to species living in habitats with more than one dimension. For example, the analogue of equation 6.35 for a two-dimensional habitat is

$$f_{st} \approx \frac{1}{1 + 8\pi\delta\sigma^2/[-\ell n\,(2m_\infty)]} \qquad (6.36)$$

For the same values of δ, σ, and m_∞, the one-dimensional f_{st} will be much larger than the two-dimensional f_{st}, as illustrated in Figure 6.6. The reason for this is that any single population or geographical point is genetically interconnected with more distinct populations or individuals as dimensionality increases. For example, in the discrete two-dimensional stepping-stone model in which demes are placed at intersections in a lattice, there are four adjacent demes to any one deme (ignoring edge effects) whereas there are only two adjacent demes to any one deme in the one-dimensional case (Figure 6.5). Hence, as dimensionality increases, the homogenizing effects of gene flow become more powerful because more local

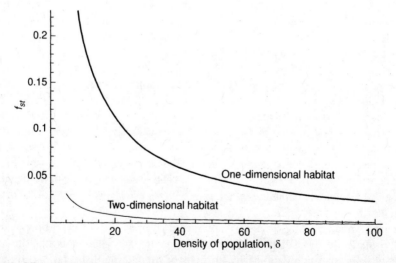

Figure 6.6. Effect of dimensionality of habitat upon population subdivision as measured by f_{st} in gene flow model over continuous habitat. The dispersal variance σ^2 is fixed at 1, and the long-distance dispersal parameter m_∞ is fixed at 0.01. Density δ is allowed to vary from 5 to 100 for both one-dimensional (upper curve) and two-dimensional (lower curve) habitats.

Table 6.2. Density-Dependent Gene Flow in Four Colonies of Herb *Liatris aspera* Living in Continuous Two-Dimensional Habitats

Parameter	Colony			
	I	II	III	IV
Density δ, (plants/m^2)	1	3.25	5	11
Seed plus pollen dispersal variance, σ^2 (m^2)	2.38	1.83	1.51	1.35
Neighborhood area (m^2)	30	23	19	17
Neighborhood size	30	75	97	191

Source: From Levin and Kerster (1969).

demes are intermixing at the short-distance rate, resulting in smaller amounts of genetic subdivision as measured by f_{st}.

The plots of f_{st} given in Figure 6.6 show how f_{st} varies as a function of density for fixed σ and m_∞. However, in many species σ and m_∞ change along with changes in density such that the parameters δ, σ, and m_∞ are not biologically independent. For example, Levin and Kerster (1969) studied four separate colonies of the herb *Liatris aspera*. Individual plants were continuously distributed within each colony in their two-dimensional habitats (fields), but different colonies had different densities. Table 6.2 gives the densities of these plants in the four colonies, which vary over an 11-fold range. Both seed dispersal and movement of pollen by pollinators mediated gene flow in these plant populations, and both types of movements were studied to obtain an estimate of σ^2 within each colony, also shown in Table 6.2. Because the plants were distributed continuously within these colonies, there were no discrete demes within colonies to which we can assign a meaningful N_{ev}. For such continuously distributed populations, Wright (1946) proposed an alternative to a discrete deme called the **neighborhood**, that is, the subregion within the population's continuous distribution that surrounds a point in space from which the parents of individuals born near that point may be treated as if drawn at random. Assuming that dispersal is random in direction and follows a normal distribution, Wright showed that the neighborhood area is given by

$$\text{Neighborhood area for one dimension} = 2\sigma\sqrt{\pi}$$

$$\text{Neighborhood area for two dimensions} = 4\pi\sigma^2 \quad (6.37)$$

The **neighborhood size**, the continuous analogue to N_{ev}, is the neighborhood area times the density, δ.

Table 6.2 shows both the neighborhood areas and the neighborhood sizes for the four colonies of *Liatris*. As can be seen, the dispersal variance decreases with increasing density, and as a consequence the neighborhood area also decreases as density increases. The reason for the decrease in neighborhood area with increasing density in *Liatris* was due almost entirely to the gene flow caused by pollinators. The pollinators tended to fly from one plant to its nearest-neighbor, but of course the nearest neighbor distance between plants decreased with increasing plant density. As a consequence, the neighborhood area became smaller as density increased. Thus, even though the density increased 11-fold in going from colony I to IV, the neighborhood size, the continuous analogue of the variance effective size, only increased 6-fold (Table 6.2). Hence, actual population sizes or densities are not necessarily reliable indicators of variance effective or neighborhood sizes, which are functions of population size/density, dispersal, and the interactions between them.

Another interesting implication of the isolation-by-distance model is that the degree of genetic differentiation between two demes or two points on a geographical continuum should increase with increasing separation—either the number of "steps" (in the stepping stone model) or the geographical distance (in the continuous distribution models). A **population genetic distance** measures the degree of genetic differentiation between two populations. There are several types of population genetic distances, and Box 6.1 shows Nei's population genetic distance, one of the more commonly used measures. However, the distance measure most convenient for isolation-by-distance models is the pairwise f_{st}. A **pairwise** f_{st} is an f_{st} calculated from equation 6.26 and applied to just two populations at a time. The total population now used to calculate H_t refers just to the two populations of interest and all other populations in the species are ignored. Note that this population genetic distance (and all others as well) is biologically quite distinct from the molecule genetic distances discussed in Chapter 5. A molecule genetic distance ideally measures the number of mutations that occurred between two DNA molecules during their evolution from a common ancestral molecule. A pairwise f_{st} is a function of allele frequencies between two demes, and mutation is not even necessary for this population genetic distance to take on its maximum value. For example, suppose an ancestral population were polymorphic for two alleles, A and a, at a locus. Now assume that the ancestral population split into two isolates, with one isolate becoming fixed for A and the other for a. Then, the pairwise f_{st} for these two isolates would be 1 (and the Nei's distance in Box 6.1 would be infinite), the maximum value possible, even though not a single mutation occurred. Population genetic distances should never be confused with molecule genetic distances.

In the isolation-by-distance models, let $f_{st}(x)$ be the pairwise f_{st} between two populations x steps apart or x geographical units apart. Malécot (1950) has shown that

$$f_{st}(x) = \frac{e^{-x\sqrt{2m_\infty/m_1}}}{1 + 4N_{ev}\sqrt{2m_1 m_\infty}} \qquad m_\infty << m_1 \qquad (6.38)$$

for the discrete, one-dimensional stepping-stone model and for the continuous habitat models

$$f_{st}(x) \approx \begin{cases} \dfrac{e^{-x\sqrt{2m_\infty/\sigma^2}}}{1 + 4\delta\sigma\sqrt{2m_\infty}} & \text{for one-dimensional habitats} \\[2ex] \dfrac{e^{-x\sqrt{2m_\infty/\sigma^2}}}{\left\{1 + 8\pi\delta\sigma^2/[-\ell n(2m_\infty)]\right\}\sqrt{x}} & \text{for two-dimensional habitats} \end{cases} \qquad (6.39)$$

In general, the isolation-by-distance models can be approximated by an equation of the form (Malécot 1950)

$$f_{st} = ae^{-bx}x^{-c} \qquad (6.40)$$

where $c = \frac{1}{2}$(dimensionality of the habitat -1) and a and b are estimated from the $f_{st}(x)$ data. The parameter c can also be estimated from the observed pairwise population genetic distances when it is not obvious what the dimensionality of the habitat may be. For example,

BOX 6.1 NEI'S (1972) POPULATION GENETIC DISTANCE

Consider two populations, 1 and 2, scored for allelic variation at a locus. Let p_{1i} and p_{2i} be the frequencies of the ith allele in populations 1 and 2, respectively. Then the probability of identity by state of two genes chosen at random from population 1 is $j_1 = \sum p_{1i}^2$, where the summation is taken over all alleles. Similarly, the probability of identity by state of two genes chosen at random from population 2 is $j_2 = \sum p_{2i}^2$. The probability of identity by state between two genes, one chosen at random from population 1 and one chosen at random from population 2, is $j_{12} = \sum p_{1i} p_{2i}$. Nei (1972) defined the normalized genetic identity by state between these two populations as

$$I_{12} = \frac{j_{12}}{\sqrt{j_1 j_2}}$$

Note that this measure of identity between the two populations ranges from 0 (when the two populations share no alleles in common, thereby making $j_{12} = 0$) to 1 (when the two populations share all alleles in common and at the same allele frequencies, thereby making $j_{12} = j_1 = j_2$).

In contrast to an identity measure, a distance measure should get larger as the two populations share less and less in common in terms of alleles and their frequencies. Nei mathematically transformed I_{12} into a distance measure by taking the negative of the natural logarithm of identity:

$$D_{12} = -\ell n(I_{12})$$

This population genetic distance ranges from zero when the populations share all alleles in common and at the same allele frequencies ($I_{12} = 1$) to infinity when the populations share no alleles in common ($I_{12} = 0$).

When data from multiple loci exists, Nei (1972) recommended that the j's be averaged over all loci, including monomorphic loci (loci with only one allele, thereby ensuring that all j's are 1 at monomorphic loci). These average j's are then used to calculate an overall identity, which is then transformed to yield an overall genetic distance. Hillis (1984) pointed out that averaging the j's across loci can sometimes lead to distances that make little sense biologically. For example, suppose that two populations are fixed for the same allele at one locus (and hence all the j's are 1 and $I_{12} = 1$), and at a second locus they share no alleles, but each population is polymorphic for two alleles each with a frequency of 0.5. At this second locus, $j_{12} = 0$ and $I_{12} = 0$ because they share no alleles, but $j_1 = j_2 = 0.5$. Hence, across both loci, the average $j_{12} = \frac{1}{2}(1 + 0) = 0.5$ and the average j_1 equals the average j_2 equals $\frac{1}{2}(1 + 0.5) = 0.75$. Using these average values of the j's, I_{12} is calculated to be $0.5/0.75 = 0.667$ and $D_{12} = 0.41$.

Now consider another case in which the first locus is polymorphic with both populations sharing two alleles and with each allele in each population having a frequency of 0.5. In this case, all the j's are 0.5 in value, and I_{12} for this locus has a value of 1. At the second locus in this case, each population is fixed for a different allele, so $j_{12} = 0$ and $j_1 = j_2 = 0.5$. In this second case, the average $j_{12} = \frac{1}{2}(0.5 + 0) = 0.25$ and the average j_1 and average j_2 are both $\frac{1}{2}(0.5 + 1) = 0.75$. Using these average j's, $I_{12} = 0.25/0.75 = 0.333$ and $D_{12} = 1.1$. In this case and in the case given in the previous paragraph, the two populations are both completely identical at the first locus and completely different at the second, yet they have very different genetic distances (0.41 versus 1.1) as originally defined by Nei (1972). Hillis (1984) points out that situations like this are likely to occur when different loci have different overall rates of evolution. To make the population genetic distance measure robust to this heterogeneity in evolutionary rates across loci, Hillis (1984) recommends that the I's be averaged across loci, not the j's. For example, in the first case where the populations are identical at the monomorphic locus and different at the polymorphic one, the average $I_{12} = \frac{1}{2}(1 + 0) = 0.5$, yielding a population genetic distance of 0.69. In the second case where the populations are identical at the polymorphic locus and different at the monomorphic one, the average $I_{12} = \frac{1}{2}(1 + 0) = 0.5$, yielding the same population genetic distance of 0.69. Therefore, it is better to average the I's and not the j's in calculating this type of population genetic distance from multilocus data when there is rate heterogeneity across loci.

a species may be distributed over a two-dimensional habitat, but more constraints on movement may occur in one direction than another. Hence, noninteger dimensions between 1 and 2 are biologically meaningful.

Equation 6.40 has been applied to human data. Figure 6.7 (redrawn from Cavalli-Sforza et al. 1994) shows the human pairwise f_{st} on a global scale along with the isolation-by-distance curve fitted from equation 6.40. As can be seen, the patterns of genetic differentiation between human populations fit well to the isolation-by-distance model. This

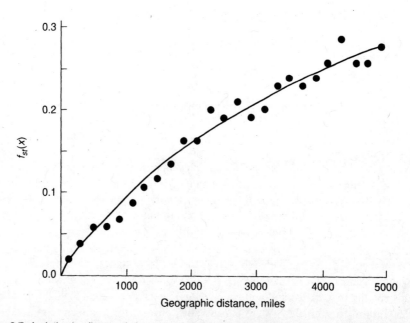

Figure 6.7. Isolation by distance in human populations on global scale. The x axis gives the geographical distance in miles between pairs of human populations. The y axis gives the pairwise $f_{st}(x)$, where x is the geographical distance in miles. The solid line gives the predicted curve under a model of isolation by distance. Modified with permission from Fig. 2.9.2 in L. Cavalli-Sforza et al., *The History and Geography of Human Genes.* Copyright ©1994 by Princeton University Press.

good fit is not unexpected. Table 6.1 shows that humans in the Ina Valley in Japan display an isolation-by-distance pattern, and other studies have shown that isolation by distance holds for human gene flow on many different geographical scales and virtually all studied populations (Lasker and Crews 1996; Santos et al. 1997).

IMPACT OF POPULATION SUBDIVISION UPON GENETIC VARIATION IN TOTAL POPULATION

So far we have only considered the effects of genetic drift at the local level. For example, the variance and inbreeding effective sizes invoked in all the models given above refer to *local* demes, not to the total population. Because these local populations are genetically interconnected, it is of great interest to see what impact limited gene flow and population subdivision have upon the evolutionary properties of the reproductive community as a whole. The mathematics for this problem can become somewhat complicated, so only some basic results are given here. More details and results are given in Crow and Maruyama (1971) and Maruyama (1972, 1977).

Consider first the impact of population subdivision on the long-term inbreeding effective size of the total reproductive community, N_{efT}. A long-term effective size assumes a reference population in the distant past and that sufficient time has elapsed for an equilibrium to have been established among the relevant evolutionary forces (here, genetic drift, mutation, and gene flow under neutrality). If we also assume random mating within each local deme, then the average probability of identity by descent among the individuals of this subdivided population is F_{st} (e.g., equation 6.15). Using the basic definition of an equilibrium inbreeding effective size, we have

$$\bar{F}_{eq,T} = F_{st} = \frac{1}{1 + 4N_{efT}\mu} \tag{6.41}$$

where μ is the mutation rate. Equation 6.41 yields the long-term inbreeding effective size of the total population to be

$$N_{efT} = \frac{1}{4\mu}\left(\frac{1 - F_{st}}{F_{st}}\right) \tag{6.42}$$

Note that the total inbreeding effective size is a decreasing function of F_{st}. Hence, the more subdivided a population is, the higher the chance of identity by descent under random mating at the local level and thus the smaller the total inbreeding effective size. Consider the special case of an equilibrium island model with an infinite number of local demes. Under this model, each local deme has an inbreeding effective size of N_{ef}, but because there is an infinite number of local demes, the total population size is also infinite. If there were no population subdivision, the total inbreeding effective size should also be infinite. But when there is subdivision, we can substitute equation 6.16 into the F_{st} term of equation 6.42 to yield

$$N_{efT} = N_{ef}\left(\frac{\mu + m}{\mu}\right) \tag{6.43}$$

When there is no gene flow ($m = 0$) and each local deme is an isolate, the total inbreeding effective size is simply the local inbreeding effective size, as expected. This observation has important implications in conservation biology. When a species becomes fragmented into completely isolated subpopulations, pedigree inbreeding will accumulate due to drift at a rate determined by the population sizes within the fragments, *not* the total species population size. Thus, the total inbreeding effective size of a fragmented species can be much smaller than the census size.

When there is gene flow, $(\mu + m)/\mu > 1$, and the total inbreeding effective size is larger than the local inbreeding effective size under the island model. Crow and Maruyama (1971) show that in general N_{efT} is larger than N_{ef} when $m > 0$ but smaller than the total population size. Note also that the larger the gene flow rate m, the larger the total inbreeding effective size. Given that the mutation rate μ is typically small, even a small degree of gene flow can result in a total inbreeding effective size that is many orders of magnitude larger than the total inbreeding effective size under complete isolation. For example, let $\mu = 10^{-5}$ and $m = 0.01$; then from equation 6.43 we have that $N_{efT} = 1001 N_{ef}$. Hence, achieving even low levels of genetic interchange among fragmented subpopulations is often a high conservation priority.

Although the total inbreeding effective size tends to decrease with increasing population subdivision (Crow and Maruyama 1971), the opposite is generally true for the total variance effective size. The total variance effective size measures how much variance in allele frequency is induced by genetic drift for the population *as a whole* and not for each local deme. For example, Wright (1943) considered a finite island model in which an otherwise ideal population is subdivided into n local populations, each of ideal size N, with a gene flow rate of m. The total population size in this model is nN, and if the population were panmictic, its variance effective size would be nN. Wright showed that the total variance effect size in this case is

$$N_{evT} = \frac{nN}{1 - f_{st}} \quad (6.44)$$

When the population is indeed panmictic with no subdivision, $f_{st} = 0$ and the total variance effective size equals the census size, as expected for this idealized population. However, note that when the population is subdivided due to restricted gene flow, $f_{st} > 0$ so $1 - f_{st} < 1$ and $N_{evT} > nN$. Thus a subdivided population has a total variance effective size *larger* than the census size! Indeed, if gene flow were completely eliminated and each population became an isolate, we would expect f_{st} to eventually approach 1 as each isolate became fixed for a new mutational lineage. Note that as f_{st} approaches 1, the total variance effective size goes to infinity even though the total census size is finite.

By contrasting equations 6.42 and 6.44, we see that increasing population subdivision reduces total inbreeding effective size but increases total variance effective size. This paradoxical effect of genetic subdivision is also seen with the rate of loss of alleles due to drift at the level of the total reproductive community. For example, the rate of loss of alleles per generation due to drift in a two-dimensional continuous model of isolation by distance in an otherwise idealized population of total size N_T is expected to be $1/(2N_T)$ if $\delta\sigma^2 \geq 1$ (Maruyama 1972). That is, with sufficient gene flow ($\delta\sigma^2 \geq 1$), the total reproductive community is losing its allelic variation at the same rate as a single panmictic deme of idealized size N_T. However, when gene flow is decreased such that $\delta\sigma^2 < 1$, then the rate of loss of allelic variation becomes $\delta\sigma^2(2N_T) < 1/(2N_T)$. Thus, a continuous population of total size N_T with restricted gene flow has lower overall rates of loss of allelic variation than a panmictic population of total size N_T.

The above models raise an important question: How can population subdivision increase the amount of pedigree inbreeding while simultaneously slowing down the rate of loss of genetic variation and reducing the variance in allele frequencies at the total population level? This paradox is resolved when we recall that any evolutionary force, including genetic drift, can have an impact *only when there is genetic variation*. As a species becomes subdivided due to restricted gene flow, more and more of its local demes can become temporarily fixed for a particular allele. Such fixation increases the overall level of homozygosity due to identity by descent in the total population. However, the subset of the local demes that are fixed for a single allele at a given time is immune to evolutionary changes caused by drift because they have no genetic variation. Because of the random nature of drift, it is unlikely that the local demes that are fixed for some allele are all fixed for exactly the same allele. Thus at any given time at the total population level, genetic variation is preserved as fixed differences between local areas that are immune to genetic drift. These fixed differences cannot be lost due to drift, at least temporarily, and thus the overall rate of loss of genetic variation is reduced relative to the total panmictic case. Moreover, the overall allele frequencies in the subset of the total population that is fixed for some allele cannot change due to drift, so the overall variance in allele frequencies is reduced to the total panmictic case.

The protection of global genetic variation by local fixation is illustrated by populations of eastern collared lizards (*Crotaphytus collaris collaris*) living in glades in the Missouri Ozarks (Templeton et al. 1990b). Ozark glades are barren rocky outcrops, usually with a southern or southwesterly exposure on a ridge top that creates a desert or prairielike microhabitat. Desert- and prairie-adapted plants and animals (such as prickly pear cacti, scorpions, tarantulas, and collared lizards) invaded the Ozarks about 8000 years ago during the Xerothermic maximum (the period of maximum warmth in our current interglacial period). Increasing rainfall at the end of the Xerothermic about 4000 years ago allowed forests to expand into the Ozarks (Mondy 1970). As a result, the glades became desert and prairielike habitat islands in an ocean of forest. Until European settlement, frequent fires occurred that maintained the forest surrounding the glades as an open woodland. With European settlement, rounds of clear cutting occurred throughout most of the Ozarks and fires were suppressed as a new forest grew in, particularly after the mid-twentieth century. This new forest was an oak–hickory forest with a dense, woody understory and a forest floor covered with leaf litter that prevented many of the woodland grasses and perennials from growing (and thereby reducing the insect communities associated with these woodland forest-floor plants). Collared lizards will disperse several kilometers through an open woodland that has abundant insect prey items on the forest floor, but they are reluctant to move even as little as 50 meters through a forest with a dense understory (Templeton et al. 2001). Consequently, gene flow between glade populations in areas of fire suppression had virtually been stopped. A single isolated glade can support relatively few lizards, so glade population sizes typically are between 10 and 30 adults, with only exceptional glades having larger local sizes (Sexton et al. 1992 and unpublished data). Small local population sizes coupled with little or no gene flow have resulted in extreme genetic fragmentation and differentiation between even neighboring glades, including fixation of different alleles (Figure 6.8).

The effects of this fragmentation can easily be seen in the f_{st} statistics based on microsatellite loci for collared lizards from the northeastern Ozarks, the area most affected by

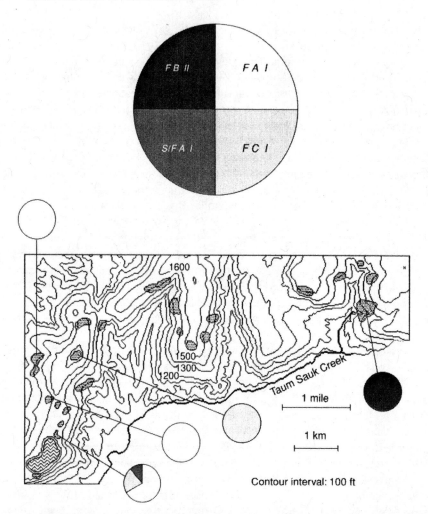

Figure 6.8. Topographical map showing distribution of genetic variation found in glade populations of *C. collaris* on ridge system between Taum Sauk Mountain on right and Profitt Mountain on left in Missouri Ozarks. Glades are indicated by stippled areas, and a large reservoir on Profitt Mountain is indicated by wavy lines. Lizards sampled from these glades were scored for genetic variation at the *malic dehydrogenase* locus (*Mdh*), which had two alleles (*S* and *F*), and for restriction site map variation in mitochondrial DNA (with three haplotypes being found in these populations: *A*, *B*, and *C*) and in nuclear ribosomal DNA (rDNA, with two haplotypes being found in these populations: *I* and *II*). The gene pools of local glade populations are shown by a pie diagram, with the legend to the pie diagram indicated at the top of the figure in the order of *Mdh* genotypes, mtDNA genotypes, and rDNA genotypes.

fire suppression, versus populations from the central range in western Texas and Oklahoma, an area not influenced by forested habitat (Hutchison 2003). In the northeastern Ozarks, $f_{st} = 0.40$ with the most distant populations being only 90 km from one another. In contrast, in western Texas and Oklahoma, $f_{st} = 0.10$ even though the most distant populations are nearly 700 km apart. The pairwise f_{st} values also show a dramatic difference (Figure 6.9). In western Texas and Oklahoma, the f_{st} values gradually rise with increasing geographical distance, reaching a value averaging around 0.2 for the longest distances. This is the classic

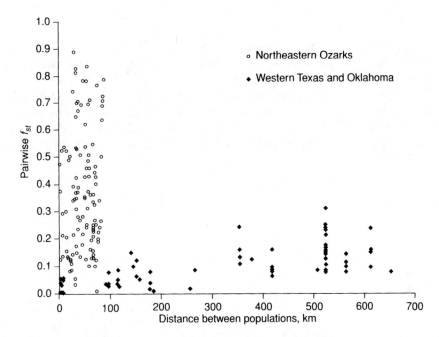

Figure 6.9. Plot of pairwise f_{st} versus geographical distance for populations of collared lizards from southwestern United States and for populations from northeastern Ozarks. Based on data from Hutchison (2003).

pattern for isolation by distance, and it shows that collared lizards disperse well over relatively long distances in the absence of forests. The situation dramatically changes in the northeastern Ozarks. Here the geographical scale is less than 90 km, a scale over which the west Texas and Oklahoma lizards show little differentiation (most f_{st}'s are less than 0.05 for distances less than 100 km in Texas and Oklahoma). In the Ozarks the f_{st}'s take on their full range of values and show no significant correlation with geographical distance. This pattern is consistent with much gene flow in the northeastern Ozarks in the past resulting in little to no differentiation among the ancestral northeastern Ozark populations, followed by extreme and sudden fragmentation into small populations that rapidly differentiated from a common ancestral gene pool due to strong genetic drift.

Although the fragmentation of the Ozark populations has led to much loss of genetic variation within local demes, there is still much genetic variation in the Ozark populations as a whole. Consider sampling genes from the current Ozark lizard populations shown in Figure 6.8. If you sampled two genes at random from the same glade population, chances are they would be identical, and repeated sampling from the same glade would reveal little or no genetic variation. These results would reflect the small inbreeding and variance effective sizes at the local glade level. Because almost all individuals on all glades are homozygous and probably identical by descent, averaging over all individuals across glades would still result in high levels of identity by descent and homozygosity, reflecting a small total inbreeding effective size. However, now consider sampling two genes at random from all the populations shown in Figure 6.8 combined together. The sampled genes now would most often come from different glades, and they would have a high probability of being different because different glades are fixed for different alleles. Thus, much genetic variation would be evident despite a high average level of identity by descent. Even though there is little genetic variation within any one glade population in the northeastern Ozarks, there is much genetic variation in the glade populations considered in toto (Hutchison 2003).

Consider now following these populations through time. Most of the populations shown in Figure 6.8 are already fixed for an allele at the loci being scored. If these populations continue to be isolated from one another, they will simply remain fixed for the same alleles, or occasionally quickly go to fixation for a new mutant allele arising in their local gene pool. Thus, the status quo of a high rate of fixed differences among glade populations should persist as long as gene flow levels are extremely low. This means that genetic variation will also persist at the total population level in a way that would be largely unaffected by drift. Remember that genetic drift cannot cause any evolution in a glade population fixed for a particular allele. Because genetic drift is not operating in those glade populations that are fixed and different glade populations are fixed for different alleles, the global population loses its genetic diversity at a greatly reduced rate, reflecting a large total variance effective size.

The effect of genetic subdivision on increasing total variance effective sizes can be quite large. For example, in an island model in which the effective number of migrants ($N_{ev}m$) is 0.5 (the exchange of one effective individual every other generation), the rate of loss of allelic variation in the total population will be half of what it would have been under panmixia, corresponding to a doubling of the total variance effective size over total census size (equation 6.44). As we saw in Chapter 4, we have to be very careful when dealing with effective sizes to define both what type of size we are measuring and what our reference population is in time. The results given here indicate that we also need to be careful in defining our spatial reference. The impact of an evolutionary force upon local demes can be just the opposite of its impact on a collection of local demes.

The opposing effects of population subdivision upon individual pedigree inbreeding levels and total population levels of genetic variation can create difficult choices for conservation biologists. As noted earlier, a priority in many conservation management programs is to establish gene flow between fragmented isolates. Such gene flow has many beneficial effects from a population genetics perspective: It increases local inbreeding and variance effective sizes, it increases local levels of genetic variation and hence local adaptive flexibility, and it reduces the overall level of pedigree inbreeding, thereby minimizing the dangers of inbreeding depression. However, a goal of many management programs is also to maintain high levels of genetic variation in the total population for long periods of time. This goal is most easily achieved by fragmenting the population into isolates (Chesser et al. 1980). Thus, there is a trade-off, and management decisions require a careful assessment of what the priorities are for any given endangered species. There are no easy, universal answers.

USING SEQUENCE OR RESTRICTION SITE DATA TO MEASURE POPULATION SUBDIVISION

In our previous models of population subdivision, any pair of homologous genes was classified into one of two mutually exclusive categories: The pair was identical by descent or not (for defining F_{st}) or the pair was heterozygous or not (for defining f_{st}). However, with the advent of restriction site and DNA sequence data, we can refine this categorization by the use of a molecule genetic distance. Suppose, for example, that we have sequence data on a 10-kb locus, and a pair of genes at this locus are examined that differ by only a single nucleotide site. Now consider another pair of genes, but this time differing by 20 nucleotide sites. In the models used previously in this chapter, both of these pairs of genes would be placed into the same category: They are not identical by descent at this locus or they are heterozygous at this locus. Intuitively, the second pair is much less identical than the first pair or, alternatively, the first pair has much less nucleotide heterozygosity than the second pair. These quantitative differences in nonidentity or in the amount of heterozygosity at the molecular level are ignored in all of our previous formulations. Taking into account these quantitative differences can increase power and sensitivity for detecting population subdivision and restricted gene flow (Hudson et al. 1992). Accordingly, there have been several proposed alternatives to f_{st} that incorporate the amount of difference at the molecular level between heterozygous pairs of alleles (that is, a molecule genetic distance). Among these are N_{st}, which quantifies the amount of molecular heterozygosity by the average number of differences between sequences from different localities (Lynch and Crease 1990), and K_{st}, which uses the average number of differences between sequences randomly drawn from all localities (Hudson et al. 1992). In general, we will let Φ_{st} designate any f_{st}-like statistic that uses a molecule genetic distance instead of heterozygosity as its underlying measure of genetic diversity (Excoffier et al. 1992).

Recall that we can define multiple f statistics to partition genetic variation at different biological levels in a hierarchy of individuals and populations. For example, by using the relationship $1 - f_{it} = (1 - f_{st})(1 - f_{is})$, we could partition genetic diversity as measured by heterozygosity in the Yanomama Indians into a portion due to heterozygosity within local demes $(1 - f_{is})$ and a portion due to heterozygosity between local demes $(1 - f_{st})$. The same types of hierarchical partitions can be made using Φ statistics instead of f statistics. A partition of genetic variation that substitutes a molecule genetic distance for heterozygosity is called an **analysis of molecular variance (AMOVA)**.

These newer measures of population structure are essential when dealing with data sets with extremely high levels of heterozygosity, as are becoming increasingly common with DNA sequencing. When there are high levels of allelic or haplotypic heterozygosity in all populations, both H_T (the expected heterozygosity if the total population were mating at random) and H_S (the actual heterozygosity in the total population) are close to 1 in value. When all heterozygosities are high, the f_{st} value calculated from equation 6.26 approaches zero regardless of the values of the underlying evolutionary parameters (such as N_{ev}, m, or μ). Thus, the all-or-nothing nature of traditional heterozygosity measures can induce serious difficulties when dealing with highly variable genetic systems. For example, recall from Chapter 1 the survey of 9.7 kb of the human *lipoprotein lipase* locus (*LPL*) in which 88 haplotypes were found in a sample of 142 chromosomes coming from three human populations: a population in Rochester, Minnesota; a population of African-Americans from Jackson, Mississippi; and a population of Finns from North Karelia, Finland. Regarding each haplotype as an allele, the value of f_{st} was 0.02 (Clark et al. 1998), which was not significantly different from zero. This seemingly indicates that these populations display no significant genetic differentiation. However, using the nucleotide differences at the sequence level as the molecular genetic distance, Φ_{st} was calculated to be 0.07, a value significantly different from zero. Therefore, these populations did indeed show significant genetic differentiation, but these differences were not detected by the traditional f_{st} using haplotypes as alleles and heterozygosity as a qualitative measure of genetic differentiation. In general, when high levels of genetic variation are encountered, a quantitative scale of differences between alleles is preferable to a qualitative one.

MULTIPLE MODES OF INHERITANCE AND POPULATION STRUCTURE

Most of the theory presented in this chapter up to now has focused upon autosomal loci. However, we can study DNA regions with different modes of inheritance that are found in the same individuals. For example, many animals, including humans and fruit flies in the genus *Drosophila*, have DNA regions with four basic modes of inheritance:

- Autosomal nuclear DNA regions, with diploid, biparental inheritance
- X-linked nuclear DNA regions, with haplo–diploid, biparental inheritance
- Y-linked nuclear DNA regions, with haploid, uniparental (paternal) inheritance
- Mitochondrial DNA, with haploid, uniparental (maternal) inheritance

These different modes of inheritance interact strongly with the balance between genetic drift and gene flow that shapes genetic population structure. In our previous models, we assumed an autosomal system with diploid, biparental inheritance. As a consequence, an effective population size (of whatever type) of N_e individuals corresponds to an effective sample of $2N_e$ autosomal genes. As we saw earlier, the force of drift in this case is given by $1/(2N_e)$. In general, the force of drift is given by 1 divided by the effective number of gene copies in the population. An autosomal nuclear gene is only a special case in which the effective number of gene copies is $2N_e$.

The effective number of gene copies is not the same when we examine other modes of inheritance. For example, in a species with a 50–50 sex ratio (such as humans or *Drosophila*), N_e individuals corresponds to an effective sample of $\frac{3}{2}N_e$ X-linked genes. Because different modes of inheritance have different numbers of genes per individual, the force of drift is expected to vary across the different genetic systems that are coexisting in the same individuals. Indeed, we saw in Chapter 5 that the strength of drift as measured by coalescence times varies as a function of mode of inheritance (Figure 5.10). Now we note that variation in the force of drift directly induces variation in the *balance* of drift and gene flow. Therefore, we do not expect different DNA regions imbedded within the same populations to display the same degree of genetic subdivision, even when the *rate* of gene flow is identical for all DNA regions. For example, if dealing with a haploid genetic system in a population of size N_{ev}, equation 6.28 becomes

$$f_{st} = \frac{1}{1 + 2N_{ev}m} \qquad (6.45)$$

Note that the term $4N_{ev}$ in equation 6.28 has been replaced by $2N_{ev}$ in equation 6.45. This change by a factor of 2 is due to the fact that a diploid system in a population with variance effective size N_{ev} actually is based upon an effective sample of $2N_{ev}$ genes, whereas the haploid system only has an effective sample of N_{ev} genes. If we also stipulate that the haploid system has uniparental inheritance in a species with a 50–50 sex ratio, only half the individuals sampled can actually pass on the haploid genes, reducing the effective size by another factor of 2 (assuming the sexes have equal variance effective sizes). Thus, for haploid, uniparental systems like mitochondrial or Y-chromosomal DNA in humans or *Drosophila*, the effective size for a haploid, uniparental locus is approximately $\frac{1}{2}N_{ev}$, so equation 6.45 becomes

$$f_{st} = \frac{1}{1 + N_{ev}m} \qquad (6.46)$$

Similarly, adjusting for the number of genes per individual, the analogue of equation 6.28 for an X-linked locus in a species with a 50–50 sex ratio is

$$f_{st} = \frac{1}{1 + 3N_{ev}m} \qquad (6.47)$$

Hence, as we go from autosomal to X-linked to haploid uniparental systems, we expect to see increasing amounts of genetic subdivision for the same rate of gene flow (m). This expected variation in the balance of drift and gene flow presents an opportunity to gain more insight into population structure by simultaneously studying several inheritance systems rather than just one.

For example, populations of the fruit fly *D. mercatorum* were studied that live in the Kohala Mountains on the northern end of the Island of Hawaii (DeSalle et al. 1987). Samples were taken from several sites in the Kohalas (Figure 6.10), but the bulk of the collection came from site IV and the three nearby sites B, C, and D. We will consider this as a two-population system by pooling the nearby sites B, C, and D and contrasting them with site IV. An isozyme survey of nuclear autosomal genes yields an f_{st} of 0.0002 for this contrast, which is not significantly different from zero. Given the sample sizes and using equation 6.28, this nonsignificant f_{st} is statistically incompatible with any $N_{ev}m$ value less than 2. Hence, the nuclear autosomal system tells us that $N_{ev}m > 2$. This is not surprising, as

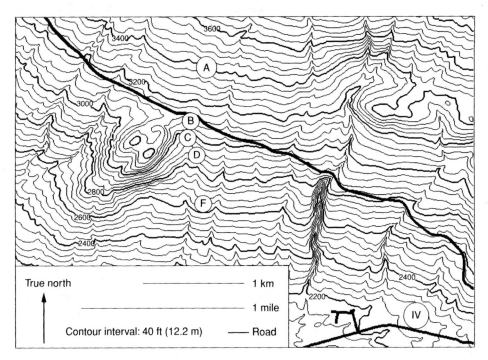

Figure 6.10. Topographic map of collecting sites for *D. mercatorum* in Kohala Mountains near town of Kamuela (also known as Waimea) on Island of Hawaii. A transect of collecting sites on the slopes of the Kohalas are indicated by the letters A, B, C, D, and F and a site in the saddle at the base of the Kohalas is indicated by IV.

sites IV and B, C, and D are less than 3 km apart, with the intervening area consisting of inhabited area for this species. Although the nuclear f_{st} tells us that there is significant gene flow between these two localities, we cannot distinguish between $N_{ev}m$ of 3, or 30, or 300, or 3000 with the nuclear autosomal isozyme data alone.

The same individual flies were also scored for restriction site polymorphisms for mtDNA, yielding a statistically significant f_{st} of 0.17. Using equation 6.46, this signficant f_{st} implies $N_{ev}m \approx 5$. This value of $N_{ev}m \approx 5$ from the mtDNA data is consistent with the inference that $N_{ev}m > 2$ from the isozyme data. Thus, there is no biological contradiction for these samples in yielding a nonsignificant f_{st} for isozymes and a significant f_{st} for mtDNA. This population of flies is also polymorphic for a deletion of ribosomal DNA on the Y chromosome (Hollocher and Templeton 1994). The Y-DNA f_{st} was 0.08 and significantly different from zero. Using equation 6.46 for the Y-DNA data yields an estimate of $N_{ev}m \approx 12$, a result also compatible with the isozyme data. Hence, the nuclear autosomal system told us very little about the possible values for $N_{ev}m$ other than that the value is greater than 2, but by combining all the data, we know that $N_{ev}m$ in this population is around 5–12 and not in the hundreds or thousands or more.

Although both of the haploid elements were individually consistent with the isozyme results, note that the f_{st} calculated from the Y-DNA is less than half that from the mtDNA. This difference was statistically significant even though both Y-DNA and mtDNA are haploid, uniparental systems. To understand why these two haploid, uniparental systems could yield significantly different results, recall that equation 6.46 was applied to both of them under the assumptions that the variance effective sizes of both sexes were equal and gene flow was equal for both sexes. In most nonmonogamous species, the variance in offspring number is greater in males than in females, resulting in males having a smaller variance effective size than females (see equations 4.31 and Chapter 5). This in turn, through equation 6.46, would imply that the f_{st} for the Y-DNA should be larger than that for mtDNA—exactly the *opposite* of the observed pattern. Therefore, the typical pattern in variance of male versus female reproductive success and effective sizes cannot explain the observed results. We therefore turn our attention to sex-specific influences on the rate of gene flow.

Direct studies on dispersal in this species reveal identical dispersal behaviors in males and females (Johnston and Templeton 1982), but this does not mean that m is identical in males and females. All dispersal in this species occurs during the adult phase, and almost all adult *D.mercatorum* females are inseminated by one to three males. The females have a special organ for storing sperm and can retain viable sperm for several days. Recall that m measures the rate of exchange of gametes, not individuals. Therefore, when a male disperses, only male gametes are potentially being dispersed. However, dispersing females carry not only their own gametes but also those of one to three males. So, even with equal *dispersal* rates for males and females, there is actually much more *gene flow* of male gametes than female gametes. This male-biased gene flow predicts a smaller f_{st} for Y-DNA versus mtDNA, as is observed. Hence, we need two gene flow rates, m_f for female gametes and m_m for male gametes. The overall gene flow rate for an autosomal locus is just the average of the two sexes as they are in a 50–50 ratio; that is, $m = \frac{1}{2}(m_f + m_m)$. Assuming the variance effective sizes are the same for both sexes, the nuclear isozyme results imply that $N_{ev}m = \frac{1}{2}N_{ev}(m_f + m_m) > 2$; the mtDNA implies that $N_{ev}m_f = 5$; and the Y-DNA implies that $N_{ev}m_m = 12$. All systems together therefore yield $N_{ev}m = \frac{1}{2}N_{ev}(m_f + m_m) = \frac{1}{2}(5 + 12) = 8.5$. Thus, by combining the results of several genetic systems with different modes of inheritance, we can gain more insight into population structure, including sex-specific differences.

A FINAL WARNING

The statistics F_{st} and f_{st} (and related measures such as N_{st}, K_{st}, or Φ_{st}) can be measured from genetic survey data. In contrast, measuring gene flow directly from dispersal studies is often difficult. The theory developed in this chapter shows that even rare exchanges between populations can result in much effective gene flow and have major consequences for population structure. Long-distance dispersal is the most difficult type of dispersal to study, but as we learned, its genetic impact can be great even when exceedingly rare. Moreover, dispersal of individuals is not the same as gene flow, as shown by the *D. mercatorum* example. Therefore, gene flow, as opposed to dispersal, is more accurately measured from genetic survey data than from direct observations on the movements of plants and animals. However, estimating gene flow from F or f and related statistics *assumes* that the underlying cause of differentiation among demes or localities is due to the equilibrium balance of genetic drift and recurrent gene flow. Such estimates of gene flow are sometimes of dubious biological validity when forces other than recurrent gene flow and drift influence the spatial pattern of genetic variation and/or when drift and gene flow may not be in an equilibrium balance.

For example, suppose a species is split into two large subpopulations that have no genetic interchange whatsoever (that is, $m = 0$). Equation 6.28 then tells us that $f_{st} = 1$. However, this is an equilibrium prediction, but historical events could have occurred that placed the populations far out of equilibrium or that created large temporal fluctuations in the amount of gene flow or drift. When equilibrium is disrupted, it takes time for the relationship shown in equation 6.28 to become reestablished. If the subpopulations are large, it would take many generations after the cessation of gene flow before we would actually expect to see $f_{st} = 1$. Until that equilibrium is achieved, $f_{st} < 1$ so that using the equilibrium equation in this case would incorrectly indicate that $m > 0$ when in fact $m = 0$.

On the other hand, suppose a species recently expanded its geographical range into a new area from a small and genetically homogeneous founder population. The local demes formed in the new geographical range of the species would display much genetic homogeneity for many generations because of their common ancestry regardless of what $N_{ev}m$ value is established in the newly colonized region (Larson 1984). Once again, if we sampled shortly after such a range expansion event, we would mistakenly infer high values of m regardless of what the current values of m were. Suppose now that an event occurred that restored gene flow among previously long-isolated populations. If m were small, f_{st} would decline only slowly, so the observation of a nonequilibrium f_{st} in this case would imply less gene flow than is actually occurring at present. These hypothetical examples show that f_{st} is an effective indicator of gene flow over evolutionary time *only if* your population is in equilibrium and has not been influenced by recent historical events. This is a big IF, so we will address how to separate current population structure from historical events in the next chapter.

7

GENE FLOW AND POPULATION HISTORY

Premise 1 from Chapter 1—that DNA can replicate—implies that genes have an existence in space and time that transcends the individuals that temporarily bear them. In Chapter 6 we saw how we can study and measure the pattern of genetic variation over space. Genetic variation is ultimately created by the process of mutation (premise 2), and when a new genetic variant is first created, it is confined to one point in space. An initial mutant gene can only spread through space with the passing of generations and many DNA replication events. The spread of genes through space therefore takes time. During this time, the spread of the gene through space is influenced in part by evolutionary forces such as gene flow. Our first model of gene flow in Chapter 6 had dispersal occurring at a rate m every generation. Under this model, gene flow would influence the spread of the gene every generation. Suppose that genetic exchange among populations only occurred every tenth generation. Still, any genetic variants that persist for hundreds or more of generations (as many do in large populations) would have their spatial distribution influenced by multiple occurrences of gene flow. Gene flow in these models is therefore **recurrent**; that is, forces or events that occur multiple times during the time from the present to the coalescence time to the common ancestral molecule for all the homologous molecules in a sample. The time to the most recent common ancestral molecule is a natural time period for defining recurrence because this is the time period in which the current array of genetic variation at the locus of interest has been shaped and influenced by evolutionary forces. Once we go back in the past beyond the most recent common ancestral molecule, there is no genetic variation observable in the coalescent process, so there is no potential for observing the effects of evolutionary forces such as gene flow.

Besides recurrent forces or events, unique events such as colonizing a new area or having the population split into two or more isolates by a climatic change could also have influenced the current spatial pattern of a genetic variant. **Historical events** are events that occurred only once or at most a few times during the time from the present to the coalescence time to

---

Population Genetics and Microevolutionary Theory, By Alan R. Templeton
Copyright © 2006 John Wiley & Sons, Inc.

the common ancestral molecule for the sample of homologous DNA molecules. In general, both recurrent and historical events influence how genes spread through space and time.

When we measure a pattern of spatial variation through a statistic such as f_{st}, we often do not know how much the measured value has been influenced by recurrent evolutionary forces such as gene flow, genetic drift, and system of mating and how much it has been influenced by historical events such as **population fragmentation** (the split of a population into two or more subpopulations with no genetic interchange), **range expansion** (the expansion of populations of a species into new geographical areas), and bottleneck or founder events. As indicated in Chapter 6, equilibrium equations relating to the balance of recurrent evolutionary forces such as drift, gene flow, and/or mutation must be used cautiously when interpreting genetic observations. Such equations were derived under the *assumption* of a long-standing balance of recurrent evolutionary forces that are unchanging through time and uninfluenced by any historical events. Such an assumption may not be true for many real populations. To avoid making such an assumption, we need to go beyond just describing the current spatial pattern of genetic variation and instead investigate the spread of genes through both space and time together.

In rare instances, such space/time investigations can be done directly by actually sampling ancient populations and assaying their DNA. However, the vast majority of methods used to study genetic variation over space and time depend upon some sort of indirect inference of the past based upon samples of current populations. So the question becomes, can surveys of current genetic variation yield information about past patterns of gene flow and history?

A partial answer to this question was already given in Chapter 2. We saw in that chapter that the single-locus Hardy–Weinberg model has no historical effects: It only takes one generation of random mating for the deme to achieve its equilibrium state regardless of the starting conditions or past history. However, we saw that this property of an instant equilibrium was lost as soon as we went to two or more loci or polymorphic sites. The two-locus Hardy–Weinberg equilibrium is achieved only gradually through time and indeed may never be achieved during the time of a species' existence for closely linked polymorphic sites. The initial conditions (the exact genetic background upon which a mutation occurred, the gamete frequencies at that time, etc.) influence the dynamics of the two-locus Hardy–Weinberg model for many, many generations. Thus, when we move from a one-locus model to a two-locus model, *history* can have an impact upon the current patterns of genetic variation. Because history influences multilocus patterns, studies on multilocus or multisite genetic variation can in principle contain information about that history. As a consequence, most methods used in population genetics to recover history and to reconstruct the existence of genes through both space and time depend either upon multilocus data and/or multisite data within a locus or DNA region. This chapter will examine some of these methods, with an emphasis on how multi–polymorphic locus/site data are used to make inferences about the fate of demes and genes over both space and time.

DIRECT STUDIES OF SPACE AND TIME

The most straightforward way of inferring past genetic states is to sample extinct populations directly. With current DNA technologies, this is a difficult but not always impossible task. For example, Paxinos et al. (2002) scored haplotype variation in present-day and ancient populations of the nene, the Hawaiian goose (*Branta sandvicensis*). The nene was once widespread throughout the Hawaiian Islands but had gone extinct on all but the Island of Hawaii by the time of Captain Cook's arrival in 1778. The population on the Island of Hawaii further declined in the 1800s, and fewer than 30 individuals existed by the middle of the twentieth century. Conservation efforts since then have resulted in an increase in both nene population size and geographical distribution. Their genetic survey of mtDNA in 26 present-day nene revealed only a single haplotype, designated *RH* (reference haplotype). Given that mtDNA is usually highly polymorphic, this complete lack of diversity is startling. Paxinos et al. hypothesized that this lack of diversity was due to a bottleneck effect (Chapter 4), and perhaps due to the historically documented bottleneck that occurred in the twentieth century. To test the hypothesis that the current pattern of genetic variation in nene mtDNA was due to a past bottleneck event, they also scored mtDNA variation in 14 museum specimens collected between 1833 and 1928, 16 bones from archaeological middens dated to 160–500 years before present, and 14 bones from paleontological sites dated from 500–2540 years before present. Figure 7.1 summarizes their results. Starting with the oldest paleontological samples, several haplotypes are found, including *RH*, which was the most common haplotype in this earliest sample. Because *RH* is the most common haplotype in the paleontological sample and because of its central position in the mtDNA haplotype tree of the variants found at that time, *RH* is the most likely ancestral haplotype

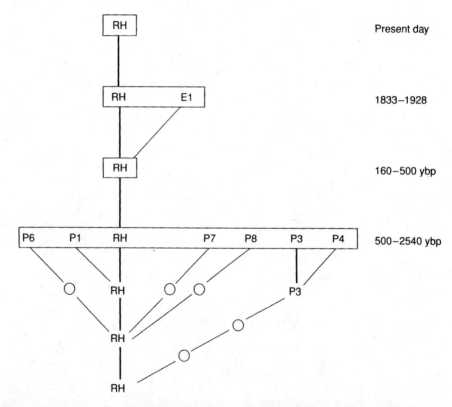

Figure 7.1. Nene mtDNA haplotype tree through time. Thin lines indicate a single mutational change, and open circles represent an intermediate haplotype form not found in the sample. Thick lines indicate a DNA lineage that left descendants at a later time without any mutational change. The boxes surround the haplotypes found during the time periods indicated on the right (ybp = years before present). Based on data from Paxinos et al. (2002).

for the haplotype tree (Castelloe and Templeton, 1994) and is drawn as such in Figure 7.1. A dramatic reduction in genetic diversity occurs in going from the paleontological to the archaeological samples: All haplotype variation is lost except *RH*. A single museum specimen has a haplotype other than *RH*, and this variant is only a single mutational step derivative of *RH*. Hence, there was a dramatic reduction of nene mtDNA diversity in the past, but it is not due to the bottleneck effect that occurred in the twentieth century. Rather, the genetic diversity had already been lost sometime before the time period of 160–500 years before present. Paxinos et al. (2002) note that there was a phase of human population growth on the Hawiian Islands between 350 and 900 years ago, and during this human growth phase, the nene went extinct on Kauai and at least five of the nine large ground-dwelling Hawaiian birds also went extinct at this time. Given the data summarized in Figure 7.1, it is more likely that the nene suffered its diversity-depleting bottleneck during this time of human population size expansion rather than during the more recent, historically documented bottleneck of the twentieth century.

LINKAGE DISEQUILIBRIUM AND MULTILOCUS ASSOCIATIONS

As shown in Chapter 2, both past history as well as recurrent evolutionary forces influence the genetic parameter of linkage disequilibrium, D. The creation of disequilibrium is influenced by historical factors such as the exact haplotype background upon which a new mutation originated or historical admixture between two previously genetically differentiated subpopulations (Chapter 2). Recurrent processes such as the system of mating and the amount of recombination then influence the decay of linkage disequilibrium over time—a decay that can be quite slow for closely linked genes or sites. This slow decay means that the signature of the historical events that initially created linkage disequilibrium is often apparent for many generations.

For example, in Chapters 3 and 6 we saw that the current African American gene pool has been greatly shaped by historical admixture between previously differentiated European and African populations (see Figure 6.2). In human populations without evidence of historical admixture, most of the significant linkage disequilibrium disappears between marker pairs greater than 2 cM (Wilson and Goldstein 2000). However, in African Americans, significant linkage disequilibrium extends up to 30 cM (Lautenberger et al. 2000), a result consistent with their known population history of admixture. An even more dramatic example of ancient admixture affecting current linkage disequilibrium is provided by the Lemba, a southern African group who speak a variety of Bantu languages but claim Jewish ancestry (Wilson and Goldstein 2000). According to their oral traditions, the Lemba are descended from a group of Jewish males who centuries ago came down the eastern coast of Africa by boat. Many were lost at sea, but the remainder interbred with local Bantu women, thereby establishing the ancestors of the current Lemba, who are now found mostly in South Africa and Zimbabwe (Thomas et al. 2000). Consistent with this oral tradition of historical admixture, the modern-day Lemba have significant linkage disequilibrium over intervals up to 19–24 cM. In contrast, modern-day populations of Bantus and Jews with no evidence of historical admixture have significant linkage disequilibrium only up to 1–6 cM.

Identifying specific populations that have high levels of linkage disequilibrium due to historical admixture has practical importance in human genetics because the linkage disequilibrium allows one to localize disease-associated genes with a bank of genetic markers scattered throughout the genome (Lautenberger et al. 2000). However, the history that makes these populations so valuable in disease association studies means that any detected associations between a marker and a disease are specific to the population being studied and cannot be generalized to all of humanity. Other human populations that had different histories will not show the same marker associations, as already illustrated by the example of using color blindness as a marker for G6PD deficiency in the Italian population west of the Apennines versus the population on the island of Sardinia (Chapter 4). Associations detected through linkage disequilibrium are due to history, not cause and effect, and therefore cannot be generalized across populations with different histories.

The fact that linkage disequilibrium is often population specific can also be a serious limitation to using linkage disequilibrium as a tool for investigating the general influence of historical events upon a species' population structure. Often, we are interested in the history of the entire species, or at least that of many different subpopulations living in a certain geographical area. Many interesting historical events (such as past fragmentation or colonization events) are expected to influence primarily the amount of genetic differentiation between subpopulations rather than the amount of linkage disequilibrium within any one subpopulation. Linkage disequilibrium can also be created by pooling genetically distinct subpopulations, as shown in Chapter 3. Hence, disequilibrium within and among subpopulations detects different aspects of evolutionary history. This in turn leads to the problem of how to interpret linkage disequilibrium; is it due to historical admixture or current pooling of differentiated demes?

Another limitation of using linkage disequilibrium to investigate population structure and history is that disequilibrium is inherently a two-locus (or two-site) statistic, whereas most genetic surveys assay multiple polymorphic loci or sites. You can always calculate all the pairwise disequilibrium between the multiple markers, but the resulting large matrix of correlated numbers is a very cumbersome investigative tool. Consequently, other methods of multilocus association are often used, particularly when the desired inference concerns not a single deme but a collection of many demes.

One method used to measure multilocus associations between demes is **principal-component analysis**, a statistical procedure used to simplify multivariate data with a minimal loss of information that has many uses in data analysis. Multivariate data can always be plotted in a multidimensional space, with each dimension corresponding to one of the variables being measured. A principal-component analysis rotates the axes of the original coordinate system through this multidimensional space such that one rotated axis, called the first principal component, captures the maximum amount of variation possible in a single dimension. The second principal component is the rotated axis constrained to be perpendicular (and thereby statistically independent) to the first principal component that captures the maximum amount of the remaining variation, and so on until an entire new set of perpendicular rotated axes has been defined.

To illustrate how principal-component analysis is specifically applied as a measure of multilocus association, consider the data set given in Table 7.1 on the frequencies of the $D-$ allele at the human *Rh* blood group locus and the frequencies of the O allele at the human *ABO* blood group locus in five human populations (data from Cavalli-Sforza et al. 1994). No linkage disequilibrium exists between these unlinked loci *within* these populations. However, if we amalgamate these five human populations into a single pooled population (weighting each population equally), the $D-$ allele has a frequency of 0.146, the O allele of 0.720, and the combination of $D-$ and O has a gamete frequency of 0.096. The linkage disequilibrium in the amalgamated population is given by the difference

Table 7.1. Allele Frequency Data on Five Human Populations for *O* Allele at *ABO* Blood Group Locus and *D−* Allele at *Rh* Blood Group Locus

Population	Frequency of O	Frequency of $D-$
Africa	0.69	0.20
Asia	0.60	0.15
Europe	0.65	0.36
America	0.90	0.02
Australia	0.76	0.00

Source: From Cavalli-Sforza et al. (1994).

between the observed gamete frequency and the product of the two allele frequencies: $0.096 - 0.105 = -0.009$. This nonzero value of linkage disequilibrium in the pooled population tells us little beyond the fact that some of the populations being pooled differed in their allele frequencies for $D-$ and O.

Now consider plotting the allele frequencies at these two loci against one another, as shown in Figure 7.2. These original axes correspond to the frequencies of the $D-$ and O alleles, respectively, in this two-dimensional space. The first principal component is defined to be the line through this two-dimensional space that minimizes the sum of the perpendicular distances of each population's data point from the line, as is shown in Figure 7.2. The projections of the original allele frequencies onto this line can always be expressed as a linear combination of the original allele frequencies; that is, $ap_{D-} + bp_O$, where p_{D-} and p_O are the frequencies of the $D-$ and O alleles, respectively, from a particular population and a and b are the weights assigned to a particular allele by minimizing perpendicular distances (many computer programs exist to calculate these weights). The second principal component is then defined as a line perpendicular to the first principal component that minimizes the distances of the data points to this second line. Because the example shown in Figure 7.2 has only two dimensions (one corresponding to each of the two alleles being considered), there is only one way for the second principal component to be drawn in this case, as shown in Figure 7.2. Therefore, we do not need to worry about the minimization criterion for the second principal component in this case. However, when one has multiple alleles per locus and multiple loci, a total of $\sum(n_i - 1)$ dimensions are needed in such a plot, where n_i is the number of alleles at locus i and the summation is over all loci. In these higher dimension cases, the minimization criterion is needed to calculate all but the very last of the $\sum(n_i - 1)$ principal components.

When the frequencies of alleles at different loci are correlated due to differentiation among demes, most of the variation in how the populations differ genetically will be found in the first principal component. There is progressively lesser amounts of variation in the amount of genetic differentiation as you go to the second, third, etc., principal components. This is also shown in Figure 7.2, which shows the projections of the original data upon the first and second principal components. As can be seen, the populations are much more spread out on the line corresponding to the first principal component than they are on the line corresponding to the second principal component. This reflects the fact that the first principal component captures more of the genetic differentiation among the populations than the second. In general, if there are strong multilocus associations among the populations, most of the information about genetic differentiation should be contained in just the first

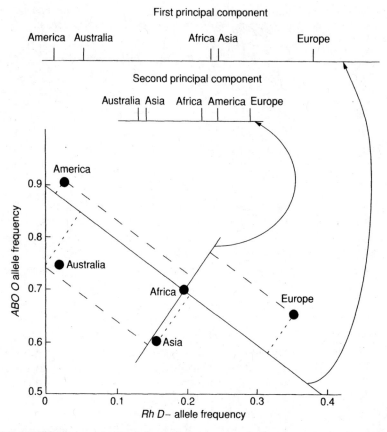

Figure 7.2. Principal-component analysis of allele frequency data shown in Table 7.1 for five human populations scored for *ABO O* allele and *Rh D−* allele. The two perpendicular lines representing the first and second principal components are shown. Dotted lines show the perpendicular projections of each point corresponding to a population's allele frequencies upon the first principal-component line; dashed lines show the perpendicular projections onto the second principal component. At the top of the diagram the principal components are redrawn, showing the points of intersection of the perpendicular projections from each population. Modified with permission from Fig. 1.13.1 in L. Cavalli-Sforza et al., *The History and Geography of Human Genes*. Copyright © 1994 by Princeton University Press.

few principal components. For example, Ammerman and Cavalli-Sforza (1984) present data on 39 independent alleles at 9 loci sampled in human populations in Europe (later expanded to 95 alleles). The first principal component accounts for 27% of the total genetic differentiation among these European populations, and the second 18%. Thus, only two linear combinations of the original allele frequencies account for nearly half of the total genetic differentiation observed with 95 independent alleles.

When Ammerman and Cavalli-Sforza (1984) plotted the values of the first principal component upon a map of Europe, they discovered that the values formed a sequence of bands running as arcs across Europe that gradually changed in value as one went from the Middle East to northwestern Europe (Figure 7.3). This same spatial pattern in Europe is observed in the archeological record for the spread of agriculture out of the Middle East. Indeed, the correlation coefficient between the values of the first principal component and the archeological dates of arrival of agriculture is 0.89, an extremely high value. Ammerman and

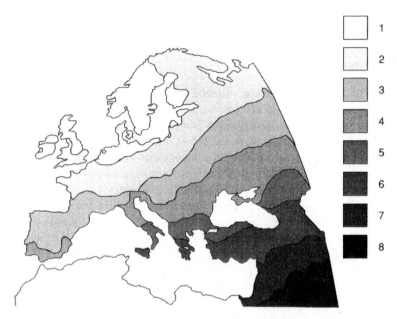

Figure 7.3. Distribution of first principal component of 95 allele frequencies across Europe. The scale of 1–8 is a ranking of the values of the principal component. Modified with permission from Fig. 5.11.1 in L. Cavalli-Sforza et al., *The History and Geography of Human Genes.* Copyright © 1994 by Princeton University Press.

Cavalli-Sforza (1984) therefore concluded that the first principal component corresponds to a genetic cline established by early agriculturalist spreading out of the Middle East, spreading both their genes and agricultural system. Interestingly, the strong clinal pattern found in the first principal component is not easily seen in the plots of the individual allele frequencies over Europe. Therefore, concentrating the information from many alleles into a single principal component revealed a genetic signature of an old historical event that was not readily evident from any single locus.

There are multiple ways of actually doing the formal statistical analysis of how the various principal components overlay upon geographical maps, and there is controversy over which methodologies are most appropriate (Sokal et al. 1999a,b; Rendine et al. 1999). However, one area of agreement is that these techniques should not be used for historical inferences without corroboration from other kinds of evidence (Rendine et al. 1999). The reason for this restriction on the use of principal components is that the patterns revealed by principal components (and other statistics, such as linkage disequiilibrium, f_{st}, or Φ_{st}) are often compatible with many sources of biological causation. For example, if a deme is found to have extensive linkage disequilibrium, is that due to an historical admixture event or to an historical bottleneck event? Quite often, it is impossible to discriminate among such alternatives with just the information provided by linkage disequilibrium alone (or, e.g., principal components). Consequently, these procedures are best used to confirm prior hypotheses (such as the Lemba coming from an admixed ancestral population or the spread of agriculture in Europe being also a spread of peoples). Fortunately, there are other analyses that can infer historical events and recurrent forces of population structure without a priori expectations.

POPULATION GENETIC DISTANCES, POPULATION TREES, AND GEOGRAPHICAL PLOTS

When Is a "Tree" a Tree?

The concept of a population genetic distance was introduced in Chapter 6, and we saw how such genetic distances could be plotted over geographical distance to test the fit to a model of recurrent gene flow constrained by isolation by distance (Figure 6.7). Population genetic distances can also be used to test the fit of a model of historical fragmentation followed by complete genetic isolation. Populations that were fragmented in the past and then remain genetically isolated inevitably diverge from one another over time due to genetic drift and mutation. This was shown experimentally in Chapter 4 with Buri's experiments on 107 isolated subpopulations fragmented from a common ancestral stock of *Drosophila melanogaster* (Figure 4.3). The population genetic distances among isolated subpopulations increase with increasing time since the fragmentation event as the different subpopulations acquire different alleles by mutation and diverge in allele frequencies by drift. If different subpopulations split from one another at different times, the population genetic distances should ideally reflect the order of their splitting in time; that is, the older the split, the larger the population genetic distance.

In Chapter 5 we saw that molecule genetic distances can be used to estimate an evolutionary tree for haplotypes through such algorithms as neighbor joining. Similarly, population genetic distances can be used to estimate an evolutionary tree of populations by use of the same algorithms. Recall from Chapter 5 that recombination can undercut the very idea of a haplotype tree itself. When recombination occurs, a single haplotype can come to bear different DNA segments that had experienced different patterns of mutation and coalescence in the past. Thus, there is no single evolutionary history for such recombinant haplotypes. When recombination is common and uniform in a DNA region, the very idea of a haplotype tree becomes biologically meaningless. Similarly, an evolutionary tree that portrays the historical relationships among populations can be estimated *only if the populations are isolates*. The analogue of recombination in this case is gene flow or admixture. If genetic interchange occurs among members of the populations, then populations cannot be placed into an evolutionary tree that depicts common ancestry and divergence under isolation as the sole agents of current evolutionary relatedness.

Algorithms such as neighbor joining generate population trees from pairwise population genetic distances regardless of whether or not the populations are exclusively related to one another through common ancestral populations and divergence under strict genetic isolation. For example, Table 7.2 shows a matrix of pairwise population genetic distances (pairwise f_{st}'s in this case) based on 100 DNA polymorphisms from Bowcock et al. (1991). Assuming for the moment that human populations are true isolates, Figure 7.4 shows the population tree estimated from this population genetic distance data (Cavalli-Sforza et al. 1994). This and most other human population genetic distance trees have their deepest divergence between Africans and non-Africans. Under the assumption that human populations are genetic isolates, this divergence reflects the accumulated population genetic distance since non-Africans split off from Africans. Just as a molecular clock can be calibrated for molecular genetic distances, population genetic distances can also be calibrated under the assumption that populations are isolates. When this is done, the "split" between Africans and non-Africans is estimated to have occurred around 100,000 years ago (Cavalli-Sforza et al. 1994; Nei and Takezaki 1996). The estimated ages of other splits between human populations are also shown in Figure 7.4.

Table 7.2. Pairwise f_{st}'s among Five Human Populations Based on 100 DNA Polymorphisms (±standard errors)

	Zaire Pygmies	Melanesians	Chinese	Europeans
Central African Republic Pygmies	0.043 ± 0.011	0.242 ± 0.031	0.235 ± 0.032	0.141 ± 0.022
Zaire Pygmies		0.265 ± 0.034	0.235 ± 0.033	0.171 ± 0.019
Melanesians			0.171 ± 0.019	0.148 ± 0.019
Chinese				0.093 ± 0.016

Source: From data in Bowcock et al. (1991) as presented in Cavalli-Sforza et al. (1994).

However, are human populations truly related to one another only by a series of historical splits followed by genetic isolation? As shown in Chapter 6, nonzero population genetic distances can also arise and persist between populations with recurrent gene flow as long as that gene flow is sufficiently restricted relative to genetic drift. Thus, the mere existence of a nonzero population genetic distance between two populations does *not* imply that a split ever occurred between them. Therefore, the population genetic distances shown in Table 7.2 could be measuring the amount of gene flow between the populations and not the time since a split because no split ever occurred! In this case, there is no true evolutionary tree of populations. Thus, population genetic distances can be interpreted either as indicators of time after a split in the context of an evolutionary tree or as indicators of gene flow.

Fortunately, these two interpretations of population genetic distance can be distinguished. If human populations were truly related to one another by past fragmentation events followed by genetic isolation, then the resulting population genetic distances should satisfy several constraints. For example, under the evolutionary tree interpretation, all non-African

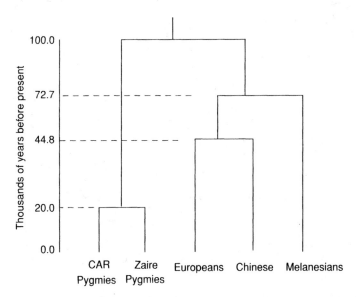

Figure 7.4. Population tree estimated from pairwise f_{st} data given in Table 7.2 (CAR = Central African Republic). Modified with permission from Fig. 2.4.4 in L. Cavalli-Sforza et al., *The History and Geography of Human Genes*. Copyright © 1994 by Princeton University Press.

human populations split from Africans at the same time (Figure 7.4). Therefore all population genetic distances between African and non-African populations have the same expected value under the tree interpretation. The constraints on population genetic distances imposed by an evolutionary tree are collectively called **treeness**. When population genetic distances instead reflect the amount of genetic interchange, treeness constraints are no longer applicable. Therefore, how well the population genetic distances fit treeness provides a means of discriminating between these two interpretations of population genetic distances.

For example, the tree shown in Figure 7.4 implies that the population genetic distances between the African (Pygmy) populations to the Melanesian, Chinese, and European populations should all be equal under a molecular clock *if the tree model of historical fragmentation and isolation is true*. This expected population genetic distance between African and non-African populations can be estimated by the average distance between these populations, which is 0.213 using the data in Table 7.2. However, Table 7.2 also reveals that these distances are not equal. Instead, the European population is much closer in population genetic distance to the African populations than are the Chinese and Melanesian populations. Treeness also implies that the population genetic distance between the Melanesians and the Europeans should be the same as the population genetic distance between the Melanesians and the Chinese, which is estimated by the average of these two distances as 0.160. The expected population genetic distances between these human populations under the assumption that they are related to one another by a series of splits followed by genetic isolation is shown in Table 7.3.

One method of quantifying how well a matrix of population genetic distances fits treeness is the **cophenetic correlation**—the correlation between the observed population genetic distances to the expected values generated by the estimated "tree." The word tree is put in quotation marks here because we want to test the hypothesis that an evolutionary tree of populations exists. At this point, we need to remain open to the possibility that there is no tree of populations at all. Therefore, the tree estimated from the population genetic distances by using some algorithm such as neighbor joining is only a vehicle to test treeness. The estimated tree should never be accepted as an actual evolutionary tree until treeness has been tested. Because the tree is estimated from the population genetic data itself, a positive cophenetic correlation is always expected, even for randomly generated matrices of population genetic distances. Moreover, the tree does not constrain all population genetic distances, so some distances will always fit exactly and elevate the correlation. For example, the genetic distance between the two Pygmy populations and the genetic distance between Europeans and Chinese are identical in Tables 7.2 and 7.3 because the tree imposes no constraint on these distances. The cophenetic correlation for this human example (the standard correlation between the values given in Table 7.2 to their counterparts in Table 7.3)

Table 7.3. Expected Pairwise f_{st}'s among Five Human Populations[a]

	Zaire Pygmies	Melanesians	Chinese	Europeans
Central African Republic Pygmies	0.043	0.213	0.213	0.213
Zaire Pygmies		0.213	0.213	0.213
Melanesians			0.160	0.160
Chinese				0.093

[a] Under the assumption That the populations are related to another through a series of splits followed by genetic isolation as depicted in the evolutionary tree in Figure 7.4.

is 0.87. In general, any cophenetic correlation less than 0.8 is regarded as a poor fit (Rohlf 1993). Hence, the 0.87 value calculated for the tree in Figure 7.4 would not reject treeness. However, when the number of populations is small (say, less than 10), the proportion of the distances that are expected to fit exactly is rather large (2 out of 10 for this human example), which can greatly inflate the cophenetic correlation. Therefore, when dealing with a small number of populations, it is best to exclude the distance pairs unconstrained by treeness. Excluding the two unconstrained distances from the analysis of the tree in Figure 7.4, the cophenetic correlation drops to 0.53, indicating that the tree is not a true evolutionary tree of these human populations. Using several larger human data sets, the cophenetic correlations ranged from 0.45 to 0.79 (Templeton 1998a), thereby implying that human populations in general are *not* genetic isolates related to one another in a treelike fashion due to historical splits and divergences as implied by Figure 7.4.

Gene Flow versus Admixture

The rejection of treeness among populations is usually due to some sort of genetic interchange among the populations, but treeness can be violated either by recurrent gene flow or by rarely occurring, historical episodes of admixture (Cavalli-Sforza et al. 1994). If the lack of treeness is due to rare episodes of admixture, then the tree would still be yielding biologically meaningful information about population history. For example, Cavalli-Sforza et al. (1994) argue that human populations could have truly split from one another as shown in Figure 7.4 and that the fragmented populations subsequently represented distinct evolutionary lineages that had been "long separated" from one another. The lack of treeness in the data matrix shown in Table 7.2 is primarily caused by the genetic distances between Africans and Europeans being too small. This in turn could be explained by Europeans arising from just a single admixture event between ancestral African and Chinese populations occurring in Europe about 30,000 years ago (Cavalli-Sforza et al. 1994). In contrast, the same pattern could be explained by an isolation-by-distance model in which most humans live and breed in the area near their birth, and those that do move generally move only to nearby geographical areas. The European population is geographically closer to the African populations than are the Chinese and Melanesian populations, so the observed pattern of smaller African–European population genetic distances and larger African–Asian distances makes sense under a model of recurrent flow constrained by isolation by distance. Using a fuller data set, Bowcock et al. (1991) concluded that the tree estimated from their population genetic distances is not a real tree of populations at all, but rather that the underlying genetic distance data reflect genetic interchange among the populations in a manner consistent with isolation by distance. However, they also pointed out that if you are willing to invoke admixture events as needed, any pattern of population genetic distances lacking treeness can also be explained by a finite number of admixture events between historical isolates. Can we discriminate between recurrent genetic interchange among populations versus nonrecurrent historical admixture events?

Fortunately, rare episodes of admixture can be discriminated from some models of recurrent gene flow such as isolation by distance. One approach is to perform finer scale geographical sampling. For example, if sub-Saharan Africans truly split from non-Africans 100,000 years ago as shown in Figure 7.4, then any sub-Saharan population should be equally distant from any European population. Admixture events should cause only local deviations from this pattern, as illustrated earlier by the Lemba. However, the genetic distances between northern sub-Saharan African populations and Europeans are smaller than those between southern African populations and Europeans (Cavalli-Sforza et al. 1994), as predicted under isolation by distance. Indeed, as shown in Figure 6.7, the human population genetic distance data fit well to a recurrent gene flow model with isolation by distance. To explain the results shown in Figure 6.7 by admixture requires many admixture events with a timing and geographical placement that fit the expectations of isolation by distance. This is possible but not plausible. Thus, with finer geographical sampling, the admixture model becomes increasingly implausible as a general explanation of human population genetic distances.

Serre and Pääbo (2004) presented an analysis showing the importance of fine-scale geographical sampling in interpreting human population structure. In many human studies, it is commonplace to sample individuals from "populations" defined by shared cultural and linguistic attributes that are also often correlated with geography or geographical origins. The individuals thus sampled are parts of predetermined categories, and these categories or groups of them often serve as the units in the analysis of population structure (e.g., an AMOVA). Such analyses typically find that humans are subdivided into genetically distinct groups, and frequently such results have been interpreted to justify the concept of race in humans (Risch et al. 2002; Rosenberg et al. 2002). However, an isolation-by-distance model with no genetic clusters at all will still have the *appearance* of genetic clusters if individuals are sampled by populations that are separated by geographical sampling gaps. To investigate the possible role of sampling strategy, Serre and Pääbo (2004) drew samples out of a human cell line database that had been collected worldwide with each cell line scored for 377 autosomal microsatellites. The original cell lines had been sampled with the traditional population sampling scheme of human genetics, and when Serre and Pääbo drew out subsamples with such population sampling, they obtained the typical result of humans being subdivided into discrete genetic clusters (Rosenberg et al. 2002). Serre and Pääbo next drew out a similarly sized subsample of individual cell lines based on geographical location rather than population affiliation with the goal of obtaining a more uniform geographical coverage of the human species without major geographical gaps. The distinct genetic clusters of humanity disappeared with uniform geographical sampling. Serre and Pääbo (2004) now observed gradients of allele frequencies that extend worldwide instead of discrete clusters. This study serves as a warning about the importance of sampling in studies of population structure and evolutionary history. The sampling process itself can create genetic patterns that are easily confused with biological reality. In particular, fine geographical sampling without large spatial gaps is essential for discriminating past fragmentation (with or without subsequent admixture) from recurrent gene flow restricted by isolation by distance.

There is another way of making use of finer geographical sampling to discriminate between isolation by distance and historical admixture events. If past isolation truly existed between two populations, the populations would show allele frequency differences at many loci, as shown in Chapter 4. When such previously isolated demes come into contact, admixture results in intermediate allele frequencies in the area of geographical contact. Such intermediate allele frequencies should be created simultaneously for *all* differentiated loci because the admixture event itself creates extensive linkage disequilibrium for *all* loci showing allele frequency differences between the original populations (Chapter 6). This in turn results in a strong geographical concordance in the allele frequency changes over geographic space for all genetic systems. In contrast, isolation by distance does not result in such a strong spatial concordance across loci. Mutations arise at different times and places and spread out gradually from their point of geographical origin under isolation by distance. This variation in space and time in the origin of alleles creates spatial patterns in allele frequencies that may vary from locus to locus. There can be constraints on the patterns of movement that influence the spatial patterns for all loci under recurrent gene flow (such as a low pass in a mountain range for a species that cannot tolerate high altitudes), and multilocus analysis can often provide some insight into these gene flow constraints. Nevertheless, the amount of variation across loci in spatial patterns of allele frequencies is far greater under recurrent gene flow constrained by geographical distance than under an historic episode of admixture of previous isolates, which causes allele frequency shifts in many loci at the same point in time and space.

Consider again human populations. Figure 7.5 shows the allele frequencies at the *ABO* blood group locus for 20 human populations, including three African and 17 non-African. Figure 7.5 shows some clustering of populations with similar allele frequencies at this locus, but none of these clusters corresponds to African populations. Instead, each of the three African populations is found in a different cluster! Thus, what is supposed to be the deepest and oldest split among human populations (Figure 7.4) is not apparent at this particular locus. This conclusion is not unique to the *ABO* locus. Indeed, the lack of concordance in the geographical distribution of different genetic traits in humans has been one of the primary traditional arguments against the validity of human races (Futuyma 1986). This lack of concordance across genetic systems favors the hypothesis of recurrent gene flow constrained by isolation by distance as playing the major role in recent human evolution.

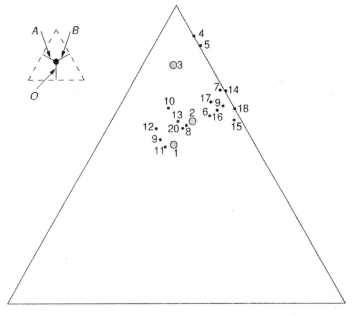

Figure 7.5. Triallelic frequencies of *ABO* blood group locus allele frequencies for human populations. Each point represents a population with the perpendicular distances from the point to the sides of the triangle representing the frequencies of the *A*, *B*, and *O* alleles, as indicated in the small triangle at the top left in the diagram. Populations 1–3 are African and are shown by large, shaded circles. All non-African populations are indicated by small, closed circles. Populations 4–7 are Amerindian, 8–13 are Asian, 14 and 15 are Australian aborigine, and 16–20 are European. Adapted from Fig. 67 in A. Jacquard, *Structures Génétiques des Populations*. Copyright © 1970 by Masson & Cie with permission from Masson S.A.

By testing the fit of population genetic distance data to treeness, testing the fit to isolation-by-distance models (given an adequate sampling design), and examining concordance across loci, we have a suite of tools for discriminating between a series of historical splits followed by divergence versus cases in which populations have experienced various types of genetic interchange, including historical admixture events and recurrent gene flow. You should never simply convert population genetic distances into trees without an assessment of treeness and isolation by distance. A population tree need not exist at all when the genetic relationships among intraspecific populations are dominated by gene flow patterns rather than historic splits.

HAPLOTYPES

Haplotypes are defined by the genetic state of many polymorphic sites within a single DNA molecule. Haplotypes are usually defined for relatively small genetic regions, and many of these regions have low rates of or even no recombination. Haplotypes in such DNA regions therefore retain their historical configuration of nucleotide sites for many generations or even throughout their entire evolutionary existence in time. The persistence of their original evolutionary state through time serves as a valuable tool for tracing their fate through space.

Of particular utility for gene flow studies are those haplotypes that are rare in the species as a whole. After a new haplotype has been created by the process of mutation, it exists initially only as a single copy at a single point in space. Many successful DNA replication events are needed before this newly arisen haplotype can become common and spread through space. As a consequence, many newly arisen haplotypes tend to be rare in the species as a whole (even though they may be common in some demes) and are restricted in their geographical range. When individuals or populations bearing such globally rare haplotypes move through space, their movements are traced by these rare haplotypes (Richards et al. 2000). Finding such globally rare, geographically restricted haplotypes is no longer difficult. When genetic diversity is scored at the DNA sequence level, haplotypes with these properties are frequently found in most species. At first this may seem counterintuitive. Obviously, by definition, any particular rare haplotype is rare, but what we mean is that the category of rare haplotypes is collectively common. Large samples of genes in most species will reveal one or more rare haplotypes. This is shown for the coalescent model given in Box 7.1. Consequently, finding recently arisen and globally rare haplotypes is not difficult in most species, and such haplotypes provide a powerful marker for recent movements through space of individuals or populations (Richards et al. 2000).

For example, let us reconsider the admixture between European, African, and Amerindian populations in the New World. Hammer et al. (1998) surveyed a nonrecombining region of the Y chromosome. Twelve mutational events were found in this sample, and the haplotype associated with the oldest inferred mutation is widespread throughout the world. However, as one progresses through time, the haplotypes created by more recent mutations become increasingly rare in the global human species and become more geographically restricted. As a consequence, Y-chromosomal haplotypes were identified that are almost completely restricted to Europeans, others that are almost completely restricted to sub-Saharan Africans, and yet others that are restricted to Amerindians. As mentioned in Chapter 6, modern-day Brazilians represent a recent and ongoing admixture of peoples from these three geographical regions. Carvalho-Silva et al. (2001) surveyed 200 unrelated Brazilian males and 93 unrelated Portuguese males (Portugal had been the colonial power in Brazil) for this

BOX 7.1 AGES OF MUTATIONS IN THE n-COALESCENT

Consider the coalescent process with mutation associated with a sample of n genes in a diploid population of inbreeding effective size $N_{ef} = N$. As shown in Box 5.1, the expected time between the $k - 1$ coalescent event and the kth event is

$$E(\text{time between } k - 1 \text{ and } k \text{ coalescent events}) = \frac{4N}{(n - k + 1)(n - k)} \quad (7.1)$$

If mutations occur at a rate μ per generation in every DNA lineage, the expected number of mutations that occur in any given lineage between the $k - 1$ coalescent event and the kth event is

$$E(\text{number of mutations between } k - 1 \text{ and } k \text{ coalescent events})$$
$$= \frac{4N\mu}{(n - k + 1)(n - k)}$$
$$= \frac{\theta}{(n - k + 1)(n - k)} \quad (7.2)$$

Between the $k - 1$ coalescent event and the kth event there exist $n - k + 1$ DNA lineages, each at risk for mutation. Hence, the total number of mutations that are expected to be observed between the $k - 1$ coalescent event and the kth event in a sample of n genes, say S_k, is

$$S_k = (n - k + 1) \frac{\theta}{(n - k + 1)(n - k)} = \frac{\theta}{n - k} \quad (7.3)$$

Over the entire coalescent process in which the n sampled genes trace back to a single ancestral molecule, the expected number of mutational events is, summing equation 7.3 over all $n - 1$ coalescent events,

$$S = \sum_{k=1}^{n-1} S_k = \theta \sum_{k=1}^{n-1} \frac{1}{n - k} \quad (7.4)$$

For a large sample of genes, the sum of $(n - k)^{-1}$ from $k = 1$ to $k = n - 1$ converges to 2, so the total number of mutations expected in the n-coalescent process under neutrality is 2θ when n is large. Note that the last term in this sum (corresponding to $k = n - 1$) is 1. Recall also from Box 5.1 that the expected time it takes between the next to last ($k = n - 2$) and last ($k = n - 1$) coalescent event is half of the total expected coalescent time for the entire sample of n genes. Thus, for large n, half of the expected mutations occur in the more ancient half of the total coalescent process and half of the mutations are expected to occur in the more recent half (each of these two equal time intervals should experience θ mutational events). Now, the next to the last term in the sum of $(n - k)^{-1}$ in equation 7.4 (corresponding to $k = n - 2$) equals $\frac{1}{2}$. As just mentioned above, the sum up to and including the $n - 2$ term is about 1 for large n, so the number of mutations expected to occur between the $n - 3$ to $n - 2$ coalescent events is $\theta/2$, and thus $\theta/2$ are expected

> to occur between the present and the $n-3$ coalescent event. As shown in Box 5.1, both of these expected time intervals are equal in duration. Continuing with this argument, it is evident that the expected number of mutations is uniformly distributed across time. Thus, in a large sample of genes, any single mutation that occurred during the coalescent process is as likely to be recent as ancient. If either we sequence a DNA region with a high mutation rate and/or sequence a large number of nucleotides to ensure that θ for the total sequenced region is large, we should encounter many recently arisen mutational events and their associated haplotype states. Collectively, young haplotypes are common in large samples with large θ even though any particular young haplotype is expected to be globally rare.

paternally inherited, Y-chromosomal region. Using these geographically specific Y-chromosomal haplotypes, they estimated that about 97.5% of the Y chromosomes found in present-day Brazilians were of European origin, 2.5% of African origin, and 0% of Amerindian origin. These results are quite divergent from the admixture studies based on nuclear genes described in Chapter 6 that indicated that present-day Brazilians have received significant genetic input from all three geographical regions. The discrepancies between these two types of inheritance systems can be reconciled if most of the admixture was mediated by European men mating with African and Amerindian women. Concordant with this hypothesis, Carvalho-Silva et al. (2001) report that more than 60% of the maternally inherited mitochondrial haplotypes found in "white" Brazilians (Chapter 6) are of African or Amerindian origin. Note that by simultaneously studying globally rare and geographically restricted Y-chromosomal and mitochondrial haplotypes, Carvalho-Silva et al. (2000) could conclude not only that present-day Brazilians represent an admixed population but also specifically that the Brazilians are an admixture of genes of New World, European, and sub-Saharan African origin and that the admixture was primarily associated with European men mating with African and Amerindian women. This example illustrates the detailed inferences about recent population history and gene flow that are possible from haplotypes, particularly when coupled with genetic surveys on elements with different inheritance systems.

To reinforce this conclusion, consider again the Lemba of southern Africa. The nuclear linkage disequilibrium studies show that the Lemba are an admixed population, a conclusion consistent with their oral history of an ancient Jewish–Bantu admixture. However, their oral history further specifies that the admixture was due to Jewish men mating with Bantu women. Linkage disequilibrium of autosomal genes per se sheds no light on this detail. However, using Y-chromosomal haplotypes of restricted geographical distribution, Thomas et al. (2000) showed that about two-thirds of the Lemba Y chromosomes have a Middle Eastern origin and one-third a Bantu origin. Moreover, one particular Y-chromosomal haplotype is found in frequencies ranging from 0.100 to 0.231 in various Jewish populations but is very rare or absent from most other human populations. Yet the frequency of this globally rare haplotype is 0.088 in the Lemba of southern Africa, consistent with a genetic interconnection between the Lemba and Jews of Middle Eastern origin. In contrast to the Y haplotypes, there is no evidence of Semitic admixture with the maternally inherited mtDNA (Soodyall 1993)—a pattern also consistent with the oral traditions of the Lemba that the original admixture involved Jewish men and Bantu women.

These studies illustrate the richness of detailed inference that haplotype studies can provide about recent movements of individuals and populations through space. However, what about more ancient movements? Rare haplotypes are generally young in an evolutionary sense, so they offer little to no insight into older movements and historical events. We will now see that much more information can be obtained about past history and population structure by studying all the haplotypes and not just the rare haplotypes.

USING HAPLOTYPE TREES TO STUDY POPULATION HISTORY AND PAST POPULATION STRUCTURE

Haplotype trees (Chapter 5) provide information about a haplotype's existence through evolutionary time. Can the temporal information inherent in a haplotype tree be used to shed light upon the spatial distribution of current haplotypic variation? To answer this question, consider the study of Templeton and Georgiadis (1996) on mtDNA restriction site variation in Eastern African populations of buffalo (*Syncerus caffer*) and impala (*Aepyceros melampus*). The F-statistic estimator of Davis et al. (1990) yields an F_{st} of 0.08 for the buffalo and an F_{st} of 0.10 for the impala. Both of these F_{st} values are significantly different from zero, but they are not significantly different from each other. Moreover, in both species most of the geographical sites surveyed are relatively close together in Kenya and Tanzania, but one site (Chobe) is far to the south. In both species, the Chobe samples had many haplotypes not found in the other locations, and it was the Chobe samples that were responsible for the significant F_{st}'s in both cases. Hence, this F-statistic analysis implies that both species are equally subdivided, have comparable rates of gene flow, and display restricted gene flow primarily between the Chobe versus the Kenya/Tanzania localities. Now, let us add on information from the mtDNA haplotype trees for these two species with seemingly similar patterns of spatial variation as measured by F statistics.

Figure 7.6 shows the haplotype tree estimated from the mtDNA and indicates the haplotypes found only in Chobe in both species. In the buffalo, the Chobe haplotypes are scattered throughout the haplotype tree; in the impala, the Chobe haplotypes are tightly clustered within the tree. Although both species show the same degree of spatial subdivision as measured by F_{st}, they have obviously achieved this degree of subdivision in very different fashions through time. Clearly, the use of haplotype trees allows a finer discrimination of biological pattern than an F-statistic analysis. The reason for this is straightforward: By using a haplotype tree, you examine a spatial/temporal pattern of genetic variation whereas you examine only the current spatial pattern with the F statistic.

The scattered spatial/temporal pattern found in the buffalo (Figure 7.6) indicates recurrent genetic interchange between Chobe and the more northern populations throughout the time period from the coalescence of the sampled mtDNA haplotypes to the present. The impala pattern is more difficult to interpret. Such a strong evolutionary clustering of haplotypes in a geographical region, particularly when the haplotype clusters are separated by a long branch length with missing intermediates, is often interpreted as evidence of a past fragmentation event (Avise 1994). However, because impala are found in intermediate geographical locations that were not sampled, it is possible that this pattern arose from recurrent gene flow under an isolation-by-distance model (Chapter 6) such that geographically intermediate populations would fill in the missing haplotype nodes and show a gradual shift from one cluster of haplotypes to the other, just as happened with the human data in the analysis of Serre and Pääbo (2004). Indeed, a rigorous quantitative analysis of these data (of a type to be discussed below) reveals that the sparseness of sampling prevents one from distinguishing between isolation by distance versus fragmentation of the Chobe population from the Kenyan/Tanzanian populations of impala (Templeton and Georgiadis 1996).

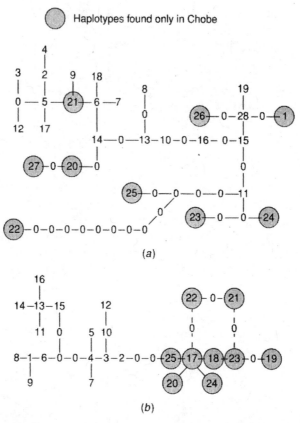

Figure 7.6. Unrooted mtDNA haplotype trees for two species of ungulates: (*a*) the buffalo (*S. caffer*) and (*b*) the impala (*A. melampus*) sampled from same geographical sites over eastern Africa. Each line represents a single restriction site change in the haplotype tree, with dashed lines indicating ambiguity under statistical parsimony. Haplotypes are designated by numbers, and a 0 indicates an internal node in the haplotype tree that was not found in the sample of current haplotypic variation. The haplotypes found in the distant, southern location of Chobe are shaded.

The analysis of Templeton and Georgiadis (1996) on the impala illustrates the dangers of making biological inferences by a visual inspection of how geography overlays upon a haplotype tree. Such visual inferences make no assessment of adequate sample sizes for statistical significance or adequate sampling of geographical locations for distinguishing among potential causes of geographical associations. One method for addressing these difficulties is a **nested-clade phylogeographic analysis** (Templeton et al. 1995) that quantifies the associations between geography, time, and the haplotype tree. As we saw earlier concerning rare haplotypes, information about the age of a haplotype can be used as a powerful tool to investigate the fate of that haplotype over space. A haplotype tree contains much information about the relative ages of all haplotypes and not just the rare, newly arisen tips of the haplotype tree. One method of obtaining temporal information from the haplotype tree is to use a molecular clock to estimate the ages of the various haplotypes in a rooted tree. This seemingly simple approach is often not easy to implement because of the difficulties associated with obtaining a good calibration for the clock, possible deviations from the clock, and the inherent large evolutionary stochasticity associated with the coalescent process itself. We will make use of such detailed temporal reconstruction later in this chapter, but nested-clade analysis extracts much temporal information from a haplotype tree without the need for a molecular clock or age estimates.

Nested-clade analysis first uses the haplotype tree to define a series of hierarchically nested clades (branches within branches). Such nested hierarchies are commonly used in comparative evolutionary analyses of species or higher taxa but can also be applied to the haplotype variation found within a species if that variation can be placed into a haplotype tree (Templeton et al. 1987a). Temporal information is captured by the nested hierarchies because the age of a specific haplotype or clade of haplotypes must be less than or equal to the age of the clade within which it is nested. If the haplotype tree is rooted, then we can define a set of nested hierarchies in which each clade is strictly younger than the clade within which it is nested. In either case, a nested hierarchy of clades captures much of the relative temporal information inherent in a haplotype tree that does not depend upon a molecular clock.

Templeton et al. (1987a) and Templeton and Sing (1993) proposed a set of rules to produce a nested series of haplotypes and clades. To achieve the first level of nesting, one starts at the tips of the haplotype tree. Recall from Chapter 5 that a tip simply refers to a haplotype that is connected to the tree by only one branch. For example, Figure 7.7 shows a

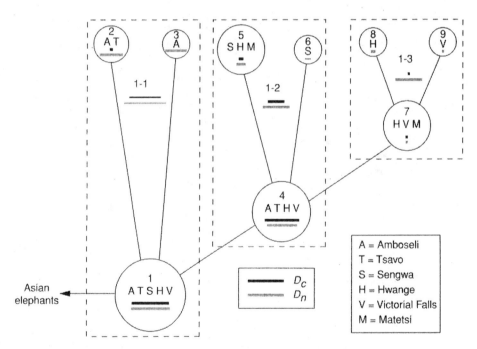

Figure 7.7. Rooted mtDNA haplotype trees for African elephants sampled from six geographical sites over eastern Africa. Each line represents a single restriction site change in the haplotype tree. Each circle represents a haplotype, with the number in the circle being the haplotype designation. Below the number, letters indicate the sampling sites at which that particular haplotype was found. Below that are bars that indicate the clade distance (black) and nested-clade distance (grey), as given in Table 7.4. The length of the bars in the legend box corresponds to a distance of 1000 km. The relative sizes of the circles represent the number of different sites at which a haplotype was present. The dashed-line rectangles enclose the haplotypes nested together into one-step clades using the nesting rules of Templeton et al. (1987a) and Templeton and Sing (1993). Three one-step clades result, designated by $1 - x$, where x is a number designating a specific one-step clade.

simple haplotype tree for mtDNA from elephants obtained by mapping restriction enzyme cut sites (N. J. Georgiadis and A. R. Templeton, unpublished data). The tree consists of eight haplotypes found in savanna elephant populations from eastern Africa, with an Asian elephant haplotype used as an outgroup to root the tree. Notice that haplotypes 2, 3, 5, 6, 8, and 9 are all connected to the tree by a single branch and hence are tips. In contrast, haplotypes 1, 4, and 7 have more than one branch connecting to them in the tree and therefore represent interior nodes of the tree. These haplotypes are therefore called interior haplotypes. To create the first level of nested haplotypes, move one mutational step from the tips into the interior and place all haplotypes that are interconnecting by this procedure into a single clade. For example, moving one mutational step in from tip haplotype 9 connects you to interior haplotype 7 (Figure 7.7). Similarly, moving in one step from tip haplotype 8 also connects you to interior haplotype 7. Therefore, haplotypes 7, 8, and 9 are grouped together into a "one-step clade" which is designated as clade 1-3 in Figure 7.7. Similarly, moving in one mutational step from tip haplotypes 5 and 6 connects to interior haplotype 4, which are now nested together into one-step clade 1-2; and moving in one mutational step from tip haplotypes 2 and 3 connects to interior haplotype 1, which defines clade 1-1 (Figure 7.7). In this small haplotype tree, all haplotypes are now found in one of the one-step clades, but in larger haplotype trees, there may be many interior haplotypes that are more than one mutational step from any tip haplotype. In those cases, the initial set of one-step clades are pruned off the haplotype tree, and the same nesting procedure is then applied to the more interior portions of the pruned tree. Additional rounds of pruning and nesting are repeated as needed until all haplotypes have been placed into one-step clades. Additional nesting rules are needed in case some haplotypes or clades are left stranded between two other nesting categories. Such a situation is shown in Figure 7.8. In cases such as these, the stranded haplotype or clade (haplotype C in Figure 7.8) should be nested with either the A-B clade or the D-E clade. To determine which, first see if one clade has haplotypes that are mutationally closer to the stranded clade than the other. For example, suppose in Figure 7.8 that haplotypes A, B, C, and E are actually present in the sample of genes but that D is an inferred intermediate node that is not actually present in the sample. In such a case, haplotype C is mutationally closer to clade 1-1 than to clade 1-2, and therefore C should be nested into clade 1-1. When the stranded clade is mutationally equidistant to its nesting alternatives, the stranded clade should be nested with the nesting clade that has the smallest number of observations as this nesting maximizes statistical power (Templeton and Sing 1993).

The second level of nesting uses the same rules, but the rules are now applied to one-step clades rather than haplotypes and result in "two-step clades." For example, for the elephant mtDNA tree shown in Figure 7.7, clade 1-3 is a tip one-step clade. Moving into the interior one mutational step from 1-3 connects to the interior clade 1-2. Therefore, clades 1-2 and 1-3 are nested together into a two-step clade. However, in this case clade 1-1 is left unnested.

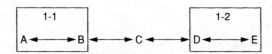

Figure 7.8. Simple haplotype network consisting of five haplotypes (A through E). The standard nesting rules start at the tips and go one mutational step inward, so haplotypes A and B are nested together into one-step clade 1-1 and haplotypes D and E are nested together into one-step clade 1-2. This leaves haplotype C unnested between clades 1-1 and 1-2.

In order not to leave any clade out of the nested design at this level, clade 1-1 is nested together with 1-2 and 1-3 to form a single two-step clade that contains all the African elephant haplotypes as there is no other possible clade with which to nest it. In the case of a larger haplotype network, this nesting procedure is repeated using two-step clades as its units, and so on until a nesting level is reached such that the next higher nesting level would result in only a single clade spanning the entire original haplotype network. In Figure 7.7, the procedure ends at the two-step level as there is only a single two-step clade.

Nested haplotype trees such as that shown in Figure 7.7 contain much temporal information. For example, within clade 1-3, haplotypes 8 and 9 are younger than haplotype 7. Note that we do not know if haplotype 8 is younger than haplotype 9 or vice versa, but we do know that both of these tip haplotypes must be younger than the interior haplotype to which they are connected. Even if the tree were unrooted, coalescent theory predicts that tips are highly likely to be younger than the interiors to which they are connected (Castelloe and Templeton 1994), so both rooted and unrooted trees contain temporal information in their nested-clade hierarchies. This temporal information extends to the higher nesting levels. For example, defining the age of a clade as the age of its oldest member, we know that clade 1-3 is younger than clade 1-2 because 1-3 is the tip relative to the 1-2 interior. In this manner, turning a haplotype tree into a series of nested clades captures much information about relative temporal orderings, although some aspects of time are left undefined. Nevertheless, even this partial information about temporal ordering can be used to analyze the spread of haplotypes and clades through space and time in a manner that does not depend upon a molecular clock or dating.

A nested-clade analysis also requires the spatial distribution of haplotypes and clades of haplotypes to be quantified. The geographical data are quantified in two main fashions (Templeton et al. 1995). The first is the **clade distance**, D_C, which measures how widespread the clade is spatially. When measuring spatial spread with geographical distances, the clade distance is determined by calculating the average latitude and longitude for all observations of the clade in the sample, weighted by the local frequencies of the clade at each location. This estimates the geographical center for the clade. Next, the great circle distance (the shortest distance on the surface of a sphere between two points on the surface) from a location containing one or more members of the clade to the geographical center is calculated, and these distances are averaged over all locations containing the clade of interest, once again weighted by the frequency of the clade in the local sample. Sometimes geographical distance is not the most appropriate measure of space. For example, suppose a sample is taken of a riparian fish species. Because rivers do not flow in straight lines and because the fish are confined in their movements to the river, the geographical distance between two sample sites on the river is not relevant to the fish; rather, the important distance in this case is the distance between the two points going only along the river. In cases such as these, the investigator should define the distances between any two sample points in the most biologically relevant fashion, and the clade distance is now calculated as the average pairwise distance between all observations of the clade, once again weighted by local frequencies.

Table 7.4 shows the geographical clade distances calculated for all the haplotypes and clades for elephant mtDNA shown in Figure 7.7. Haplotype 3 has a clade distance of 0 km, indicating that this haplotype was found in only a single location. Haplotype 2 has a clade distance of 81 km, indicating the close geographical proximity of the two locations (Amboseli and Tsavo, both located in Kenya) in which this haplotype is found. Haplotype 1 has a clade distance of 1021 km, reflecting the fact that it is found at high frequency over a large geographic area extending from Kenya to Zimbabwe. Table 7.4 gives the clade

Table 7.4. Clade and Nested-Clade Distances for Haplotypes and One-Step Clades Shown in Figure 7.7 and Their Old Minus Average Young Clade and Nested-Clade Distances in Nested Group

	Within One-Step Clades			Within Total Tree		
Haplotypes	Number in sample	D_c	D_n	One-Step Clades	D_c	D_n
1	35	1021^{L^a}	1027^{L^a}			
2	20	81^{S^a}	657^{S^a}			
3	1	0	601	1-1	884	1173^{L^a}
Old–young		944^{L^a}	373^{L^a}			
4	11	959^{L^a}	832^{L^a}			
5	16	114	249^{S^b}			
6	3	0	156^{S^b}	1-2	460^{S^a}	768^{S^a}
Old–young		862^{L^a}	598^{L^a}			
7	27	47	47			
8	1	0	126			
9	1	0	68	1-3	49^{S^a}	759^{S^c}
Old–young		47	−50		626^{L^a}	409^{L^a}

[a] Significant at 0.1% level.
[b] Significant at 5% level.
[c] Significant at 1% level.

Note: Significance relative to the null hypothesis of no geographical associations of haplotypes or clades within a nesting clade is indicated as determined by 1000 random permutations using the program GEODIS, with L and S designating significantly large and small, respectively.

distances for all the other haplotypes in Figure 7.7 and for all the one-step clades formed from them. It is also important to note that the clade distances, being an average weighted by local clade frequencies, are a function of both how geographically widespread a clade is and also the frequencies of the clade across this geographical range. For example, note in Table 7.4 that clade 1-2 has a clade distance of 460 km whereas clade 1-1 has a clade distance of 884 km. Yet, as shown in Figure 7.7, members of clade 1-2 are found at all six sampling sites in eastern Africa, whereas clade 1-1 is found in only five. The reason for this apparent discrepancy is that the five sites at which 1-1 is found cover nearly the same geographical distance as the six sites at which 1-2 is found and because clade 1-1 is found in high frequency in both Kenya and Zimbabwe, whereas clade 1-2 is common only in one area, Zimbabwe. Therefore, on the average, clade 1-2 is much more geographically concentrated than clade 1-1.

The second measure of geographical distribution of a haplotype or clade is the **nested-clade distance**, D_n, which quantifies how far away a haplotype or clade is located from those haplotypes or clades to which it is most closely related evolutionarily, that is, the clades with which it is nested into a higher level clade. For geographical distance, the first step in calculating the nested-clade distance is to find the geographical center for all individuals bearing not only members of the clade of interest but also any other clades that are nested with the clade of interest at the next higher level of nesting. For example, to calculate D_n for haplotype 1 nested in 1-1 in Figure 7.7, one first finds the geographical center of haplotypes 1, 2, and 3 pooled together (the haplotypes nested within 1-1). This is the geographical center of the nesting clade. The nested-clade distance is then calculated as the average distance that an individual bearing a haplotype from the clade of interest lies from the geographical center of the nesting clade. Once again, all averages are weighted by local frequencies. When the investigator defines the distances between sample locations, the nested-clade distance is the average pairwise distance between an individual bearing a haplotype from the clade of interest to individuals bearing any haplotype from the nesting clade that contains the clade of interest. In those cases where some other distance is used (as in our fish example), the nested-clade distance is defined as the average pairwise difference between all copies of the clade of interest to all copies of clades in the same nesting group, including itself.

Table 7.4 also gives the geographic nested-clade distance for all the elephant mtDNA haplotypes and the one-step clades into which they are nested. For example, the nested-clade distance of haplotype 1 is 1027 km, which is nearly identical to its clade distance of 1021 km. This means that the geographical center of haplotype 1 is almost the same as the geographical center for clade 1-1, resulting in very little difference between these two distance measures. In contrast, the nested-clade distances for haplotypes 2 and 3 are 657 and 601 km, respectively, indicating that the geographical centers of these two haplotypes are located far away from the geographical center of clade 1-1. In contrast, the clade distances for haplotypes 2 and 3 are 81 and 0 km, respectively. Thus, although these two haplotypes are far away from the geographical center of clade 1-1, all individuals bearing these haplotypes are found close together, with bearers of haplotype 3 being confined to a single sampling location (corresponding to a clade distance of zero).

Just as we saw above that it is dangerous to make biological inferences from a visual overlay of geography upon a haplotype tree, it is equally dangerous to make biological inferences from just the observed values of quantitative distance measurements such as those given in Table 7.4. For example, haplotype 3 has a clade distance of 0 km. However, from Table 7.4 we also see that haplotype 3 was only observed once in the entire sample. Hence, haplotype 3 has to have a clade distance of zero because a single copy of this haplotype must by necessity occur at only one location. Consequently, this small clade distance is without statistical significance. In contrast, tip haplotype 2 has a clade distance of 81 km, but this is based on 20 observations of this haplotype (Table 7.4). Thus, we can be much more confident that haplotype 2 has a highly restricted geographical range than haplotype 3. Our degree of confidence in the numbers presented in Table 7.4 can be quantified by testing the null hypothesis that the haplotypes or clades nested within a high-level nesting clade show no geographical associations given their overall frequencies. This null hypothesis is tested by randomly permuting the observations within a nesting clade across geographical locations in a manner that preserves the overall clade frequencies and sample sizes per locality (Templeton et al. 1995). After each random permutation, the clade and nested-clade distances can be recalculated. By doing this a thousand or more times, the distribution of these distances under the null hypothesis of no geographical associations for a fixed frequency can be simulated. The observed clade and nested-clade distances can then be contrasted to this null distribution, and we can infer which distances are statistically significantly large and which are significantly small. Table 7.4 also shows the significantly (at the 5% level) small and large clade and nested-clade distances as determined by the computer program GEODIS (Posada et al. 2000).

Because our biological interest in haplotype trees centers around how space and time are associated, some statistical power can be enhanced within a nesting clade by taking the average of the clade and nested-clade distances for all the tips pooled together and subtracting the tip average from the corresponding average for the older interiors. The average interior-tip difference still captures the temporal contrast of old versus young within a nesting clade but often has greater power to reject the null hypothesis of no geographical association. The interior minus tip clade and nested-clade average distances are also shown in Table 7.4.

Statistical significance is not the same as biological significance. Statistical significance tells us that the measures we are calculating are based upon a sufficient number of observations that we can be confident that geographical associations exist with the haplotype tree. However, statistical significance alone does not tell us how to interpret those geographical associations. To arrive at biological significance, we must examine how various types of recurrent gene flow or historical events can create specific patterns of geographical association.

Expected Patterns under Isolation by Distance

As detailed in Chapter 6, isolation by distance creates associations between genetic variation and geography. Because gene flow restricted by isolation by distance implies only limited movement by individuals during any given generation, it takes time for a newly arisen haplotype to spread geographically. Obviously, when a mutation first occurs, the resulting new haplotype is found only in its area of origin. With each passing generation, a haplotype lineage that persists has a greater and greater chance of spreading to additional locations via restricted gene flow. Hence, the clade distances should increase with time under a model of restricted gene flow. One of the more common types of restricted gene flow is isolation by distance (Wright 1943). Under recurrent gene flow restricted through isolation by distance, the spread of a haplotype through space occurs via small geographical movements in any given generation, resulting in a strong correlation between how widespread a haplotype (or clade) is (as measured by D_c) and its temporal position in the haplotype tree. The older the haplotype, the more widespread it is expected to be. Moreover, newer haplotypes are found within the geographical range of the haplotype from which they were derived (taking into account sampling error), and since geographical centers move slowly under this model, the clade and nested distances should yield similar patterns of statistical significance.

The expectations under isolation by distance are illustrated by the mtDNA of elephants in eastern Africa (Figure 7.7 and Table 7.4). As can be seen from Figure 7.7, the geographical range of a haplotype consistently increases as we go from younger to older haplotypes no matter where we are in the haplotype tree. The distances given in Table 7.4 quantify this visual pattern. Starting with haplotypes nested within clade 1-1, we see that the clade distance of the older interior haplotype 1 is significantly large, whereas the clade distance of the younger tip haplotype 2 is significantly small. Overall, the older interior clade distance is significantly larger than the average of the younger tip clade distances. This same pattern of statistical significance is observed in a parallel fashion with the nested-clade distances. Thus, both clade and nested-clade distances within clade 1-1 show a significant decline as one goes from older to younger haplotypes, as expected under isolation by distance. Moreover, this same pattern is repeated within clade 1-2. (Because tip haplotypes 8 and 9 were each only observed one time, there is no statistical significance to the patterns observed within clade 1-3.) As we go to the one-step clades nested within the single two-step clade for the elephant mtDNA tree, we also see both clade and nested-clade distances declining with declining age in a statistically significant fashion. Hence, the pattern of statistically significant results obtained with elephant mtDNA implies that savanna elephants over eastern Africa were genetically interconnected by recurrent gene flow that is restricted by geographical distance.

Expected Patterns under Fragmentation

Historical events can also create strong associations between haplotypes and geography. One such event is past fragmentation followed by complete or nearly complete genetic isolation. Genetic isolation means that haplotypes or clades that arose after fragmentation but in the same isolate will show concordant restricted spatial distributions that correspond to the geographical area occupied by the isolates in which they arose. Genetic isolation also means that fragmented populations behave much as separate species, but with the barrier to gene flow being only geographical in this case. Thus, the relationship of haplotype trees to geographically fragmented populations is like that shown in Figure 5.15 for the relationship between haplotype trees and species, but now with the additional restriction that isolates (the "species" shown in Figure 5.15) inhabit different geographical areas.

Not all types of fragmentation are covered by nested-clade analysis. For example, nested-clade analysis is not applicable to the case of **microvicariance**, in which an ancestral population is fragmented into numerous local isolates such that there is no to little correlation of geographical distance with genetic isolation (Templeton et al. 1995). However, nested-clade analysis is applicable to **allopatric fragmentation**, in which an ancestral population is split into two or more isolates such that each isolate consists of geographically contiguous subpopulations occupying an area that is geographically separated from the areas occupied by other isolates. Only allopatric fragmentation is considered in this chapter.

Figure 5.15 reveals that several patterns are possible under fragmentation depending on the time of fragmentation relative to the time scale of the coalescent process. If the fragmentation event lasts longer than the typical coalescent time, the haplotype tree will develop monophyletic clades of haplotypes that mark the isolates (Figure 5.15a). If the fragmentation event is much older than the coalescent time, many mutations should accumulate, resulting in clades that mark the different isolates being interconnected with branch lengths that are much longer than the average branch length in the tree. However, not all cases of fragmentation and isolation are marked by strict monophyly of haplotype tree clades (Figures 5.15a and b), and strict monophyly can also be destroyed by subsequent admixture events (Templeton 2001). Therefore, a strict monophyletic correspondence of clades with geography is a strong but not a necessary indicator of fragmentation given adequate geographical sampling. Regardless of whether there is monophyly or not, haplotypes or clades that arose after the fragmentation event cannot spread beyond the confines of the isolate in which they arose. This means that the clade distance cannot increase beyond the geographical ranges of the fragmented isolates. Even if this clade had been introduced to another isolate by some rare admixture or dispersal event, the frequency of the clade in the other isolate will generally be rare as long as isolation is the norm. Such rarely occurring admixture events therefore have little impact on clade distance.

Although the magnitude of the clade distance is severely restricted by fragmentation, the same is not true for the nested-clade distances. The nested-clade distances can suddenly become much larger than the clade distances when the nesting clade contains haplotypes or clades found in other isolates, either because the nesting clade is older than the fragmentation event itself and therefore has descendants in more than one isolate or because the nesting clade has members in more than one isolate due to the sorting of ancestral polymorphic lineages (Figures 5.15a and b). These sudden increases of nested-clade distance relative to clade distance generally occur in the higher nesting levels under fragmentation because the discrepancies generally arise within those nesting clades containing ancestral polymorphic lineages that predate the fragmentation event (Figure 5.15). In contrast, under restricted

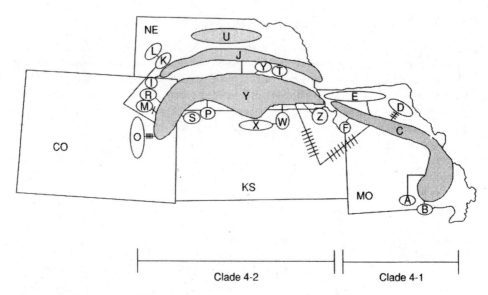

Figure 7.9. Rough geographical overlay of *A. tigrinum* mtDNA haplotype tree upon sampling locations found in Missouri (MO), Kansas (KS), Nebraska (NE), and Colorado (CO). Haplotypes are designated by letters and are enclosed within a shape that indicates their rough geographical range. Many of these haplotypes are found as polymorphisms in the same ponds but are shown as nonoverlapping in this figure for ease of pictorial representation. Consequently, the indicated geographical distributions are only approximate. A line without any tick marks indicates a single restriction site change on that branch in the haplotype tree. Lines with tick marks indicate multiple restriction sites changes, with the number of tick marks indicating the number of restriction site differences.

gene flow, as clades get older and older, the clades tend to become increasingly uniform in their spatial distributions. This means that the clade and nested-clade distances tend to converge with increasing age of the clade under restricted but recurrent gene flow models. Indeed, this is the pattern observed for the elephant mtDNA (Figure 7.7). Thus, a large increase in nested-clade distance over clade distance as one goes backward in time is yet another signature of past fragmentation.

Figure 7.9 provides an example of these signatures of past fragmentation. This figure shows a mtDNA haplotype tree from the tiger salamander, *Ambystoma tigrinum*, roughly overlaid upon the geographical area sampled (Templeton et al. 1995). There is prior evidence that these salamanders were split into an eastern and western group (formally recognized as different subspecies) during the last glaciation, which created inhospitable conditions in the upper Great Plains. After the end of the Ice Age, these two groups of salamanders have expanded their ranges, coming into contact in northwestern Missouri in historic times. Thus, this mtDNA tree can illustrate the signature of a past fragmentation event. As can be seen, there is one branch in the tree that consists of 14 mutational steps that is much longer than any other branch in the tree. Moreover, this long branch separates the mtDNA tree into two clades that have strong geographical associations. Clade 4-1 [using the nested terminology in Templeton et al. (1995)], consisting of the haplotypes labeled A through F in Figure 7.9, is found in Missouri. In contrast, clade 4-2 (consisting of haplotypes I through Z) is found from the front range of the Rocky Mountains, across the Great Plains, and just barely overlaps clade 4-1 in the northwestern corner of Missouri. Note that *within* clades 4-1 and 4-2 there is a tendency for geographical range to increase as one goes from tips to interiors, just like the elephants in eastern Africa. A nested-clade analysis reveals that much of the pattern of significant clade and nested-clade distances within 4-1 and within 4-2 are concordant with the expectations of isolation by distance. However, this increase of geographical range with time comes to an abrupt halt for clades 4-1 and 4-2. The accumulation of mutations on the long branch connecting these two clades indicates the passage of much time, but this time has passed without any increase in geographical range. In contrast, under an isolation-by-distance model, the geographical range of haplotypes and haplotype clades should increase with increasing time into the past.

Another way of having a clade not increase its geographical distribution with increasing time into the past is when there is sufficient gene flow such that the clade has spread over the entire geographical range of the species. This results in a loss of significant geographical associations at the deepest clade levels because there is sufficient gene flow and time to have homogenized the geographical distributions of these older clades. However, Figure 7.9 shows its strongest geographical associations at the deepest levels—just the opposite of the expectations for spatial homogenization via gene flow. This is yet another signature of an historical fragmentation event.

As mentioned above, a sudden increase in nested-clade distance without a corresponding increase in clade distance is one signature of fragmentation. This signature is also present in the salamander data. The clade distance of haplotype C in Figure 7.9, the oldest haplotype within 4-1, is 191 km and its nested-clade distance is 189 km. Likewise, the clade distance of haplotype Y, the oldest haplotype within 4-2, is 208 km and its nested-clade distance is also 208 km. The clade distance of clade 4-1 as a whole is 201 km and that of clade 4-2 is 207 km, indicating no further increase in geographical range for these higher level clades. Thus, within clades 4-1 and 4-2, a pattern of isolation by distance prevails that has reached its maximum geographical extent by the time we encounter the oldest haplotypes within each of these clades. The failure of the clade distances of the four-step clades to increase indicates the abrupt halt of recurrent gene flow. In contrast to the stability of the clade distances at the four-step level, the nested-clade distances of 4-1 and 4-2 increase significantly to 607 and 262 km, respectively, indicating that both of these four-step clades are marking fragmented populations located in different geographical areas. Thus, there are a variety of observable patterns that can discriminate fragmentation from isolation by distance.

Expected Patterns under Range Expansion

Another type of historical event that can create strong geographical associations is range expansion (including colonization). When range expansion occurs, those haplotypes found in the ancestral populations that were the source of the range expansion will become widespread geographically (large clade distances). This will sometimes include relatively young haplotypes or clades that are globally rare and often restricted just to the ancestral area (recall our discussion of rare haplotypes). However, some of those young, rare haplotypes in the ancestral source population can be carried along with the population range expansion, resulting in clade distances that are large for their frequency. Moreover, some haplotypes or clades that arise in the newly colonized areas (and being new mutations, tend to be tips) may have small clade distances but will often be located far from the geographical center of their ancestral range, resulting in large nested-clade distances.

An example of a well-documented range expansion is the movement of humans into the New World from Asia. Figure 7.10 shows a portion of a nested analysis of some

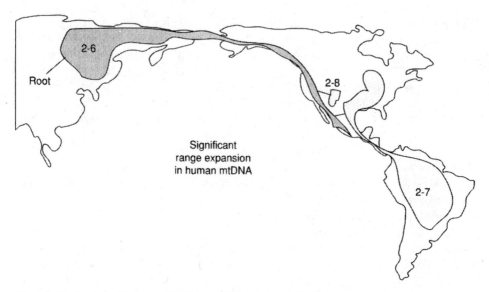

Figure 7.10. Portion of human mtDNA tree from Torroni et al. (1992) as nested in Templeton (1998c). The rough geographical ranges of three two-step clades nested within a single three-step clade are shown. Clade 2-6 is the oldest clade as inferred from outgroup rooting and is indicated by darker shading. The geographical ranges of the two tip clades, 2-7 and 2-8, are indicated by lighter shading.

human mtDNA clades that yields a significant signature of range expansion (Templeton 1998c) based upon a survey of mtDNA genetic variation in Torronni et al. (1992). This figure indicates the rough geographical distributions of three two-step clades of mtDNA haplotypes that are nested together into a single three-step clade. The oldest clade in this group, 2-6, is found primarily in northeastern Asia, but also in North and Central America. This peculiar geographical distribution caused by the human range expansion into the Americas results in a significant reversal of the clade versus nested-clade distances, with the clade distance being significantly small whereas the nested-clade distance is significantly large. This reversal reflects the fact that almost all copies of this haplotype are found in a relatively restricted region of Asia, but the range expansion caused the geographical center of the three-step clade within which 2-6 is nested to shift to the east, yielding a significantly large nested-clade distance. The mutation defining clade 2-7 apparently occurred during the expansion into the Americas, and as a result this clade was carried along with the expansion to yield a widespread distribution within the Americas. Its clade distance of 2431 km is larger than the clade distance of 2103 km for 2-6 even though 2-7 is younger than 2-6. This illustrates a reversal of the expectation under isolation by distance. The mutation defining clade 2-8 occurred after the expansion, being limited to a portion of North America with a significantly small clade distance of 644 km. However, its nested-clade distance is very large (3692 km), indicating that this relatively rare and spatially restricted haplotype clade arose far from the geographical origins of its closest evolutionary neighbors. Note in this case that it is the younger clade that has the small clade distance and large nested-clade distance. In contrast, under fragmentation, it was the older clade in the salamanders that had the small clade distance and large nested-clade distance.

These patterns associated with range expansion are distinct from those generated by either isolation by distance or fragmentation. These and the other patterns described above are

Table 7.5. Expected Patterns under Isolation by Distance, Allopatric Fragmentation, and Range Expansion

Isolation by Distance	Allopatric Fragmentation	Range Expansion
D_c tends to increase with increasing age, interior status, and nesting level.	D_c can abruptly stop increasing with increasing age, interior status, and nesting level.	Some tip or young clades can have significantly large D_c's when their ancestral clades do not.
D_n tends to increase with increasing age, interior status, and nesting level and converges with D_c with increasing nesting level.	D_n can abruptly increase with increasing age, interior status, and nesting level while D_c does not.	Some tip or young clades can have significantly large D_n's and significantly small D_c's
There is an increasing overlap of spatial distributions of clades with increasing age, interior status, and nesting level.	Older, higher level clades can show nonoverlapping or mostly nonoverlapping spatial distributions. Such clades are often connected by long branches in the tree.	Clades with the patterns described in the rows above are found in the same general area that is geographically restricted relative to the species' total distribution.

summarized in Table 7.5. As can be seen from this table, restricted gene flow, fragmentation, and range expansion can all be distinguished by a detailed examination of the patterns formed by significantly small or large clade and nested-clade distances. However, sometimes a range expansion is accompanied by a fragmentation event. For example, a species colonizes a new area during a favorable climatic period (a range expansion), but subsequent climatic change isolates the colony from the ancestral range (allopatric fragmentation). In such cases, the signature is a mixture of the fragmentation and range expansion patterns (Templeton 2004a).

Multiple Patterns in Nested-Clade Analysis

No single test statistic discriminates between recurrent gene flow, past fragmentation, and past range expansion in the nested-clade analysis. Rather, it is a pattern formed from several statistics that allows discrimination. Also, as indicated in the above discussion, many different patterns can sometimes lead to the same biological conclusion. Finally, as pointed out with the impala example (Figure 7.6), sometimes the pattern associated with significant clade and nested-clade distances is an artifact of inadequate geographical sampling. In light of these complexities (which reflect the reality of evolutionary possibilities and sampling constraints), an inference key is provided as an appendix to Templeton et al. (1995), with the latest version being available at http://darwin.uvigo.es/ along with the program GEODIS for implementing the nested-clade phylogeographic analysis.

The use of this inference key not only protects against making biological inference affected by inadequate geographical sampling but also is essential in searching out multiple, overlaying patterns within the same data set. For example, in the analysis of mtDNA restriction site variation in the salamander *A. tigrinum* given in Templeton et al. (1995), an historical fragmentation event is inferred between two named subspecies, as previously discussed. However, the "stretched-out" distributions of many of the haplotypes shown in Figure 7.9 also reflect the pattern associated with range expansion within each subspecies, with the western subspecies expanding to the east and the eastern subspecies expanding from southeastern Missouri into northwestern Missouri. Because the clades yielding the inference of range expansion are nested within clade 4-1 and within clade 4-2 (the clades marking the fragmentation event), these range expansions must have occurred after the fragmentation event. This inference is consistent with historic records that indicate that it was only recently that these two populations came into contact in northwestern Missouri. This pattern of independent range expansion within each subspecies is overlaid upon a pattern of isolation by distance occurring within each subspecies, as previously noted. Thus, the present-day spatial pattern of mtDNA variation found in these salamanders is due to the joint effects of fragmentation, range expansion, and gene flow restricted by isolation by distance. There is nothing about the evolutionary factors of restricted gene flow, fragmentation events, or range expansion events that makes them mutually exclusive alternatives. One of the great strengths of the nested-clade inference procedure is that it explicitly searches for the combination of factors that best explains the current distribution of genetic variation and does not make a priori assumptions that certain factors should be excluded. Moreover, by using the temporal polarity inherent in a nested design (or by outgroups when available), the various factors influencing current distributions of genetic variation are reconstructed as a dynamic process through time. Hence, nested-clade analysis does not merely identify and geographically localize the various factors influencing the spatial distribution of genetic variation, rather it brings out the dynamical structure and temporal juxtaposition of these evolutionary factors.

INTEGRATING HAPLOTYPE TREE INFERENCES ACROSS LOCI OR DNA REGIONS

No single locus or DNA region can capture the totality of a species' population structure and evolutionary history. Therefore, ideally many loci or gene regions should be examined. Moreover, Templeton (1998c, 2004a) examined several examples where the nested-clade analysis was applied to data sets for which much prior information existed about the organisms' recent evolutionary history. These examples reveal that the nested-clade inference key works well, but sometimes mistakes are made, mostly the failure to detect an event. The processes of mutation and genetic drift, which shape the haplotype tree upon which the nested analysis is based, are both random processes, so sometimes the expected pattern will not arise just by chance alone. Moreover, haplotype trees only contain the branches in the coalescent process that are marked by a mutational change, so any haplotype-tree-based analysis can detect only those events marked by a mutation that occurred in the right place in time and space. The occurrence of such a mutation is to some extent random, and therefore we expect that a nested-clade analysis will miss some events or processes just by chance alone due to the random nature of the mutational process. One method of overcoming these problems is **cross validation**, in which an inference must be confirmed by another data set or by subsampling the original data set. In experimental science, cross validation can be achieved by replicating the experiment. However, a particular gene has only one, unreplicated evolutionary history. What we can do is look at other genes scattered over the genome that should all share a common history to some extent; that is, we replicate evolutionary history across genes within the genome. By looking at multiple DNA regions rather than just one, we can not only cross validate our inferences but also obtain a more complete evolutionary history because multiple loci are collectively more likely than any single gene to contain some mutations at the right place in time and space to mark past events or processes.

Combining and cross validating inferences across loci can be complicated by certain sampling features that sometimes imply that all genes do *not* completely share a common evolutionary history. One of the most important sampling features for inference in a nested-clade analysis is the spatial distribution of the samples relative to the distribution of the species. Obviously, inferences are confined to those geographical areas that are sampled. If all loci or DNA regions were surveyed from the same set of sampled individuals, all regions would share this geographical constraint. However, in some cases, different laboratories study different loci or regions, each with different sampling designs. For example, Templeton (2005) subjected 25 different regions of human DNA to nested-clade analysis, including mtDNA, Y-DNA, 11 X-linked regions, and 12 scattered autosomal regions. These data sets came from many different laboratories [see the references in Templeton (2002b, 2005)] and did not cover the same geographic areas. For example, the nested-clade analysis of three DNA regions detected range expansions and restricted gene flow occurring in the Pacific region (Australia and the islands of the South Pacific). Many of the other data sets did not detect these events and gene flow patterns for the simple reason that they did not include populations from the Pacific region.

Another constraint on nested-clade inference is the number of locations sampled, which varied from 4 to 35 in the human data sets analyzed by Templeton (2002b, 2005). If the sampling is too geographically sparse, it is impossible to make many inferences. For example, one of the data sets was the mtDNA data of Torroni et al. (1992), a part of which was considered above and illustrated in Figure 7.10. Within clade 2-7 in that figure, the nested analysis detected significant range expansion within the Americas. Other parts of that data set also detected range expansion, long-distance dispersal, and isolation by distance within the Americas. None of the other human data sets analyzed detected these events. The reason was that Torroni et al. (1992) sampled many locations within the Americas, covering North, Central, and South America. Most of the other data sets had none or only one Native American population in their sample, thereby making it impossible to infer range expansion or isolation by distance *within* the American continents. Obviously, the more locations sampled within an area, the finer the spatial scale of possible inference.

A third aspect of sampling is sample size, which varied from 35 to 1544 in the human data sets analyzed by Templeton (2002b, 2005). Since biological inference is confined to statistically significant geographic associations, the number of individuals sampled can have a major effect on the number and type of inferences made. Generally, the larger the sample size, the more power in making inferences. Also, older historical events and processes that influence many populations and haplotypes are generally easier to detect than more recent events or forces because they generally affect a greater portion of the sample. For example, the smallest sample in this analysis was the X-linked *PDHA1* locus. The nested-clade analysis of this locus could only detect a statistically significant signal at the highest level of nesting, a contrast that makes use of all of the 35 individuals sampled.

The DNA region sampled also determines important properties for inference. First, the ability to detect events with haplotype trees depends upon having mutations occurring at the right time and place. For example, only the nested analysis of the mtDNA detected a fragmentation event between Native American populations and Old World populations. This event was too recent (perhaps only 14,000 years) to have much chance of being marked by appropriate mutations in the other genetic systems examined due to their slower rates of accumulating mutations. Thus, any specific haplotype tree may not detect certain events or processes due to inadequate genetic resolution (Templeton 1998c, 2004a). Second, genetic resolution also constrains what is meant by a "recurrent" evolutionary process. When gene flow restricted by isolation by distance is inferred, it does not necessarily mean that gene flow occurred every generation. Rather, such an inference only means that gene flow occurred recurrently relative to the mutation rate at that locus, and this can vary from locus to locus. Hence, the potential for a locus to detect a recurrent process can vary across loci. Third, each locus can potentially sample different periods of time. Any particular haplotype tree contains spatial/temporal information only up to but excluding the final coalescent event to the most recent common ancestral molecule. The reason for this is that at the final coalescent event there is only one lineage of DNA and therefore no chance for genetic variation to exist. Without genetic variation, no evolutionary force can produce observable effects. Therefore, even though this ancient population was undoubtedly variable and evolving, the coalescent filter through which we examine this ancient population lacks variation, and therefore we can see no evolutionary effects. This also influences what we see as recurrent versus historical. Recurrence only means the process occurred repeatedly upon the time scale of coalescence, which varies from locus to locus (Figure 5.10).

With regard to the temporal sampling period imposed by ultimate coalescence, it is important to note that different genetic systems have different expected coalescent times as a function of their pattern of inheritance (Chapter 5). In general, the expected coalescent time for a large sample of genes is 2 times the inbreeding effective number of *genes* in the population (not individuals). For an autsomal locus in a population with inbreeding effective size N_{ef} individuals, each individual bears two genes at this locus, so the inbreeding effective number of genes (as opposed to individuals) is $2N_{ef}$. Hence, the expected coalescence time is $4N_{ef}$, as previously derived in Box 5.1. Similarly, as also shown in Chapter 5, the expected coalescent time for an X-linked DNA region is $3N_{ef}$. Assuming a human population with an equal sex ratio and equal inbreeding effective sizes of both sexes, the expected coalescence time for the unisexual haploid elements (mtDNA and Y-DNA) is N_{ef}, as also shown in Chapter 5. As a result, mtDNA and Y-DNA tend to detect recent events and processes in a species' evolutionary history, while X-linked and autosomal genes primarily detect older events and processes.

The temporal sampling period can also vary greatly from gene to gene even when the genes share the same pattern of inheritance because the variance in coalescence time is much larger than the mean coalescence time (Chapter 5). Indeed, this is the case for the 25 human DNA regions included in this nested-clade phylogeographic analysis (Figure 5.10).

The variation in the temporal period being sampled by a particular locus creates a problem in integrating inferences across loci. When the nested-clade analysis is applied to a single locus, the nesting hierarchy itself gives us the relative temporal sequence of the events and gene flow patterns that we are inferring, as already mentioned for the *A. tigrinum* results. However, there is no such simple temporal ordering across loci, even if they have the same pattern of inheritance. For example, 15 human DNA regions yield as their oldest inference in the nested-clade analysis a range expansion out of Africa into Eurasia or an Old World range expansion of ambiguous origin but concordant with an out-of-Africa expansion (Templeton 2005). Although this is the oldest historical event detected within each of these 15 loci, is it the same historical event or not? That is, was there just one expansion of humans out of Africa or up to 15 separate ones detected by these loci? In order to address this question, we need to place a common time scale upon all loci.

The molecular clock method of Takahata et al. (2001) was described and used in Chapter 5 to estimate the time of the most recent common ancestral haplotype (TMRCA) for the

25 DNA regions subjected to nested-clade analysis by Templeton (2005). The results using a 6-MYA calibration for the divergence of humans and chimpanzees (Haile-Selassie 2001; Pickford and Senut 2001) are shown in Figure 5.10. Errors in the calibration point would shift the estimated times for all loci by an equal proportional amount, so such errors do not affect cross validation across loci but only affect absolute dating.

The procedure of Takahata et al. (2001) can also be used to date any monophyletic clade in a rooted haplotype tree by calculating the ratio of the average nucleotide differences within the monophyletic clade, say k, to one-half the average nucleotide difference between an outgroup species and the species of interest, in this case chimpanzees and humans. The age of a significant inference (fragmentation, range expansion, or gene flow) emerging from the nested-clade analysis is estimated as the age of the youngest monophyletic clade that contributes in a statistically significant fashion to the inference, reflecting the fact that this age represents the time in the population's history during which all haplotype lineages affected by the event or process were present but more derived haplotype lineages not affected by the event had not yet evolved (Templeton 2002b).

The method of Takahata et al. (2001) estimates the mean age of a monophyletic clade but not the error associated with the coalescent process, which can be quite large (Chapter 5). There are many ways to estimate the age of an allele or haplotype clade and its associated coalescent error (Slatkin and Rannala 2000; Rannala and Bertorelle 2001), but a simple heuristic method (Templeton 1993) gives results comparable to more complicated procedures. Assuming an unfragmented population, Tajima (1983) showed that given k, the average pairwise number of nucleotide differences among present-day haplotypes across the node in the tree to be aged, the expected time to coalescence, T, is given as

$$T = \frac{\theta(1+k)}{2\mu(1+n\theta)} \tag{7.5}$$

with a variance

$$\sigma^2 = \frac{\theta^2(1+k)}{4\mu^2(1+n\theta)^2} \tag{7.6}$$

where n is the number of nucleotides that were sampled, μ is the mutation rate, and θ is the expected nucleotide heterozygosity (this is a different but equivalent parameterization of that given by Tajima). In this case, we estimate T using the procedure of Takahata et al. (2001). Note that equation 7.5 can be substituted into equation 7.6 to yield

$$\sigma^2 = \frac{T^2}{1+k} \tag{7.7}$$

The above variance reflects the fact, as shown in Chapter 5, that the variance of the coalescent process is proportional to the square of the mean (in this case the mean is T). It also reflects the fact that our ability to obtain accurate estimates from a molecular clock depends upon the mutational resolution, in this case measured by k. If very few mutations have occurred since an event happened, equation 7.7 tells us that we will not be able to estimate the time of the event accurately using this phylogenetic approach based upon a molecular clock.

Kimura (1970) has shown that the overall distribution of time to coalescence is close to a gamma distribution (Appendix 2). The mean calculated from the procedure of Takahata et al. (2001) and the variance calculated from equation 7.7 can be used to estimate the gamma probability distribution of the haplotype clade that marks the inference of interest. In this manner, the ages of the inferred events and processes can be regarded as random variables rather than known constants.

These equations do assume that the population is not fragmented. As mentioned above, the only significant fragmentation event detected in these nested-clade analyses was between Native Americans and Old World (Africa and Eurasia) populations. This fragmentation event was detected only with mtDNA and therefore was not cross validated by any other DNA region. Nevertheless, to be cautious, equation 7.7 will only be applied to events occurring in Old World human populations as they show no evidence for any persistent fragmentation on the time scales measured by these DNA regions (Figure 5.10).

We can use these gamma distributions to cross validate inferences across loci. For example, let us return to the problem posed by the 15 DNA regions that yield an inference of range expansion out of Africa or an inference of a range expansion of ambiguous origin but concordant with an out-of-Africa expansion. All 15 of these DNA regions could be marking a single out-of-Africa range expansion, but at the other extreme each DNA region could be detecting a different out-of-Africa expansion. If all events are in fact the same event, then they should be temporally concordant in addition to being concordant in type (range expansion) and location (out of Africa into Eurasia). Temporal concordance can be tested with the gamma distributions, which are shown in Figure 7.11 for all 15 inferred out-of-Africa expansion events. Templeton (2004b) derived a maximum-likelihood ratio test (see Appendix 2) of the null hypothesis that there is only one event (all gamma distributions share a common T) versus the alternative of multiple events (each locus or DNA region is

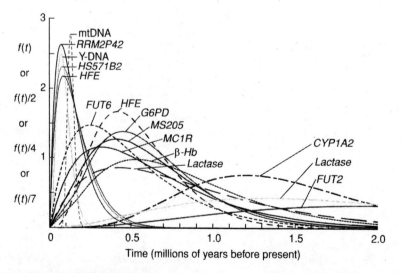

Figure 7.11. Gamma distributions for ages of range expansion events involving Africa and Eurasia, all of which are out-of-Africa events when geographical origin is unambiguous. The x axis gives the age in millions of years before present, and the y axis gives the gamma probability distribution, $f(t)$, that was fitted to the data from a particular locus or DNA region. Because the probability mass is so concentrated close to the y axis for several genes, the gamma distribution was divided by 7 for mtDNA, by 4 for Y-DNA, and by 2 for *HFE, HS571B2*, and *RRM2P4* to yield a better visual presentation. The age distributions fall into three clusters, shown by thin lines, medium lines, and thick lines, respectively [see Templeton (2005) for details on the genes].

associated with a unique T, say t_i for locus i) given by

$$G = -2\sum_{i=1}^{j}(1+k_i)\left(1 - \frac{t_i}{\hat{T}} + \ln t_i - \ln \hat{T}\right) \tag{7.8}$$

where j is the number of loci that detected the event, t_i is the time estimated for the event from locus i, k_i is the average nucleotide divergence at locus i used to estimate t_i, and

$$\hat{T} = \frac{\sum_{i=1}^{j} t_i(1+k_i)}{\sum_{i=1}^{j}(1+k_i)} \tag{7.9}$$

is the maximum-likelihood estimator of the time of the event under the null hypothesis that all loci are detecting a single event. The test statistic G is asymptotically distributed as a chi square with $j-1$ degrees of freedom. Small values of G favor the hypothesis of a single event, whereas large values favor the hypothesis of many distinct events. Applying test 7.8 to the human data yields $G = 102.54$ with 14 degrees of freedom with a p value of 3.89×10^{-15}. Therefore, the hypothesis of a single out-of-Africa expansion event is strongly rejected.

An inspection of Figure 7.11 reveals that the time distributions for the 15 events cluster into three distinct groupings. Accordingly, the null hypotheses of temporal concordance *within* each of these three groupings were tested, and in all cases there was strong concordance within ($p = 0.95$ for the most recent expansion out of Africa dated to 130,000 years ago, $p = 0.51$ for the middle expansion dated to 651,000 years ago, and $p = 0.62$ for the oldest expansion dated to 1.9 MYA). Pooling together the inferences from j homogeneous loci also results in a gamma distribution with mean and variance (Templeton, 2004a)

$$\text{Mean} = \hat{T} = \frac{\sum_{i=1}^{j} t_i(1+k_i)}{\sum_{i=1}^{j}(1+k_i)} \tag{7.10}$$

$$\text{Var}(\hat{T}) = \frac{\sum_{i=1}^{j}(1+k_i)^2 \text{Var}(t_i)}{\left(\sum_{i=1}^{j}(1+k_i)\right)^2} = \frac{\sum_{i=1}^{j}(1+k_i)t_i^2}{\left(\sum_{i=1}^{j}(1+k_i)\right)^2} \tag{7.11}$$

Combining equation 7.7 with equation 7.11, the effective number of informative mutations about the age of the event based on pooled data, k_{eff}, is given by

$$k_{\text{eff}} = \frac{\hat{T}^2}{\text{Var}(\hat{T})} - 1 \tag{7.12}$$

The log likelihood ratio test given in equations 7.8 and 7.9 can be used to test the null hypothesis that two or more times based on pooled data are the same time by using equations 7.10 and 7.12 instead of t_i and k_i for individual loci. The log likelihood ratio test of temporal homogeneity of the most recent and middle out-of-Africa expansion events yields a chi-square statistic of 40.84 with one degree of freedom with a p value of 1.66×10^{-10}. Hence, the null hypothesis of temporal concordance is strongly rejected, and the first two clusters shown in Figure 7.11 define two distinct out-of-Africa expansion events. The log likelihood ratio test of homogeneity of the middle and oldest out-of-Africa expansion events also rejects temporal concordance with a chi-square statistic of 8.85 with one degree of freedom and a *p* value of 0.0029. Hence, all three clusters shown in Figure 7.11 identify separate events that are cross validated by multiple loci. All of these expansion events are concordant with the fossil and archaeological record (Templeton 2005). The 1.9-million-year expansion event corresponds to when the fossil record indicates that ancient humans, called *Homo erectus*, first spread out of Africa into Eurasia (Gabunia et al. 2000). The second expansion corresponds to when a tool-making culture known as the Acheulean spread throughout Eurasia after first appearing in Africa. The most recent expansion corresponds to a time period in the fossil record during which many anatomically modern traits began to expand out of Africa after appearing earlier in African fossils.

Given the fossil and genetic evidence that human populations lived both in Africa and in Eurasia at least by 1.9 MYA, there is also the possibility that these human populations experienced recurrent gene flow in addition to the genetic interconnections fostered by range expansions out of Africa. Indeed, multiple inferences of gene flow were obtained from the nested-clade analysis of these 25 genes (Templeton 2005). Figure 7.12 shows the gamma distributions just for the inferences of restricted gene flow between African and Eurasian populations that were sufficiently old to date with a molecular clock. The nested-clade analysis also indicates that this gene flow was restricted by isolation by distance (gene flow due to long-distance dispersal only appears in the more recent portions of the human haplotype trees). This means that the genetic interchange between Africa and Eurasia, at least until recently, was mostly due to short-distance movements between neighboring demes (other than the three major population-level expansions). Moreover, there were no

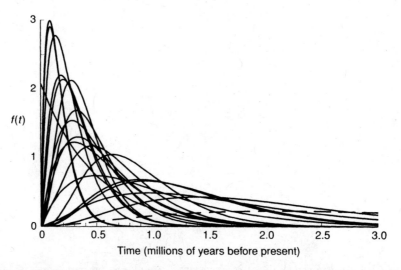

Figure 7.12. Distributions for ages of youngest clade contributing to significant inference of restricted gene flow, primarily with isolation by distance. The *x* axis gives the age in millions of years before present, and the *y* axis gives the gamma probability distribution, $f(t)$. The genes or DNA regions yielding these distributions are, as ordered by their peak values of $f(t)$ going from left to right: Xq13.1, *MSN/ALAS2, HFE, FIX, HFE, G6PD, bHb, ECP, RRM2P4, EDN, PDHA1, CYP1A2, FUT2, FUT6, FUT6, FUT2, CYP1A2, CCR5*, and *MX1* [see Templeton (2005) for details on the genes]. The curve for *MX1* is shown as a dashed line to emphasize its outlier status (Templeton 2002b). Several other inferences of restricted gene flow that were too recent to date phylogenetically are not shown and not used in the analyses.

major, persistent barriers (at least on a time scale of thousands to tens of thousands of years) between human demes in Africa and Eurasia, so copies of genes could spread throughout Africa and Eurasia via gene flow, using local demes as stepping stones to cross vast distances over many generations (the stepping-stone model of Chapter 6).

Because gene flow is a recurrent evolutionary force, there is no expectation of different inferences of restricted gene flow from the various genes to be temporally concordant, in contrast to historical events such as rapid range expansions. Here, we can cross validate the inference of gene flow in a given time period by quantifying the amount of overlap of the gamma distributions in that time period. A glance at Figure 7.12 reveals that the inferences of restricted gene flow form a continuum across the Pleistocene (which started about 2 MYA), in great contrast to the clustering shown by the 15 inferences of range expansion events (Figure 7.11). The one exception is the gamma distribution associated with the gene *MX1*. This locus is also an extreme outlier in TMRCA (Figure 5.10). Because this is the only gene that indicates gene flow in the Pliocene, this inference is not cross validated and therefore discarded. The continuum defined by the remaining genes is expected if gene flow restricted by isolation by distance were a recurrent evolutionary force throughout the Pleistocene, with no lengthy interruptions. Hence, the factors that allowed *H. erectus* to spread out of Africa nearly 2 MYA also allowed humans to go back and forth between African and Eurasia, at least in the stepping-stone sense.

The probability of the time at which recurrent gene flow commenced among populations in a geographical area can also be estimated from data on multiple DNA regions (Templeton 2004b). Let $f_i(t)$ be the gamma distribution for recurrent gene flow at time t as inferred by gene region i. Then, the probability that gene flow is *not* occurring by time t as inferred from DNA region i, say $F_i(t)$, is

$$F_i(t) = 1 - \int_t^\infty f_i(t)\,dt = \int_0^t f_i(t)\,dt \qquad (7.13)$$

The probability that gene flow is occurring by time t based on information from j loci with overlapping distributions for this inference is 1 minus the probability that gene flow is not occurring at any of the j loci; that is,

$$\text{Prob(gene flow by } t) = 1 - \prod_{i=1}^{j} \int_0^t f_i(t)\,dt \qquad (7.14)$$

Applying equation 7.14 to the distributions shown in Figure 7.12 (excluding *MX1*), the probability of recurrent gene flow constrained by isolation by distance for human populations in Africa and Eurasia reaches 0.95 by 1.46 MYA. Thus, recurrent gene flow between African and Eurasian populations goes back to the lower Pleistocene and was established at or shortly after the initial spread of *H. erectus* out of Africa.

Another issue that has been very controversial in the human evolution literature is whether or not the out-of-Africa range expansions were also replacement events in which the expanding African population drove all the Eurasian populations to complete genetic extinction. The alternative is that the expanding African populations interbred with at least some of the Eurasian populations they encountered, and hence these expansion events enhanced genetic interconnections between Africa and Eurasia through admixture.

The earliest expansion event represents the initial spread of the human lineage into Eurasia, so there could be no interbreeding or replacement, only colonization of a previously uninhabited area. However, the two more recent out-of-Africa expansion events did involve people moving out of Africa into Eurasia, which was then inhabited. Perhaps the Acheulean peoples, with their superior tools, drove the earlier Eurasian inhabitants to extinction, completely replacing them. Alternatively, the expanding Acheulean peoples could have interbred with the Eurasian populations. The Acheulean replacement hypothesis can be tested by noting that if complete replacement had occurred, there would be no genetic signatures of events or genetic processes in Eurasia that would be older than this expansion event (Templeton 2004b). This prediction stems from the simple fact that the coalescent can only have information on events and processes that affected past populations that left genetic descendants in present-day populations (Chapter 5). The Acheulean replacement hypothesis can therefore be tested by testing the null hypothesis that the gamma distribution marking the Acheulean expansion occurred at a time that was not significantly different than other Eurasian events or processes with older estimated times. To be conservative in the definition of "older," an event or process will only be regarded as older than the Acheulean expansion if its estimated age falls in the older 1% tail of the pooled gamma distribution that describes the Acheulean expansion. This 1% cutoff is calculated from the pooled Acheulean gamma distribution to be 1.0476 MYA. The nested-clade analyses identified five events or processes with estimated times older than 1.0476 MYA: the first out-of-Africa expansion at 1.9 MYA and four inferences of restricted gene flow dating from 1.25 to 3.4 MYA (Templeton 2005). The oldest of these inferences of restricted gene flow is based on the *MX1* gene that is an outlier. Excluding *MX1*, the chi square is 10.37 with four degrees of freedom, yielding $p = 0.0346$. Hence, the null hypothesis of Acheulean replacement is rejected at the 5% level of significance. The expansion of people from Africa about 0.6–0.8 MYA was therefore marked both by bringing a new culture to and by interbreeding with Eurasian populations.

We can similarly test whether or not the most recent out-of-Africa range expansion was a replacement event. The older 1% tail of the gamma distribution that describes this recent out-of-Africa expansion occurs at 0.1774 MYA. There are two older out-of-Africa expansion events and 16 inferences of restricted gene flow involving Eurasian populations that have estimated ages older than 0.1774. Excluding the inference of restricted gene flow associated with the outlier gene *MX1*, the chi square is 118.18 with 17 degrees of freedom, which yields a p value of less than 10^{-17}. Hence, the genetic data overwhelmingly reject the recent out-of-Africa replacement hypothesis.

The inferences of ancient and recurrent gene flow punctuated by major population movements out of Africa coupled with interbreeding are consistent with the fossil record. Many fossil traits display a pattern of first appearing in Africa and then spreading throughout Eurasia (Stringer 2002), whereas other traits display a pattern of regional continuity (Wolpoff et al. 2000; Wu 2004). These two patterns have sometimes been regarded as mutually exclusive alternatives in the human evolution literature. However, both patterns can be simultaneously true under a model of genetically interconnected populations and no total replacement. As long as there is genetic interchange among populations, the Mendelian mechanisms of recombination and assortment allow different traits influenced by different genes to have different evolutionary fates. Some traits could have spread due to the joint actions of gene flow, admixture, and natural selection (as will be discussed in Chapter 12), whereas other traits may not have spread as rapidly or not at all due to a lack of selection or due to local selective pressures (see Chapter 14). Recurrent gene flow and admixture therefore provide the genetic interconnections that explain *all* of the fossil trait patterns during this time period as well as current distributions of genetic variation in humans (Figure 6.7). This model of gene flow and interbreeding also explains why current genetic variation in human populations does not fit an evolutionary tree model in which different human populations are treated as distinct "branches" on an evolutionary tree. Although the human evolutionary genetic literature is filled with portrayals of human populations as branches on a tree, none of these population trees actually fit the genetic data when tested, as previously noted. Instead, patterns of genetic differentiation among current human populations fit an isolation-by-distance model much better than a tree model (Eller 1999, 2001; Templeton 1998a, 2003).

The nested-clade analyses also reveal many cross-validated recent range expansion events that represent range extensions of the human species into northern Eurasia, the Pacific region, and the Americas. There is also a cross-validated range expansion out of Asia into previously occupied regions that occurred after the last out-of-Africa expansion event that is detected with Y-DNA and β-globin. Interestingly, this recent out-of-Asia event is not detected by mtDNA despite good genetic resolution and sampling. This implies that this recent out-of-Asia expansion may have been primarily male mediated and therefore must have been characterized by interbreeding between the expanding male population with local populations in order to leave a genetic signature.

All of the inferences made from nested-clade analyses of these 25 loci can be combined into a single model for recent human evolution (Figure 7.13). What is most dramatic from Figure 7.13 is how incomplete our view of human evolution would be if based upon just one locus or DNA region. As more DNA regions are examined, more details and insights into the recent evolution of humans are sure to follow.

IMPLICATIONS OF STUDIES ON HISTORICAL EVENTS AND PAST POPULATION STRUCTURE

In this chapter we have reviewed a handful of the approaches being used and developed to investigate the recent evolutionary history of a species from molecular genetic survey data. Such studies have many implications for both basic and applied research. Historical events such as fragmentation, colonization of new areas, and founder events often serve as important components of the process of speciation (Templeton 1981). The ability to reconstruct these events therefore adds a powerful analytical tool to studies on the origin of species and indeed upon the very meaning of species (Templeton 1998b, 1999c, 2001; Templeton et al. 2000c).

Understanding a species' recent past is also important in conservation biology. For example, the nested-clade analysis of the buffalo and impala (Figure 7.5) revealed that the buffalo has a population structure dominated by recurrent gene flow throughout eastern Africa, whereas the impala is prone to fragmentation (Templeton and Georgiadis 1996). Nested-clade analyses of other ungulates living in this same landscape revealed that savanna food specialists, such as the impala, are prone to fragmentation across nonsavanna habitat barriers, whereas feeding generalists such as the buffalo or elephant are not fragmented genetically by these same habitat barriers. Therefore, the genetic population structures of the species that inhabit this African savanna landscape are determined by an interaction between the species' ecological niches and habitat patchiness at the landscape level. This information can be used in management decisions for conservation. Wise management

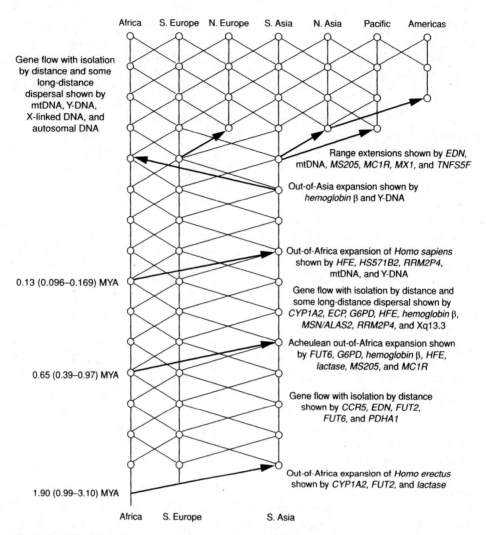

Figure 7.13. Recent human evolution as reconstructed from all cross-validated inferences made using nested-clade analyses on 25 DNA regions. Major expansions of human populations are indicated by arrows. Genetic descent is indicated by vertical lines and gene flow by diagonal lines.

decisions require more than just knowledge of the current spatial genetic pattern; they also require knowledge of the underlying processes that created the pattern. For example, suppose a manager wanted to translocate individuals from one part of the species' range to another where the native population had been extirpated or reduced to low numbers. There is no genetic rationale for not proceeding with translocation programs in the buffalo or elephant because they have recurrent genetic exchange throughout the studied range. On the other hand, the long-term fragmentation observed in the impala raises the possibility of local adaptation, so translocation should be avoided or undertaken in a cautious manner.

The observed and statistically significant patterns of gene flow and fragmentation revealed by mtDNA in these African species were not predictable from short-term studies on female dispersal behavior. For example, buffalo females are strongly philopatric and hence were thought to have little gene flow over the sample area compared to the seemingly more mobile impala. However, just the opposite occurred. This perhaps is not too surprising. As shown in Chapter 6, gene flow is not the same as dispersal, and even relatively rare events can have a major impact on gene flow patterns. Rare events are precisely the type of events that are not easily observable, much less quantifiable, by traditional dispersal studies but can leave a genetic signature.

Knowledge of recent human evolutionary history has clinical applications. As mentioned before, relatively isolated founder populations of humans have great value in studies on disease association, as do admixtures of previously differentiated populations. Moreover, disease association studies that are done without knowledge of population subdivision and history can often lead to spurious results, as discussed in Chapter 4. Therefore, a knowledge of humanity's past is essential for understanding current disease associations (Templeton 1999a). Knowledge of our evolutionary history has social implications as well. For example, "racial" categories in humans are not biological units because humans are not fragmented into distinct lineages (Figure 7.13, with the possible exception of Amerindians, who are no longer genetically isolated and whose fragmentation event inferred from mtDNA has not yet been cross validated with any other locus). Hence, the differences among human populations are dominated by geography and not racial categories (Figure 6.7), genetic differences are discordant with racial categories (Figure 7.5), and the overall level of differentiation among human populations is modest when compared to most other species (Templeton 1998a). Yet ideas of race being true biological entities have had and continue to have many social and economic impacts upon various human populations.

GENOTYPE AND PHENOTYPE

8

BASIC QUANTITATIVE GENETIC DEFINITIONS AND THEORY

In Chapter 1 we introduced the three premises upon which population genetics is founded. In Chapters 2–7, we explored the roles of premise 1 (DNA replicates) and premise 2 (DNA mutates and recombines) on the fate of genes through space and time. Many powerful evolutionary mechanisms were uncovered during this exploration of premises 1 and 2, but our discussion of evolutionary mechanisms remains incomplete until we weave the third premise into this microevolutionary tapestry. The third premise is that the information encoded in DNA interacts with the environment to produce phenotypes (measurable traits of an individual). Premise 1 tells us that genes have an existence in time and space that transcends the individual. This transcendent behavior of genes does not imply that individuals are not important. The evolutionary fate of genes *does* depend on the individuals that carry the genes. DNA cannot replicate except through the vehicle of an individual living and interacting with its environment. Therefore, how an individual interacts with the environment plays a direct role in the ability of DNA to replicate. As pointed out in Chapter 1, the fact that DNA replication is sensitive to how an individual interacts with its environment is the basis of natural selection and adaptive evolution. Premise 3 says that you inherit a response to an environment, not traits per se. Thus, the environmental context in which individuals live and reproduce cannot be ignored if we want a full understanding of evolution. In this chapter and the following two, we will lay the foundation for understanding the relationship between genotype and phenotype, a relationship that is essential to understand before turning our attention to natural selection and adaptive evolution in the final chapters of this book. We will also explore how the genetic variation that arises in a population via premise 2 influences phenotypic variation in the population.

Our basic approach to modeling evolution in the previous chapters was to start at a particular stage in the life history at one generation and then continue through the organisms' life cycle until we reach the comparable stage at the next generation. We had to specify the genetic architecture and the rules of inheritance in going from individual genotypes to

Population Genetics and Microevolutionary Theory, By Alan R. Templeton
Copyright © 2006 John Wiley & Sons, Inc.

gametes, and we had to specify the rules of population structure to go from gametes to diploid individuals in the next generation. However, up to now, we ignored the fact that after fertilization an individual zygote develops in and reacts to its environment to produce the traits that characterize it at different stages of life, including its adult traits. We now extend our models to include how individuals within each generation develop phenotypes in the context of an environment and how phenotypes are transmitted from one generation to the next. This problem of phenotypic transmission is far more complicated than Mendelian transmission of genes, as we shall see in this chapter.

SIMPLE MENDELIAN PHENOTYPES

Mendel was able to work out many of the rules for inheritance by focusing upon phenotypes that were primarily determined by a single locus and that had a simple mapping from genotype to phenotype in the environment in which the organisms normally lived and developed. Such simple genotype/phenotype systems are still the mainstay of introductory genetics textbooks because they illustrate Mendelian inheritance in a straightforward manner. A deeper examination of these "simple" Mendelian systems typically reveals more complexity in the relationship between genotype and phenotype than is usually presented in the textbooks. These simple Mendelian systems therefore provide an excellent vehicle for illustrating the issues we must deal with in going from genotype to phenotype.

Consider sickle cell anemia—the workhorse example of a simple Mendelian trait found in most genetic textbooks. Sickle cell anemia is a form of hemolytic anemia (that is, the red blood cells tend to lyse) that can lead to a variety of deleterious clinical effects and early death. Sickle cell anemia is commonly presented as a single-nucleotide trait in which one nucleotide change in the sixth codon of the gene coding for the β chain of hemoglobin produces the S allele (with valine at the sixth position) from the more common A allele (with glutamic acid at the sixth position). This single-amino-acid substitution in turn changes the biochemical properties of the resulting hemoglobin molecule, which in turn is typically presented as leading to the phenotype of sickle cell anemia in individuals who are homozygous for the S allele. Thus, sickle cell anemia is typically presented as a single-locus, autosomal genetic disease and the S allele is said to be recessive to A. Note that the word "recessive" refers to the mapping of genotype to phenotype, in this case meaning that the genotype homozygous for the S allele has the phenotype of sickle cell anemia, whereas AA and AS individuals do not have this phenotype.

Recessiveness and allied concepts in Mendelian genetics, such as dominance, codominance, **epistasis** (when the phenotype is influenced by interactions between two or more genes), and **pleiotropy** (when a single genotype influences many different traits), are not innate properties of an allele. Instead, such words apply to the genotype-to-phenotype relationship in the context of a particular environment. To illustrate this, we will now explore many different phenotypes associated with the S and A alleles and examine the genotype-to-phenotype relationship in the context of an environment.

Phenotype of Electrophoretic Mobility

Appendix 1 outlines the genetic survey technique of protein electrophoresis. Under some pH conditions, glutamic acid (associated with the A gene product) and valine (associated with the S gene product) will have a charge difference. Hemoglobin normally exists as a

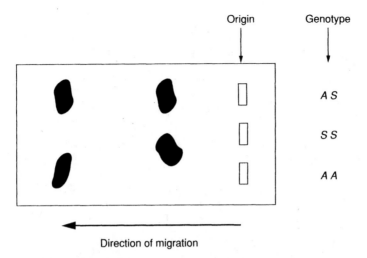

Figure 8.1. Protein electrophoresis of genotypic variation associated with *A* and *S* alleles at hemoglobin β-chain locus in humans. The buffer environment is such that the hemoglobin molecule is disassociated into its component α and β chains and the β chains have a charge difference depending upon the amino acid they have at the sixth position of the amino acid chain.

tetramer consisting of two α-globin chains and two β-globin chains. Under the appropriate buffer conditions, the hemoglobin tetramer can be disassociated into its component polypeptide chains. When protein electrophoresis is performed on blood samples in the proper pH and buffer environment, the β-hemoglobin chains display different electrophoretic mobility phenotypes in individuals differing in their genotypes with respect to the *S* and *A* alleles, as shown in Figure 8.1 (the α-globin chains, which have a distinct electrophoretic mobility from both types of β chains, are not shown). As can be seen in that figure, the *SS* and *AA* homozygous genotypes each produce a single band, but with distinct phenotypes of electrophoretic mobility. However, note that the electrophoretic phenotype of the *AS* heterozygotes is the sum of the phenotypes of the two homozygotes. Thus, the phenotype associated with each allele in homozygous conditions is fully expressed in the heterozygote as well. In this case, we would say that the *A* and *S* alleles are *codominant*.

Phenotype of Sickling

Sickle cell anemia gets its name from the fact that the red blood cells (the cells that carry the hemoglobin molecules) will distort their shape from their normally disk-shaped form to a sickle shape (Figure 8.2) under the environmental conditions of a low partial pressure of molecular oxygen (O_2). Hemoglobin is the molecule that transports oxygen from the lungs to the tissues throughout our bodies. When the oxygen is released by the hemoglobin molecule due to a low partial pressure of oxygen in the ambient environment, an allosteric change occurs in the hemoglobin molecule. This three-dimensional change causes the valine in the β^S-globin to protrude outward, where it can stick into a pocket in the three-dimensional structure of an α chain on an adjacent hemoglobin molecule. The hemoglobin molecules are tightly packed together in the red blood cells, making such a joining of β^S-globin to α-globin likely. Indeed, long strings of these joined hemoglobin molecules can assemble, and these strings in turn distort the shape of the red blood cell, leading to the trait of sickling

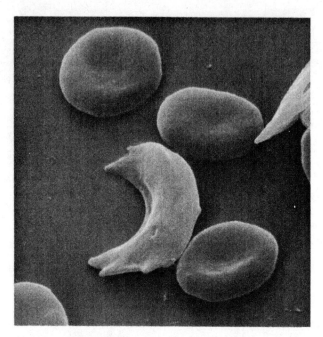

Figure 8.2. Red blood cells showing sickle cell shape (center) and normal, disk shape. Copyright © 2006 by Photo Researchers. Reprinted by permission of Eye of Science/Photo Researchers, Inc.

under environmental conditions of a low partial pressure of oxygen. Such environmental conditions occur in the capillaries when oxygen is taken up by a peripheral tissue. Such conditions also occur at high altitude, or during pregnancy (fetal hemoglobin has a higher oxygen affinity than adult hemoglobin, allowing the fetal blood to take oxygen from the mothers blood across the placenta). Because both *AS* and *SS* genotypes have β^S-globin chains in their red blood cells, both of these genotypes show the sickling trait under the appropriate environmental conditions. Therefore, with respect to whether red blood cells sickle or not, the *S* allele is *dominant*.

Phenotype of Sickle Cell Anemia

In *SS* homozygotes, there are no β^A-globin chains to disrupt the strings of joined hemoglobins, so the strings tend to be longer in *SS* individuals than in *AS* individuals. As a consequence, the distortion of the normal shape of the red blood cell tends to be more severe in *SS* individuals than in *AS* individuals under the environmental conditions of a low partial pressure of oxygen. Indeed, the distortion can be so severe in *SS* individuals that the red blood cell ruptures, losing its hemoglobin molecules and leading to anemia. The highly distorted red blood cells also pass poorly through the narrow capillaries, which is one of the places with the appropriate environmental conditions to induce sickling. The anemia and the inability of the distorted cells to move easily through the capillaries can lead to wide spectrum of phenotypic effects (Figure 8.3) known collectively as sickle cell anemia.

As Figure 8.3 shows, sickle cell anemia is actually a complex clinical syndrome with multiple phenotypic effects (pleiotropy) and much variation in expression from one individual

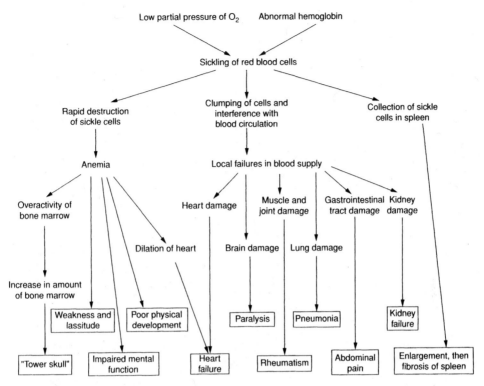

Figure 8.3. Sickle cell anemia syndrome. Modified from Fig. 12.4 in J. V. Neel and W. J. Schull, *Human Heredity* (1954). Copyright ©1954 by the University of Chicago Press.

to the next. Indeed, the symptoms vary from early-childhood death to no clinical symptoms at all (the absence of clinical symptoms in some individuals will be discussed in Chapter 12 but is due in part to epistasis with other loci). However, regardless of the exact degree of expression, the clinical syndrome of sickle cell anemia is found only in *SS* individuals. Therefore, with respect to the phenotype of sickle cell anemia, the *S* allele is *recessive*.

Phenotype of Malarial Resistance

The *AS* and *SS* genotypes show resistance to falciparum malaria (Friedman and Trager 1981), one of the most lethal forms of malaria in humans. The malarial parasite enters the red blood cells of its host. A cell infected by the falciparum but not by the other malarial parasites develops knobs on its surface, which leads to its sticking to the endothelium of small blood vessels. In such sequestered sites, sickling takes place because of the low oxygen concentration in *AS* and *SS* individuals. The infected red cell is also more acidic than the uninfected cell, a pH environment that enhances the rate of sickling. The spleen generally removes the red blood cells with the distorted sickle morphology before the parasite can complete its life cycle, leading to the phenotype of malarial resistance. Obviously, this phenotype can only be expressed under the environmental conditions of malarial infection. Because the genotypes *AS* and *SS* show this type of resistance to malaria, the *S* allele is *dominant* for the phenotype of malarial resistance.

Phenotype of Health (Viability)

The A and S alleles influence how healthy an individual is, particularly with regard to viability, the ability of the individual to stay alive in the environment. First consider an environment that does not have falciparum malaria. In such an environment, SS individuals have a substantial chance of dying before adulthood due to hemolytic anemia and other complications, as shown in Figure 8.3. This is particularly true in areas that have poor health care. Because AA and AS individuals do not suffer from sickle cell anemia, the S allele is *recessive* for the phenotype of health in a nonmalarial environment.

Now consider health in a malarial environment. The SS individuals have poor health because they suffer from sickle cell anemia, but the AA individuals also have poor health because of falciparum malaria and have a high probability of childhood death. However, the AS individuals do not suffer from sickle cell anemia, and they have some resistance to malaria. Therefore, in an environment with falciparum malaria, the S allele is *overdominant* with respect to the phenotype of health because the AS heterozygotes have superior viability to either homozygote class.

Note that the S allele can be dominant, recessive, codominant, or overdominant depending upon which phenotype is being measured and the environment in which the measurement is made. Although we frequently use such expressions as a "dominant allele" or "recessive allele," such expressions are merely a linguistic short-hand for describing the genotype–phenotype relationship in a particular environmental context. Dominance, recessiveness, and so on, are NOT intrinsic properties of an allele. Context is always important when dealing with the relationship between genotype and phenotype.

NATURE VERSUS NURTURE?

Does nature (the genotype) or nurture (the environment) play the dominant role in shaping an individual's phenotype? From premise 3, we can see that this is a false issue. Phenotypes emerge from the *interaction* of genotype and environment. It is this interaction that is the true causation of an individual's phenotype, and it is meaningless to try to separate genotype and environment as distinct causes for the individual's phenotype. However, in population genetics, we are often concerned with a population of individuals with much phenotypic variability. Accordingly, in much of population genetics our concern centers on causes of phenotypic *variation* among individuals within the deme rather than the causation of any single individual's phenotype. Causes of phenotypic variation in a population are quite distinct from causes of individual phenotypes, and the nature/nurture issue is limited only to causes of variation. We will illustrate these statements by considering yet another simple Mendelian genetic disease: phenylketonuria (PKU).

The enzyme phenylalanine hydroxylase catalyzes the amino acid phenylalanine to tyrosine and is coded for by an autosomal locus in humans. Several loss-of-function mutations have occurred at this locus (Scriver and Waters 1999), and homozygosity for loss-of-function alleles is associated with the clinical syndrome known as phenylketonuria. Let k designate the set of loss-of-function alleles and K be the set of functional alleles at this locus. Because kk homozygotes cannot catalyze phenylalanine, they have a buildup of phenylalanine, a common amino acid in most foods. The degradation products of phenylalanine, such as phenylketones, also build up in kk homozygotes. The phenylketones are typically found at high levels in the urine of the kk homozygotes, an easily scored phenotype that gives the syndrome its name. However, there are other phenotypes associated with this syndrome. For example, *kk* homozygotes tend to have a lighter skin color than most individuals that share their ethnic background because one of the main pigments in our skin, melanin, is synthesized from tyrosine, which cannot be produced from phenylalanine in *kk* homozygotes. However, the reason why PKU has attracted much attention is the tendency for *kk* homozygotes to suffer from mental retardation. As with sickle cell anemia, there is tremendous heterogeneity in the phenotype of mental ability among *kk* homozygotes that in part is due to epistasis with other loci (Scriver and Waters 1999). However, we will ignore epistasis for now and just treat PKU as a single-locus, autosomal recessive genetic disease.

The primary source of phenylalanine is our diet. The *kk* homozygotes typically have normal mental abilities at birth. While in utero, the *kk* homozygote is not eating but is obtaining its nutrients directly from the mother. Typically, the mother is a carrier of PKU with the genotype *Kk*, which means that she can catalyze phenylalanine to tyrosine. After birth, the *kk* homozygote cannot metabolize the phenylalanine found in a normal diet, and mental retardation will likely soon develop. If a baby with the *kk* genotype is identified soon after birth and placed on a diet with low phenylalanine, the baby will usually develop a normal level of intelligence. Thus, the same *kk* genotype can give radically different phenotypes depending upon the dietary environment. Because of the responsiveness of the *kk* genotype to environmental intervention, many countries require genetic screening of all newborns through a simple urine test to detect the *kk* homozygotes (Levy and Albers 2000). The PKU screening program has been successful in greatly reducing the incidence of mental retardation due to *kk* genotypes.

Individuals who are *kk* are generally advised to maintain a low-phenylalanine diet throughout their life. However, phenylalanine is such a common component of most protein-bearing foods that such diets are highly restrictive and more expensive than normal diets. Moreover, the beneficial effects of the low-phenylalanine diet are strongest in children. Once the brain has fully developed, *kk* individuals often do not perceive much of an impact of diet on their mental abilities. As a result, compliance with the diet tends to drop off with age. Note that the nature of the interaction of genotype (*kk*) with environment (the amount of phenylalanine in the diet) shifts with ontogeny (development) of the organism. Thus, genotype-by-environment interactions are not static even at the individual level.

Prior to the successful screening program, few *kk* women reproduced due to their severe retardation, but with treatment, many *kk* women married and had children. However, many of these women were now eating a normal diet and hence had high levels of phenylalanine and its degradation products in their blood. The developing fetus, usually with the genotype *Kk*, which typically develops normally, was now exposed to an in utero environment that inhibited normal brain development. Such *Kk* children of *kk* mothers on a normal diet were born with irreversible mental retardation. Is the phenotype of mental retardation due to nature or nurture in this case? Obviously, both are important. One cannot predict the phenotype of an individual on the basis of genotype alone; the genotype must be placed in an environmental context before prediction of phenotype is possible. Thus, what is inherited here is not the *trait of mental retardation* but rather the *response* to the dietary and maternal environments.

Now consider the disease of scurvy. Ascorbic acid (vitamin C) is essential for collagen synthesis in mammals. Most mammals can synthesize ascorbic acid, but all humans are homozygous for a nonfunctional allele that prevents us from synthesizing ascorbic acid. As a result, when humans eat a diet lacking vitamin C, they begin to suffer from a collagen deficiency, which leads to skin lesions, fragile blood vessels, poor wound healing, loss of teeth, and eventually death if the vitamin-deficient diet persists too long. Thus, humans have an inherited response to the dietary environment that can lead to the disease of scurvy.

Both scurvy and PKU therefore have a similar biological causation at the individual level. Both diseases result from the way in which an individual homozygous for a loss-of-function allele responds to a dietary environment. Yet, PKU is typically said to be a "genetic" disease, whereas scurvy is said to be an "environmental" disease. Phenylketonuria is considered a *genetic* disease because although the disease arises from the interaction of both genes and environment, the environmental component of the interaction is nearly universal (phenylalanine is in all normal diets) whereas the genetic component of the interaction, the *kk* genotype, is rare. As a consequence, when PKU occurs in a human population, it is because the person has the *kk* genotype since virtually all of us have a diet that would allow the PKU response given a *kk* genotype (at least until the screening program). Hence, the phenotype of PKU is *strongly associated* with the *kk* genotype in human populations. The condition of scurvy is also the result of an interaction between genes and environment, but in this case the genetic component of the interaction is universal in humans. However, the environmental component of the interaction of having a diet without sufficient amounts of ascorbic acid is rare. Therefore, the phenotype of scurvy is *associated* with a diet deficient in vitamin C in human populations.

When we ask the question, what causes PKU or scurvy, our answer is that an interaction of genes with environments is the cause of these diseases. However, when we ask the question, what causes some people to have PKU or scurvy and others not, we conclude that genetic variation is the cause of phenotypic variation for PKU whereas environmental variation is the cause of phenotypic variation for scurvy. As the PKU/scurvy example illustrates, the interaction of genes with environment creates a confoundment between frequency and apparent causation in a *population* of phenotypically variable individuals. When causation at the individual level arises from an interaction of components, then the *rarer* component at the level of the population is the one with the *stronger* association with phenotypic variation. Scurvy is an environmental disease because the dietary environment is rare but the genotypic component is common; PKU is a genetic disease because the dietary environment is common but the genotypic component is rare.

The dependency of causation of variation upon frequency in a population is illustrated by a hypothetical example in Table 8.1 in which a disease arises from the interaction of two independently varying components. In particular, the disease only occurs when the first component has state A1 and the second component has state B1. In this population, state A1 is relatively common, having a frequency of 0.9, and state B1 is relatively rare, having a frequency of 0.1. As shown in Table 8.1, the frequency of the disease in the population is given by the product $(0.9)(0.1) = 0.09$, and the remaining 91% of the population has no disease. Now suppose that a survey is done in this population on component A. Given that an individual has state A1, then that individual will only show the disease when the

Table 8.1. Hypothetical Disease Arising from Interaction of Two Factors

	B1 (0.1)	B2 (0.9)
A1 (0.9)	DISEASE (0.09)	No disease (0.81)
A2 (0.1)	No disease (0.01)	No disease (0.09)

Note: Component A has two trait states in the population, A1 with frequency 0.9 and A2 with frequency 0.1. Component B has two trait states in the population, B1 with frequency 0.1 and B2 with frequency 0.9.

individual also has trait B1, but we assume that trait B is not being monitored. Therefore, the probability of the disease in individuals with state A1 is the same as the frequency of trait B1, that is, 0.1. Hence, the frequency of the disease in individuals with trait A1 is just slightly above the overall incidence of the disease in the general population of 0.09. Thus, such a study would conclude that A1 is at best a minor cause of variation in the disease. In contrast, suppose a survey is conducted on component B but not A. The frequency of the disease in individuals with trait B1 is equal to the probability that the individual also has trait A1, that is, 0.9. Hence, trait B1 is strongly associated with the disease in this population. Knowing that a person has trait B1, we would conclude that he or she has a 10-fold higher risk for the disease than the general population incidence of 0.09, in great contrast to the trivial apparent effect of trait A1. However, we know in this hypothetical example that A1 and B1 are equally important in actually causing the disease in any individual. Although A1 and B1 jointly cause the disease in affected individuals, A1 is not a good predictor of which individuals are at high risk for the disease whereas B1 is a good predictor of disease risk. Cause of variation for a disease risk in a population is therefore a different concept than the cause of the disease in an individual.

When dealing with populations, we must therefore change our focus from *causation of phenotypes* to *causes of phenotypic variation*. Nature and nurture can never be separated as a cause of a phenotype (premise 3), but nature and nurture can be separated as causes of variation in a population, the focus of the remainder of this chapter.

FISHERIAN MODEL OF QUANTITATIVE GENETICS

The phenotypes considered in the previous section fell into discrete categories. Such discrete phenotypes were used by Mendel in his studies that uncovered the basic rules of inheritance. However, many phenotypes display continuous variation, for example, height or weight in humans. One of the major problems after the rediscovery of Mendelism at the beginning of the twentieth century was how to reconcile the discrete genotypes of Mendelian genetics with the continuous, quantitative phenotypes that often were of more practical utility in agriculture and medicine. We will now examine a model developed by R. A. Fisher (1918) that extends Mendelian genetics to cover quantitative phenotypes.

Fisher and others realized that there were two major ways in which discrete genotypes could map onto continuous phenotypes. First, there could be variation in the environment that interacts with genotypes to produce a range of phenotypes in individuals who share a common genotype that overlaps with the range of phenotypes associated with other genotypes (Figure 8.4). Second, the phenotypic variation could be associated with genetic variation at many loci. As the number of loci increases, the number of discrete genotypes becomes so large that the genotypic frequency distribution approximates a continuous distribution, so even a simple genotype-to-phenotype mapping would produce a nearly continuous phenotypic distribution (Figure 8.5). In general, Fisher regarded most phenotypic variation as arising from *both* underlying environmental and multilocus genotypic variation.

Just as we characterized our population of genotypes by genotype frequencies, we will characterize our population of phenotypes by phenotype frequencies. For quantitative phenotypes, a continuous probability distribution is used to describe the phenotype frequencies rather than the discrete probabilities that we had used in previous chapters to describe the genotype frequencies. Under the Fisherian model in which many variable factors, both genetic and environmental, contribute to phenotypic variation, we would expect many

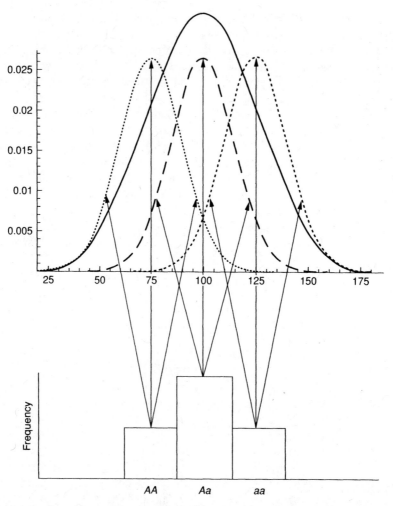

Figure 8.4. Continuous phenotypic distribution produced by interactions between genetic variation at single locus with two alleles (*A* and *a*) with environmental variation. The locus is assumed to have an allele frequency of 0.5 and obeys Hardy–Weinberg, with the height of the histogram at the bottom of the figure indicating the genotype frequencies. The arrows coming from a genotype block in the histogram indicate that the same genotype can give rise to many phenotypes depending upon the environment. A thin solid line indicates the phenotypic distribution arising from the interaction of genotype *AA* with environmental variation, a thin dashed line the phenotypic distribution associated with *Aa*, and a thin dotted line with *aa*. The thick solid line indicates the overall phenotypic distribution in the population that represents a mixture of the three genotypic specific distributions as weighted by the genotype frequencies.

phenotypic distributions to fit a normal distribution due to the central limit theorem (see Appendix 2). This normal expectation is often found empirically to be a good approximation to the population distribution of many quantitative traits. Moreover, even when a trait does not fit a normal distribution, mathematical transformations can often be used to "normalize" the trait. Therefore, in this chapter we will assume that any quantitative trait is distributed normally in the population.

The normal distribution has many optimal properties, the most important of which for the current discussion is that only two parameters, the mean (population average) and the

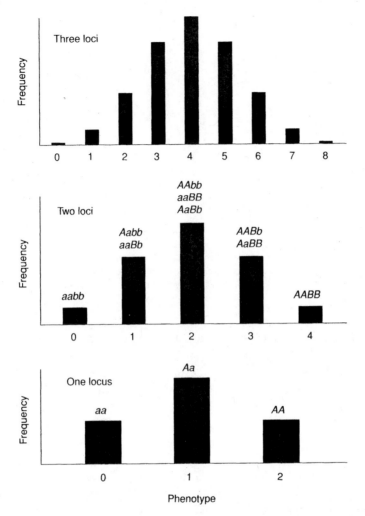

Figure 8.5. Approximate continuous phenotypic distribution produced by increasing number of loci affecting phenotypic variation. A simple genotype-to-phenotype model is assumed in which each allele indicated by a lowercase letter contributes 0 to the phenotype, and each allele indicated by a capital letter contributes +1, with the overall phenotype simply being the sum over all alleles and all loci. At the bottom of the figure, the phenotypic distribution associated with a one-locus, two-allele model with equal allele frequencies is shown, the middle panel shows the phenotypic distribution associated with a two-locus, two-allele model with equal allele frequencies, and the top panel shows the phenotypic distribution associated with a three-locus, two-allele model with equal allele frequencies (the genotypes associated with the phenotypic categories are not indicated in that case). As the number of loci increases, the phenotypic distribution approximates more and more that of a continuous distribution.

variance (the average squared deviation from the mean), fully describe the entire distribution. In particular, the normal distribution is symmetrically centered about the mean, μ as shown in Figure 8.6. The variance, σ^2, describes the width of the distribution about the mean (Figure 8.6). When the phenotypes of a population follow a normal distribution, the individuals may show any phenotypic value, but the phenotypes most common in the population are those found close to the central mean value. The frequency of a phenotype drops off with

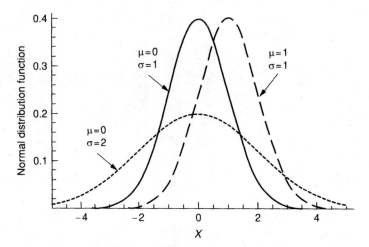

Figure 8.6. Three normal distributions that show role of mean and variance upon shape and position of distribution. The x axis gives the possible values for a random variable, X, and the y axis is the corresponding value of the normal distribution function for the specified mean and variance.

increasing deviations from the mean, with the rate of dropoff of the frequency depending upon the variance. Because it takes only two numbers, the mean and the variance, to fully describe the entire phenotypic distribution, Fisher's model of quantitative variation has two basic types of measurements: those related to mean or average phenotypes and those related to the variance of phenotypes. Both types of measurements will be considered, starting with the measurements related to the mean.

QUANTITATIVE GENETIC MEASURES RELATED TO MEAN

The most straightforward measure related to the mean is the mean phenotype itself, μ. (All the quantitative genetic definitions used in this chapter are summarized in Table 8.2.) The **mean** μ is the average phenotype of all individuals in the population. Alternatively, μ can be thought of as the mean phenotype averaged over all genotypes and all environments. Let $P_{ij,k}$ be the phenotype of a diploid individual with genotype ij (with ij referring to gamete i and gamete j that came together to form genotype ij) living in a specific environment k in a population of size n. Then the mean phenotype μ is given as

$$\mu = \sum_{ij} \sum_{k} \frac{P_{ij,k}}{n} \tag{8.1}$$

where the summation is over all genotypes ij and over all environments k, which is equivalent to summing over all n individuals. For example, Sing and Davignon (1985) scored the phenotype of the amount of cholesterol (measured in milligrams, mg) found in a deciliter (dl) of blood serum in 102 Canadian men (Table 8.3). Summing the values of total serum cholesterol over all 102 men and dividing by 102 yields a mean phenotype of $\mu = 174.2$ mg/dl.

To investigate the role of genetic variation as a source of phenotypic variation, it is necessary to relate mean phenotypes to genotypes. This is done through the **genotypic**

Table 8.2. Quantitative Genetic Definitions

Name	Symbol	Mathematical Definition/Meaning
Phenotype of a diploid individual	$P_{ij,k}$	Value of trait for individual with genotype ij living in specific environment k
Population size	n	Size of population (or sample) being measured
Mean	μ	Average phenotype of all individuals in population, $\mu = \sum_{ij} \sum_k P_{ij,k}/n$
Genotype number	n_{ij}	Number of individuals in population with genotype ij
Genotypic value of genotype ij	G_{ij}	Mean phenotype of all individuals sharing genotype ij, $G_{ij} = \sum_k P_{ij,k}/n_{ij}$
Genotypic deviation of genotype ij	g_{ij}	Deviation of genotypic value of genotype ij from mean of total population, $g_{ij} = G_{ij} - \mu$
Environmental deviation	e_k	Residual phenotypic deviation inexplicable with genetic model being used, $e_k = P_{ij,k} - ij$
Genotype frequency of ij	t_{ij}	Frequency of unordered genotype ij
Average excess of gamete type i	a_i	Average genotypic deviation of all individuals who received copy of gamete type i, $a_i = \dfrac{t_{ii}}{p_i} g_{ii} + \sum_{j \neq i} \dfrac{\frac{1}{2} t_{ij}}{p_i} g_{ij}$
Average effect of gamete type i	α_i	Slope of least-squares regression of genotypic deviations against number of gametes of type i borne by genotype
Breeding value or additive genotypic deviation of genotype ij	g_{aij}	Sum of average effects of gametes borne by individuals with genotype ij, $g_{aij} = \alpha_i + \alpha_j$
Dominance deviation of genotype ij	d_{ij}	Difference between genotype's genotypic deviation and its additive genotypic deviation in one-locus model, $d_{ij} = g_{ij} - g_{aij}$
Phenotypic variance	σ_P^2	Variance of phenotype over all individuals in population, $\sigma_P^2 = \sum_{ij} \sum_k (P_{ij,k} - \mu)^2 / n$
Genetic variance	σ_g^2	Variance of genotypic deviations, $\sigma_g^2 = \sum_{ij} t_{ij} g_{ij}^2$
Environmental variance	σ_e^2	Variance left over after explaining as much as possible of phenotypic variance with genetic variance, $\sigma_e^2 = \sum_k r_k e_k^2$, where r_k is the frequency of environment k.
Broad-sense heritability	h_B^2	Ratio of genetic variance to phenotypic variance, $h_B^2 = \sigma_g^2 / \sigma_P^2$
Additive genetic variance	σ_a^2	Variance in additive genotypic deviations, $\sigma_a^2 = \sum_{ij} t_{ij} g_{aij}^2$
(Narrow-sense) heritability	h^2	Ratio of additive genetic variance to phenotypic variance $h^2 = \sigma_a^2 / \sigma_P^2$
Dominance variance	σ_d^2	Residual left over after subtracting off additive genetic variance from genetic variance, $\sigma_d^2 = \sigma_g^2 - \sigma_a^2$

Table 8.3. Genotypes at *ApoE* Locus in Population of 102 Canadians, and Their Numbers, Frequencies, and Genotypic Values and Deviations for Phenotype of Total Serum Cholesterol

Genotype	$\varepsilon 2/\varepsilon 2$	$\varepsilon 2/\varepsilon 3$	$\varepsilon 2/\varepsilon 4$	$\varepsilon 3/\varepsilon 3$	$\varepsilon 3/\varepsilon 4$	$\varepsilon 4/\varepsilon 4$	Sum or Mean
Number	2	10	2	63	21	4	102
Frequency	0.020	0.098	0.020	0.618	0.206	0.039	1.000
G_{ij} (mg/dl)	136.0	161.4	178.1	173.8	183.5	180.3	174.2
g_{ij} (mg/dl)	−38.2	−12.8	3.9	−0.4	9.3	6.1	0.0

Note: All means are weighted by genotype frequencies.

value of genotype ij, G_{ij}, the mean phenotype of all individuals sharing genotype ij. Letting n_{ij} be the number of individuals in the population with genotype ij, then

$$G_{ij} = \sum_k \frac{P_{ij,k}}{n_{ij}} \tag{8.2}$$

The frequency of genotype ij in the population is given by n_{ij}/n, so equation 8.1 can be rewritten as

$$\mu = \sum_{ij}\sum_k \frac{P_{ij,k}}{n} = \sum_{ij}\frac{n_{ij}}{n}\sum_k \frac{P_{ij,k}}{n_{ij}} = \sum_{ij}\frac{n_{ij}}{n} G_{ij} \tag{8.3}$$

Hence, the overall mean phenotype is the average of the genotypic values as weighted by their genotype frequencies.

Sing and Davignon (1985) also used protein electrophoresis on blood samples from these 102 Canadians to score genetic variation at the autosomal *Apoprotein E (ApoE)* locus (see Chapter 5), a locus that codes for a protein that can form soluble complexes with lipids such as cholesterol so that they may be transported in the serum. The ApoE protein also binds with certain receptors on cells involved in the uptake of cholesterol from the blood. Protein electrophoresis reveals three alleles that code for different amino acid sequences, labeled $\varepsilon 2$, $\varepsilon 3$, and $\varepsilon 4$, which in turn define six possible genotypes. Table 8.3 gives these six genotypes, along with the number of individuals bearing each of these six genotypes and the genotype frequencies. Table 8.3 also gives the mean phenotypes (genotypic values) of total serum cholesterol found for each of the six genotypes that are obtained by adding the phenotypes of all individuals that share a particular genotype and then dividing by the number of individuals with that genotype. When these genotypic values are multiplied by their respective genotype frequencies and the products summed over all genotypes, the overall mean phenotype of 174.2 is obtained. This method of calculating μ results in the same value obtained by summing the phenotypes of all individuals in the sample and dividing by the sample size, as expected from equation 8.3. Figure 8.7 shows the normal distributions fitted to the genotypic values and phenotypic variances within each genotype class. As can be seen, the phenotypic distributions of the different genotypes are centered at different places, reflecting the differences in genotypic values given in Table 8.3.

Fisher's focus was upon genetic variation being associated with phenotypic variation. He therefore was not concerned with the actual mean phenotype of the total population. Fisher therefore calculated the overall mean phenotype μ only so that he could discard it in order

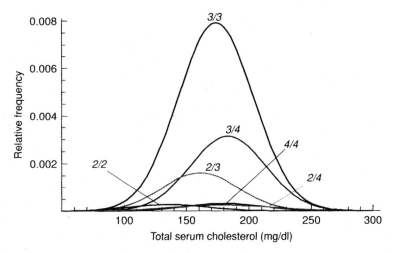

Figure 8.7. Normal distributions with observed means and variances for six genotypes defined by $\varepsilon 2$, $\varepsilon 3$, and $\varepsilon 4$ alleles at *ApoE* locus in population of 102 Canadians. The area under each curve is proportional to the genotype frequency in the population. Based on data from Sing and Davignon (1985).

to focus upon phenotypic differences associated with genotypic differences. He eliminated the effect of the overall mean phenotype simply by subtracting it from a genotypic value to yield the **genotypic deviation, g_{ij}, of genotype ij**, given by the equation

$$g_{ij} = G_{ij} - \mu \qquad (8.4)$$

Equation 8.3 shows that the average genotypic value (when weighted by the genotypic frequencies) is μ, and using equation 8.4, the average genotypic deviation (when weighted by the genotype frequencies) is

$$\text{Average}(g_{ij}) = \sum_{ij} \frac{n_{ij}}{n}(G_{ij} - \mu) = \sum_{ij} \frac{n_{ij}}{n} G_{ij} - \mu \sum_{ij} \frac{n_{ij}}{n} = \mu - \mu = 0 \qquad (8.5)$$

Hence, the average genotypic deviation of the population is always zero. This is shown in Table 8.3 for the *ApoE* example. All of Fisher's subsequent statistics related to genes and genotypes are based upon the genotypic deviations and hence are mathematically invariant to μ. In this manner, Fisher could focus exclusively upon the phenotypic *differences* between genotypes in the population irrespective of the value of the actual mean phenotype.

Fisher realized that not all differences in phenotypes among individuals are due to genotypes. He therefore gave a simple model of an individual's phenotype as

$$P_{ij,k} = \mu + g_{ij} + e_k = G_{ij} + e_k \qquad (8.6)$$

where e_k is the **environmental deviation**. Fisher's choice of terminology for e_k was unfortunate and has led to much confusion. The term e_k is simply whatever one needs to add on to the genotypic deviation of an individual in order to obtain the value of that individual's phenotype. This is shown graphically in Figure 8.8. Any factor that causes an

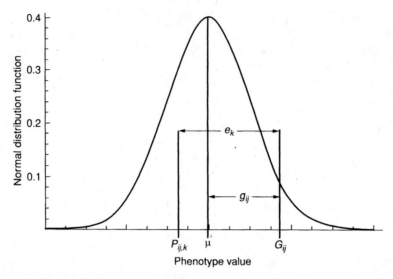

Figure 8.8. Pictorial representation of how phenotype of an individual with genotype ij is partitioned into genotypic deviation g_{ij} and environmental deviation e_k, which is number that needs to added to genotypic value G_{ij} to produce individual's phenotypic value $P_{ij,k}$.

individual's phenotype to deviate from its genotypic mean (the genotypic value in Fisher's terminology) therefore contributes to e_k. For example, many genes are known to affect cholesterol levels in addition to *ApoE* (Sing et al. 1996; Templeton 2000). Therefore, some of the phenotypic variation that occurs within a particular genotypic class (as shown by the curves in Figure 8.7) is due to genetic variation at other loci, yet these genetic factors would be treated as "environmental" deviations in a genetic model that incorporates only the variation at the *ApoE* locus. Moreover, an individual's phenotype emerges from how the genotype responds to the environment, making the roles of genotype and environment biologically inseparable at the individual level. However, equation 8.6 gives the appearance that an *individual's* phenotype $P_{ij,k}$ can be separated into a genetic component (g_{ij}) and an environmental component (e_k). This is a false appearance. Note that μ, g_{ij}, and G_{ij} are all some sort of average over many individuals. Thus, equation 8.6 is not really a model of an individual's phenotype; rather, it is a population model of how an individual's phenotype is placed in a population context. Because e_k is just whatever we need to add on to G_{ij}, a quantity only defined at the level of a population, e_k is also only defined at the level of a population. There is no way of mathematically separating genes and environment for a specific individual; equation 8.6 only separates causes of variation at the *population level*. Equation 8.6 separates phenotypic variation into two causes: the phenotypic variation due to genotypic variation in the population and whatever is left over after the genotypic component has been considered. Therefore, e_k is more accurately thought of as a residual population deviation that is inexplicable by the genetic model being used. It does not truly measure environmental causes of variation (although it is certainly influenced by them). However, Fisher's terminology is so ingrained in the literature that we will continue to call e_k the environmental deviation, although keep in mind that it is really only the residual deviation inexplicable with the genetic model being used.

Fisher defined e_k as a residual term because this results in some useful mathematical properties. Recall that μ can be interpreted as the mean phenotype averaged over all genotypes and all "environments" (actually residual factors). Hence, averaging both sides of equation 8.6 across both genotypes and environments yields

$$\sum_{ij}\sum_{k}\frac{P_{ij,k}}{n} = \sum_{ij}\sum_{k}r_k\frac{n_{ij}}{n}G_{ij} + \sum_{ij}\sum_{k}r_k\frac{n_{ij}}{n}e_k$$

$$\mu = \sum_{ij}\frac{n_{ij}}{n}G_{ij}\sum_{k}r_k + \sum_{ij}\frac{n_{ij}}{n}\sum_{k}r_k e_k$$

$$\mu = \mu + \sum_{k}r_k e_k \Rightarrow \sum_{k}r_k e_k = 0 \tag{8.7}$$

where r_k is the frequency of residual factor k in the population. The term $\Sigma r_k e_k$ is the average environmental deviation in the population, and equation 8.7 shows that it must be zero. Therefore, by using the device of deviations from the mean, Fisher has eliminated the mean μ from all subsequent calculations. Subtracting off μ also results in both the genetic factors and residual factors having a mean of zero.

So far, all of the parameters of Fisher's model simply give an alternative description of the phenotypes of individuals. Fisher also addressed the more complicated problem of how phenotypes are transmitted from one generation to the next. This is a difficult problem because under Mendelian genetics the genotypes are not directly transmitted across generations; rather, haploid gametes transmit the genetic material. The essential problem faced by Fisher was that although the gametes represent the bridge across generations, the haploid gametes do not normally express the phenotypes found in diploid individuals. Therefore, to crack the problem of the role of genetic variation on *transmission* of phenotypic variation, Fisher had to come up with some way of assigning a phenotype to a haploid gamete even though only diploid individuals actually express the phenotype. Indeed, the core to understanding Fisher's model is to *think like a gamete!* That is, one must look at the problem of the transmission of phenotypes from one generation to the next from the gamete's perspective. Fisher came up with two ways of quantifying phenotypic transmission from a gamete's point of view.

Fisher's first measure of the phenotype associated with a gamete is the **average excess of gamete type** i, a_i, which is the average genotypic deviation of all individuals who received a copy of gamete type i. Consider drawing a *known* gamete of type i from the gene pool. Then draw a second gamete according to the normal rules defined by population structure. Then, the average excess of gamete type i is the expected phenotype of the genotypes resulting from pairing the known gamete i with the second gamete minus μ. The average excess mathematically is called a conditional expectation or conditional average. Specifically, the average excess of gamete type i is the conditional average genotypic deviation given that at least one of the gametes an individual received from its parents is of type i.

To calculate a conditional expected genotypic deviation, we first need the conditional genotype frequencies given that one of the gametes involved in the fertilization event is of type i. For the one-locus model, let p_i be the frequency of gamete i in the gene pool. As seen in previous chapters, p_i represents the probability of drawing at random a gamete of type i out of the gene pool. Now, let t_{ij} be the frequency of diploid genotype ij in the population, where j is another possible gamete that can be drawn from the gene pool. Gamete j is drawn from the gene pool with the rules of population structure that are applicable to the population under study. Note, we are not assuming random mating or Hardy–Weinberg in

266 BASIC QUANTITATIVE GENETIC DEFINITIONS AND THEORY

this model. As shown in Appendix 2, the conditional probability of event A given that B is known to have occurred is the probability of A and B jointly occurring divided by the probability of B. Therefore, the probability of an individual receiving one gamete of type i and one gamete of type j (thereby resulting in genotype ij) given that one of the gametes is known to be of type i is t_{ij}/p_i. There is one complication with this equation for conditional genotype frequencies. In population genetics it is customary to interpret t_{ij} as the frequency of the unordered genotype ij, that is, the frequency of ij plus the frequency of ji. We saw this convention in Chapter 2 in deriving the Hardy–Weinberg genotype frequencies for a model with two alleles, A and a. As shown in Table 2.2, there are actually two ways of obtaining the Aa heterozygote, Aa with genotype frequency pq and aA with genotype frequency qp, for a total unordered frequency of $2pq$. When *given* one of the gamete types, there are no longer two ways of drawing a heterozygous combination. Now there is only one way of getting an ij heterozygote ($j \neq i$); the other gamete *must* be of type j. Therefore, when using unordered genotype frequencies, the conditional genotype frequencies given one gamete is of type i are

$$\text{Prob}(ii \text{ given } i) = t(ii|i) = \frac{t_{ii}}{p_i}$$

$$\text{Prob}(ij \text{ given } i) = t(ij|i) = \frac{\frac{1}{2}t_{ij}}{p_i} \quad \text{when } j \neq i \tag{8.8}$$

With these conditional genotype frequencies given gamete i, we can now calculate the average genotypic deviation given gamete i, a_i, as

$$a_i = \frac{t_{ii}}{p_i}g_{ii} + \sum_{j \neq i} \frac{\frac{1}{2}t_{ij}}{p_i}g_{ij} = \sum_j t(ij|i)g_{ij} \tag{8.9}$$

where g_{ij} is the genotypic deviation of the genotype ij.

For example, consider the *ApoE* example given in Table 8.3. There are three possible gamete types in this population, corresponding to the three alleles $\varepsilon 2$, $\varepsilon 3$, and $\varepsilon 4$. Using the genotype frequencies in Table 8.3, the allele frequencies for these three gametes are calculated to be $p_2 = 0.078$, $p_3 = 0.770$, and $p_4 = 0.152$ for the $\varepsilon 2$, $\varepsilon 3$ and $\varepsilon 4$ alleles, respectively. Now suppose we know an individual has received the $\varepsilon 2$ allele. Then we know that this individual must be genotype 2/2, 2/3, or 2/4. The frequencies of these genotypes are 0.020, 0.098, and 0.020, respectively from Table 8.3. Using equations 8.8, the conditional frequencies of these three genotypes given an $\varepsilon 2$ allele is $0.020/0.078 = 0.256$ for genotype 2/2, $\left(\frac{1}{2}\right)(0.098)/0.078 = 0.628$ for genotype 2/3, and $\frac{1}{2}(0.0196)/0.078 = 0.128$ for genotype 2/4. These three conditional genotype frequencies sum to 1 (with a slight deviation due to rounding error in this case) because they define the set of mutually exclusive and exhaustive genotypes in which a gamete bearing an $\varepsilon 2$ allele can come to exist. From Table 8.3 we can also see that if an $\varepsilon 2$ allele is combined with another $\varepsilon 2$ allele to form the 2/2 genotype, then it has the genotypic deviation of -38.2. Likewise, if an $\varepsilon 2$ allele is coupled with an $\varepsilon 3$ allele to form a 2/3 genotype, it has a genotypic deviation of -12.8; and if an $\varepsilon 2$ allele is coupled with an $\varepsilon 4$ allele to form the 2/4 genotype, it has a genotypic deviation of 3.9. Therefore, the average excess of a gamete bearing an $\varepsilon 2$ allele is, using equation 8.9,

$$a_2 = (0.256)(-38.2) + (0.628)(-12.8) + (0.128)(3.9) = -17.2$$

Similarly, one can calculate the average excess of the $\varepsilon 3$ allele to be 0.1 mg/dl and that of the $\varepsilon 4$ allele to be 8.1 mg/dl. Thus, on the average, individuals who received an $\varepsilon 2$ allele will have a total serum cholesterol level that is 17.2 mg/dl lower than the overall population mean, individuals who received an *ε3* allele are close to the overall population mean (only 0.1 mg/dl above μ on the average), and individuals who received an *ε4* allele will have a total serum cholesterol level that is 8.1 mg/dl higher than the overall population mean. In this manner, we have assigned a phenotype of cholesterol level to a gamete even though that gamete never actually displays a cholesterol level of any sort. Basically, the average excess is answering the question of what phenotype a gamete is expected to have once fertilization has occurred using the rules of population structure for that population. The average excess looks at phenotypic variation from the gamete's perspective, not that of the diploid individual. Note also that the value of the average excess is explicitly a population-level parameter: It not only depends upon the average phenotype of individuals with a given genotype but also is a function of gamete and genotype frequencies. Thus, any factor that changes population structure or gamete frequencies can change the average excess of a gamete even if the relationship between genotype and phenotype is unchanging.

In the above calculations we made no assumption about population structure, simply taking the observed genotype frequencies as is. However, the calculations of the average excess are simplified for the special case of random mating. Under Hardy–Weinberg, $t_{ii} = p_i^2$, where p_i is the frequency of allele i, and $t_{ij} = 2p_i p_j$ when $j \neq i$. Hence, from equation 8.8, the conditional frequency of genotype ij given i is simply p_j. This makes sense for random mating. Given one gamete-bearing allele i, the probability of drawing at random a second gamete for a fertilization event of type j is simply its allele frequency, p_j. Therefore, for the special case of random mating in a Hardy–Weinberg population, equation 8.9 simplifies to

$$a_i = \sum_j p_j g_{ij} \tag{8.10}$$

where the summation is over all alleles j in the gene pool, including i.

As an example, we will redo the *ApoE* example, but this time assuming the population is in Hardy–Weinberg. Table 8.4 shows the expected Hardy–Weinberg genotype frequencies, which are not significantly different from the observed. Hence, random mating is an appropriate model for this population. Table 8.4 assumes that the genotypic values are unchanged; that is, the same genotypes have the same mean phenotypes as given in Table 8.3. However, because the genotype frequencies have changed somewhat, so has the mean μ of total serum cholesterol, and therefore the genotypic deviations are slightly different from those given in Table 8.3. These differences serve as a reminder that Fisher's quantitative genetic framework is applicable only to populations and not to individuals: The same mapping of individual genotypes to individual phenotypes (which are the same for Tables 8.3 and 8.4) can result in different quantitative genetic parameters for populations differing in their gene pool or population structure. Under random mating, the probability of an *ε2* allele being coupled with another *ε2* allele is $p_2 = 0.078$, of being coupled with an *ε3* allele is $p_3 = 0.770$, and of being coupled with an *ε4* llele is $p_4 = 0.152$. Using the genotypic deviations in Table 8.4 and equation 8.10, the average excess of a gamete bearing an *ε2* allele is

$$a_2 = (0.078)(-38.6) + (0.770)(-13.2) + (0.152)(3.5) = -12.6 \text{ mg/dl}$$

The values for the average excess of the other alleles are also shown in Table 8.4. The differences with the values calculated from the data in Table 8.3 are minor, as expected

268 BASIC QUANTITATIVE GENETIC DEFINITIONS AND THEORY

Table 8.4. Genotypes at *ApoE* Locus in Population of 102 Canadians and Their Expected Hardy–Weinberg Frequencies and Values of Various Quantitative Genetic Parameters for Phenotype of Total Serum Cholesterol under Assumption of Random Mating

Genotype	$\varepsilon 2/\varepsilon 2$	$\varepsilon 2/\varepsilon 3$	$\varepsilon 2/\varepsilon 4$	$\varepsilon 3/\varepsilon 3$	$\varepsilon 3/\varepsilon 4$	$\varepsilon 4/\varepsilon 4$	Sum or Mean
Hardy–Weinberg frequency	0.006	0.120	0.024	0.593	0.234	0.023	1.000
G_i (mg/dl)	136.0	161.4	178.1	173.8	183.5	180.3	174.6
g_i (mg/dl)	−38.6	−13.2	3.5	−0.8	8.9	5.7	0.0
Gametes	$\varepsilon 2$			$\varepsilon 3$		$\varepsilon 4$	Sum or mean
Frequency	0.078			0.770		0.152	1.000
a_i (mg/dl) = α_i	−12.6			−0.3		8.0	0.0
Genotype	$\varepsilon 2/\varepsilon 2$	$\varepsilon 2/\varepsilon 3$	$\varepsilon 2/\varepsilon 4$	$\varepsilon 3/\varepsilon 3$	$\varepsilon 3/\varepsilon 4$	$\varepsilon 4/\varepsilon 4$	Sum or mean
g_{ai} (mg/dl)	−25.3	−12.9	−4.7	−0.6	7.7	16.0	0.0
d_i (mg/dl)	−13.3	−0.3	8.2	−0.2	1.2	−10.3	0.0

Note: All means are weighted by genotype frequencies or allele frequencies.

given that the actual population is not significantly different from a Hardy–Weinberg population. As before, a gamete bearing the $\varepsilon 2$ allele tends to lower total serum cholesterol relative to μ in those individuals who receive it, a gamete bearing the $\varepsilon 3$ allele tends to produce individuals close to the overall population mean, and a gamete bearing the $\varepsilon 4$ allele tends to elevate cholesterol levels relative to μ in those individuals who receive it.

Fisher also defined a second measure that assigns a phenotype to a haploid gamete. The **average effect of gamete type *i***, α_i, is the slope of the least-squares regression of the genotypic deviations against the number of gametes of type i borne by a genotype (Figure 8.9.) To calculate the average effect, let

$$Q = \sum_i \sum_j t_{ij}(g_{ij} - \alpha_i - \alpha_j)^2 \tag{8.11}$$

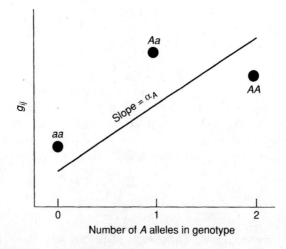

Figure 8.9. Average effect of allele *A* for one-locus, two-allele model.

Then solve for $\partial Q/\partial \alpha_i = 0$ simultaneously for every α_i. This solution minimizes the value of equation 8.11, which in turn means that this solution corresponds to the values of α_i and α_j that minimize the squared deviation from the genotypic value g_{ij} weighted by the genotype frequency. Using the criterion of a squared deviation, the average effects are those values assigned to a haploid gamete that best explain the genotypic deviations of diploid genotypes.

The average excess has the straightforward biological meaning of being the average genotypic deviation that a specific gamete type is expected to have, but the biological meaning of the average effect is less apparent. To gain insight into the biological meaning of the average effect, consider the model in which the genotype frequencies can be described in terms of allele frequencies and the system-of-mating inbreeding coefficient f (Chapter 3) or f_{it} (Chapter 6) that incorporates both deviations from random mating and population subdivision. Then the general multiallelic solution to the simultaneous equations defined by 8.11 yields (Templeton 1987c)

$$\alpha_i = \frac{a_i}{1+f} \tag{8.12}$$

Equation 8.12 reveals some important biological insights about average excesses and average effects. First, we can see that the relationship between these two measures of phenotype assigned to gametes is affected by system of mating and population subdivision, as both factors influence f (or f_{it}). *This sensitivity to population-level factors emphasizes that the phenotypes assigned to gametes depend upon the population context and not just the relationship between genotype to phenotype.* Second, we can see that the average excess and average effect always agree in sign. If the average excess for gamete i indicates that it increases the phenotype above the population average, then the average effect will also indicate the same, and likewise for lowering the phenotype below the population average. Moreover, the average excess is zero if and only if the average effect is zero. Finally, for the special case of Hardy–Weinberg, $f = 0$ and the average excess and average effect are identical. Thus, the average effect is assigning a phenotype to a gamete in a manner proportional to and sometimes identical to average excess.

So far, we have examined the parameters Fisher used to describe the phenotypes of diploid individuals (genotypic value and genotypic deviations) and to assign phenotypes to haploid gametes (average excesses and effects). However, to address the problem of the transmission of phenotypes from one generation to the next, we need to look forward to the next generation of diploid individuals. In doing so, we will assume a constant environment over time. Without this assumption, any change of environments or frequencies of residual factors in general can alter the relationship of genotype to phenotype at the individual level and thereby change all population-level parameters.

Because an individual genetically only passes on gametes to the next generation, Fisher argued that an individual's phenotypic "breeding value" is determined by the individual's gametes and not the individual's actual phenotype. In particular, Fisher defined the **breeding value or additive genotypic deviation of individuals with genotype ij, g_{aij}**, as the sum of the average effects of the gametes borne by the individual; that is, $g_{aij} = \alpha_i + \alpha_j$. Thus, an individual's breeding value depends solely upon the phenotypic values assigned to his or her gametes (which are population-level parameters) and *not* to the individual's actual phenoype.

Table 8.4 gives the additive genotypic deviations for the six genotypes at the *ApoE* locus assuming random mating. To calculate these, all we had to do is add up the average effects (which in this case are the same as the average excesses because of random mating) of the two alleles borne by each genotype. Note that the additive genotypic deviations are not the same as the genotypic deviations. For example, the genotypic deviation of the 2/4 genotype is 3.5 mg/dl, indicating that individuals with this genotype tend to have a slightly elevated serum cholesterol level relative to the average for the population. However, the additive genotypic deviation for the 2/4 genotype is −4.7 mg/dl, indicating that on the average the gametes they pass on to their children will result in a lower than average cholesterol level in their children. As this example shows, when we calculate an additive genotypic deviation, we are not asking about the phenotype of the individual but rather are looking ahead to the next generation and asking about the average phenotype of that individual's offspring. The gametes, not the individual's intact genotype, determine this future phenotypic impact, and hence the individual's breeding value is determined by what his or her gametes will do in the context of a reproducing population.

Fisher measured the discrepancy between the genetic deviation and the additive genotypic deviation by the **dominance deviation of genotype ij, d_{ij}**, the difference between the genotype's genotypic deviation and its additive genotypic deviation; that is, $d_{ij} = g_{ij} - g_{aij}$. Note that the dominance deviation is yet another residual term; it is the number that you have to add on to the additive genotypic deviation to get back to the genotypic deviation ($g_{aij} + d_{ij} = g_{ij}$). The dominance deviations for the *ApoE* genotypes under the assumption of random mating are given in Table 8.4.

It is unfortunate that Fisher named the genotypic residual term d_{ij} the "dominance" deviation. Dominance is a word widely used in Mendelian genetics and describes the relationship between a specific set of genotypes to a specific set of phenotypes. Dominance as used in Mendelian genetics is not a population parameter at all. However, the dominance deviation of Fisher is a population-level parameter; it depends upon phenotypic averages across individuals, allele frequencies, and genotype frequencies. It has no meaning outside the context of a specific population. A nonzero dominance deviation does require some deviation from codominance in a Mendelian sense, but the presence of Mendelian dominance (or recessiveness) does not ensure that a nonzero dominance deviation will occur.

In this chapter, we have developed only the one-locus version of Fisher's model, but in multilocus versions there is yet another residual term called the epistatic deviation (which we will examine in Chapter 10). Once again, the population-level residual term of epistatic deviation should never be confused with epistasis as used in Mendelian genetics. Mendelian epistasis is necessary but not sufficient to have an epistatic deviation (Cheverud and Routman 1995). Because the epistatic deviation is the last residual to be calculated, it is often very small, with most of the effects of Mendelian epistasis having already been incorporated into the additive and dominance deviations. Thus, the lack of an epistatic deviation per se tells one nothing about the amount of Mendelian epistasis affecting the phenotype of interest, just as the lack of a dominance deviation per se tells one nothing about the amount of Mendelian dominance affecting the phenotype of interest.

QUANTITATIVE GENETIC MEASURES RELATED TO VARIANCE

A normal phenotypic distribution is completely defined by just two parameters, the mean and the variance. Up to now, we have discussed Fisher's quantitative genetic parameters that are related to mean phenotypes or deviations from means. A full description of the phenotypic distributions now requires a similar set of parameters that relate to the variance of the phenotype.

The most straightforward of these measures is the **phenotypic variance**, which is the variance of the phenotype over all individuals in the population:

$$\sigma_P^2 = \sum_{ij}\sum_{k} \frac{(P_{ij,k} - \mu)^2}{n} \tag{8.13}$$

For example, the phenotypic variance of the Canadian sample is 732.5 mg²/dl². The phenotypic variance is nothing more than the variance of the phenotypic distribution. Hence μ and σ_P^2 provide a complete description of the overall phenotypic distribution under the assumption of normality.

Substituting equation 8.6 into 8.13 and noting that the summation over all n individuals is the same as summing over all genotypes and residual factors as weighted by their frequencies, we obtain

$$\sigma_P^2 = \sum_{ij}\sum_{k} t_{ij} r_k (g_{ij} + e_k)^2$$

$$= \sum_{ij}\sum_{k} t_{ij} r_k \left(g_{ij}^2 + 2 g_{ij} e_k + e_k^2\right)^2$$

$$= \sum_{ij} t_{ij} g_{ij}^2 \sum_k r_k + 2 \sum_{ij} t_{ij} g_{ij} \sum_k r_k e_k + \sum_{ij} t_{ij} \sum_k r_k e_k^2$$

$$= \sum_{ij} t_{ij} g_{ij}^2 + \sum_k r_k e_k^2 \tag{8.14}$$

because the t_{ij}'s and the r_k's must sum to 1 as they define probability distributions over genotypes and residual factors, respectively, and because the mean genotypic deviation and the mean environmental deviation are both zero, as shown previously. Equation 8.14 shows that the phenotypic variance can be separated into two components. Recalling that all variances are expected squared deviations from the mean (Appendix 2) and that both g_{ij} and e_k have means of zero, we can see that both of the components in equation 8.14 are themselves variances. The first component is the **genetic variance**, σ_g^2, the variance of the genotypic deviations or equivalently the average genotypic deviation squared:

$$\sigma_g^2 = \sum_{ij} t_{ij} g_{ij}^2 \tag{8.15}$$

Fisher called the second component of equation 8.14 the **environmental variance**, σ_e^2, the variance of the environmental deviations or equivalently the average environmental deviation squared:

$$\sigma_e^2 = \sum_k r_k e_k^2 \tag{8.16}$$

Although σ_e^2 is called the environmental variance, it is really the residual variance left over after explaining a portion of the phenotypic variance with the genetic model.

With these definitions, equation 8.14 can now be expressed as

$$\sigma_P^2 = \sigma_g^2 + \sigma_e^2 \qquad (8.17)$$

that is, the total phenotypic variation in the population can be split into a component due to genotypic variation and a component not explained by the genotypic variation being modeled or monitored.

We can use equation 8.15 to calculate the genetic variance in the Canadian population for the phenotype of total serum cholesterol, assuming Hardy–Weinberg, from the data given in Table 8.4 as

$$\begin{aligned}\sigma_g^2 &= (0.006)(-38.6)^2 + (0.12)(-13.2)^2 + (0.024)(3.5)^2 \\ &\quad + (0.593)(0.8)^2 + (0.234)(8.9)^2 + (0.023)(5.7)^2 \\ &= 50.0\,\text{mg}^2/\text{dl}^2\end{aligned}$$

The residual variance can now be calculated by solving equation 8.17 for σ_e^2 to obtain $\sigma_e^2 = 732.5 - 50.0 = 682.5\,\text{mg}^2/\text{dl}^2$.

The proportion of the total phenotypic variation that can be explained by genotypic variation is the **broad-sense heritability**, h_B^2, the ratio of the genetic variance to the phenotypic variance:

$$h_B^2 = \frac{\sigma_g^2}{\sigma_P^2} \qquad (8.18)$$

For example, in our Hardy–Weinberg Canadian population, the broad-sense heritability of the phenotype of total serum cholesterol is $50.0/732.5 = 0.07$. This means that 7% of the phenotypic variance in cholesterol levels in this population can be explained by the fact that there is genotypic variation in this population at the *ApoE* locus with respect to the alleles $\varepsilon 2$, $\varepsilon 3$, and $\varepsilon 4$.

Broad-sense heritability refers to the contribution of genotypic variation to phenotypic variation and not the contribution of gametic variation to phenotypic variation. Broad-sense heritability is therefore not useful in modeling phenotypic evolution across generations or the transmission of phenotypes from one generation to the next. To address these issues, it is necessary to have variance components that are related to the phenotypic impact of gametes. Fisher created such a measure with the **additive genetic variance**, σ_a^2, the variance in the additive genotypic deviations (breeding values) or, equivalently, the average additive genotypic deviation squared:

$$\sigma_a^2 = \sum_{ij} t_{ij} g_{aij}^2 \qquad (8.19)$$

For example, using the Hardy–Weinberg frequencies in Table 8.4 for the t_{ij}'s and using the additive genotypic deviations in that table, the additive genetic variance (often just called the additive variance) can be calculated to be $44.5\,\text{mg}^2/\text{dl}^2$.

The proportion of the transmissible phenotypic variation is measured by the **narrow-sense heritability (often just called heritability)**, h^2, which represents the ratio of the additive genetic variance to the phenotypic variance:

$$h^2 = \frac{\sigma_a^2}{\sigma_P^2} \tag{8.20}$$

Returning to the Canadian example given in Table 8.4, the heritability of total serum cholesterol is $44.5/732.5 = 0.06$. As we saw above, genetic variation at the *ApoE* locus for the alleles *ε2*, *ε3* and *ε4* could account for 7% of the phenotypic variation. The heritability tells us that the transmission of the gametes at this locus can explain 6% of the phenotypic variance in the next generation.

The additive genetic variance is always less than or equal to the genetic variance, so as he did with his mean-related measures, Fisher defined the **dominance variance** to be the residual genetic variance left over after subtracting off the additive genetic variance, that is,

$$\sigma_d^2 = \sigma_g^2 - \sigma_a^2 \tag{8.21}$$

For example, the dominance variance for the Hardy–Weinberg Canadian population in Table 8.4 is $50.0 - 44.5 = 5.5$ mg^2/dl^2. Similarly, when multilocus models are used, there is another residual variance term called the epistatic variance (Chapter 10). As with the mean-related measures, the dominance and epistatic variances refer to residual variance components in a population and should not be confused with dominance and epistasis as used in Mendelian genetics.

We now have the basic parameters and definitions of Fisher's model of quantitative genetics. In the next two chapters we will see how to apply these concepts to two major cases: the first when the gene loci underlying the quantitative variation are not measured and the second when genetic variation at causative loci or loci linked to causative loci are being measured.

9

QUANTITATIVE GENETICS: UNMEASURED GENOTYPES

In the previous chapter the basic definitions of Fisher's quantitative genetic model were introduced through the worked example of how genetic variation at the *ApoE* locus influences the phenotype of serum cholesterol level in a human population. Because the *ApoE* genotypes of the individuals in this population were known, the average phenotype of a genotype was easy to estimate since all we had to do was take the average cholesterol values of those individuals sharing a common genotype. Likewise, all the other parameters of Fisher's model could easily be estimated in this case by making use of the data on *ApoE* genotype. However, when Fisher first developed his theory, very few genes were identifiable in most species, and the genes underlying most quantitative traits were unknown. As a result, in the vast majority of cases it was impossible to measure the genotypes associated with the phenotypes of interest. Modern genetics gives us the ability to measure many genotypes related to quantitative traits, as already illustrated by the *ApoE* example in the previous chapter. However, even today the genotypes underlying most quantitative traits in most species are not known or measurable. Therefore, the methods used in the previous chapter to estimate the parameters of Fisher's model cannot be used if we cannot assign an individual to a specific genotype category.

Fisher realized that to make his model useful and practical, it would be necessary to estimate at least some of the parameters even when no genotypes were known. Fisher therefore provided statistical methods for estimating the parameters of a genetic model when no direct genetic information is available. This is a remarkable accomplishment that made his model of immediate practical utility in medicine and agriculture, a utility that is enhanced when we can measure the genotypes, as will be shown in the next chapter.

How can we do genetics when we do not measure genotypes? Fisher and others came up with three basic approaches:

- Correlation between relatives

Population Genetics and Microevolutionary Theory, By Alan R. Templeton
Copyright © 2006 John Wiley & Sons, Inc.

- Response to selection
- Controlled crosses

All of these approaches require some sort of information about the genetic relationships among individuals. In the first approach, we must gather information about the pedigree relationships between individuals whose phenotypes are being scored. In the second approach, we must measure the phenotypes of the individuals from a population being selected for some trait values and then measure the phenotypes in their offspring. With the last approach, we make crosses, such as the classical Mendelian crosses of F_1, F_2 and backcrosses. In every case, information is being generated about the genetic relationships among a group or groups of individuals. Each method uses this information about genetic relationships as a proxy for the unmeasured genotypes. We will now examine how this is accomplished for these three major unmeasured genotype approaches.

CORRELATION BETWEEN RELATIVES

Fisher (1918) used the correlation among relatives as his primary tool to calculate quantitative genetic variances and related parameters. To see how he did this, we first need to introduce the idea of a **phenotypic covariance**, which measures the extent to which a phenotype in one individual deviates from the overall population mean in the same direction and magnitude as the phenotype of a second individual. Suppose that we measure the phenotypes of paired individuals, say X and Y, such that x_i and y_i are the phenotypes for a specific pair i. We saw in Chapter 8 that the mean μ measures the central tendency of these phenotypes and that the variance σ^2 measures how likely they are to deviate from the mean. When we measure the phenotype as a sample of paired observations (x_i, y_i) with $i = 1, \ldots, n$, where n is the number of pairs sampled, we can also measure how the phenotype "covaries" between individuals. That is, if x_i is a phenotypic trait value larger than average, does this give you any information about y_i? The extent of the tendency of the paired observations to covary is measured by the covariance of x and y (Appendix 2), that is, the expected (average) value of the product of the difference between the phenotype of individual x_i and the mean phenotypic value of the population from which individual x_i is sampled times the difference between the phenotype of individual y_i from the mean value of the population from which individual y_i is sampled:

$$\text{Cov}(X,Y) = \sum_{i=1}^{n} \frac{(x_i - \mu_x)(y_i - \mu_y)}{n} \quad (9.1)$$

where Cov(X,Y) is the covariance between the paired variables X and Y. Equation 9.1 measures how the phenotypes of X and Y covary. For example, suppose there is tendency for X and Y to have similar values relative to their respective means. This implies that if X is larger than its mean, then Y also tends to be larger than its mean. Thus, both $x_i - \mu_x$ and $y_i - \mu_y$ tend to be positive together, so their product is likewise positive. The variables X and Y covarying together also implies that if X is smaller than its mean, then Y does likewise. In that case, both $x_i - \mu_x$ and $y_i - \mu_y$ tend to be negative together, so their product still tends to be positive. When such a trend exists, positive terms dominate in averaging the products, resulting in an overall positive covariance. Thus, a *positive covariance* means that the traits of the two individuals tend to deviate from their respective means in the same direction. In contrast, a *negative covariance* indicates that as one individual's phenotype deviates in one direction from its mean, the other individual's phenotype tends to deviate in the other direction with respect to its mean. This results in many products between a positive number and a negative number, yielding a negative product and an overall negative average. When the covariance is zero, positive and negative products are both equally likely and cancel one another out on the average. This means that there is no association between the trait of X and the trait of Y in the populations being studied.

There is one special case of covariance that will be useful in subsequent calculations; namely, that the covariance of a variable with itself is simply the variance of the variable. This is shown by

$$\text{Cov}(X,X) = \sum_{i=1}^{n} \frac{(x_i - \mu_x)(x_i - \mu_y)}{n} = \sum_{i=1}^{n} \frac{(x_i - \mu_x)^2}{n} = \sigma^2 \qquad (9.2)$$

Fisher (1918) related his genetic model to covariances by letting X be the phenotype of one individual and Y being that of a second (often related) individual. He then used principles of Mendelian genetics to predict how these phenotypes should covary within classes of paired individuals defined by a particular type of genetic relatedness. Let us first consider a sample of paired individuals from a population. Using the model given in Chapter 8 (equation 8.6), we can describe the phenotypes of an individual from the population as follows:

$$P_i = \mu + g_{ai} + g_{di} + e_i \qquad (9.3)$$

where P_i is the phenotype of individual i, μ is the mean phenotype in the population for all individuals, g_{ai} is the additive genotypic deviation for individual i, g_{di} is the dominance deviation for individual i, and e_i is the environmental deviation for individual i (we have dropped the multiple subscript convention used in Chapter 8 that indexes both genotypes and environments because we are now assuming we cannot measure the phenotypes of genotypes, only individuals). We can now express the covariance of the phenotypes of a pair of individuals from the population with mean μ, say xi and yi, as

$$\text{Cov}(P_{xi}, P_{yi}) = (P_{xi} - \mu)(P_{yi} - \mu) = (g_{axi} + g_{dxi} + e_{xi})(g_{ayi} + g_{dyi} + e_{yi})$$
$$= g_{axi}g_{ayi} + g_{dxi}g_{ayi} + e_{xi}g_{ayi} + g_{axi}g_{dyi}$$
$$+ g_{dxi}g_{dyi} + e_{xi}g_{dyi} + g_{axi}e_{yi} + g_{dxi}e_{yi} + e_{xi}e_{yi} \qquad (9.4)$$

Recall from Chapter 8 that Fisher defined his environmental deviation as a residual term relative to the genotypic deviation and similarly defined the dominance deviation as a residual term relative to the additive genotypic deviation. Mathematically, this means that all the unlike deviation parameters in equation 9.4 have a covariance of zero. Thus, all the cross-product terms of unlike deviations must average to zero by definition. Hence, equation 9.4 simplifies to

$$\text{Cov}(P_{xi}, P_{yi}) = g_{axi}g_{ayi} + g_{dxi}g_{dyi} + e_{xi}e_{yi} \qquad (9.5)$$

Because all of these deviation parameters by definition have a mean of zero, equation 9.5 can also be expressed as

$$\text{Cov}(P_{xi}, P_{yi}) = \text{Cov}(g_{axi}, g_{ayj}) + \text{Cov}(g_{dxi}, g_{dyj}) + \text{Cov}(e_{xi}, e_{yi}) \qquad (9.6)$$

Fisher assumed that each individual experienced an independent "environment" (that is, residual factors) so the last term in equation 9.6 is zero by assumption. This is a critical assumption because it means that the covariance in phenotypes among individuals is only a function of covariances among genetic contributors to the phenotype, that is, the first two terms in equation 9.6. The assumption that different individuals have environmental deviations with zero covariance can be violated, and when that happens the simple Fisherian model being presented here is no longer valid. When correlations exist in the environmental deviations of different individuals, either more complicated models are needed that take into account the correlations [see Lynch and Walsh (1998)] or special sampling designs are needed to eliminate such correlations. For example, one class of paired individuals frequently used in human genetic studies of quantitative traits is identical twins. Such paired individuals share their entire genotype, but they also frequently share many common environmental factors since much of the human environment is defined by our families, schooling, and so on, which are typically also shared by identical twins. To eliminate this environmental covariance, some studies go to great expense and labor to find identical twins who were separated shortly after birth and reared apart, in the hope that this will eliminate or at least reduce the environmental covariance. However, even in these studies significant covariance can still exist for some phenotypes because of other shared environmental features, such as sharing the same womb before birth (Devlin et al. 1997). When a positive covariance exists between the environmental deviation terms among related individuals, the simple Fisherian model will inflate the genetic component of phenotypic variation. This is a serious error that can lead to spurious biological conclusions, but for the purposes of this book, we will assume the ideal case of no covariance between the environmental deviations of any two individuals in the population.

Now we will consider specific classes of pairs of individuals. First of all, consider the class of mating pairs drawn from a random-mating population. Random-mating means that there is no covariance between the phenotypes of the mating individuals, so equation 9.6 must be zero in this case. Moreover, random mating means that the mating individuals do not share any more genes in common than expected by chance alone, so the covariance of all the genetically related parameters of the mating individuals (the first two terms of equation 9.6) must also be zero. Hence, under random mating, the covariance between parental phenotypes from mating pairs is zero because the covariance between all the terms in equation 9.6 is also zero.

Now consider the covariance between a parent (indicated by the subscript p) and an offspring (indicated by the subscript o). For this pair, equation 9.6 becomes (with the assumption of no covariance between environmental deviations)

$$\text{Cov}(P_p, P_o) = \text{Cov}(g_{ap}, g_{ao}) + \text{Cov}(g_{dp}, g_{do}) \qquad (9.7)$$

Recall from Chapter 8 that the additive genotypic deviations are the only aspects of phenotype that can be passed on from parent to offspring through a gamete. Because the dominance deviation is the residual genetic component that *cannot* be passed on from generation to generation, the covariance between the dominance deviations of parent and offspring must be zero by definition. Therefore, a phenotypic correlation between parent and offspring is attributed solely to the additive genotypic component of equation 9.7: $\text{Cov}(g_{ap}, g_{ao})$. However, a parent passes on only one gamete to an offspring. The other gamete comes from the other parent, and it is assumed to be unknown. Recall that the additive genotypic deviation of an individual is the sum of the average effects of two gametes. However, because the genotypes are not measured, we cannot split the additive genotypic deviation of the parent into two gametic components, as we did in Chapter 8. However, we do know under Mendelian genetics that one-half of the genes of the parent are passed on to the offspring. Therefore, under the supposition of Mendelian inheritance, our best estimate of the average effect of the gamete passed onto the offspring o from parent p is $\alpha_p = \frac{1}{2}g_{ap}$. Of course, the offspring received another gamete from the other parent, say parent m, and let the average effect of this gamete be α_m. Therefore, we have that the additive genotypic deviation of the offspring is the sum of the two average effects:

$$g_{ao} = \tfrac{1}{2}g_{ap} + \alpha_m \tag{9.8}$$

Substituting equation 9.8 into 9.7 and recalling that $\text{Cov}(g_{dp}, g_{do}) = 0$ by definition, we have

$$\text{Cov}(g_{ap}, g_{ao}) = \text{Cov}(g_{ap}, \tfrac{1}{2}g_{ap} + \alpha_m) = \text{Cov}(g_{ap}, \tfrac{1}{2}g_{ap}) + \text{Cov}(g_{ap}, \alpha_m) \tag{9.9}$$

Under random mating, there is no covariance between the average effects of gametes from one parent with that of the other. Therefore, in a random-mating population, $\text{Cov}(g_{ap}, \alpha_m) = 0$, and equation 9.9 simplifies to

$$\text{Cov}(g_{ap}, g_{ao}) = \text{Cov}(g_{ap}, \tfrac{1}{2}g_{ap}) = \tfrac{1}{2}\text{Cov}(g_{ap}, g_{ap}) = \tfrac{1}{2}\text{Var}(g_{ap}) = \tfrac{1}{2}\sigma_a^2 \tag{9.10}$$

since the covariance of any variable with itself is the variance of the variable (equation 9.2) and the variance of g_{ap} is by definition the additive genetic variance σ_a^2 (equation 8.19). With equation 9.10, Fisher has managed to relate a fundamental quantitative genetic parameter, the additive genetic variance, to a phenotypic covariance that can be calculated simply by measuring the phenotypes of parents and their offspring. No actual genotypes need to be measured to estimate this fundamental genetic parameter.

Fisher also related other genetic parameters to the correlation between the phenotypes of classes of individuals. A correlation is a covariance that has been standardized by the product of the standard deviations of the two variables being examined (Appendix 2). In particular, the correlation between X and Y is

$$\rho_{XY} = \frac{\text{Cov}(X,Y)}{\sqrt{\text{Var}(X) \times \text{Var}(Y)}} \tag{9.11}$$

The correlation coefficient measures associations among variables on a -1 to $+1$ scale, regardless of the original scale of variation.

Substituting equation 9.10 into the numerator of equation 9.11, we have that the correlation between the phenotypes of parents and offspring is

$$\rho_{po} = \frac{\frac{1}{2}\sigma_a^2}{\sqrt{\text{Var}(P_p) \times \text{Var}(P_o)}} \tag{9.12}$$

Assuming that the phenotypic variance, σ_p^2, is constant for the parental and offspring generations, then $\text{Var}(P_p) = \text{Var}(P_o) = \sigma_p^2$, and equation 9.12 becomes

$$\rho_{po} = \frac{\frac{1}{2}\sigma_a^2}{\sigma_p^2} = \frac{1}{2}h^2 \tag{9.13}$$

where h^2 is the heritability of the trait (equation 8.20). In this manner, Fisher could estimate the important quantitative genetic parameter of heritability as twice the observable phenotypic correlation between parents and offspring. Note that equation 9.13 allows us to estimate that portion of the phenotypic variance that is due to genotypic variation that can be transmitted to the next generation through gametes even though not a single genotype contributing to the phenotypic variance is being measured or is even known.

Fisher also refined this estimate based upon parent/offspring correlations. Any offspring has two parents, and equation 9.13 makes use of information on only one parent. If the phenotype of both parents is known, then we can calculate the average phenotype of the two parents, known as the **midparent value**, as $\frac{1}{2}P_m + \frac{1}{2}P_f$, where the subscript m denotes the phenotype of the mother and f the phenotype of the father. When we have information about both parents, equation 9.8 becomes

$$g_{ao} = \frac{1}{2}g_{am} + \frac{1}{2}g_{af} \tag{9.14}$$

The covariance between the midparent value and the offspring is therefore

$$\text{Cov}(\text{midparent, offspring}) = \text{Cov}\left(\frac{1}{2}g_{am} + \frac{1}{2}g_{af}, \frac{1}{2}g_{am} + \frac{1}{2}g_{af}\right)$$
$$= \frac{1}{4}\text{Var}(g_{am}) + \frac{1}{4}\text{Var}(g_{af}) = \frac{1}{2}\sigma_a^2 \tag{9.15}$$

This is the same result as when we used just one parent (equation 9.10). However, the total phenotypic variance of the midparents is only half the original phenotypic variance because it is a variance of an average of two values. Therefore,

$$\rho_{\bar{p}o} = \frac{\frac{1}{2}\sigma_a^2}{\sqrt{\frac{1}{2}\sigma_p^2 \times \sigma_p^2}} = \frac{\sqrt{\frac{1}{2}}\sigma_a^2}{\sigma_p^2} = \sqrt{\frac{1}{2}}h^2 \tag{9.16}$$

Another statistical method for measuring the association between the phenotypes in the parental generation with the phenotypes in the offspring generation is the least-squares regression coefficient (Appendix 2) of offspring phenotype on midparent value. Fisher showed that in general the least-squares regression coefficient between X and Y is related to the correlation coefficient between X and Y as follows:

$$b_{YX} = \rho_{XY}\sqrt{\frac{\sigma_Y^2}{\sigma_X^2}} \tag{9.17}$$

where b_{YX} is the regression coefficient of Y on X. Putting equation 9.16 into 9.17, we have

$$b_{o\bar{p}} = \rho_{\bar{p}o}\sqrt{\frac{\sigma_o^2}{1/2\sigma_p^2}} = \sqrt{\frac{1}{2}h^2}\sqrt{\frac{1}{\frac{1}{2}}} = h^2 \tag{9.18}$$

Hence, another way of estimating the narrow-sense heritability of a trait is to calculate the regression coefficient of offspring phenotype on midparent value, as illustrated in Figure 9.1.

The above equations show how data on the phenotypes of parents and offspring, with no genotypic data, can be used to estimate the additive genetic variance and the heritability of the trait. This makes sense, because the phenotype of a parent and offspring are genetically linked only by what is transmissible through a gamete, and that is exactly what the additive genetic variance and heritability were designed to measure. Fisher also showed that estimates of the other genotypic components of variance are possible by looking at the phenotypic correlation of other types of genetic relatives. For example, consider the covariance between full sibs (two individuals with the same mother and father). Full sibs receive half of their genes from the common mother and half from the common father. Accordingly, equation 9.14 is applicable to both of the full sibs, and likewise equation 9.15 describes their covariance for their additive genotypic deviations, which is $\frac{1}{2}\sigma_a^2$. Recall from Chapter 8 that the motivation for developing the additive genotypic deviation was the fact that parents pass on to their offspring only a haploid gamete and not a diploid genotype. The dominance deviation was defined as that residual of the *genotypic* deviation that is left over after taking out the effects transmissible through a haploid gamete. However, full sibs share not only half of

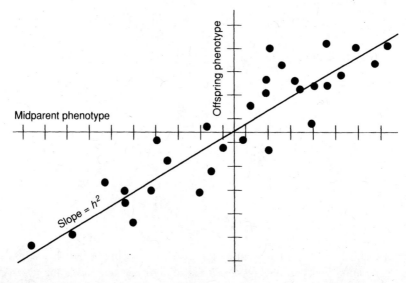

Figure 9.1. Hypothetical example of regression of offspring phenotype upon midparent value. The slope of the resulting regression line is the heritability of the trait, h^2.

their genes but also some of the same *genotypes*. In particular, under Mendelian inheritance, full sibs share exactly the same genotype at one-quarter of their autosomal loci; that is, both sibs receive identical alleles from both parents with probability $(\frac{1}{2})(\frac{1}{2}) = 1/4$. Therefore, full sibs share one quarter of their dominance deviations, the part of the residual genotypic deviations due to single-locus diploid genotypes. That is, $\text{Cov}(g_{ds1}, g_{ds2}) = 1/4\, \sigma_d^2$, where $s1$ and $s2$ index a pair of full sibs. Moreover, full sibs will share some multilocus genotypes as well and therefore some of the epistatic deviations. This will not be explicitly modeled here [for a fuller treatment, see Lynch and Walsh (1998)]. Hence, equation 9.7 when applied to full-sib pairs become:

$$\text{Cov(full sibs)} = \tfrac{1}{2}\sigma_a^2 + \tfrac{1}{4}\sigma_d^2 \qquad (9.19)$$

Dividing equation 9.19 by the phenotypic variance yields the phenotypic correlation between full sibs:

$$\rho_{s1,s2} = \frac{\text{Cov(full sibs)}}{\sigma_p^2} = \frac{\tfrac{1}{2}\sigma_a^2 + \tfrac{1}{4}\sigma_d^2}{\sigma_p^2} = \tfrac{1}{2}h^2 + \tfrac{1}{4}\frac{\sigma_d^2}{\sigma_p^2} \qquad (9.20)$$

Recall from equation 9.13 that the phenotypic correlation between a parent and offspring is half the heritability. Therefore, if $\sigma_d^2 > 0$, then the phenotypic correlation between full sibs is expected to be greater than that of a parent and offspring under Mendelian inheritance. For example, consider the phenotype of systolic blood pressure in humans (Miall and Oldham 1963). The correlation between parent and offspring in one human population was measured to be 0.237. This implies that the heritability of systolic blood pressure in this population is $2(0.237) = 0.474 = h^2$. In this same population, the phenotypic correlation between full sibs is 0.333. From equation 9.20, the difference between the phenotypic correlation of full sibs from that of parent and offspring should be $\tfrac{1}{4}(\sigma_d^2/\sigma_p^2)$. Therefore, we can estimate the portion of the total phenotypic variance that is due to dominance deviations as $\sigma_d^2/\sigma_p^2 = 4(0.333 - 0.237) = 0.384$. Also note that adding σ_d^2/σ_p^2 to the heritability provides an estimate of the broad-sense heritability, which in this case is $h_B^2 = 0.474 + 0.384 = 0.858$. Thus, 85.8% of the variation in systolic blood pressure is attributable to genotypic differences in this population, with 14.2% of the phenotypic variance due to "environmental" deviations. Moreover, the genetic proportion in turn can be split into an additive part that is transmissible from parent to offspring (47.4%) and a nontransmissible portion of 38.4%. Thus, by studying more than one class of relatives, a fuller description of the contribution of genetic variation to phenotypic variation is possible even though not a single genotype is actually measured or known.

The unmeasured approach using phenotypic correlations can be extended to many other classes of relatives and can accommodate some shared environmental influences, such as shared nongenetic maternal effects (Lynch and Walsh 1998). This approach therefore can provide much insight into the genetic contributions to phenotypic variation for many traits in any species for which it is possible to know the genetic relatedness among individuals. This approach is limited to the estimation of those quantitative genetic parameters related to the variance and does not provide estimators for the quantitative genetic parameters related to the mean. When genotypes cannot be measured, even this partial answer is better than no answer at all.

Indeed, frequently the primary genetic parameter that a geneticist wants to estimate is the heritability. As we will see in the next section and in Chapter 11, it is the heritability that determines the response to selection, either artificial or natural. As we have seen, heritability can be estimated even when no genotypes are being examined. Because heritability is such an important concept in quantitative and evolutionary genetics, it is critical to draw a distinction between heritability and inheritance, two words that are sometimes confused with one another. **Inheritance** describes the way in which genes are passed on to the next generation and how the specific genotypes created after fertilization develop their phenotypes. For example, the mental retardation associated with phenylketonuria is said to be *inherited* as a single-locus, autosomal recessive trait under a normal dietary environment. The terms "single-locus, autosomal" tells us that the genes that are passed on from generation to generation that are associated with the phenotype of PKU-induced mental retardation are found at just one locus on an autosome, and hence Mendel's first law of equal segregation is required to understand and model the passage of PKU from generation to generation. The word "recessive" tells us that an individual must be homozygous for a specific allele in order to display the PKU trait in a normal dietary environment. The inheritance of traits is the primary focus of Mendelian genetics, and it requires some knowledge of the underlying genetic architecture. The concept of inheritance is applied to specific individuals in specific family contexts.

In contrast, **heritability** measures that portion of the parental phenotypic variance that is passed on to the offspring generation through gametes. Heritability requires no knowledge of the underlying genetic architecture. Heritability is defined not for individuals, but rather for a population. Heritability is explicitly a function of allele and genotype frequencies, whereas inheritance is not. If any of these population-level attributes are altered, heritability can change.

Inheritance is necessary but not sufficient for heritability. If we know a trait is heritable, then we know that there are genetic loci affecting phenotypic variation, and these loci will have a pattern of inheritance, whether we can observe it or not. On the other hand, knowing that a trait is inherited only tells us that genetic variation affects the phenotype in some individuals, but without knowing the frequencies of the genotypes, we can say nothing of heritability. For example, PKU is inherited as a single-locus, autosomal recessive trait under normal dietary environments (Chapter 8), but is PKU heritable? Recall that the p allele class is very rare in the human population and that a random-mating model seems to be appropriate for this locus. If we assign a phenotypic value of 1 to the "normal" genotypes under this pattern of recessive inheritance (PP or Pp) and a value of 0 to the affected genotype (pp) in a normal dietary environment, then the average excess/average effect of the PKU allele p in a random-mating population in which q is the frequency of allele p is

$$a_p = (1-q)[1-(1-q^2)] + q[0-(1-q^2)] = q^2 - q^3 - q + q^3 = -q(1-q)$$

(9.21)

As q goes to zero, so does $-q(1-q)$. Therefore, because q is very small in human populations, the average excess/average effect of the p allele is close to zero. Similarly, the average excess/average effect of the P allele is also close to zero. Therefore, PKU is *not heritable even though it is inherited!* Another way of showing the lack of heritability of PKU is to consider the phenotypic correlation between parents and offspring. For a trait like PKU that is due to a rare, autosomal recessive allele in a random-mating population, almost all babies born with PKU came from parents who are heterozygotes; that is, PKU children come from $Kk \times Kk$ crosses. However, the phenotype of Kk individuals is 1.

Therefore, PKU children with the phenotype of 0 come from normal parents with phenotype 1, and normal children came from normal parents with phenotype 1. This means that there is no correlation between the phenotypes of the parents and the phenotypes of the children. Once again, we conclude that PKU is not a heritable trait. Indeed, many genetic diseases in humans are inherited as autosomal recessives for a rare allele. Very few of these genetic diseases that afflict humankind are heritable, although all are clearly inherited. Inheritance and heritability should never be equated.

RESPONSE TO SELECTION

Another way of estimating heritability with phenotypic data alone is through the response to selection. This approach is widely used in agriculture. Keeping the environmental factors as constant as possible across the generations, plant and animal breeders select some segment of the parental population on the basis of their phenotype (e.g., the cows that produce the most milk and the plants with the highest yields) and breed these selected individuals only. Obviously, the point of all this is to alter the phenotypic distribution in the next generation along a desired direction. The Fisherian model given above allows us to predict what the response to selection will be. From equation 9.18 or Figure 9.1, we see that the offspring phenotype can be regressed upon the midparent values. Now suppose that only some parental pairs are selected to reproduce. Assume that the selected parents have a mean phenotype μ_s that differs from the mean of the general population, μ, as shown in Figure 9.2. The **intensity of selection** $S = \mu_s - \mu$ is the mean phenotype of the selected parents minus the overall mean of the total population (the selected and nonselected individuals). We can then use the regression line to predict the phenotypic response of the offspring to the selection, as shown in Figure 9.2, to be

$$R = Sh^2 \qquad (9.22)$$

where h^2 is the heritability of the phenotype being selected and **response to selection R** the is measured by the mean phenotype of the offspring (μ_o) minus the overall mean (selected and nonselected individuals) of the parental generation, μ; that is, $R = \mu_o - \mu$.

Equation 9.22 shows that a population can only respond to selection (R) if there is both a selective force (S) and heritability (h^2) for the trait. If there is no additive genetic variation, heritability is zero, so there will be no response to selection no matter how intense the selection may be. Hence, the only aspects of the genotype–phenotype relationship that are important in selection (and hence in agricultural breeding programs and, as we will see in Chapter 11, adaptation) are those genetic aspects that are additive as a cause of phenotypic variation.

If the heritability of a trait is known, say from studies on the phenotypic correlations among relatives, then equation 9.22 allows the prediction of the response to selection. However, if the heritability is not known, then the heritability can be estimated by monitoring the intensity and response to selection as $h^2 = R/S$. For example, Clayton et al. (1957) examined variation in abdominal bristle number in a laboratory population of the fruit fly *Drosophila melanogaster*. They first estimated the heritability of abdominal bristle number in this base population through offspring–parent regression (equation 9.18) to be 0.51. In a separate experiment, Clayton et al. selected those flies with high bristle number to be the parents of a selected generation. The base population had an average of 35.3 bristles per fly, and the selected parents had a mean of 40.6 bristles per fly. Hence, the intensity of

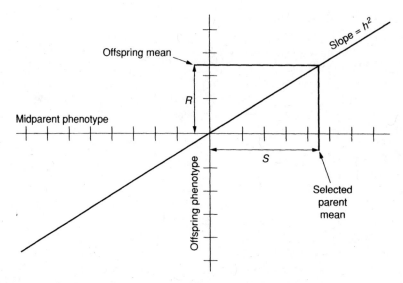

Figure 9.2. Response (R) to selection as function of heritability, h^2, and intensity of selection (S). The y axis is drawn to intersect the x axis at μ, the overall mean phenotype of the parental generation (including both selected and nonselected individuals).

selection is $S = 40.6 - 35.3 = 5.3$. The offspring of these selected parents had an average of 37.9 bristles per fly, so the response to selection is $R = 37.9 - 35.3 = 2.6$. From equation 9.22, the heritability of abdominal bristle number in this population is now estimated to be $R/S = 2.6/5.3 = 0.49$, a value not significantly different from the value estimated by offspring–parent regression.

PROBLEM OF BETWEEN-POPULATION DIFFERENCES IN MEAN PHENOTYPE

Fisher designed all the quantitative genetic parameters in his model (Chapter 8) for a single population sharing a common gene pool and characterized by a single system of mating for the phenotype of interest. Moreover, in order to estimate some of these parameters when genotypes are unmeasured, Fisher also had to assume that the environment (actually residual factors) was constant across the generations in the sense that the probability distribution of the environmental deviations was unchanging with time and all individuals had an independent environmental deviation (this assumption can be relaxed in more complicated models). However, a frequent problem in quantitative and population genetics occurs when we want to compare two populations with distinct phenotypic distributions. Because two different populations may have distinct gene pools and may live in a different range of environmental conditions, the biological meaning of changes in the phenotype from one to the other are difficult to evaluate. Do the populations differ because they have different allele frequencies but have identical genotype-to-phenotype mappings and the same environment? Or do they differ because they have completely different allelic forms? Or do they differ because they have the same alleles and the same allele frequencies but differ in the probabilities of various environmental or other residual factors? Or do they differ because of a combination of different allele frequencies, different alleles, and different environments? The Fisherian model outlined in Chapter 8 does not address any of these questions because it was designed to be applicable only to a single population and does not contain one statistic related to the comparison of different populations.

The inability of the Fisherian model to compare populations is particularly evident when we want to understand a difference in mean phenotype between two populations. The first step in Fisher's model is to subtract off the mean of the population from all observations. All of Fisher's quantitative genetic parameters are defined in terms of deviations from the mean and hence are mathematically invariant to the overall mean value of the population. Consequently, none of Fisher's quantitative genetic measures such as broad-sense or narrow-sense heritability have anything to do with the mean phenotype of the population. Nevertheless, sometimes the argument is made that because a trait is heritable within two different populations that differ in their mean trait value, then the average trait differences between the populations are also influenced by genetic factors (e.g., Herrnstein and Murray 1994). Because heritability is a within-population concept that refers to variances and not to means, such an argument is without validity. Indeed, heritability is irrelevant to the biological causes of mean phenotypic differences between populations. To see this, we will consider four examples.

First, in Chapter 8 we examined the role of the amino acid replacement alleles at the *ApoE* locus in a Canadian population of men from the mid-1980s upon the phenotype of total serum cholesterol level. The mean phenotype in that population was 174.2 mg/dl. Hallman et al. (1991) studied the role of the same *ApoE* polymorphisms in nine different human populations, whose mean total serum cholesterol levels varied from 144.2 mg/dl (Sudanese) to 228.5 mg/dl (Icelanders). These mean differences in cholesterol levels span a range of great clinical significance, as values above 200 mg/dl are considered an indicator of increased risk for coronary artery disease. Hence, these nine populations are greatly different in their phenotypic distributions in a manner that is highly significant both statistically and biologically. Despite these large differences in mean total serum cholesterol levels, a Fisherian analysis of the *ApoE* polymorphism *within* each of these populations results in estimates of the average excesses and effects, heritabilities, and so on, that are statistically indistinguishable, as illustrated in Figure 9.3 for the average excesses for the Sudanese versus Icelanders. How can such large mean phenotypic differences between populations be totally invisible to the Fisherian model? It is known that the phenotype of total serum cholesterol is strongly influenced by many environmental factors, such as diet, exercise regimens, alcohol consumption, smoking, and so on. The populations examined by Hallman et al. (1991) differ greatly in these environmental variables, so it is not surprising that they also differ in their mean phenotypes. The homogeneity of the results of the Fisherian analyses of the *ApoE* polymorphism arises from two factors:

- The populations all have similar allele frequencies in their respective gene pools; for example, the frequencies of the $\varepsilon 2$, $\varepsilon 3$, and $\varepsilon 4$ alleles in the Sudanese with the lowest average total serum cholesterol are 0.081, 0.619, and 0.291, respectively, and in the Iceland population with the highest average total serum cholesterol they are 0.068, 0.768, and 0.165.
- The genotypes all map onto phenotypes in a similar manner *relative to their population means*; that is, a genotype that tends to be above the mean in one population tends to be above the mean and by the same relative amount in all populations.

The main impact of the environmental factors in this case is to shift the genotypic values (the mean phenotype of a genotype) up or down by approximately the same amount in all

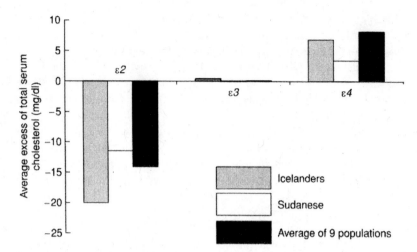

Figure 9.3. Average excesses of $\varepsilon 2, \varepsilon 3$ and $\varepsilon 4$ alleles at *ApoE* locus on phenotype of total serum cholesterol in Sudanese and Icelanders and average over nine populations of humans. No significant differences exist in the average excess values across populations. Data from Hallman et al. (1991).

genotypes. Therefore, although environmental factors are clearly having a large effect on μ, the mean phenotype of the population, and upon the genotypic values G_{ij}, there is hardly any effect at all upon the genotypic deviations, $G_{ij} - \mu$. As shown in Chapter 8, all of the important parameters in the Fisherian quantitative genetic model are ultimately functions of the genotypic deviations, so if the genotypic deviations are the same and the genotype frequencies are similar, the quantitative genetic inferences will also be similar, as indeed they are in this case. This example shows that two populations can differ greatly in their mean phenotypes even though the populations are genetically homogenous. This example also illustrates that differences in environment can contribute to significant differences in mean phenotypes even though the phenotype is influenced by genetic variability in all populations to the same quantitative degree. Hence, the Fisherian parameters are irrelevant to and tell us nothing about the mean phenotypes of populations.

Our second example relates to the phenotype most abused by the spurious argument that high heritabilities imply that mean population differences are due to genetic differences between the populations: the intelligence quotient score (IQ) designed to measure general cognitive ability in humans. One of the earliest studies on the heritability of IQ is that of Skodak and Skeels (1949), a study that is still cited by those such as Herrnstein and Murray (1994), who make the argument that differences in IQ scores between populations are genetically based. The study of Skodak and Skeels is frequently cited in this context because they concluded that IQ has a high heritability. However, a closer examination of their results actually illustrates the inapplicability of heritability to mean phenotypic differences between populations. As mentioned earlier, one great complication in human studies is the fact that environmental variables are often correlated among relatives due to shared family environmental effects. Skodak and Skeels attempted to eliminate or at least reduce these environmental correlations by examining the IQ scores of a population of adopted children and comparing the adopted children both to their biological mothers and to their adoptive mothers. Figure 9.4 shows the normal curves obtained with the observed means and variances of these three populations (adopted children, biological mothers, adoptive mothers). As shown in Figure 9.4, the adoptive mothers had a much higher average IQ

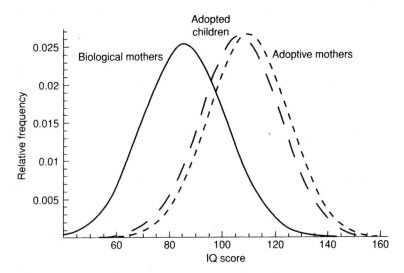

Figure 9.4. Normal distributions of IQ scores in biological mothers (solid line), adopted children (long-dash line), and adoptive mothers (short-dashed line) from study of Skodak and Skeels (1949).

score (110) than the biological mothers (86). These two populations of mothers also differed greatly in socioeconomic status, a strong indicator of many environmental differences in human populations.

Skodak and Skeels also measured the correlation of the phenotypic scores between the adoptive children and their biological and adoptive mothers. There was no significant correlation between the IQ score of the children and that of their adoptive mothers, but there was a highly significant correlation of 0.44 between the IQ scores of the children and their biological mothers. From equation 9.13, this implies that the heritability of IQ is 0.88. Skodak and Skeels therefore concluded that genetic variability is the major contributor to variation in IQ scores *within* the population of adopted children. But it would be a mistake to infer from these data that IQ is genetically *determined* at the individual level or that environmental factors play no role in an individual's IQ score.

The average IQ score for the general population is adjusted to have a mean of 100. The subpopulation of mothers willing to give up their children for adoption are not randomly drawn from this general population, but rather are drawn disproportionately from the lower socioeconomic sectors of the general population. The highly selected nature of the population of biological mothers is reflected in the fact that they have a mean IQ of 86, nearly a full standard deviation below the general mean. The selective intensity in this case is $S = (86 - 100) = -14$, showing that the biological mothers are 14 IQ points below the mean of the general population. Using a heritability of 0.88, the response in IQ score of the children of the biological mothers should be, using equation 9.22, $R = (-14)(0.88) = -12.32$. Therefore, the expected average IQ of the adoptive children should be, under the assumption of a constant environment across the generations, $100 + R = 87.68$. However, as Figure 9.4 shows, the mean IQ score of the adopted children was 107—nearly 20 IQ points higher than that predicted from equation 9.22 and statistically indistinguishable from the mean IQ of the adoptive mothers! Why such a discrepancy? The adoption agencies at that time also highly selected the adoptive families, placing the children in families with higher than average socioeconomic status. The highly selective nature of the families into which the children were placed is reflected in the average IQ scores of the adoptive mothers, which deviates from the general mean of 100 but in a direction opposite that of the biological mothers. Hence, Skodak and Skeels concluded that the environments associated with the adoptive families strongly contributed to a significant increase in IQ scores (well over a standard deviation) for the adoptive children. Thus, two major conclusions emerge out of the study of Skodak and Skeels:

- IQ has a high heritability within the population of adopted children and genetic variation is the primary cause of phenotypic variation in IQ score in these children.
- The environment strongly influences the IQ scores of these children.

For those who do not understand the concept of heritability, these two conclusions may seem contradictory, but they are not. Heritability has nothing to do with the mean phenotype. If the altered socioeconomic environments into which these children were placed had a uniform and highly beneficial impact upon all of them, the mean IQ score could rise substantially (as it did) but without altering the strong positive correlation between IQ scores of biological mothers and adopted children. As shown explicitly in equation 9.1, the correlation coefficient measures the association between the paired observations *relative to their respective population means*. Consequently, the high heritability of IQ in this study indicates that biological mothers who had an IQ below the mean of 86 had a strong tendency to have a child with an IQ below the children's mean of 107. Likewise, a biological mother with an IQ above the mean of 86 had a strong tendency to have a child with an IQ above the children's mean of 107. The actual mean values themselves of 86 and 107 are subtracted off each observation in the very first step of calculating the correlation coefficient (equation 9.1) and are thereby mathematically irrelevant to the calculation of the correlation coefficient and of the heritability. It is only the *deviations* from the mean that influence heritability, not the mean value itself. Therefore, even if the heritability were 1, the environment could still be a major determinant of the mean phenotype and thereby a major determinant of the actual phenotypic values of each individual.

In the previous two examples, the environment had a strong impact upon the phenotype, but the environmental variation simply shifted the phenotypes up or down in a similar manner for all individuals irrespective of their genotype. The interpretation of phenotypic differences between populations is even more complicated when different genotypes respond to the environment in different, nonuniform fashions. For example, different populations of the plant *Achillea* (a yarrow) live at different altitudes and show large phenotypic differences in size and growth form, as well as much variation within a population (Clausen et al. 1958). The bottom panel of Figure 9.5 shows some of the variation found in five individual plants collected along an elevation gradient in California. An entire adult plant can be grown from a cutting in this species, which clones the entire intact genotype of that individual. Cuttings from these five plants were then grown at low-, middle-, and high-elevation sites, with the results shown in the upper three panels of Figure 9.5. The responses to environmental variation were substantially different across these five fixed genotypes. For example, the genotype that was the tallest at low elevation is of medium height at the middle elevation and dies at the highest elevation. In contrast, the shortest plant at the low-elevation site becomes even shorter at the middle-elevation site but then grows much taller at the high-elevation site. This heterogeneity in genotypic response to variation in elevation illustrates the **norm of reaction**, that is, the phenotypic response of a particular genotype to an environmental factor. In the *ApoE* example, all genotypes shared a common norm of reaction to the environmental

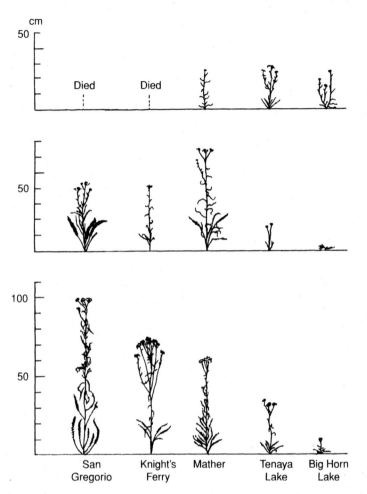

Figure 9.5. Variation in growth form and height as function of elevational environmental variation in five cloned genotypes of plant *Achillea* (yarrow) grown in low elevation (bottom panel), midlevel elevation (middle panel), and high elevation (top panel). The original locations of the source of the clones are indicated by the names given below the bottom panel and are ordered from lowest elevation to highest in going from left to right. Based on data From Clausen et al. (1958).

variation that existed among the nine populations sampled. However, for the plant *Achillea*, there is considerable genotypic variation in the norm of reaction itself. In this sense, the norm of reaction can be regarded as a phenotype itself (albeit a difficult one to measure as it requires measurements upon common genotypes in different environments)—a phenotype that can also show genotypic variability and heritability in some populations for some traits (height in *Achillea*) but not others (serum cholesterol as influenced by *ApoE* in humans).

The fourth and last example involves the differences in male head shape between two species of Hawaiian *Drosophila*, *D. silvestris* and *D. heteroneura* (Figure 9.6). As can be seen from that figure, the head shapes are extremely different, with *D. heteroneura* having a hammer-head shape, whereas *D. silvestris* has a rounded head that is typical of most *Drosophila* species. Indeed, there is no phenotypic overlap between the male head shapes of these two species. Moreover, there is little phenotypic variation within some laboratory

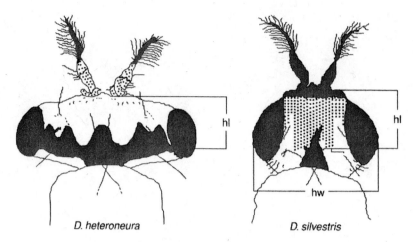

Figure 9.6. Male shapes in two species of Hawaiian *Drosophila*, *D. heteroneura* (left) and *D. silvestris* (right). Shape is measured by the ratio of head length (hl) to head width (hw) or various mathematical transformations of that ratio. From Fig. 1 in Val (1977) Copyright ©1977 by The Society for the Study of Evolution.

strains of each species for male head shape (Val 1977; Templeton 1977), and the little variation that exists in these laboratory strains seems to be attributable to environmental variation (Templeton 1977a). Therefore, the heritability of head shape is small or even zero *within* these strains of both species. *Drosophila heteroneura* and *D. silvestris* are interfertile, and the genetic basis of their head shape differences can therefore be studied directly, as will be detailed in the next section. These crosses clearly indicate that the head shape differences between these strains of different species are largely genetic. In this case, heritability of head shape differences is small *within* these laboratory strains of *D. heteroneura* or *D. silvestris*, yet the head shape differences *between* the strains of *D. heteroneura* and *D. silvestris* are primarily genetically based.

Collectively, these four examples show that there is *no relationship* between heritability within a population and the biological basis of phenotypic differences between populations. Traits can have high heritabilities within a population yet large phenotypic differences between may be primarily due to environmental factors (e.g., IQ in the study of Skodak and Skeels 1949); or traits can have low or no heritability within a population yet large phenotypic differences between may be primarily due to genetic differences (e.g., head shape in *D. heteroneura* and *D. silvestris*). This is not a surprising conclusion, because the entire Fisherian model given in Chapter 8 was designed to investigate causes of variation in deviations from the overall population mean within a single population. The Fisherian model given in Chapter 8 was not designed to study between-population differences. We therefore need a new quantitative genetic approach to study between-population differences. An unmeasured genotype approach to this problem is given in the following section.

CONTROLLED CROSSES FOR ANALYSES OF BETWEEN-POPULATION DIFFERENCES

As shown in the previous section, the traditional quantitative genetic model is inapplicable to the study of between-population differences. Moreover, with unmeasured genotypes, there is a serious confoundment between environmental and genetic factors when comparing differences between populations. This problem can be circumvented by finding or creating a hybrid or admixed population and then studying the distribution of phenotypes in these hybrid and other crosses under a common environmental regimen. Hybridization and admixture occur naturally in many circumstances. Such mixed populations provide a valuable opportunity for studying the genetic basis of phenotypic differences between the original parental populations. However, the uncontrolled nature of such natural crosses poses difficulties. If the admixture was very recent or ongoing and it is possible to identify individuals as being members of the traditional Mendelian categories of F_1, F_2, backcross, and so on then the approaches to be discussed in this section are applicable. In many cases of natural hybridization or admixture, it is impossible to accurately classify individuals or otherwise quantify the degree of admixture for an individual without the use of genetic markers. Those cases in which admixed populations are used in conjunction with genetic markers will be discussed in the next chapter on measured genotype approaches. In this chapter, we limit the discussion to the third major way of performing unmeasured genotype analyses: *controlled crosses between populations or the use of ongoing admixture between populations in which individuals can be unambiguously classified into traditional Mendelian cross categories*. The approach of controlled or known crosses has the ability to address the genetic basis of phenotypic differences between populations.

We will only deal with the simplest case in which the two populations have fixed genetic differences for the genes affecting the trait of interest, although more complicated situations can be handled (Lynch and Walsh 1998). In this simple case, the original parental lines can be regarded as homozygous stocks with no internal genetic variation. This also means that the trait has no heritability within either parental line for the reason that there is no genetic variation within lines. Consider now the simplest situation in which the phenotypic differences between the populations are due to a single locus in which individuals from the two parental populations (P_1 and P_2) are homozygotes and the F_1 between them are heterozygotes:

Population	P_1	F_1	P_2
Genotype	AA	Aa	aa
Genotypic value	G_{AA}	G_{Aa}	G_{aa}

As with the Fisherian model for a single population, our focus is not upon the genotypic values per se, but rather upon the phenotypic differences between the two parental populations. The difference in mean phenotypes between the parental populations is $G_{AA} - G_{aa}$, and as long as we preserve this difference, we can rescale the genotypic values to any mathematically convenient value, just as the traditional one-population Fisherian analysis is mathematically invariant to the overall population mean μ, as shown in the previous section. A convenient transformation of the original genotypic values is based upon the midparental value of the two parental populations, $m_p = \frac{1}{2}(G_{AA} + G_{aa})$. Then, let

$$a = \text{new genotypic value of } AA = G_{AA} - m_p = \frac{1}{2}(G_{AA} - G_{aa})$$

$$d = \text{new genotypic value of } Aa = G_{Aa} - m_p$$

$$-a = \text{new genotypic value of } aa = G_{aa} - m_p = \frac{1}{2}(G_{aa} - G_{AA}) = -\frac{1}{2}(G_{AA} - G_{aa})$$

Note that the difference in mean phenotypes between the parental populations is now $a - (-a) = 2a = (G_{AA} - G_{aa})$. Thus, the original mean phenotypic differences are preserved under this transformation, but now we have the mathematical advantage of describing the three original genotypic values by just two parameters, a and d, thereby simplifying our model.

Let σ_e^2 be the phenotypic variance within population P_1. This variance is assigned a subscript of e to emphasize that it must be an environmental variance because by assumption there is no genetic variation within P_1; all individuals are homozygous AA. Now assume that environmental variation is identical in both parental populations and all crosses between them (the "common garden" aspect of this design) and that all genotypes respond to this variation with identical phenotypic variances. The phenotypic environmental variance σ_e^2 is applicable to the individuals within population P_2 (aa) and the heterozygotes Aa as well. Thus, σ_e^2 has the biological interpretation in this model as being the within-genotype phenotypic variance. With these assumptions, the various traditional Mendelian crosses have phenotypic means and variances as given in Table 9.1 (Cavalli-Sforza and Bodmer 1971).

Because we have assumed that the original parental strains are homozygous, the first cross in Table 9.1 that has any genotypic variation within it is the F_2. As can be seen from Table 9.1, the phenotypic variance in the F_2 can be partitioned into a portion reflecting the mean differences between the genotypes as weighted by their F_2 frequencies and σ_e^2, the within-genotype variance. Moreover, note that the between-genotype component of the phenotypic variance can be divided into two terms, one that is solely a function of a and one that is solely a function of d. The portion that is a function of a is defined to be the additive genotypic variance and the portion that is a function of d is defined to be the dominance variance. Note that despite the identical terminology, the additive and dominance variances given in Table 9.1 are not in general the same as the additive and dominance variances given in Chapter 8. In Chapter 8, the additive and dominance variances are defined for a population with arbitrary genotype frequencies, but in Table 9.1 the additive and dominance variances are defined specifically for an F_2 population. The system of mating in this case is highly nonrandom and fixed by the investigator. Because of this the genotype frequencies are the same as Mendelian probabilities; that is, the genotype frequencies of AA, Aa, and aa are in a 1 : 2 : 1 ratio. Hence, the additive and dominance variances here are strictly limited to a manipulated F_2 cross and have no generality to any other population.

Although Table 9.1 was derived only for a one-locus model, as long as each locus contributes to the phenotype in a completely additive fashion with no epistasis, then a has the biological interpretation of being the sum of all the homozygote effects in the P_1 and d the sum of all the heterozygote effects in the F_1, and all else in Table 9.1 remains the same. Hence, Table 9.1 gives the expected phenotypic means and variances when the genetic component of the phenotypic difference between the original two parental populations is due to fixed differences at one locus or many loci with additive gene action. By performing these crosses and measuring the phenotypic variances and means, one can estimate the underlying genetic parameters of this idealized genetic model.

For example, do the male head shape differences between *D. heteroneura* and *D. silvestris* (Figure 9.6) have a genetic basis? To answer this question, we first need a measure of head shape. Val (1977) measured the head length (hl) and head width (hw) (Figure 9.6) in individuals from laboratory strains of both species and their F_1, F_2 and backcross populations. Head length and width are a function of both head shape and head size. In order to measure only head shape and not size, Templeton (1977a) transformed the original measurements to

Table 9.1. Expected Means and Variances for One-Locus Model of Between-Population Phenotypic Differences

Population or Cross	Genotype Frequency			Phenotypic Mean	Phenotypic Variance
	AA	Aa	aa		
P_1	1	—	—	a	σ_e^2
P_2	—	—	1	$-a$	σ_e^2
F_1	—	1	—	d	σ_e^2
F_2	$\tfrac{1}{4}$	$\tfrac{1}{2}$	$\tfrac{1}{4}$	$\tfrac{1}{4}(a) + \tfrac{1}{2}(d) + \tfrac{1}{4}(-a) = \tfrac{1}{2}d$	$\tfrac{1}{4}(a - \tfrac{1}{2} + \tfrac{1}{2}(d - \tfrac{1}{2}d)^2 + \tfrac{1}{4}(-a - \tfrac{1}{2}d)^2 + \sigma_e^2$ $= \tfrac{1}{2}a^2 + \tfrac{1}{4}d^2 + \sigma_e^2 = \sigma_a^2 + \sigma_d^2 + \sigma_e^2$
BC_1	$\tfrac{1}{2}$	$\tfrac{1}{2}$	—	$\tfrac{1}{2}(a) + \tfrac{1}{2}(d) = \tfrac{1}{2}(a + d)$	$\tfrac{1}{2}[a - \tfrac{1}{2}(a + d)]^2 + \tfrac{1}{2}[d - \tfrac{1}{2}(a + d)]^2 + \sigma_e^2$ $= \tfrac{1}{4}a^2 + \tfrac{1}{4}d^2 - \tfrac{1}{2}ad + \sigma_e^2$
BC_2	—	$\tfrac{1}{2}$	$\tfrac{1}{2}$	$\tfrac{1}{2}(d) + \tfrac{1}{2}(-a) = \tfrac{1}{2}(d - a)$	$\tfrac{1}{2}[d - \tfrac{1}{2}(d - a)]^2 + \tfrac{1}{2}[-a - \tfrac{1}{2}(d - a)]^2 + \sigma_e^2$ $= \tfrac{1}{4}a^2 + \tfrac{1}{4}d^2 + \tfrac{1}{2}ad + \sigma_e^2$
Sum of BC_1 and BC_2				$\tfrac{1}{2}(a + d) + \tfrac{1}{2}(d - a) = d$	$\tfrac{1}{2}a^2 + \tfrac{1}{2}d^2 + 2\sigma_e^2 = \sigma_a^2 + 2\sigma_d^2 + 2\sigma_e^2$

Note: The between-population additive and dominance variances are defined in this model to be $\sigma_a^2 = \tfrac{1}{2}a^2$ and $\sigma_d^2 = \tfrac{1}{4}d^2$, respectively. All other parameters are defined in the text.

Table 9.2. Quantitative Genetic Analysis of Male Head Shape Differences between *D. silvestris* and *D. heteroneura* using Model in Table 9.1

Population or Cross	N	Mean θ	Var($\times 10{,}000$)	Expected Variance
D. silvestris	20	1.157	1.21	σ_e^2
D. heteroneura	20	1.234	0.66	σ_e^2
F_1	37	1.197	0.67	σ_e^2
F_2	71	1.207	1.59	$\sigma_a^2 + \sigma_d^2 + \sigma_e^2$
BC_1	141	1.187	1.55	
BC_2	123	1.224	1.21	Sum $= \sigma_a^2 + 2\sigma_d^2 + 2\sigma_e^2$

Source: From Templeton (1977a).

polar coordinates. The angle $\theta = \arctan(hl/hw)$ measured in radians is a convenient measure of head shape because it depends only upon the relative proportions of the original head measurements through their ratio hl/hw and is therefore mathematically invariant to their absolute size. Table 9.2 shows the results obtained by using this transformation on the results obtained by Val (1977).

One simple way of estimating the parameters of this model is to equate the expectations with the observed values (Cavalli-Sforza and Bodmer 1971). When multiple observations can be related to the same parameter, an average of the relevant observed values is taken weighted by their sample sizes (for variances, the sample size is adjusted for the estimation of the mean by subtracting 1; see Appendix 2). For example, the phenotypic variances of the two parental strains and the F_1 should all equal σ_e^2. Therefore, letting V_i be the observed estimated phenotypic variance in strain or population i (letting $i = s$ designate *D. silvestris* and $i = h$ *D. heteroneura*), the estimate of σ_e^2; is

$$\hat{V}_e = \frac{(n_s - 1)V_s + (n_h - 1)V_h + (n_{F_1} - 1)V_{F_1}}{(n_s - 1) + (n_h - 1) + (n_{F_1} - 1)} = 0.000081 \tag{9.23}$$

Similarly, the additive and dominance variances can be estimated by

$$\hat{V}_a = 2V_{F_2} - V_{BC_1 + BC_2} = 0.000042 \tag{9.24}$$

and

$$\hat{V}_d = V_{F_2} - \hat{V}_a - \hat{V}_e = 0.000036 \quad \text{or}$$
$$\hat{V}_d = \tfrac{1}{2} V_{BC_1 + BC_2} - \tfrac{1}{2}\hat{V}_a - \hat{V}_e = 0.000036 \tag{9.25}$$

Hence, the total genotypic variance in this case is $0.000042 + 0.000036 = 0.000078$. Given that the estimated environmental variance is 0.000081, this means that about half of the variability in male head shape between these species is attributable to genetic differences between the species.

More sophisticated statistical and genetic models can be used for control crosses and natural hybrids with individuals of known cross type [e.g., Wu 2000; also see Lynch and Walsh (1998)]. However, the important message is that to study the genetic basis of between-population differences, genetic crosses in a common environment are essential. It is impossible to infer the biological basis of interpopulation differences in phenotypes from observations on the phenotypes of populations that potentially live in different environments. It is essential to control the environment and place all genotypes into identical environments (as was done, for example, in the studies on the plant *Achillea*) or to do crosses, the traditional mainstay of Mendelian genetic analysis. The traditional definitions of genotypic variance, heritability, and so on, as given in Chapter 8, are irrelevant to the problem of between-population phenotypic differences, and arguments based on such intrapopulation parameters have no biological validity.

BALANCE BETWEEN MUTATION, DRIFT, AND GENE FLOW UPON PHENOTYPIC VARIANCE

When genotypes are observed, we can measure the balance between the evolutionary forces of mutation, drift, and gene flow upon genetic variation (Chapters 2–7). However, in this chapter we are assuming that only phenotypic variation is observable, not genotypic variation. Nevertheless, with some additional assumptions, it is still possible to make some predictions about the balance of these evolutionary forces upon quantitative genetic variation that contributes to phenotypic variation.

Suppose there is phenotypic variation for a neutral trait. By a neutral trait, we mean that all individuals make the same average genetic contribution to the next generation regardless of an individual's trait value. Moreover, suppose phenotypic variation for this trait is influenced by environmental variation in a manner that is homogeneous throughout the species' range and affects within-genotype phenotypic variance to the same extent in all genotypes (just as assumed in the previous section). With these assumptions, first consider the balance between mutation and drift upon the genotypic variance in a single, undivided population. Because phenotypes are being measured and not genotypes, the mutation rate μ is not directly relevant; rather, only the extent to which new phenotypic variation is being produced by mutation is relevant. Therefore, let σ_m^2 be the mutational additive variance, that is, the additive genetic variance in the phenotype that is created by mutation every generation. This mutational additive variance is the analogue of the mutation rate in this model of phenotypic evolution. Just as we assumed in previous chapters that μ was constant over the generations, we make the same assumption about mutational additive variance. Lynch and Hill (1986) then showed that the equilibrium additive genetic variance is given by

$$2N_{e\lambda}\sigma_m^2 \qquad (9.26)$$

where $N_{e\lambda}$ is the eigenvalue effective size that measures the rate at which alleles become fixed or lost (Chapter 4). As we saw in Chapter 4, the balance of mutation and drift upon various measures of genetic variation depends upon the product of an effective population size times the mutation rate (reflecting the balance of the power of mutation divided by the power of drift, which is inversely proportional to an effective population size). Equation 9.26 says much the same at the phenotypic level. Additive genetic variance is created by an amount σ_m^2 per generation by the action of mutation, and it is lost at a rate $1/(2N_{e\lambda})$ due to the action of drift per generation. The equilibrium additive genetic variance is just the ratio of the rate of creation divided by the rate of loss.

Whitlock (1999) extended this basic model to incorporate various models of population subdivision and gene flow. In dealing with subdivided populations, we are concerned both with the amount of additive genetic variance available within the local demes, designated by V_{within}, and the additive genetic variance found among the demes, V_{among}. When population subdivision is measured by F_{st} (Chapter 6), then the equilibrium additive genetic variances are

$$V_{\text{within}} = 2N_{e\lambda}(1 - F_{st})\sigma_m^2 \qquad V_{\text{among}} = 4N_{e\lambda}F_{st}\sigma_m^2 \qquad (9.27)$$

where $N_{e\lambda}$ is the eigenvalue effective size for the species as a whole and not the local demes. Recall that F_{st} itself is a function of the inbreeding effective size and the pattern and amount of gene flow (Chapter 6). Hence, a great deal of biological complexity is implicit in equations 9.27.

Equations 9.27 show that the amount of additive genetic variance within a local deme increases with decreasing F_{st}, converging to the single, undivided population result (equation 9.26) when $F_{st} = 0$. As shown in Chapter 6, this means in turn that as local inbreeding effective sizes become larger and/or as there is more gene flow between demes, more and more additive genetic variance becomes available within the local demes. In contrast, the additive genetic variance among demes is an increasing function of F_{st}, so this component of additive genetic variance is augmented by smaller local inbreeding effective sizes and less gene flow. As subdivision becomes increasingly extreme, more and more of the additive genetic variance shifts from the within component to the among, until at $F_{st} = 1$ there is no additive genetic variance within demes and all additive genetic variance exists due to fixed genetic differences among demes.

The total additive genetic variance in the species is $V_{\text{within}} + V_{\text{among}}$, which from equation 9.27 is

$$2N_{e\lambda}\sigma_m^2(1 + F_{st}) \qquad (9.28)$$

Whenever a population is subdivided due to local drift and restricted gene flow, $F_{st} > 0$ and equation 9.28 will be greater than equation 9.26, the total additive genetic variance in a species consisting of a single, undivided population. Recall from Chapter 6 that a subdivided population has a total variance effective size *larger* than an undivided population with the same census size. The same is true for eigenvalue effective size (Maruyama 1972). Consequently, for the same number of individuals, $N_{e\lambda}$ is generally going to be larger for a subdivided population than for an undivided one. Moreover, population subdivision converts more of what would be nonadditive genetic variance in an undivided population into additive variance (Goodnight 1995). These impacts of population subdivision further increase the total additive genetic variance. Hence, population subdivision not only causes a partitioning of additive genetic variance within and among demes, it also increases the total amount of additive genetic variance found in the species as a whole given the same mutational input per generation. As we will see in Chapter 12, this impact of subdivision upon additive variance can have important adaptive consequences.

10

QUANTITATIVE GENETICS: MEASURED GENOTYPES

As seen in the previous two chapters, Fisher developed a framework for analyzing the genetic contributions to phenotypic variability even when no genotypes were measured. In the same year that Fisher published his paper (1918), Payne (1918) studied the genetic basis of scutellar bristle number in *Drosophila melanogaster* by crossing two lines with different bristle numbers and followed their response to selection. Payne made his genetic inferences about the underlying genetic contribution to the phenotype of scutellar bristle number by monitoring the fate of genetic differences at several visible genetic markers scattered through the genome that displayed fixed differences between the original parental strains. By following these markers and overlaying bristle number data on the marker genotype data, Payne inferred that several loci affect bristle number. This was perhaps the first example of using measured genotypes to study the genetics of a multilocus quantitative trait.

Although the measured genotype approach is as old as the Fisherian unmeasured approach, the measured approach has only become widespread recently. There are two primary reasons for this new excitement over an old approach.

- Molecular genetics has revealed many polymorphic loci scattered throughout the genome of most species. As a result, genetic markers are now available in virtually any organism, and the scoring of the markers is more amenable to automation than the visible markers used by Payne. Moreover, much is known about the biochemical and/or physiological functioning of some of these genetic markers, which can allow the formulation of direct hypotheses about specific genotypes influencing specific phenotypes.
- Computing power is now available to implement intensive calculations.

This combination of molecular genetic and computer technology allows studies on quantitative traits that were impractical earlier and that extend the applicability of measured genotype approaches in quantitative genetics to many more phenotypes and organisms.

Population Genetics and Microevolutionary Theory, By Alan R. Templeton
Copyright © 2006 John Wiley & Sons, Inc.

There are two main types of measured genotype approaches, which in turn can be subdivided into subtypes. First are **marker loci studies** that measure genotypes at loci that are not known to directly affect the phenotype of interest but that may display indirect associations with those loci that do influence the phenotypic variation. Payne's study is an example of this approach. He had no reason to propose that the visible markers he was monitoring were in any way related to the phenotype of scutellar bristle number. However, in the context of a cross between two strains and subsequent reproduction, he reasoned that any marker located on the same chromosome as a locus actually affecting scutellar bristle number would display phenotypic associations through genetic linkage. Marker loci studies can be subdivided into three major subtypes on the basis of the primary factor that is thought to lead to the indirect associations between genetic variation at marker loci and the phenotypic variation of interest:

- *Markers of Interpopulation Disequilibrium.* As seen in Chapters 3 and 7, past or ongoing admixture events between genetically differentiated populations induce extensive linkage disequilibrium. If the parental populations also display phenotypic differences that have a genetic basis, then marker loci that have large differences in allele frequency in the parental populations are expected to show indirect phenotypic associations in the admixed population through linkage disequilibrium with the loci actually contributing to the phenotypic differences between the parental populations. For recent or ongoing admixture, physical linkage is not needed for "linkage" disequilibrium. However, the rate of decay of the linkage disequilibrium is a function of recombination frequency, so such studies make use of indirect associations induced by temporal proximity to an admixture event and/or physical proximity on a chromosome. This type of marker study is appropriate for studying the genetics of between-population phenotypic differences.
- *Markers of Intrapopulation Disequilibrium.* Linkage disequilibrium occurs within a single population as well. Indeed, as shown in Chapter 2, the very act of mutation creates disequilibrium. Disequilibrium induced by mutation generally decays very rapidly due to recombination under random mating and is therefore found between polymorphic sites with a low recombination rate. As shown in Chapter 4, linkage disequilibrium can extend across larger segments of DNA if recent founder or bottleneck effects had occurred. As with markers of interpopulation disequilibrium, this approach makes use of a population's past evolutionary history, but because the linkage disequilibrium induced by founder and bottleneck effects is generally less extreme than that induced by admixture, physical linkage is generally required as well to detect indirect phenotypic associations. This type of marker study is appropriate for studying the genetics of the phenotypic diversity found within a single population.
- *Markers of Linkage.* Linkage disequilibrium is a population concept, but linkage can affect the cosegregation of genetically variable loci within a family or set of controlled crosses even in the absence of linkage disequilibrium between the loci at the population level. Therefore, studies of marker segregation in families with known pedigrees or controlled or known crosses provide a method for detecting indirect associations that does not depend upon a population's evolutionary history or upon linkage disequilibrium in the population. This approach is appropriate for studying the genetics of both intra- and interpopulational phenotypic differences.

All of these marker approaches share the common theme of indirect associations influenced by physical linkage between markers and causative loci. Moreover, they are not mutually exclusive. For example, often marker linkage studies are performed in populations with a recent evolutionary history of a founder event or in admixed populations.

The second major type of a measured genotype approach is the **candidate locus study**, in which prior information is used to implicate specific loci as likely to be associated with the phenotype of interest. The candidate locus approach is appropriate for studying the genetics of both intra- and interpopulational phenotypic differences. This approach can and often is integrated with the marker loci approach. In choosing markers for a marker association study, an investigator can include genetic markers at known candidate loci. Also, when marker studies reveal a segment of the genome that appears to contribute to phenotypic variation, that segment can be screened in detail for potential candidate loci, which then become the focus of subsequent studies.

The measured genotype approaches offer greater insight into the underlying genetic architecture of phenotypic variation than is possible with unmeasured approaches. Moreover, they allow aspects of the genotype-to-phenotype mapping to be studied and estimated that are inaccessible to the unmeasured approaches. For example, we saw in Chapter 8 that a measured genotype approach using the candidate *ApoE* locus allowed the estimation of genotypic values, a simple quantitative genetic parameter that is difficult or impossible to estimate with unmeasured genotype approaches. Also, interactions among factors, both genes interacting with genes and genes interacting with environments, are much easier to study with measured than with unmeasured genotype approaches. Finally, both types of measured genotype approaches offer the promise of actually identifying the specific loci that contribute to phenotypic variation in the population, thereby opening up many more research possibilities. We now will discuss in more detail these measured genotype approaches in quantitative genetics.

MARKER LOCI

Markers of Interpopulation Disequilibrium

The study of phenotypic associations with genetic markers in recently admixed populations is the measured genotype analogue to the use of controlled or known crosses to study between population differences with the unmeasured approaches (Chapter 9). The advantage of the measured approach over the unmeasured approach is that it can be applied to controlled or known crosses and also to uncontrolled cases of natural hybridization or admixture. With uncontrolled admixture events, we often can identify an admixed population but cannot characterize the cross status of specific individuals. To implement the unmeasured approach illustrated by Table 9.2 in the previous chapter, it was essential to classify individuals as coming from specific crosses such as F_1, F_2, various backcrosses, and so on. After a few generations of admixture, such a simple categorization is usually impossible, and the unmeasured genotype approach described in Chapter 9 cannot be used. However, molecular markers can be used to estimate the degree of admixture for particular individuals within a recently or ongoing admixed population.

Because admixture creates linkage disequilibrium, even between unlinked loci (equation 6.3), phenotypic associations with the measured markers sometimes reflect only the shared history of admixture and not physical linkage. Nevertheless, linkage disequilibrium induced by admixture will generally decay at a rate determined by the recombination frequency, so genetic linkage can have an observable impact on the magnitude of the association, thereby allowing inference on chromosomal location of potentially causative loci. Therefore, physical linkage between markers and causative loci are not necessary in admixture studies, but additional genetic insights are possible when markers are sufficient in number to ensure close physical linkage between at least some markers and the causative loci.

The act of measuring genotypes does not necessarily avoid the problem of spurious associations that can be created when populations and environments are nonrandomly associated. Hence this approach is best when it can be demonstrated that the individuals from the admixed population also live in a common environment. Under these circumstances, such studies can study the genetic basis of between-population differences in a manner that eliminates or reduces the confoundment that arises when the parental populations have both genetic and environmental differences.

One early study of this type is that of Scarr et al. (1977), who used measured genotypes to mark individual admixture history but not physical linkage. As shown in Chapters 3, 6, and 7, African Americans represent a population strongly influenced by admixture between humans originally from Europe and from sub-Saharan Africa. Scarr et al. (1977) scored 14 loci, most with multiple alleles, in a sample of African Americans from the city of Philadelphia, a sample of West Africans, and a sample of Europeans and European Americans. The latter two samples were chosen to represent the original parental populations involved in the admixture, and the 14 loci were chosen because their genotype frequencies in Africans and Europeans were known to be very different. An *individual* in the admixed population of African Americans was then scored for his or her genotypes at these 14 loci and an "odds" ratio for ancestry would be calculated for the individual as follows:

$$\text{Odds} = \log\left[\frac{\prod_{i=1}^{14} A_i}{\prod_{i=1}^{14} B_i}\right] \tag{10.1}$$

where A_i is the genotype frequency in West African populations of the genotype observed in this particular individual at locus i and B_i is the genotype frequency for the same genotype in Europeans.

Socioeconomic status is an important indicator of the human environment, and this environmental indicator varies significantly between European and West African populations and between European American and African American populations. Consequently, any phenotypic difference between European American and African American populations is potentially confounded by both genetic and environmental differences. In order to use admixture as a tool for investigating genetic contributions to phenotypic variation, it is necessary that the individuals from the African American population experience the same environment or the same range of environments in a statistically independent fashion regardless of their degree of admixture. Therefore, as a control for environmental influences, Scarr et al. showed that there was no correlation between the ancestral odds scores calculated from the genetic markers using equation 10.1 with indicators of socioeconomic status *within* the African-American population.

The next step of the analysis is to look for associations between phenotypes and the ancestral odds scores calculated from the marker loci. Scarr et al. looked at several phenotypes. For example, skin color reflectance through a red filter off the forehead showed a significant correlation of 0.27 with the ancestral odds coefficient. This implies that the mean skin color difference observed between Europeans and Africans has a genetic basis. Scarr et al. also measured cognitive skills through the use of five standardized tests and combined the cognitive test results into a single phenotypic measure through a principal-component analysis. Although the European Americans and African Americans differed significantly

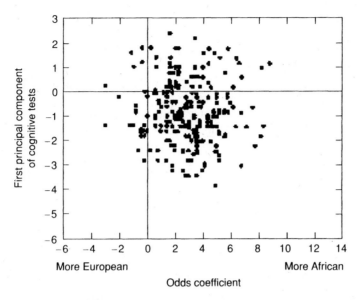

Figure 10.1. Plot of measure of cognitive ability (first principal component score derived from five standard tests of cognitive ability) versus degree of European and African ancestry, as estimated from marker loci, in population of African Americans. From Fig. 3 in Scarr et al., *Human Genetics* 39:69–86 (1977). Copyright © 1977 by Springer. Reprinted with permission of Springer Science and Business Media.

in their mean phenotypes of these test scores, no significant correlation was found between any of these test phenotypes with the ancestral odds ratio within African Americans. Figure 10.1 shows the results they obtained for the phenotype of the first principal component of all the cognitive test scores, which yields a statistically insignificant correlation coefficient of −0.02. Hence, in contrast to the skin color differences, there is no detectable genetic contribution for the differences between European and African Americans with respect to cognitive abilities. Recall from Chapter 9 that cognitive test score results (e.g., IQ tests) often have high heritabilities, and such studies have been done for both European American and African American populations (Herrnstein and Murray 1994). Yet Scarr et al. (1997) could find no evidence for any genetic contribution in their study. There is no contradiction here: Heritability refers to within-population variation and, as shown repeatedly in Chapter 9, heritability is biologically and mathematically irrelevant to between-population mean differences. In contrast, studies within an admixed population focus upon the differences *between* the parental populations, and none were detected in this study.

The study of Scarr et al. only used the genetic markers as indicators of ancestral history. To also use the genetic markers as indicators of linkage, many more marker loci are required to ensure close physical linkage of the unknown causative loci to one or more markers. In particular, one or more markers need to be well within the recombination range associated with the large magnitudes of linkage disequilibrium found in admixed populations. This range in turn is a function of the degree and timing of admixture and the magnitude of the initial genetic differentiation between the parental populations (McKeigue 1998). Thus, there is no one minimal genetic map resolution that will be sufficient for all populations, but from the human examples described in Chapter 7, a resolution no less than 20–30 cM is generally required. On the basis of computer simulations of admixture events 2–10 generations ago (the time scale most relevant for human studies), McKeigue (1998) recommended 2000 marker loci for human studies. However, not just any polymorphic marker suffices for this purpose. Successful mapping using admixed populations requires that the marker loci display strong differentiation in allele frequencies in the parental populations (McKeigue 1998), just as was required in the study of Scarr et al. (1977).

Studies using fewer markers have already proven successful. For example, Zhu et al. (2005) scored 269 microsatellite loci in three populations: an admixed population of African Americans, a population of European Americans, and a population of Africans from Nigeria. One can then use equation 6.4 to estimate the degree of admixture. Equation 6.4 calculates the amount of European input into the African American gene pool, but in this case, Zhu et al. (2005) were interested in calculating the extent of African input into the African American gene pool. Accordingly, equation 6.4 is modified to be

$$M = \frac{p_A - p_E}{p_W - p_E}$$

$$= \frac{\text{change in allele frequency in African Americans from Europeans}}{\text{initial difference in allele frequency between West Africans and Europeans}}$$

(10.2)

The value of M in equation 10.2 when averaged over all marker loci was 0.81 in the African American population, a result virtually identical to the studies mentioned in Chapter 6.

As pointed out in describing the work of Scarr et al. (1977), admixture mapping requires that the degree of admixture be estimated for each *individual* in the admixed population. Zhu et al. (2005) obtained this individual estimate by modifying equation 10.2 instead of using equation 10.1. In particular, the degree of African ancestry for a particular individual i was measured by

$$M_i = \frac{1}{L} \sum_{j=1}^{L} \left(\frac{p_j - p_E}{p_W - p_E} \right) \quad (10.3)$$

where p_j is the allele frequency most common in Africans at locus j in individual i and L is the number of marker loci ($n = 269$ in this case). Because p_j is measuring the frequency of an allele in an individual, this frequency can only take on the values of 0, 0.5, or 1. However, by averaging these locus-specific values of admixture over a large number of markers, an accurate characterization of the extent of African ancestry for individual i can be obtained.

African Americans as a group suffer from earlier and more severe hypertension (high blood pressure) than most other groups in America. Zhu et al. therefore split their African American sample into those individuals with high blood pressure (cases) and those with normal blood pressure (controls). If there were genetic differences between Europeans and Africans that influenced risk for hypertension, then those markers most closely linked to these hypertensive genes should show either increased African ancestry if the African alleles at the hypertensive loci tended to increase blood pressure in African Americans or decreased African ancestry if the European alleles tended to increase blood pressure in African Americans. Zhu et al. therefore estimated the amount of excess African ancestry for marker locus l, $\Delta \Pi_l$, as

$$\Delta \Pi_l = \frac{1}{n} \sum_{i=1}^{n} (q_{il} - M_i) \quad (10.4)$$

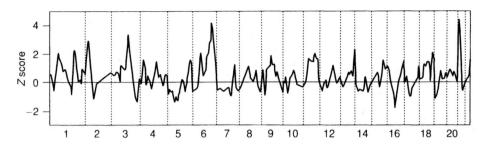

Figure 10.2. Plot of Z scores for 269 microsatellite marker loci against their position in human genome. The numbers on the x axis indicate the number of the human chromosome. Dashed lines separate different chromosomes. The width between dashed lines reflects the length of the chromosome in centimorgans. Adapted with permission from X. Zhu et al., *Nature Genetics* 37:177 (2005). Copyright © 2005 by Macmillan Publishers.

where q_{il} is the amount of admixture estimated for individual i based on marker locus l, M_i is the amount of admixture estimated for individual i based upon all the marker loci (equation 10.3), and n is the number of individuals in the sample. The excess African ancestry is then transformed with a normalized score (Appendix 2) as

$$Z_l = \frac{\Delta \Pi_l}{\text{std}(\Delta \Pi_l)} \qquad (10.5)$$

where $\text{std}(\Delta \Pi_l)$ is the standard deviation (Appendix 2) of the excess African ancestry at marker l across individuals. These Z scores were calculated for all 269 markers in the hypertensive African Americans, and Figure 10.2 shows a plot of these scores against the location of the markers in the human genome. Two regions of the human genome on chromosome 6 and chromosome 21 have Z scores in excess of 4 and are highly statistically significant. When these analyses were repeated on the African American controls, no peaks were found. Both of the Z scores in the hypertensive sample are positive, indicating that the African alleles in these two chromosomal regions are predisposing individuals to increased risk for high blood pressure.

Markers of Intrapopulation Disequilibrium

Another critical aspect of a population's recent evolution for understanding and interpreting present-day associations between markers and phenotypes is its demographic history of founder and bottleneck events. Populations with a recent history of founder or bottleneck effects are particularly valuable in quantitative genetic studies for several reasons:

1. Founder and bottleneck effects augment the magnitude and physical extent of likely linkage disequilibria, thereby creating strong associations between markers and phenotypes.
2. The founder or bottleneck event reduces the overall levels of genetic variation, thereby resulting in a more uniform genetic background. This uniformity is particularly important for traits whose underlying genetic architecture is characterized by epistasis. A more uniform genetic background converts much of the epistasis, which can appear erratic when the actual causative genes are not being measured, into more stable marginal effects attributable to the remaining polymorphic loci (Cheverud and Routman 1996; Cheverud et al. 1999; Goodnight 1995, 2000).

3. A founder event can increase the frequencies of some genes associated with a phenotype of interest. It is therefore easier to sample individuals with a phenotype of interest in such founder populations, and the alleles underlying this phenotype are upon a more homogeneous genetic background.

The founder populations most likely to have strong associations with noncausative markers are those that have experienced little or no subsequent gene flow with other populations after the founder event. Postfounder reproductive isolation allows the persistence of the linkage disequilibrium patterns, commonness of a phenotype of interest, and the genetic background homogeneity induced by the original founder or bottleneck event. Moreover, sampling individuals from a relatively isolated population makes it less likely to inadvertently sample individuals from more than one local deme in a subdivided population. Pooling individuals from genetically differentiated demes creates linkage disequilibrium in the total sample that is a function only of the allele frequency differences between demes and is independent of actual linkage relationships (equation 6.3). Disequilibrium that is independent of true linkage undercuts the fundamental biological premise of the marker association analysis and can create spurious associations. Obviously, pooling of individuals from genetically differentiated demes should be avoided when population subdivision is known, but often the subdivision was not known or taken into account during the sampling stage of a study. There are methods for trying to detect or adjust for such cryptic subdivision after the fact (Pritchard et al. 2000; Devlin et al. 2001), but pooling is best avoided in the first place. Sampling from a relatively isolated founder population is one effective method of minimizing the dangers of cryptic population subdivision in a sample. For all these reasons, the identification and characterization of relatively genetically isolated populations that have undergone recent founder or bottleneck effects have become a high priority in quantitative genetic studies (Peltonen et al. 2000, Peltonen 2000).

The basic idea of the intrapopulational disequilibrium approach is to score the population for a large number of markers and test the marker genotypes for phenotypic associations. If a phenotypic association is detected, it is attributed not to the marker genotypes themselves but rather to linkage disequilibrium between the marker locus with a nearby **quantitative trait locus, or QTL**, a locus whose genetic variation significantly contributes to some of the phenotypic variation observed in the population. Since it is not known a priori how many QTLs may exist or their chromosomal locations, it is critical to have sufficient marker loci scattered throughout the genome to ensure likely linkage disequilibrium between at least one of the markers and any given QTL. Even for founder populations, sufficient coverage often requires a marker density of less than 5 cM (Ober et al. 2001).

This approach is most straightforward when the relevant phenotypic variation is primarily due to a single locus, as for example a Mendelian disease in humans. The value of isolated founder populations is illustrated in such a case by the first successful example of positional cloning of the gene of a human genetic disease, Huntington's chorea. Huntington's chorea is inherited as an autosomal dominant genetic disease and is associated with a late age-of-onset degeneration of the central nervous system (usually after 40 years of age) that ultimately causes death. The cloning of this disease gene depended first upon identifying a relatively isolated founder population that had a high frequency for this disease. Such a population was found in the region of Lake Maricaibo in Venezuela (Gusella et al. 1983). A restriction fragment length polymorphism (RFLP) was found to be associated with the disease in this population, which indicated that the disease locus was nearby on the tip of the short arm of chromosome 4 of the human genome. It must be emphasized that the association of

Huntington's disease with this RFLP is not universally true for all human populations; it is applicable primarily to that one founder population. However, the known chromosomal location of the RFLP was a universal indicator of the approximate chromosomal location of the locus for Huntington's disease and allowed the eventual identification and cloning of that disease gene.

Linkage disequilibrium often allows the mapping of the disease gene to an interval of just 1–2 cM (Peltonen et al. 2000). As pointed out in Figure 2.5 for the sickle cell allele, when a disease allele arises by mutation it occurs on a specific haplotype background that can be stable for many generations in a small region about the original mutation. Often disease mutations are recurrent, thereby occurring on a variety of haplotype backgrounds in the species as a whole. This indeed is what Figure 2.5 shows for the sickle cell allele. However, in any one founder population, it is likely that most or even all of the copies of the disease allele trace back to a recent common ancestral DNA molecule through the coalescent process, thereby making the haplotype background associated with the mutation far more uniform in a specific founder population. Therefore, once a general region of the genome has been indicated by disequilibrium with single markers, an examination of haplotype associations within this region can often refine the location of the disease gene to 50–200 kb, greatly facilitating the targeting of physical cloning and sequencing efforts (Peltonen 2000).

Difficulties arise with this marker approach when the genetic architecture is more complex. Suppose, for example, that many genes contribute to the phenotypic variation but in a manner characterized by extensive epistasis. Many of the epistatic QTLs could be overlooked (Cheverud et al. 1996) if phenotypic associations are only examined one marker at a time. Thus, the genetic model used constrains what is discovered; we tend to see only what we look for, particularly when it comes to epistasis (Frankel and Schork 1996).

Another difficulty is in choosing the marker loci. The analysis depends upon linkage disequilibrium, so any factor that tends to reduce disequilibrium undermines the power of this approach. As pointed out in Chapter 5, not all marker pairs display significant disequilibrium, sometimes even over less than 2 kb (e.g., Figure 5.18). Therefore, it is doubtful if a single marker allele would always display disequilibrium with a QTL. Moreover, when using haplotypes in small DNA regions, the linkage disequilibrium that exists is not well correlated with physical distance, as also pointed out in Chapter 5. This means that when attempting fine-scale mapping of a QTL, a significant phenotypic association with a marker is not a reliable guide to the actual location of the QTL.

Another problem with markers is the danger of selecting a marker at a highly mutable site. Frequent mutation at a site weakens linkage disequilibrium because it places the same genetic state on a variety of chromosomal backgrounds. It is commonplace in much of the evolutionary literature to assume the infinite-sites model (which does not allow multiple hits) for nuclear DNA. At first glance, the infinite-sites model seems reasonable. For example, a genetic survey of the human gene *lipoprotein lipase* (*LPL*) discussed in Chapter 5 revealed 88 polymorphic sites out of 9734 in the sequenced region—a figure that seems to be well below saturation. Moreover, almost all these 88 polymorphic sites have only two alternative nucleotide states, which seemingly further bolsters the argument against multiple mutational hits at the same nucleotide sites. However, these arguments are based upon the premise that mutations are equally likely to occur at all sites and, given a mutation, that any of the three nucleotide states are equally likely to arise. Neither of these premises is justified in this case, as pointed out in Chapter 5. For example, about a third of all mutations in human nuclear DNA are transitions from 5-methylcytosine to thymine that occur exclusively at

CpG dinucleotides, a combination markedly underrepresented in human DNA. Mutational hotspots have also been reported for mononucleotide-repeat regions, DNA polymerase α arrest sites, and other rarely occurring sequence motifs in human DNA. The pattern of site polymorphism in the *LPL* gene parallels the results of these mutation studies, with 9.6% of the nucleotides in CpG sites being polymorphic, 3.3% of the nucleotides in mononucleotide runs of length 5 or greater, 3.0% of the nucleotides within 3 bp of the polymerase α arrest site motif of TG(A/G)(A/G)GA, and 0.5% at all other sites (Templeton et al. 2000a). Altogether, almost half of the polymorphic sites in the sequenced portion of the *LPL* gene were from one of these three highly mutable classes and therefore would be less than ideal choices for a genetic marker, and an analysis of the haplotype trees estimated for this region (corrected for recombination) show that nucleotides at these highly mutable sites have indeed experienced multiple mutational changes in the evolutionary history of this DNA region. Unfortunately, the only consideration in finding markers is often that they are highly polymorphic in all populations, a procedure that biases in favor of sites with high mutability. One solution to the problem of multiple genetic backgrounds for mutable marker loci is to choose populations with extreme founder or bottleneck events in their recent evolutionary history. This greatly reduces the chances of multiple mutational hits at the marker locus in this particular population. Hence, an understanding of the recent evolutionary history of a population is critical in designing and executing this type of measured genotype study (Templeton 1999a).

Markers of Linkage. Physical linkage between two loci affects their pattern of cosegregation in a pedigree or set of controlled crosses regardless of whether or not there is linkage disequilibrium between the two loci in the population as a whole. Consequently, if one of these loci is a measured marker locus and the other is an unmeasured polymorphic QTL that contributes to phenotypic variation, then the pattern of segregation at the marker locus in the pedigree or controlled crosses will be associated with phenotypic differences among individuals within the pedigree or set of crosses. The strength of this association depends upon two factors:

- Magnitude of phenotypic impact of QTL
- Amount of recombination between marker locus and QTL

An experimenter has no control over the first of these factors, but the second can be controlled to some extent by the choice of the number and the genomic locations of the marker loci. Ideally, enough markers should be studied to cover the entire genome. Minimally, this means markers every 20 cM, which implies that any QTL will be ≤ 10 cM from a marker (and therefore double cross overs will not be important).

In humans, this linkage mapping approach is used with pedigree data. Because the associations are detected through linkage within a pedigree, there is no need for linkage disequilibrium in the population as a whole. Nevertheless, such linkage mapping studies in humans are still primarily and most powerfully executed in founder populations. The linkage disequilibrium induced by founder events is no longer directly relevant in a pedigree study, but founder populations are still the preferred objects of study because of their greater uniformity of genetic background, the greater incidence of phenotypic variants that exist in some founder populations, and the lessening of the chances for artifacts due to cryptic population subdivision. Moreover, although population-level linkage disequilibrium is not strictly required for this approach, its presence can make informative patterns of cosegregation within a pedigree much more likely. Therefore, linkage disequilibrium in the population as a whole enhances the power of linkage association studies within pedigrees (Ober et al. 2001). Finally, individuals from founder populations in humans are more likely to share many common environmental variables than individuals from most other human populations, thereby resulting in a more uniform environmental background as well.

An example of this approach is given by Ober et al. (2000, 2001), who measured 20 quantitative traits associated with asthma, diabetes mellitus, cardiovascular disease, hypertension, and autism in a population of Hutterites. The Hutterites are a recent, religiously defined founder population who practice a communal farming life-style which attenuates many sources of environmental variation. The phenotypes chosen for study have high incidences in this population, as expected from a severe founder event. In particular, 11% of the population have asthma, 28% of the Hutterites over 30 have diabetes or impaired glucose tolerance, and 34% have hypertension. Over 500 marker loci have been scored in about 700 living Hutterites that come from a single, 13-generation pedigree. This marker density yields a 9.1-cM map overall, with greater marker density in some regions of the genome. The simplest approach to detect QTLs is to model each theoretical QTL as a two-allele locus with an additive contribution to the phenotype and look for associations between such hypothetical QTLs and each marker. Because of the large number of statistical tests that are not independent because of linkage, the analysis is complicated by the need to correct for these multiple comparisons, but procedures exist for doing so (e.g., Cheverud 2001). Ober et al. (2000) found evidence for 23 QTLs influencing the phenotype of asthma susceptibility in the Hutterites.

Instead of using a single-site analysis of association, it is also possible to combine information from two or more sites into an integrated analysis of linkage association. One of the most common approaches of this type is **interval mapping**, in which a QTL is hypothesized to lie between two adjacent markers and the likelihood of the QTL being at various intermediate positions between the flanking markers is statistically evaluated. For example, consider the simple case of a controlled cross design in which two inbred strains are crossed to produce an F_1, which in turn is then backcrossed to one of the parental stains. Let the marker alleles from one parental strain be designated by capital letters and by small letters for the other strain, and assume that the backcross was to this latter strain. Now consider two marker loci, say locus A and locus B, that are adjacent to one another on the chromosome map with a recombination frequency of r. Now we hypothesize a QTL, say locus X, in between these two marker loci. Assume that one parental strain was originally fixed for the X allele and that the other strain is fixed for the x allele at this hypothetical QTL. Assume that the genotypic value of Xx is G_{Xx} for the phenotype being measured in the backcross progeny and G_{xx} for those backcross individuals with genotype xx. Assuming that the markers have a recombination frequency of r that is sufficiently small that double crossovers can be ignored (this assumption can be dropped, but then one needs a mapping function), the recombination frequency between marker A and QTL X is r_x and the recombination frequency between marker B and the QTL is $r - r_x$, as shown in Figure 10.3. Given these assumptions, Table 10.1 shows the expected phenotypic means for the observed marker genotypes as a function of the hypothesized phenotypic effects of the QTL at the hypothesized map position of r_x from marker locus A. Similar models exist for F_2 crosses and pedigree data, although they are generally more complex. The important point is that the fit of the observed phenotypes to the expected phenotypes depends on both the hypothesized phenotypic impact of locus X and its chromosomal position. This fit is typically measured by a likelihood ratio test (Appendix 2) of the hypothesis that the QTL

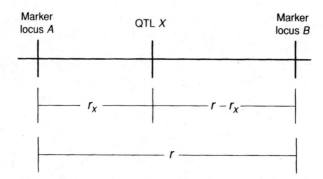

Figure 10.3. Hypothetical QTL (X) located between two adjacent marker loci, A and B. The recombination frequency between A and B is r, and that between locus A and the QTL is r_x.

is at position r_x ($G_{Xx} \neq G_{xx}$ in the backcross model shown in Table 10.1) versus the null hypothesis of no phenotypic effect at position r_x ($G_{Xx} = G_{xx}$ in the backcross model). The **LOD score** is the logarithm to the base 10 of the likelihood ratio of a QTL at a specific position in the genome. Because many such tests are performed and they are not independent, various procedures are used to determine which LOD scores are statistically significant at the level of the entire genome (e.g., Cheverud 2001; Lander and Kruglyak 1995).

The procedure of interval mapping has been extended to make use of information from several markers at once and not just the two flanking markers. An example of this type of study is given by Van Eerdewegh et al. (2002), who used a multisite mapping procedure to scan the human genome for QTLs that contribute to risk for asthma in 460 Caucasian families from the United Kingdom and the United States. Only one region of chromosome 20p13 had significant LOD scores for the combined phenotype of asthma and bronchial hyperresponsiveness, as shown in Figure 10.4. The 23 QTLs affecting asthma in the Hutterites by Ober et al. (2000) were not replicated in this study. These results indicate that the genetic factors influencing risk of asthma in the general Caucasian population are different than those in the Hutterite population. This illustrates an important point; all results of any marker association study are strictly limited to the populations being examined. Because of the possibility of differences in allele frequencies between populations and genetic and

Table 10.1. Observed and Expected Phenotypic Means of Measured Marker Genotypes at Two Adjacent Loci in Backcross Experiment

Observed Marker Genotype	Observed Average Phenotype	Possible Genotypes When QTL Is Included	Expected Frequency Assuming Model in Figure 10.3	Expected Phenotype Given QTL with Genotype Values of G_{Xx} and G_{xx}
AB/ab	G_{AB}	AXB/axb	$\frac{1}{2}(1-r)$	$G_{Xx} = G_{AB}$
Ab/ab	G_{Ab}	AXb/axb	$\frac{1}{2}(r - r_x)$	$[(r - r_x)G_{Xx} + r_x G_{xx}]/r$
		Axb/axb	$\frac{1}{2}r_x$	
aB/ab	G_{aB}	aXB/axb	$\frac{1}{2}r_x$	$[r_x G_{Xx} + (r - r_x)G_{xx}]/r$
		axB/axb	$\frac{1}{2}(r - r_x)$	
ab/ab	G_{ab}	axb/axb	$\frac{1}{2}(1-r)$	$G_{xx} = G_{ab}$

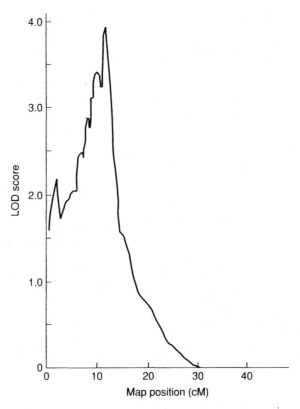

Figure 10.4. Genome scan for asthma and bronchial hyperresponsiveness in 460 Caucasian families. The LOD score is plotted on the *y* axis, and the *x* axis corresponds to the map position in centimorgans. Only the portion of chromosome 20 that had a significant LOD score peak is shown. Adapted with permission from P. van Eerdewegh et al., *Nature* 418:426–430 (2002). Copyright © 2002 by Macmillan Publishers.

environmental background effects, there is no a priori expectation that the genetic loci contributing to phenotypic variation in one population should be the same as those contributing to phenotypic variation in another population.

The significant region shown in Figure 10.4 spans 4.28 cM, corresponds to 2.5 Mb of DNA region, and contains 40 genes (Van Eerdewegh et al. 2002). Hence, although QTL refers to a quantitative trait *locus*, genome scans do not actually identify loci but rather chromosomal regions. It is possible that more than one locus in this region is influencing the phenotype of interest, but their effects are not separable in the QTL genome scan. This possibility is illustrated by the study of Steinmetz et al. (2002) on the genetic basis of phenotypic variation in the ability to grow under high temperatures in two yeast strains of *Saccharomyces cerevisiae*. They performed a high-resolution genome scan of the haploid offspring of hybrids between these two strains using 3444 markers with an average interval of 1.2 cM. They found two significant QTL regions, one of which corresponded to an interval of 32 kb. The entire 32-kb interval was then sequenced in 12 haploid strains, six each of opposite phenotypes for growth under high temperatures. This information when combined with additional genetic tests allowed them to identify three tightly linked QTLs in this interval. No one locus was either necessary or sufficient for growth at high temperatures;

rather, the phenotypic effects of this single-QTL region emerge from the joint effects of all three loci with complex cis and trans effects among them. This shows that the acronym QTL is a misleading one; a better acronym would be **QTR**, a quantitative trait region. Hence, from now on, QTR will refer to studies that only identify chromosomal regions that could contain more than one locus and QTL for those studies that have actually refined the region to a single locus or theoretical models that *assume* a locus.

Just as mapping studies can make use of more than one marker at a time to detect associations, it is also possible to use more sophisticated genetic models of how QTRs contribute to phenotypic variation. Many models treat each QTR as a single locus with two alleles making its own additive contribution to the phenotype. This is a highly simplified genetic architecture, and more complicated mappings of genotype to phenotype can be used. For example, van der Knaap et al. (2002) crossed a strain of cultivated tomato with highly elongated fruit to a wild tomato relative that produces nearly spherical fruit. Such crosses and their subsequent F_2's and backcrosses are commonly used in linkage mapping approaches in laboratory or agricultural populations, and indeed the controlled nature of these crosses and the frequent choice of marker loci with fixed genetic differences between the original parental strains often increase genetic power relative to the uncontrolled pedigrees used in human studies. In the case of the tomatoes, 97 markers showing fixed differences between the original tomato strains were used to cover the entire tomato genome with an average interval of 12.8 cM. Both single-site and interval mapping approaches were used to analyze 85 F_2 plants assuming an additive QTR model. The interval mapping analysis revealed four QTRs influencing fruit shape, each on a different chromosome. After identifying these four QTRs, van der Knaap et al (2002) looked for epistasis between them for the phenotype of fruit shape and found strong, highly significant interactions. They therefore concluded that the four QTRs can be viewed as functionally related in that they control one aspect of fruit shape: the degree of eccentricity. Thus, the original genetic architecture assumed for the initial mapping was shown to be significantly incomplete.

One difficulty with the approach of van der Knaap et al. (2002) is that all the QTRs examined for epistasis were identified solely on the basis of their additive effects on the phenotype. However, suppose the phenotypic contribution of two loci was primarily through their interactions with little marginal or additive effect at a single locus. In such a situation a mapping analysis that examined only one marker site at a time or even one map interval at a time would not detect the QTRs in the first place. Therefore, just to incorporate pairwise epistasis into the genetic architecture requires a more refined analysis that examines pairs of markers or intervals. Such an analysis is computationally and analytically much more difficult and moreover generally requires larger sample sizes. Larger sample sizes are required because in a single-locus, two-allele model, the sample is subdivided into two (backcrosses) or three (F_2's) genotypic categories. However, with just pairwise epistasis, there are four genotypes in a backcross and nine F_2 genotypic categories, so the same amount of data is now distributed over more genotypic categories, with a resulting reduction in statistical power. Despite these challenges, epistasis can be evaluated (Templeton et al. 1976; Cheverud and Routman 1995; Jannink and Jansen 2001; Marchini et al. 2005).

The importance of searching for epistasis is illustrated by the work of Peripato et al. (2004) on the phenotype of litter size in mice. They performed a genome scan on the litter size of 166 females from an F_2 cross between two inbred strains of mice. Standard interval mapping that assumes a single QTL identified two significant regions on chromosomes 7 and 12, respectively. These two QTRs account for 12.6% of the variance in litter size. They next reanalyzed the same data with a genomewide epistasis scan based on a two-locus interaction model and found eight epistatic QTRs on chromosomes 2, 4, 5, 11, 14, 15, and 18 that explained 49% of the variance in litter size. Note that the regions found by standard interval mapping that assumes no epistasis were *not* involved at all in the epistatic component of the genetic architecture. Thus searching for significant QTRs under the assumption of no epistasis and then looking for epistasis between the identified QTRs (as done in the tomato study of van der Knaap et al. 2002) can miss most if not all of the epistasis that contributes to phenotypic variance. Note also that the single-locus model genome scan only explained 12.6% of the phenotypic variance, whereas the two-locus epistatic model genome scan explained 49% of the phenotypic variance. This indicates the importance of epistasis as a component of the genetic architecture of litter size in this mouse population.

These experiments with mice reveal a major limitation of the measured marker approach to quantitative genetics. Because the loci directly affecting the trait are not known a priori, a model for their phenotypic effects must always be invoked. Even simple models of single-locus, two-allele additive phenotypic effects allow us to detect many loci contributing to phenotypic variation, but the mouse experiments illustrate that what we see is dependent upon what we look for. The model of genotype-to-phenotype mapping imposes a major constraint upon all measured marker approaches and ensures that such approaches will give an incomplete and sometimes misleading picture of the genetic architecture underlying the phenotypic variation. Even when epistasis is specifically looked for between two genome regions, additional epistasis that depends upon three or more genes may go undetected. Indeed, two- and three-locus epistasis was detected in the *Drosophila* experiments by Templeton et al. (1976), and the three-locus epistasis was not predictable from the two-locus interactions. Since most of the models used to search for epistasis are limited to interactions between two QTLs, we still tend to underestimate the significance of epistasis as a contributor to phenotypic variance and obtain only an incomplete view of genetic architecture. Also, recall the distinction between a QTR and a QTL. Almost all studies in this area identify QTRs which could hide strong interactions among tightly linked loci, as shown by the work of Steinmetz et al. (2002). Indeed, this is an exceedingly likely possibility because the genomes of many species often have families of functionally related genes close together (Cooper 1999). All of these properties put together indicate that marker association studies are a flawed vehicle for illuminating genetic architecture, but nevertheless such approaches are yielding much more insight into genetic architecture than the unmeasured genotype approaches discussed in the previous chapter.

Direct comparisons between the unmeasured genotype approach and the marker linkage mapping association studies are possible because both approaches require either pedigree data or controlled crosses. Consequently, the same data can be analyzed using both measured and unmeasured genotype approaches. For example, Ober et al. (2001) not only analyzed several phenotypes showing variation in the Hutterite pedigrees with a marker association approach but also used the pedigree data to estimate heritabilities with the unmeasured approach outlined in Chapter 9. They found no correlation between the heritabilities of these traits with the strength of association found in the QTR mapping study, indicating that the two approaches were detecting different genetic contributors to phenotypic variation. This result is not surprising. As pointed out in Chapter 9, the unmeasured genotype approaches detect the *heritable* genetic contributors to phenotypic variation. In contrast, marker association studies detect *inherited* genetic contributors to phenotypic variation. For example, marker association studies generally work best when the phenotype is primarily determined by a single Mendelian locus of large effect, as is the case for many autosomal recessive genetic diseases. Yet, as pointed out in Chapter 9, such genetic diseases are generally *not heritable*.

Therefore, these two quantitative genetic approaches are not designed to detect the same components of genetic variation contributing to phenotypic variation. The lack of correlation between heritability and QTRs in studies such as Ober et al. (2001) is surprising only to those who confuse inheritance with heritability.

CANDIDATE LOCI

The function of many genes is known. In such cases, it is sometimes possible to identify genes whose known function might contribute directly to a phenotype of interest. For example, the gene *ApoE* discussed in Chapter 8 codes for a protein that combines with insoluble lipids to form soluble apolipoproteins that can be transported in blood serum. One of the apolipoproteins that contains the ApoE protein is HDL (high-density lipoprotein), the second most important contributor to total serum cholesterol. In addition, the ApoE protein binds to the low-density lipoprotein (LDL) receptor, thereby competitively inhibiting the binding and uptake of LDL in peripheral cells and having an impact on LDL levels, the major component of total serum cholesterol. Thus, the known function of the protein coded for by the *ApoE* locus relates directly to the phenotype of total serum cholesterol, as well as to both HDL and LDL levels. Therefore *ApoE* is said to be a candidate locus for the phenotype of total serum cholesterol and its major subcomponents.

Candidate loci can also be identified when no a priori knowledge of their function exists. When a QTR study indicates a region of the genome that influences a phenotype, it is sometimes feasible to identify all the genes in that region. For example, Steinmetz et al. (2002) sequenced the entire 32-kb QTR for growth under high temperatures in yeast, as described in the previous section, and found 20 genes in that region. Using additional markers, they further restricted the QTR to a region containing 6 genes. These 6 genes were then regarded as candidate loci for the phenotype of growth under high temperatures. Another method of identifying candidate loci is through cDNA microarray technology (Appendix 1). Huang et al. (2002) used this technology to match the transcriptional activity of a large number of genes with biomechanical and physiological functions related to a study of blood vessels in the lung in response to pulmonary hypoxic hypertension. They identified the genes whose expression was strongly correlated with each physiological function. For example, they found one set of genes whose expression was correlated with the remodeling of arterial wall thickness and another set of genes whose integral of activity over time best fit the blood vessel wall thickness change. These sets of genes then become candidates for the physiological and biochemical phenotypes to which they are highly correlated.

Candidate Locus Analyses with Alleles, SNPs, and Haplotypes

Once one or more candidate loci have been identified, genetic variation is scored at the candidate loci and tested for association with phenotypic variation. With low-resolution genetic screens such as protein electrophoresis, such tests of association are relatively straightforward by measuring the phenotypic means and variances of the genotypes at the candidate locus. For example, *ApoE* is a candidate locus for the phenotype of total serum cholesterol. In Chapter 8, we showed how all of the quantitative genetic parameters in Fisher's original model could be estimated for the measured genotypes at this locus. Hence, that worked example of *ApoE* (Sing and Davignon 1985) is a classic measured genotype study of a candidate locus.

This simple approach can be complicated by two interrelated factors: high levels of genetic variation and linkage disequilibrium. As the genetic resolution of survey techniques increased, high levels of genetic variation became commonly observed. This is illustrated by studies on another candidate gene for cholesterol related phenotypes, the *LPL* locus. Recall from Chapter 1 the genetic survey of Nickerson et al. (1998) on 71 individuals from three populations that were sequenced for a 9.7-kb region within this locus, representing only about a third of the total gene. Eighty-eight polymorphic sites were discovered, and 69 of these sites had their phases determined to define 88 distinct haplotypes (Clark et al. 1998). Thus, using only a subset of the known polymorphic sites in just a third of a single gene, a sample of 142 chromosomes reveals 88 "alleles" or haplotypes, which in turn define 3916 possible genotypes—a number considerably larger than the sample size of 71 people. Indeed, virtually every individual in the sample had a unique genotype, and there were more haplotypes or alleles than individuals. The simple candidate approach illustrated by Sing and Davignon's (1985) work on *ApoE* cannot be applied when almost all the possible genotypes have no observations in the sample. Obviously, increased genetic resolution poses some daunting statistical challenges.

One approach for dealing with large amounts of genetic variation is to analyze each polymorphic site separately, effectively treating each polymorphic nucleotide site as a separate locus. However, polymorphic sites within a candidate locus are, virtually by definition, tightly linked and often show strong linkage disequilibrium, as is the case for *LPL*. As a consequence, the multiple single-site tests are not independent from one another, making statistical and biological interpretation difficult. Most commonly, such single-site approaches are interpreted as a mini–linkage analysis with each polymorphic site being treated as a "marker." Those markers that show the strongest phenotypic associations are sometimes regarded as being physically closest to the causative site or being the causative site. This, however, often leads to incorrect biological inference, because linkage disequilibrium is often uncorrelated with physical distance in small chromosomal regions.

To illustrate the difficulties in biological interpretation that arise from high levels of genetic variation coupled with extensive linkage disequilibrium, consider again from Chapter 5 the genetic survey of Fullerton et al. (2000). They sequenced 5.5 kb of the *ApoE* region in 96 individuals, revealing 23 SNPs. Figure 10.5 shows the distribution of these SNPs over the sequenced region. This figure also shows the physical positions of the two amino acid changing mutations that define the three major amino acid sequence alleles at this locus, $\varepsilon 2$, $\varepsilon 3$, and $\varepsilon 4$. As shown in Chapters 8 and 9, these alleles are associated with large differences in the phenotype of total serum cholesterol in human populations. Moreover Stengård et al. (1996) showed that men bearing the $\varepsilon 4$ allele have a substantial increase in risk of mortality from coronary artery disease (CAD), as shown in Figure 10.6. Because the protein product of the $\varepsilon 4$ allele has many altered biochemical properties relative to the protein products of the other alleles, such as its binding affinity to the LDL receptor protein, it is reasonable to assume that the amino acid replacement mutation that defines the $\varepsilon 4$ allele is the causative mutation of some of this clinically important phenotypic variation. Such will be assumed here. The mutation defining the $\varepsilon 4$ allele is the boxed mutation at position 3937 in Figure 10.5.

All studies of candidate loci only sequence a small portion of the chromosome. Suppose, hypothetically, that instead of sequencing the 5.5 kb shown in Figure 10.5 the sequencing had been terminated near the beginning of exon 4, as shown by the gray boxed region in Figure 10.5. Because position 3937 is in exon 4, this position would not have been sequenced in this hypothetical study. Suppose further that no prior knowledge existed about

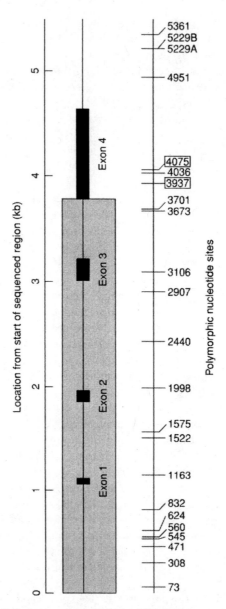

Figure 10.5. Physical position of polymorphic SNPs found by Fullerton et al. (2000) in 5.5-kb region of *ApoE* gene. The two amino acid replacement polymorphic sites that determine the $\varepsilon 2$, $\varepsilon 3$, and $\varepsilon 4$ alleles are indicated by boxed numbers. A larger box shaded with gray encloses the portion of this region that was "sequenced" in a hypothetical study.

the importance of the amino acid replacement mutations or even their very existence. Hence, the investigators in this hypothetical study would only have available the SNPs in the large box shown in Figure 10.5. Suppose now that each of these available SNPs would be tested one by one for associations with CAD in a study similar to that of Stengård et al. (1996). Because the causative mutation at site 3937 is not included in this hypothetical study, any associations detected with the available SNPs would be due to linkage disequilibrium with

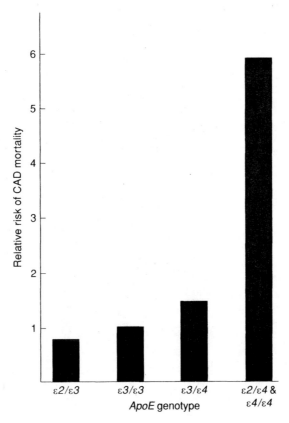

Figure 10.6. Relative risk of mortality due to coronary artery disease as function of *ApoE* genotype in longitudinal study of Finnish men. Based on data from Stengård et al. (1996).

the ε4 allele. Interestingly, the ε4 allele (the SNP at site 3937 in Figure 10.5) does not show any significant disequilibrium with the SNP at site 3701, the SNP that lies closest to 3937 in the region sequenced in our hypothetical study. In contrast, there is significant disequilibrium between the SNPs at sites 3937 and 832. Accordingly, this hypothetical study would find an association between the SNP at site 832 and CAD mortality. Site 832 lies in the 5′ enhancer region of this gene, so our hypothetical investigators could easily create a plausible story about how a 5′ regulatory mutation was "causing" the observed associations with CAD mortality, even though the causative mutation was actually located near the opposite 3′ end of their sequenced region.

Why does this hypothetical single SNP study associate the 5′ regulatory end of the gene region with CAD mortality but find no association with a SNP only a little more than 200 bp away from the causative mutation? The answer to this was already discussed in Chapters 2 and 5; when dealing with DNA regions showing little or no recombination (as is the case for the *ApoE* region, Chapter 2), the primary cause of the pattern and magnitude of linkage disequilibrium is evolutionary history. On the larger scale of centimorgan intervals used in linkage disequilibrium mapping in marker association studies, linkage disequilibrium is used as a proxy for physical location. But this is a dangerous assumption on the small physical scales of thousands of base pairs typically used in candidate locus studies. This is shown for the *ApoE* example in Figure 10.7. This figure displays the *ApoE* haplotype tree

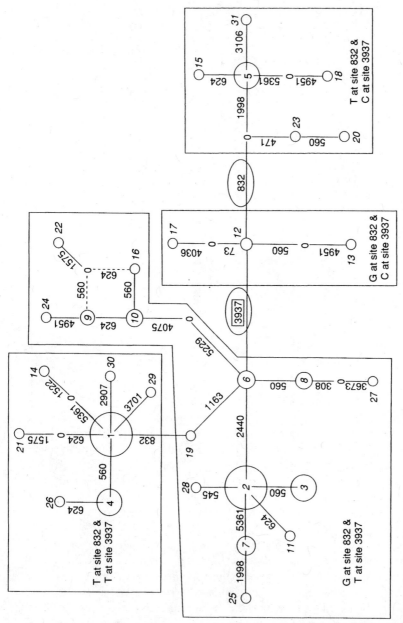

Figure 10.7. Statistical parsimony network based on 23 variable sites (numbered by their nucleotide position in reference sequence) in 5.5-kb segment of *ApoE* gene (modified from Fullerton et al. 2000). Circles designate the haplotypes, each identified by its haplotype number (1–31), either inside or beside the circle. The relative sizes of the circles indicate the relative frequencies of each haplotype in the sampled population. A "0" indicates an inferred intermediate haplotype that was not found in the sample. Each line represents a single mutational change, with the number associated with the line indicating the nucleotide position that mutated. A solid line is unambiguous under the principle of statistical parsimony, whereas dashed lines represent ambiguous inferences under statistical parsimony. The two sites showing significant disequilibrium are enclosed by ovals, and the tree is subdivided by the state of these two sites.

as estimated in Figure 5.19. A glance at this figure shows why sites 832 and 3937 show significant disequilibrium; the mutations at these sites are located next to one another in the evolutionary tree of haplotypes. As a result, most haplotypes today fall into just two genetic states for these sites, resulting in significant disequilibrium. Indeed, if it were not for homoplasy at site 832, the disequilibrium would have been even stronger. Because no recombination is detected in this region, recombination is irrelevant and linkage disequilibrium primarily reflects *temporal* proximity in evolutionary history rather than *physical* proximity on the DNA molecule. In general, linkage disequilibrium is strongest between markers that are old (near the root of the tree) and that are on nearby branches in the haplotype tree (temporal proximity). Physical proximity is irrelevant in the absence or near absence of recombination. Consequently, a candidate gene study should not be regarded as a mini-QTL marker association study; the biological meaning of linkage disequilibrium can be extremely different at the physical scales of a marker study measured in centimorgans versus a candidate locus study covering only a few thousand base pairs. If the candidate gene region has much recombination between its polymorphic sites, then the approaches described in the previous section are applicable. However, when recombination is rare or even totally absent, other approaches are required.

One way to eliminate the problems caused by disequilibrium is to organize all the genetic variation in the DNA region into haplotypes and then use haplotypes as the units of analysis. This method transforms the problem into a one-locus, multiallelic problem. Several studies have reported finding significant phenotypic associations with haplotypes that are not detectable or are much weaker when each SNP is analyzed separately (Balciuniene et al. 2002; Knoblauch et al. 2002; Martin et al. 2000; Seltman et al. 2001; Van Eerdewegh et al. 2002; Zaykin et al. 2002). For example, Drysdale et al. (2000) examined 13 SNPs in the human beta-2-adrenoceptor gene, a candidate locus for asthma because it codes for a receptor protein on the bronchial smooth muscle cells in the lung that mediates bronchial muscle relaxation. Drysdale et al. could find no association between any of the 13 SNPs and asthma, but when the SNPs were phased (Appendix 1) to produce 12 haplotypes, they did find significant associations between haplotypes at this locus and asthma. This result is not surprising: SNPs do not affect phenotypes in isolation; rather they only affect phenotypes in the context of the genetic state of all the other SNP states that are found in a particular individual. Haplotypes recover some of this context, and when the SNP is removed from this context by analyzing it in isolation, biological information is lost. Placing the SNP back into the context of a haplotype recovers that lost information, as the studies mentioned above show.

Analyzing genotypes for which haplotypes are treated as alleles works well as long as the number of distinct haplotypes or alleles is not too large (Seltman et al. 2001), as was the case in Chapter 8 for the measured genotype analysis of *ApoE* using the 3 major alleles defined by protein electrophoresis or even the 12 haplotypes in the asthma study as only 3 haplotypes tended to be common in most human populations. In cases such as these in which the genotypic diversity is low, the problem of detecting associations between genotype and phenotype can be approached using standard statistical tests. For example, Sing and Davignon (1985) used a standard analysis of variance (ANOVA, see Appendix 2) to test the null hypothesis that there is no difference in the genotypic values of total serum cholesterol described in Chapter 8. Hence, the heterogeneity in the genotypic values illustrated in Chapter 8 is statistically significant, leading to the biological conclusion that the protein electrophoretic variation detected at the *ApoE* locus is significantly associated with phenotypic variation in the trait of total serum cholesterol.

Candidate Locus Analyses Using Evolutionary History

Placing individual SNPs into the context of a haplotype increases biological information, although when haplotype diversity is high, statistical power can be eroded. In such cases, even extremely large samples only provide sparse coverage of the possible genotypic space defined by the haplotypes. This sparseness results in low statistical power (Seltman et al. 2001; Templeton 1999a). The number of distinct alleles or haplotypes is often quite large relative to the sample size when DNA sequencing is used to survey genetic variation in a candidate region, as illustrated by *ApoE* (Figure 10.7) or *LPL* with its 88 haplotypes detected in a sample of 71 individuals (Chapter 5). Just as SNPs can be placed into the context of a haplotype to increase the level of biological information, so can haplotypes be placed into their evolutionary context through the device of a haplotype tree, at least when recombination is sufficiently rare in the candidate region. The evolutionary historical context of haplotypes can provide additional biological information and a solution to the problem of large amounts of haplotype diversity.

A simple example of the use of a haplotype tree is provided by the study of the *ApoAI–CIII–AIV* gene cluster by Haviland et al. (1995). This is a tightly linked cluster of three genes, each coding for a type of apoprotein involved in lipid transport and metabolism. Seven haplotypes (numbered *1–7*) were found in a restriction site survey of this candidate region in a population of 167 French Canadians scored for total serum cholesterol and other lipid phenotypes. One hundred and forty-seven of these individuals had their haplotypes determined unambiguously, and of these, 140 individuals bore at least one copy of haplotype *6* and therefore only one haplotype state varied among these 140 individuals. Thus, the seven genotypes in this subset of 140 (the *6/6* homozygote and the six *6/-* heterozygous genotypes) differ by only one haplotype. The haplotype tree, shown in Figure 10.8, is a simple star in which the six rarer haplotypes radiate from the central and common haplotype *6*, with all six of the tip haplotypes differing from haplotype *6* by only a single restriction site change. The basic goal in a candidate gene study is to show significant phenotypic differences between the genotypes being measured. In this example, there are 21 different paired contrasts among the seven genotypes found in these 140 individuals. However, there are only six degrees of freedom in the ANOVA (appendix 2) for genotypic effects in this subset of seven distinct genotypes. In general, there are $n - 1$ degrees of freedom in a haploid analysis of n haplotypes and $\frac{1}{2}n(n - 1)$ pairwise contrasts, so the discrepancy between degrees of freedom

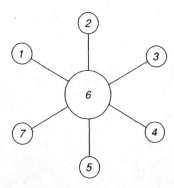

Figure 10.8. Unrooted haplotype tree for restriction site variation at *ApoAI–CIII–AIV* region. Based on data in Haviland et al. (1995).

and the number of genotypic contrasts gets worse with increasing haplotype diversity. One of the basic principles of the comparative method of evolutionary biology (Chapter 1) is that the most meaningful contrasts are between evolutionary neighbors. Applying this principle, the evolutionarily relevant contrasts are those between haplotype *6* versus the six others. It would make no sense evolutionarily, for example, to contrast haplotype *1* with haplotype *3*, as the closest evolutionary neighbor of both of these haplotypes is haplotype *6*. Accordingly, only the six paired contrasts of evolutionary neighbors were performed by Haviland et al. (1995) (haplotype *6* versus the other six haplotypes), each using one of the available degrees of freedom. By concentrating the available statistical power into these six contrasts rather than spreading it over 21 contrasts, significant associations were discovered between the haplotypes and the phenotypes of total cholesterol, LDL cholesterol, serum levels of the ApoB protein in males, and the phenotypes of HDL cholesterol and serum levels of the Apo AI protein in females. In contrast, no significant phenotypic associations were discovered using the same haplotypes as the treatment effects with a one-way ANOVA (see Appendix 2) that does not use evolutionary principles to concentrate statistical power upon the most relevant comparisons.

Most haplotype trees have a more complex topology than the simple star shown in Figure 10.8. Regardless of topology, any haplotype tree of n haplotypes will have $n-1$ connections in a fully resolved tree, so in theory all haplotype trees could define contrasts across their branches that would fully use all available $n-1$ degrees of freedom and not squander any statistical power upon evolutionarily irrelevant contrasts. Statistical problems can still exist because many haplotypes, particularly those on the tips of the haplotype tree, are expected to be quite rare in the population (Castelloe and Templeton 1994), resulting in little power for some contrasts. This problem can be diminished by pooling haplotypes into larger categories, particularly tip haplotypes.

One method of using the haplotype tree to pool is the nested design already discussed in Chapter 7 for the nested-clade analysis of geographic data. The premise upon which a nested analysis is based is that any mutation having functional significance will be imbedded in the historical framework defined by the haplotype tree and therefore whole branches (clades) of this tree will show similar functional attributes. Nesting has several advantages. First, nesting categories are determined exclusively by the evolutionary history of the haplotypes and not by a phenotypic preanalysis, thereby eliminating a major source of potential bias. Second, the clades define a nested design that makes full and efficient use of the available degrees of freedom. Nesting performs only evolutionarily relevant contrasts and does not squander statistical power on less informative or redundant constrasts. Third, statistical power has been enhanced by contrasting clades of pooled haplotypes instead of individual haplotypes, thereby directly addressing the problem of too much diversity eroding statistical power.

Figure 10.9 shows the haplotype tree and nested clades of the haplotypic variation found at the *alcohol dehydrogenase (Adh)* locus in 41 homozygous strains of the fruit fly. *D. melanogaster* (Aquadro et al. 1986). Table 10.2 shows the results of a nested ANOVA of the phenotype of alcohol dehydrogenase (Adh) activity using the nested design shown in Figure 10.9 (Templeton et al. 1987a). As can be seen from Table 10.2 or Figure 10.10, four significant effects are identified, which subdivide the haplotypic variation into five functional allelic classes with respect to Adh activity. Figure 10.11 gives the actual mean phenotypes of all the homozygous strains used in this study, grouping these strains by which one of the five functional clades they bore. As can be seen, the strongest effect in Table 10.2 is associated with the difference between clade 3-1 versus clade 3-2. The actual

320 QUANTITATIVE GENETICS: MEASURED GENOTYPES

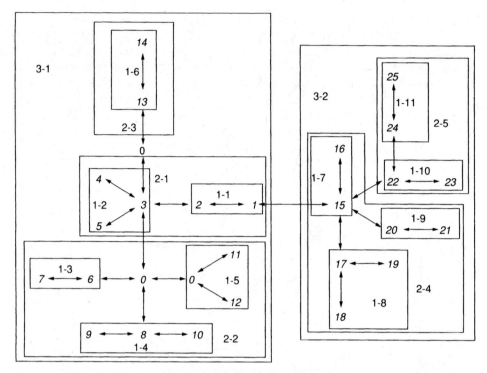

Figure 10.9. Unrooted haplotype tree for restriction site variation at *D. melanogaster Adh* region along with nested design (from Templeton et al. 1987a). Haplotypes are indicated by numbers, with a "*0*" indicating an inferred intermediate not observed in the sample. Each arrow indicates a single mutational change. Haplotypes are nested together into one-step clades, indicated by "1-#," which in turn are nested together into two-step clades, indicated by "2-#," which in turn are nested together into three step clades, 3–1 and 3–2.

distribution of Adh activity is bimodal (Figure 10.11), and this clade contrast captures most of that bimodality. This bimodality is also easily detected by a standard one-way ANOVA using the haplotypes as treatments. However, the nested analysis also detected phenotypic heterogeneity *within* both clades 3-1 and 3-2. Within clade 3-1, the nested analysis in Table 10.2 identifies clade 1-4 as having a significantly different Adh activity than the other clades nested within clade 3-1. As can be seen from Figure 10.11, the four strains that constitute clade 1-4 have the four highest Adh activities within the lower Adh activity mode. Thus, within the lower activity mode, there is a strong association between evolutionary relatedness (being in clade 1-4) and phenotypic similarity. This significant effect is only detectable using evolutionary information. The same is true for the additional significant phenotypic associations found within clade 3-2 (Figure 10.11). Haplotype *23* occupies the lower tail of the upper Adh activity mode, and haplotype *19* is found in the lower Adh activity mode. This latter type of phenotypic association is invisible to an analysis that does not take evolutionary history into account. The Adh activity associated with haplotype *19* is not at all unusual, falling close to the mean of the lower mode. What makes it significant is that haplotype *19* is evolutionarily associated with clade 3-2, the clade that otherwise is associated with the upper Adh mode. Hence, this is a phenotypic reversion from high Adh activity to low. Without using evolutionary history, only the difference between the

Table 10.2. Nested-Clade Analysis of Adh Activity at *Adh* Locus

Source	Sum of Squares	Degrees of Freedom	Mean Square	F Statistics
Three-step clades	138.33	1	138.33	366.50[a]
Two-step clades	0.88	3	0.29	0.78
One-step clades				
Within 2–1	1.50	1	1.50	3.98
Within 2–2	5.74	2	2.87	7.61[b]
1–4 vs. 1–3			Bonferroni Significance	<0.05
1–4 vs. 1–5			Bonferroni Significance	<0.10
1–3 vs. 1–5			Bonferroni Significance	>0.50
Within 2–4	5.38	2	2.69	7.12[b]
1–7 vs. 1–8			Bonferroni Significance	<0.50
1–7 vs. 1–9			Bonferroni Significance	>0.50
Within 2–5	0.31	1	0.31	0.82
Zero-step clades				
Within 1–1	0.18	1	0.18	0.47
Within 1–2	1.26	2	0.63	1.66
Within 1–3	0.01	1	0.01	0.04
Within 1–4	0.80	2	0.40	1.06
Within 1–5	0.15	1	0.15	0.39
Within 1–6	0.39	1	0.39	1.02
Within 1–7	0.00	1	0.00	0.00
Within 1–8	10.49	2	5.25	13.90[b]
17 vs. 18			Bonferroni Significance	>0.50
17 vs. 19			Bonferroni Significance	<0.01
Within 1–9	1.66	1	1.66	4.39
Within 1–10	4.14	1	4.14	10.97[b]
Within 1–11	0.06	1	0.06	0.16
Error	6.04	16	0.38	

[a] Significant at 0.1% level.
[b] Significant at 1% level.
Source: From Templeton et al. (1987a).
Note: The sums of squares for the one-step and zero-step clade (haplotype) levels are decomposed into their independent components nested within the next higher level. When a clade with a significant association contained three or more possible contrasts, each contrast was tested and subject to a Bonferroni correction to localize the specific branch with the strongest effect.

upper and lower activity modes is detectable, creating two functional allelic categories. With evolutionary history, five functional allelic categories are detectable. As this Adh example shows, pooling haplotypes according to their evolutionary history has enhanced the statistical power to detect phenotypic associations.

Nested designs based on haplotype trees, such as that shown in Figure 10.9, obviously produce a statistical design for haplotypes, not genotypes. This is not a problem when dealing with a haploid genetic element, such as mitochondrial DNA, or homozygous strains, as in the Adh example, as each individual is genetically characterized by a single haplotype. However, when dealing with diploid genotypes in natural populations, a single individual can bear two distinct haplotypes. In the diploid case, the phenotypic data (from the individuals) do not in general overlay upon a haploid statistical design. This problem can be solved by

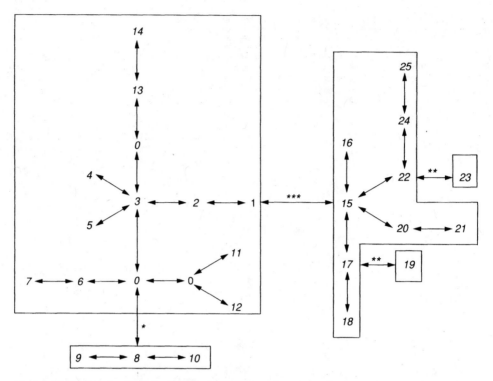

Figure 10.10. Functional allelic classes of *Adh* haplotypic variation for phenotype of Adh activity as inferred from nested-clade analysis shown in Table 10.2.

the traditional solutions of quantitative genetics of assigning a phenotype to a haplotype (Chapter 8). For example, Templeton et al. (1988) used the average excess to assign a "phenotype" to a haplotype or clade of haplotypes in a nested-clade analysis of restriction site variation at the human *ApoAI–CIII–AIV* region with the log of the serum triglyceride level. Because the individuals are diploid, the haplotypes do not define independent phenotypic observations, so the statistical significance of these average excesses was determined by a nonparametric nested permutational analysis. A significant phenotypic association was detected that was undetectable with a standard nonevolutionary analysis. Alternatively, Hallman et al. (1994) used the average effect to assign phenotypes to haplotypes and clades of haplotypes and then used likelihood ratio tests (Appendix 2) to identify clades of haplotypes in the *apoprotein B* (*ApoB*) region that explained about 10% of the genetic variance and 5% of the total variance in HDL cholesterol and triglyceride levels. None of the sites used to define these *ApoB* haplotypes showed any significant phenotypic associations when tested individually. Evolutionary analyses provided greater statistical power than nonevolutionary alternatives in both of these diploid examples.

Another method of using haplotypes and haplotype trees in phenotypic association tests is through the **evolutionary tree, transmission disequilibrium test (ET-TDT)**. The TDT is applied to family data consisting of offspring characterized by some phenotype (often a disease) and their parents and tests for genotype–phenotype association by examining for deviations from the null hypothesis of equal probabilities of all alleles or haplotypes being transmitted to the affected offspring from a heterozygous parent (Spielman et al. 1993). The ET-TDT uses nested clades of haplotypes as defined by Templeton et al. (1987a) to focus

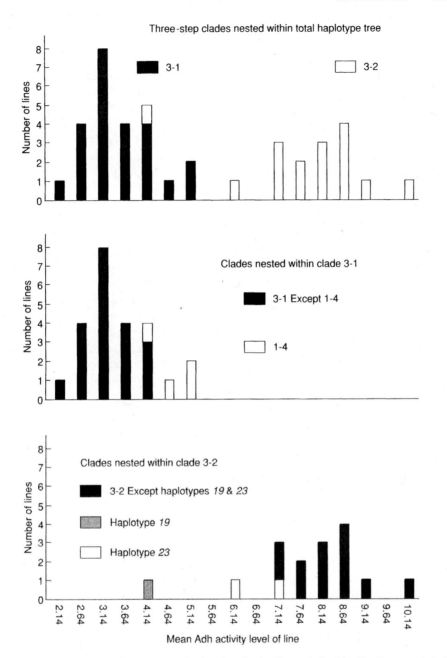

Figure 10.11. The *Adh* activity histograms for functionally significant clades identified by nested-clade analysis.

statistical power upon only the evolutionarily relevant contrasts of transmission probabilities (Seltman et al. 2001). Given n haplotypes, the standard TDT tests the null hypothesis that all n haplotypes have the same transmission probabilities. The ET-TDT, however, tests the null hypothesis that tip haplotypes (or higher level clades of tip haplotypes, using the same nesting rules discussed in Chapter 7) are transmitted with the same probability as the interior haplotype (or nested clade of haplotypes) to which they are evolutionarily connected in the haplotype tree. Just as illustrated by Haviland et al. (1995), this produces a series of ordered contrasts that makes efficient use of the available degrees of freedom.

As shown in Chapter 8, average excesses and average effects do not measure all of the genotypic deviations, and the TDT, with its focus on transmitted gametes, is applicable only to family data. As a consequence, there could be associations between genotypes and phenotypes that would be missed by an analysis based exclusively upon haplotypes. Haplotype trees can still be used to study phenotypic associations with *genotypes* and not just *haplotypes*. One such evolutionary method is tree scanning (Templeton et al. 2005). **Tree scanning** partitions the haplotype tree into two or more mutually exclusive and exhaustive clades and then treats each clade as an "allele" in a genotypic analysis of phenotypic associations. A tree scan starts with all possible biallelic partitions of the haplotype tree. A biallelic partition is created by cutting a branch in the haplotype tree and then grouping together all the haplotypes on one side of the cut into allelic class A and all the haplotypes on the other side of the cut into allelic class a. Figure 10.12 shows one such partitioning of the *Adh* haplotype tree. For an autosomal locus, all individuals can now be classified into the genotypic categories of AA, Aa, and aa. The null hypothesis is then tested that all these genotypes have the same genotypic value (Chapter 8). The strength of association can be measured when the null hypothesis is rejected by the percent of the phenotypic variance explained by the genetic variance this partition generates. The scan is accomplished by repeating this procedure over all $n - 1$ branches in a tree of n haplotypes. These multiple tests are also nonindependent because the allelic partitions produced by cutting branches overlap to some extent in the haplotypes they put into a specific allelic category, with the degree of overlap being maximal

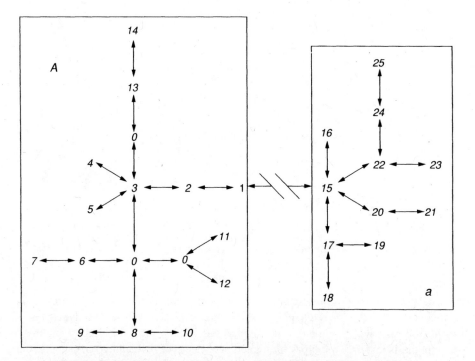

Figure 10.12. Example of split in *Adh* tree dividing into two "alleles," *A* and *a*.

for adjacent branches in the tree. Tree scanning therefore corrects all the significance levels for multiple testing with a procedure that can accommodate correlated tests.

In analogy with a genome scan that uses marker loci to mark physical sections of the chromosome (Figure 10.3), the tree scan uses mutations to mark temporal sections of the haplotype tree, except the haplotype tree can have multiple branches from a single node and therefore is not one dimensional like a genome scan. Just as adjacent markers in a genome scan yield highly correlated phenotypic associations, so do adjacent branches in a tree scan. Even when each test result is corrected for multiple, correlated testing, the phenotypic associations can spill over from one branch to adjacent branches. Sometimes only one branch is significant after correcting for multiple tests, but often there is a peak branch of association that falls off with other branches as they become located farther and farther away in the haplotype tree from the peak branch, just as genome scans produce peaks of phenotypic association that fall off with physical distance on the chromosome (Figure 10.4).

Figure 10.13 shows the tree scan results for testing for associations between the *Adh* haplotype tree and the phenotype of Adh activity. As can be seen, the partition that splits

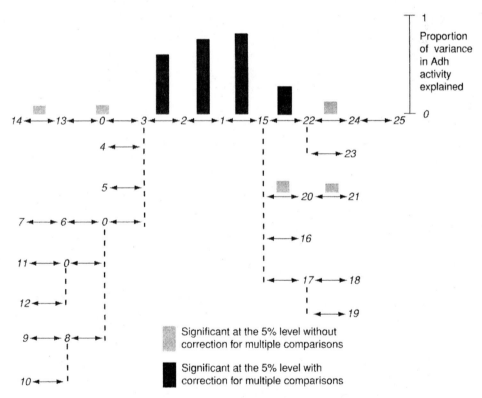

Figure 10.13. Results of initial tree scan on *Adh* haplotype tree. The haplotype tree shown in Figure 10.9 has been redrawn to show mutational distances only on the horizontal axis. Each horizontal double-headed arrow indicates a single mutational change. Dashed lines are drawn where branching occurs in the haplotype tree. The dashed lines do not represent any mutational change and only represent topological connectors to a common node. The proportion of the total phenotypic variance of *Adh* activity that was explained by a partitioning at a specific mutational branch is indicated by the solid bar adjacent to the evolutionary branch that defined the partition only for those partitions that were significant at the 5% level, both with and without correcting for multiple tests.

the tree between haplotypes *1* and *15* explains 81% of the phenotypic variance, and the amount of phenotypic variance falls off for branches farther and farther away from this peak branch. Thus, this initial tree scan detects only one peak. The conservative interpretation of this result is that one branch is associated with a significant shift in phenotypic effects at the genotypic level, and that branch is most likely the branch connecting haplotypes *1* and *15*. Note that this is the same branch identified by the nested-clade analysis as having the strongest phenotypic effect. The other branches with significant phenotypic associations shown in Figure 10.13 are most likely correlated effects from the *1–15* branch, as the various bipartitions are not independent of one another.

Although the tree scan did successfully identify the branch with the strongest phenotypic association found by the nested-clade analysis, the other branches associated with significant phenotypic transitions in the nested-clade analyses (Table 10.2 or Figure 10.11) were undetected. The strong phenotypic effect associated with the *1–15* branch obscures all of these more minor phenotypic associations because of the nonindependence of the contrasts in a biallelic partition. A second round of tree scanning can correct this problem. The branches associated with significant peaks of explained phenotypic variance in the first round are fixed as being cut to create two or more "alleles," and then a second round of tree scanning now cuts another single branch to subdivide one of the first round alleles into two allelic categories. For the Adh example, the results of the first round of tree scanning lead to cutting the *1–15* branch, dividing the tree into two "alleles" called *A* and *a* (Figure 10.14). The second round of tree scanning cuts, one by one, all the remaining branches, thereby creating a series of three-allele genotype systems. The second round of tree scanning identifies a second significant peak within allele *A* on the branch connecting an internal node to haplotype *8*, thereby identifying two of the four phenotypic associations previously discovered by nested-clade analysis.

The tree scan analysis is consistent with the nested-clade analysis but appears to have lower power. However, this appearance may be misleading because the two analyses differ in other ways besides how they use the haplotype tree. In the original nested-clade analysis, there was no correction for multiple testing. Without such correction, the tree scan identifies a third peak on the branch connecting haplotypes *17* and *19* (Figure 10.14), once again corresponding to a branch found in the nested-clade analysis (Figure 10.9). Another difference between the nested-clade analysis and the tree scan is the nature of the statistical test used for evaluating phenotypic associations. The nested-clade analysis evaluates statistical significance through a parametric nested analysis of variance (NANOVA), but the tree scan evaluates significance through a nonparametric permutation test (Templeton et al. 2005). In general, nonparametric tests are less powerful than appropriate parametric alternatives. However, the permutation test is applicable to diploid data with heterozygotes whereas the NANOVA is not. Thus, for most data sets, we will have to use permutation testing even though it is less powerful. When the nested-clade analysis of Adh activity is redone using a nonparametric permutation test instead of NANOVA, the significant effect associated with haplotype *23* (Figure 10.9) is lost (Templeton et al. 1988) and only three branches are significant—the same three found by tree scanning. Hence, when comparable statistical tests and criteria are used, the nested-clade analysis and the tree scan analysis of the *Adh* data set yield identical inferences at similar *p* levels.

Tree scanning was also applied to the *ApoE* haplotype tree (Figure 10.7, but the tree used in the tree scan had more haplotypes and branches because it was based on a larger sample) for several lipid-related traits (Templeton et al. 2005). As expected, the tree scan identified each of the branches that define the three electrophoretic alleles ($\varepsilon 2$, $\varepsilon 3$, and $\varepsilon 4$) as

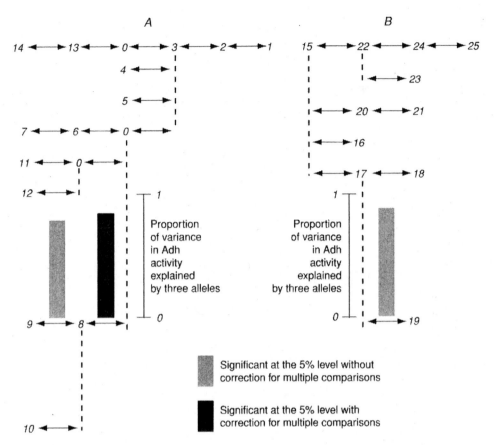

Figure 10.14. Results of second round of tree scanning on *Adh* haplotype tree. The first round resulted in splitting the tree into two allelic categories, *A* and *a*. The proportion of the total phenotypic variance of *Adh* activity within these two initial allelic classes that is explained by cutting an additional branch to create a three-allele system is indicated by the solid bar adjacent to the evolutionary branch that defined the partition for those partitions that were significant at the 5% level, both with and without correcting for multiple tests.

being functionally different alleles for lipid traits. In addition, the tree scan found significant partitions *within* each of the three electrophoretic alleles for one or more lipid-related traits, indicating that there is more functionally significant variation at this locus than previously thought. Each SNP (Figure 10.5) was also analyzed for phenotypic associations, but the SNP analysis detected fewer associations with lower statistical significance (Templeton et al. 2005). Using evolutionary history, here in the form of a haplotype tree, obviously helped in detecting current phenotypic associations. But why is cutting each branch one at a time and looking for phenotypic associations more powerful than examining each SNP one at a time for phenotypic associations?

One answer to this question is homoplasy, the occurrence of multiple mutations at the same nucleotide site. Under the infinite-sites model (Chapter 5), each mutation occurs only once in evolutionary history, and there should be little difference between examining single branches (most often defined by one mutation) versus single SNPs for phenotypic associations. However, the infinite-sites model is often violated, and this is certainly the case for *ApoE* (Figure 5.19). For example, tree scanning found significant associations of HDL cholesterol levels with a branch defined by a mutation at nucleotide site 560, but SNP 560 had no significant associations. As shown in Figure 5.19 or 10.7, site 560 mutated multiple times in the evolutionary history of the *ApoE* gene, so a SNP analysis will pool together into a single "allele" many disjoint portions of the haplotype tree, thereby violating the fundamental premise of the comparative method. In contrast, cutting branches places all of these multiple mutational events into a different allelic category in the tree scan analysis, and the evolutionary/haplotype context of mutations at site 560 apparently matters phenotypically. The greater power of the tree scan shows that evolutionary history can be a powerful tool in understanding the current phenotypic significance of genetic variation.

There are other ways of using haplotype trees for phenotypic analyses of genetic variation at candidate loci. Haplotype trees can extend beyond a single species, and combined interspecific/intraspecific haplotype trees can provide important clues as to the phenotypic significance of genetic variation within a species. For example, Knudsen and Miyamoto (2001) have developed a test for detecting evolutionary rate shifts in synonymous and nonsynonymous mutations at specific protein positions. As will be shown in Chapter 12, such tests can be used to detect the presence of natural selection at the protein level as it has operated over long periods of evolutionary time. Amino acid positions with an excess of synonymous over nonsynonymous mutational substitutions tend to be under conservative selection in which selection favors the status quo. Amino acid substitutions at conserved positions tend to be deleterious and as such are preferentially associated with disease phenotypes when they appear as intraspecific variants. On the other hand, directional selection causes a shift toward nonsynonymous substitutions over synonymous. Mutations at those positions that have been subjected to directional selection also tend to have deleterious phenotypic consequences when they mutate in a species that was influenced by directional selection in its evolutionary past. For example, familial Mediterranean fever is an autosomal recessive disease characterized by recurrent attacks of fever and inflammation (Schaner et al. 2001). The disease is caused by mutations in the gene encoding the pyrin protein 2–4, with most of the disease-causing mutations being in the ret finger protein domain. This domain was sequenced in 20 primate and two nonprimate mammalian species, and the sequence data were used to construct an interspecific haplotype tree. Considerable heterogeneity in rates of synonymous and nonsynonymous substitutions was discovered consistent with episodes of directional selection that affected some clades and not others. In particular, the evolutionary branch leading to humans, chimpanzees, and gorillas bears the signature of directional selection. Interestingly, when these amino acid positions that were positively selected in our ancient ancestors mutate in contemporary human populations to an ancestral state, the phenotypic impact is often familial Mediterranean fever. These disease mutations also often represent the wildtype amino acid found in nonhominoid primate species. This illustrates again that single-amino-acid changes must be placed in their evolutionary context for accurate phenotypic predictions. This realization has led to the development of "The Adaptive Evolution Database," which identifies evolutionary branches with deviant normalized ratios of nonsynonymous to synonymous substitutions (Liberles et al. 2001). Using a combination of biochemical and evolutionary considerations, Sunyaev et al. (2001) have estimated that about 20% of the common human nonsynonymous SNPs damage the protein and that the average human carries 10 such damaging SNPs, usually in heterozygous form.

CANDIDATE LOCI AND GENETIC ARCHITECTURE

The candidate locus approach tends to provide an incomplete picture of the genetic architecture of a trait because only those loci for which a priori information exists are studied. Because our prior information may be incomplete (or incorrect), some loci that influence the phenotype of interest may not be included in the study. However, for the subset of the loci included in a candidate study, the candidate locus approach has many advantages for revealing some aspects of genetic architecture, particularly pleiotropy and interaction effects.

Pleiotropy

Generally, the knowledge of the biochemical or physiological function of a candidate gene suggests several potential phenotypes, not just one. Therefore, a candidate locus approach often suggests a hypothesis for pleiotropy. For example, we saw in Chapter 8 that the three major electrophoretic alleles at the *ApoE* locus in humans contributed significantly to the phenotype of total serum cholesterol. Given the central role that the ApoE protein plays in lipid metabolism, it is also likely that this genetic variation influences other lipid-related phenotypes. Boerwinkle et al. (1987) tested the pleiotropic hypothesis by examining the simultaneous effects of the *ApoE* electrophoretic polymorphism on several lipid phenotypes. The genetic variation at the *ApoE* locus did indeed display significant associations with several lipid-related phenotypes, a few of which are given in Table 10.3. Note that the pattern of pleiotropy between traits can vary across genotypes as well. For example, in Table 10.3, the traits of total serum cholesterol and serum β-lipoprotein are positively correlated for all three alleles. However, the $\varepsilon 2$ allele has opposite effects (a negative correlation) on the traits of total serum cholesterol and serum triglyceride, as does the $\varepsilon 3$ allele, but the $\varepsilon 4$ allele elevates both traits, resulting in a positive correlation. Hence, the correlation between phenotypic traits in a pleiotropic system is itself subject to genetic variation. Both the traits themselves and their pattern of correlation are inherited in this example.

The pleiotropic effects of *ApoE* extend beyond lipid phenotypes related to CAD. First, the role of cholesterol as a precursor for steroid hormones suggests that the same genes that predispose humans to CAD may also play a role in reproduction. Indeed, studies in humans and other mammals indicate that genes affecting cholesterol metabolism also affect ovarian function, the menstrual cycle, and fertility. For example, the $\varepsilon 2$, $\varepsilon 3$, and $\varepsilon 4$ alleles at the *ApoE* locus in humans not only influence cholesterol levels and CAD incidence but also are associated with differences in male fertility (Gerdes et al. 1996).

Table 10.3. Patterns of Pleiotropy for Three Lipid Phenotypes as Measured by Average Excesses of Three Common Electrophoretic Alleles at *ApoE* Locus in Humans

Phenotype	Average Excess for Allele (mmol/l)		
	$\varepsilon 2$	$\varepsilon 3$	$\varepsilon 4$
Total serum cholesterol	−0.52	0.047	0.26
Serum β-lipoprotein	−0.98	0.063	0.61
Serum triglyceride	0.11	−0.031	0.074

Source: From Boerwinkle and Utermann (1988).

Second, lipids are also a critical component of cell membranes, including those of neurons. Many genes involved in CAD risk also have pleiotropic effects on brain development and functioning. As already mentioned, the electrophoretic variants at the *ApoE* locus are associated with both CAD and Alzheimer's disease. In addition, *ApoE* genetic variants are associated with peripheral arterial disease (Resnick et al. 2000), synapse development by neurons (Mauch et al. 2001), susceptibility to and progression of multiple sclerosis (Schmidt et al. 2002), the optic neuropathy of primary open-angle glaucoma (Copin et al. 2002), cognitive functioning in healthy adults (Flory et al. 2000), the risk of dementia and peripheral neuropathy after infection with HIV-1 (Corder et al. 1998), the age of onset of Huntington's disease (Kehoe et al. 1999), age-related macular degeneration (Klaver et al. 1998), the age of onset of schizophrenia (Martorelli et al. 2001), risk for fetal iodine deficiency disorder (Wang et al. 2000), risk for (Tang et al. 2002) and age of onset for (Zareparsi et al. 2002) Parkinson's disease, susceptibility to declines in cognitive performance after head trauma (Mahley and Rall 2000), response to specific types of viral infections and malarial resistance (Wozniak *et al.* 2003), response to therapeutic drugs (Siest et al. 2004), risk for discontinuing the use of statins (a type of cholesterol-lowering drug) (Maitland-van der Zee et al. 2003), and obesity (Long et al. 2003). As this list shows, pleiotropy can be extensive.

Epistasis

The prior knowledge used in candidate locus studies commonly identifies several candidate loci that are biochemically or physiologically interrelated. As a consequence, the candidate locus approach often suggests possible interactions among several candidate genes in producing the phenotype of interest.

For example, we have already seen how the electrophoretic alleles at the *ApoE* locus are associated with phenotypic variation in total serum cholesterol levels (Chapter 8). The two major cholesterol components of total serum cholesterol are HDL and LDL. The genetic variation at the *ApoE* locus is associated with variation in both of these subcomponents as well. Since the ApoE protein is found in HDL particles, the effect on HDL is not surprising. However, the ApoE protein is not found in LDL particles, yet the major phenotypic impact of genetic variation at *ApoE* seems directed at this lipoprotein component. The LDL particles bind to another protein called the LDL receptor protein coded for by the unlinked *LDLR* autosomal locus. This binding is the first stage of the removal of LDL particles from the serum and their uptake by peripheral cells. Thus, the amount of binding has a strong influence on the serum levels of LDL. The ApoE protein also binds to the LDL receptor and competitively inhibits the binding of LDL. Moreover, the various forms of ApoE coded for by the different electrophoretic alleles show different binding affinities. Therefore, a candidate mechanism for the association of *ApoE* on LDL levels is through the binding of ApoE to LDL receptors. This in turn suggests the possibility of epistasis between the *ApoE* and *LDLR* loci with respect to the phenotypes of LDL and total serum cholesterol levels.

Pedersen and Berg (1989, 1990) indeed found such epistasis for these cholesterol phenotypes. Recall from Chapter 8 that there are three common electrophoretic alleles at the *ApoE* locus and that the $\varepsilon 4$ allele is associated with elevated total and LDL cholesterol levels. Pedersen and Berg discovered two major alleles at the *LDLR* locus (A_1 and A_2) that modulate the phenotypic effect associated with the $\varepsilon 4$ allele at *ApoE*. In particular, they found that the $\varepsilon 4$ allele is associated with the elevation of cholesterol levels only in individuals that are homozygous $A_2 A_2$ at the *LDLR* locus. Hence, the high positive average excess of the $\varepsilon 4$ allele shown in Chapter 8 is actually due to epistasis between the $\varepsilon 4$ allele and the A_2 allele.

The epistasis between *ApoE* and *LDLR* also reveals the complexity behind another aspect of genetic architecture: the distinction between major and minor loci that is made in much of the evolutionary genetic literature. In some cases, this distinction seems to have a biological basis. For example, as mentioned above, the electrophoretic alleles at *ApoE* also influence the age of onset of Huntington's disease (Kehoe et al. 1999). As discussed in Chapter 9, Huntington's disease is an autosomal dominant associated with the H allele at the *huntingin* locus on the tip of the short arm of chromosome 4. Homozygosity or heterozygosity for the H allele is necessary for the development of Huntington's disease, although it is not sufficient. The age of onset of this neurodegenerative disease can be so late that some individuals bearing the H allele die of other causes before they manifest any sign of Huntington's disease. As a consequence, not all people bearing the H allele have Huntington's disease, so H is necessary but not sufficient for this disease. In contrast, *ApoE* variants influence the age of onset of the disease *only* in those individuals who also bear the H allele. The *ApoE* locus is neither sufficient nor necessary for the development of Huntington's disease but rather is only a modifier of phenotypic effect given H, the *necessary* genetic element. It seems reasonable to call the *huntingin* locus the major locus and *ApoE* the minor or modifier locus.

In the case of the cholesterol phenotypes, neither *ApoE* nor *LDLR* are necessary or sufficient for elevated cholesterol levels, yet they both clearly influence cholesterol levels. In cases such as this, the term *major locus* generally means that the genetic variation at the major locus explains a larger portion of the total phenotypic variance than any other locus, whereas a *minor locus* explains only a small portion of the total phenotypic variance. The identification of such major and minor loci is biologically problematic because of the confoundment between frequency and apparent causation that was discussed in Chapter 8. To illustrate this, Table 10.4 presents a hypothetical set of genotypic values for the *ApoE/LDLR* genotypes that modifies the single-locus *ApoE* genotypic values for total serum cholesterol presented in Table 8.3 to incorporate the finding of Pedersen and Berg that genotypes bearing the $\varepsilon 4$ allele are associated with elevated serum cholesterol levels relative to genotypes bearing the $\varepsilon 3$ instead of $\varepsilon 4$ when the individual is also $A_2 A_2$ at *LDLR*.

We need to know the genotype frequencies to perform a quantitative genetic analysis of the data given in Table 10.4. In all cases, we assume a Hardy–Weinberg model with no linkage disequilibrium between the unlinked loci of *ApoE* and *LDLR*, assumptions that fit real human populations. Therefore, all we need to specify the genotype frequencies are the allele frequencies. Assume first that the allele frequencies at *ApoE* are 0.078 for $\varepsilon 2$, 0.770 for $\varepsilon 3$, and 0.152 for $\varepsilon 4$, corresponding to the observed allele frequencies in the study by Sing and Davignon (1985). Next assume that the allele frequencies at *LDLR* are 0.22 for A_1 and

Table 10.4. Hypothetical Genotypic Values for Phenotype of Total Serum Cholesterol as Function of Genetic Variation at Two Loci, *ApoE* and *LDLR*[a]

		ApoE genotypes					
		$\varepsilon 2/\varepsilon 2$	$\varepsilon 2/\varepsilon 3$	$\varepsilon 3/\varepsilon 3$	$\varepsilon 2/\varepsilon 4$	$\varepsilon 3/\varepsilon 4$	$\varepsilon 4/\varepsilon 4$
LDLR genotypes	A_1/A_1 or A_1/A_2	136	161	174	161	174	174
	A_2/A_2	136	161	174	189	185	185

[a] Consistent with results of Pedersen and Berg (1989).

Table 10.5. Two- and Single-Locus Genotype Frequencies for Genetic Variation at *ApoE* and *LDLR* Loci under Assumption of Two-Locus Hardy–Weinberg Equilibrium

		ApoE genotypes						Sum
		$\varepsilon 2/\varepsilon 2$	$\varepsilon 2/\varepsilon 3$	$\varepsilon 3/\varepsilon 3$	$\varepsilon 2/\varepsilon 4$	$\varepsilon 3/\varepsilon 4$	$\varepsilon 4/\varepsilon 4$	
LDLR genotypes	A_1/A_1 or A_1/A_2	0.002	0.047	0.232	0.009	0.092	0.009	0.392
	A_2/A_2	0.004	0.073	0.361	0.014	0.142	0.014	0.608
	Sum	0.006	0.120	0.593	0.024	0.234	0.023	1

Note: Allele frequencies at *ApoE* are 0.078 for $\varepsilon 2$, 0.770 for $\varepsilon 3$, and 0.152 for $\varepsilon 4$, and the allele frequencies at *LDLR* are 0.22 for A_1 and 0.78 for A_2. The column sums of the two-locus genotype frequencies give the marginal genotype frequencies for the genotypes at the *ApoE* locus, and the row sums give the marginal genotype frequencies for the genotypes at the *LDLR* locus.

0.78 for A_2, corresponding to the observed allele frequencies by Pedersen and Berg (1989). Table 10.5 shows the two-locus and marginal single-locus genotype frequencies under these assumptions. We can also calculate the marginal genotypic values by making use of the concept of conditional probability (Appendix 2). For example, the frequency of the genotype $\varepsilon 3/\varepsilon 4$ and $A_1/$- (where the dash refers to either the A_1 or A_2 allele) in Table 10.5 is 0.092, and the frequency of the genotype $\varepsilon 3/\varepsilon 4$ and A_2/A_2 is 0.142. Adding these two numbers gives the marginal frequency of the genotype $\varepsilon 3/\varepsilon 4$ to be 0.234, as shown in Table 10.5. However, now consider the conditional probability of the genotype $\varepsilon 3/\varepsilon 4$ and $A_1/$- given the genotype $\varepsilon 3/\varepsilon 4$. As shown in Appendix 2, such a conditional probability is the probability (which is the genotype frequency in this case) of the genotype $\varepsilon 3/\varepsilon 4$ and $A_1/$- divided by the probability of the genotype $\varepsilon 3/\varepsilon 4$; that is, $0.092/0.234 = 0.393$. Similarly, the conditional probability of the genotype $\varepsilon 3/\varepsilon 4$ and A_2/A_2 given the genotype $\varepsilon 3/\varepsilon 4$ is $0.142/0.234 = 0.607$. Now, we can calculate from Table 10.4 the marginal genotypic value of the genotype $\varepsilon 3/\varepsilon 4$ as the conditional expectation (recall Chapter 8) of the genotypic value given the marginal genotype is $\varepsilon 3/\varepsilon 4$ as $(0.393)(174) + (0.607)(185) = 180.692$. The other marginal genotypic values for the single-locus genotypes at both the *ApoE* and *LDLR* loci can be calculated in a similar fashion, and you should do so as an exercise. Once these marginal genotypic values have been calculated, they can be combined with the marginal genotype frequencies to perform single-locus analyses of the effects of *ApoE* and *LDLR* considered separately upon the phenotype of total serum cholesterol using the equations given in Chapter 8 (once again, you should do so as an exercise). The marginal single-locus analysis of *ApoE* yields average excesses (and average effects since we are assuming Hardy–Weinberg genotype frequencies) of -12.385 mg/dl for $\varepsilon 2$, -0.021 mg/dl for $\varepsilon 3$, and 6.461 mg/dl for $\varepsilon 4$. These values in this hypothetical population are close to the values observed in the real population in Table 8.3. The marginal genetic variance for *ApoE* is calculated as 41.0 mg^2/dl^2 for this hypothetical population, split into an additive variance of 36.6 and a dominance variance of 4.4. Using a total phenotypic variance of 732.5 mg^2/dl^2 (the same as that given in the real population described in Chapter 8), this yields a broad-sense heritability attributed just to *ApoE* of 0.056 and a heritability of 0.050. Thus, in this hypothetical population, the electrophoretic alleles at *ApoE* account for 5.6% of the total phenotypic variance.

In contrast, the same type of marginal quantitative genetic analysis on *LDLR* yields average excesses/average effects of -2.125 for A_1. bearing gametes and 0.599 for A_2. The genetic variance for *LDLR* is just 2.907, split into an additive variance of 2.548 and dominance variance of 0.359, yielding a broad-sense heritability of 0.004 and a heritability of 0.003. Thus, *LDLR* genetic variation explains only 0.4% of the phenotypic variance. Thus, there seems to be no doubt that *ApoE* is the major locus in this case, and *LDLR* is the minor locus. Indeed, the marginal effects of *LDLR* are so minor that in most samples such figures would have little chance of statistical significance. From a mechanistic sense, these conclusions might seem perplexing. In this hypothetical population, we *know* that the elevation in cholesterol level is due to the *interaction* of the $\varepsilon 4$ allele with the A_2 allele. Given this knowledge, the large positive marginal average excess of 6.461 for the $\varepsilon 4$ allele makes sense, but given that the A_2 allele is just as important mechanistically in causing elevated cholesterol, why is the marginal average excess of the A_2 allele just 0.359—an effect so small as to be undetectable in most samples of real populations?

One possible explanation is that this is an artifact of studying a two-locus system as two single-locus systems. However, we can apply the quantitative genetic framework outlined in Chapter 8 to the two-locus system with a little additional effort. First, we can couple the two-locus genotypic values in Table 10.4 with the two-locus genotype frequencies in Table 10.5 to calculate the two-locus genetic variance to be 52.833. Note that the two-locus genetic variance is greater than either single-locus genetic variance and indeed is greater than their sum. Hence, we really did lose biological information when we treated this two-locus system as two one-locus systems. We can now decompose this two-locus genetic variance into additive and dominance subcomponents using the numbers we have already calculated in the marginal analyses. By definition, the average effect of a two-locus gamete is the sum of the average effects of the two single-locus average effects. For example, the average effect of a gamete bearing the $\varepsilon 4$ allele and the A_2 allele is $6.461 + 0.599 = 7.060$. Similarly, one can calculate all the other two-locus average effects and with these numbers calculate additive genotypic deviations and the additive genetic variance, which is 39.166 in this case. The dominance deviations for any two-locus genotype are calculated by first adding the two marginal genotypic deviations of each single-locus genotype that contributes to the two-locus genotype and then subtracting off the additive genotypic deviation for that two-locus genotype (Falconer and Mackay 1996). Once the dominance deviations have been calculated, the two-locus dominance variance is calculated in the standard fashion (the average over two-locus genotypes of the two-locus genotype's dominance deviations squared) to be 4.731. Note that the sum of the additive and dominance variances is 43.897, which is less than the two-locus genetic variance of 52.833. In the single-locus analyses given in Chapter 8, the additive and dominance variances always summed to the genetic variance. This occurred because in the single-locus model only two genetic parameters are needed to describe completely the genotypic deviations, the average effects and the dominance deviations. As we saw in Chapter 8, the dominance deviation is also the residual of the genetic variance once the additive variance has been subtracted off—there is simply nothing more left to fit in a one-locus model. This is not the case in the two-locus model because of epistasis. By definition, the **epistatic or interaction variance of the phenotype** is the residual of the genetic variance after the additive and dominance variance components have been subtracted. In our hypothetical example, this means that the epistatic variance is $52.833 - 43.897 = 8.936$. Note that even though epistasis has a strong effect on genotypic values (Table 10.4), the epistatic variance is much smaller in this case than the additive variance (8.936 versus 39.166). The two-locus analysis does not add much beyond the marginal single-locus analysis of *ApoE* alone, as shown graphically in Figure 10.15*a*.

Thus, it makes little difference whether or not you analyze this system one locus at a time or take into account the two-locus genotypic values. In either case, *ApoE* emerges as the major locus, and the elevated cholesterol levels observed in many individuals are attributed

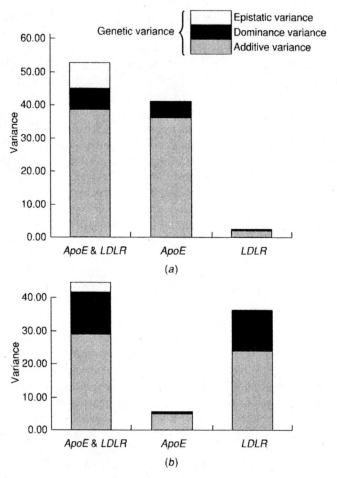

Figure 10.15. Decomposition of genetic variance into additive, dominance, and epistatic components and into marginal single-locus components for genotypic values given in Table 10.4 and assuming total phenotypic variance of 732.5. (a) ApoE allele frequencies are 0.078, 0.770, and 0.152 for $\varepsilon 2$, $e3$, and $\varepsilon 4$ alleles, respectively, and LDLR allele frequencies are 0.22 and 0.78 for A_1 and A_2 alleles, respectively. (b) ApoE allele frequencies are 0.02, 0.03, and 0.95 for $\varepsilon 2$, $e3$, and $\varepsilon 4$ alleles, respectively, and LDLR allele frequencies are 0.5 and 0.5 for A_1 and A_2 alleles, respectively.

primarily to the additive effect of the $\varepsilon 4$ allele with epistasic variance playing at most a minor role. Now let us redo these calculations for a second hypothetical population with identical genotypic values (Table 10.4) and identical total phenotypic variance (732.5). All that will be changed are the allele frequencies and nothing else. In particular, the ApoE allele frequencies are 0.02, 0.03, and 0.95 for the $\varepsilon 2$, $e3$, and $\varepsilon 4$ alleles, respectively, and the LDLR allele frequencies are 0.5 and 0.5 for the A_1 and A_2 alleles, respectively. The results are summarized in Figure 10.15b. As before, epistatic variance makes very little contribution to phenotypic variance, but in great contrast to Figure 10.15a, LDLR is now clearly the major locus and ApoE is the minor locus. Given that the relationship between genotype and phenotype has *not* been altered (Table 10.4), how could this be?

The answer to this question lies in the confoundment between frequency and apparent causation as discussed in Chapter 8. Recall from Chapter 8 that additive, dominance, and (now) epistatic variance refers to a population of phenotypically variable individuals and these variance components are explicitly a function both of how individual genotypes map to phenotypes and of the genotype frequencies in the population. Recall also from Chapter 8 that Mendelian epistasis is necessary for there to be an epistatic variance but that Mendelian epistasis contributes both to the dominance and additive variances in addition to the epistatic variance and Mendelian dominance contributes to the dominance and additive variances (Cheverud and Routman 1995). Thus, Fisher's additive variance includes the effects of additive gene action plus some effects of Mendelian dominance and epistasis, whereas Fisher's epistatic variance includes only a fraction of the effects of Mendelian epistasis. The epistatic variance is therefore an inappropriate tool for measuring the importance of epistasis in genetic architecture. Additivity in Fisher's model tells one little about the underlying genetic architecture. In particular, we can see from our hypothetical examples that interaction effects tend to be additively attributed to the rarer component of the interaction system. Table 10.4 shows that elevated cholesterol levels are due to an interaction between the $\varepsilon 4$ and A_2 alleles. In the population illustrated in Figure 10.15a, the frequency of $\varepsilon 4$ is much rarer than the frequency of A_2 (0.152 vs. 0.78, respectively). In Figure 10.15b these relative frequencies are reversed (0.95 for $\varepsilon 4$ and 0.5 for A_2). In both cases, the major locus is the locus with the rarer allele in the $\varepsilon 4-A_2$ interaction system and the minor locus is the one with the more common allele. A major locus can become a minor locus simply by shifting allele frequencies and without any change in the relationship between genotypes and phenotypes. Thus, the distinction between a major and minor locus is strictly in the context of a particular population and says nothing per se about the importance of a specific locus as a contributor to an individual's phenotype. It cannot be emphasized enough that medical applications of genetics are primarily directed at individuals and not populations. The candidate locus approach provides medically useful information for individuals concerning epistasis in contrast to the Fisherian model.

The confoundment of frequency and apparent causation in interaction systems provides an explanation for the nonreplicability found in many human genetic studies of phenotypic association. For example, Hegele et al. (2001) found a significant association of a promoter sequence variant in the gene encoding cholesterol 7-α hydroxylase with reduced plasma HDL cholesterol levels in a Canadian Hutterite population. However, this same variant had no significant association with any lipid trait in a Canadian Oji-Cree population and no significant association with HDL cholesterol but a significant elevation of LDL cholesterol in an Inuit population. The frequency of this variant was 0.708 in the Hutterites, 0.466 in the Oji-Cree, and 0.490 in the Inuit. These inconsistencies could be explained by the allele frequency differences coupled with epistasis at other loci (also potentially showing allele frequency differences) or interactions with environmental variables. This type of inconsistency is not rare in human studies. Indeed, "major" genes found in one study commonly vanish in other studies at a rate that cannot be explained by just statistical sampling alone (Crow 1997). Often the populations in disease association studies are sampled from one or a few large kindreds or from isolated founder populations. Consequently, we expect large differences in allele frequencies among these human populations. If the phenotypic associations are not caused solely by a major gene but rather are due to the interactions of many genes and environments, we expect such vanishing genes. Consequently, the lack of replicability in many human genetic studies is to be expected and is an indicator of the importance of interactions in producing human phenotypes.

Because candidate loci studies generally involve only a small number of loci, it is often feasible to examine all pairwise combinations for epistasis regardless of marginal effects.

For example, Nelson et al. (2001) surveyed a human population for genetic variation at 18 markers in six candidate gene regions for lipid phenotypes: *ApoAI–CIII–AIV, ApoB, ApoE, LDLR, LPL,* and *Pon-1*. The same individuals were also scored for several lipid-related traits. Without invoking any prior model of gene action that favors marginal or interaction effects, Nelson et al. exhaustively searched for all marginal and pairwise phenotypic associations with the markers. Because of the large number of statistical tests required in such an exhaustive search, those results significant at the 1% level were subjected to a 10-fold cross validation to guard against spurious results. Marginal and epistatic terms were then estimated only after significant and validated genetic markers had been identified. Nelson et al. found that the validated pairs of loci that explained the greatest amount of phenotypic variability show epistasis for most traits examined. Moreover, in many cases there were no significant marginal effects at all! An example of this is shown in Figure 10.16. This study shows a severe flaw and bias in the traditional method of building multilocus models by first starting with marginal or additive effects and then adding on interaction effects. If Nelson et al. (2001) had taken such an approach, they would have failed to detect those markers that explained the greatest amount of phenotypic variation for most of the lipid traits studied. Thus, the Fisherian approach that first looks for a marginal effect before examining for the presence of epistasis is biased and may miss much of the underlying genetic architecture.

Candidate loci approaches are also biased; indeed, the very information used to pick the candidate loci in the first place is a source of bias. By confining inference to candidate loci, epistasis involving one or more noncandidate loci will always be missed. Thus, although

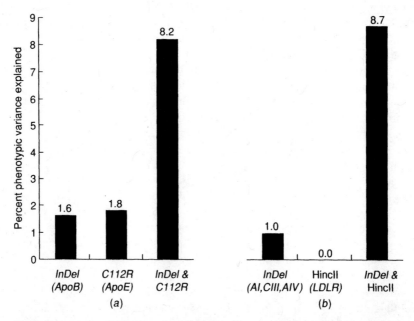

Figure 10.16. Amount of variability in natural logarithm of serum triglyceride (ln Trig) levels explained by genotypic classes. The height of each bar indicates the percent of the total phenotypic variance of ln Trig levels that is explained by two markers together or each marker separately. (*a*) In females, the insertion/deletion polymorphism at the *ApoB* locus and the *C112R* polymorphism at the *ApoE* locus explain the largest significant and validated proportion of ln Trig variability. (*b*) In males, the insertion/deletion polymorphism at the *ApoAI–CIII–AIV* gene complex and the *Hinc*II restriction site polymorphism at the *LDLR* locus explain the largest significant and validated proportion. Based on data from Nelson *et al.* (2001).

the candidate locus approach can yield much information about the contributions of the candidate loci to epistasis and genetic architecture in general, there is never a guarantee that all relevant loci are included in the candidate set in the first place. One method of choosing candidate loci that reduces this bias is the use of microarray data. Such an approach not only identifies candidate loci, as mentioned earlier, but also can measure the coexpression of candidate genes in a quantitative fashion, which suggests candidate interactions among genes as well (de la Fuente et al. 2002).

Pleiotropy and Epistasis

A gene can have both pleiotropic and epistatic effects, For example, we have already seen that the genetic variation at the *ApoE* locus affects many phenotypes and displays many epistatic interactions. The complexity of this genetic architecture centered on *ApoE* is greatly augmented by the fact that the epistatic interactions displayed by *ApoE* differ greatly for the various pleiotropic phenotypes. Thus, we have already seen strong epistasis between *ApoE* and *LDLR* for the phenotype of total serum cholesterol, epistasis between *ApoE* and *ApoB* for the phenotype of ln Trig in women, and epistasis between *ApoE* and *Hd* for the phenotype of age of onset in Huntington's disease. Other traits associated with *ApoE* also display epistasis, each with a different set of loci. For example, the elevated marginal risk for Alzheimer's disease associated with the $\varepsilon 4$ allele is due in part to epistasis with the *LDLR* locus (Cheng et al. 2005), just as for the phenotype of elevated total serum cholesterol. However, risk for Alzheimer's disease is also elevated in those individuals bearing the $\varepsilon 4$ at *ApoE* that also bear the *A* allele at the α-1-antichymotrypsin locus (Kamboh et al. 1995). These and other studies on *ApoE* have revealed an expanding web of pleiotropic effects and epistatic interactions. Similar webs of pleiotropy and epistasis have been revealed by candidate loci studies on other systems, including those initially regarded as "simple" Mendelian traits (Templeton 2000). Thus, the genetic architectures of many traits are interwoven by a complex pattern of shared loci through epistasis and pleiotropy. As a consequence, it is difficult to alter just one trait in isolation of all other traits.

This interweaving of genetic architectures through pleiotropy and epistasis has great evolutionary significance, as we will see in later chapters dealing with natural selection. This interweaving of genetic architectures also has clinical significance. For example, two different loci, *BRCA1* and *BRCA2*, have been implicated in increased risk for breast and ovarian cancer, but as with most traits there is much heterogeneity in these risks from one population to another. For example, *BRCA2* is predictive of cancer risk in Ashkenazi Jewish females but not in all populations. Levy-Lahad et al. (2001) discovered that Ashkenazi Jews also have a high frequency of a 5' SNP at the *RAD51* locus. The protein products of *BRCA1*, *BRCA2*, and *RAD51* function in conjunction with one another in DNA repair inside the cell. The 5' SNP at the *RAD51* locus was found to interact with genetic variation at the *BRCA2* locus in this Ashkenazi population to influence cancer risk, but not with the *BRCA1* locus. Thus, just as the marginal effects of a single locus often gave a biased and sometimes misleading view of genetic architecture in the examples discussed above, clinical information based upon single-locus marginal effects is likewise limited. Better *individual*-level clinical prediction is possible with increased knowledge of genetic architecture and the role of epistasis.

Gene–Environment Interactions

Genes interact not only with one another to influence phenotypic variation but also with environments to produce phenotypes. Indeed, this is one of the fundamental premises of evolutionary genetics and provides the basis for all adaptive evolution (Chapters 1 and 11). The prior knowledge used to identify a candidate locus commonly identifies environmental factors as well that may influence the candidate biochemical or physiological process. As a consequence, the candidate locus approach often suggests possible interactions between genotypes and specific environmental variables.

Studies on *ApoE* can illustrate such candidate gene-by-environment interactions. We have already seen that the electrophoretic variants at the *ApoE* locus influence many serum lipid phenotypes. Such phenotypes are also influenced by our diet. A large study was done on young children randomized into control and dietary intervention groups (Rask-Nissila et al. 2002). The intervention group was placed on a low-saturated-fat, low-cholesterol diet. Many lipid phenotypes were then tracked over time in these two groups. This study revealed that children with the genotypes $\varepsilon 2/e3$ or $\varepsilon 3/e3$ were more responsive to this dietary intervention for non-HDL cholesterol levels than children with other genotypes at the *ApoE* locus.

Many drugs have also been developed to lower cholesterol levels, and *ApoE* variants interact with the drug environment as well. For example, inhibitors of the enzyme HMG CoA reductase (statins) are widely used to lower cholesterol levels, but there is considerable interindividual variation. Vohl et al. (2002) found a significant association between the *ApoE* polymorphisms and LDL cholesterol response to the drug simvastatin, but this response was modulated by epistasis with the *LDLR* locus. Fenofibrate is another drug used to control lipid levels. Brisson et al. (2002) found that people bearing the $\varepsilon 2$ allele were more responsive to fenofibrate for many lipid parameters, but this response was modulated by epistasis with the *LPL* locus and with the *PPARα* locus (fenofibrate is an agonist of the protein product of this last locus). Note that these studies also reveal the importance of pleiotropy and epistasis in modulating genetic responsiveness to the environment.

These and many other studies reveal that a single genotype can give rise to a variety of phenotypes depending upon environmental conditions (**phenotypic plasticity**) and that different genotypes will display different phenotypic responses to the same environmental change (**norms of reaction**). These features of gene-by-environment interaction will be important later in this book for understanding many features of natural selection. As the drug response studies given in the paragraph above reveal, phenotypic plasticity and norms of reaction also have great practical and clinical importance. A widespread appreciation for the importance of these quantitative genetic concepts is only recent in human genetics. This appreciation was slow in developing because drug and food metabolism is sufficiently redundant and complex as to mask the effects of single-locus variants when studied with classic Mendelian approaches (Marshall 1997). However, there are now a plethora of candidate loci for many clinical phenotypes, so the increased feasibility and power of measured genotype approaches are revolutionizing the field of dietary and drug therapies. Indeed, "pharmacogenomic" companies have been founded in recent years with the expressed purpose of using genetic knowledge in predicting drug response.

Genetic Architecture: Is the Whole Greater Than the Sum of the Parts?

One of the greatest strengths of the candidate locus approach is that it not only focuses attention upon particular loci that are likely to contribute to phenotypic variation but also suggests potential interaction effects. In the previous section, we have seen the impact of pleiotropy, epistasis, and environmental interactions in shaping genetic architecture. These studies suggest that there will be limits to reductionism and stress the importance of context dependency; in other words, the whole is greater than the sum of the parts.

We end this chapter with an example that directly relates to this issue. We have seen many studies in this chapter that reveal that the electrophoretic variation at the *ApoE* locus is associated with large effects upon serum cholesterol level and upon the incidence of CAD. Many studies have also shown that high cholesterol level itself is an indicator of risk for CAD. This suggests a simple, reductionistic model for the impact of *ApoE* on CAD; namely, *ApoE* variation influences cholesterol levels, which in turn influence CAD risk. This simple, linear hypothesis can be portrayed as follows:

$$ApoE \text{ variation} \Rightarrow \text{cholesterol-level variation} \Rightarrow \text{variation in CAD risk}$$

Such simple, linear models of causation abound in science and often have proven to be quite useful. But the complexities in genetic architecture revealed in the studies mentioned in this chapter suggest that such a simple model may be inappropriate.

If the model given above is true, cholesterol levels should explain the effects of *ApoE* on CAD risk. If true, such a model would have great clinical implications in risk assessment for CAD, the most common cause of death in the developed countries. If all the effects of *ApoE* on CAD are modulated through cholesterol levels, there is no point to measuring genotypes for clinical predictions because cholesterol level is "closer" to the disease endpoint in the linear chain of causation given in the model portrayed above.

Sing et al. (1995) tested this simple model with a retrospective endpoint analysis. A population of older adults from Rochester, Minnesota, was studied in which 50% of them suffered from CAD. The study then asked about what variables were predictive of whether or not an individual suffered from CAD in this population. Individuals were classified by several criteria. One criterion was *ApoE* genotype. Because of sample size considerations, only the three most common genotypes were considered: $\varepsilon 4/\varepsilon 3$, $\varepsilon 3/\varepsilon 3$, and $\varepsilon 3/\varepsilon 2$. The population was also divided into three equal-sized samples based on serum cholesterol level: the highest third (high cholesterol), the middle third (medium cholesterol), and the lowest third (low cholesterol). The overall odds of having CAD in this population are 1 (50–50), so deviations above or below 1 are indicators of how these categories influence CAD incidence.

The left-hand side of Figure 10.17 shows how the odds vary as a function of cholesterol-level tertile. As expected, individuals with high cholesterol had slightly more than a twofold increase in CAD incidence, whereas those with low cholesterol had slightly more than a twofold decrease in CAD incidence. The results for *ApoE* genotype are also not particularly surprising (middle of Figure 10.17), with the $\varepsilon 4/\varepsilon 3$ having slightly more than a twofold increase in CAD odds compared to the $\varepsilon 3/\varepsilon 3$ genotype with slightly more than a twofold decrease. Given that $\varepsilon 4$ tends to elevate cholesterol level and that both *ApoE* and cholesterol level explain a comparable range of odds, the simple model of linear causation provides a plausible explanation for these marginal results. This plausibility is tested directly by cross-classifying people both by cholesterol tertile and *ApoE* genotype, thereby subdividing the population into nine categories. The right-hand side of Figure 10.17 shows the results. A glance at the now jagged graph immediately disproves the simple linear causation chain. There is no simple mapping from the marginal odds of each factor to their two-way interaction. For example, notice that some individuals with high cholesterol level (those with the $\varepsilon 3/\varepsilon 3$ genotype) fall within the normal range of incidence of CAD and indeed have lower observed incidences of CAD than other individuals with either average or even low cholesterol levels (those with the $\varepsilon 4/\varepsilon 3$ genotype). Cholesterol level per se obviously does not "cause" CAD in either a necessary or sufficient sense. The same is true for genetic factors. Note from Figure 10.17 that the highest incidence by far of CAD is found in bearers

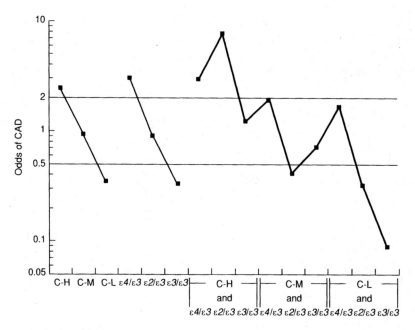

Figure 10.17. Odds of CAD in Rochester, Minnesota, population as function of cholesterol-level tertiles (H, high tertile; M, middle tertile; L, low tertile), three most common *ApoE* genotypes ($\varepsilon 4/\varepsilon 3$, $\varepsilon 3/\varepsilon 3$, and $\varepsilon 3/\varepsilon 2$), and their pairwise combinations. The odds of CAD in the overall population is 1. Solid lines indicate the "normal range" that lies between twofold above and twofold below the population average (from Sing et al. 1995).

of the $\varepsilon 2$ allele who have high cholesterol and *not* the bearers of the $\varepsilon 4$ allele, which elevates the marginal levels of cholesterol and CAD risk. In contrast to high-cholesterol individuals, the $\varepsilon 2$ allele is associated with the lowest (among genotypes) CAD incidence in individuals with medium cholesterol levels and intermediate (among genotypes) incidence in individuals with low cholesterol. The $\varepsilon 3$ allele, which does not appear to have much impact on cholesterol levels in the marginal analysis (Chapter 8), emerges as the allele associated with the lowest overall incidence of CAD (the $\varepsilon 3/\varepsilon 3$ genotype in people with low cholesterol). Finally, the $\varepsilon 4$ allele, which raises cholesterol levels marginally, is only associated with the highest incidence of CAD among genotypes when cholesterol levels are medium to low. It is obvious that there is no universal "defective" allele in this situation: Which genotype is "best" is highly context dependent and can vary over an individual's own life span as cholesterol level can change due to life alterations in life style, diet, or drugs.

Such context dependency is the norm in a system in which true causation arises from interactions rather than single factors. The resulting complexity and context dependency may at first be discouraging to those who want to apply genetic knowledge. But Figure 10.17 also reveals a beneficial side to embracing the context dependency of interaction systems. Note that the scale for the odds of CAD in Figure 10.17 is logarithmic. Both cholesterol levels and *ApoE* genotypes vary over about a 5-fold range when considered in isolation, but together the odds ratios vary a 100-fold. Obviously, there is much more information in the two pieces put together than in either separately. Learning how to put the pieces back together is a major challenge in modern biology. This challenge illustrates the limits of reductionism, but the potential benefits are great. The whole really is greater than the sum of the parts.

NATURAL SELECTION AND ADAPTATION

11

NATURAL SELECTION

The basic premises of population genetics are as follows (Chapter 1):

1. DNA can replicate.
2. DNA can mutate and recombine.
3. Phenotypes emerge from the interaction of DNA and environment.

As seen in Chapters 2–7, the first two premises result in genetic variation that exists in both time and space in reproducing populations. The third premise focuses upon how this genetic variation is translated into phenotypic variation within a population at a given point in time and space, as discussed in Chapters 8–10. A phenotype is simply any measurable trait of an individual (or sometimes another level of biological organization, as we will see in later chapters). Among the traits that can be measured are the following:

- Being alive versus being dead: the phenotype of **viability** (the ability of the individual to survive in the environment)
- Given being alive, having mated versus not having mated: the phenotype of **mating success** (the ability of a living individual to find a mate in the environment)
- Given being alive and having mated, the number of offspring produced: the phenotype of **fertility** (capable of having offspring in the environment) and **fecundity** (the number of offspring the fertile, mated, living individual can produce in the environment)

As mentioned in Chapter 1, these three phenotypes play a special role in microevolutionary theory and are called fitness components. **Fitness components** collectively determine the chances of an individual passing on its DNA in the context of the environment. Each of these three fitness components is necessary for DNA to be passed on from one generation to the next in a sexually reproducing population. Note the conditional nature of the later two fitness components; these phenotypes cannot be displayed unless the previous requirements

---

Population Genetics and Microevolutionary Theory, By Alan R. Templeton
Copyright © 2006 John Wiley & Sons, Inc.

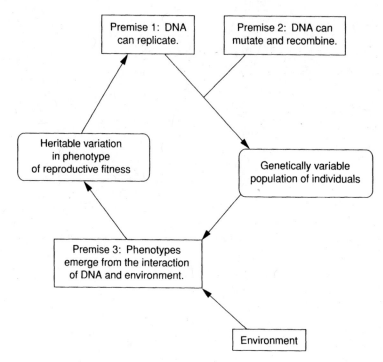

Figure 11.1. Integration of three fundamental premises of population genetic theory through phenotype of reproductive fitness.

for passing on DNA have been met. Thus, the successful passing on of DNA to the next generation requires that the individual be viable, mated, *and* fertile. Any one fitness component is necessary but, by itself, insufficient for successful DNA replication. Therefore, these three fitness components are combined into a collective phenotype called **reproductive fitness**, which measures the amount of successful DNA replication of an individual. Reproductive fitness (usually just called "fitness" in the evolutionary literature) turns premise 1 into reality (offspring who receive the replicated DNA) and unites premise 3 with premise 1, as shown in Figure 11.1.

The unification of premises shown in Figure 11.1 reveals that the probability of successful DNA replication can be influenced by how the genotype interacts with the environment. In a population of genetically diverse individuals (arising from premise 2), it is possible that some genotypes will interact with the environment to produce more or fewer acts of successful DNA replication than other genotypes. This means that genetic variation can contribute to variation in the phenotype of reproductive fitness in the population. When different genotypes respond to their environments to yield different reproductive fitness phenotypes, the environment influences the relative chances of the various genotypes for passing on their DNA to the next generation, a phenomenon called **natural selection**. Those individuals with higher than average phenotypes of reproductive fitness make a higher than average genetic contribution to the next generation's gene pool, and hence their genes tend to be overrepresented in the next generation. As long as the phenotype of reproductive fitness is heritable (Chapters 8 and 9), the unevenness of genetic contributions to the next generation induced by natural selection will alter the gene pool. In particular, natural selection tends to alter the gene pool in a manner that increases the average reproductive fitness of the population. This increase in reproductive fitness caused by natural selection is called **adaptation**. Because natural selection is modulated by how genotypes respond to their environment to produce the phenotype of reproductive fitness, adaptation is *always* relative to an environment. **Adaptive traits** are the attributes that aid organisms in living, mating, and reproducing in specific environments. Because selection and adaptation are so central to much of evolution, the remainder of this book will concentrate on natural selection and adaptation and the interactions of natural selection with other evolutionary forces.

Natural selection completes a circle among premises 1, 2, and 3 (Figure 11.1), but natural selection is *not* a tautology. Genes cannot replicate themselves; genes can only replicate through an individual that is viable, mated, and fertile in the context of the environment in which the individual develops and lives. Selection is not a tautology because a context (environment) outside the genetic system plays a direct role in determining which DNA replicates through premise 3 (Figure 11.1). Natural selection therefore arises from the *interaction* of genotypes with environment and is not self-contained within genotypes. Natural selection is based on differential DNA replication *in the context of the environment in which individuals live, mate, and reproduce*.

Although fitness is defined as an individual phenotype, in practice it is only measured and applied to groups of individuals. Population genetics differs greatly from evolutionary ecology in the groups of individuals to which fitness is assigned. In evolutionary ecology, fitness is usually assigned to a group of individuals that share some common phenotype for a trait other than fitness itself (e.g., foraging strategy, cryptic coloration pattern, etc.). In population genetics, fitness is typically assigned to all individuals sharing a common genotype. This is the convention that will be used in this book except when we use the unmeasured genotype approach in a quantitative genetic model of natural selection. Hence, in terms of quantitative genetic terminology, *the fitness that is used in population genetics is the genotypic value of the individual fitnesses of a group sharing a common genotype*. Fitness as a genotypic value is therefore a measure of a genotype's collective genetic contribution to the next generation and not necessarily the contribution of any specific individual with that genotype. This shift from fitness being an individual phenotype to fitness being a genotypic value is justified in evolutionary theory because evolution is defined in terms of genetically based changes that occur across generations. As shown in Chapter 8, average excesses and average effects measure that aspect of phenotypic variation that can be transmitted from one generation to the next through a Mendelian gamete, and both of these measures are functions solely of genotypic values and *not* each individual phenotype. This does not mean that individual phenotypes are not important; obviously, it is the individual phenotypes that determine the average phenotypic mean of a genotypic group. However, once we know the genotypic value, nothing is gained by specifying the phenotypes of each and every individual when our focus is on the next generation. Therefore, it is only the genotypic values of fitness that are important in natural selection and evolution. Hereafter, whenever the word "fitness" is used, it will refer to the genotypic value of a genotypic class and no longer to specific individuals.

FUNDAMENTAL EQUATION OF NATURAL SELECTION: MEASURED GENOTYPES

We will first explore the consequences of natural selection through a simple extension of the one-locus, two-allele model introduced in Chapter 2. As before, we need to go through a complete generation transition, ending up in the next generation at a comparable point to the starting generation. In the models used in Chapter 2 we went from parental genotypes to gametes via Mendelian probabilities, then to offspring genotypes through the mechanisms of uniting gametes (population structure). To include natural selection, we need to expand this model by explicitly modeling the life stages measured by the fitness components. In this chapter, we will assume each of the fitness components can be measured as a constant probability that is assigned as a genotypic value to a particular group of individuals sharing a common genotype.

Figure 11.2 presents the basic one-locus, two-allele model that allows us to incorporate natural selection. As we did in Chapter 2, we assume the life stages are discrete with no overlapping of generations. We start with a gene pool at the initial generation. Our genetic model considers only variation at a single autosomal locus with two alleles, A and a, so the initial gene pool is completely characterized by specifying the frequency of the A allele to be p (the frequency of a is q). We initially consider only a single, isolated local population. This means that we can ignore gene flow as a component of population structure. We also assume that the population is infinite in size, so we can equate probabilities to their frequencies. This means that we can ignore genetic drift as a component of population structure. However, we make no assumptions at this point about the deme's system of mating. Accordingly, the results to be given apply both to random-mating and non-random-mating populations. Under these assumptions, the deme's system of mating will determine the probabilities of the various combinations of gamete pairs to yield the zygotic genotype frequencies, z. These zygotic genotype frequencies represent the frequencies of the possible genotypes at the moment of fertilization. In the models given in Chapter 2 where there was no natural selection, the zygotic genotypic frequencies were equated to the adult genotype frequencies, reflecting the assumption that all zygotes were equally viable.

We now deviate from the models of Chapter 2 by assuming that the individuals develop phenotypes in an environment that is constant through time. The first phenotypic attribute that we focus upon is viability. We assume that each genotype has a genotypic value of ℓ_{ij} that measures the probability of a zygote with genotype ij living to adulthood in the environment in which it grows and develops. The frequency of an adult with genotype ij is proportional to the product of its initial zygotic frequency, z_{ij}, times the probability that it lives to adulthood, ℓ_{ij} (Figure 11.2). Because some or all of the ℓ_{ij} can be less than 1 (not every zygote lives to be an adult), these products do not define a probability distribution. Hence, it is necessary to divide the products $z_{ij}\ell_{ij}$ by the average viability, $\bar{\ell} = z_{AA}\ell_{AA} + z_{Aa}\ell_{Aa} + z_{aa}\ell_{aa}$, to obtain the adult genotype frequencies, $z_{ij}\ell_{ij}/\bar{\ell}$, which sum to 1.

In Chapter 2 we also assumed that all adults are mated, or at least that the same proportion of adults are mated in each genotype category. We now assume that each genotype has a genotypic value of m_{ij} that measures the probability of an adult with genotype ij successfully finding a mate in the environment (Figure 11.2). The frequency of a mated adult with genotype ij is proportional to the product of its adult genotypic frequency times the probability that it finds a mate, m_{ij}. These products are proportional to $z_{ij}\ell_{ij}m_{ij}$. Because some or all of the m_{ij} can be less than 1 (not every adult finds a mate), these products do not define a probability distribution. Hence, it is necessary to divide the products $z_{ij}\ell_{ij}m_{ij}$ by the average product of viability with mating success, $\overline{m\ell} = z_{AA}m_{AA}\ell_{AA} + z_{Aa}m_{Aa}\ell_{Aa} + z_{aa}m_{aa}\ell_{aa}$, to obtain the mated adult genotype frequencies, $z_{ij}\ell_{ij}m_{ij}/\overline{m\ell}$, which sum to 1.

In Chapter 2 we assumed that all mated adults contributed equally to the next generation's gene pool. However, we now assume that mated adults interact with their environment to produce the phenotype of the number of gametes passed on to the next generation (that is,

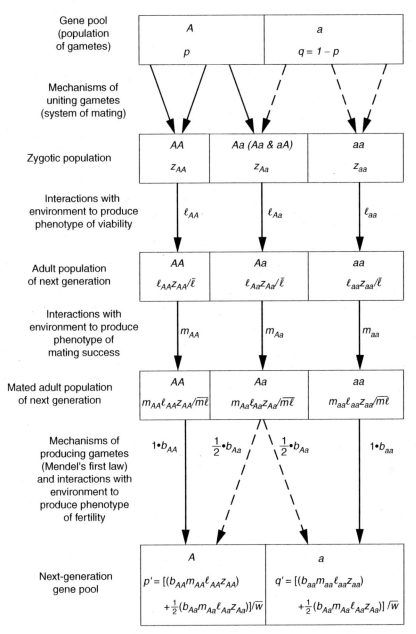

Figure 11.2. Derivation of impact of natural selection upon allele frequencies for single autosomal locus with two alleles, A and a. The numerators of the genotype frequencies of the adult population are divided by the average zygote-to-adult viability, $\bar{\ell} = z_{AA}\ell_{AA} + z_{Aa}\ell_{Aa} + z_{aa}\ell_{aa}$, to ensure that the adult genotype frequencies sum to 1. Similarly, the numerators of the genotype frequencies of the mated adult population are divided by $\overline{m\ell} = z_{AA}m_{AA}\ell_{AA} + z_{Aa}m_{Aa}\ell_{Aa} + z_{aa}m_{aa}\ell_{aa}$. In order for the allele frequencies in the next generation's gene pool to sum to 1, the portions indicated in brackets are divided by the average fitness, $\bar{w} = z_{AA}b_{AA}m_{AA}\ell_{AA} + z_{Aa}b_{Aa}m_{Aa}\ell_{Aa} + z_{aa}b_{aa}m_{aa}\ell_{aa}$.

gametes that are actually involved in fertilization events). This phenotype is designated by b_{ij}, the genotypic value for the number of successful gametes produced by mated adults with genotype ij. When interactions between the genotypes and their environment cause these b's to vary across genotypes, the contribution of mated adults with genotype ij to the next generation's gene pool is proportional to b_{ij}. The b's in turn are multiplied by the appropriate Mendelian probabilities to determine the proportional contribution of mated adults to a specific gamete type. For example, the gametic contribution of mated adult heterozygotes, b_{Aa} is split 50–50 between the A and a gamete types due to Mendel's first law (Figure 11.2).

We have now gone through a complete life cycle, going from gene pool to gene pool (Figure 11.2). Accordingly, we can now address the question of whether or not evolution occurred. From Figure 11.2, we see that the frequency of the A allele in the next generation's gene pool, p', is

$$p' = \frac{(b_{AA}m_{AA}\ell_{AA}z_{AA}) + \frac{1}{2}(b_{Aa}m_{Aa}\ell_{Aa}z_{Aa})}{(b_{AA}m_{AA}\ell_{AA}z_{AA}) + (b_{Aa}m_{Aa}\ell_{Aa}z_{Aa}) + (b_{aa}m_{aa}\ell_{aa}z_{aa})} \quad (11.1)$$

where the denominator ensures that the gamete frequencies sum to 1. Notice that the fitness components of viability, mating success, and fertility only influence the next generation's gene pool through their product. The product $b_{ij}m_{ij}\ell_{ij}$ represents the probability that a zygote with genotype ij will live to adulthood and mate ($m_{ij}\ell_{ij}$) times the number of successful gametes it contributes to the next generation given that it is alive and mated (b_{ij}). Accordingly, $b_{ij}m_{ij}\ell_{ij}$ represents the average reproductive contribution (number of successful gametes) of a *zygote* with genotype ij. Hence, $b_{ij}m_{ij}\ell_{ij}$ is the genotypic value of the phenotype of overall reproductive success throughout the entire lifespan measured from fertilization; that is, $b_{ij}m_{ij}\ell_{ij}$ is the **fitness** of genotype ij. We therefore define $w_{ij} = b_{ij}m_{ij}\ell_{ij}$ as the genotypic value of the phenotype of fitness for genotype ij. In this simple model, the three fitness components of viability, mating success, and fertility combine in a simple multiplicative fashion to yield the phenotype of reproductive fitness. This is not always true in more complicated models of natural selection, but fitness is always some function of viability, mating success, and fertility.

In terms of the phenotype of fitness, equation 11.1 can be expressed as

$$p' = \frac{w_{AA}z_{AA} + \frac{1}{2}w_{Aa}z_{Aa}}{w_{AA}z_{AA} + w_{Aa}z_{Aa} + w_{aa}z_{aa}} \quad (11.2)$$

The denominator in equation 11.2, $w_{AA}z_{AA} + w_{Aa}z_{Aa} + w_{aa}z_{aa}$, is the mean or average value of the phenotype of fitness in the zygotic population. Therefore, let $\bar{w} = w_{AA}z_{AA} + w_{Aa}z_{Aa} + w_{aa}z_{aa}$. By subtracting the starting allele frequency, $p = z_{AA} + 1/2 z_{Aa}$, from equation 11.2, we find that the change in allele frequency over this generation is

$$\begin{aligned} \Delta p = p' - p &= \frac{w_{AA}z_{AA} + \frac{1}{2}w_{Aa}z_{Aa}}{\bar{w}} - \left(z_{AA} + \tfrac{1}{2}z_{Aa}\right) \\ &= \frac{z_{AA}(w_{AA} - \bar{w}) + \frac{1}{2}z_{Aa}(w_{Aa} - \bar{w})}{\bar{w}} \end{aligned} \quad (11.3)$$

Recall from Chapter 8 that the genotypic deviation of a phenotype is the genotypic value minus the overall population mean. Hence, the terms $w_{ij} - \bar{w}$ in the numerator of equation 11.3 are the genotypic deviations for the phenotype of fitness. Recall also from Chapter 8 (equation 8.8) that the conditional frequency of an A-bearing gamete being found in genotype AA is z_{AA}/p and in genotype Aa is $1/2 z_{Aa}/p$. Hence, equation 11.3 can be expressed as

$$\Delta p = \frac{p}{\bar{w}} \left[\frac{z_{AA}(w_{AA} - \bar{w}) + \frac{1}{2} z_{Aa}(w_{Aa} - \bar{w})}{p} \right] \tag{11.4}$$

The portion of equation 11.4 in the brackets is the average excess (Chapter 8) of the phenotype of fitness for gametes bearing the A allele. Thus, the change in allele frequency of A due to natural selection is

$$\Delta p = \frac{p}{\bar{w}} a_A \tag{11.5}$$

where a_A is the average excess of fitness of the A allele. Although the above derivation is only for a two-allele model, equation 11.5 is also true for a locus with multiple alleles. In the multiple allelic case, equation 11.5 shows how any one of the alleles changes across a generation due to the action of natural selection.

Equation 11.5 is the **fundamental equation of natural selection for a measured genotype**, which states that the change in allele frequency is proportional to the average excess of fitness for that allele. Because the average excess is the only part of equation 11.5 that can vary in sign or take on the value of zero (except for the trivial case $p = 0$), our answer to the question about whether evolution occurs or not through natural selection is determined solely by the *average excess of fitness*: If A has a positive average excess of fitness, the A allele increases in frequency; if A has a negative average excess of fitness, it decreases in frequency; and if A has a zero average excess of fitness, its frequency is unaltered by natural selection. Equation 11.5 reveals that any evolution at this locus induced by natural selection is due solely to the *transmission of fitness phenotypes through gametes*. Genotypes and their mean fitnesses are relevant only in how they contribute to the gametic measure of average excess. This means that only the *heritable* component of fitness can yield evolutionary change under natural selection. Therefore, the simplistic definition of natural selection as "survival of the fittest" is wrong when "the fittest" refers to individuals and their phenotypes. Rather, natural selection favors gametes with positive average excesses for fitness. Therefore, to truly understand natural selection as an evolutionary force, we must understand the quantitative genetic concept of average excess and we must think in terms of gametes and not individuals.

SICKLE CELL ANEMIA AS EXAMPLE OF NATURAL SELECTION

One of the textbook classics of natural selection, selection at the human β-Hb locus that codes for the β chain of adult hemoglobin, illustrates the fact that selection is not survival of the fittest and the importance of understanding natural selection in terms of average excesses. Although this story is usually told starting in Africa, we will start the story thousands of kilometers away, in Southeast Asia.

The people living in the coastal areas of Southeast Asia and the islands that now make up Indonesia were perhaps the best sailors of ancient times (Wiesenfeld 1967). About 2000 years ago, these people established a colony on the island of Madagascar, which lies just off the eastern coast of Africa and is many thousands of kilometers away from modern-day Indonesia. The colonists brought not only their language (the language of the present-day people of Madagascar has its closest affinities to the language of the Ma'anyan of Southeast Borneo) but also their agricultural system. Agriculture originated at several locations and times during human history, and the system brought by these ancient colonists is called the Malaysian agricultural complex. The Malaysian agricultural complex includes many root and tree crops such as yams and bananas, plants that are adapted to wet, tropical conditions. In contrast, the cereal-based agricultural system that arose in the Middle East is centered on crops such as wheat and barley that do not do well under wet, tropical conditions. Consequently, the cereal-based agricultural system penetrated in only a limited way into the wet, tropical portions of Africa. However, shortly after the Malaysian agricultural complex was introduced to Madagascar, Bantu-speaking peoples on mainland Africa also took it up. The Bantus rapidly expanded into much of wet, tropical Africa with this new agricultural complex and in so doing greatly changed the environment in which they lived. In particular, this new slash-and-burn agricultural system created breeding sites and optimal habitat for the mosquito *Anopheles gambia*, the primary vector for transmitting the deadly malaria parasite *Plasmodium falciparum*. Moreover, the agricultural system allowed human populations to become much more dense in this area. These two factors combined to allow falciparum malaria to become a sustained epidemic disease in this part of Africa and a major source of mortality, particularly in children. We now examine how these Bantu populations adapted to their new environment, and in particular to malaria, through natural selection operating upon genetic variation at the human β-chain hemoglobin autosomal locus (β-Hb).

In Chapter 8, we noted that the sickle cell allele (S) at this locus is a dominant allele for the phenotype of malarial resistance. In contrast, another allele at this locus, C (a mutation in the same codon as S), is a recessive allele for malarial resistance (Cavalli-Sforza and Bodmer 1971). [Modiano et al. (2001) suggest that C is not completely recessive, but for now, we will treat it as a recessive allele before turning our attention to the possibility that it is not.] Moreover, both the S and C alleles are associated with hemolytic anemia, another source of mortality. Hence, these alleles are expected to influence the phenotype of viability in both the premalarial and the malarial environment of wet, tropical Africa. Cavalli-Sforza and Bodmer (1971) estimated these viabilities for a West African population under malarial conditions, as shown in Table 11.1. Table 11.1 also shows the estimated viabilities under nonmalarial conditions by removing the beneficial effects of the malarial-resistant phenotypes. Table 11.1 gives what is known as **relative fitness**, in which the fitness of one genotype is set to 1 and all other fitnesses are measured relative to this standard. Such a rescaling of fitness values has no effect on equation 11.5 because the impact of natural selection depends only upon $(w_{ij} - \bar{w})/\bar{w}$, which is invariant to scale transformations. Hence, we can scale the fitnesses for mathematical or biological convenience without altering the evolutionary impact of natural selection through equation 11.5.

Table 11.1 makes explicit that fitness phenotypes emerge from interactions between genotypes and environments. The relative fitnesses of the various genotypic classes are altered in going from a nonmalarial to a malarial environment. The AS and CC genotypes will be our focus with regard to the human adaptation to the malarial environment because both have malarial resistance without serious hemolytic anemia. Interestingly, Table 11.1

Table 11.1. Phenotypic Attributes and Relative Fitnesses (Viabilities) of Six Genotypes Formed by A, S, and C Alleles at β-Hb Locus in Humans in Wet, Tropical Africa

Genotype	Phenotypic Attributes	Fitness in Nonmalarial Environment	Fitness in Malarial Environment
AA	Malarial susceptibility	1.00	0.89
AS	Malarial resistance	1.00	1.00
SS	Hemolytic anemia	0.20	0.20
AC	Malarial susceptibility	1.00	0.89
SC	Hemolytic anemia	0.71	0.70
CC	Malarial resistance	1.00	1.31

Note: The fitness of the *AS* heterozygote is set to 1. The malarial fitnesses are estimated from data given in Cavalli-Sforza and Bodmer (1971).

indicates that the *CC* genotype actually has much greater viability in a malarial environment than the *AS* heterozygote or any other genotype, a result consistent with an estimated 93% reduction in risk of clinical malaria in *CC* homozygotes (Modiano et al. 2001). If natural selection were truly survival of the fittest, a simple glance at Table 11.1 tells us that human populations should adapt to malaria by becoming fixed for the *C* allele so that everyone in the population would have the fittest genotype of *CC*. However, equation 11.5 tells us that natural selection is not survival of the fittest individuals, but rather *natural selection favors those gametes with positive average excesses of fitness*. Although the italicized phrase is less dramatic than survival of the fittest, it is the key to truly understanding how natural selection operates. From a gamete's point of view, the situation is not nearly so simple.

To examine the adaptive situation from a gamete's point of view, we first have to consider in which genotypes a gamete is likely to be found. This in turn depends upon the allele frequencies in the initial gene pool at the time of the transition from a nonmalarial to a malarial environment and upon population structure. Concerning the initial gene pool, Table 11.1 reveals that the *S* allele is strongly deleterious in a nonmalarial environment. However, because its deleterious effects are recessive, it would have little heritability in a random-mating population as long as it is rare (Chapter 9). Therefore, we can safely assume that the *S* allele either was not present in the initial Bantu populations and arose through mutation after the environmental transition or was present but in very low frequency. The *C* allele is neutral relative to the *A* allele in Table 11.1 under nonmalarial conditions. However, this allele is extremely rare in humans as a whole, so we can also reasonably assume that it also was either absent or rare in the initial expanding Bantu populations. Thus, we assume that the initial gene pool was $p_A \approx 1$, $p_S \approx 0$ and $p_C \approx 0$ for the frequencies of the *A*, *S*, and *C* alleles, respectively. Concerning population structure, we will assume a random-mating population of infinite size (no genetic drift) and with no gene flow with other populations.

Now consider the fitness situation from a gamete's point of view after the transition to a malarial environment has occurred. First, consider the situation from a *C* gamete's perspective, given that this is the gamete associated with the fittest genotype. The average genotypic deviation in fitness that a *C* gamete will display (that is, its average excess, equation 8.10) under our assumptions is

$$a_C = p_A(0.89 - \bar{w}) + p_S(0.70 - \bar{w}) + p_C(1.31 - \bar{w}) \qquad (11.6)$$

Under random mating, the initial population consists almost entirely of *AA* homozygotes with only the rare occurrence of any other genotypes. As a result, the average fitness of the initial population is approximately the average fitness of *AA*, which is 0.89 under malarial conditions (Table 11.1). Hence, under our initial conditions, equation 11.6 is approximately

$$a_C \approx 1(0.89 - 0.89) + 0(0.70 - 0.89) + 0(1.31 - 0.89) = 0 \qquad (11.7)$$

Equation 11.7 reveals that although the *C* allele has the best beneficial fitness effects at the genotypic level ($w_{CC} = 1.31$), at the gametic level *C* has virtually no benefit at all. The reason is straightforward if you take a gamete's perspective. Given a *C* gamete, the initial allele frequencies with *C* extremely rare and random mating ensure that this gamete is almost always going to be found in an *AC* genotype, which has the same fitness as the *AA* homozygote in a malarial environment. Hence, the *C* gamete has no selective advantage in this initial population living in a malarial environment because it is almost never expressed in a homozygous state, the only state in which it has an individual fitness advantage.

Now let us look at the situation from the perspective of an *S*-bearing gamete. The average excess of fitness of an *S* gamete and its approximate initial value is

$$a_S = p_A(1 - \bar{w}) + p_C(0.70 - \bar{w}) + p_S(0.2 - \bar{w}) \approx 1(1 - 0.89) + 0(0.70 - 0.89)$$
$$+ 0(0.2 - 0.89) = 0.11 \qquad (11.8)$$

Coupling equation 11.8 with equation 11.5, we see that *S*-bearing gametes are strongly favored by natural selection in this initial response to a malarial environment. Once again, the reason is straightforward from a gamete's perspective: Almost all copies of *S* alleles will find themselves in *AS* individuals in the initial population, and these *AS* individuals have a higher fitness than the predominant *AA* individuals. Therefore, the initial evolutionary response to a malarial environment induced by natural selection is to increase rapidly the frequency of the *S* allele, whereas the *C* allele frequency is hardly changed at all.

As the generations pass, the *C* gametes find themselves in a gene pool with many *A* gametes and an increasing number of *S* gametes. Therefore, as natural selection drives up the frequency of the *S* allele, it soon becomes likely that a *C* allele will be paired under random mating not only with an *A* allele to form the malarial-susceptible *AC* genotype with a relative fitness of 0.89 but also with an *S* allele to form the anemic *SC* genotype with an even worse fitness of 0.71. Because *C* was not initially favored by natural selection and therefore remains rare, a *C* allele is still almost never paired with another *C* allele under random mating to form the *CC* genotype, the fittest genotype ($w_{CC} = 1.31$). For example, suppose that $p_A = 0.95$, $p_S = 0.05$, and $p_C \approx 0$ after several generations of adapting to malaria. Then, with these new allele frequencies, we have $\bar{w} = 0.90$ and

$$a_C \approx 0.95(0.89 - 0.90) + 0.05(0.70 - 0.90) + 0(1.31 - 0.90) = -0.02 \qquad (11.9)$$

The negative average excess of *C* tells us that natural selection is operating to eliminate *C*-bearing gametes in the *context of this population in a malarial environment*. Under these population and environmental conditions, the *C* allele does not join with another *C* allele often enough for natural selection to favor it on the average. In contrast, the *S* allele still has a positive average excess of 0.06 when $p_A = 0.95$, $p_S = 0.05$, and $p_C \approx 0$. Accordingly, the *S* allele is still increasing in frequency. The more *S* increases in frequency, the more the average excess of C decreases. From the perspective of a C-bearing gamete, that gamete is increasingly likely to be found only in genotypes with lower than average fitness (AC genotypes that are susceptible to malaria and SC genotypes that suffer from hemolytic anemia). Therefore, using equation 11.5, we find that *natural selection will decrease the frequency of the C allele until it is eliminated from the population!* This example shows that evolution by natural selection is *not* survival of the fittest. Even though the CC genotype is the most fit genotype ($w_{CC} = 1.31$) and even though the S allele is associated with the genotype with the lowest fitness ($w_{SS} = 0.2$), it is the S allele that increases in frequency under natural selection while the C allele frequency decreases.

Once the C allele is eliminated by natural selection, the system becomes a one-locus, two-allele model with three genotypes: AA, AS, and SS. The average excess of the S allele is now simplified to

$$a_S = (1 - p_S)(1 - \bar{w}) + p_S(0.2 - \bar{w}) \qquad (11.10)$$

Figure 11.3 shows a plot of equation 11.10 against p_S over the allele frequency range of 0–0.25. As noted above, the average excess of S is positive when p_S is small, and hence the frequency of the S allele will increase in frequency due to natural selection. However, Figure 11.3 shows that once the frequency of the S allele gets above 0.12, its average excess for fitness becomes negative, implying that natural selection will cause the frequency of S to decline. Therefore, regardless of the initial frequency of S in this two-allele population, natural selection will cause the population to evolve until $p_S = 0.12$. At $p_S = 0.12$, the average excess of fitness of the S allele is zero, so equation 11.5 tells us that there is no evolution due to natural selection at this allele frequency. Hence, $p_S = 0.12$ represents a stable equilibrium; when the allele frequency drops below this frequency, natural selection operates to bring it back up to $p_S = 0.12$ (Figure 11.3), and when the allele frequency exceeds this frequency, natural selection operates to decrease it back down to $p_S = 0.12$. This is an example of a **balanced polymorphism**, in which natural selection favors an intermediate allele frequency due to a stable balance of positive and negative fitness contributions to the average excesses of the polymorphic alleles. Looking at equation 11.10, we can see the first term of the average excess $(1 - \bar{w})$ represents the positive impact of the S allele when in a heterozygote, whereas the second term $(0.2 - \bar{w})$ represents the negative impact of the S

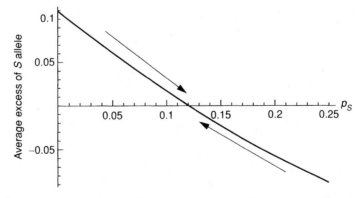

Figure 11.3. Plot of average excess of S allele for phenotype of fitness in malarial environment over allele frequency range of 0–0.25 in random-mating population polymorphic for A and S alleles. Arrows show the direction of allele frequency change induced by natural selection using equation 11.5.

allele when homozygous. When $p_S = 0.12$, the weights given to the positive and negative contributions to the average excess are exactly balanced, resulting in an average excess of zero and a stable polymorphism.

We can use equation 11.5 to find the general solution to a balanced polymorphism for the two-allele case (A and a) when there is a trade-off between the positive-fitness impact of an allele when heterozygous versus a negative-fitness impact when homozygous. Equation 11.5 will be zero (no evolutionary change) only when the average excess of the A allele is 0. One can always set $a_A = 0$ and solve for p to get the equilibrium solution, but there is an easier method. If a selective equilibrium exists, it means that the average excesses of all the alleles present in the equilibrium population must have an average excess of fitness of zero; otherwise equation 11.5 implies that some alleles will be changing in frequency and the population is therefore not in equilibrium. In the two-allele model being considered now, this means that at equilibrium $a_A = a_a = 0$; that is,

$$a_A = p(w_{AA} - \bar{w}) + (1 - p)(w_{Aa} - \bar{w}) = p(w_{Aa} - \bar{w}) + (1 - p)(w_{aa} - \bar{w}) = a_a \tag{11.11}$$

The $\bar{w}$ terms on the two sides of equation 11.11 cancel out, leaving

$$pw_{AA} + (1 - p)w_{Aa} = pw_{Aa} + (1 - p)w_{aa}$$
$$(1 - p)(w_{Aa} - w_{aa}) = p(w_{Aa} - w_{AA}) \tag{11.12}$$

Without loss of generality, we can set the fitness $w_{Aa} = 1$ and measure all other fitnesses relative to it. The fitness terms in parentheses in equation 11.12 thus become $(1 - w_{aa})$ and $(1 - w_{AA})$. These two parenthetical terms are called **selection coefficients** and measure the fitnesses of the genotypes as a deviation from the reference genotype that was assigned a fitness of 1. In this special case, the reference genotype is the heterozygote Aa. Let s be the selection coefficient of the AA homozygote such that $s = 1 - w_{AA}$ and let t be the selection coefficient of the aa homozygote such that $t = 1 - w_{aa}$. In terms of selection coefficients, the equilibrium equation 11.12 can be solved for the equilibrium allele frequency as

$$(1 - p_{eq})t = p_{eq}s \qquad p_{eq} = \frac{t}{s + t} \tag{11.13}$$

At this equilibrium frequency, equation 11.5 predicts no further evolution because the average excesses of fitness are all zero. For the A and S example, $s = 1 - w_{AA} = 0.11$ and $t = 1 - w_{SS} = 0.8$ using the fitness for the AA and SS genotypes under malarial conditions as given in Table 11.1. Hence, using equation 11.13, the equilibrium frequency of the A allele in this special case is $t/(s + t) = 0.8/(0.11 + 0.8) = 0.88$, so the equilibrium frequency of the S allele is 0.12, the same result shown graphically in Figure 11.3. Hence, equation 11.5 reveals that the key to understanding both the evolutionary dynamics and the equilibrium caused by natural selection is the average excesses of fitness of the various gamete types. The equilibrium defined by equation 11.13 also tells us that natural selection is not the same as evolution. Equation 11.5 shows that natural selection can cause evolution, but equation 11.13 and Figure 11.3 show that natural selection can prevent evolution. At the equilibrium defined by equation 11.13, natural selection operates to maintain the population at a stable allele frequency and can counteract any perturbation caused by another evolution force, such as genetic drift. Hence, selection plays a dual role in evolution; it is the force responsible for adaptive change in evolution, and it is also the force that can prevent evolution to preserve an adaptive status quo.

Because the process of mutation is random and the chance of a new mutant persisting past the first few generations is primarily due to chance even in large populations (Chapter 5), it is likely that some Bantu populations may not have had either the C or S alleles when the transition to a malarial environment occurred. Suppose that a Bantu population existed in which, just by chance, there were no S alleles in the gene pool but there were A and C alleles. As before, we assume that at the time of the environmental transition the frequency of A was close to 1 and the frequency of C close to 0. Because the malarial-resistant phenotype associated with the C allele is recessive, the average excess of fitness of C in this initial population will be close to zero but not exactly zero. For example, suppose $p_A = 0.999$ and $p_C = 0.001$. Then using equation 11.6 (with $p_S = 0$), the average excess of the A allele is -0.0000004 and the average excess of the C allele is 0.00042. As we saw earlier, when S is present, even though extremely rare, it has an average excess of about 0.1, which is many orders of magnitude greater than the average excess of C. This meant that the initial response to selection in a malarial environment was dominated by a rapid increase in the frequency of S. But now we are assuming that S is not present at all in the population, so these more modest average excesses determine the response to natural selection. In particular, we can see that selection will cause a small increase in the initial frequency of C at the expense of a small decline in the frequency of A. However, as C slowly becomes more and more common, a C-bearing gamete becomes increasingly likely to be found in the high-fitness CC homozygote under random mating. Hence, a positive-feedback loop is created in which selection increases the frequency of the C allele, which in turn increases the average excess of the C allele, which in turn allows selection to increase the C allele even more.

The C allele may have gotten a head start in its increase in frequency just by genetic drift alone in at least some populations. From Table 11.1, we can see that the C allele is expected to be neutral relative to the A allele in the absence of the S allele under the premalarial environment. It is therefore possible that the C allele could have drifted to relatively high frequencies in some Bantu populations even before the transition to malarial conditions. For example, if the initial frequency of the C allele were 0.02, its initial average excess in fitness once the transition to a malarial environment had occurred would be 0.008, still a small value but one that would result in a much larger increase in allele frequency due to natural selection than at $p_C = 0.001$. Regardless of the exact initial allele frequency, the C allele eventually becomes common and then its average excess begins to decline (Figure 11.4), reflecting the fact that there is decreasing amounts of phenotypic variance in fitness in the population as more and more of the population consists of a single genotypic class, CC. Unlike the plot of the average excess for the S allele (Figure 11.3), note that the average excess of the C allele under malarial conditions is never negative (Figure 11.4). Using equation 11.5, this means that selection will always increase the frequency of C until $p_C = 1$; that is, the population is fixed for the C allele, everyone is a CC homozygote, and the average excess is zero because there is no genetic variation.

Because the average excess of the C allele is so small when starting from a population in which it is rare, equation 11.5 implies that it will take hundreds of generations to go to fixation. For example, Figure 11.5 shows the plot of the frequency of the C allele versus number of generations that is obtained by iterating equation 11.5 and starting with an initial set of allele frequencies with $p_A = 0.98$ and $p_C = 0.02$. As can be seen, the increase in allele frequency is modest during the first 100 or so generations, but after that the C allele rapidly goes to fixation. During this extended time period before fixation, the β-Hb locus

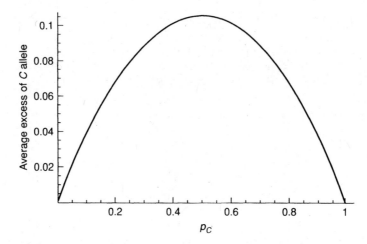

Figure 11.4. Plot of average excess of C allele for phenotype of fitness in malarial environment over allele frequency range of 0–1 in random-mating population polymorphic for A and C alleles.

is polymorphic for the A and C alleles. However, this is not a balanced polymorphism with a stable equilibrium, but rather a **transient polymorphism**; that is, the population is polymorphic during the time period in which selection operates to increase the frequency of the favored allele to ultimate fixation.

The situation described above in which the S allele was not present appears to have occurred in human populations inhabiting the region of the upper Volta rivers in Western Africa. In the center of this malarial region, the frequency of the C allele is about 0.15 and the S allele is virtually absent, just the opposite of the situation found in most other malarial regions of Africa. Indeed, a frequency of C of about 0.15 is what one would expect after about 2000 years of selection in the scenario illustrated in Figure 11.5. Eventually, if

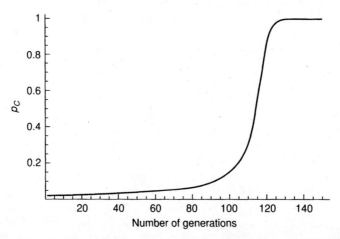

Figure 11.5. Plot of frequency of C allele (p_C) as function of number of generations in malarial environment starting with random-mating population polymorphic for A and C alleles with initial frequencies of 0.98 and 0.02, respectively.

the malarial conditions persist, we would expect this population to go to fixation for the C allele, the only stable selective equilibrium allowed by the average excess of C in such an environment (Figure 11.4).

We saw earlier that natural selection is *not survival of the fittest* when "fittest" refers to those individuals or even genotypes with maximal fitness. By examining the two alternative routes of adapting to malaria at the hemoglobin β-chain locus illustrated by the A/S balanced polymorphism versus fixation of the C allele, we can now see that natural selection is *not survival of the fittest population* either. Consider a population that adapts to malaria through the A/S polymorphism. The average fitness of such a population has increased from 0.89 before the adaptation to 0.90 after the adaptation. Moreover, $p_A^2 = (0.88)^2$ or 77% of the population has a genotype still susceptible to malaria (AA), $p_S^2 = (0.12)^2$ or 2% of the population suffers from hemolytic anemia, and only $2p_A p_S$ or 21% of the population is resistant to malaria and not suffering from anemia. In contrast, if a population had adapted to malaria through the fixation of the C allele, the average fitness would have increased from 0.89 to 1.31 and everyone in the population would be resistant to malaria at a level even greater than that of AS heterozygotes, and no one would suffer from severe hemolytic anemia. Despite these population- and individual-level advantages to fixing the C allele, an initial random-mating population with both the S and C alleles present but rare is driven by natural selection to the A/S balanced polymorphic solution and the elimination of the C allele. Natural selection favors neither the fittest individuals nor necessarily the fittest population. Rather, equation 11.5 tells us that *natural selection always favors the fittest gamete, not the fittest individual or the fittest population*.

The West African populations with high frequencies of the C allele are not completely isolated genetically from the surrounding populations that have a high frequency of the S allele. Given that the C allele seems to allow a superior adaptation to malaria for both individuals and populations, why does the C allele not spread throughout Africa? To answer this question, we must first recognize that the "superiority" of C at the individual and population levels is irrelevant. We must answer this question only from a gamete's perspective. Consider a C-bearing gamete introduced by gene flow into a population characterized by an A/S balanced polymorphism (recall, this means that $p_A = 0.88$, $p_S = 0.12$, and $\bar{w} = 0.90$). Such a C allele will have an average excess of approximately (using the malarial fitnesses in Table 11.1)

$$a_C \approx 0.88(0.89 - 0.90) + 0.12(0.70 - 0.90) = -0.03 \qquad (11.14)$$

Hence, in the context of a population already adapted to malaria through the A/S balanced polymorphism, a C allele is strongly deleterious and would be rapidly eliminated. Likewise, consider an S-bearing gamete introduced by gene flow into a population that has adapted to malaria through the fixation of C. Such a gamete will almost always find itself in an SC genotype, and hence the average excess of the introduced S-bearing gamete is $0.70 - 1.31 = -0.61$. The negative average excesses of the gametes introduced by gene flow from a population displaying one mode of malarial adaptation to a population showing the other indicate that from the gamete's perspective these two adaptive solutions are mutually exclusive. In this example, natural selection prevents the spread of alleles from one population into the other.

The mutually exclusive nature of these two adaptive solutions to malaria explains another perplexing feature of these malarial adaptations: The frequencies of S and C are negatively correlated in the malarial regions of western Africa (Figure 11.6). Since both the S and C

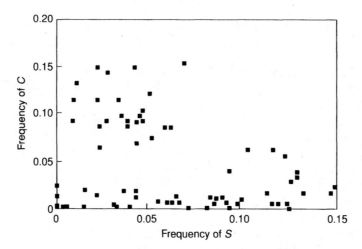

Figure 11.6. Frequencies of *S* and *C* alleles in 72 West African populations. Reprinted with permission from Fig. 4.12 in L. L. Cavalli-Sforza and W. F. Bodmer, *The Genetics of Human Populations*. Copyright ©1971 by W. H. Freeman and Co.

alleles are alleles associated with malarial resistance, it may at first glance seem reasonable to expect them both to be favored by natural selection in a malarial environment and therefore to have a positive correlation with one another. However, natural selection is *not* survival of the fittest trait either. Obviously, malarial resistance is the fittest trait to have in a malarial environment, but selection operates only from the gamete's point of view, which includes a weighted average of the effects of that gamete in all possible genotypes and over the entire constellation of traits (including hemolytic anemia in this case) associated with these genotypes. The negative-fitness interaction between the *S* and *C* alleles when put together in *SC* heterozygotes ensures that these two adaptive strategies are negatively correlated from the gametic perspective. Only the gametic perspective determines the pattern observed in the actual populations (Figure 11.6). Thus, two alleles associated with malarial resistance are negatively correlated in a malarial environment!

The contrast in the gene pools of the upper Volta populations versus the Bantu populations in the remaining malarial regions of sub-Saharan Africa illustrates that evolution induced by natural selection is very much an historical process; two populations can (and apparently did in this case) adapt to the same environment in completely different fashions depending upon their initial historical states. *The course of evolution is always constrained by what genetic variation is available for natural selection to work upon and by the initial state of the gene pool.*

ADAPTATION AS POLYGENIC PROCESS

One way humans adapted to malaria in wet, tropical Africa was through the A/S polymorphism at the β-Hb locus, an adaptation that confers malarial resistance to only about 20% of the population. In general, many loci can contribute to how an individual responds to its environment in determining the phenotype of fitness. As a consequence, adaptation to a new environment is usually a polygenic process, with genetic variation at many loci responding to natural selection. This is certainly the case for malarial adaptation in human populations. We will now examine a few of the other loci involved in this adaptation.

Glucose-6-phosphate dehydrogenase (G6PD) is a cytoplasmic enzyme coded for by the X-linked *G6PD* locus. The G6PD enzyme is essential for a cell's capacity to withstand oxidant stress (Ruwende and Hill 1998), as mentioned in Chapter 4. Over 300 independent mutations have been identified that lead to a common phenotype of a deficiency of the enzyme G6PD. G6PD deficiency affects over 400 million persons worldwide, mostly living in malarial regions. This geographical distribution implies that G6PD deficiency has risen in frequency through natural selection with malaria as the selective agent. This hypothesis is supported by data from in vitro studies that demonstrate impaired growth of *P. falciparum* parasites in G6PD-deficient red blood cells. In wet, tropical Africa, the allele A^- is the most common allele at the *G6PD* locus that results in an enzyme deficiency. Studies conducted in East and West Africa provide strong evidence that the A^- allele is associated with a significant reduction in the risk of severe malaria for both *G6PD* A^- female heterozygotes and male hemizygotes. The effect of female homozygotes for A^- on severe malaria remains unclear but can probably be assumed to be similar to that of comparably deficient male hemizygotes.

If a person does not have the A/S genotype at the β-*Hb* locus in a malarial region, the person could still be protected against malaria through the *G6PD* A^- allele. Thus, the Bantu-speaking peoples adapted to malaria as they expanded into wet, tropical Africa with the Malaysian agricultural complex not only by the C and S alleles at the β-*Hb* locus but also by the A^- allele at the *G6PD* locus. However, as with the S allele, there are fitness trade-offs associated with the A^- and other G6PD-deficient alleles. Although reducing the red blood cell's capacity to withstand oxidant stress provides resistance to the malarial parasite by causing premature lysis of infected cells, it also makes the red blood cells subject to lysis when exposed to oxidizing agents, thereby resulting in hemolytic anemia. For example, about 25% of the people with G6PD deficiency experience an adverse reaction to eating fava beans, a bean containing strong oxidizing agents. This sensitivity, known as favism, can result in death. Thus, the adaptation to one environmental agent (malaria) is counterbalanced by the interactions with other environmental agents (oxidizing agents), and these two antagonistic effects are weighted through the average excess to determine the frequency of the G6PD-deficient alleles. Such deficient alleles vary tremendously in frequency from one human population to the next as a function of malaria and oxidizing agents in the diet or environment, ranging from 0.001 in nonmalarial regions such as Japan or northern Europe to 0.62 in Kurdish Jews living in historical malarial regions (Ruwende and Hill 1998).

There are even more loci involved in malarial adaptation in humans. The first genetic syndrome suggested to be a malarial adaptation in humans was thalassemia (Haldane 1949). All normal hemoglobins are tetramers of two pairs of unlike globin chains (Weatherall 2001). Adult hemoglobin (HbA) has two α chains and two β chains, so HbA can be symbolized as $\alpha_2\beta_2$. To obtain efficient production of HbA, it is necessary that α chains should be produced in about the same quantity as β chains. We have already discussed the β-*Hb* locus on chromosome 11 that encodes for the β chain. As shown in Figure 2.5, the β-*Hb* locus normally exists as a single copy of the gene per chromosome. In contrast, the α-*Hb* locus that codes for the α chain generally exists as two tandem copies (α-*Hb*1 and α-*Hb*2) on chromosome 16. Normally, the rates of transcription and translation at these three loci are adjusted to produce an equal balance of α and β chains in the adult human. Thalassemia occurs when that balance is disrupted, and a large number of mutations at either the α-*Hb* loci and the β-*Hb* locus can lead to thalassemia. For example, some 80 different deletions and point mutations can induce α-thalassemia, including a deletion of one or both of the duplicated copies of α-Hb, a frame shift mutation in one of the copies or any point mutation that results in an inactive α chain. Over 200 mutations have been discovered that induce β-thalassemia, including deletions of all or part of the β-Hb gene, point mutations that inactivate the production of functional β chains, promoter mutations that lower the amount of transcription, mutations that alter posttranscriptional processing such as mutations that create or destroy exon/intron splice sites (including both noncoding mutations in the introns and silent mutations in the exons), and 3' mutations that alter the signal for the 3' cleavage site of the primary transcript (followed normally by polyadenylation). All of these mutations can result in an imbalance between α and β chains that in turn results in a spectrum of phenotypic effects ranging from no clinical symptoms to lethal anemia but also to resistance to the malarial parasite. In general, the severity of the thalassemia depends upon the degree of imbalance. Homozygotes for mutations inducing thalassemia generally have deleterious clinical symptoms, whereas heterozygotes generally have mild to no deleterious clinical symptoms.

Thalassemic heterozygotes appear to have an advantage against *P. falciparum* malaria as well (Weatherall 2001). As a consequence, the global distribution of thalassemia is largely coincident with the distribution of falciparum malaria. For example, the African populations that we discussed earlier with respect to the S and C alleles at the β-Hb locus and the A^- allele at the $G6PD$ locus also have a high frequency of β-thalassemia due to 5' point mutations and a high frequency of α-thalassemia due to a deletion of one of the copies of α-Hb. Hence, the adaptation to malaria in these Bantu-speaking populations involves changes in allele frequencies at α-Hb, β-Hb, and $G6PD$. However, the list does not end here, as malarial-resistant variants have been associated also with the *ApoE* locus (Woznaik et al. 2003), the *acid phosphatase 1* locus (Bottini et al. 2001a), the *major histocompatibility complex* loci (Chapter 1), the *tumor-necrosis factor-*α locus, the *intercellular adhesion molecule 1* locus, and others (Weatherall 2001).

Different human populations have adapted to malaria using different combinations of these alleles and loci. We have already pointed out how some western African populations adapted to malaria through β-HbS and others through β-HbC, and different alleles representing independent mutation events are in high frequencies in different human populations living in malarial regions across the globe at the other loci involved in malarial adaptation. This pattern illustrates three general points about adaptation through natural selection:

1. *The course of adaptation is always constrained by the available genetic variation.* Whether or not a human population adapts to malaria through S or C at the β-Hb locus depends upon which variants were available to the population due to mutation and/or gene flow. Natural selection is only selection on *existing* variation that arises by mutation and spreads by gene flow. Natural selection acts upon genetic variation but does *not directly create* it.

2. *Even uniform selective pressures produce divergent adaptive responses because selection operates upon variation whose creation and initial frequencies are profoundly influenced by random factors such as mutation and drift.* The randomness of the mutation and recombination processes that create variation in the first place and the randomness of drift influencing the frequencies of newly created variation cause much variation among populations in the state of their initial gene pools and array of available genetic variation. This heterogeneity can lead to diverse selective outcomes among different populations adapting to the same environmental influence.

The diversity in selective outcomes in turn means that even though we can explain adaptation in terms of natural selection, it is difficult to predict exactly how any population will adapt to an altered environment.

3. *Adaptation generally involves many loci with different biochemical, cellular, or developmental functions.* This is certainly true for malarial adaptation in humans, as shown above, and seems to be a general property of most adaptive responses that have been investigated in detail. However, in most cases we do not know the specific loci involved in adaptation nearly as well as we do in the case of malarial adaptation in humans. Indeed, in many cases, the genes underlying adaptive traits are not known at all.

Up to now, we have taken a measured genotype approach to the study of natural selection and adaptation, focusing upon genotypic values of specific, measured genotypes and average excess of gametes bearing specific alleles. To broaden our insights into adaptive evolution through natural selection, we must extend our models to include the unmeasured genotype approaches that have been applied to many other polygenic traits.

FUNDAMENTAL THEOREM OF NATURAL SELECTION: UNMEASURED GENOTYPES

Equation 11.5 is the fundamental equation describing natural selection for a measured genotype at a single locus. Given that many adaptive traits are polygenic but with the underlying loci being unknown, Fisher (1930) elucidated many properties of natural selection using the theory of the quantitative genetics of unmeasured genotypes (Chapter 9). His central result is summarized in an equation that is the unmeasured genotype analogue of the measured genotype equation 11.5. Fisher called this equation the **fundamental theorem of natural selection**, which describes how natural selection operates upon the phenotype of fitness when fitness is regarded as a heritable but genetically unmeasured trait. There are many ways of deriving this theorem, and we will present only one of the simplest.

Because we assume that no genotypes are being measured, we must focus exclusively upon phenotypes. This means that we can no longer use the primary definition of fitness in population genetics as a genotypic value for a measured genotype; rather, we now use the evolutionary ecology definition of fitness as a measure of average reproductive success of a phenotypic class of individuals rather than a genotypic class. Therefore, let x be the phenotypic value of some trait for an individual in a population (which we assume is a continuous, quantitative trait), and let $f(x)$ be the probability distribution that describes the frequencies of x in the population. The mean phenotype is then

$$\mu = \int_x x f(x) \, dx \qquad (11.15)$$

where the integration is over all possible values of x. We now assign a fitness value, $w(x)$, to those individuals sharing a common *phenotypic* value x. The mean or average fitness of the population is

$$\bar{w} = \int_x w(x) f(x) \, dx \qquad (11.16)$$

In equation 11.16 we average the fitness over all phenotypic classes, and the $\bar{w}$ appearing in the denominator of the measured genotype equation 11.5 is the average fitness over all genotypic classes. Regardless of whether we average over all phenotypes or over all genotypes, we are always calculating the average fitness over all *individuals* in the population. Hence, $\bar{w}$ is the same in equations 11.5 and 11.16.

As we saw in Chapter 9, selection can alter the mean phenotype of the population of individuals who successfully contribute gametes to the next generation. The fitness function $w(x)$ weights each phenotypic class x by its overall gametic contribution to the next generation. Therefore, the frequency of selected individuals is proportional to $w(x)f(x)$. However, $w(x)f(x)$ does not in general define a probability distribution, but $w(x)f(x)/\bar{w}$ does integrate to 1 and defines the probability distribution of the selected individuals. Hence, the mean phenotype of the selected individuals is

$$\mu_s = \frac{\int_x x w(x) f(x) \, dx}{\bar{w}} \tag{11.17}$$

Now let h^2 be the heritability of the trait. As shown in Chapter 9, the phenotypic response to selection is given by $R = h^2 S$, where $S = (\mu_s - \mu)$, $R = (\mu_o - \mu)$, and μ_o is the phenotypic mean of the offspring of the selected parents.

Fitness is just a phenotype, so Fisher considered what happens when the phenotype of interest is fitness itself. When $x = w$, $w(w) = w$ by definition, and $\mu = \bar{w}$. Equation 11.17 then becomes

$$\mu_s = \frac{\int_w w \times w f(w) \, dw}{\bar{w}} = \frac{\int_w w^2 f(w) \, dw}{\bar{w}} \tag{11.18}$$

Expressing w^2 as $(w^2 - \bar{w}^2) + \bar{w}^2$, equation 11.18 now takes the form

$$\mu_s = \frac{\int_w \left[(w^2 - \bar{w}^2) + \bar{w}^2\right] f(w) \, dw}{\bar{w}} = \frac{\int_w (w^2 - \bar{w}^2) f(w) \, dw + \bar{w}^2 \int_w f(w) \, dw}{\bar{w}} \tag{11.19}$$

Because $\bar{w} = \int_w w f(w) \, dw$, the term $\int_w (w^2 - \bar{w}^2) f(w) \, dw$ in equation 11.19 can be expressed as $\int_w (w - \bar{w})^2 f(w) \, dw$. Also, $\int_w f(w) \, dw = 1$ by definition. Hence, equation 11.19 simplifies to

$$\mu_s = \frac{\int_w (w - \bar{w})^2 f(w) \, dw + \bar{w}^2}{\bar{w}} = \frac{\sigma^2 + \bar{w}^2}{\bar{w}} \tag{11.20}$$

where $\sigma^2 = \int_w (w - \bar{w})^2 f(w) \, dw$ and is the variance in the phenotype of fitness in the population. The standard quantitative genetic measure S of the intensity of selection is for the special case of $x = w$ (so that $\mu = \bar{w}$):

$$S = \mu_s - \mu = \frac{\sigma^2 + \bar{w}^2}{\bar{w}} - \bar{w} = \frac{\sigma^2 + \bar{w}^2 - \bar{w}^2}{\bar{w}} = \frac{\sigma^2}{\bar{w}} \tag{11.21}$$

When $x = w$, the response to selection, R, is $\Delta \bar{w}$. Hence, the standard quantitative genetic equation for predicting the response to selection is, for the special case when the phenotype is fitness itself,

$$R = h^2 S$$
$$\Delta \bar{w} = \left(\frac{\sigma_a^2}{\sigma^2}\right)\left(\frac{\sigma^2}{\bar{w}}\right) = \frac{\sigma_a^2}{\bar{w}} \qquad (11.22)$$

where σ_a^2 is the additive genetic variance for the phenotype of fitness. Equation 11.22 is **Fisher's fundamental theorem of natural selection** and states that the change in average fitness is proportional to the additive genetic variance in fitness. Equation 11.22 is the unmeasured genotype analogue of equation 11.5 . Both equations 11.5 and 11.22 have many implications for our understanding of natural selection.

SOME IMPLICATIONS OF FUNDAMENTAL EQUATIONS OF NATURAL SELECTION

Equations 11.5 and 11.22 offer many common insights into natural selection and its attributes:

1. *Natural selection can only operate when there is genetic variation associated with phenotypic variation for fitness in the population.* If there were no genetic variation influencing the phenotype of fitness, fitness could not possibly be heritable and therefore equation 11.22 implies no evolution due to natural selection. Likewise, without genetic variation associated with phenotypic variation of fitness, there can be no nonzero average excesses of fitness, and therefore equation 11.5 also implies no evolution by natural selection.

2. *The only fitness effects that influence the response to natural selection are those transmissible through a gamete.* Equation 11.5 shows that there is no response to selection unless the average excess for fitness is not zero, and equation 11.22 shows that there is no response to selection unless the additive genetic variance for fitness is not zero. The additive genetic variance can only be nonzero if the average effects are nonzero, and the average effects are nonzero if and only if the average excesses are nonzero (Chapter 8). Hence, both equations demonstrate that the response to natural selection depends upon how individual fitness variation is funneled through a gamete and transmitted to the next generation. Natural selection can only be understood by taking a gamete's perspective.

3. *The adaptive outcome represents an interaction of fitness variation with population structure.* As shown in Chapter 8, both the average excesses and average effects are explicitly functions of the rules by which gametes are paired together to form genotypes, that is, population structure. Since it is average excesses and average effects that exclusively determine the response to natural selection (equations 11.5 and 11.22, respectively), population structure plays a direct role in determining the adaptive response to an environment. This reinforces the fact that natural selection cannot be survival of the fittest. The impact of the fitness of an individual upon the response to selection can only be evaluated in the context of a population structure. Even complete knowledge of the fitness of every individual in a population is insufficient to determine the response to selection.

4. *Selective equilibria can only occur when all the average excesses and all the average effects are zero, that is, when all gametes have the same average fitness impact.* Equation 11.5 implies that evolution will occur whenever a gamete has a nonzero average excess, and equation 11.22 implies that evolution will occur whenever the additive variance for fitness is nonzero. Evolution due to natural selection stops only when there is no heritability for fitness. This in turn means that at a selective equilibrium there is no correlation between the fitness of parents and the fitness of their offspring even when there is genetic variance in the phenotype of fitness. For example, consider a random-mating human population in a malarial environment that is at selective equilibrium for the A/S polymorphism at the β-Hb locus. Three genotypes are present in the equilibrium population, and their fitnesses differ greatly (Table 11.1). However, individuals with genotype AA will produce some children with the AA genotype who have a fitness lower than the population average and some children with the AS genotype who have a fitness higher than the population average. Likewise, individuals with the AS genotype will produce some children with lower than average fitnesses (AA and SS) and some children with higher than average fitnesses (AS), as do SS individuals who will have both SS and AS children. Thus, every genotype in the equilbrium population produces some children with higher than average fitness and some with lower than average fitness. The equilibrium occurs when the balance between the higher than average offspring and the lower than average offspring is the same for each possible parental genotype. Thus, a selective equilibrium does not necessarily imply that there is no genetic variance for the phenotype of fitness, only that there is no *additive* genetic variance.

This prediction of Fisher's fundamental theorem and equation 11.5 has some interesting implications for interpreting the evolutionary meaning of the heritability of a trait in a natural population. If the heritability is low, this could mean either that there is little genetic variation for the trait or that the trait is closely related to fitness in a population at or near a selective equilibrium. If the heritability is high, this implies that the trait is not fitness or that the population is far from equilibrium. For example, Merilä and Sheldon (1999) reviewed the literature and found that fitness-related traits do indeed tend to have lower heritabilities than traits not clearly related to fitness. However, they also discovered that fitness-related traits tend to have higher, not lower, genetic variances. The reason for this seeming paradox is that a greater proportion of the genetic variance was nonadditive for fitness-related traits. Thus, the low heritability of fitness-related traits was due not to a lack of genetic variation at equilibrium but rather to that variation being balanced in such a way as to result in low heritabilities.

5. *Natural selection acts to increase the average fitness of a population on a per-generational basis.* Because the additive genetic variation must be greater than or equal to zero, equation 11.22 implies that $\Delta \bar{w} \geq 0$ under natural selection. Because $\bar{w}$ is the same for both the measured and unmeasured genotype approaches, equation 11.22 also implies that the allele frequency changes induced by natural selection as described by equation 11.5 can never decrease average fitness. Because average fitness can only increase or stay the same under natural selection, the selective equilibria discussed under point 4 must always correspond to an average fitness optimum. Wright (1932) expressed this idea in terms of an **adaptive surface or landscape**, which is a plot of the average fitness as a function of the gamete frequencies for a given population structure. The fundamental theorem of natural selection tells us that natural selection causes populations to always "climb up" in this surface until they reach a "peak" which corresponds to a selective equilibrium. One cannot climb down from the peak in this adaptive surface because to do so would cause $\Delta \bar{w} < 0$,

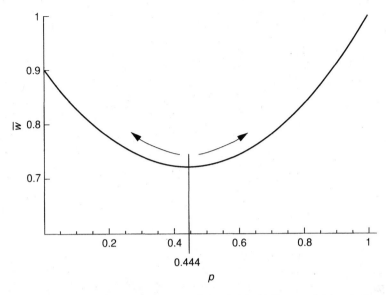

Figure 11.7. Adaptive surface for one-locus, two-allele model with fitness of $AA = 1$, $Aa = 0.5$, and $aa = 0.9$. The surface represents a plot of $\bar{w}$ under the assumption of random mating against p, the frequency of the A allele. A line at $p = 0.444$ indicates the minimum value of $\bar{w}$. Arrows indicate the direction of evolution induced by natural selection on either side of this minimum value.

which violates equation 11.22. This inability of natural selection to allow a climb downward in a fitness surface leads to the sixth major feature that emerges from these fundamental equations.

6. *Natural selection only takes populations to local adaptive solutions and not necessarily to the adaptive state with the highest average fitness and indeed may operate to prevent an adaptive state with higher average fitness from evolving.* To see this, consider a simple one-locus, two-allele (A and a) model in which the genotype AA has a fitness of 1, Aa a fitness of 0.5, and aa a fitness of 0.9. Under random mating, $\bar{w} = p^2(1) + 2pq(0.5) + q^2(0.9) = p + 0.9(1 - p)^2$. Figure 11.7 shows a plot of $\bar{w}$ versus p, the adaptive surface, for this case. This fitness surface has two peaks, corresponding to $p = 0$ with $\bar{w} = 0.9$ and to $p = 1$ with $\bar{w} = 1.0$. The "valley" between these two peaks occurs at $p = 0.444$ (Figure 11.7). Equation 11.22 tells us that if we start at any $p < 0.444$, we must go to the peak with $\bar{w} = 0.9$ because that is the direction of increasing average fitness. Equation 11.5 tells us the same thing because the average excess of the A allele is negative for any nonzero $p < 0.444$. For example, if the initial allele frequency were $p = 0.4$ then the average excess of the A allele is -0.024. Hence, the population will evolve toward smaller values of p and will continue to evolve under natural selection until $p = 0$. However, if the population starts at $p = 0.45$, the average excess of A is 0.003, so p increases under natural selection and will continue to do so until $p = 1$. Note that the highest average fitness on the entire surface is $\bar{w} = 1$. However, when p starts out at 0.4, natural selection takes the population farther and farther away from this global optimum and instead takes the population to the lower peak of $\bar{w} = 0.9$. In this case, natural selection prevents the globally optimal solution from evolving.

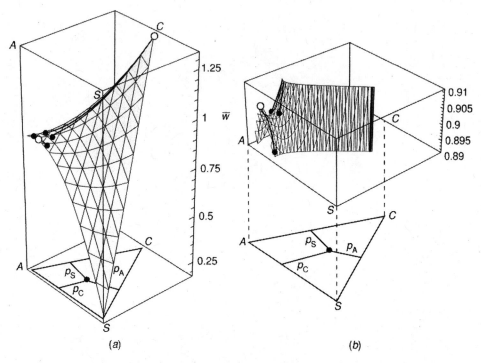

Figure 11.8. Adaptive surface defined by A, S, and C alleles at β-Hb locus in random-mating population using fitnesses given in Table 11.1 for malarial environment. The gene pool space is shown by the triangle near the bottom, with the allele frequencies given by the perpendicular distances from the point to the sides of the triangle and with the vertices associated with fixation of a particular allele labeled by the letter corresponding to that allele. The vertical axis gives $\bar{w}$ as a function of these three allele frequencies. (a) Entire adaptive landscape. Two peaks exist in this adaptive surface, as indicated by large white dots at $p_A = 0.88$, $p_S = 0.12$, and $p_C = 0$ with $\bar{w} = 0.90$ and at $p_A = 0$, $p_S = 0$, and $p_C = 1$ with $\bar{w} = 1.31$. Small black dots indicate the initial state of the gene pool for four populations: one starting at $p_A = 0.95$, $p_S = 0.025$, and $p_C = 0.025$; a second at $p_A = 0.85$, $p_S = 0.025$, and $p_C = 0.125$; a third at $p_A = 0.85$, $p_S = 0.125$, and $p_C = 0.025$; and a fourth at $p_A = 0.75$, $p_S = 0.125$, and $p_C = 0.125$. Black lines from these small dots plot the evolutionary trajectory across generational time as defined by equation 11.5. Because the saddle separating these two adaptive peaks is shallow, part (b) expands that part of the adaptive landscape that contains the saddle and the polymorphic A/S peak.

We already saw another example of this with the S and C alleles at the β-Hb locus. Figure 11.8 shows a plot of $\bar{w}$ in a malarial environment as a function of the frequencies of alleles A, S, and C in a random-mating population. This adaptive landscape has two peaks, one at the A/S polymorphic solution with $\bar{w} = 0.90$ and the other at fixation of C with $\bar{w} = 1.31$. The A/S peak is rather low, being barely above the $\bar{w}$ of 0.89 that corresponds to fixation of the A allele. In contrast, the C peak is high, clearly indicating that it is the global optimum. However, the two peaks are distinct enough to have a shallow saddle between them. How a population evolves due to natural selection on this landscape depends upon which side of the saddle it is initially located, as illustrated in Figure 11.8. As long as the initial frequency of C is low and S is present in the random-mating population, natural selection takes the population to the A/S peak and eliminates the C allele, thereby preventing the optimal adaptation from evolving. In this sense, natural selection can be a *hindrance* to adaptation, even though natural selection is the one evolutionary force that is necessary for adaptation.

7. *Natural selection generally does not optimize, even in a local sense, any individual trait other than fitness itself, even if the trait contributes to fitness in a positive fashion.* The peak climbing feature of natural selection locally optimizes the average phenotype of fitness. Frequently, however, we are not examining fitness directly; rather we are examining some other trait that may contribute to fitness. When studying such traits, it is commonplace in much of the evolutionary literature to equate adaptation to a process that evolves the "optimal" trait value that maximizes the fitness associated with that trait. Crow and Nagylaki (1976) showed that this is not true, but their proof is rather complex. Here, we use an alternative approach. Let x be some phenotype related to fitness through the function $w(x)$. Suppose we are at a selective equilibrium ($\Delta \bar{w} = 0$) at a local peak and the phenotypic distribution of the trait at equilibrium is given by $f_{eq}(x)$. Then, the average fitness at equilibrium is $\bar{w}_{eq} = \int_x w(x) f_{eq}(x) dx$ and the average value of trait X at equilibrium is $\bar{x}_{eq} = \int_x x f_{eq}(x) dx$. The fitness associated with the equilibrium trait x value is $w(\bar{x}_{eq})$. Equation 11.22 tells us that $\bar{w}_{eq}$ is a local maximum, so $\bar{x}_{eq}$ is the optimal value of trait X only if $w(\bar{x}_{eq})$ is also the local maximum; that is, if $\bar{w}_{eq} = w(\bar{x}_{eq})$. To see if $\bar{x}_{eq}$ is the optimal value of the trait, we use Taylor's theorem from calculus to expand $w(x)$ about the point $\bar{x}_{eq}$,

$$w(x) \approx w(\bar{x}_{eq}) + w'(\bar{x}_{eq})(x - \bar{x}_{eq}) + \tfrac{1}{2} w''(\bar{x}_{eq})(x - \bar{x}_{eq})^2 \qquad (11.23)$$

where $w'(\bar{x}_{eq})$ is the first derivative of $w(x)$ evaluated at $x = \bar{x}_{eq}$ and $w''(\bar{x}_{eq})$ is the second derivative of $w(x)$ evaluated at $x = \bar{x}_{eq}$. We now take the average value of both sides of equation 11.23 by integrating across the equilibrium probability distribution of the trait, noting that $w'(\bar{x}_{eq})$, $\bar{x}_{eq}$, and $\tfrac{1}{2} w''(\bar{x}_{eq})$ are all constants:

$$\bar{w}_{eq} = w(\bar{x}_{eq}) \int_x f_{eq}(x) dx + w'(\bar{x}_{eq}) \int_x (x - \bar{x}_{eq}) f_{eq}(x) dx$$
$$+ \tfrac{1}{2} w''(\bar{x}_{eq}) \int_x (x - \bar{x}_{eq})^2 f_{eq}(x) dx \qquad (11.24)$$

noting that $\int_x (x - \bar{x}_{eq}) f_{eq}(x) dx = \int_x x f_{eq}(x) dx - \bar{x}_{eq} \int_x f_{eq}(x) dx = \bar{x}_{eq} - \bar{x}_{eq} = 0$ and that $\int_x (x - \bar{x}_{eq})^2 f_{eq}(x) dx$ is the phenotypic variance in trait X at equilibrium. Let this variance be designated by $\sigma^2_{eq}(x)$, then equation 11.24 reduces to

$$\bar{w}_{eq} = w(x_{eq}) + \tfrac{1}{2} w''(x_{eq}) \sigma^2_{eq}(x) \qquad (11.25)$$

Therefore, $\bar{x}_{eq}$ only represents an optimal value of trait X that maximizes $w(x)$ when

$$w''(x_{eq}) \sigma^2_{eq}(x) = 0 \qquad (11.26)$$

Equation 11.26 tells us that natural selection only optimizes the value of a trait that contributes to fitness but is not fitness itself under two conditions:

1. The trait has no phenotypic variance at equilibrium [$\sigma^2_{eq}(x) = 0$] or
2. The trait is related to fitness in a *strictly linear* fashion at equilibrium [$w''(\bar{x}_{eq}) = 0$].

If there is nonlinearity and phenotypic variation, the trait value that maximizes fitness is not the trait value that will evolve due to natural selection. Equation 11.26 tells us that natural selection does not optimize individual traits that contribute to fitness except under highly unusual circumstances. Selection works on the whole of the individual as measured by one and only one phenotype—fitness—and all other traits are only indirectly selected and are not individually optimized.

8. *The process of adaptation can result in the evolution of some seemingly nonadaptive traits.* Equation 11.26 looks at only one trait and its mapping onto fitness through $w(x)$. In general, many traits contribute to fitness, not just one. Consider the case in which two traits, say X and Y, contribute to fitness such that $w(x, y)$ is the fitness of those individuals with trait values x and y for the two traits, respectively. Then, the two-dimensional analogue of equation 11.26 is

$$\frac{\partial^2 w(x_{eq}, y_{eq})}{\partial x^2}\sigma^2_{eq}(x) + 2\frac{\partial^2 w(x_{eq}, y_{eq})}{\partial x\, \partial y}\text{Cov}_{eq}(x, y) + \frac{\partial^2 w(x_{eq}, y_{eq})}{\partial y^2}\sigma^2_{eq}(y) = 0 \quad (11.27)$$

Equation 11.27 shows that generally the same restrictive conditions needed to optimize an individual trait (equation 11.26) are necessary for *each* of the two traits contributing to fitness for both traits to be optimized. However, equation 11.27 indicates that the single-trait optimizing conditions are not sufficient for multitrait optimization. In the two-trait case, optimization of the traits also requires that the covariance between X and Y at equilibrium must be zero. Pleiotropy can create a covariance between two traits, and moreover the covariance can be negative. A negative covariance can lead to the evolution of a trait through natural selection that, when considered by itself, is deleterious at the individual level. We have already seen this with sickle cell anemia. The trait of hemolytic anemia is correlated with fitness in a malarial environment, but there is nothing optimal about having severe hemolytic anemia! Yet natural selection has increased the frequency of hemolytic anemia as part of the adaptive response to malaria. The deleterious trait in this case arises from pleiotropy: The *S* allele affects the traits of both malarial resistance, which increases fitness in a malarial environment, and hemolytic anemia, which decreases fitness. This is an example of **antagonistic pleiotropy**, in which the same allele is associated with traits that have opposite effects on fitness. Antagonistic pleiotropy ensures that natural selection will produce many nonoptimal individual trait values that can have high genetic variances even at selective equilibrium (Merilä and Sheldon 1999).

The correlations between traits induced by pleiotropy are sometimes called **developmental constraints**. Such constraints are often presented as an impediment to adaptive evolution in which the optimal ideal cannot be achieved; that is, with antagonistic pleiotropy, it is impossible for all traits to be at the values that maximize their individual contributions to overall fitness. However, Wagner (1988) has shown that in some circumstances adaptive evolution *requires* developmental constraints. He considered the case of the evolution of a complex adaptation in which many individual trait states must be simultaneously true to achieve high fitness. Such complex adaptations result in adaptive surfaces that have peaks or ridges that are sharply defined and fall off steeply. He modeled the evolution of a population upon such an adaptive surface under the assumption that there were no developmental constraints; that is, each trait was genetically independent from all other traits with no pleiotropy whatsoever. Wagner found that natural selection became increasingly ineffective in taking populations to the peaks or ridges of this landscape as the number of component traits increased. Similar results have been obtained by Orr (2000). The reason for this is that the peak or ridge associated with a complex adaptation is found in only a small portion of the possible genetic space. Otherwise, the landscape is "flat," implying through equation 11.22 that no evolution by natural selection will occur unless the population's initial state happened to be in that small portion of the adaptive surface on a "slope" of the peak or ridge. Although developmental constraints could obviously prevent some adaptations from evolving, without such constraints it would be unlikely for natural selection to cause the evolution of complex structures that require much functional integration of many individual traits. Thus, Wagner (1988) concluded that developmental "constraints" are necessary to retain adaptive versatility in the face of functional interdependencies and high phenotypic complexity.

An example of a complex adaptation is provided by the work of Ferea et al. (1999) on the adaptation over more than 250 generations of three independent cultures of the yeast *Saccharomyces cerevisiae* to a novel environment characterized by continuous aerobic growth in glucose-limited media. Several hundred genes evolved significantly altered expression that increased fitness by altering the regulation of central metabolism. Moreover, the altered expression profiles were similar in all three independently evolved strains. The speed and similarity of these parallel adaptations are difficult to reconcile with the idea that each of these several hundreds of genes independently evolved altered expression profiles that resulted in a new but adaptive and functionally integrated central metabolism. In a separate experiment, Cavalieri et al. (2000) found a single locus in this yeast species that significantly altered transcription for 378 genes (6% of the genome). Indeed, molecular genetic studies on the basis of transcriptional regulation and signaling pathways during development ensure that pleiotropy and epistasis, the genetic underpinnings of developmental constraints, are ubiquitous (Gibson 1996; Guss et al. 2001). Perhaps some aspects of the altered central metabolism found in the three evolved strains of yeast described by Ferea et al. (1999) are individually suboptimal, but it is doubtful that this threefold parallel, adaptive evolution of altered central metabolism could have occurred *without* developmental constraints. Pleiotropy can therefore be both an adaptive constraint and an adaptive facilitator.

9. *The course of adaptive evolution is strongly influenced by genetic architecture.* Point 8 shows the important role of pleiotropy and epistasis in shaping adaptive evolution and that both pleiotropy and epistasis are critical components of genetic architecture. Recall from Chapter 2 that genetic architecture refers to the number of loci and their genomic positions, the number of alleles per locus, the mutation rates, and the mode and rules of inheritance of the genetic elements. As explained in Chapter 9, the mode of phenotypic inheritance includes such factors as dominance/recessiveness, pleiotropy, and epistasis. Genetic architecture is a strong determinant of the shape of an adaptive surface and hence the outcome of evolution via natural selection. For example, the A/S polymorphic peak in the adaptive surface shown in Figure 11.8 arises from the pleiotropic effects of the S allele on the traits of hemolytic anemia and malarial resistance. Similarly, the saddle between the A/S peak and the C peak in Figure 11.8 arises from the negative phenotypic interaction between the S and C alleles to cause hemolytic anemia in S/C heterozygotes; that is, S is codominant relative to C for the phenotype of hemolytic anemia. Finally, the extension of the slopes of the A/S peak to that part of the adaptive surface associated with the S and C alleles being rare depends critically upon S being a dominant allele for viability relative to A in a malarial environment whereas C is recessive to A (Table 11.1). All of these features of genetic architecture help shape the adaptive surface and hence the adaptive outcome under natural selection. Even slight changes in genetic architecture can drastically alter the course of adaptive evolution. For example, the data of Cavalli-Sforza and Bodmer (1971) indicate C is recessive to A with respect to the trait of viability, but a larger recent study indicates that CC homozygotes have a 93% reduction in risk of clinical malaria in western Africa

while AC homozygotes have a 29% reduction (Modiano et al. 2001). This study indicates that the C allele is not completely recessive to the A allele for the phenotype of malarial resistance. Although malaria risk reduction is not the same as viability, these data indicate that the C allele may also not be completely recessive to the A allele for the phenotype of viability in a malarial environment. Suppose the AC heterozygotes have a slight viability advantage over AA homozygotes in a malarial environment, say a fitness of 0.93 relative to AS instead of the fitness of 0.89 of AA. Such a small difference could not be detected with the sample sizes used by Cavalli-Sforza and Bodmer (1971) in calculating the relative viabilities shown in Table 11.1. Although letting C have such a slight phenotypic effect in AC heterozygotes is a seemingly minor change in the genetic architecture of this system, this slight change has major effects on the average excess of C when it is rare. For example, assuming as we did in equation 11.7 that we start with a random-mating population close to fixation for the A allele but with both S and C present at low frequencies, the average excess of C is now

$$a_C \approx 1(0.93 - 0.89) + 0(0.70 - 0.89) + 0(1.31 - 0.89) = 0.04 \qquad (11.28)$$

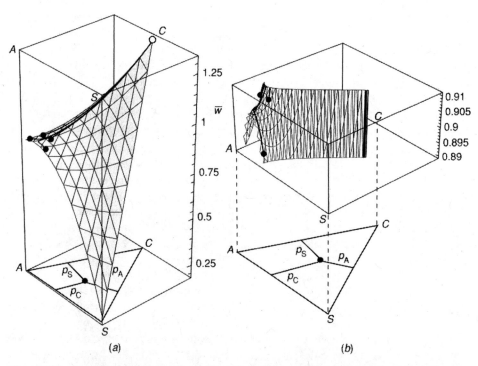

Figure 11.9. Adaptive surface defined by A, S, and C alleles at β-Hb locus in random-mating population using fitnesses given in Table 11.1 for malarial environment except for AC, which is now assigned fitness of 0.92 to reflect alternative genetic architecture in which C is not completely recessive to A for phenotype of viability. All the graphing conventions used in Figure 11.8 are used in this figure. Only one peak exists in this adaptive surface, as indicated by the large white dot at $p_A = 0$, $p_S = 0$, and $p_C = 1$ with $\bar{w} = 1.31$. Small black dots indicate the initial state of the gene pool for the same four populations given in Figure 11.8. Black lines from these small dots plot the evolutionary trajectory across generational time as defined by equation 11.5. (a) Entire landscape and (b) same expanded portion as Figure 11.8.

Thus, unlike the situation described in Table 11.1 and equation 11.7, the C allele now increases in frequency due to natural selection even when extremely rare. Moreover, even if the C allele were introduced via mutation or gene flow into a population at the A/S selective equilibrium, its average excess would now be

$$a_C \approx 0.88(0.93 - 0.90) + 0.12(0.70 - 0.90) + 0(1.31 - 0.90) = 0.002 \qquad (11.29)$$

The C allele now has a positive average excess for fitness and would increase in frequency due to natural selection, in great contrast to its selective elimination under these conditions with a recessive genetic architecture. The impact of this change in genetic architecture is also evident in the adaptive landscape, as shown in Figure 11.9. This slight change in genetic architecture now makes the adaptive surface have only a single peak, so that now all populations that have the C allele present will evolve to fixation for C regardless of the initial frequency of C or S.

Equations 11.5 and 11.22 can teach us much about natural selection and the adaptive process. Natural selection is the only evolutionary force that can cause adaptive responses to the environment over the course of evolution. However, as we also saw, sometimes natural selection can be a force that prevents adaptive breakthroughs. The course of adaptation is determined not exclusively by natural selection but also through the interaction of natural selection with other factors such as initial conditions, population structure, and genetic architecture. Therefore, *natural selection is necessary for adaptation, but it is not sufficient to determine the course of adaptation.* We will see this even more clearly in the next chapter, which considers in more detail how natural selection interacts with other evolutionary forces during the process of adaptive evolution.

12

INTERACTIONS OF NATURAL SELECTION WITH OTHER EVOLUTIONARY FORCES

The fundamental equation of natural selection for a measured genotype was derived in Chapter 11 as

$$\Delta p = \frac{p}{\bar{w}} a_A \qquad (12.1)$$

where p is the frequency of the allele A whose evolution is being monitored through Δp, $\bar{w}$ is the average fitness of the population, and a_A is the average excess of fitness of the A allele. As shown in Chapter 11, the average excess, a_A, drives the evolutionary changes of the frequency of the A allele induced by natural selection when $a_A \neq 0$. However, as shown in Chapters 2–7, population structure and many other factors can induce evolutionary change as monitored through Δp. What happens when both natural selection and other evolutionary forces are simultaneously influencing evolutionary changes at a locus? When all evolutionary forces induce only small changes in allele frequency when considered one by one, then the total change in allele frequency can be approximated by the sum of the changes induced by each force alone; that is,

$$\Delta p = \frac{p}{\bar{w}} a_A + \Delta p(\text{other}) \qquad (12.2)$$

where the first term on the right-hand side of equation 12.2 is equation 12.1 and represents the evolutionary changes induced by natural selection and the second term, $\Delta p(\text{other})$, represents the evolutionary changes induced by factors other than natural selection. When $\Delta p(\text{other}) \neq 0$, the evolutionary dynamics change from what is expected from natural selection alone, and moreover the evolutionary equilibrium is altered as well. Under natural selection alone as described by equation 12.1, potential equilibria occur only at allele frequencies such that the average excess of fitness is zero, and the stability of any equilibrium is

---

Population Genetics and Microevolutionary Theory, By Alan R. Templeton
Copyright © 2006 John Wiley & Sons, Inc.

determined by the signs that the average excess takes above and below any potential equilibrium frequency. However, under equation 12.2, the potential equilibria of the evolutionary process are defined by

$$p_{eq} = -\bar{w}\frac{\Delta p(\text{other})}{a_A} \tag{12.3}$$

Notice that now the potential equilibria are determined by the *balance* of the other evolutionary forces relative to natural selection as measured by the ratio $\Delta p(\text{other})/a_A$.

Equation 12.3 also continues a theme seen throughout Part I; evolutionary outcomes are determined not by one evolutionary force in isolation but rather by their balance (relative strengths) when considered together. Indeed, equation 12.3 has a remarkable implication for the role of evolutionary factors other than natural selection. Whenever we are near a selective equilibrium, we know that a_A has to be close to zero (as shown in Chapter 11). This means that the ratio $\Delta p(\text{other})/a_A$ tends to be amplified in magnitude as the population gets closer and closer to its selective equilibrium, unless $\Delta p(\text{other})$ also tends toward zero at the same equilibrium. Thus, other evolutionary forces that seemingly have only minor effects as agents of evolutionary change when considered by themselves can sometimes have major evolutionary effects when combined with natural selection. Equation 12.3 makes it clear that natural selection does not always *overwhelm* other forces of evolutionary change but rather *interacts and amplifies* their evolutionary importance. In this chapter we will investigate the implications of equations 12.2 and 12.3 for how natural selection is balanced against other evolutionary forces.

INTERACTION OF NATURAL SELECTION WITH MUTATION

Mutation is the source of new allelic variation and directly changes allele frequencies by converting a copy of one allele (thereby decreasing its frequency) into another type of allele (thereby increasing its frequency). Because mutation rates are generally very small, mutation by itself seems to be a minor cause of change in allele frequencies. But Equation 12.3 has already warned us that seemingly minor forces can have major evolutionary implications when combined with natural selection, as we will now see is the case for natural selection combined with mutation.

Consider a model of recurrent mutation in which one allele, say A, repeatedly mutates into another allelic class, say a, at rate μ per generation but a does not mutate back to A. This is a realistic mutational model for many genetic diseases. For example, we mentioned in Chapter 11 that the common phenotype of a deficiency of the enzyme G6PD is due to over 300 independent mutations at the molecular level (Ruwende and Hill 1998). Many of these mutations are loss-of-function mutations of some sort that can usually only be rectified by an exact reverse mutation. Thus, an allele coding for a functional G6PD enzyme has many ways of mutating into a nonfunctional G6PD-deficient allele, but a G6PD-deficient allele is extremely unlikely to mutate back into a functional *G6PD* allele. A similar situation exists for the thalassemias, with some 80 different deletions and point mutations at the duplicated α-globin locus inducing α-thalassemia and over 200 different mutations at the β-globin locus inducing β-thalassemia (Weatherall 2001). A general model for many human genetic diseases is that the allelic class A describes those copies of the gene that code for a functional protein, and the allelic class a describes those copies of the gene that code for a nonfunctional or deficient protein or even the complete absence of the protein. The recurrent, one-way mutational model is applicable to such genetic diseases.

If we let q be the frequency of the a allelic class and $p = 1 - q$ be the frequency of the A allelic class, then mutation alone induces the following evolutionary change:

$$\Delta q(\text{mutation}) = p\mu \tag{12.4}$$

Equation 12.4 reflects the fact that each copy of the A allele has a probability μ of becoming an a allele each generation. Because μ is generally a very small number, the evolutionary change induced by mutation is going to be minor over any one generation, but equation 12.4 has only one equilibrium at $q_{eq} = 1$ (that is, all copies of the functional allele are eventually made nonfunctional by the operation of mutation over a large number of generations).

Now suppose there is a deleterious fitness consequence to bearing a nonfunctional allele such that the fitness of AA is 1 (our relative standard fitness), the fitness of Aa is $1 - hs$, and the fitness of aa is $1 - s$, where $s > 0$ is the selection coefficient against aa homozygotes and $0 \leq h \leq 1$ is a measure of dominance ($h = 0$ means that A is completely dominant over a for the phentoype of fitness, and $h = 1$ means that a is completely dominant over A for the phenotype of fitness, with intermediate values of h reflecting an intermediate deleterious effect on the heterozygotes). The average excess of fitness for the a allele under random mating is (equation 8.10)

$$a_a = p(1 - hs - \bar{w}) + q(1 - s - \bar{w}) \tag{12.5}$$

The average excess of allele a is negative for all $q > 0$, so equation 12.1 implies that the only selective equilibrium is $q_{eq} = 0$ (that is, all copies of the genes at this locus are from the functional allelic class). Note that the selective equilibrium and the mutational equilibrium are complete opposites. If we assume s is much larger than μ, then we can assume that selection will overwhelm mutation when far from the selective equilibrium (which ensures a negative value of a_a that is large in magnitude). But when close to the selective equilibrium, even the weak force of recurrent mutation can have a major effect.

Given that the population is close to the selective equilibrium, then $p \approx 1$ and equation 12.4 becomes $\Delta q(\text{mutation}) \approx \mu$. Also, $\bar{w} \approx 1$ near selective equilibrium because almost all individuals in this randomly mating population are AA homozygotes. Hence, equation 12.5 simplifies to $a_a \approx p(-hs) + q(-s) = -hs + qs(h - 1)$. From equation 12.3, the equilibrium between selection and mutation is given by

$$q_{eq} = \frac{\mu}{hs + q_{eq}s(1 - h)} \tag{12.6}$$

since $\bar{w} \approx 1$. For the special case of a recessive genetic disease, $h = 0$ and equation 12.6 can be rearranged as $q_{eq}^2 = \mu/s$. Hence, the equilibrium allele frequency for an autosomal, recessive deleterious allele is

$$q_{eq} = \sqrt{\frac{\mu}{s}} \tag{12.7}$$

Note that the equilibrium frequency of a is explicitly a balance between mutation and selection, in this case as measured by the square root of the ratio of mutation rate divided by the selection coefficient. As an example of this balance, let the a allele be completely lethal when homozygous ($s = 1$) and the mutation rate be 10^{-6}. Then, the equilibrium frequency of the a allele from equation 12.7 is 0.001—some three orders of magnitude greater than the mutation rate! Because the square root of any number much smaller than 1 is going to be much larger than the original number, equation 12.7 implies that recessive, deleterious alleles will accumulate in randomly mating populations until they reach frequencies several orders of magnitude greater than their mutation rates. The reason for the ineffectiveness of natural selection in eliminating these recessive, deleterious mutations is that selection can only eliminate a recessive, deleterious allele when it is in a homozygote, which is a rare event under random mating for a rare allele. Almost all copies of the recessive, deleterious alleles are in heterozygotes where they are protected from selection. As a consequence, the average excess (equation 12.5) is dominated by the heterozygote genotypic deviation (which is close to zero) and the trait has very low heritability (as shown in Chapter 9). The fundamental theorem of natural selection (Chapter 11) predicts that selection is ineffective on this recessive, deleterious trait as long as it is rare, thereby allowing the seemingly weak force of mutation to have a major impact on the equilibrium allele frequency. Recall from Chapter 3 that the average human from the United States and Europe bears between five and eight lethal equivalents (Stine 1977) and that 1.25% of all live births in this area are affected by a severely deleterious genetic disease, most of which are autosomal recessives (Czeizel 1989). Thus, although rare at any one locus, recessive deleterious alleles are collectively common in human populations from Europe and the United States. One reason for this terrible load of deleterious alleles in humans lies in the implications of equation 12.7.

Now consider a deleterious allele that has some effect on the heterozygotes such that $h > 0$. Given that q_{eq} is small because a is deleterious, even small values of h will often yield an hs term that is much larger than $q_{eq}s(1 - h)$. Accordingly, a good approximation to equation 12.6 is now

$$q_{eq} = \frac{\mu}{hs} \qquad (12.8)$$

For example, consider the previous case in which $s = 1$ and $\mu = 10^{-6}$, but now let $h = 0.01$. Thus, the fitness of the heterozygote is reduced by just 1% relative to the fitness of AA, a decrease so slight that it would be difficult to detect in most studies. Despite this tiny heterozygous effect, equation 12.8 now predicts an equilibrium frequency of 0.0001—an order of magnitude less than the equilibrium for this case under complete recessiveness. Thus, a fitness decrease of 0.01 in the heterozygote yields a 10-fold reduction in the frequency of this deleterious allele from the recessive case. Indeed, note from equation 12.8 that as long as the heterozygote has a 1% fitness decline ($hs = 0.01$), it really does not make much difference how deleterious this allele is in the homozygotes. In terms of the equilibrium between selection and mutation, it makes little evolutionary difference whether the fitness of the homozygote is reduced 1% or 100% (lethal); the 1% decline in the heterozygote dominates the evolutionary balance. The average excess provides the reason why even slight effects in heterozygotes dominate the evolutionary balance of mutation and selection against deleterious alleles; newly arisen mutations in a randomly mating population exist almost exclusively in heterozygotes and the homozygote effect is therefore relatively unimportant. This illustrates that in measuring the balance between mutation and selection, we must always look at selection from the gamete's perspective as quantified through the average excess.

INTERACTION OF NATURAL SELECTION WITH MUTATION AND SYSTEM OF MATING

The system of mating can also interact with natural selection to influence evolutionary dynamics and equilibria. However, the interaction of selection with system of mating is more direct than with many other evolutionary forces because system of mating directly influences the average excess (Chapter 8). To illustrate this direct interaction between selection and system of mating, consider the mutation/selection balance models of the previous section but now allowing deviations from random mating as measured by f (Chapter 3). The average excess associated with a deleterious allele now becomes

$$a_a = p(1 - hs - \bar{w})(1 - f) + [f + q(1 - f)](1 - s - \bar{w}) \quad (12.9)$$

Note that equation 12.5 is a special case of equation 12.9 when $f = 0$ (random mating). Assuming, as before, that $\bar{w} \approx 1$, the average excess of the deleterious allele a is approximately $-fs + (1 - f)[-hs + qs(h - 1)]$, and when dealing with a recessive allele ($h = 0$) the average excess further simplifies to $-fs - (1 - f)qs$. At equilibrium, we expect q_{eq} to be very small, so even modest deviations from random mating with $f > 0$ can ensure that the fs term dominates over the $(1-f)q_{eq}s$ term. Hence, at equilibrium in an inbreeding population ($f > 0$), the average excess of the deleterious a allele becomes $-fs$, and the equilibrium allele frequency defined by equation 12.3 is

$$q_{eq} = \frac{\mu}{fs} \quad (12.10)$$

In the example of a recessive lethal allele ($s = 1$) with a mutation rate of 10^{-6}, recall that equation 12.7 yields an equilibrium frequency of 0.001 in a randomly mating population. Now consider the evolutionary impact of a seemingly minor deviation from random mating with $f = 0.01$ (a 1% deviation from Hardy–Weinberg genotype frequencies). With this modest f, equation 12.10 yields an equilibrium allele frequency of 0.0001. Thus, a 1% deviation from random mating is amplified into a 10-fold reduction in the frequency of a recessive, lethal allele.

Equation 12.10 tells us that populations that systematically deviate from random mating with an inbreeding system of mating accumulate far fewer deleterious recessive alleles than would a randomly mating population. For example, the Tamils of India have strongly favored first-cousin matings for centuries. The Tamils display no detectable inbreeding depression (Rao and Inbaraj 1980), in great contrast to the human populations in Europe and the United States who avoid inbreeding as a system of mating and have an average of five to eight lethal equivalents per person (Stine 1977).

Because the average excess is a direct function of f, deviations from random mating will influence the outcome of natural selection on *all* selected genetic systems, not just selection against rare, deleterious alleles. To illustrate this pervasive impact of system of mating upon adaptive outcomes, let us return to the example from Chapter 11 of the Bantu-speaking peoples adapting to a malarial environment at the Hb-β locus with respect to the A, S, and C alleles. In Chapter 11, we only considered the case of a randomly mating population. Now we assume a hypothetical inbreeding population with $f > 0$. As before, consider an initial gene pool with $p_A \approx 1$, $p_S \approx p_C \approx 0$ for the frequencies of the A, S, and

C alleles, respectively, and with fitnesses as given in Table 11.1 for the malarial environment. With inbreeding, the average excesses of the S and C alleles are (from equation 8.9)

$$a_S = p_A(1-f)(1-\bar{w}) + [f + p_S(1-f)](0.2 - \bar{w}) + p_C(1-f)(0.70 - \bar{w})$$
$$a_C = p_A(1-f)(0.89 - \bar{w}) + p_S(1-f)(0.70 - \bar{w}) + [f + p_C(1-f)](1.31 - \bar{w})$$
(12.11)

Under the initial conditions, almost everyone in the population is AA, so the average fitness is close to 0.89, the fitness of the AA genotype under malarial conditions. Hence, the initial response to selection for the S and C alleles is determined by their initial average excesses of

$$a_S = p_A(1-f)(0.11) + [f + p_S(1-f)](-0.69) + p_C(1-f)(-0.19)$$
$$\approx (1-f)(0.11) - f(0.69) = 0.11 - 0.8f$$
$$a_C = p_A(1-f)(0) + p_S(1-f)(-0.19) + [f + p_C(1-f)](0.42) \approx 0.42f$$
(12.12)

using $p_A \approx 1$, $p_S \approx p_C \approx 0$ in the initial gene pool. Recall that the initial average excesses in the random-mating case were 0.11 for S and 0 for C (equations 11.6 and 11.8 in Chapter 11), which is a special case of 12.12 when $f = 0$. As equation 12.12 reveals, inbreeding decreases the average excess of the S allele and increases the average excess of the C allele relative to their values under random mating. When $f > 0.11/0.8 = 0.14$, equations 12.12 reveal that natural selection *always* operates to eliminate the S allele and favor the C allele. However, even when selection under inbreeding initially favors an increase in both the S and C alleles (when $0 < f < 0.11/0.8$), the adaptive topography is altered by inbreeding to decrease the "height" of the A/S polymorphic peak relative to the random-mating case. This alteration broadens the conditions under which the population will evolve toward fixation of C under natural selection. For example, Figure 12.1 shows the adaptive topography with the fitnesses given in Table 11.1 for a malarial environment with $f = 0.04$, as well as the evolutionary trajectories over this landscape for the same set of initial conditions considered in Figure 11.8. In contrast to Figure 11.8, all initial populations adapt to malaria in this case through the fixation of the C allele and none remain polymorphic for sickle cell.

These results make sense if you take the gamete's perspective as measured by the average excess. An S allele in an inbreeding population is more likely to be coupled with another S allele, so the deleterious fitness effects of the SS homozygote become more important and the beneficial effects of the AS heterozygote become less important under inbreeding from an S-bearing gamete's perspective. In contrast, the highly beneficial fitness effects of the CC homozygote become more important under inbreeding from the perspective of a C-bearing gamete. The contrast in evolutionary outcomes between Figures 11.8 versus 12.1 reveals that *adaptation cannot be explained only in terms of the fitnesses of genotypes*, which are identical in these two cases. Even complete knowledge of the fitnesses of every individual is inadequate for predicting the impact of natural selection. We need in addition the information about population structure that directly affects the average excesses of fitness. Hence, our fundamental equation of natural selection for measured genotypes (equation 12.1) reveals

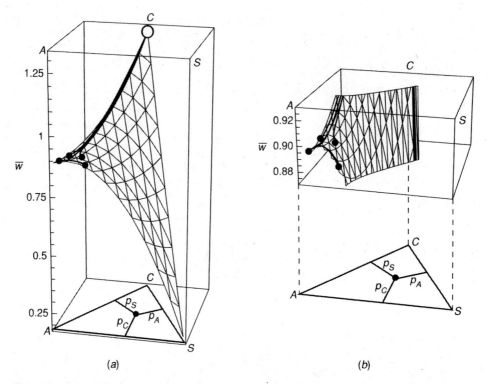

Figure 12.1. Adaptive surface defined by A, S, and C alleles at β-Hb locus in inbreeding population with $f = 0.04$ using fitnesses given in Table 11.1 for malarial environment. The gene pool space is shown by the triangle near the bottom, with the allele frequencies given by the perpendicular distances from a point to the sides of the triangle and with the vertices associated with fixation of a particular allele labeled by the letter corresponding to that allele. The vertical axis gives $\bar{w}$ as a function of these three allele frequencies. One peak exists in this adaptive surface corresponding to fixation for C, as indicated by the white dot. Black dots indicate the initial state of the gene pool for four populations: one starting at $p_A = 0.95$, $p_S = 0.025$, and $p_C = 0.025$; a second at $p_A = 0.85$, $p_S = 0.025$, and $p_C = 0.125$; a third at $p_A = 0.85$, $p_S = 0.125$, and $p_C = 0.025$; and a fourth at $p_A = 0.75$, $p_S = 0.125$, and $p_C = 0.125$. Black lines from these dots plot the evolutionary trajectory across generational time as defined by equation 12.1. (a) Entire adaptive landscape and (b) same expanded portion as Figure 11.8b.

that the outcome of adaptive evolution represents an *interaction between fitness differences and population structure*. Once again, we see the fallacy in the statement that "natural selection is survival of the fittest."

INTERACTION OF NATURAL SELECTION WITH GENE FLOW

Gene flow can influence the adaptive course of natural selection both by determining what genetic variation is available within a deme's gene pool and by directly altering allele frequencies (Chapter 6). To see this, consider the simple model from Chapter 6 of symmetrical gene flow at rate m between two demes, 1 and 2. The change in allele frequency at an autosomal locus in deme 1 is given by $\Delta p_1 = -m(p_1 - p_2)$ where p_i is the frequency of the allele in deme i (equation 6.2). Combining the effects of selection and gene flow, equation 12.2 becomes

$$\Delta p_1 = \frac{p_1}{\bar{w}} a_A - m(p_1 - p_2) \qquad (12.13)$$

Suppose that $a_A > 0$ in deme 1; that is, natural selection favors an increase in the frequency of allele A. However, if p_1 were initially zero, this adaptive course could never start. Suppose further that the A allele is present in population 2. Once A is introduced into deme 1 through gene flow, natural selection can now operate to increase its frequency. As we saw in Chapter 11, natural selection does not create genetic variants but only operates upon the genetic variation available in the gene pool. Gene flow can be an important source of such variation.

The sign of the component of allele frequency change induced by gene flow in equation 12.13 is determined solely by the initial difference in allele frequencies among the demes, in this case $p_1 - p_2$. The sign of the selective component of equation 12.13 is determined solely by the average excess of fitness, which in turn is a function of within-deme allele frequencies, system of mating, and genotypic deviations of fitness in the deme. As a result, there is no biological necessity for the selective and gene flow components to have the same sign. In some cases, selection and gene flow can operate in the same direction, allowing an allele to increase (or decrease) in frequency more rapidly than possible through either selection or gene flow alone; in other cases selection and gene flow will be in opposite directions and the evolutionary outcome will depend upon their balance. To illustrate these diverse outcomes, we return to the example of African Americans, a population influenced by asymmetric gene flow involving demes originally derived from Europe and West Africa (Chapter 6). In this case, the impact of asymmetric gene flow over several generations was measured by (equation 6.4)

$$M = \frac{p_A - p_W}{p_E - p_W}$$
$$= \frac{\text{change in allele frequency in African Americans from West Africans}}{\text{initial difference in allele frequency between Europeans and West Africans}}$$

$$(12.14)$$

Because M is standardized for the initial allele frequency difference, M should be identical for all alleles at all polymorphic loci showing any initial difference in allele frequency between Europeans and West Africans *if* gene flow were the only evolutionary force operating.

Adams and Ward (1973) estimated M for alleles at several loci in African Americans from Claxton, Georgia (Table 12.1), and found significant heterogeneity across alleles. Such heterogeneity implies that selection may have altered the allele frequency dynamics at some loci. A majority of the alleles yielded an estimate of M of about 0.11, and combining the estimates across all alleles as weighted by the variances of the estimators yields an overall M of 0.13. The most straightforward interpretation of these results is that these alleles are neutral or nearly neutral in this population and are reflecting the overall impact of asymmetric gene flow that has resulted in about 13% admixture. However, there are many allelic outliers from this apparently neutral background M of 0.13. For example, the A allele at the *ABO* blood group locus shows a slightly negative M, but not significantly different from zero. Based on this allele alone, there seems to be no admixture at all. Waterhouse and Hogben (1947) have implicated the A allele as a cause of fetal wastage (spontaneous abortions) of

Table 12.1. Admixture Estimates (M) for Several Loci in African Americans from Claxton, Georgia

Allele	M	Possible Explanation
R_o	0.107 ⎫	Alleles at several blood group loci that may be neutral and
R_1	0.110 ⎪	that may reflect the overall impact of asymmetric gene
r	0.117 ⎪	flow resulting in about 11% admixture in this African
Fy^a	0.108 ⎬	American population
P	0.092 ⎪	
Jk^a	0.164 ⎭	
A	−0.037	Maternofetal incompatibility at the ABO blood group locus may select against European alleles
R_2	0.446 ⎫	Unknown
T	0.466 ⎭	
Hp^1	0.619 ⎫	Alleles at loci implicated with malarial adaptation;
$G6PD\ A^-$	0.395 ⎬	selection may occur against African alleles in
β-$Hb\ S$	0.614 ⎭	nonmalarial environment

Source: From Adams and Ward (1973).
Note: The estimates were significantly heterogeneous across loci. The last column gives possible explanations for that heterogeneity.

AO fetuses arising from matings between women with blood type O (genotype OO) with men of blood type A (genotypes AA or AO) (this type of selection will be examined in more detail in the next chapter). Bottini et al. (2001b) have shown increased recurrent spontaneous abortion in matings between women of blood type B (genotype BB or BO) with men of blood type A. West African populations have a higher frequency of the O and B alleles and a lower frequency of the A allele than European populations (Adams and Ward 1973), and much of the original admixture is known historically to be between men of European ancestry with women of African ancestry. This allele frequency and mating pattern would result in preferential elimination of A-bearing fetuses, thereby reducing the flow of A alleles from the European American population into the African American population.

In contrast to the A allele, there are several alleles at other loci that have high values of M, about 0.4 or above. A majority of these alleles have been implicated in malarial adaptation, and moreover these same alleles are often associated with deleterious effects in a nonmalarial environment (as already discussed for the $G6PD\ A^-$ and β-$Hb\ S$ alleles). Selection for malarial resistance was much reduced in North America relative to West Africa, so it is reasonable to conclude that selection would be occurring against the antimalarial alleles in the North American environment. Thus, selection would favor an increase in alleles of European ancestry at these loci, thereby accentuating the effects of asymmetric gene flow. Overall, the results shown in Table 12.1 illustrate that *natural selection interacts with gene flow as a selective filter that blocks or retards the flow of some genes from one population to another but accentuates the spread of others.*

INTERACTION OF NATURAL SELECTION WITH GENETIC DRIFT

Genetic drift induces random changes in allele frequency, so the second term in equation 12.2 when applied to drift is expected to vary in sign at random from generation to generation.

Hence, unlike gene flow, there is no consistent enhancement or retardation of the effects of selection with genetic drift. To R. A. Fisher, this meant that genetic drift would play little role in shaping adaptive outcomes unless the population size is extremely small or, as we will see below, in dealing with the fate of a newly arisen mutation. To Fisher, selection (the first term in equation 12.2) provides a consistent direction to evolution or maintains an equilibrium, whereas drift causes random deviations from the selective trajectory that tend to cancel each other out in the long run. Under this view, drift has little long-term importance in adaptive evolution. As we will see later in this chapter, Sewall Wright disagreed with this point of view and contended that drift could under some conditions strongly influence the course of adaptive evolution through its interactions with natural selection. However, we begin this section by examining the fate of newly arisen, selected mutations—an area in which both Fisher and Wright agreed that drift plays an important evolutionary role. Drift is always strong when dealing with newly arisen mutant alleles because, regardless of the total population size, the mutant is originally present as a single copy and hence is subject to maximal sampling error.

To see how selection and drift interact to determine the fate of newly arisen alleles, recall from Chapter 5 that the probability of a newly arisen, neutral, autosomal allele surviving in a population and going to fixation is $1/(2N)$, its initial allele frequency as a new mutation in a population of size N. Let this probability of fixation of a neutral mutation be given by $u_0 = 1/(2N)$. Natural selection changes this probability. For example, suppose initially that the population is fixed for allele a at an autosomal locus in a finite randomly mating deme of size N but mutation creates a single copy of the new allele A. Suppose further that A is a favorable allele under the existing environmental conditions such that the relative fitnesses are 1 for aa, $1 + s$ for Aa, and $1 + 2s$ for AA where the selection coefficient s is greater than zero. In this case a_A is greater than zero for every $p > 0$, so under natural selection alone (equation 12.1) we would expect the favorable A allele to go to fixation. However, the second term of equation 12.2 can take on negative values by chance due to drift, and when the A allele is still rare in the population, the drift term can be sufficiently negative to lower the frequency of A, even to the point of loss, contrary to the predictions of the fundamental theorem of natural selection. In particular, the probability of fixation of the favorable allele A is not 1, as expected from selection alone, or u_0, as expected from drift alone, but rather (Crow and Kimura 1970) is

$$u = \frac{1 - e^{-2(N_{ev}/N)s}}{1 - e^{-4N_{ev}s}} \tag{12.15}$$

where N_{ev} is the variance effective size. For simplicity, let $N_{ev} = N$; then equation 12.15 becomes

$$u = \frac{1 - e^{-2s}}{1 - e^{-4Ns}} \tag{12.16}$$

If we take the limit of equation 12.16 as s goes to zero (that is, the A allele approaches neutrality), we get (using Taylor's theorem) that $u \approx 2s/(4Ns) = 1/(2N)$. Consequently, the neutral fixation probability is a special case of equation 12.16 when the selection coefficient becomes very small. As selection becomes stronger, drift and natural selection interact with one another to influence the evolutionary fate of the favorable allele, as indicated by the product $4Ns$ appearing in the denominator of equation 12.16. In this simple codominant

Table 12.2. Fixation Probabilities of Selectively Favorable Mutant Allele with $s = 0.01$ as Function of Population Size N

N	u	$u_0 = 1/(2N)$	Percent Change in Selected Case from Neutral Case
10	0.061	0.050	22%
50	0.023	0.010	129%
100	0.020	0.005	300%
∞	0.020	0.000	—

Note: The fixation probabilities of a neutral mutant in a population of the same size are also shown, as well as the percent increase in the fixation probability in the selected case compared to the neutral case, $100(u - u_0)/u_0$.

fitness model, the strength of natural selection as measured by the average excess is proportional to s when the allele is rare, and the strength of drift is, of course, proporational to $1/N$. Hence, the product Ns is the ratio of the strength of selection divided by the strength of drift. As we have seen many times before, it is the relative strengths of two evolutionary forces that must be taken into account to determine their joint evolutionary impact.

Table 12.2 shows the values of u and u_0 for $s = 0.01$ and for various values of N. This table shows that as N gets larger (selection becomes stronger relative to drift), selection is able to cause ever increasingly large deviations from the expectations under pure genetic drift (neutrality). Note, however, that the probability of fixation of this favorable mutant stabilizes at 0.020. This result may seem surprising given that selection favors the fixation of the A allele, yet even in an effectively infinite population the favored allele is much more likely to be lost than fixed. This is not just limited to the case of $s = 0.01$. In general, as N goes to infinity, the denominator in equation 12.16 goes to 1, so the fixation probability is approximately $1 - e^{2s}$ in large populations. When s is small, $1 - e^{2s} \approx 2s$. This approximation predicts that the fixation probability of a favorable allele with $s = 0.01$ should be 0.02 in a large population, as is indeed found to be the case in Table 12.2. Note that in this case the *favorable* allele will be lost 98% of the time. Hence, even in a large population, the most likely fate of a selectively favorable allele is to be lost due to genetic drift. This occurs because genetic drift overpowers selection in the first few generations during which it is likely that only one or a handful of copies of the favorable allele exist in the population.

Because the direction of genetic drift is random, it is even possible for genetic drift to overpower natural selection and lead to the fixation of a deleterious mutant. Suppose now that A is a deleterious allele under the existing environmental conditions such that the relative fitnesses are 1 for aa, $1 - s$ for Aa, and $1 - 2s$ for AA where the selection coefficient s is greater than zero. In this case a_A is less than zero for every $p > 0$, so under natural selection alone (equation 12.1) we would expect the deleterious A allele to be lost from the population. However, with drift, the fixation probability for the deleterious mutant is $u = (e^{2s} - 1)/(e^{4Ns} - 1)$, which is positive for all finite values of N. Thus, there is a finite chance of a *deleterious* allele going to fixation in a finite population, in violation of Fisher's fundamental theorem of natural selection. Table 12.3 shows these fixation probabilities for $s = 0.01$ as a function of population size. As the strength of genetic drift decreases relative to selection with increasing population size, the deviations from neutrality are accentuated in magnitude. However, unlike the case for an advantageous mutant, as population size becomes large, the results converge to that expected from selection alone (elimination of

Table 12.3. Fixation Probabilities of Selectively Deleterious Mutant Allele with $s = 0.01$ as Function of Population Size N

N	u	$u_0 = 1/(2N)$	Percent Change in Selected Case from Neutral Case
10	0.041	0.050	-18%
50	0.003	0.010	-68%
100	0.00038	0.005	-92%
∞	0.000	0.000	—

Note: The fixation probabilities of a neutral mutant in a population of the same size are also shown as well as the percent increase in the fixation probability in the selected case compared to the neutral case, $100(u - u_0)/u_0$.

the deleterious allele). This occurs because the primary effect of genetic drift in a large population is to eliminate rare alleles. For an advantageous allele, this effect of drift was in opposition to selection, but in the case of a deleterious allele, both selection and drift tend to reinforce one another in eliminating the newly arisen rare allele.

Although Fisher and Wright agreed that drift can interact with selection to have a large impact on newly arisen mutations, Fisher looked upon drift as primarily limiting the amount of genetic variation that would effectively enter a gene pool. However, once a new variant had achieved an appreciable frequency, Fisher regarded the adaptive process as being dominated by natural selection. Wright, in contrast, felt that adaptively important interactions between natural selection and genetic drift could emerge when multiple selective equilibria existed with unequal average fitness [multiple "peaks" of unequal height in the adaptive surface metaphor introduced by Wright (1932)]. There are two aspects to the adaptive importance of genetic drift when multiple selective peaks exist: first, the role of genetic drift in determining the initial conditions upon which selection acts and, second, the role of genetic drift in allowing populations to evolve from one selective peak to another in violation of Fisher's fundamental theorem (called peak shifts).

To illustrate the potential importance of genetic drift in influencing the initial conditions under which selection operates, let us return to the hemoglobin β-chain locus with the three alleles A, S, and C. Figure 12.2 shows the adaptive surface under the nonmalarial environmental fitness given in Table 11.1 in a randomly mating population. This represents the adaptive surface in West Africa before malarial conditions were created by the introduction of the Malaysian agricultural complex into that area. Instead of a peak, the adaptive landscape in this nonmalarial environment has a "ridge" connecting the points corresponding to fixation of the A allele and fixation of the C allele. This reflects the fact that in the nonmalarial environment the A and C alleles are neutral with respect to one another. Figure 12.2 shows the adaptive trajectories from the same four initial conditions illustrated in Figure 11.8 under a malarial environment. In the nonmalarial environment, adaptive evolution simply results in the elimination of the deleterious S allele (or its near loss, as selection becomes ineffective against this deleterious recessive allele as it becomes rare) and takes the population to a point on the A/C adaptive ridge. However, once on top of this ridge, there is no longer any evolution induced by natural selection. The ridge surface is a neutral one influenced only by genetic drift. Hence, populations are free to evolve toward any point on this ridge according to the random dictates of drift. However, once the environment changes to a malarial one (Figure 11.8), none of the populations shown in Figure 12.2 would be upon an adaptive peak. However, the two populations close to fixation of A would evolve under

384 INTERACTIONS OF NATURAL SELECTION WITH OTHER EVOLUTIONARY FORCES

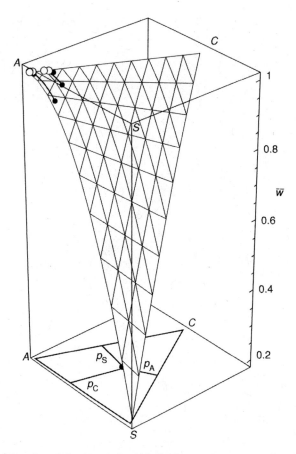

Figure 12.2. Adaptive surface defined by A, S, and C alleles at β-Hb locus in randomly mating population using fitnesses given in Table 11.1 for nonmalarial environment. All the graphing conventions used in Figure 12.1 are used in this figure. An adaptive ridge exists in this adaptive surface connecting the A and C vertices, corresponding to all possible A/C polymorphic states with no S alleles. Black dots indicate the initial state of the gene pool for four populations: one starting at $p_A = 0.95$, $p_S = 0.025$, and $p_C = 0.025$; a second at $p_A = 0.85$, $p_S = 0.025$, and $p_C = 0.125$; a third at $p_A = 0.85$, $p_S = 0.125$, and $p_C = 0.025$; and a fourth at $p_A = 0.75$, $p_S = 0.125$, and $p_C = 0.125$. Black lines from these dots plot the evolutionary trajectory across generational time as defined by equation 12.1, ending at the selective equilibria points indicated by white dots.

natural selection toward the A/S polymorphic adaptive peak, whereas the two populations that had higher initial frequencies of C would evolve toward the fixation for C adaptive peak. Evolution upon a ridge of neutrality governed by genetic drift could produce local populations with a variety of initial frequencies of the A and C alleles, which in turn would lead to highly divergent adaptive outcomes once the environment changed. Such neutral ridge evolution may be a contributor, perhaps even the determining factor, to the divergent adaptive trajectories observed in West African populations today in which some increase the frequency of S and others the frequency of C (Figure 11.6). In this manner, *genetic drift can create a diversity of initial conditions for allelic subsets that are neutral under one environmental condition that strongly influence the adaptive outcome when the environment is altered.* This adds to the difficulty in predicting the course of adaptive evolution, but it also increases the diversity of adaptive responses shown by populations to altered environments.

The other adaptively important interaction between selection and drift is a peak shift. Recall that Fisher's fundamental theorem of natural selection requires populations to evolve under selection only in a manner that increases average fitness. In terms of the metaphor of an adaptive surface, this means that it is impossible for natural selection to cause a population to evolve "downhill." However, genetic drift can cause populations to evolve in a manner that violates the fundamental theorem, as we already saw with the loss of favorable alleles or the fixation of deleterious alleles in finite populations. This ability of genetic drift to cause violations of the fundamental theorem creates a finite chance that a population will evolve downhill and cross a selective "valley" to reach the slopes of a different selective peak. Once in the domain of this new selective peak, natural selection makes it more likely to evolve toward the top of the new peak rather than return to the original peak. To see this, consider first a simple two-peak adaptive surface defined by a one-locus, two-allele system in a randomly mating population with the following fitnesses: AA with a fitness of 1; Aa with a fitness of 0.9; and aa with a fitness of 0.95. The adaptive surface for this example is plotted in Figure 12.3 as a function of p, the frequency of the A allele. This adaptive surface has two adaptive peaks, one at $p = 0$ with $\bar{w} = 0.95$ and the second at $p = 1$ with $\bar{w} = 1$. The average excess of fitness of the A allele in this case is $a_A = p(1) + (1-p)(0.9) - \bar{w}$, and the average excess of a is $a_a = p(0.9) + (1-p)(.95) - \bar{w}$. At the polymorphic equilibrium, $a_A = a_a$, so $p + (1-p)(0.9) = p(0.9) + (1-p)(0.95)$, which yields $p_{eq} = \frac{1}{3}$. This polymorphic equilibrium point is also indicated in Figure 12.3. However, as can be seen from Figure 12.3, this equilibrium point is at a fitness valley, not a peak. Therefore, depending upon which side of $p = \frac{1}{3}$ a population initially lies, it should either go to fixation of the

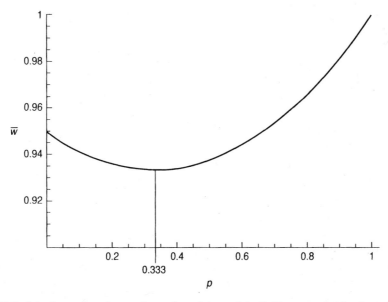

Figure 12.3. Adaptive surface for one-locus, two-allele model with fitnesses of $AA = 1$, of $Aa = 0.9$, and of $aa = 0.95$. The surface represents a plot of $\bar{w}$ under the assumption of randomly mating against p, the frequency of the A allele. A line at $p = 0.333$ indicates the minimum value of $\bar{w}$ and an unstable equilibrium.

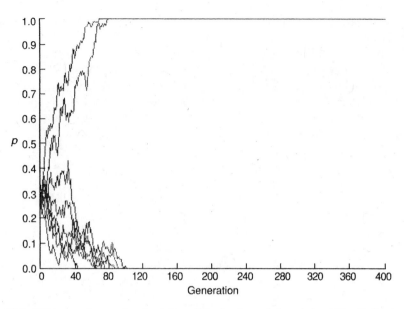

Figure 12.4. Eight runs of computer simulation of idealized population of size 100 with fitness model described in Figure 12.3, all starting with initial condition of $p = 0.3$.

A allele (initial $p > \frac{1}{3}$) or the a allele (initial $p < \frac{1}{3}$). Suppose now that a population was at a selective equilibrium with $p = 0.3$ but the environment shifts to create a new adaptive surface like that shown in Figure 12.3. According to equation 12.1 and the fundamental theorem of natural selection, this population should evolve toward loss of the A allele and go to the lower peak shown in Figure 12.3. It is impossible under natural selection alone for such a population to evolve toward fixation of A and thereby achieve the higher fitness peak. However, Figure 12.4 shows the results of eight runs of a computer simulation of an idealized, randomly mating population of size 100 with these fitnesses and an initial allele frequency of 0.3. As can be seen, two of the eight runs evolved to the peak defined by $p = 1$: a clear violation of the fundamental theorem of natural selection! Hence, *drift interacts with selection to allow evolutionary outcomes that would be prevented by selection alone. In particular, drift plus selection allows for peak shifts that would be prevented by natural selection in large populations.*

Although drift allowed peak shifts for some of the runs of this computer simulation, six of the eight runs evolved to the peak defined by $p = 0$, thereby obeying the fundamental theorem of natural selection (Figure 12.4). In general, the direction of evolution is biased in favor of the direction initially favored by natural selection. Nevertheless, the fact that some runs went to $p = 0$ and others to $p = 1$ shows that genetic drift permits populations to "explore" the adaptive surface more thoroughly than by natural selection alone. Note also that it is impossible to predict the outcome of any given population in the runs shown in Figure 12.4. We can only describe what should happen if the evolutionary "experiment" were to be repeated many times. If the species of interest were subdivided into several small subpopulations with the same initial allele frequencies, we would have multiple replications of the experiment. Each of the subpopulations would evolve, and some would go to the higher peak even if the initial conditions made that violate Fisher's fundamental theorem. Moreover, when we have multiple subpopulations, another evolutionary force can enter: gene flow. Indeed, Wright considered gene flow a critical force for modulating the adaptive interactions between selection and drift, as we will now see.

INTERACTIONS OF NATURAL SELECTION, GENETIC DRIFT, AND GENE FLOW

Wright (1932) felt that in a population with a large effective size there would be virtually no chance for a peak shift. Hence, the population would evolve toward the nearest local peak, and then selection would maintain the population on that local peak, even if a vastly superior adaptive alternative existed. We have already seen an example of this with the β-Hb locus in human populations living in the malarial regions of Africa; most such populations have evolved toward the A/S polymorphic peak even though the fixation for the C peak seems to be a far superior adaptive outcome in terms of average fitness (Figure 11.8), percentage of the population protected against malaria, degree of protection against malaria, and eliminating deleterious side effects such as hemolytic anemia from the population. Because selection could keep populations on inferior adaptive peaks, Wright regarded natural selection as a potential *impediment* to adaptation when it dominated over genetic drift.

On the other hand, an isolated population of small variance effective size is also not optimal for adaptive evolution. Genetic drift is powerful in such a population, but the primary manifestation of the power of drift is the rapid loss of genetic variation. Without genetic variation, there is no evolution of any sort. What is needed is to simultaneously have demes of small variance effective size but with access to much genetic variation. As shown in Chapters 6 and 9, population subdivision can both induce small variance effective sizes at the local deme level yet maintain *higher* levels of genetic diversity and additive genetic variance at the global level than an equally sized panmictic species (see equations 6.44 and 9.28). What is needed is the right amount of gene flow between the local demes: too much, and the population becomes effectively panmictic and selection prevents genetic drift from allowing populations to explore the adaptive surface; too little, and the local demes generally have little genetic variation and hence low adaptive potential. Under an island model of population subdivision, Wright (1932) argued that an Nm term (the product of the local variance effective size and the migration rate) of about 1 provided the right balance. This conclusion has been supported by Barton and Rouhani (1993), who showed that peak shifts are most likely if Nm is slightly below 1.

To illustrate peak shifts in a subdivided population, a computer simulation was run using the same fitness parameters as that given in Figure 12.3 and with the local variance effective size being 100, as in Figure 12.4, but now with an island model of gene flow with $m = 0.01$. Hence, $Nm = 1$ in these simulations. As before, we assume that the population had adapted to a previous environment with an equilibrium $p = 0.3$. More of the local demes will tend to go to the $p = 0$ peak than the $p = 1$ peak upon the environmental change, so we will assume that the average frequency of the A allele in the global population is 0.1. Note that 0.1 is well below the threshold of $p = \frac{1}{3}$ that separates the domain of the lower peak from that of the higher. Hence, gene flow in these simulations acts as a directional force to bring local populations to the lower fitness peak. Figure 12.5 shows the evolution of eight local demes, each starting with $p = 0.3$ but receiving input at rate $m = 0.01$ every generation from the global gene pool with $p = 0.1$. As can be seen, two of the eight local demes still evolved toward the higher peak (although gene flow now prevents fixation of A). Obviously,

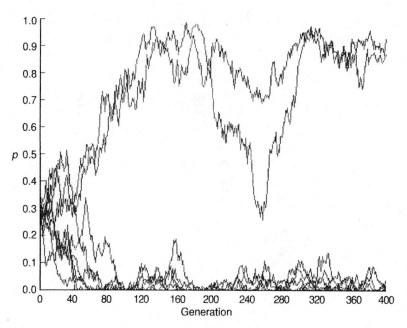

Figure 12.5. Eight runs of computer simulation of idealized population of size 100 with fitness model described in Figure 12.3, all starting with initial condition of $p = 0.3$, and part of subdivided global population with overall frequency of 0.1 for A allele with subdivision corresponding to island model with $Nm = 1$.

gene flow is not making peak shifts impossible despite the fact that gene flow is consistently pulling down the frequency of A in these simulations.

We need to keep in mind the balances of the evolutionary forces relative to one another to understand why gene flow is not preventing shifts to the higher allele frequency peak even though gene flow is a *directional* force against shifts to the higher allele frequency peak in this case. Gene flow in one sense is helping genetic drift to explore the adaptive surface because *gene flow helps maintain local genetic diversity*. This allows both selection and drift to operate at the local level. As long as there is genetic variation within the deme, even the demes that are near the lower frequency peak still have a finite chance of undergoing a peak shift (Figure 12.5). Without gene flow, once fixed on the lower peak, there is no possibility of a peak shift for an isolated deme in the absence of mutation (Figure 12.4). Moreover, the amount of gene flow is still sufficiently low ($Nm = 1$) that genetic drift can overcome the biases of both selection and directional gene flow to cause a peak shift. Once a deme drifts into the domain of the higher peak, there is a shift in the balance of selection versus the bias of gene flow and the force of genetic drift that favors selection. Recall from our earlier models that what is important is not the magnitude of selection or drift but rather their ratio (measured by Ns in the models of newly arisen mutations). Similarly, the ability of selection to overcome a gene flow bias depends upon the relative magnitudes of the selective intensities to the gene flow rate. When selection is stronger in keeping a deme close to the higher peak, the same amount of gene flow or drift is less important *relative* to selection on the higher peak than on the lower. Thus, for the same amount of gene flow and drift, it is generally more likely to shift from a lower peak to a higher peak than the opposite. *Although peak shifts are random at the local deme level, the global probabilities of peak shifts over many demes are biased in favor of higher peaks.* Therefore, even though "random" genetic drift is the mechanism for exploring the adaptive surface, the demes preferentially end up on the higher peaks (Barton and Rouhani 1993; Templeton 1982).

The random exploratory mechanism of genetic drift allows the demes to explore more than one adaptive solution. The shift in the balance of selection to drift makes it likely that the demes will remain at the better adaptive solutions that they encounter. Even though natural selection is the only evolutionary force necessary for adaptation, Wright (1932) argued that adaptive evolution is more effective when natural selection is *not* in sole control. Indeed, Wright felt that natural selection could *prevent* adaptive evolution by keeping a population on a local but not globally optimal adaptive peak. Thus, natural selection is necessary to explain adaptation, but it is not sufficient in Wright's view.

There is another shift in the balance of gene flow relative to selection and drift as this evolutionary process in a subdivided population continues. We assumed in the previous simulations that most populations rather rapidly shifted from $p = 0.3$ toward the lower peak, resulting in a global allele frequency of 0.1 for the A allele. However, because gene flow now maintains local genetic diversity and hence local genetic drift and because the shift in the balance of selection to drift and gene flow creates a bias in favor of the higher peak shown in Figure 12.3, more and more local demes will evolve toward the domain of the higher peak as time progresses. This in turn will increase the global allele frequency of the A allele. As this occurs, gene flow shifts from being an evolutionary force biased against the higher peak to a force that is more neutral and finally to one that is biased in favor of the higher peak. For example, suppose the global allele frequency has reached a value of 0.3 as more and more demes shifted toward the higher peak. This global allele frequency is still below the $\frac{1}{3}$ threshold, but now gene flow is only slightly biased against the higher peak. Figure 12.6 shows the results of computer simulations that are identical to the simulations

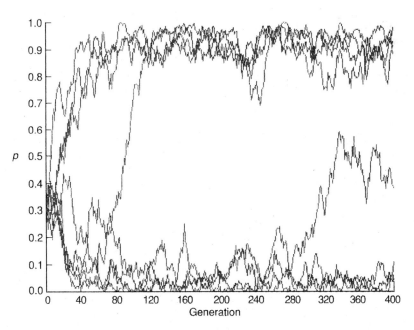

Figure 12.6. Eight runs of computer simulation of idealized population of size 100 with fitness model described in Figure 12.3, all starting with initial condition of $p = 0.3$, and part of subdivided global population with overall frequency of 0.3 for A allele with subdivision corresponding to island model with $Nm = 1$.

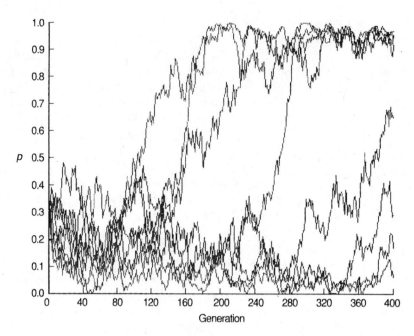

Figure 12.7. Eight runs of computer simulation of idealized population of size 100 with fitness model described in Figure 12.3, all starting with initial condition of $p = 0.3$, and part of subdivided global population with overall frequency of 0.5 for A allele with subdivision corresponding to island model with $Nm = 1$.

shown in Figure 12.5 except for a global allele frequency of 0.3 instead of 0.1. In this case, four of the eight subpopulations have gone to the higher peak, and another is well on the way! This is the result of the shift in the balance of gene flow to selection and drift, with gene flow now favoring peak shifts by placing populations into the fitness valley where drift is most effective. Thus, even more demes go toward the higher peak, and the global allele frequency increases even more. Figure 12.7 shows the simulated results obtained when the global allele frequency has reached 0.5. At this point, gene flow has shifted toward a bias favoring the higher peak. It is now virtually inevitable for all the local demes to evolve toward the higher peak (Figure 12.7).

Wright argued that gene flow might be even a more powerful force in bringing more and more demes to a superior adaptive peak when there is an interaction between selection and the *amount* of gene flow. His argument applies to the case in which the superior adaptive peak results in a higher absolute fitness (recall that the adaptive surface is defined from relative fitnesses, so this is not necessarily the case). For example, in the β-Hb A, S, and C example, the superior adaptive peak is associated with fixation for C (Figure 11.8), and populations on this peak are characterized by 100% of the individuals having resistance to malaria that is superior to the resistance of A/S individuals, who represent only a minority of the individuals in populations on the lower peak. Hence, the absolute average viability of individuals (their ability to survive in a malarial environment) is indeed superior for those populations on the higher peak in this example. Wright felt that such a situation would often be true, so that those demes on the higher peaks would have more offspring than demes on lower peaks. As a result, the same migration *rate* per offspring would result in those demes on the higher peaks producing a greater number of migrants going out to other demes. This increase in the absolute numbers of migrants biases the evolutionary process to favor the spread of those alleles associated with the higher peaks throughout the species.

Wright (1931, 1932) called the above model of adaptive evolution the **shifting balance theory**, in which shifting balances between the relative strengths of selection, drift, and gene flow allow local demes in a subdivided population to explore the adaptive surface, then preferentially evolve toward the higher peaks in this surface, and ultimately draw other demes toward the higher peaks via asymmetric gene flow. From the onset, the shifting balance theory has been controversial (Coyne et al. 2000; Goodnight and Wade 2000; Whitlock and Phillips 2000). Nevertheless, shifting balance was not empirically tested until 60 years after it was proposed. This long gap between hypothesis and experiment stems in large part from the difficulty and labor involved in testing the shifting balance theory. Such tests require multiple coexisting demes and an overall large population size and manipulation and/or monitoring of many evolutionary parameters such as gene flow, drift, and selection. Wade and Goodnight (1991) provided these first empirical tests using laboratory populations of the flour beetle *Tribolium castaneum* and demonstrated that shifting balance can work under the appropriate circumstances. However, this does not resolve the question of how often the appropriate circumstances occur. Even though shifting balance is possible, it may play little role in adaptive evolution if those circumstances rarely occur. We will therefore examine some of the critical requirements of shifting balance to gauge how likely this mode of adaptive evolution may be.

Population Subdivision

Population subdivision with Nm close to 1 was a critical requirement for shifting balance according to Wright. Many species do have subdivided populations. Even humans have an Nm close to 1 (Chapter 6). Our own species is instructive about the potential for shifting balance. Although our overall Nm is close to 1, there is much spatial heterogeneity over our geographical distribution for the amount of local population subdivision. This is often true of other species. For example, the eastern collared lizard (*Crotaphytus collaris collaris*) shows extreme population subdivision with little to no gene flow among isolated demes in the northeastern Ozarks (a central highland area in North America), gene flow constrained by isolation by distance in the southwestern Ozarks, and less extreme isolation by distance in Texas (Hutchison and Templeton 1999). For many species the necessary degree of subdivision may exist only in a part of the species' range. Therefore, the entire species does not have to be highly subdivided for shifting balance to occur. Barton and Rouhani (1993) point out that if Nm varies sufficiently gradually from place to place, then a superior fitness peak can be established in the geographical regions where $Nm \approx 1$ and can then spread through the rest of the range by the synergistic effects of selection and gene flow.

Besides spatial heterogeneity in the degree of population subdivision, there can also be temporal heterogeneity in subdivision. For example, collared lizards in the northeastern Ozarks are currently highly fragmented with virtually no gene flow and therefore little chance for shifting balance due to a lack of genetic variation in local demes. However, this lack of current gene flow has only been recently created by the suppression of forest fires in the area, primarily since the 1950s in much of the Ozarks. When fire regimens are reestablished, a population structure characterized by some gene flow rapidly emerges (Templeton et al. 2001). Consequently, there may be only occasional times in a species' history when $Nm \approx 1$, but when these episodes occur, a species could experience adaptive breakthroughs associated with going to a superior peak that will profoundly influence its future evolutionary fate.

Subsequent work (Peck et al. 1998, 2000; Slatkin 1981) has also indicated that the gene flow/population subdivision conditions that are required for shifting balance may not be so restrictive as originally envisioned by Wright (1931, 1932). For example, consider a type of population structure known as a **metapopulation** in which the species is subdivided into many local demes that are subject to local extinction and recolonization. Such a population structure can promote local deme differentiation and peak shifts, particularly when recolonization is associated with repeated founder events. Metapopulations can also help spread a superior adaptation throughout the local demes, particularly if most recolonization occurs from just one or a few nearby demes and if the less fit demes are the ones more likely to go extinct and the more fit demes contribute disproportionately to the pool of colonists (McCauley 1993; Wade and Goodnight 1998).

Genetic Architecture

Shifting balance also requires multiple selective equilibria, that is, multiple adaptive peaks that are separated by either fitness valleys or fitness ridges in Wright's metaphor. The shape of the adaptive surface depends in part upon the underlying genetic architecture of fitness. Recall from Chapter 10 that the genetic architecture refers to the number of loci and their linkage relationships and the numbers of alleles per locus that contribute to a trait, along with the mapping of genotype onto phenotype (dominance, recessiveness, pleiotropy, epistasis, etc.). The multiple-peak adaptive surface required for shifting balance arises in part from a genetic architecture characterized by strong interactions between genes (either between alleles at the same locus or between alleles at different loci, that is, epistasis) and/or pleiotropy. For example, the two-peak surface shown in Figure 11.8 for the β-Hb A, S, and C example in a malarial environment arises in part from the pleiotropic effects associated with these alleles on the traits of malarial resistance and hemolytic anemia and in part from the interactions between alleles (S is dominant to A for the trait of malarial resistance and recessive for the trait of anemia, C is recessive to A for the trait of malarial resistance and codominant with S for the trait of anemia). One implication of a multiple-peak adaptive surface is that the fitness effects of an allele are highly context dependent. For example, in Figure 11.8 the S allele is a beneficial allele in the context of a randomly mating population close to fixation for the A allele but is a deleterious allele in the context of a randomly mating population close to fixation for the C allele.

Fisher rejected Wright's view of genetic architecture and its resulting rugged adaptive landscapes and counteracted it with his own visual metaphor: the adaptive hypersphere. In Fisher's metaphor, the degree of adaptation of a population is represented by its closeness to a fixed point in a multidimensional space that corresponds to a single optimal adaptive type, in contrast to the multipeak adaptive landscapes of Wright. A second point in Fisher's hyperspace represents the population's current average adaptive phenotype. In contrast to the rugged adaptive surfaces of Wright, Fisher envisioned a smooth, continuous relationship between the optimal point and other points in this adaptive space such that the degree of adaptation of any population was a simple decreasing function of its distance from the optimal point. Hence, all the points closer to the optimal type than the current population are more fit than the current population. The points corresponding to better adapted populations relative to the current population are therefore found in a hypersphere whose center is at the optimum point and whose radius is the distance between the optimum and the current

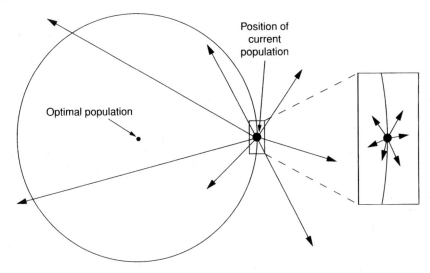

Figure 12.8. Fisher's adaptive target. The point in the center of the circle corresponds to a population that is adapted to the environment in the optimal fashion. The point on the circle corresponds to the level of adaptation of the current population, with increasing distance from the optimal point corresponding to decreased levels of adaptation. The circle encloses the area within which a population would be better adapted than the current population. Fixation for a mutation is indicated by an arrow starting at the current population and ending at a new point in space that is random in direction and magnitude from the current population. On the right is an expanded section of the circle in the vicinity of the current population to illustrate the increased chances of mutations with small phenotypic effect being selectively advantageous.

population. When dealing with two dimensions, this hypersphere becomes a simple circle or "target" (Figure 12.8). Fisher felt that natural selection would ensure that the population was near the optimal point. The only reason why populations would not be exactly at the optimal point is because the appropriate mutations had not yet occurred. Thus, the random mechanism responsible for exploring Fisher's adaptive space is mutation, with genetic drift playing no role whatsoever.

Fisher represented mutation as a vector coming from the current population that is random in both direction and magnitude. Fixation for this mutation would move the population to the point in the adaptive space indicated by the end of the mutational arrow. Fisher then calculated the probability of such a random vector having its end point land within the adaptive target, that is, closer to the optimum than the starting point. Fisher showed that this probability is zero whenever the magnitude of the vector exceeds the diameter of the hypersphere (see Figure 12.8). As the size of the fitness effect associated with a mutation declines, the probability of a mutation of random direction resulting in a favorable change increases and ultimately reaches a limit of $\frac{1}{2}$ for mutations of very small effect (Figure 12.8). Fisher therefore concluded that mutations of very small effect would provide the primary raw material for adaptive change since they are more likely to be advantageous. Fisher assigned each mutation a single vector to represent its fitness effects. The phenotypic effect is determined at the moment of mutation and remains constant throughout evolutionary history. Crow (1957) provided a rationale for Fisher's assumption by arguing that the genetic background is constantly changing in large, random-mating populations such that those mutations that have a consistent advantageous phenotypic effect regardless of genetic background will be the ones most likely utilized by natural selection to build adaptive traits. Therefore, in the Fisherian model, adaptation occurs by the accumulation of many small mutational steps toward an optimum, with the underlying genetic architecture of adaptive traits being due to a large number of loci with each locus having functional alleles with small, additive phenotypic effects.

Wright agreed with Fisher that adaptive traits generally have a multilocus genetic architecture. Unlike Fisher, Wright did not model the phenotypic effects as intrinsic, constant attributes of a specific mutation, but rather Wright regarded the phenotypic effects of a mutation as highly context dependent because of interallelic interactions, epistasis, and pleiotropy. Indeed, such interactions ensure that only rarely could we regard an allele as being intrinsically of major or minor effect (the critical distinction in Fisher's metaphor). As with the *ApoE/LDLR* example in Chapter 10, alleles of major or minor effect arise out of contextual interactions and can change dramatically as the genetic background is altered. In contrast to the fixed-length vectors in Fisher's metaphor (Figure 12.8), the magnitude of the phenotypic effects associated with a specific mutation or allele can be dramatically altered in an interaction system (Figures 10.15–10.17). Such interactions, particularly epistasis, make the shifting balance process far more probable and adaptively important (Bergman et al. 1995; Brodie 2000; Goodnight 1995, 2000). Consequently, one critical difference between Fisher and Wright is their opposing views of genetic architecture.

Until recently, we had little insight into the genetic architecture of most quantitative traits. Recall from Chapter 9 that the classical, unmeasured genotype analysis of quantitative genetic traits provides little information about genetic architecture and the role of epistasis and other interaction effects. Indeed, the Fisherian unmeasured genotype analysis is biased against the detection of epistasis. The Fisherian quantitative genetic measures were designed to first fit the "additive" component, then fit the "dominance" component, and finally, attribute what is left over to the "epistasis" component. This creates the artifact that epistasis seems weak compared to additive effects (Cheverud and Routman 1995; Goodnight 1995). For example, recall from Chapter 10 that the extensive epistasis between the *ApoE* and *LDLR* loci for the phenotype of serum cholesterol level is translated into "additive variance" at the *ApoE* locus by the Fisherian quantitative genetic measures (Figure 10.15A).

The dominance of Fisher's quantitative genetic paradigm that places little importance upon epistasis and other interaction effects has more to do with mathematical and statistical convenience than with biological reality. All that we know of biological systems—from the control of gene expression, to biochemical pathways, to developmental processes, to physiological regulation, and so on—indicates that interactions are the norm. Indeed, the idea that a single locus can have a marginal effect that is invariant to the remainder of its biological context is implausible in the extreme, yet Fisher's metaphor is based upon this unlikely premise. Measured genotype approaches are now allowing us to detect epistasis and other interactive effects (Chapter 10), and these interactions are found to be the norm in genetic architecture, not the exception (Templeton 2000). It is therefore not surprising that Frankel and Schork (1996) concluded that "where complex genetic traits loom, epistasis is not far behind."

Indeed, measured genotype approaches are finding extensive epistasis in the genetic architecture of "simple" Mendelian traits that had traditionally been regarded as examples of single-locus genetic architectures. By definition, a single-locus genetic trait is free of epistasis. But are such simple traits truly single-locus traits, and if not, is there any role for epistasis? Consider sickle cell anemia—the workhorse example of a simple Mendelian trait found in most genetic textbooks. Up to now, sickle cell anemia has been treated as a single-locus trait in this book. Indeed, sickle cell anemia is commonly presented as a single-nucleotide trait due to one A-to-T nucleotide change in the second position of the sixth codon of the β chain of hemoglobin (the S allele) that is said to "cause" sickle cell anemia when homozygous. However, when individuals who are homozygous for the S allele are examined, tremendous heterogeneity in clinical severity is revealed (Odenheimer et al. 1983; Sing et al. 1985). Epistasis is a major determinant of this heterogeneity. For example, the S allele is found in high frequency in certain Greek populations, but clinical manifestations are mild. Studies (e.g., Berry et al. 1992) have shown that persistence of fetal hemoglobin into the adult can ameliorate the severity of sickle cell anemia. Fetal hemoglobin is coded for by the tandemly duplicated γ loci (called A and G), which are closely linked to the β locus (Figure 2.6). Normally the γ genes are turned off after birth and the β gene is activated, leading to the transition from fetal to adult hemoglobin. Several mutations in or near the γ loci can cause persistence of fetal hemoglobin and are found in Greek populations (Berry et al. 1992; Patrinos et al. 1996).

Sickle cell is also common in certain populations in Saudi Arabia that live in historic malarial regions (el-Hazmi and Warsy 1996). An XmnI polymorphic site 5' to the $G\gamma$ locus (Figure 2.6) and a HindIII polymorphic site in the $G\gamma$ locus is associated with persistence of fetal hemoglobin and is in disequilibrium with the sickle cell allele in some Arabian populations. This same haplotype is also found in populations from India, who likewise have mild clinical symptoms with homozygous SS and have persistence of fetal hemoglobin (Ramana et al. 2000). Note that having persistence of fetal hemoglobin at the $G\gamma$ locus increases the fitness of homozygotes for the S allele at the β-Hb locus because that genetic combination results in sickle cell homozygous individuals that display mild clinical symptoms (el-Hazmi et al. 1992; Ramana et al. 2000). However, persistence of fetal hemoglobin is expected to decrease the ability of the blood to deliver oxygen to the peripheral tissues because fetal hemoglobin has a higher binding affinity for oxygen [this allows the developing fetus to take oxygen from the mother's blood across the placenta (Giblett 1969)]. The disequilibrium found in both Greek and Arab populations therefore favors the high-fitness combination at the expense of the low-fitness combination. This nonrandom association between alleles at different loci is therefore presumably due to natural selection operating upon a multilocus, epistatic genetic architecture. Hence, the sickle cell phenotype is not really a single-locus phenotype but rather in some human populations is a **coadapted gene complex**, in which the frequencies of alleles at different loci are mutually adjusted with respect to one another by natural selection favoring epistatic combinations with high fitness. In the particular case of the $G\gamma$ and β-Hb loci in Greek and Arab populations, the physical closeness of these two loci in the genome (see Figure 2.6) means that the disequilibrium built up by natural selection is not readily dissipated by recombination. The resulting genetic stability of this combination of closely linked alleles means that the combination itself approximates an "allele" in its inheritance pattern. Closely linked loci with coadapted combinations of alleles in linkage disequilibrium are called **supergenes** (Kelly 2000). Accordingly, the DNA region around the β-Hb locus (Figure 2.6) is really a supergene in the selective context of a malarial environment. The behavior of this region as a supergene also means that the adaptive topographies shown in Figures 11.8, 11.9, 12.1, and 12.2 are actually only three-dimensional projections of a higher dimensional adaptive landscape. For example, the frequency of the S allele as shown in these previous adaptive landscapes is actually the frequency of the superallele S that is not linked to the $G\gamma$ allele that causes persistence of fetal hemoglobin. To accommodate the Greek and Arab populations into the adaptive surface, the adaptive surface would now need a fourth dimension that measures the frequency of the superallele that has *S* and the *Gγ* allele that causes persistence of fetal hemoglobin on the same chromosome. In other words, at the *Gγ/β-Hb* supergene level there are four alleles: *A, C, S-no persistence of fetal hemoglobin*, and *S-persistence of fetal hemoglobin*. By ignoring the epistasis between the *Gγ* and *β-Hb* loci and focusing only on the *β-Hb* locus (the norm in most textbooks), we would mistakenly conclude that Greek, Arab, and most Bantu populations living in malarial environments had adapted to malaria by being on the same *A/S* polymorphic adaptive peak. However, when we consider the four-dimensional adaptive surface associated with the *Gγ/β-Hb* supergene (which unfortunately cannot be adequately depicted in a two-dimensional figure), we find that many of the Greek and Arab populations that are polymorphic for sickle cell anemia are actually in a different portion of the adaptive surface than the Bantu populations that are polymorphic for sickle cell anemia. Thus, just considering this one small DNA region alone, different human populations have adapted to malaria by evolving toward three different parts of the fitness surface (increasing the frequency of *C* as in Upper Volta, the *A/S-no persistence of fetal hemoglobin* polymorphism as in many Bantu populations, and the *A/S-persistence of fetal hemoglobin* polymorphism as in some Greek and Arab populations). Obviously, there is no single adaptive target for selection to hit, and different human populations have evolved different adaptive strategies in response to a malarial environment.

Even our four-dimensional fitness surface with its added adaptive option is an oversimplification. As pointed out in Chapter 11, several other loci are involved in malarial adaptation, and these and other loci show epistasis with one another and with sickle cell. For example, there is also evidence for an X-linked locus as a contributor to the persistence of fetal hemoglobin in Arabian populations (el-Hazmi et al. 1994a). The protein haptoglobin forms complexes with hemoglobin and is thereby the major determinant of hemoglobin excretion (Giblett 1969). Given this direct physiological interaction, it is not surprising that genetic variants of haptoglobin also display epistasis with the *S* allele (Giblett 1969). Epistatic interactions for clinical severity of anemia have also been reported between the *S* allele and α-thalassemia (caused by genetic variation at the α-*Hb* locus) and G6PD deficiency (Chapter 11) (el-Hazmi et al. 1994b). G6PD deficiency in turn displays epistasis with thalassemia (caused by mutations at the α- and β-globin loci) (Siniscalco et al. 1966).

G6PD deficiency achieves malarial resistance by reducing the ability of the malarial parasite to use the oxidative shunt pathway in the parasitized red blood cell (Giblett 1969)—a different molecular mechanism than that associated with the *S* allele. Accordingly, G6PD deficiency has a different set of clinical effects than the *S* allele, even though epistasis partially interweaves the two systems. Some bearers of G6PD deficiency are extremely sensitive to environmental oxidizing agents such as fava beans (Chapter 11). Sensitivity to fava beans (favism) can cause death through hemolytic crisis. However, not all carriers of G6PD deficiency are susceptible to favism, and some of this heterogeneity is due to epistasis between the X-linked *G6PD* locus and the two autosomal globin loci associated with thalassemia (Siniscalco et al. 1966). There is also epistasis for favism between G6PD deficiency and the locus for red cell acid phosphatase (*ACP-1*) (Bottini et al. 1971; Palmarino et al. 1975; Bottini et al. 1995), yet another locus that has a direct effect on malarial resistance (Bottini et al. 2001a). As a consequence, the joint allele frequencies at these interacting loci are a complex function of the presence of malaria and the ingestion of fava beans (Palmarino et al. 1975).

The *ACP-1* locus not only interacts with *G6PD* for favism and is involved with malarial adaptation but also influences fetal growth, birth weight in human males, and body mass (Amante et al. 1990; Greene et al. 2000). The effects of acid phosphatase on intrauterine growth and neonatal survival themselves are subject to epistatic modification with the adenosine deaminase locus (*ADA*) (Gloria-Bottini et al. 1989b) and with the maternal–fetal incompatibility reactions modulated by the *ABO* blood group locus (Lucarini et al. 1995). These epistatic interactions in turn interact with the diabetic status of the individual (Gloria-Bottini et al. 1989a; Bottini et al. 1991). Diabetes is a complex genetic system influenced by many loci with epistasis (Gloria-Bottini et al. 1989a; Bennett and Todd 1996) and also shows interactions with other systemic diseases, such as rheumatoid arthritis [another epistatic trait (Cornelis et al. 1998)] and coronary artery disease (yet another epistatic trait, as shown in Chapter 10). This linking of systems through epistasis could continue, but it should be clear by now that simple Mendelian systems are in reality merely low-resolution projections of complex systems involved in an expanding web of pleiotropic effects and epistatic interactions. This web of epistasis was revealed in the case of sickle cell anemia because it has such strong clinical relevance.

Frankel and Schork (1996) point out that when the tools to investigate epistasis are available and used, epistasis and pleiotropy are almost always found, even for simple Mendelian traits. The Fisherian view of many genes of intrinsic, small, additive effect appears to have little relevancy for any genetic system that is examined in detail. Consequently, when it comes to genetic architecture, Wright was right and Fisher was off target. Computer simulations by Bergman et al. (1995) indicate that the evolutionary advantage of the shifting balance process is an increasing function of the amount of epistasis. Hence, the extensive amount of epistasis that is being found in measured genotype approaches indicates that the conditions for shifting balance are broader than previously thought. Genetic architecture therefore does not appear to be a factor that would limit the operation of Wright's shifting balance process between selection, genetic drift, and gene flow.

INTERACTIONS OF NATURAL SELECTION, GENETIC DRIFT, AND MUTATION

The balance between genetic drift and mutation was shown in Chapter 5 to influence the rates of mutational substitution (rate of fixation due to drift times the rate of input of new mutations, or $1/2N \times 2N\mu = \mu$) and the expected level of polymorphism (which is a function of twice the mutation rate divided by the force of drift, or $2\mu \div 1/(2N) = 4N\mu = \theta$). In Chapter 7, the balance of drift and mutation as measured by θ was shown to influence many properties of the coalescent process under the assumption of neutrality, such as the expected number of mutations that occur between coalescent events and over the entire haplotype tree, which are proportional to θ. Natural selection interacts with drift and mutation to alter all of these balances and expectations. These alterations provide some simple methods of detecting natural selection.

After Kimura proposed the neutral theory (Chapter 5), there were extensive efforts made to test it. However, the basic parameters of the neutral theory—μ, the mutation rate to *neutral* alleles, and various types of effective population size—are difficult to estimate. As a result, many of the resulting data sets were interpreted as supporting neutrality by Kimura and other "neutralists" and interpreted as supporting selection by "selectionists." A test statistic was needed that did not depend upon estimating any of these parameters.

Maynard Smith (1970) proposed testing the neutral hypothesis through its fundamental prediction that the neutral mutation rate both determines the rate of interspecific divergence and influences the amount of intraspecific polymorphism (Kimura 1968a,b). As we saw in Chapter 5, the rate of interspecific divergence under neutrality is proportional to μ, and the expected heterozygosity is $\theta/(1 + \theta) \approx \theta$ for small θ. Hence, the ratio of intraspecific polymorphism to interspecific divergence is proportional to $\theta\mu = 4N$, where N is the inbreeding effective size. Hence, by looking at the relative proportion of intraspecific polymorphism to interspecific divergence, we end up with a statistic that does not depend upon the inestimable neutral mutation rate. Moreover, if all types of mutations and all loci are evolving under the neutral model, this ratio should be $4N$ for *all* mutations and loci. Therefore, one could in theory test neutrality by testing for homogeneity across mutational types and/or loci in the amount of intraspecific polymorphism relative to interspecific divergence, thereby eliminating any need to estimate N. Unfortunately, in 1970, when Maynard Smith published this paper, such a simple homogeneity test was not possible because the primary data on intraspecific polymorphism came from protein electrophoresis and the primary data on interspecific divergence came from amino acid sequencing. These two techniques are not comparable in how they measure genetic diversity, making the ratio of intraspecific polymorphism to interspecific divergence difficult to estimate. This situation changed with the advent of haplotype trees, which as pointed out in Chapter 5 can include both intra- and interspecific branches in a common evolutionary tree.

Like so many other analyses involving haplotype trees, the testing of neutrality through a simple contingency test of homogeneity was first implemented using a combined inter/intraspecific evolutionary tree of inversion variation in *Drosophila* (Templeton 1987b). At the time, there were few other data sets that combined intra- and interspecific data into a single evolutionary tree. However, DNA sequencing of homologous genes within and between species became increasingly common, and now there are many intra/interspecific data sets available for which the contingency test approach is applicable. McDonald and Kreitman (1991) performed the first contingency test of neutrality with DNA sequence data. Templeton (1987b) had suggested that the mutations found on the intra/interspecific tree be divided into two topological categories that reflect their position in the tree: polymorphic (those mutations located on branches contained within a species) and fixed (those mutations located on branches that interconnect species) (see Figure 12.9 for an example). The rationale for this choice was to test the basic prediction of the neutral theory about the relationship between interspecific divergence and intraspecific polymorphism. If selection were present, this choice of tree positions should be sensitive to selection. The fixed differences have obviously gone to fixation within a species and have persisted through time; therefore, they are of proven evolutionary success. The polymorphic class is not yet proven, containing some mutations that will eventually go to fixation and others that will be lost. Hence, if selection biases the probabilities of loss and fixation, as we have seen that it does (Tables 12.2 and 12.3), these two topological categories should be differentially affected. McDonald and Kreitman (1991) used the same tree positional categories of fixed and polymorphic that had been defined by Templeton (1987b). One can also just look at the number of fixed and polymorphic nucleotide sites without going through the intermediate step of estimating a haplotype tree, and this is commonly done. However, using only sites and not a haplotype tree makes the analysis much more sensitive to the infinite-sites model, which as pointed out in Chapter 5 is frequently violated even intraspecifically but is violated even more commonly for interspecific comparisons. By constructing a haplotype tree, one can obtain an estimate of homoplasy (multiple mutations to the same state at the same site and therefore violations of the infinite-sites model), and therefore mutational counts based on tree branches are more robust to deviations from the infinite-sites model (Chapter 5) than are counts based on fixed and polymorphic sites. Moreover, with some procedures such as

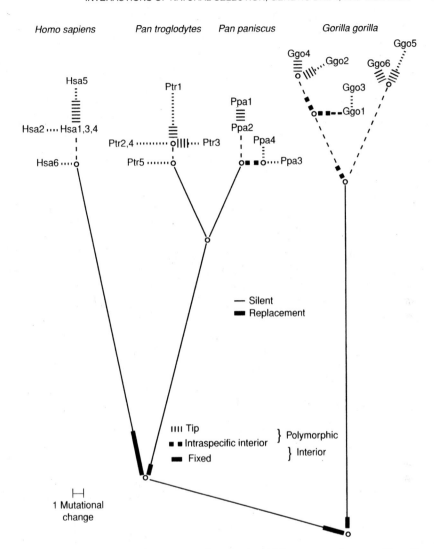

Figure 12.9. Intra/interspecific haplotype tree for mitochondrial gene *cytochrome oxidase II* gene in humans (*Homo sapiens*), two species of chimpanzees (*Pan troglydytes* and *Pan paniscus*), and the gorilla (*Gorilla gorilla*). Thick lines indicate replacement mutations and thin silent mutations. Solid lines indicate fixed differences between species, dashed lines the intraspecific interior branches, and dotted lines the tip branches. The polymorphic branches refer to the combined class of tip and intraspecific interior, and the interior branches refer to the combined class of fixed and intraspecific interior. Small circles indicate interior nodes in the tree that are not represented by any current haplotype. Haplotypes are named with the abbreviation of the species in which they were found and a number indicating the sample identity. Three samples in humans (1, 3, and 4) all had the same haplotype. Modified from Templeton (1996b).

statistical parsimony, it is possible to further correct for the undercounting of mutations that are caused by deviations from the infinite-sites model (Templeton 1996b).

McDonald and Kreitman (1991) chose mutational categories that should also be sensitive to the presence of selection. They were analyzing protein-coding regions, so they divided their mutations into two categories: synonymous mutations that do not alter the amino acid

Table 12.4. Contingency Analysis of Silent/Replacement Mutations versus Polymorphic/Fixed Tree Topological Categories for Mitochondrial *COII* Gene

	Silent	Replacement
Polymorphic	42	14
Fixed	113	8

Source: From Templeton (1996b).
Note: The probability under the null hypothesis of homogeneity is determined by Fisher's exact test (FET). FET probability under the null hypothesis of homogeneity: 0.001.

sequence (due to the redundancy of the genetic code) and replacement (or nonsynonymous) mutations that do alter the amino acid sequence of the resulting protein. If the DNA region being examined contains non-protein-coding regions, the category silent (synonymous plus noncoding mutations) can be used to contrast with replacement mutations. Kimura (1968a) and other neutralists had strongly argued that silent mutations were far more likely to be neutral than replacement ones, so this categorization should also be sensitive to selection operating at the level of the amino acid sequence of the protein.

The tree topological categories and the mutational categories together define a two-dimensional contingency table. The hypothesis of neutrality implies homogeneity within this contingency table, so now the test of neutrality is a straightforward contingency test of homogeneity (Templeton 1987b), for which there are many well-known statistical options. As an example, Figure 12.9 shows the estimated combined intra- and interspecific haplotype tree for mitochondrial *cytochrome oxidase II* (*COII*) DNA sequences from several hominoid primates, including humans, as well as the distribution of silent and replacement substitutions upon all branches (Templeton 1996b). As can be seen, the replacement substitutions are primarily found in the intraspecific portions of the haplotype tree. When the mutational numbers are counted and placed in a contingency table (Table 12.4), this visual impression is confirmed by the Fisher's exact test (Appendix 2), which reveals a highly significant deviation from homogeneity/neutrality. In particular, you can see from either Figure 12.9 or Table 12.4 that the replacement mutations are disproportionately found in the polymorphic part of the haplotype tree, that is, the portion of the tree of unproven evolutionary success. Under the assumption that the silent substitutions are more likely to be neutral, this implies that selection at this gene is conservative and preferentially eliminates amino acid changes.

This simple contingency test can be extended to yield even more insight into the nature of selection. For example, the motivation for subdividing the mutations into replacement and silent categories was the strong a priori belief that these two classes of mutations would differ considerably with respect to natural selection and therefore would be sensitive indicators to selection when present. In many cases, we have additional a priori information to make even finer distinctions among mutations. In the case of the COII protein, we know that the molecule is split into two halves with drastically different biochemical functions. On the N-terminal side of the central aromatic domain, the polypeptide is hydrophobic and is found in association with the transmembrane portion of the cytochrome oxidase complex. The C-terminal side of the molecule is hydrophilic and protrudes into the cytosol. It contains the Cu_A site, crucial for the transfer of electrons to O_2 and for the cytochrome *c* binding site. Because of the extreme difference in the biochemical role played by these two regions, mutations are categorized as N-terminal mutations or C-terminal mutations, yielding a total

Table 12.5. Contingency Analysis of *COII* Data with Four Mutational Categories and Three Tree Topological Categories

	N-Terminal Silent	N-Terminal Replacement	C-Terminal Silent	C-Terminal Replacement
Tip	8	2	12	7
Interior	10	3	12	2
Fixed	60	6	53	2

Source: From Templeton (1996b).
Note: Because some of the cells have few observations, an exact permutation test was used. Permutational probability under the null hypothesis of homogeneity: 0.000.

of four mutational categories: N-terminal silent, N-terminal replacement, C-terminal silent, and C-terminal replacement.

We can also refine our categorization of the topological positions in the haplotype tree. For example, in Chapter 7 we noted how we could distinguish between tip branches (those that lead to a single haplotype) and interiors branches (those that interconnect internal nodes). All fixed branches are interiors, but so are some of the polymorphic branches. Hence, we can subdivide the tree positions into three types: fixed, intraspecific interior, and tip. Recall that the motivation for subdividing the tree into the fixed and polymorphic classes was the a priori belief that these categories should be differentially affected by selection when selection was present. Splitting the intraspecific polymorphic class into tip and intraspecific interior also makes sense in potential for being sensitive to selection. Interior haplotypes tend to be older than tip haplotypes (Chapter 7) and have given rise to one or more descent haplotypes. Hence, mutations on interior branches also have more evolutionary success on the average than mutations on tip branches. Table 12.5 shows a new contingency analysis on *COII* that makes use of these more refined mutational and tree topological positions. The evidence for selection still remains strong, but now we can see that the excess of replacement substitutions is primarily found on the tip branches in the C-terminal portion of the molecule. Hence, the selection-conserving amino acid sequence is primarily directed to that portion of the molecule involved in electron transport and cytochrome *c* binding.

Both the fixed and intraspecific interior categories have demonstrated some degree of evolutionary success in terms of temporal persistence and leaving descendants whereas the tip branches have not. Therefore, pooling tips with intraspecific interiors to form the polymorphic class will tend to dilute the signal for selection. If any classes are to be pooled, a stronger signal for selection will be achieved by pooling fixed and intraspecific interior into the interior class and contrasting it with tips. Such pooling is particularly important when trying to detect weak selection. For example, although silent substitutions are regarded as more likely to be neutral in general compared to replacement substitutions, silent mutations can have some functional consequences and hence could be under selection to some extent. For example, if the pools of the tRNAs that share a common amino acid differ in concentration within a cell, a silent substitution could affect the rate of translation and thereby potentially have a selectable phenotypic effect. Such selection can result in **codon bias**, the preferential use of some codons over others among a synonymous set. For example, Llopart and Aguade (1999) found such codon bias in the *RpII215* gene in several species of *Drosophila*. Using these data, they could divide the synonymous mutations into two categories: transitions from unpreferred to preferred codons and transitions from

Table 12.6. Contingency Analyses of Preferred (u → p) and Unpreferred (p → u) Synonymous Substitutions in *Drosophila Rpll215* Gene

Substitution	Fixed	Polymorphic	Interior	Tip
u → p	8	11	16	3
p → u	5	27	14	18
	FET probabilty: 0.0497		FET probabilty: 0.0074	

Source: From Llopart and Aguade (2000).

preferred to unpreferred. To see if natural selection was influencing this codon bias, they then performed contingency tests using first the fixed-versus-polymorphic categories and second the tip-versus-interior (both inter- and intraspecific in this case) categories (Llopart and Aguade 2000). Their results are summarized in Table 12.6. As can be seen, the evidence for selection is only marginally significant when the fixed-versus-polymorphic categories are used, but it became very strong when contrasting the interior-versus-tip categories.

Another advantage of using tips versus interiors is that it allows the contingency test approach to be used on data sets that contain only intraspecific observations, which is a common occurrence in population genetics. For example, Markham et al. (1998) sequenced the HIV-1 glycoprotein 120 gene in HIV clones extracted from 15 subjects positive for the AIDS causing virus HIV-1. This gene influences the mode of entry of HIV virions into human cells and is a target of the immune system, so there are many potential selective forces that could operate (Templeton et al. 2004). The polymorphic and fixed classes cannot be applied to these data, but tips versus interior can, as shown in Table 12.7. A highly significant contingency chi square is obtained in this case, resulting in a strong rejection of neutrality. Note that in Table 12.7 there are proportionately more replacement mutations in the interior class compared to the tips. This implies that natural selection is favoring amino acid changes in this protein that is both a target for the host's immune system and influences the host cell types that the HIV-1 virus can infect. This shows that the contingency test approach can not only detect selection but also discriminate between different types of selection: selection for conserving amino acid sequence in Table 12.4 and selection favoring amino acid change in Table 12.7.

All of the above contingency tests are based upon the prediction of a homogeneous balance of mutation and drift over mutational categories if all such categories are neutral. As mentioned at the beginning of this section, such a homogeneous balance is also expected across loci if the variation at all loci is neutral. Therefore, Hudson et al. (1987) suggested that neutrality could be tested by testing for homogeneity across two or more loci in the balance

Table 12.7. Contingency Analyses of Silent versus Replacement Mutations in Tip versus Interior Branches for Haplotype Trees of HIV-1 Glycoprotein 120 Sequences in 15 HIV-Positive Subjects

	Silent	Replacement
Tip	273	640
Interior	110	421

Note: Based on data from Markham et al. (1998). The abbreviation df stands for degrees of freedom, and p is the probability of the null hypothesis being true given the data. Contingency chi square = 14.06, df = 1, $p = 0.0002$.

of fixed interspecific differences to intraspecific polymorphisms. If one rejects homogeneity across loci with this HKA test (named after the authors, Hudson, Kreitman, and Aguade), then one or more loci in the sample are not neutral or closely linked to a selected locus. However, the HKA test does not identify which loci are not evolving under the neutral model. In contrast, the contingency tests discussed above typically choose mutational categories such that when homogeneity is rejected, some insight is obtained about the mutational types being selected and the nature of the selection (e.g., the contrast of Tables 12.4 and 12.7).

As shown in Chapters 5 and 7, the balance of mutation and drift under neutrality is frequently measured by the parameter $\theta = 4N\mu$. Another category of tests of neutrality is based upon different methods of estimating θ. For example, equation 5.7 shows that the expected heterozygosity (actually, the expected non–identity by descent) for neutral alleles is $\theta/(1+\theta) \approx \theta$ for θ small. When measuring heterozygosity at the nucleotide level from sequence data, a molecular genetic distance is substituted for the concept of non–identity by descent (Chapter 6). In particular, let π_{ij} be the number of nucleotide differences between sequences i and j grom a sample of n genes. There are $n/(n-1)/2$ unordered pairs (ij is pooled with ji) in a sample of n sequenced genes, so an estimate of θ at the nucleotide level based upon the average number of nucleotide differences between all pairs of n sequences is

$$\Pi = \frac{\sum_{i=1}^{j} \sum_{j=2}^{n} 2\pi_{ij}}{n(n-1)} \qquad (12.17)$$

where Π is the estimator of θ based on average nucleotide heterozygosity. Under neutrality, θ is also related to S, the expected number of mutational events over the time from the present to coalescence back to a single ancestral molecule (see equation 7.4). Under the infinite-sites model in which each mutation occurs at a different nucleotide site, S also has the biological interpretation as being the number of nucleotide sites showing genetic variation in the sample (called the number of segregating sites). Consequently, we can estimate θ under the infinite-sites model by solving equation 7.4 for θ as

$$\Theta = \frac{S}{\sum_{k=1}^{n-1} [1/(n-k)]} = \frac{S}{\sum_{i=1}^{n-1} (1/i)} \qquad (12.18)$$

where S is the observed number of segregating sites in the sample of n sequences, $i = n - k$, and Θ is the estimator of θ based upon the number of segregating sites. (Note, sometimes θ is estimated on a per-nucleotide basis, in which case both equations 12.17 and 12.18 are divided by L, the length of the sequenced region in nucleotides.)

Under neutrality, both Π and Θ should be estimating the same parameter. Therefore, Tajima (1989a) suggested a test for neutrality based on the standardized difference between Π and Θ:

$$D = \frac{\Pi - \Theta}{\sqrt{\text{Var}(\Pi - \Theta)}} \qquad (12.19)$$

where $\text{Var}(\Pi - \Theta)$ is an estimate of the sampling variance of the difference beween Π and Θ that Tajima obtained under the infinite-sites model. Tajima showed that equation 12.19 does not have a zero expectation if natural selection occurs. For example, suppose a mutation arose that was favored by natural selection and rapidly went to fixation. Fixation of the selected site would also increase the frequencies of variants at other polymorphic sites that were in linkage disequilibrium with the selected site, a phenomenon known as hitchhiking. Indeed, in the extreme case of no recombination (such as with mtDNA), fixation of the selected site causes fixation of all other sites as well, resulting in what is called a selective sweep. Such selective sweeps and hitchhiking effects due to positive, directional selection favoring a new mutant cause a reduction of variation in the selected DNA region. After the selective sweep, each new mutation creates a new segregating site under the infinite-sites model, so Θ recovers rapidly from the sweep. However, these new mutations are initially rare and contribute little to average heterozygosity, which is most sensitive to haplotypes with intermediate gamete frequencies. As a result, Π will be smaller than Θ after a selective sweep, converging only slowly at a rate determined by the mutation rate. Hence, if a selective sweep has occurred in the recent past in the DNA region sequenced, D should be negative ($\Pi < \Theta$). In contrast, if natural selection favors the maintenance of polymorphisms, then there will be haplotypes at intermediate frequencies beyond neutral expectations. Haplotypes with intermediate frequencies contribute much to average heterozygosity but not to the number of segregating sites, so D should be positive ($\Pi > \Theta$). Thus, not only do significant differences between Π and Θ indicate selection but also the sign of the difference indicates the nature of selection.

There are difficulties with Tajima's D statistic as a test for neutrality. First, D is sensitive not only to departures from neutrality but also to departures from demographic stability, in contrast to the contingency and HKA tests (Hudson 1993). For example, Tajima (1989b) showed that a population bottleneck or founder event also reduces variation by causing fixation at many sites, and once again as variation is re-created by mutation, the number of segregating sites increases more rapidly than average heterozygosity. Therefore, population bottlenecks and founder events will also cause $D < 0$, thereby mimicking the effects of a selective sweep. When the null hypothesis is rejected with the D test, it means either that neutrality is rejected and/or that demographic stability is rejected. Since few real populations display long-term demographic stability, this is a serious confoundment.

Fay and Wu (2000) have suggested a way of overcoming this confoundment of selection with demography by devising yet a third estimator of θ, called θ_H. Under neutrality and the infinite-sites model, another unbiased estimator of θ is given by

$$\theta_H = \sum_{i=1}^{n-1} \frac{2S_i i^2}{n(n-1)} \qquad (12.20)$$

where S_i is the number of derived variants found i times in the sample of n sequences. By a derived variant, Fay and Wu are referring to the nucleotide state at the polymorphic sites that represents a mutation after the most recent common ancestral molecule. Outgroup data are used to determine whether a state is derived or ancestral, so equation 12.20 requires and makes use of more information than is contained in either equations 12.18 or 12.19. Note also that those variants that are in high frequency contribute heavily to equation 12.20 (the i^2 term), whereas haplotypes of intermediate frequency contribute heavily to equation 12.18. Therefore θ_H gives added weight to derived variants that are in high frequency, but ancestral states of high frequency are given little weight. Under neutral coalescence, ancestral states tend to be more common on the average (Castelloe and Templeton 1994). Fay and Wu (2000) showed that population growth after a bottleneck does not tend to make derived variants common, so that an excess of derived haplotypes at high frequency is a unique pattern associated with recurrent directional selection. Fay and Wu measure this potential excess by the statistic $H = \Pi - \theta_H$. They found that only a few high-frequency-derived variants are needed to detect directional selection with this statistic since not many are expected under neutrality.

The H statistic of Fay and Wu shows that information about the relative age of a variant when coupled with frequency data can be an indicator of positive selection. Others have also proposed tests of neutrality based upon indicators of haplotype age (Nielsen 2001a; Slatkin 2000). These methods are particularly strong when trying to infer selection upon a class of alleles defined by a common functional feature. For example, recall from Chapter 11 that over 300 independent mutations have been identified that lead to a common phenotype of a deficiency of the enzyme G6PD, coded for by an X-linked locus. Tishkoff et al. (2001) estimated the evolutionary history of the *G6PD* DNA region with three highly variable microsatellite repeats (Appendix 1) within 19 kb of the *G6PD* locus in sub-Saharan African populations and populations around the Mediterranean. These populations contain two of the G6PD-deficient alleles, A^- (found in sub-Saharan Africans, as mentioned in Chapter 11) and *Med*, found in the Mediterranean populations. They found that the A^- and *Med* alleles had low levels of microsatellite variation compared to the nondeficient alleles of *G6PD*, indicating that they were of recent origin. An estimated haplotype network also indicated that the deficient alleles were of recent origin. The fact that these two independent and relatively new G6PD-deficient alleles went to high frequency in their respective populations is a strong indicator of selection favoring the phenotypes associated with the deficient alleles in these regions in the recent past.

Although the Fay and Wu H statistic solves the confoundment of demography and selection inherent in the Tajima D statisic, both the H and D statistics share another source of potential error in interpreting the results of these test statisics: the infinite-sites model. As pointed out in Chapter 5, serious deviations from the infinite-sites model can occur (e.g., see Figure 5.19 or 5.21), and moreover, different regions of the genome can deviate from the infinite-sites model, but in extremely different ways from one another (recall the contrast of *LPL* and *ApoE* in Chapter 5). Palsbøll et al. (2004) showed that estimates of θ are extremely sensitive to the underlying model of mutation. They estimated θ from fin whale mtDNA sequence data under the standard infinite-sites model using a Bayesian procedure (Appendix 2) that gives the entire posterior distribution of the estimate of θ, as shown in Figure 12.10. They then used the program ModelTest (Posada and Crandall 1998) to search over 52 possible mutation models. The results indicated that a different mutation model should be used with their data that allows the possibility of multiple mutations at the same site. Figure 12.10 also shows the posterior distribution of the estimate of θ under this alternative mutation model. As can be seen, the two distributions are quite distinct, showing little overlap. It must be emphasized that both θ distributions were estimated from the *same* data; the differences are due *entirely* to the assumed mutation model. Consequently, a significant deviation from the infinite-sites estimator of θ could be due to selection, demography (for Tajima's D), or the underlying mutation model. The H statistic is likewise sensitive to deviations from the infinite-sites model (Baudry and Depaulis 2003). Tajima's D statistic and similar statistics should *not* be applied to sequence data when programs like ModelTest or the existence of homoplasy in the haplotype tree indicate deviations from the infinite-sites model.

All of the studies discussed in this section reveal that natural selection alters the balance of mutation and genetic drift. Molecular genetics provides much information about how selection interacts with mutation and drift, and these interactions can be detected through a variety of statistical tests, some of which are surprisingly simple. These tests also frequently

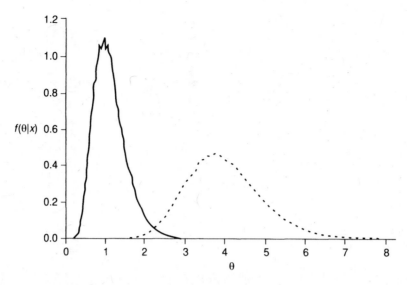

Figure 12.10. Posterior distributions of estimator of θ under infinite-sites model (solid line) and mutation model fitted to data that allows multiple mutations at same site (dotted line). The data (indicated by *x* in the posterior distribution) are mtDNA sequences from the fin whale *Balaenoptera physalus*. Modified from Fig. 1a in Palsbøll et al. (2004). Copyright © 2004 by The Society for the Study of Evolution.

identify specific mutational categories, protein regions, or allelic classes that have been subject to selection and the nature of that selection (purifying or conservative, directional, and balancing). With the availability of whole genome data, it is now possible to scan entire genomes to identify those genes subject to selection during the evolution of a species, including humans (Akey et al. 2004; Bamshad and Wooding 2003; Clark et al. 2003; Fay et al. 2001; Storz et al. 2004; Sunyaev et al. 2003; Vallender and Lahn 2004). Such scans are finally addressing the issues debated between the classical and balanced schools and later between the neutralists and the selectionists (Chapter 5). For example, Smith and Eyre-Walker (2002) concluded that 45% of all amino acid substitutions have been fixed by natural selection in *Drosophila*, and Fay et al. (2001) estimated that only 20% of amino acid substitutions in humans are neutral.

These tests for selection at the molecular level also provide a powerful tool for identifying previously undetected genes within the genome (Nekrutenko et al. 2001), identifying those regions in the genome that are functionally important (Nielsen 2001b), and predicting the functional consequences of mutations at specific sites within a protein (Chasman and Adams 2001; Knudsen and Miyamoto 2001; Miller and Kumar 2001; Sunyaev et al. 2001). For example, Miller and Kumar (2001) show that mutations associated with genetic disease in humans are primarily due to amino acid replacement mutations at sites shown to be under purifying or conservative selection. Thus, evolutionary analyses of the interaction of selection, mutation, and drift provide a useful tool for understanding and predicting patterns of human disease mutations.

13

UNITS AND TARGETS OF SELECTION

Most genetic models used in this book have been single-locus models. This emphasis upon single-locus models is typical of much of population genetics. One of the main reasons for the dominance of single-locus models is mathematical tractability. Such a mathematical rationale does not necessarily justify biologically the dominance of single-locus models. The biological adequacy of a single-locus model in describing evolution is a troubling issue because qualitatively new biological features can emerge as soon as we go beyond the single-locus models. For example, we saw in Chapter 2 that the important conclusion of "no evolution" under the assumptions of the single-locus Hardy–Weinberg model is not necessarily true for a two-locus model.

This reliance upon single-locus models greatly bothered Ernst Mayr (1959, 1970), one of the major architects of the neo-Darwinian theory of evolution. He called such single-locus population genetic models "beanbag genetics" in which each "bean" (locus) is studied independently and then added together in a beanbag to reconstruct the whole. Mayr pointed out that the individual is the "target of selection" and not a single locus. It is the individual that lives or dies, mates or fails to mate, is fertile or sterile. Any single locus may contribute to the individual's fitness, but because of epistasis and coadapted complexes, that single-locus contribution must be placed into the context of the genotype as a whole. Mayr was strongly influenced by the earlier work of Chetverikov (1926), who suggested that the individual is not divisible into discrete traits coded for by individual genes, but instead the individual is a product of the genotype as a whole. Mayr called this idea the "unity of the genotype."

Mayr (1970) pointed out an important complication to the concept of the unity of the genotype as applied to evolution, namely, that the individual's unified genotype is broken apart in meiosis and fertilization. As a result, although there is the *unity* of the genotype at the individual level, there is *no continuity* of an individual's genotype across generations. Indeed, we now know that in many species each individual genotype is a unique event, never to be replicated in the history of the species (Chapter 1). Evolutionary predictions

---

Population Genetics and Microevolutionary Theory, By Alan R. Templeton
Copyright © 2006 John Wiley & Sons, Inc.

can only be made when there is genetic continuity over space and time (premise 1 in Chapter 1). An allele at a single locus displays such continuity (Chapter 1) even though the genetic background upon which that allele can be placed is unique for each individual and is constantly changing across generations. Therefore, Mayr realized that despite the individual being the target of selection, the individual is not the meaningful unit for measuring the evolutionary response to natural selection. Instead, Mayr argued that fitness must be used in a "statistical" sense at the level of a reproducing population in the context of its gene pool.

This same idea was expressed somewhat differently in Chapter 11. In that chapter, the fundamental equation of natural selection for a measured genotype and the fundamental theorem of natural selection for unmeasured genotypes both reveal that the only fitness effects that influence the response to natural selection are those transmissible through a gamete. We showed repeatedly in Chapters 11 and 12 that understanding the evolutionary response to natural selection requires taking the gamete's perspective, not the individual's. This is also why we measured fitness in Mayr's statistical sense as a genotypic value, the mean phenotype of a genetically defined group of individuals that automatically averages over all genetic backgrounds that are not used to define the group.

The gametic perspective for the response to natural selection potentially undermines Mayr's use of the unity of the genotype as an argument against beanbag genetics. Just as meiosis and fertilization thoroughly break down and rearrange the individual's total genotype, meiosis and fertilization also ensure that the meaningful genetic unit for predicting the response to selection is something far less than an individual's genotype at all loci in the genome. Could this genetic unit be as small as a single locus? The pervasive occurrence of epistasis noted by Mayr and pointed out in Chapter 10 does not necessarily imply that a single locus is not a meaningful genetic unit for the response to natural selection. Recall from Chapter 10 the example of epistasis between the *ApoE* and *LDLR* loci for the phenotype of total serum cholesterol. Despite strong epistasis in this case, it made little difference whether or not this was treated as a two-locus system or a one-locus system in the context of a particular gene pool (Figure 10.15). As pointed out in Chapter 10, epistasis contributes to the additive (i.e., beanbag) variance; indeed in many possible gene pools virtually all of the epistasis appears as an additive effect associated with just one locus (Figure 10.15). In the case given in Figure 10.15a, selection on the phenotype of total serum cholesterol would induce a response at the *ApoE* locus and virtually none at *LDLR*. In that case, the selective response could be predicted quite well just using the *ApoE* locus alone in the context of the gene pool with allele frequencies close to those observed in the study populations of Pederson and Berg (1989) (Figure 10.15a). Mayr correctly argued that the genetic response to selection must be placed in the context of the gene pool, but once placed in such a gene pool context, it is possible that virtually all of the epistasis found in a multigenic complex may be statistically allocated to the average excesses or effects of single loci. Thus, the unity of the genotype does not necessarily undermine the single-locus approach of beanbag genetics.

Mayr's discussion of these issues brings into focus two separate but often confused issues in population genetics. The first is the **unit of selection**, the level of genetic organization that allows the prediction of the genetic response to selection. As shown in Chapter 11, fitnesses in population genetics are assigned to a genotypic class of individuals rather than individuals themselves (the statistical fitness of Mayr). What was not addressed in Chapter 11 was the level of genetic organization that defines these genotypic classes; the genotypic classes can be single-locus genotypes, two-locus genotypes, and so on. The unit of selection is the level of genetic organization to which a fitness phenotype can be assigned that allows the response to selection to be accurately predicted. This means that the unit of selection must have genetic continuity across the generations. For example, if the unit of selection is a multilocus unit, then the combinations of alleles across the loci in this complex must recur in evolutionary time and not be unique events. The requirement for continuity over the generations limits units of selection to a number of loci that must be orders of magnitude less than the total number of loci in the whole genome, at least in outbreeding populations.

The second issue raised by Mayr is the **target of selection**, the level of biological organization that displays the phenotype under selection. Mayr only discussed one target of selection, the individual. Individuals were the target of selection in the models given in Chapters 11 and 12. However, we shall see in this chapter that fitness phenotypes can be assigned to biological levels both below and above the level of the individual.

Units and targets of selection should never be confused. The unit of selection is always some level of *genetic* organization that recurs over time and space. A target of selection is some level of *biological* organization that displays a *phenotype* that influences the probability of the unit of selection's recurrence over time and space. Sometimes the unit and target of selection can be the same. For example, a transposable element is a level of genetic organization, but it also displays a phenotype (transposition) that influences its chances for recurrence and is therefore a target of selection. In general, however, units and targets of selection are different entities. In this chapter we will discuss both units and targets of selection, starting with the problem of the unit of selection.

UNIT OF SELECTION

In Chapter 11, we derived the fundamental equation for natural selection for a measured genotype (equation 11.5) using a single-locus genetic architecture. This equation can be generalized to genotypes defined by two or more loci. We consider now a two-locus, two-allele model in a random-mating population of the sort used in Chapter 2. As shown in Chapter 10 (e.g., Table 10.4), genotypic values can be assigned to genotypes defined by two loci. In our current model we will let $w_{im/jn}$ be the genotypic value of the fitness phenotype of individuals (our target of selection) sharing the two-locus genotype im/jn, where im denotes the gamete type received from one parent and jn the gamete type from the other parent. Hence, ij denotes the genotype at the first locus (with alleles A or a) and mn the genotype at the second locus (with alleles B and b). We also assume that cis and trans double heterozygotes have the same fitness; that is, $w_{AB/ab} = w_{Ab/aB}$. Using these two-locus genotypic fitnesses as weights in the same manner done in deriving equation 11.4 but with a two-locus genotype model like that given in Figure 2.4 from Chapter 2, the change in gamete frequencies can be derived (a useful exercise) as follows:

$$\Delta g_{AB} = \frac{g_{AB}}{\bar{w}} a_{AB} - rD \frac{w_{AB/ab}}{\bar{w}}$$

$$\Delta g_{Ab} = \frac{g_{Ab}}{\bar{w}} a_{Ab} + rD \frac{w_{AB/ab}}{\bar{w}} \qquad (13.1)$$

$$\Delta g_{aB} = \frac{g_{aB}}{\bar{w}} a_{aB} + rD \frac{w_{AB/ab}}{\bar{w}}$$

$$\Delta g_{ab} = \frac{g_{ab}}{\bar{w}} a_{ab} - rD \frac{w_{AB/ab}}{\bar{w}}$$

where g_{im} is the frequency of gamete type im in the gene pool, r is the recombination frequency between the two loci, D is the linkage disequilibrium between the loci in the gene pool, and a_{im} is the average excess for fitness of the two-locus gamete im. For example, the two-locus average excess of the gamete bearing the AB allelic combination in a random-mating population is

$$a_{AB} = g_{AB}(w_{AB/AB} - \bar{w}) + g_{Ab}(w_{AB/Ab} - \bar{w}) + g_{aB}(w_{AB/aB} - \bar{w}) + g_{ab}(w_{AB/ab} - \bar{w})$$

(13.2)

Note that the first terms in equations 13.1 are similar to equation 11.5. This first term states that the response to natural selection for this two-locus system is driven by the gametic perspective, in this case, gametes defined by two loci. However, unlike equation 11.5, equations 13.1 contain a second term that reflects evolution not driven by the gamete's perspective as measured by average excess. This reflects the fact, pointed out in Chapter 2, that recombination and linkage disequilibrium are also forces of evolutionary change in a two-locus system. Hence, equations 13.1 are somewhat akin to equation 12.2, where the "other" forces influencing evolution are recombination and linkage disequilibrium. However, even this second term is influenced by natural selection because it is a product of the neutral evolutionary impact of recombination and disequilibrium revealed in Chapter 2 (rD) times a weight determined by genotypic fitnesses ($w_{AB/ab}/\bar{w}$). Note that this weighting contains the genotypic value $w_{AB/ab}$. What is a *genotypic* value doing in an equation about changes in *gamete* frequencies? The answer lies in the special role that the double-heterozygote class plays in evolutionary change driven by recombination and linkage disequilibrium. As pointed out in Chapter 2, recombination is a force for change in two-locus gamete frequencies *only* in double-heterozygotes. Therefore, *any factor that influences the frequency of double heterozygotes plays a direct role in modulating the evolutionary impact of recombination*. Among those factors is the fitness of the double-heterozygote class relative to the average fitness of the population as a whole. If the double heterozygote is more fit than the average individual, selection accentuates the effect of recombination in breaking down linkage disequilibrium; on the other hand, if the double heterozygotes are less fit than average, selection reduces the importance of recombination.

The second terms in equations 13.1 reveal that qualitatively new features, including genotypic as well as gametic measures of fitness, influence multilocus evolution. Thus, the average excess is not the sole arbiter of multilocus natural selection as it was for single-locus natural selection. This makes the problem of the unit of selection even more important, because if multilocus complexes are the true units of selection, then recombination, linkage disequilibrium, and the fitnesses of specific genotypes modulate the impact of natural selection in addition to average excess. For the two-locus model given in equations 13.1, the question of the unit of selection becomes the question of whether the course of selective response can be adequately described by looking at each locus separately (equation 11.5) or we must consider the entire two-locus complex as a unit (equations 13.1). Consider just the first locus. Let p be the frequency of the A allele at this locus. Since $p = g_{AB} + g_{Ab}$, the two-locus system described by equations 13.1 implies that the selective dynamics of this single locus is given by

$$\Delta p = \Delta g_{AB} + \Delta g_{Ab} = \frac{g_{AB}}{\bar{w}}a_{AB} + \frac{g_{Ab}}{\bar{w}}a_{Ab}$$
$$= \frac{p}{\bar{w}}\left(\frac{g_{AB}}{p}a_{AB} + \frac{g_{Ab}}{p}a_{Ab}\right)$$

(13.3)

The term g_{AB}/p is the conditional probability of an A allele being coupled with a B allele given that the gamete carries an A allele, and similarly g_{Ab}/p is the conditional probability of an A allele being coupled with a b allele given that the gamete carries an A allele. Hence, the term in parentheses in equation 13.3 is an extension of the concept of average excess (a conditional genotypic deviation given a gamete bearing the A allele) that includes the genetic background defined by the second locus (B or b). From equation 11.5, the single-locus dynamics should be given by

$$\Delta p = \frac{p}{\bar{w}} a_A \qquad (13.4)$$

This single locus will be the unit of selection rather than the entire two-locus system when

$$a_A \approx \left(\frac{g_{AB}}{p} a_{AB} + \frac{g_{Ab}}{p} a_{Ab} \right) \qquad (13.5)$$

When equation 13.5 is a good approximation, the selective dynamics are given by equation 13.4, and we do not need to place the A allele into its multilocus context to predict its response to selection.

Insight into the factors influencing the unit of selection can be achieved by a closer examination of the terms on the right- and left-hand sides of approximation 13.5. First, note that the right-hand side depends upon both single-locus allele frequencies (p) and the two-locus gamete frequencies. The left-hand side of 13.5 can be expanded for this random-mating population as (see Chapter 11)

$$a_A = p(w_{AA} - \bar{w}) + (1 - p)(w_{Aa} - \bar{w}) \qquad (13.6)$$

which is a function only of p. The single-locus allele frequency p can always be determined from the two-locus gamete frequencies ($p = g_{AB} + g_{Ab}$), but the inverse is not true due to the potential presence of linkage disequilibrium. Therefore, biological information about the state of the gene pool is lost as we go from the right-hand side of approximation 13.5 to the left-hand side. How important this loss of information is in affecting the accuracy of approximation 13.5 depends upon the magnitude of linkage disequilibrium. This in turn is a function in part of the amount of recombination between the loci, as recombination influences the rate at which D is dissipated (equation 2.7). Hence, one biological factor that influences the accuracy of approximation 13.5 is the amount of recombination. In particular, we would expect little loss of biological information to occur when recombination rates are high and linkage disequilibrium is at or close to zero. However, when recombination is low, nonzero disequilibrium is likely and the two-locus gametes have much continuity across the generations. Under these conditions, approximation 13.5 could break down and the single locus is no longer an adequate unit of selection. More generally, this insight goes back to Mayr's point that the individuals total genotype is broken apart and scrambled by meiosis and fertilization. Meiosis (including recombination and assortment) destroys the multilocus genetic continuity that can be passed on to the next generation though gametes. Hence, the *meiotic* factors that break down multilocus genotypes are a major determinant of the unit of selection.

A second major difference between the right- and left-hand sides of approximation 13.5 is that the right-hand side depends upon fitnesses assigned to two-locus genotypes (e.g., equation 13.2) whereas the left-hand side depends upon fitnesses assigned to a single-locus genotype (equation 13.6). The single-locus fitnesses are related to the two-locus fitnesses through a conditional expectation. For example, the single-locus, marginal fitness of the *AA* genotype is

$$w_{AA} = \frac{g_{AB}^2 w_{AB/AB} + 2g_{AB}g_{Ab}w_{AB/Ab} + g_{Ab}^2 w_{Ab/Ab}}{p^2} \quad (13.7)$$

The quantity g_{AB}^2/p^2 is the conditional frequency of the *AB/AB* genotype *given* that the individual is *AA* (with probability p^2), and similarly for the other two-locus genotypes that are homozygous for the *A* allele. Thus, the one-locus fitness w_{AA} is a weighted average of the fitnesses of the two-locus genotypes that are marginally homozygous *AA*. Averaging several two-locus fitnesses into a single number also represents a potential loss of biological information, and this loss of fitness information can also influence the accuracy of approximation 13.5. As equation 13.7 illustrates, we assign a single-locus fitness by averaging the fitnesses of that single-locus genotype across all the multilocus contexts in which it exists. If this context dependency upon the background genotype is important for the phenotype assigned to the single-locus genotype of interest, approximation 13.5 can break down. On the other hand, if the two-locus fitnesses can be determined in some simple fashion from the marginal single-locus fitnesses, there may be little loss of biological information and the single locus is the appropriate unit of selection. Under the ideal Fisherian model, an additive fitness deviation can be assigned to each single locus, and multilocus fitnesses can be approximated by adding the appropriate single-locus fitness deviations. As we saw in Chapters 9 and 10, epistasis can contribute to both additive and nonadditive phenotypic deviations. Hence, the presence of epistasis for the phenotype of fitness can undermine approximation 13.5, although not always (recall the example of epistasis between *ApoE* and *LDLR* shown in Figure 10.15 in which the additive terms dominate). Therefore, *epistasis for fitness* is another biological factor influencing the unit of selection.

Another simple way of predicting multilocus fitnesses from single-locus fitnesses occurs when the fitness effects of each locus are statistically independent and the fitnesses are rescaled probabilities (such as the viability fitness components in Chapter 11). Under these conditions, the multilocus fitnesses are the products of the single-locus fitness. Multiplicative fitnesses do generate deviations from additivity. However, under the ideal Fisherian conditions in which all fitness deviations are small (Chapter 12, Figure 12.8), multiplicative effects can be adequately approximated by additive effects. This approximation breaks down as the intensity of selection increases. Hence, a third factor influencing the unit of selection is the *intensity* or *magnitude* of selection.

Overall, epistasis and strong selection favor the buildup of multilocus adaptive complexes. These complexes are then partially broken down during the process of meiosis and the transmission of genetic material to the next generation through the gametes. The unit of selection emerges from the balance between natural selection working on epistatic systems to build up multilocus complexes versus the factors of meiosis and fertilization that break apart these same complexes. We have already seen the importance of this balance in Chapter 12 in our examination of the sickle cell polymorphism in malarial regions. Is the unit of selection in this case the hemoglobin β-chain locus or is it the multilocus β–γ gene region (Figure 2.6). As noted in Chapter 12, there is strong epistasis for fitness between the γ loci and the β locus in malarial regions, given the appropriate allelic variation at these loci. Moreover, there is little recombination between them. Hence, as concluded in Chapter 12, a meaningful unit of selection in this case is not the β locus, but rather the multilocus

β–γ **supergene**, a multilocus complex of tightly linked genes with extensive epistasis that approximates a single locus with respect to its patterns of Mendelian segregation.

Despite the importance of the concept of the unit of selection in population genetics, there have been few experiments investigating the unit of selection. One of the few experiments was conducted on the fruitfly *Drosophila mercatorum*. This fly is normally a sexually reproducing species, like most other species of *Drosophila*. However, when virgin females are isolated from some strains, they lay unfertilized eggs that can successfully develop into viable adult females (Carson 1967), a phenomenon known as parthenogenesis (virgin birth). In *D. mercatorum*, the unfertilized eggs undergo a normal meiosis (and hence segregation and recombination) followed by the mitotic duplication of a haploid nucleus. Two such haploid nuclei fuse to form a diploid nucleus, which then undergoes development (Figure 13.1). The resulting adults are all female (as they have two X chromosomes) and are homozygous at all loci. Because the females are diploid and retain normal meiosis, they can reproduce sexually if given the opportunity to mate with a male. Because there is no recombination in males in this species and visible markers exist for the chromosomes, it is possible to breed a male that is homozygous for the same autosomes as a parthenogenetic strain and has its X chromosome from the parthenogenetic strain, with only a single Y chromosome being introduced from a sexual strain (Templeton 1983b). Using such males, two different parthenogenetic strains, called S and O, were crossed, and the resulting F_1 females were isolated as virgins to produce a parthenogenetic population of totally homozygous F_2 females. One advantage of using parthenogenesis is that the number of genotypes in a two-locus model is reduced from the normal F_2 number of nine genotypic categories to just four homozygous genotypes. Indeed, even a three-locus model in this parthenogenetic F_2 has fewer genotypic categories (eight) than the two-locus F_2 model with sexual reproduction. Moreover, because each individual is a doubled haploid gamete, the fitness of a class of individuals sharing the same haploid genotype is equivalent to the fitness of the gamete they inherited; that is, average excesses of fitness are measured directly through individual fitnesses in this system.

Templeton et al. (1976) used four isozyme loci and two visible genetic markers for coarse coverage (the average map distance between markers was 35 cM) of the *D. mercatorum* X chromosome and one autosome that together represent about 60% of the genome to perform a low-resolution QTL scan (Chapter 10) for the phenotype of egg-to-adult viability. They then examined all possible single markers, paired markers, and three-way marker associations with the phenotype of viability (probability of surviving to adulthood) in flies from several different parthenogenetic F_2 populations. In addition to the normal F_2 (the offspring of the F_1 females), Templeton et al. (1976; Templeton 1979a) extended this breeding scheme to make female flies that were hybrids for specific chromosomes from different parental parthenogenetic strains but homozygous for other parthenogenetic chromosomes, as shown in Figure 13.2. A female fly that was hybrid for two complete S and O parental genomes produced only an average of 1.63 parthenogenetic offspring (the $SO_{100\%}$ line in Figure 13.2). In contrast, females flies that were hybrid for only two of the four major chromosomes of the *mercatorum* genome produced an average of 5.36 progeny per female (the $SO_{60\%}$ line in Figure 13.2), and females that were hybrid for only one chromosome produced 10.25 progeny on the average (the $SO_{40\%}$ line in Figure 13.2). Hence, the absolute fitnesses of these hybrid flies under parthenogenesis varied by almost an order of magnitude. The visible and isozyme markers used to test for fitness associations were located on the X chromosome and the large metacentric autosome. Note from Figure 13.2 that the X-linked markers were hybrid in two of the lines (the $SO_{100\%}$ and $SO_{60\%}$ in Figure 13.2) and the metacentric markers

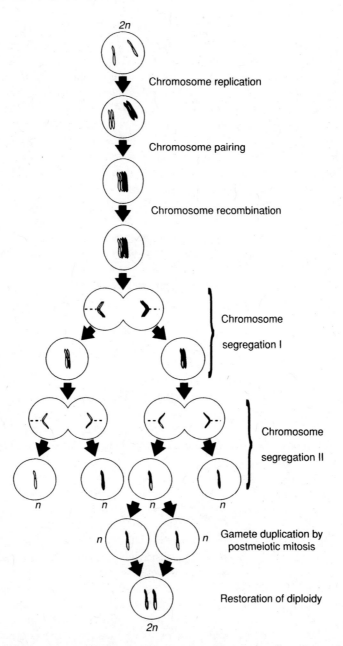

Figure 13.1. Parthenogenesis in *D. mercatorum*. A normal diploid (2*n*) oocyte undergoes normal meiosis, with the two meiotic segregation divisions producing four haploid nuclei. Three of the haploid nuclei become polar bodies, and one becomes the egg nucleus, as in normal meiosis. However, after meiosis, the haploid egg nucleus undergoes mitosis to make a duplicate copy, the process called gamete duplication. These two haploid copies of a single product of meiosis then fuse to restore diploidy, and development of a new fly proceeds from this diploid nucleus.

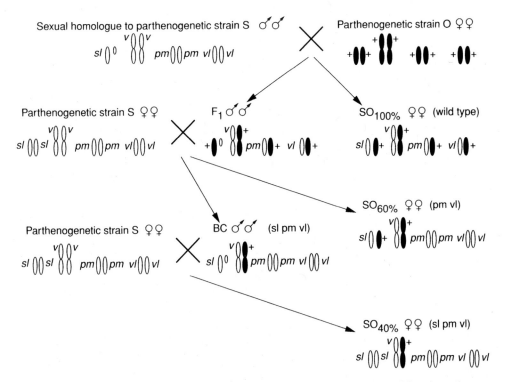

Figure 13.2. Breeding scheme in *D. mercatorum*. The two parental strains are an all-female parthenogenetic strain designated by O and males from a sexual homologue to a second parthenogenetic strain designated by S. A pictorial representation of the *mercatorum* genome is shown below each strain, with the first pair on the left indicating the XX or XY chromosome pairs, and the next three pairs indicating the major autosomes, with one metacentric pair and two acrocentric pairs. A small dot chromosome pair is not shown. White indicates chromosomes from the S strain and black from the O strain. The visible markers on the major chromosomes are also shown, with *sl* being *spotless* at the base of supraorbital bristle, *v* being *vermillion* eye color, *pm* being *plum* eye color, and *vl* being *veinless*, the failure of the posterior wing vein to reach the margin of the wing. The three types of females used in the unit of selection experiments are designated by SO_x with x being 100%, 60%, or 40%, the percentage of the total genotype that is hybrid between S and O.

were hybrid in all three lines. Hence, the fitnesses associated with the X-linked and between X-linked and autosomal markers could be studied at two different levels of selective intensity (1.63 and 5.36 progeny per female) and the fitnesses associated with the autosomal markers could be studied at all three levels of selective intensity. This sexual/parthenogenetic system is ideal for empirically investigating the unit of selection because normal meiosis generates much genetic variation in flies that are hybrid for one or more chromosomes, gamete fitness can be measured directly as individual fitness, epistasis could be easily monitored, and the intensity of selection (absolute number of progeny capable of surviving to adulthood) could be experimentally manipulated.

The fitnesses (egg-to-adult viability) associated with the markers were estimated for all one-, two-, and three-locus genotypes. Little selection was detected at the single-locus level. For example, in the parthenogenetic F_2 from the $SO_{100\%}$ F_1 adults (with the strongest intensity of selection), the single-marker analysis detected significant viability differences at only one of the six markers—a 25% reduction in viability associated with homozygosity

Table 13.1. Observed Viabilities Associated with Four Homozygous Marker Genotypes Associated with Marker Pair of *EstB* (with Alleles *B1* and *B2*) and *vermillion* Eyes (with Alleles v^+ and *v*) in Parthenogenetic F$_2$ Offspring of Two Parthenogenetic Strains of *D. mercatorum*

	$B1\,v^+$	$B1\,v$	$B2\,v^+$	$B2\,v$
Relative viability	1.000	0.520	0.616	0.832

Source: From Templeton et al. (1976).
Note: The genotypes of the two parental strains are $B1\,v^+$ and $B2\,v$, respectively. Because all the flies are homozygous, the genotypes are indicated by the allelic state of just one chromosome. The viability of the genotype $B1\,v^+$ is set to 1 and all other viabilities are measured relative to this genotype.

for the recessive allele at one of the visible markers. Such viability reductions for recessive visible markers are commonplace in *Drosophila* studies. Hence, from the single-site analysis there was little evidence for genetic variability for the phenotype of viability.

This situation changed greatly at the two- and three-locus levels. For example, significant viability effects were detected with 7 of the 15 pairs of markers in the $SO_{100\%}$ experiment, involving both linked and unlinked markers. For example, two of the markers used were an eye color marker (*v* for vermillion eyes) and an isozyme esterase marker (*EstB*) found on the opposite ends of the metacentric autosome. Neither of these markers showed any significant association with viability when considered one at a time, but when analyzed as a pair, highly significant effects on viability were detected (Table 13.1). The pattern of fitnesses in Table 13.1 indicates strong epistasis. First, there are no selective effects associated with either marker considered by itself. Hence, the whole is more than the sum of the parts. Second, if fitness effects were additive across loci, the recombinant genotypes should be intermediate between the two parental genotypes, but as can be seen, the parental genotypes had high fitness whereas the recombinants had low fitness. The fitnesses shown in Table 13.1 therefore indicate the presence of strong epistasis for fitness. If epistasis had been examined only for those markers showing significant marginal effects, no epistasis would have been detected at all in these experiments. However, by examining all pairwise combinations of markers, extensive genetic variation for the phenotype of viability was detected. These experiments were also extended to all three-locus comparisons, and 13 of the 20 three-locus combinations showed significant viability effects in the $SO_{100\%}$ line. For example, the two-locus analysis detected no significant epistasis between the marker pair *vermillion* and *xanthine dehydrogenase* (an isozyme locus), yet this pair of markers displays significant epistasis when coupled with *Est B*. Thus, even the pairwise analysis had not detected all the epistasis that was actually present.

Although the qualitative patterns of fitness at the two- and three-locus levels indicate the presence of fitness epistasis (e.g., Table 13.1), a quantitative analysis of the role of epistasis was also performed. The two- and three-locus fitnesses were reparameterized into single-locus additive deviations and multilocus epistatic deviations (no dominance deviations are needed because the parthenogenetic offspring are homozygotes). The two-locus fitness partitioning is shown in Table 13.2. The fitnesses were then estimated under the constraint of no epistasis; that is, $\varepsilon = 0$. The fit of the nonepistatic model to the unconstrained fitness model was tested using a likelihood ratio test. If the nonepistatic fitness model explained the observed genotype frequencies in the parthenogenetic F$_2$ (that is, the response to selection) by this likelihood ratio test criterion, then the individual loci were considered additive units of selection. Another alternative fitness model is the multiplicative model, shown for the two-locus case in Table 13.3. These multiplicative fitness components were also estimated

Table 13.2. Two-Locus Fitness Model Applied to Parthenogenetic, Homozygous Offspring of Hybrid and Partial-Hybrid *D. mercatorum* Females

		Alleles at second locus	
		B	b
Alleles at first locus	A	$\mu + \alpha + \beta + \varepsilon \equiv 1$	$\mu + \alpha - \beta - \varepsilon$
	a	$\mu - \alpha + \beta - \varepsilon$	$\mu - \alpha - \beta + \varepsilon$

Source: From Templeton et al. (1976).
Note: The parameter μ represents the overall average fitness, α and β represent the additive fitness effects of the A and B alleles (with $-\alpha$ and $-\beta$ being the additive fitness effects of the a and b alleles), and ε is the epistatic term between the loci. The relative fitness of the AB/AB homozygous genotype is set to 1, reflecting the fact that there are only three statistically independent fitness parameters to be estimated.

under maximum likelihood and tested against the unconstrained model using a likelihood ratio test. If the multiplicative model explained the observed genotype frequencies well, then the individual loci were considered multiplicative units of selection. More complicated models are possible with the three-locus system, but they will not be discussed here.

When selection was at its most intense level in the parthenogenetic offspring of the $SO_{100\%}$ hybrid females, extensive epistasis was found between markers on the same chromosome arm, on the same chromosome but different arms, and on different chromosomes. This epistasis was so strong that the response to selection usually could not be predicted from the additive ($\varepsilon = 0$) or the multiplicative models. Hence, when the selective pressures were intense, the unit of selection was the highest multilocus complexes observable, spanning both linked and unlinked loci. As the selective intensity decreased by lowering the levels of hybridity of the female parent (Figure 13.2), the fitness effects involving unlinked markers tended to be adequately predicted by additive or multiplicative fitness models, but the fitness effects involving two or more markers on the same chromosome could not be broken down into smaller predictive units of selection. At the lowest intensity of selection, only those markers closely linked on the same chromosome arm behaved as multiloci units of selection and all markers on different chromosome arms could be treated as additive or multiplicative units of selection. Overall, the unit of selection changed from multichromosomal complexes to chromosomes to chromosome arms as selective intensity dropped. The greater the intensity of selection, the broader was the unit of selection. The *D. mercatorum* experiments illustrate that *the unit of selection is an emergent property of the balance*

Table 13.3. Two-Locus Multiplicative Fitness Model Applied to Parthenogenetic, Homozygous Offspring of Hybrid and Partial-Hybrid *D. mercatorum* Females

		Alleles at second locus	
		B	b
Alleles at first locus	A	1	$1 \cdot m_2 = m_2$
	a	$1 \cdot m_1 = m_1$	$m_1 \cdot m_2$

Source: From Templeton et al. (1976).
Note: The marginal fitnesses of the AA and aa genotypes at the first locus are 1 and m_1, respectively, and the marginal fitness of the BB and bb genotypes at the second locus are 1 and m_2, respectively. These marginal fitnesses then determine the two-locus genotype fitness in the multiplicative fashion shown.

between selection operating to build up epistatic gene complexes and recombination operating to break down such complexes. In particular, these experiments illustrate that the unit of selection increases in breadth (over larger recombinational distances) as the intensity of selection increases.

These experimental conclusions are reinforced by the theoretical work of Michalakis and Slatkin (1996) on a two-linked-locus model of natural selection starting in a population initially fixed for the *a* and *b* alleles, respectively, at the two loci. Suppose mutations can occur to yield the *A* and *B* alleles at these two loci such that fixation for *A* and *B* takes the population to the highest adaptive peak, but negative epistasis exists in this case such that the *aB* and *Ab* gametes are associated with deleterious fitness consequences. They showed that selection and drift can interact to cause a shift from the lower to higher adaptive peak in this case, and this is a special case of the shifting balance theory (Chapter 12). However, they also investigated the impact of selective intensity and amount of recombination upon the evolutionary trajectory taken during this peak shift. When selection was intense and recombination low, populations generally remained at the *ab* peak until mutations had occurred to produce the double-mutant gamete *AB*. Then, natural selection would drive the *AB* gamete to fixation. In this case, the unit of selection was the two-locus supergene. In contrast, when selective intensities were weaker and/or recombination stronger, Michalakis and Slatkin found that other evolutionary trajectories to the new adaptive peak usually occurred. For example, we saw in Chapter 12 that even a deleterious mutant has a finite probability of fixation in a local population with small variance effective size. Once fixed for one of the mutants at one of the loci, selection will now favor in this new genetic background the fixation of the mutant allele at the other locus once it is created by mutation or enters the population via gene flow. In general, Michalakis and Slatkin found that the evolutionary trajectories were such that the population will be monomorphic for one of the two loci most of the time, which implies that at any given time the unit of selection is a single locus because only one locus is responding to selection most of the time. Moreover, although epistasis was critical in defining the fitness surface and driving the evolutionary process, there would be little opportunity to detect the presence of epistasis in these evolutionary transitions because, as shown in Chapter 10, epistasis is not apparent when a critical allele at one locus is very common. This theoretical work shows that epistasis is necessary, but not sufficient, to result in multilocus units of selection.

The *D. mercatorum* experiments and theoretical work of Michalakis and Slatkin (1996) illustrate the importance of recombination in influencing the unit of selection. As we previously noted in Chapter 3, the *effective* amount of recombination is a function not only of genetic architecture but also of any factor that influences the frequency of double (or higher) heterozygotes in the population. These population-level factors include system of mating (Chapter 3) and population subdivision (Chapter 6). In the case of selected complexes of genes, this also includes the fitnesses of the double (or higher) heterozygotes (see equations 13.1). Because the unit of selection emerges from the balance of selection on the whole genotype versus effective recombination, many population factors are expected to influence the unit of selection. For example, in populations with an inbreeding system of mating ($f > 0$), there are fewer double heterozygotes and therefore less "effective" recombination. Thus, the unit of selection should be broader in an inbreeding population than in a population with $f \leq 0$. Such large, multilocus units of selection have indeed been demonstrated in plant populations with much selfing (Jain and Allard 1966; Weir et al. 1974).

In a highly outcrossing species, such as humans, effective recombination is high, so units of selection will tend to be small, perhaps just single genes much of the time, despite the existence of extensive epistasis (Chapter 10). One potential exception to this in humans, and many other species as well, stems from the fact that functionally related genes tend to be found in tightly linked gene clusters, as we have seen for globin genes (Figure 2.6) and apoprotein genes (Figures 2.7 and 2.8). These clusters have the potential for much epistasis because the genes often belong to the same functional pathway and also have tight linkage. The combination of a high potential for epistasis and tight linkage implies that many of these clusters might constitute supergene units of selection. Singer et al. (2005) tested this hypothesis by screening the human genome for genes with similar expression patterns via microarray analysis. After eliminating duplicate genes (such as the duplicate α-Hb loci discussed in Chapter 11), they discovered that clusters of coexpressed genes tend to be more tightly spaced than genes in general. By doing an analysis of chromosome breakpoints during mammalian evolution, they also discovered that coexpressed clusters of genes have significantly fewer breakpoints than expected by chance, indicating that these clusters are being held together by natural selection. This work implies that many of these expression clusters define supergene units of selection in the human genome, as we already saw was the case for the human globin gene cluster as discussed in Chapter 12.

The unit of selection arises from the interaction of selective intensity and genetic architecture (primarily recombination and epistasis) in the context of population properties (e.g., system of mating, variance effect size, degree of subdivision). The unit of selection is *not* just a function of the unity of the genotype; rather, the unit of selection is a dynamic compromise between selection building up coadapted complexes and *effective* recombination breaking them down. Thus the unit of selection can change as the population evolves or experiences altered demographic conditions.

TARGETS OF SELECTION BELOW LEVEL OF INDIVIDUAL

In Chapters 11 and 12, fitness phenotypes were always regarded as properties of individual organisms. However, even in those chapters, we saw that the fitness values of specific individuals were important only through their contribution to genotypic values; that is, the average fitness phenotype for a group of individuals sharing a common genotype. Consequently, to study natural selection, all we need is a group that shares a common genetic state. We can then assign a "genotypic value" or average fitness deviation to that group. There is nothing about the theory of natural selection or the concept of average excess that requires that the biological units that are pooled together and assigned an average fitness value have to be individual organisms. As long as the group we define has some genetic identity that displays a fitness phenotype and there is variation in the population for those genetic identities, natural selection can occur. The genetic identities do not have to correspond to individual-level genotypes, although the response to selection will still be modulated by what fitness variation can be passed through the gametes produced by individuals. In this section, we will discuss targets of selection that are at levels of biological organization below the level of individuals (that is, nested within individuals).

Meiotic Drive

One target of selection below the level of diploid individuals is selection among the gametes produced by a single individual. Groups of gametes can be defined that share a common genetic state, as we saw in developing the concepts of average excess and average effect. For every locus at which the diploid individual is heterozygous, the gametes produced by that individual will constitute a genetically variable group due to Mendel's first law of segregation. For multilocus genetic states, meiosis in a single-individual diploid organism will also produce new and genetically diverse gamete types through novel combinations generated by independent assortment and recombination. Gametes also display a phenotype of great evolutionary importance—their ability to engage in a fertilization event. Therefore, the gametes produced by a single individual can be a target of selection when the gametes themselves directly express a phenotype that alters their chances of being passed into the next generation through a fertilization event.

In some species, the haploid phase of life is dominant, so the idea of a haploid expressing selectable phenotypes should not be surprising. Even though the haploid phase in species such as humans is limited to a single cell generation, this transient haploid phase can still display important phenotypes that influence the chances of being involved in a fertilization event. One important type of selectable phenotype is **meiotic drive** or **segregation distortion**, the excess recovery of one of a pair of alleles in the gametes of a heterozygous individual. Gamete types displaying meiotic drive distort their effective segregation ratios from the 50–50 meiotic segregation expected under Mendel's first law. Consider a one-locus, two-allele model with alleles A and a. Suppose the A- and a-bearing gametes display a phenotype that influences either meiosis or the chances for fertilization such that the probability of an A-bearing gamete produced by an Aa diploid heterozygote being involved in a fertilization event is k, and $1 - k$ is the probability of an a-bearing gamete from an Aa individual being involved in a fertilization event. Normal segregation is now a special case in which $k = 1 - k = \frac{1}{2}$. The population and evolutionary consequences of normal segregation when coupled with the other Hardy–Weinberg assumptions were shown in Figures 2.1 and 2.2. Figure 13.3 shows the population consequences when meiotic drive is allowed. From that figure, we can see that the frequency of the A allele before meiosis is $p = G_{AA} + \frac{1}{2}G_{Aa}$, our standard formula for relating genotype frequencies to allele frequencies. However, the frequency of the A allele in the gene pool after one round of meiosis with segregation distortion is

$$p' = G_{AA} + kG_{Aa} = G_{AA} + \tfrac{1}{2}G_{Aa} - \tfrac{1}{2}G_{Aa} + kG_{Aa} = p + G_{Aa}\left(k - \tfrac{1}{2}\right) \qquad (13.8)$$

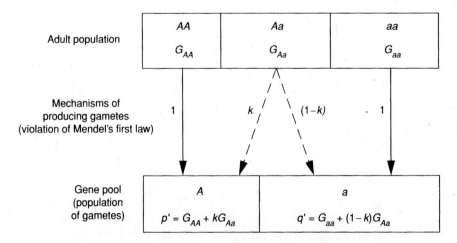

Figure 13.3. Meiotic drive at single locus with two alleles, A and a.

To see if meiotic drive is an evolutionary force, we calculate the change in allele frequency as

$$\Delta p = p' - p = G_{Aa}\left(k - \tfrac{1}{2}\right) \qquad (13.9)$$

Equation 13.9 tells us that meiotic drive is an evolutionary force that will change allele frequencies whenever $G_{Aa} > 0$ and $k \neq \tfrac{1}{2}$. The first condition relates to a specific individual-level genotype frequency ($G_{Aa} > 0$) and reflects the fact that meiotic drive can only occur in the heterozygotes because it is only in heterozygotes that there is the necessary variation in the gametic genetic states. Thus, this selection is clearly nested within certain individuals and differs fundamentally from the individual-level selection given by equation 11.5. The second condition for evolution ($k \neq \tfrac{1}{2}$) is simply that meiotic drive is occurring rather than normal 50–50 segregation. The $k - \tfrac{1}{2}$ component in equation 13.9 is a measure of the fitness deviation of the gamete; that is, the actual amount of meiotic drive relative to normal segregation. Thus, $k - \tfrac{1}{2}$ plays a role similar to that of average excess in our standard equations for natural selection at the individual level in equation 11.5. In this case, the measure of fitness deviation is assigned directly to the gamete rather than by statistical averaging over individuals. However, the common theme between equations 13.9 and 11.5 is that the response induced by natural selection is funneled through the gametes. Once again, selection operates from the gamete's perspective, and in this case the gamete has its own direct phenotype.

There are many examples of meiotic drive. For example, there are genetic diseases in humans that are associated with trinucleotide repeats (Rubinsztein 1999). One example is the fragile X syndrome, the most common form of inherited mental retardation in humans that affects about 1 in 4000–5000 males and about 1 in 7000–8000 females. This disease is associated with an X-linked locus that has tandem CGG repeats in the 5' untranslated portion of the *FMR1* gene that codes for a RNA-binding protein. There is much genetic variation in the number of repeats, and the phenotype of mental retardation is associated with alleles with more than 200 repeats. Alleles with 50 to 200 repeats are not associated with the syndrome but have an increased risk of mutating (that is, an expansion of the number of repeats) to disease alleles compared to alleles with fewer than 50 repeats. Drasinover et al. (2000) discovered that some of the intermediate alleles show segregation distortion in females heterozygous for normal and intermediate alleles. In particular, meiotic drive favored the intermediate allele with $k = 0.69$ when there were between 51 and 55 repeats and $k = 0.74$ when there were between 56 and 60 repeats. However, there was no significant meiotic drive for alleles with more than 60 repeats. This suggests that selection favors alleles in the 51–60 repeat range through meiotic drive in heterozygous females.

Another trinucleotide-repeat disease is myotonic dystrophy, the most common form of adult-onset muscular dystrophy. This autosomal disease is associated with a CTG repeat in the 3' untranslated part of the myotonin–protein kinase gene. The number of repeats normally varies from 5 to 72,000, although there is one case with a repeat number of 309,550. Alleles with more than 50 repeats are associated with the disease state. Carey et al. (1995) found meiotic drive favoring those alleles with a copy number greater than or equal to 19 in heterozygotes for one allele with fewer than 19 repeats and one allele with 19 or more repeats. The direction of this meiotic drive is unfortunate because the higher repeat number alleles are the ones associated with the disease, as is true for other trinucleotide-repeat diseases.

Equation 13.9 raises an interesting problem for alleles such as those associated with myotonic dystrophy. If an allele is favored by meiotic drive ($k > \frac{1}{2}$), then Δp in equation 13.9 is always positive for every p between 0 and 1. Meiotic drive should therefore cause the favored allele to go to fixation. If the only phenotype shown by an allele were meiotic drive, it should go to fixation after it originated by mutation. Hence, even if alleles with meiotic drive arise commonly, we do not expect to find many meiotic drive polymorphisms as such alleles should go rapidly to fixation. Yet, we do find several polymorphisms showing meiotic drive. Moreover, as with the myotonic dystrophy example, it is often the rarer allele that is favored by meiotic drive. Why do these meiotic drive alleles stay rare but polymorphic? The answer to this question lies in the observation that *a single unit of selection can have more than one target of selection*; that is, the same genetic unit that is responding to selection can have phenotypic manifestations at more than one level of biological organization. For the myotonin–protein kinase gene, the same allelic class favored by meiotic drive at the level of haploid gametes within certain heterozygous individuals tends to affect the phenotype at the level of the diploid individual in a highly deleterious fashion. This suggests that selection operating upon different targets of selection in an antagonistic fashion can maintain polymorphisms in the genetic unit of selection.

The t complex in mice provides an example of antagonistic selection for different targets of selection (Redkar et al. 2000; Schimenti 2000). The t complex is located in a 20-cM region of chromosome 17 of the mouse genome that constitutes about 1% of the mouse genome. The t complex normally differs from the non-t state in this region by four different inversions that suppress most but not all recombination in this region. A small central region is not spanned by the inversions. This region contains a large number of candidate genes for sperm motility, capacitation (a process which includes changes in sperm membranes, motility, and metabolism that is essential for subsequent fertilization), binding to the zona pellucida of the oocyte, binding to the oocyte membrane, and penetration of the oocyte. Extensive epistasis exists between subregions of this complex for many of these sperm-related phenotypes that are subject to intense selective pressures, as we will soon see. The combination of low recombination, extensive epistasis, and intense selection makes this large, 20-cM genomic region behave as a single unit of selection. For our purposes, we can treat this complex, multilocus unit of selection as if it were a single supergene with two alleles, t and T.

Male mice that are heterozygous T/t show extreme segregation distortion favoring the t allele, with k values going as high as 0.99 for some t alleles. The t/t homozygotes are frequently lethal, and if they live, t/t males are invariably sterile. Thus, t alleles are strongly favored by meiotic drive within T/t heterozygote males, but t alleles are strongly selected against at the individual level due to their lethality and sterility effects in homozygotes. Note that meiotic drive alone should result in fixation of the t allele through equation 13.9, but selection on lethality and sterility of diploid individuals should result in fixation of the T allele through equation 11.5. Natural populations of mice generally go to neither of these fixation points, instead remaining highly polymorphic even for lethal t alleles. To understand this polymorphic balance of antagonistic targets of selection, we need to see how both these phenotypic levels are filtered through the gamete in going from one generation to the next.

As shown in Chapter 11, we measure that aspect of fitness differences among individuals that is transmissible through a gamete by the average excess. However, meiotic drive alters the allele frequencies even before the fitness effects measured by the average excess occur (remember, the average excess measures the average individual fitness effects the gamete is expected to have in the next generation). In equation 13.9 and in Figure 13.3, meiotic drive would alter the allele frequency of the t allele from p to p'. However, that p' was derived under the assumption that all heterozygotes experience meiotic drive. In the case of the t complex, only male heterozygotes display meiotic drive, so only half of the heterozygotes are affected. Accordingly, p' in is this case equals $p + \frac{1}{2}G_{Tt}(k - \frac{1}{2})$. Assuming random mating, meiotic drive should change the frequency of t from p to $p' = p + pq(k - \frac{1}{2})$. Now suppose the relative individual-level fitness of T/T is 1, of T/t is $1 - s$, and of t/t is 0 (a lethal t allele). Given random draws of gametes from the gene pool already altered to p' by meiotic drive, the average excess of the t allele for individual-level fitnesses is

$$a_t = q'(1 - s - \bar{w}) + p'(-\bar{w}) \tag{13.10}$$

where $\bar{w} = q'^2 + 2p'q'(1 - s)$. Selection at the diploid level further alters the allele frequency as given by equation 11.5 to $p'' = p' + p'a_t/\bar{w}$. The total change in allele frequency can be written as

$$\begin{aligned}\Delta p = p'' - p &= p'' - p' + p' - p \\ &= p'\frac{a_t}{\bar{w}} + pq\left(k - \frac{1}{2}\right)\end{aligned} \tag{13.11}$$

Equation 13.11 clearly reflects the impacts of selection at two levels of biological organization such that one unit of selection (the t complex) has two targets of selection. The first term in equation 13.11 reflects selection at the individual level and as before is proportional to the average individual-level fitness deviation (average excess) of the gamete of interest, the t allele in this case. The second term reflects the impact of selection among gametes within male heterozygotes, that is, meiotic drive. However, one basic property is preserved in both equations 13.11 and 11.5 (the equation incorporating only individual-level selection); namely, the response to natural selection is determined by the fitness effects that are transmissible through a gamete. The first term in equation 13.11 has its sign determined by the average excess of the t allele, the phenotypic measure of the fitnesses of diploid bearers that is transmissible by a t-bearing gamete. The second term in equation 13.11 has its sign determined by k, a phenotypic measure of meiotic drive that is assigned directly to the t-bearing gamete. Hence, both terms of equation 13.11 are gametic measures of fitness phenotypes, albeit at two distinct levels of biological organization. We still need to take the gamete's perspective in order to understand natural selection. Even when natural selection involves multiple levels of selection, the response to selection is always filtered through a gamete.

Equation 13.11 also makes clear the antagonistic nature of selection at the gametic and individual levels in this case. For t alleles, $k > \frac{1}{2}$, so the second term of equation 13.11 is positive for all p between 0 and 1. At the individual level, $\bar{w}$ is a strictly decreasing function of p, so the only adaptive peak at the individual level is at $p = 0$ as the fitness surface falls with increasing p. This means that the first term in equation 13.11 is always negative for all p between 0 and 1. Hence, selection at the t complex is always going in opposite directions for the two different targets of selection. The equilibrium occurs when the magnitudes of these opposing selective forces are equal. This is shown graphically in Figure 13.4,

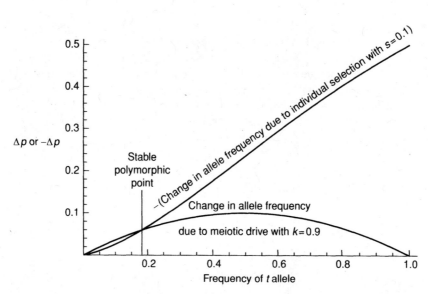

Figure 13.4. Plot of two components of change in t allele frequency from equation 13.11 with $k = 0.9$ and $s = 0.1$.

which plots the two components of selection and reveals the equilibrium as the intersection point of the second component (meiotic drive) with the negative of the first component (individual selection). Figure 13.4 also shows that this equilibrium point is stable. Note that below the equilibrium point the magnitude of meiotic drive increasing p is greater than the magnitude of individual selection decreasing p. Hence, below the equilibrium point meiotic drive overpowers individual selection and increases the frequency of the t allele. However, when above the equilibrium point, the opposite is true, so individual selection overpowers meiotic drive and decreases the frequency of the t allele. Hence, the equilibrium is selectively stable. Thus, the widespread polymorphisms of t alleles are no longer a mystery. These polymorphisms simply reflect the balance between the two targets of selection. Note that this polymorphic balance is inexplicable if one knew only of selection at just one biological level. Seemingly nonsensical evolutionary outcomes are possible with a unit of selection that has multiple targets of selection if one pays attention to only a single target of selection.

Biased Gene Conversion

Unequal gene conversion occurs in meiosis when an allele or stretch of DNA converts its homologue to its own genetic state in a heterozygous individual in a manner that results in non-Mendelian segregation ratios. Gene conversion is a common occurrence during meiosis and is strongly associated with recombination. Usually unequal gene conversion is symmetric; that is, it is equally likely for either homologous stretch of DNA to be converted. However, sometimes the genetic state of an allele or a stretch of DNA has the property of preferentially converting its homologue to its own state when gene conversion occurs. Such biased gene conversion is a phenotype with an underlying genetic basis and therefore constitutes yet another target of selection below the level of the individual.

Walsh (1983) examined a one-locus, two-allele model (A, a) of biased gene conversion. Let A and a have different phenotypes during meiosis within a heterozygous individual such that γ is the probability of an *unequal* gene conversion event and β is the conditional probability that a converts to A *given* an unequal conversion occurs. Note that the above two parameters describe the "phenotype" of gene conversion at the molecular level within heterozygous individuals. Now, $1 - \gamma$ is the probability that no unequal gene conversion event occured; that is, the probability of getting a 1:1 ratio with Mendelian segregation in a Aa heterozygote. With probability $\gamma\beta$ conversion is biased in favor of A, yielding a segregation only of A alleles in that meiotic event from an Aa heterozygote. Finally, with probability $\gamma(1 - \beta)$, conversion is biased in favor of a, yielding a segregation only of a alleles in that meiotic event from an Aa heterozygote. Hence, the overall segregation ratio from Aa heterozygotes is $\frac{1}{2}(1 - \gamma) + \gamma\beta$ A alleles to $\frac{1}{2}(1 - \gamma) + \gamma(1 - \beta)$ a alleles rather than the normal 1:1 segregation. Note that this biased segregation ratio can be expressed as $k : 1 - k$ where $k = \frac{1}{2}(1 - \gamma) + \gamma\beta$. From the standpoint of evolutionary dynamics, biased gene conversion yields the same situation as described in Figure 13.3 and in equation 13.9. Thus, if biased gene conversion is the sole target of selection, we should get fixation of A when $k > \frac{1}{2}$ and fixation of a when $k < \frac{1}{2}$.

Just like meiotic drive, the selective dynamics of a locus affected by biased gene conversion will strongly interact with any selective impact the locus may have at the level of individual fitness. Equation 13.9 describes the evolutionary dynamics of both meiotic drive and biased gene conversion in the absence of any other target of selection. Note that the dynamics depend not only upon k, the selected phenotype, but also upon G_{Aa}, the frequency of heterozygotes. This dependency is expected because selection for meiotic drive or biased gene conversion can only occur within heterozygotes. However, the dependency upon G_{Aa} also ensures that other population-level forces will influence the evolutionary process. For example, suppose we have a non-random-mating population such that $G_{Aa} = 2pq(1 - f)$ (see equation 3.1). Then any system of mating that results in a positive f will diminish the selective response to this target of selection, whereas any system of mating that results in a negative f will enhance the response. Thus, meiotic drive or biased gene conversion would be expected to play only a minor role in a population of obligate selfers because such populations have very few heterozygotes, and therefore there is little opportunity for selection within heterozygotes. In contrast, a population that is actively avoiding inbreeding would create optimal conditions for selective responses to meiotic drive and biased gene conversion. Another population-level factor is population subdivision, which can also diminish the frequency of heterozygotes through the Wahlund effect (equation 6.24). Therefore, meiotic drive and biased gene conversion are less effective in subdivided populations. If the population is further subdivided into local demes with small variance and inbreeding effective sizes, heterozygosity is further reduced and meiotic drive and biased gene conversions have less opportunity to operate. Thus, although the target of selection may be below the level of individuals, selection at the individual- and population-level forces such as system of mating, gene flow patterns, and genetic drift all strongly modulate the evolutionary dynamics of targets of selection below the level of the individual.

Transposons

Transposons are repetitive mobile sequences that are dispersed throughout the genome. There are two broad classes of transposons: DNA transposons and retrotransposons. DNA transposons generally move within the genome as pieces of DNA, cutting and pasting themselves into new genomic locations. Retrotransposons duplicate through an RNA intermediate, usually with the original transposon remaining at its original site where it is transcribed. The resulting RNA transcript is then reverse transcribed into DNA, which then can integrate into new genomic locations. In either event, this phenotype of transposition is expressed within cells and can be a target of selection, with the within-individual selection favoring those sequences that can make more copies of themselves. A transposon that can make many copies of itself and disperse throughout the genome has a much greater chance of being passed on through a gamete to the next generation than another transposon that has poor replicative abilities. For example, a transposon that exists as a single copy on an autosome will be passed on to the next generation in only half of the gametes. However, a transposon that has produced many copies that are dispersed across many locations ensures that virtually all gametes will carry multiple copies to the next generation. In this manner, those transposons most successful at the genomic level also have greater success in spreading throughout the population. As a result, the genomes of many organisms are filled with many different transposons, with the copy number of particular types of transposons sometimes going into the millions. For example, just one class of retrotransposons called human endogenous retrovirus (HERV) makes up 7% of the human genome (Prak and Kazazian 2000).

Transposons are units of selection that can have multiple targets of selection. In addition to selection within cells on their ability to transpose and replicate within genomes, transposons often affect fitness at the individual level (Cooper 1999; Kidwell and Lisch 2000; Prak and Kazazian 2000). Just as we saw for the *t* complex in mice, the evolutionary dynamics of transposons must take into account multiple targets of selection, and phrasing the evolution of transposons only as "selfish" DNA that parasitizes the genome is inappropriate and misleading. For example, different types of transposons have inserted into the promoter of the *hsp70Ba* gene that codes for the stress-inducible molecular chaperone Hsp70 in *Drosophila melanogaster* (Lerman et al. 2003). These transposon insertions underlie the natural variation found in the expression of this gene, and this in turn directly alters two components of individual fitness, inducible thermotolerance and female reproductive success. Another example is provided by the *Alu* transposon, which exists as roughly a million copies per human genome. *Alu* elements frequently insert into noncoding regions and modify the expression of nearby genes (Cooper 1999). For example, the estrogen responsiveness of the human breast cancer gene (*BRCA1*) appears to have been conferred by an *Alu* element located within the promoter region of the gene. An *Alu* sequence in the last intron of the human *CD8A* gene modulates the activity of an adjacent T-lymphocyte-specific enhancer. This particular *Alu* sequence differs at seven nucleotides from its probable source *Alu* sequence. Two of these nucleotide changes are in an area of the derived sequence that acts as a transcription factor binding site, and site-directed mutagenesis indicates that both nucleotide substitutions are necessary for this function, These results suggest that these nucleotide changes were due to selection at the individual level in this specific inserted *Alu* sequence. Thus, this *Alu* unit of selection seems to have been shaped by positive selection for its phenotypic impact at the individual level.

Indeed, selection at the individual level can sometimes co-opt completely the subsequent evolution of a transposon. One of the most startling examples of this is the jawed vertebrate immune system that mounts an antigen-specific response to infection (Agrawal 2000). Vertebrates generally have much longer generation lengths than the infectious agents that attack them, yet the vertebrate immune system effectively allows genetic diversity to be generated and selected on a rapid time scale *within* individuals. This nongermline genetic diversity can be generated because our antigen receptor genes are divided into gene segments, called V and J, and a third segment called D at some loci. DNA rearrangements, called V(D)J recombination, of these segments can be generated within the cells of our immune system. This combinatorial mechanism generates huge amounts of variation in the antigen recognition portion of the receptor, and mechanisms exist to preferentially select *at the cellular level within individuals* those combinations that are most effective in dealing with a particular infectious agent. Note that our immune response represents a type of selection at a level below the individual and involves the movement of DNA elements. These features suggest that V(D)J recombination has evolved from a transposable element, and recent studies on the molecular details of this recombination mechanism strongly indicate that this unique feature of the jawed vertebrate immune system evolved from a transposable element called the *RAG* transposon. This novel immune system, co-opted from a transposon, constitutes one of the most important adaptive breakthroughs in the jawed vertebrates, an adaptation that arose 450 MYA, and retains its critical adaptive significance to the present. Indeed, one can reasonably speculate that humans, along with many other jawed vertebrates, could have never evolved if it had not been for this *RAG* transposon.

Some transposons display a qualitatively different aspect to their evolution not seen in the other targets of selection discussed previously. In all previous cases, no matter how intense the selection is below the level of the individual, the selective response of the unit of selection was always constrained and shaped by the necessity of passing on to the next generation through a gamete. However, some transposons have the ability to "infect" a new individual in a manner independent of gametic transmission. This infectious type of transmission is called **horizontal transmission**, whereas the transmission to new individuals through a gamete is called **vertical transmission**. The ability of some transposons for horizontal transmission blurs the line between retroviruses and retrotransposons, and indeed in many cases no such line is readily discernable. This means that to some extent many retrotransposons evolve as an independent organism and to some extent as a genetic element imbedded with the genome of the host. The most dramatic cases of horizontal transmission are those in which a transposable element infects individuals from a different species. Interspecific horizontal transmission can be detected by constructing the molecular phylogeny of a transposon sequence found in many different species and comparing it to the molecular phylogeny of some single-copy gene from the same species. If all transposon transmission is vertical, then the two phylogenies should be the same. Horizontal transmission will create topological incongruence between the two phylogenies. Such topological incongruence is shown in Figure 13.5 for the *P*-element transposon found in several species of *Drosophila*. The topological incongruence shown in that figure requires a minimum of 11 horizontal transfer events among the 18 species surveyed.

Once a transposon has invaded a new species via horizontal transmission, it can rapidly spread through vertical transmission, particularly if the species has a population structure characterized by a random or outbreeding system of mating and much gene flow. For example, prior to 1949, *P* elements were generally not found in strains of *D. melanogaster* collected throughout the world (Anxolabehere et al. 1988). Starting in the 1950s, a few strains collected in the Americas and in the Pacific and Australia began to have *P* elements (Table 13.4). Over time, the incidence of strains bearing *P* elements tended to increase in these geographical areas, and moreover *P* elements spread to populations in Europe, Asia, and Africa. Thus, after the initial horizontal transfer around 1950, it took only about 20 years for *P* elements to spread throughout *D. melanogaster* on a global basis.

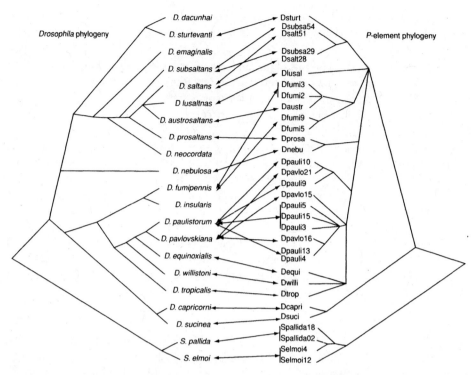

Figure 13.5. Comparison of *Drosophila* species and P-element phylogenetic histories. Double-headed arrows unite P-element clades with the *Drosophila* species from they were sampled. Modified from Fig. 4 in J. C. Silva and M. G. Kidwell, *Molecular Biology and Evolution* 17: 1542–1557 (2000). Copyright © 2000 by Oxford University Press.

By escaping the constraints of gametic transmission, some transposable elements have acquired a remarkable strategy for evolutionary success. However, their evolution as an independent infectious agent still interacts with targets of selection at and below the level of individuals after horizontal transfer has occurred. These multiple levels of selection are not mutually exclusive but rather are interactive in how they shape the response of these remarkable and highly successful units of selection.

Table 13.4. Number and Percentage[a] of Tested Strains Collected in Four Major Geographical Regions during Five Time Periods without (P_{neg}) and with (P) P Elements

Period	Americas		Europe and Asia		Africa		Orient and Australia	
	P_{neg}	P	P_{neg}	P	P_{neg}	P	P_{neg}	P
1920–1949	11 (100)	0 (0)	10 (100)	0 (0)	3 (100)	0 (0)	—	—
1950–1959	11 (85)	2 (15)	11 (100)	0 (0)	4 (100)	0 (0)	11 (92)	1 (8)
1960–1969	6 (32)	13 (68)	24 (86)	4 (14)	4 (80)	1 (20)	9 (75)	3 (25)
1970–1979	4 (8)	49 (92)	35 (51)	33 (49)	4 (68)	2 (33)	9 (39)	14 (61)
1980–1986	1 (4)	27 (96)	50 (56)	40 (44)	11 (37)	19 (63)	16 (43)	21 (57)

Source: Modified from Anxolabehere et al. (1988).
[a] Percentage are numbers in parentheses.

Unequal Exchange in Tandem Multigene Families

As transposons spread throughout a genome, they form what is known as a **multigene family**; that is, many copies of what was originally a single DNA element now coexist at different locations within a single genome. In the case of transposons, this multigene family is frequently dispersed, meaning that the copies are not necessarily found next to one another. Another major type of multigene family is a tandem family in which the copies tend to exist adjacent to one another on the same chromosome. These are not mutually exclusive categories because some multigene families consist of several dispersed clusters throughout the genome, with each cluster containing multiple tandem copies.

Within tandem families, many mechanisms of **unequal exchange** exist that allow what is originally a single copy in this family to duplicate itself and occupy more than one position on the chromosome. Among the mechanisms of unequal exchange in tandem families is gene conversion, and another common mechanism is unequal crossing over, as shown in Figure 13.6. To understand better the evolutionary dynamics of a multigene family, we need to extend our concept of genetic homology. Traditionally, **genetic homology** refers to all the copies of a gene that exist at a particular locus, literally a position in the genome. From coalescent theory (Chapter 5), we expect all such copies to be descendants of a single common ancestral gene, which ties genetic homology to the more general idea of homologous traits being traits derived from a common ancestral condition. Multigene families create problems with this traditional definition of genetic homology. As can be seen in Figure 13.6, unequal crossing over can cause a gene originally at just one position or locus to have descendant copies that occupy different positions or loci on the same chromosome. These copies are also homologous in the fundamental sense of being descendant from a common ancestor, but they violate the usual concept of genetic homology by occupying different loci or positions in the genome. To accommodate this problem, the concept of genetic homology has been extended to **orthology**, the original definition of genetic homology of all the copies of a gene occupying the same locus, and **paralogy**, sets of genes related by descent from a common ancestral gene but that occupy different loci in the genome. Unequal cross over, paralogous gene conversion, and transposition are just three mechanisms

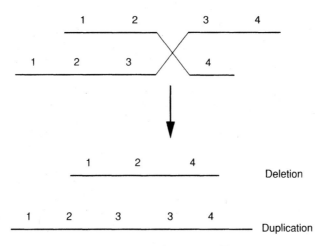

Figure 13.6. Unequal crossing over in tandem, multigene family. The numbers represent tandem copies of a repeating DNA unit.

that can generate paralogous copies of an originally orthologous set of genes. Thus, genetic elements affected by such processes represent another target of selection below the level of the individual.

As with the other targets of selection below the level of the individual, genes in tandem families are often associated with multiple targets of selection. Note from Figure 13.6 that unequal crossing over generates variation in the number of copies of the gene in the tandem family, and this is also true for other mechanisms of unequal exchange. However, there is often selection at the level of the individual for the number of copies in the multigene family. We already have seen an example of this in Chapter 11 with respect to the genes coding for the α chain of hemoglobin. Normally, this is a simple multigene family consisting of just two tandem copies ($Hb\alpha$ 1 and $Hb\alpha$ 2) on chromosome 16. As discussed in Chapter 11, it is important for the health of the adult individual that the production of α chains matches that of the β chains; otherwise an anemic condition known as thalassemia develops. In humans, this production is normally balanced when a person has two orthologous copies of the β-chain locus and the four orthologous/paralogous copies of the duplicated α-chain locus. The greater the discrepancy in the number of copies of the α-chain family from the normal diploid copy number of 4, the more clinically severe is the thalassemia (Chapter 11). Although a mild thalassemic condition can be favored by natural selection in a malarial environment (Chapter 11), natural selection at the individual level usually favors those gametes bearing just two copies of the α-chain genes because of thalassemia. Thus, although a process of reduction and increase in copy number is constantly occurring within our species, stabilizing selection ensures that most chromosome 16's in humans bear exactly two copies. Once again, to understand the evolution of even this simple tandem family, we have to integrate the effects of multiple targets of selection.

Mechanisms of unequal exchange also interact with other population-level evolutionary forces, such as genetic drift. For example, Weir et al. (1985) examined the joint effects of unequal exchange and genetic drift upon shaping the amount and pattern of genetic variation within a tandem multigene family. Recall from Chapter 5 that the overall rate of fixation of neutral, orthologous alleles at a single locus is given by $(2N\mu)[1/(2N)] = \mu$, where μ is the neutral mutation rate and N is the population size. Now consider a tandem multigene family with n tandem repeats per chromosome. We will assume that n is a constant, thereby mimicking the situation seen with the α-globin genes in which natural selection maintains a nearly constant n in the population. However, we will assume neutrality of the mutants arising within the multigene family. Because unequal exchange allows a gene to spread to paralogous positions, there are now two components to fixation: fixation of all orthologous copies at a particular locus (or in the coalescent sense, descent of all orthologous copies from a common ancestral gene) and fixation of all paralogous copies on the same chromosome to the same ancestral form (in the coalescent sense, the descent of all paralogous copies on the chromosome to a common ancestral gene). Thus, we have a combination of fixation at the population level of all orthologous copies and fixation at the level of the chromosome of all paralogous copies. However, ultimately all genes in the tandem family are homologous, so all orthologous and paralogous copies will eventually undergo fixation or coalescence to a common ancestral gene. The total number of genes in the tandem family is $2Nn$, and under neutrality all of these copies are equally likely to become the common ancestor of all future genes in this family. Thus, under neutrality, the probability of fixation of any particular gene copy in this family is $1/(2Nn)$. Retaining μ as the neutral mutation rate per locus, the rate of input of new mutations into the entire family is $2Nn\mu$. In analogy to equation 5.3, in the multigene family

$$\text{Rate of neutral evolution} = \frac{1}{2Nn} \times 2Nn\mu = \mu \qquad (13.12)$$

Equation 13.12 reveals a rather startling conclusion: The evolutionary dynamics of a multigene family under genetic drift and neutral mutation when coupled with a molecular mechanism(s) of paralogous fixation are the same as that of a single locus over long periods of evolutionary time. Thus, the entire family, regardless of n, evolves under neutrality as if it were a single locus. The evolutionary impact of unequal exchange is invisible in equation 13.12.

Equations 5.3 and 13.12 only refer to long-term evolutionary dynamics and are based on the assumption of ultimate coalescence of all orthologous and paralogous copies to a common ancestral gene, no matter how long that coalescence may take. However, on shorter time scales, the time to coalescence, in both the orthologous and paralogous senses, does matter. As we saw in Chapter 5, it takes an average of $4N$ generations to go to population fixation (coalescence to the common ancestral gene) for all orthologous copies at a particular locus. Weir et al. (1985) showed that for all orthologous *and* paralogous copies

$$\text{Expected time to fixation} = \begin{cases} 4N & \text{when } \alpha > \dfrac{1}{2N} \\ \dfrac{2}{\alpha} & \text{when } \alpha \leq \dfrac{1}{2N} \end{cases} \qquad (13.13)$$

where α is the probability of a gene converting (by any applicable molecular mechanism) a paralogous gene to its state. In the first case in equation 13.13, evolution proceeds with the same coalescent dynamics as a single-locus system. In particular, the α parameter, which measures the strength of the molecular-level forces causing paralogous exchange, has no impact on the expected time to fixation. Note that the molecular-level forces become irrelevant when α is strong relative to the population-level force of genetic drift, whose strength as we saw in Chapter 4 is measured by $1/(2N)$. When $\alpha > 1/(2N)$, a gene spreads to paralogous positions within chromosomes more rapidly than drift causes orthologous fixation between chromosomes. As a result, by the time orthologous fixation has occurred for a particular chromosome, paralogous fixation has already occurred on that chromosome. Thus, orthologous fixation is the limiting step to global fixation in the multigene family when drift is a weak evolutionary force compared to the molecular forces of paralogous conversion. Because mutations are occurring throughout this process, the rapid fixation dynamics of paralogous copies relative to population fixation of orthologous copies means that more of the genetic variation in the multigene family exists as differences between chromosomes at the population level rather than among paralogous copies within a chromosome at the genome level.

In the second case in equation 13.13, the molecular-level force of paralogous exchange is only as strong as or weaker than genetic drift at the population level. This means that orthologous fixation will tend to occur within the population at any given locus as or more rapidly than fixation of the paralogous copies within a chromosome. Hence, the time to total fixation of all orthologous and paralogous copies is limited by the rate of paralogous fixation. Now, the time to total fixation is a function of α and genetic drift does not influence this expected time. In this case, there will be much variation among the paralogous copies within a chromosome, with the level of variation increasing as α decreases.

Note that the molecular phenotype (measured by α) is important in determining the evolutionary dynamics of fixation and patterns of neutral variation *only* when the molecular phenotype is *weak* relative to the population-level evolutionary force of genetic drift. This at first may seem counterintuitive, and indeed some have verbally argued that strong molecular-level forces will override population-level processes (Dover 1982). However, the model of Weir et al. (1985) shows that just the opposite happens: Strong forces at the molecular level in this case accentuate the importance of population-level evolutionary forces. The resolution of this paradox is to focus upon genetic variation at the population level and the dynamics of a population through time—the standard focus of microevolutionary theory. A strong force that results in rapid fixation will have little impact on levels of variation over long periods of time; a weak force in contrast can make a substantial contribution to variation during that long time period. Indeed, we have seen this phenomenon before. In Chapter 5, we investigated the level of neutral genetic variation found in a population as a function of μ and genetic drift. One of the fundamental breakthroughs of Kimura's neutral theory was the realization that large amounts of genetic variation can be maintained in a population whose only population-level evolutionary force is drift. As shown by equation 5.7, the levels of neutral variation go up as drift becomes weaker and weaker (N increases). The startling conclusion of the neutral theory was that genetic drift was an important evolutionary force even in large populations, not just small populations, and indeed large populations have more genetic variation influenced by drift than a small population (Chapter 5). So saying an evolutionary force is weak is not the same as saying it is unimportant in evolution; often, it means just the opposite.

Equation 13.13 has another important implication when the molecular processes of unequal exchange are stronger than drift. Suppose a multigene family is created by a gene duplication event in an ancestral species that subsequently gives rise to two or more present-day species (Figure 13.7). If no molecular mechanisms of unequal exchange existed, a mutation could not spread from its original locus of origin to paralogous locations. Thus, the paralogous copies within a species would not coalesce until sometime *before* the original gene duplication in the ancestral species (Figure 13.7a), implying they should be very divergent from one another. In contrast, orthologous genes between the current species should coalesce to the time of the original duplication event (Figure 13.7a). Thus, without a mechanism for paralogous spread of new mutants, the orthologous comparisons *between* species should be more similar than the paralogous comparisons *within* species (Figure 13.7a). However, when the molecular forces for paralogous spread are strong, all orthologous and paralogous genes should coalesce within a species with an expected time of $4N$ generations. Generally, $4N$ is going to be much smaller than the time at which speciation occurred in the past. Because of the convergence of the entire multigene family to single-locus coalescent dynamics under strong mechanisms of unequal exchange, the orthologous comparisons between species should be much more divergent than the paralogous comparisons within species (Figure 13.7b). This pattern is called **concerted evolution** because all copies of the multigene family evolve together, in concert, sharing the same mutational substitutions that discriminate one species from another. Many multigene families do indeed display the pattern of concerted evolution, as expected from equation 13.13.

In all of the examples of targets of selection below the level of the individual, we see that it is the balance between molecular- and population-level forces that determines the evolutionary outcome. A single unit of selection can have targets of selection both at the

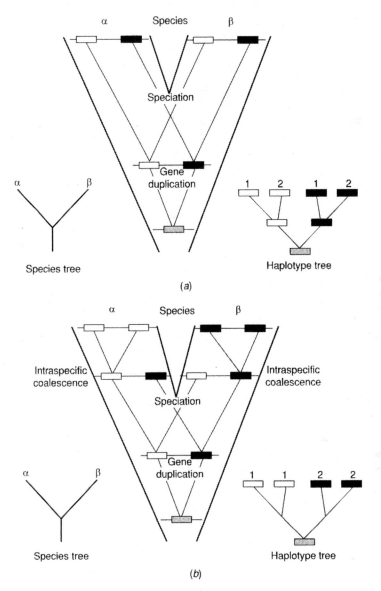

Figure 13.7. Evolution of tandem duplicated locus (*a*) without mechanisms of exchange between paralogous copies within a species and (*b*) with mechanisms of paralogous exchange. In both panels, a species tree is shown on the left and a haplotype tree on the right. The two trees are combined in the middle, with the thick lines outlining the species tree and the thin lines the haplotype tree imbedded within in it. An ancestral gene is shown in grey, which is then duplicated into two paralogous copies, shown initially in black and white, in the common ancestral species to present-day species α and β.

molecular or meiotic levels and at the individual level, and all the targets of selection must be considered to determine how the unit of selection responds. Even in those cases where the only target of selection is below the level of the individual, the within-individual processes leading to this target of selection are still constrained by and the evolutionary outcomes shaped by population-level evolutionary forces such as system of mating or genetic drift.

It is misleading and inappropriate to think of selection below the level of the individual as separate from or independent of evolution at other biological levels.

TARGETS OF SELECTION ABOVE LEVEL OF INDIVIDUAL

Many phenotypes emerge at the level of interactions between two or more individuals. If these interactions are recurrent across generations, they will have continuity over time and can be targets of selection. Targets emerging from interactions among individuals are not difficult to find. For example, of the three major fitness components outlined in Chapter 11, two in general are more appropriately assigned to an interacting pair of individuals: mating success and fertility/fecundity. An individual in a dioecious species does not truly have a phenotype of mating success or fertility; such phenotypes take on biological reality only in the context of an interaction with another individual. Another common target of selection is intraspecific competition, a phenotype emphasized by Darwin as being important in his theory of natural selection. But competition also takes on biological reality only in the context of interactions among individuals. It is always possible to assign an average phenotype to the individuals engaging in such interactions, but it is more accurate biologically to assign such phenotypes to the interacting individuals rather than to any single individual. Similarly, individuals who are relatives, particularly in species that have family structures, often interact in complex ways that can affect each other's fitnesses. As we shall see in this section, qualitatively new properties of natural selection emerge when we assign such interaction phenotypes directly to the set of interacting individuals rather than to each individual as a separate entity.

Sexual Selection

Sexual selection refers to the selection targeting the events that lead up to successful mating or its failure. Many of these events emerge from interactions among individuals and thereby constitute targets of selection above the level of the individual. Such targets under sexual selection are often split into two types:

- **intrasexual selection**, which arises out of competition between individuals of the same sex for mates, and
- **intersexual selection**, which arises out of the interactions between individuals of opposite sex that lead to mating or its failure.

As we saw in the previous section, a single unit of selection can have multiple targets of selection, and the same is true for traits under sexual selection. For example, males of the cricket, *Gryllus integer*, produce a trilled calling song that attracts females, resulting in enhanced mating success. Female crickets prefer male calling songs that are close to the mean of the population for the number of pulses per trill. Thus, the sexual selection that emerges from this male–female interaction is stabilizing upon the phenotype of number of pulses. However, gravid females of the parasitoid fly *Ormia ochracea* use this same song to localize new victims and indeed prefer the same number of pulses as the female crickets. Once parasitized, the males die in about seven days, so those males with songs that are preferred by females also have lowered viabilities, which selects for males with songs away from the average. As a consequence, the trait of male calling song has targets of selection at the level of mating pairs and at the level of individual viability. In this case, the selective targets are antagonistic and together maintain high levels of genetic variation for male calling song in this species, just as antagonistic targets of selection maintain the polymorphic t complex in mice.

Not only can a single unit of selection have multiple targets of selection, but in addition a single target of selection can induce selective responses at multiple units of selection. Targets of sexual selection frequently are aimed at multiple units of selection because intrasexual and intersexual selections are often interwoven, resulting in conflicts that direct selective responses to different units of selection in the two sexes. For example, males of the butterfly *Pieris napi* attempt to mate with many females, and male reproductive success typically increases with the number of matings obtained (Wedell 2001). Females also can mate more than once during a reproductive cycle, and if ejaculates overlap, the reproductive success of one or more of the males that mated with a multiply-mated female can be decreased relative to what it would have been with no ejaculate overlap, a phenomenon known as sperm competition and a type of male–male competition. Females of this species undergo a period of nonreceptivity for remating, and the duration of this period has a genetic component. Another component of phenotypic variation in the nonreceptive period is the degree to which the female's sperm storage organ is filled; the more full this organ, the less willing the female is to remate. Such a response by females normally ensures that they mate often enough to always have sperm available. Thus, the genetic component for the duration of the female nonreceptivity period as a response to the amount of stored sperm can serve as a unit of selection arising out of intersexual mating interactions.

The phenotypic variation in female receptivity period also creates variation in the amount of intrasexual sperm competition. Male butterflies produce two types of sperm: normal, eupyrene sperm capable of fertilization and anucleate, apyrene sperm that are incapable of fertilization. Up to 90% of the sperm can be apyrene. Such sperm can provide nutrients to the female that can affect her reproductive success, and such nonfunctional sperm also fill the sperm storage organ. These nonfunctional sperm have a major impact on decreasing the female's receptiveness to other males. Thus, the ejaculate, a major indicator of male success in intersexual selection, has also been strongly shaped by male–male intrasexual competition. Therefore, the genes underlying the production of nonfunctional sperm in males provide a second unit of selection.

In Chapter 10, we saw that when phenotypes arise from genetic architectures characterized by epistasis (interactions between loci), there is an apparent confoundment between apparent causation of phenotypic variation and frequency (recall Figure 10.15). The same is true when the phenotype of fitness emerges from the interactions of two or more individuals. In general, *selection arising out of interactions among individuals is frequency dependent when fitness is averaged over interactors in order to assign a fitness measure to individuals*.

With sexual selection, the other interactors are other individuals with a given genotype. Hence, the mating success or fertility of an individual is expected to depend upon the attributes of the individual *and* the frequencies of genotypes in the population. This dependence upon genotype frequency was shown in one of the early empirical estimates of male mating success (Ehrman 1966, 1967). Two inversions, *AR* and *CH*, behave as polymorphic alleles in the fruit fly *Drosophila pseudoobscura*. Experiments were executed to estimate the male mating success of the three genotypes *AR/AR*, *AR/CH*, and *CH/CH*.

Male mating success was estimated by the number of matings each male genotype achieved divided by the expected number of matings for that genotype under a neutral, random-mating model. Mating success was measured in many different populations over a range of genotype frequencies. Anderson (1969) found that the mating success in Ehrman's data fit well equations of the form

$$0.6 + \frac{\alpha_{ij}}{G_{ij}} \quad (13.14)$$

where G_{ij} is the frequency of genotype ij and α_{ij} is an empirically derived constant that scales the frequency-dependent mating success of males with genotype ij. Note that the mating success of any given male genotype is inversely proportional to its frequency, a result often described as a rare male mating advantage. The α constants obtained for these three inversion genotypes were 0.2 for *AR/AR*, 0.1 for *AR/CH*, and 0.05 for *CH/CH*. We will generalize these α values to x, y, and z, respectively, for the three genotypes. Note that the *AR/CH* heterozygote α value appears intermediate between the α values of the two homozygotes (codominance), and it seems like there ought to be directional selection favoring the *AR/AR* genotype because it has the largest α term. A graph of male mating success using these values also shows that for all male genotype frequencies the *AR/AR* genotype has the highest fitness and the *CH/CH* genotype has the lowest (Figure 13.8). This graph also seems to imply that there should be directional selection favoring the *AR/AR* genotype. However, not all males are in equal genotype frequencies in most populations, so this seeming superiority of *AR/AR* can be violated in many populations.

To see this, consider the following model. We assume that female fitness is not affected by this polymorphism. Thus, we set the relative fitness of all female genotypes equal to 1. We assume the three male genotype fitnesses are as given above. Then, the average female fitness is 1, and the average male fitness (noting that the three male genotype frequencies

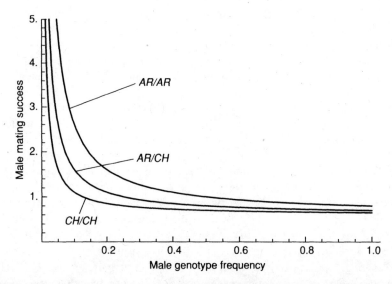

Figure 13.8. Plot of male reproductive success for three genotypes *AR/AR*, *AR/CH*, and *CH/CH* in *D. pseudoobscura* as function of male genotype frequency. Based upon the fitness model given in Anderson (1969).

must sum to 1), is

$$\bar{w}_{\text{male}} = G_{AR/AR}\left(0.6 + \frac{x}{G_{AR/AR}}\right) + G_{AR/CH}\left(0.6 + \frac{y}{G_{AR/CH}}\right)$$
$$+ G_{CH/CH}\left(0.6 + \frac{z}{G_{CH/CH}}\right) \quad (13.15)$$
$$= 0.6 + x + y + z$$

which equals 0.95 for the empirical values of x, y, and z given above. Assuming a 50–50 sex ratio, the average fitness for the population over these three genotypes is $\frac{1}{2}(1) + \frac{1}{2}(0.95) = 0.975$. Note that in this case the average fitness is a constant over all possible genotype frequencies. Recall Fisher's fundamental theorem of natural selection and the concept of an adaptive surface from Chapter 11. Because the average fitness is a constant over the entire genotypic space, there is no fitness peak and there is no possibility that selection of any sort can increase average fitness. Does this mean that there is no selection or that selection exists in this case but in a manner that violates or at least makes irrelevant Fisher's fundamental theorem?

To answer these questions, we will simplify the model by assuming Hardy–Weinberg genotype frequencies within each generation (although we do not assume constant allele frequencies across generations). Obviously, mating is not totally random in this case, but the large deviations from random-mating expectations occur only for very rare genotypes (Figure 13.8), so we can approximate the overall population genotype frequencies within each generation by the Hardy–Weinberg expectations. Despite the flat fitness surface, we can see if evolution is occurring due to sexual selection by seeing if the allele frequency is constant. Because there are no fitness differences among the females, there is no change in allele frequency in the half of the gametes contributed by females. Using the model shown in Figure 11.2 with all fitness components set to 1 except mating success where we use $m_{ij} = 0.6 + \alpha_{ij}/G_{ij}$ and using Hardy–Weinberg genotype frequencies, the allele frequency p' of AR in the male half of the gamete pool starting with an initial frequency of p is

$$p' = \frac{p^2\left(0.6 + x/p^2\right) + pq\left(0.6 + y/2pq\right)}{\bar{w}_{\text{males}}}$$
$$= \frac{0.6p + x + y/2}{0.6 + x + y + z} \quad (13.16)$$

Hence, the change in the allele frequency in the male-derived gametes is given by

$$\Delta p = p' - p = \frac{0.6p + x + y/2}{0.6 + x + y + z} - p$$
$$= \frac{x + y/2 - p(x + y + z)}{0.6 + x + y + z} \quad (13.17)$$
$$= \frac{p}{\bar{w}_{\text{males}}}\left[p\left(0.6 + \frac{x}{p^2}\right) + q\left(0.6 + \frac{y}{2pq}\right) - \bar{w}_{\text{males}}\right]$$
$$= \frac{p}{\bar{w}_{\text{males}}} a_{AR}$$

where a_{AR} is the male average excess of the AR allele for the phenotype of male mating success (fitness), as given in the brackets of equation 13.17. Note that in general $\Delta p \neq 0$, so the fitness differences associated with male mating success are indeed inducing evolution via natural selection. Hence, Fisher's fundamental theorem is violated in this case because selection is inducing evolutionary change but with no change in average fitness. Such a violation of Fisher's theorem is a general property of frequency-dependent fitness models, and this constitutes a major limitation of this theorem as frequency-dependent processes arise commonly. However, note that equation 13.17 is simply a special case of equation 11.5, so once again we see that selection is driven from the gamete's point of view even though the target of selection is at the level of interacting individuals. Hence, the measured genotype approach to selection described by equation 11.5 is applicable to a broader array of selective situations than Fisher's fundamental theorem.

Because equation 13.17 is a special case of equation 11.5 and because no allele frequency change is induced in gametes derived from females in this model, the selective equilibrium defined by the variation of male mating sense occurs when the average excess of male mating success is zero. Setting the bracketed term in equation 13.17 to zero and solving for p, the equilibrium allele frequency is

$$p_{eq} = \frac{x + y/2}{0.6 + x + y + z} \tag{13.18}$$

Using the empirically derived values given earlier, the selective equilibrium in the *D. pseudoobscura* population should be 0.71. To see if this equilibrium is stable, equation 13.17 tells us we need to look at the sign of the average excess of mating success above and below the equilibrium frequency. If $p = 0.75$, then the male $a_{AR} = -0.017$ and the AR allele decreases back to the equilibrium point. Likewise, when $p = 0.6$, the male $a_{AR} = 0.067$ and selection increases the AR allele up toward the equilibrium point. Hence, there is a stable, balanced equilibrium associated with this selection. From the gamete's point of view as measured by average excesses, there is an *adaptive* peak (Curtsinger 1984), and the gametes are on this peak when $p = 0.71$. However, this peak is not "visible" when we assign the fitnesses to the individuals, a process which results in a flat adaptive surface with a constant average fitness of 0.975. Indeed, this selective equilibrium that has no impact on average fitness seems even more peculiar when we look at the actual fitness values assigned to individuals. Substituting the Hardy–Weinberg genotype frequencies at $p = 0.71$ into equations 13.14 with the empirical α values, the equilibrium fitnesses assigned to the three male genotypes are 0.997 for *AR/AR*, 0.843 for *AR/CH*, and 1.195 for *CH/CH*. Note that at equilibrium the male mating success of the heterozygote is *lower* than that of either homozygote. Normally, this would suggest an unstable equilibrium state, yet the average excess calculations show that this is a stable equilibrium. The average fitness of the population and the fitnesses of individual genotypes are either uninformative or, worse, misleading when trying to understand the selective equilibrium and dynamics that emerge from frequency-dependent mating success. Targets of selection above the level of the individual lead to many paradoxes when we try to assign the fitnesses to individuals rather than interacting sets of individuals. The target of selection is not at the individual level in this case, and this is what results in the strange behavior of fitnesses assigned to the individual level.

Only by taking the gamete's perspective as measured by average excesses can we successfully predict selective equilibria and stability properties. Curtsinger (1984) shows that these roles of the average excess are true in general for frequency-dependent selection as well as for lower level targets of selection such as meiotic drive. Thus, the key to understanding the evolutionary response to natural selection remains in taking a gametic perspective regardless of the target of selection. When a target of selection is below the level of the individual, gametes can have selectable phenotypes directly or can be the direct bearers of the consequences of this lower level selection. Only when a DNA element can be transmitted to future generations in a vehicle other than a gamete (such as some transposons) can the importance of the gamete be circumvented. Chapters 11 and 12 showed that individual-level selection is determined by the gamete's average excess and that selective responses cannot be made just from a knowledge of individual fitnesses. Now we see that even when fitness emerges from interacting sets of individuals, the average excess still retains its selective importance.

Fertility

Fertility is a fitness component that emerges inherently from the interaction of a mating pair of individuals. A nonselfing individual cannot display this phenotype except in the context of an interaction with a mate. For example, in humans, the *ABO* blood group locus has three common alleles that determine one's ABO blood type (Chapter 7). Knowing one's ABO blood type is clinically important because transfusing donor blood bearing an antigen (A and/or B) not found in the recipient can create a potentially lethal immunological reaction in the recipient. If there is leakage of blood across the placenta, a mother can mount an immunological reaction against a developing fetus that bears antigens not found in the mother. This results in a significant increase in the rate of spontaneous abortions in those couples that have *ABO* genotypes that yield ABO maternal–fetal incompatible combinations (Takano and Miller 1972). In the ABO system, the same individual genotype can have drastically different fertilities in the context of different mates. For example, a type O woman married to a type O man would not be at risk for any ABO-induced spontaneous abortions, but a type O woman married to a type AB man would have every pregnancy at risk. Therefore, fertility is not an individual attribute but rather is an attribute of a mating pair of individuals.

To model fertility as a target of selection on mating pairs, we simplify our model as before by assuming random mating and a within-generation approximation to Hardy–Weinberg genotype frequencies. Table 13.5 gives a one-locus, two-allele (*A* and *a*) model in a random-mating population by inserting a fertility component into the original Weinberg model given in Table 2.3. Note that fertility is attributed *directly* to a mating pair rather than an individual in Table 13.5. Note further that the fertilities assigned to the true target of selection are *not* frequency dependent, but rather are constants. However, at the bottom of the table, effective fertilities (the b_{ij}'s) are assigned to the individual genotypes in order to predict the genotype frequencies in the next generation. These individual-level "fertilities" are frequency dependent. Table 13.5 makes it explicit that the appearance of frequency-dependent fitness at the level of the individual is a result of mapping the constant fitnesses of a target of selection at the level of interacting groups of individuals onto the individual level. In such a fitness mapping, the frequencies of the interactors becomes confounded with apparent individual-level fitness causation.

Once the frequency-dependent b_{ij}'s have been defined and assigned to individual-level genotypes, the selective response shown in the row of genotype frequencies in the next generation looks similar to the constant-fitness models given in Chapter 11, so once again we can use the normal average excesses of the b_{ij}'s to calculate the change in allele frequency and determine potential equilibria and their stabilities. However, a closer examination of

Table 13.5. Model of Fertility in Random-Mating Population with Two Alleles (A and a) at Autosomal Locus with Initial Frequency p of A allele

Mating Pair	Frequency	Fertility	Mendelian Probabilities of Offspring (Zygotes)		
			AA	Aa	aa
$AA \times AA$	$p^2 \times p^2 = p^4$	b_1	1	0	0
$AA \times Aa$	$2 \cdot p^2 \times 2pq = 4p^3q$	b_2	$\frac{1}{2}$	$\frac{1}{2}$	0
$AA \times aa$	$2 \cdot p^2 \times q^2 = 2p^2q^2$	b_3	0	1	0
$Aa \times Aa$	$2pq \times 2pq = 4p^2q^2$	b_4	$\frac{1}{4}$	$\frac{1}{2}$	$\frac{1}{4}$
$Aa \times aa$	$2 \cdot 2pq \times q^2 = 4pq^3$	b_5	0	$\frac{1}{2}$	$\frac{1}{2}$
$aa \times aa$	$q^2 \times q^2 = q^4$	b_6	0	0	1
Genotype Frequency in next generation			$\dfrac{p^4 b_1 + 2p^3 q b_2 + p^2 q^2 b_4}{\bar{b}}$ $= p^2 \dfrac{b_{AA}}{\bar{b}}$	$\dfrac{2p^3 q b_2 + 2p^2 q^2 b_3 + 2p^2 q^2 b_4 + 2pq^3 b_5}{\bar{b}}$ $= 2pq \dfrac{b_{Aa}}{\bar{b}}$	$\dfrac{p^2 q^2 b_4 + 2pq^3 b_5 + q^4 b_6}{\bar{b}}$ $= q^2 \dfrac{b_{aa}}{\bar{b}}$
where $b_{ij} =$			$p^2 b_1 + 2pq b_2 + q^2 b_4$	$p^2 b_2 + 2pq \left(\frac{1}{2} b_3 + \frac{1}{2} b_4\right) + q^2 b_5$	$p^2 b_4 + 2pq b_5 + q^2 b_6$

Note: Sex-dependent effects are ignored, so matings such as $AA \times Aa$ and $Aa \times AA$ are regarded as equivalent and hence their frequencies are added. The average fertility of the mating pairs as weighted by their frequencies is given by $\bar{b} = p^4 b_1 + 4p^3 q b_2 + 2p^2 q^2 b_3 + 4p^2 q^2 b_4 + 4pq^3 b_5 + q^4 b_6$.

the forms of these b_{ij}'s in Table 13.5 reveals another complexity that can arise when the target of selection is above the level of the individual. Biologically, the b_{ij}'s are the average fertilities of the *mating pairs* (the targets of selection) given that an offspring of genotype *ij* is produced. In contrast, consider the average fertilities $\bar{b}_{ij}$'s of the *individuals* with genotype *ij* under random mating:

$$\bar{b}_{AA} = p^2 b_1 + 2pq b_2 + q^2 b_3$$
$$\bar{b}_{Aa} = p^2 b_2 + 2pq b_4 + q^2 b_5 \qquad (13.19)$$
$$\bar{b}_{aa} = p^2 b_3 + 2pq b_5 + q^2 b_6$$

By comparing equations 13.19 with the equations for the b_{ij}'s in Table 13.5, there is not a single case in which $\bar{b}_{ij} = b_{ij}$. Thus, the fertilities that we assign to individual genotypes to predict the response to fertility selection are *not* the average fertilities of the individual genotypes. It is *only* the average fertilities of the mating pairs that matter, and this can involve mating pairs that do not even include the focal genotype (e.g., note in Table 13.5 that b_{AA} is a function in part of b_4, the fertility of the $Aa \times Aa$ mating pair). Unfortunately, almost all empirical measurements of fertility are based on average individual fertilities, which are not direct measures of the selective response of this target of selection.

The discrepancy between the $\bar{b}_{ij}$'s and the b_{ij}'s illustrates the difficulty in assigning an individual fitness measure when the target of selection is above the level of the individual. This problem cannot in general be addressed simply by looking at the average attribute of an individual as it interacts with other individuals in the population. Sometimes that works, but often it does not, as shown by the important fitness component of fertility. The heart of this discrepancy relates to a common theme we have noted for selection in many different contexts: the necessity of taking the perspective of a gamete in predicting the response to selection. Note that the b_{ij}'s are defined in terms of the mating pairs that yield a particular type of offspring in the *next* generation; that is, the b_{ij}'s measure fertility in a manner that requires the passage of a gamete from one generation to the next. In contrast, the $\bar{b}_{ij}$'s are the average fertilities of the genotypes in the parental generation and do not take into account what will happen in the next generation. Only the b_{ij}'s are relevant for natural selection. Note that we have a further weakening of the idea that "natural selection is survival of the fittest"—natural selection on fertility does not necessarily favor the most fertile individuals because of the discrepancy between the $\bar{b}_{ij}$'s and the b_{ij}'s. Equation 11.5 tells us that selection favors the fitter gametes, not the fittest individuals. At least when the individual was the target of selection, it was the average fitness of individuals that contributed to the average excess (Chapter 11). However, for fertility the average *individual* fertility does not contribute to the relevant average excess; rather, the b_{ij}'s are the genotypic values that predict the response to selection. Therefore, it is critical that the fitnesses first be assigned to the true target and then followed through the gametes to the next generation, as is done in Table 13.5.

As shown in equation 11.5, the ultimate gametic measure for determining selective response is the average excess. Table 13.5 shows that the selective response is driven by the average excesses of the b_{ij}'s not the $\bar{b}_{ij}$'s. Figure 13.9 shows a plot of the average excess for a special case of the model given in Table 13.5. As can be seen, the average excess of the *A* allele takes on three values of zero at intermediate p's. An examination of the sign of the average excess on either side of these three potential equilibrium points shows that only the middle one, with an equilibrium p at 0.6, is stable. Moreover, both loss ($p = 0$)

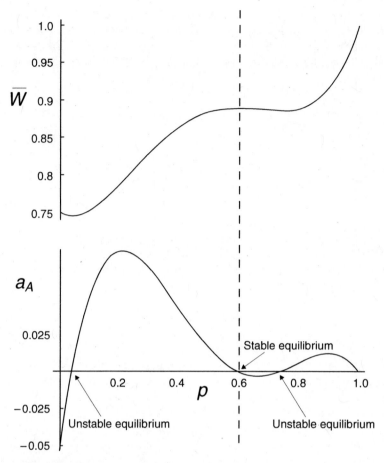

Figure 13.9. A plot of average fertility over all mating pairs and average excess across p for the special case of $b_1 = 1$, $b_2 = 0.7$, $b_3 = 0.8$, $b_4 = 1.3$, $b_5 = 0.7$, and $b_6 = 0.75$ based on the model given in Table 13.5.

and fixation ($p = 1$) are stable equilibria, for a total of three stable equilibria. In general, frequency dependent selection results in many potential equilibrium points, some of which are stable and some of which are not. Therefore, the outcome of natural selection is often sensitive to the initial conditions in frequency dependent models.

Figure 13.9 also shows another attribute of frequency dependent selection that has already been mentioned; average fitness is not necessarily maximized. The average fertility of all mating pairs (which is the same as the average fertility of all individuals) is also plotted across p in Figure 13.9, and as can be seen average fertility is maximized in this special case at $p = 1$, one of the *potential* equilibrium points. Note that average fertility is not maximized at the stable polymorphic equilibrium of $p = 0.6$, and average fertility is close to it's *minimum* value at the stable equilibrium of $p = 0$. This means that selection will often take the population *away* from the p that maximizes average fertility and instead will take the population to a much lower average fertility. Once again, selection does not favor the fittest individuals nor the fittest populations, and this is particularly true when dealing with targets of selection above the level of the individual. What remains true is that selection favors those gametes with positive average excesses, although the phenotypic deviations that go into the average excess must be defined with great care when dealing with targets above the level of the individual.

Competition

Darwin identified competition in the "struggle for existence" as a source of natural selection. Competition inherently involves interactions among individuals who are competing for some resource, and indeed we already saw a form of this in the competition for mates. When two or more individuals compete with one another, what matters is their relative competitive abilities to one another. Indeed, the very phenotype of competitive ability cannot exist at all except in the context of interacting individuals. Hence, the phenotype of competitive ability constitutes another target of selection above the level of the individual. Cockerham et al. (1972) presented a simple one-locus, two-allele model of competition involving pairs of individuals that randomly encounter one another in a random-mating population. The genotype frequencies within a generation are assumed to be in Hardy–Weinberg proportions, and the frequency of an individual competing with another individual of a specified genotype is directly proportional to the frequency of that genotype under the random-encounter assumption. A fitness effect emerges out of these competitive interactions, as given in Table 13.6.

As in the previous models, the fitness components emerging from the interaction of competing individuals are constants when assigned to the proper target of selection. To predict the evolutionary response to competitive selection, we need to put this model into the terms found in equation 11.5; that is, we need to take the gamete's perspective. Unlike fertility, Cockerham et al. (1972) show that the average genotypic values of the competitive fitness interactions that an individual experiences are the relevant projections to individual-level fitness in this case. As shown in Table 13.6, these genotypic values are frequency dependent (the w_i's in Table 13.6). Thus, competitive selection is inherently frequency dependent. As a consequence, selection under competition has the same properties we saw before: Average fitness is not maximized, Fisher's fundamental theorem is violated, projections of individual-level fitness can be misleading, and multiple equilibria exist, some stable and some not. Indeed, frequency-dependent models can yield such complex dynamics for allele frequency changes that analytical solutions are extremely difficult even for relatively simple models (Gavrilets and Hastings 1995). Thus, Cockerham et al. (1972) analyzed their model using the concept of a protected polymorphism.

Table 13.6. Competitive Model of Cockerham et al. (1972)

	AA	Aa	aa
Competing with AA	w_{22}	w_{12}	w_{02}
Competing with Aa	w_{21}	w_{11}	w_{01}
Competing with aa	w_{20}	w_{10}	w_{00}
Average competitive fitness	$w_2 = p^2 w_{22} + 2pq w_{21} + q^2 w_{20}$	$w_1 = p^2 w_{12} + 2pq w_{11} + q^2 w_{10}$	$w_0 = p^2 w_{02} + 2pq w_{01} + q^2 w_{00}$

Note: A random mating population with one locus with two alleles (A and a) with $p =$ the frequency of A. When individuals of genotype i ($i = 0$ is aa, $i = 1$ is Aa, and $i = 2$ is AA) compete with individuals of genotype j, the fitness consequence to genotype i is given by w_{ij}.

A **protected polymorphism** occurs when at least one allele at a locus is favored by natural selection when very rare and selected against when very common. The conditions for protection can be derived from either equation 11.5 or the average excess of fitness by taking the limits as the allele frequency approaches 0 and 1. For protection, the change in allele frequency predicted from equation 11.5 or from the average excess must be positive as $p \to 0$ (that is, selection causes the allele frequency to increase when it is very rare), and they must be negative as $p \to 1$ (that is, selection causes the allele frequency to decrease when it is very common). In such a case, natural selection protects the allele from going to fixation ($p = 1$) or loss ($p = 0$) and thereby favors the maintenance of a polymorphism (and in a model with no drift such as given here ensures polymorphism). For example, in Chapter 11 we discussed the balanced polymorphism of sickle cell anemia in a malarial region. As can be seen from Figure 11.3, the average excess of fitness of the S allele is positive when the frequency of the S allele is small, but it is negative when large. Hence, this balanced polymorphism is also a protected polymorphism. However, the two concepts are not identical, as a balanced polymorphism requires a stable intermediate allele frequency such that $\Delta p > 0$ when p is just below the equilibrium allele frequency and $\Delta p < 0$ when p is just above the equilibrium allele frequency (as in Figure 11.3). However, the protected polymorphism concept focuses exclusively at Δp for p close to 1 and 0; it is not concerned with potential equilibrium points and their stability. With protected polymorphisms, we look exclusively at the *instability* of the potential equilibrium points of $p = 0$ and $p = 1$.

The differences between balanced and protected polymorphisms can be illustrated with the Cockerham et al. (1972) competitive model. First, consider the conditions for protection in this model. All the information needed to calculate the average excesses and equation 11.5 is given in Table 13.6. The limits of equation 11.5 are then taken. These limits are easily determined when there is no dominance or recessiveness. When p is close to zero, the random-mating population will consist almost exclusively of aa homozygotes, so the average fitness of the population will converge to w_{00}, the fitness of aa individuals competing with other aa individuals. Similarly, the average fitness of aa individuals is likewise w_{00} because virtually all of the competitive interaction of aa individuals is with other aa individuals in such a population. As long as p is extremely small but not equal to zero, the few A alleles in this population under random mating will be found almost exclusively in Aa heterozygotes, which in turn will be competing almost always with aa individuals. Hence, the average fitness of Aa individuals for p close to zero is w_{10}. The average excess of A when rare is therefore

$$a_A = (w_{10} - w_{00}) \tag{13.20}$$

that is, only the heterozygote fitness genotypic deviation of Aa is important when $p \to 0$. When $p \to 1$, it is easier to look at the average excess of the a allele ($p \to 1$ means $q \to 0$). The average excess of the a allele under these conditions depends only upon the heterozygote fitness genotypic deviation, which is w_{12} when the population consists almost entirely of AA individuals. Similarly, the average fitness of the population converges to w_{22} when almost all individuals are AA. The average excess of the a allele when A is close to fixation is therefore

$$a_a = (w_{12} - w_{22}) \tag{13.21}$$

The conditions for protection are that equations 13.20 and 13.21 both be positive (note, if equation 13.21 is positive, then $\Delta p < 0$). Thus, in the case of no dominance or recessiveness, the polymorphism is protected whenever

$$w_{10} > w_{00} \quad \text{and} \quad w_{12} > w_{22} \tag{13.22}$$

Inequalities 13.22 mean that the polymorphism is protected whenever the rare heterozygotes are better competitors than the most common homozygote when both are competing against the most common homozygote. Note that only four of the nine competitive coefficients in Table 13.5 influence the conditions for protection. The other fitnesses coefficients obviously influence the selective dynamics of this system, but they play no role in the narrower problem of protected polymorphism.

As an example of protection, consider the competitive fitness matrix shown in Figure 13.10, along with a plot of equation 11.5 over p from 0 to 1, using the w_i's in Table 13.6 for the genotypic values in the average excess calculation. The fitness elements that contribute to inequalities 13.22 are boxed, and as can be seen, the conditions for protection are satisfied in this case. The plot of Δp against p shows this protection, as $\Delta p > 0$ when p is small and $\Delta p < 0$ when p is large. Note in this case $\Delta p = 0$ only at $p = 0.5$, and $\Delta p > 0$ when $p < 0.5$ and $\Delta p < 0$ when $p > 0.5$. Hence, this is a balanced polymorphism with a unique, stable equilibrium. Now consider the situation portrayed in Figure 13.11. Note that fitness elements for protection are identical in this case, but three of the other five elements are not. In this case, the plot of Δp against p looks the same as in Figure 13.10 when p is close to 0 or 1, reflecting the identity of the conditions for protection. However, note in this case there are three potential equilibria, with the one at $p = 0.5$ being unstable and the other two being stable. Hence, there are two local balanced polymorphic points in this case, but there is no global balanced polymorphic point because selection can drive the system to very different states depending upon the initial conditions. The selective dynamics shown in Figure 13.11 are obviously more complex than that shown in Figure 13.10, even though both are identical with respect to being protected polymorphisms. Even the simple competitive model of Cockerham et al. (1972) can result in such complex selective dynamics that in

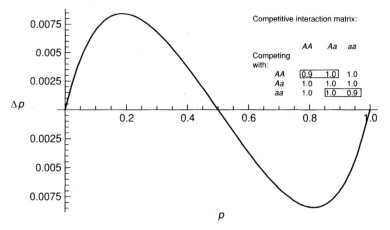

Figure 13.10. Plot of equation 11.5 (Δp) versus p for special case of competition model of frequency-dependent selection given in Table 13.6. The specific fitness values used are indicated.

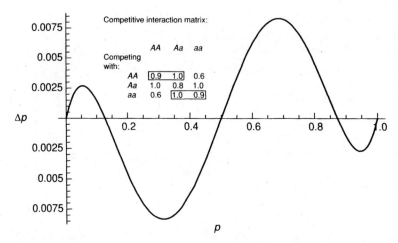

Figure 13.11. Plot of equation 11.5 (Δp) versus p for second special case of competition model of frequency-dependent selection given in Table 13.6. The specific fitness values used are indicated.

many cases we settle for knowing only the conditions for protection. Inequalities 13.22 show that most competitive parameters are not needed for protection, so the conditions for protection are very broad in this frequency-dependent competitive model. In particular, unlike the constant-fitness models of Chapter 11 and illustrated in Figure 11.3, global heterozygote superiority is not needed for polymorphism under frequency-dependent competition.

Kin/Family Selection

In both the fertility model and the competitive model, fitness phenotypes emerged from the interactions of individuals who encountered one another (or mated) at random. However, there are many targets of selection that emerge from individuals whose encounters are highly nonrandom from a genetic perspective. For example, genetically related individuals aggregate in many species, at least for some portion of their lives. This occurs in species with some sort of family structure, which places parents and offspring and/or siblings into a potentially interactive context.

Table 13.7 presents a simple one-locus, two-allele (A and a) family model in a random-mating population (Templeton 1979b). There are two simultaneous targets of selection in this model, the individual and the family. Thus, the fitnesses that we assign to individuals in the *context* of a family are constants (the w_{ij} in Table 13.7). As before, when we assign a fitness to just an individual genotype with no family context (e.g., w_{AA} in Table 13.7), the fitnesses become frequency dependent, reflecting the fact that the targets of selection include a level above the individual. Also note that the fertility selection model given in Table 13.5 is a special case of this more general model when $w_{ij} = b_j$ for all i. Hence, all the properties of the earlier frequency-dependent model are equally applicable here. Kin or family selection therefore does not maximize average fitness, evolutionary outcomes are not predictable from apparent individual-level fitness projections, and the course of evolution violates Fisher's fundamental theorem. However, as the bottom of Table 13.7 shows, just as with fertility selection, we can produce a version of equation 11.5 using, for example, the w_{AA} as genotypic values that allows us to investigate selective dynamics or at least protected polymorphisms.

Table 13.7. Model of Family Selection in Random-Mating Population with Two Alleles (A and a) at Autosomal Locus with Initial Frequency p of A allele

Mating Pair	Frequency	Mendelian Probabilities of Offspring (Zygotes) Times w_{ij}		
		AA	Aa	aa
1. $AA \times AA$	p^4	$1 \cdot w_{21}$	0	0
2. $AA \times Aa$	$4p^3q$	$\frac{1}{2} \cdot w_{22}$	$\frac{1}{2} \cdot w_{12}$	0
3. $AA \times aa$	p^2q^2	0	$1 \cdot w_{13}$	0
4. $Aa \times Aa$	$4p^2q^2$	$\frac{1}{4} \cdot w_{24}$	$\frac{1}{2} \cdot w_{14}$	$\frac{1}{4} \cdot w_{04}$
5. $Aa \times aa$	$4pq^3$	0	$\frac{1}{2} \cdot w_{15}$	$\frac{1}{2} \cdot w_{05}$
6. $aa \times aa$	q^4	0	0	$1 \cdot w_{06}$
Frequency in next generation		$\dfrac{p^4 w_{21} + 2p^3 q w_{22} + p^2 q^2 w_{24}}{\bar{w}}$ $= p^2 \dfrac{w_{AA}}{\bar{w}}$	$\dfrac{2p^3 q w_{12} + 2p^2 q^2 w_{13} + 2p^2 q^2 w_{14} + 2pq^3 w_{15}}{\bar{w}}$ $= 2pq \dfrac{w_{Aa}}{\bar{w}}$	$\dfrac{p^2 q^2 w_{04} + 2pq^3 w_{05} + q^4 w_{06}}{\bar{w}}$ $= q^2 \dfrac{w_{aa}}{\bar{w}}$
where $w_{AA} =$		$p^2 w_{21} + 2pq w_{22} + q^2 w_{24}$	$p^2 w_{12} + 2pq \left(\frac{1}{2} w_{13} + \frac{1}{2} w_{14}\right) + q^2 w_{15}$	$p^2 w_{04} + 2pq w_{05} + q^2 w_{06}$

Note: Sex-dependent effects are ignored, so matings such as $AA \times Aa$ and $Aa \times AA$ are regarded as equivalent and hence their frequencies are added. The terms w_{ij} are the fitnesses of individuals with genotype i ($2 = AA$, $1 = Aa$, and $0 = aa$) in the context of family type (mating-pair type) j. The term $\bar{w} = p^2 w_{AA} + 2pq w_{Aa} + q^2 w_{aa}$ is the average fitness of the offspring from the mating pairs as weighted by their family fitness context and Mendelian probabilities.

Darwin stated that "one of the most serious" problems to his theory of natural selection was the evolution of sterile workers in several groups of insects. In general, this leads to the problem of the evolution of **altruism**, that is, traits that appear deleterious at the individual level but beneficial for a group of individuals. The dual-target model given in Table 13.7 can accommodate these two levels of selection and can therefore be used to investigate the evolution of altruism in the context of interactions with kin in a family structure. For example, suppose a population is initially fixed for the A allele, but a new allele, a, mutates that influences an altruistic phenotype in a recessive mode. That is, within a family, any individual expressing this recessive phenotype has lower fitness than his or her sibs that do not express this phenotype, but the presence of altruists in a sibship increases the overall fitness of their nonaltruistic sibs, with the benefit increasing with an increasing proportion of altruistic individuals. This recessive case is modeled by assuming $w_{15} > w_{24} = w_{14} > w_{21} = w_{22} = w_{12} = w_{13} > w_{04}, w_{05}, w_{06}$. The equalities reflect the dominance of the nonaltruist phenotype, and the inequalities ensure that the altruist phenotype always has lower fitness than the nonaltruist in every family, but the more altruists, the higher the fitness of the nonaltruist. Because a is initially a new mutation, we start close to $p = 1$. The average excess equation is more complex in this case, but it is still a straightforward exercise to take the limit as $q \to 0$ to examine the properties of Δq when q is rare. Doing so, the conditions for $\Delta q > 0$ when q is rare is either $w_{24} > 2w_{21} - w_{04}$ or, if $w_{24} = 2w_{21} - w_{04}, 2w_{24} + w_{21} < 2w_{05}w_{15}$ (Templeton 1979b).

An examination of Table 13.7 reveals why these particular fitness components determine whether or not a increases when rare. Most families in such a population consist of family type 1, which only has AA offspring with fitness w_{21}. When q is rare, the most common family type in which the altruist phenotype will appear is family type 4, in which case the nonaltruist dominant phenotype has fitness w_{24} and the altruist phenotype has fitness w_{04}. Hence, just these three fitnesses determine the fate of the altruistic allele when rare, and the other seven components are irrelevant unless $w_{24} = 2w_{21} - w_{04}$. This equality effectively makes the contribution of family type 4 slight by itself for the selective dynamics of q when q is rare, so now the dynamics become influenced by both the first and second most common family types in which the altruist phenotype appears when q is rare, family types 4 and 5. Hence, the conditions for increase of the a allele now include the fitness of the nonaltruist phenotype in family type 5, w_{15}, and the fitness of the altruist in family type 5, w_{05}. When one of these conditions is satisfied, the altruist allele will increase in the population through its beneficial effects on kin even though the altruist itself has lower individual fitness to the nonaltruist in every family context. Similarly, by looking at Δp when p is close to zero, the fitness inequalities given above ensure that $\Delta p > 0$. Therefore, the original inequalities combined with the conditions to ensure the increase of the a allele when rare result in a protected polymorphism. If we change the inequalities to now have $w_{06} > w_{13}$ (that is, the fitness of a pure brood of altruists is higher than the nonaltruist fitness in family type 5, the most common family type in which nonaltruists exist when p is close to zero), $w_{05} > (w_{21} + w_{24})/2$, and $w_{04} > w_{21}$ (conditions that ensure that altruists are more fit on the average than nonaltruists at intermediate allele frequencies even though altruist sibs are still less fit than their nonaltruist sibs within families), then fixation of the altruist allele is the only globally stable solution (Templeton 1979b). Hence, an altruist trait can indeed evolve through the action of natural selection when the target of selection is a mixture of individual-level selection and selection among interacting related individuals in the context of a family. Hence, selection emerging from interacting biological relatives (often called kin selection) can solve Darwin's major difficulty with his theory of natural selection (Hamilton 1964).

When discussing selection below the level of the individual, we noted that a mixture of targets below the individual and at the individual level frequently occurred (e.g., the t locus in mice). Consequently, kin and related family selection models that combine individual-level and above-individual-level targets of selection should come as no surprise. We end this chapter with an example that combines targets of selection at all three levels: below, at, and above the level of the individual. In Chapter 10, we mentioned the human genetic disease gene for Huntington's chorea. Huntington's disease is inherited as an autosomal dominant genetic disease and is associated with a late age of onset degeneration of the central nervous system (usually after 40 years of age, but sometimes earlier) that ultimately causes death. After the gene was cloned (Chapter 10), it was discovered that Huntington's disease is a trinucleotide disease, in this case a CAG repeat. As mentioned earlier in this chapter, trinucleotide diseases are often targets of selection below the level of the individual, and Huntington's is no exception. Alleles with 35 or fewer CAG repeats are normal (h alleles) and are not associated with the disease. New mutations for Huntington's disease (H alleles) arise from alleles with between 29 and 35 CAG repeats that expand on transmission through the paternal germline to 36 CAGs or greater (Chong et al. 1997). Changes in CAG number occur in only 0.68% of normal chromosomes, but once the threshold of 36 is past, this increases to 70% in male germlines (Kremer et al. 1995). Once past the threshold, the mutation rate continues to increase with increasing repeat number, reaching 98% in males with at least 50 repeats. The extraordinarily high mutation rates are most consistent with a mutation process that occurs throughout germline mitotic divisions, rather than resulting from a single meiotic event (Leeflang et al., 1999). Not only is the rate of mutation nonrandom in this case, but the direction of mutation is also highly nonrandom. When mutations occur in the male germline, they are strongly biased to increase CAG number, even in normal alleles. Hence, one target of selection in this case is the genetically influenced biased mutation within male germlines. Indeed, Rubinsztein et al. (1994) showed that the observed distribution of alleles at this locus is largely explicable in terms of this nonrandom mutation within male germlines plus genetic drift at the population level.

These germline selective processes have a direct effect at the level of the individual phenotype. Because germline selection operates to increase the number of repeats, there is a preferential transition from normal to disease alleles, and once a disease allele exists, it is extremely unlikely to back mutate to a normal allele despite the extraordinarily high mutation rates. Thus germline selection is biased toward producing individuals with the disease phenotype. Moreover, the age of onset of the disease is highly correlated with repeat number (Rubinsztein et al. 1997; Figure 13.12). Interestingly, there is also epistasis with the *ApoE* locus for age of onset in men, with the $\varepsilon 2/\varepsilon 3$ genotype having a significantly earlier age of onset (Kehoe et al. 1999). This is not surprising because Huntington's disease is a neurodegenerative disease, and normal nerve functioning is highly dependent upon lipids, which in turn are influenced by *ApoE* (Chapter 8). Combining germline selection with this individual-level phenotypic consequence, most newly arisen disease alleles start with low copy numbers and therefore late ages of onset. Indeed, the age of onset is so high in individuals with low-copy-number alleles that many of the heterozygotes have completed their reproduction or die from other causes before the disease is expressed (Falush et al. 2001). Hence, new disease alleles with late age of onset are expected to be neutral at the individual level. As the alleles go through more and more male generations, germline selection causes their copy number to increase and, thereby, decreases the age of onset. As

450 UNITS AND TARGETS OF SELECTION

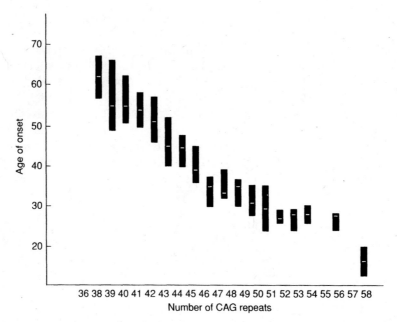

Figure 13.12. Relationship between CAG repeat number in Huntington's disease and median age of onset of disease. For each repeat number, the median age of onset is indicated by an open bar, and the 95% confidence interval for age of onset is given by a solid bar. Modified from Fig. 1 in D. C. Rubinsztein et al., *Proceedings of the National Academy of Sciences USA* 94: 3872–3876 (1997). Copyright © 1997 by the National Academy of Sciences.

can be seen from Figure 13.12, once the repeat number is greater than 46, most individuals develop the disease in their thirties or younger, the prime reproductive ages in developed countries. By the time germline selection drives the copy number above 60, virtually all disease phenotypes develop while the individual is a teenager or younger, making the disease virtually lethal in a reproductive sense (the relationship between fitness and age of onset of a deleterious genetic disease will be examined in detail in Chapter 15). Hence, the germline selection goes in a direction that creates increasingly strong individual-level selection in the opposite direction. Putting the two levels of selection together, the expected trajectory of any newly arisen disease allele is to increase repeat number during an initial phase of individual selective neutrality, and therefore the fate of the disease allele is strongly influenced by genetic drift. As the allelic lineage becomes older and older, individual selection against the alleles becomes stronger and stronger, eventually totally eliminating that lineage of disease alleles. However, the disease persists in the population as a whole because biased mutation and selection below the level of the individual is constantly re-creating new lineages of the disease allele.

However, even this is not the whole story for Huntington's disease. One common design in human genetics in estimating the fitness of a genetic disease is to use the normal sibs as controls. When this was done for Huntington's disease, a fitness *advantage* of choreic individuals was found relative to their unaffected sibs in many studies (Reed and Neel 1959; Shokier 1975; Wallace 1976). For example, Reed and Neel (1959) found that the relative fitness of affected sibs is 1.12 when the fitness of nonaffected sibs is set to 1. Hence, both germline-level selection and individual selection operating *within* families seem to favor this disease. Perhaps this within-family selection for Huntington's disease is due to more subtle neurological effects that appear before the age of onset of the disease phenotype, such as less caring about transmitting the disease to the next generation. Reproductive decisions in such families are indeed influenced by fear of transmission to potential offspring and/or social ostracism of families with Huntington's disease (Wallace 1976), so a neurological change that made an individual less caring about transmitting the disease or less sensitive to social pressure could account for the within-family advantage of Huntington heterozygotes (Neel, personal communication).

These same factors of fear of transmission and social pressures also create family-level selection. Because Huntington's disease is autosomal dominant, it tends not to skip any generations, at least once the copy number is sufficiently high. Hence, certain families are known by the general community to have this disease "run" in their families, and the individuals born into these families often know (even before the genetic basis of the disease was established) that they are at high risk, as are their potential children. Together, these social factors can and did influence *all* individuals from choreic families to lower their overall reproductive output relative to individuals from nonchoreic families. The within-family advantage to affected individuals is weaker than the family-level disadvantage (Reed and Neel 1959; Wallace 1976). This type of selective situation can be modeled as a special case of Table 13.7, with Table 13.8 giving a simplified version of Table 13.7 that is relevant when the disease alleles are very rare in the general population [a more exact treatment is given in Yokoyama and Templeton (1980)]. In this example, the relative fitness of *hh* sibs in families segregating for Huntington's disease (choreic families) is set to 1, whereas the fitness of the *Hh* sibs in these families is set to 1.12. This reflects the within-family advantage in reproductive success found in sib control studies of this disease. In contrast, *hh* offspring from $hh \times hh$ parents are assigned a fitness of 1.2, much larger than the average choreic family fitness of 1.06 in this example.

Letting $p \to 0$ ($q \to 1$) in the equations at the bottom of Table 13.8, the effective fitness assigned to *Hh* is $q^2(1.12/1.2) \to 1.12/1.2 = 0.93$ as $q \to 1$. The effective fitness assigned to *hh* is $(2pq + q^2 1.2)/1.2 \to 1.2/1.2 = 1$ as $p \to 0$ ($q \to 1$). Hence, selection near this allele frequency boundary strongly favors the *hh* genotype despite within-family, individual

Table 13.8. Model of Family- and Individual-Level Selection Associated with Huntington's Disease

Mating Pair and Number	Frequency of Mating Pair	Mendelian Probabilities of Offspring Times Fitness in Context of Family j for Each Offspring Genotype	
		Hh	hh
5. $Hh \times hh$	$4pq^3$	$\frac{1}{2} \cdot (1.12)$	$\frac{1}{2} \cdot (1)$
6. $hh \times hh$	q^4	0	$1 \cdot (1.2)$
Offspring frequency in next generation		$\frac{2pq^3(1.12)}{\bar{w}} \approx 2pq \left(q^2 \frac{1.12}{1.2} \right)$	$\frac{2pq^3(1) + q^4(1.2)}{\bar{w}} \approx q^2 \left(\frac{2pq + q^2 1.2}{1.2} \right)$

Note: The dominant allele that causes the disease is indicated by H with frequency p and the normal allele by h with frequency $q = 1 - p$. Because the H allele is generally rare in a population, the selective dynamics when p is close to zero are dominated by family types 5 and 6 in Table 13.7, so only that subset of Table 13.7 is shown in this special case. When H is rare, the average fitness is approximated by 1.2, the fitness of *hh* offspring from mating type 6, the most common mating type under these conditions.

selection favoring *Hh* over *hh*. In the case of Huntington's disease, the family-level selection strongly works against this disease despite its late age of onset in many individuals and its within-family individual-level advantage. Thus, natural selection upon this one unit of selection involves strong germline selection below the level of the individual, increasing individual selection operating in opposition to germline selection, individual-level fertility selection within families, and strong family-level selection against choreic families.

As with other frequency-dependent models of selection, selection at the level of families or kin does not maximize average fitness even locally, can result in seemingly bizarre fitnesses assigned to individuals at equilibrium, violates Fisher's fundamental theorem and Wright's peak climbing metaphor, and frequently has multiple equilibria and complex selective dynamics. Despite this complexity, one common theme emerges from this model, and indeed all the other models discussed in this chapter with the possible exception of transposons capable of horizontal transmission: All targets of selection have their selective impact filtered by a unit of selection that *must be transmitted through a gamete* to the next generation. Hence, although many of the features of natural selection discussed in Chapter 11 fail to hold true for targets of selection below and above the level of the individual, the importance of the gametic perspective on natural selection remains. To understand natural selection on most targets of selection, it is essential to examine the process from the gamete's point of view. Natural selection favors those gametes with above average fitness effects, regardless of whether or not those effects are direct phenotypes (e.g., meiotic drive) or average excesses statistically assigned to a gamete from targets of selection at or above the level of the individual. The key to understanding natural selection is therefore to *take a gametic perspective*.

14

SELECTION IN HETEROGENEOUS ENVIRONMENTS

Premise 3 (Chapter 1) states that phenotypes emerge out of a genotype-by-environment interaction. In Chapter 11 we saw that natural selection arises from this premise when there is genetic variation in the population to produce heritable variation in the phenotype of fitness. There is often variation not only in genotypes but also in environments. We have already seen this before. For example, in Chapter 8, Fisher's basic quantitative genetic model has an environmental deviation that is modeled as a random variable assigned independently to each individual. Thus, there is both genetic and environmental variation in Fisher's model. We have also seen examples of environmental variation that is not random for each individual. For example, we discussed how the environment changed in wet, tropical Africa after the introduction of the Malaysian agricultural complex and how this environmental change altered the phenotype of fitness associated with genetic variation at the human β-globin locus. This chapter will focus on how natural selection operates when there is both genetic and environmental heterogeneity influencing the interaction of genotypes and environments in producing the phenotype of fitness.

Just as we modeled genotypic variation to examine natural selection, we will now need to model environmental heterogeneity in order to study how populations adapt to changing environments. In particular, we now consider two dimensions of environmental heterogeneity. One dimension refers to the physical source of environmental heterogeneity: *spatial versus temporal*. Environments can vary over space and over time. For example, some regions of the world have environmental conditions conducive to the existence of the malarial parasite whereas others do not. Therefore, there is spatial heterogeneity at any given time for humans living in malarial and nonmalarial geographical areas. However, as seen with the introduction of the Malaysian agricultural complex in wet, tropical Africa, there is also temporal heterogeneity in malarial versus nonmalarial environments. The same region in space can be a nonmalarial environment during some times and a malarial environment at other times.

Population Genetics and Microevolutionary Theory, By Alan R. Templeton
Copyright © 2006 John Wiley & Sons, Inc.

The other dimension is **environmental grain**, how organisms perceive environmental heterogeneity. Species differ in size, their ability to move, and generation times. This creates differences in scale in the perception of environmental heterogeneity, both spatially and temporally. For example, a grizzly bear can move over large areas throughout its life and thereby experience a variety of different environments that vary spatially. In contrast, a tree cannot move through space and hence must deal with the particular spatial environment in which it originally germinated. Thus, spatial variation is experienced differently by grizzly bears and by trees. As another example, humans are a long-lived species, and as such we experience seasonal variation within our lives. In contrast, the zebra swallowtail butterfly (*Eurytides marcellus*) has two generations a year (spring and summer) in Missouri. The environmental conditions during these two seasons are quite distinct and interact with the swallowtail's basic developmental program to yield two distinct forms. The spring form is smaller and less melanic and has proportionally shorter swallowtails (extensions from the hind wings) than the summer form. In this case seasonal variation is not experienced by individual butterflies, but is only experienced across the generations in this species. Thus, the temporal variation of the seasons is experienced very differently by humans and by zebra swallowtails.

The grain of the environment can take on two extreme values. A **fine-grained environment** occurs when the individual experiences environmental heterogeneity within its own lifetime, and a **coarse-grained environment** occurs when an individual remains in a single environment throughout its lifetime, but the environment varies between demes occupying different spatial locations or across generations. The individual perceives the coarse-grained environment as a constant, but from the gamete's perspective, the gametes of the individual may experience a different environment either spatially (via gene flow) or temporally (across the generations). Since the gamete's perspective is the one that counts with respect to natural selection and adaptation, coarse-grained heterogeneity can have a great evolutionary impact even though no individual experiences this heterogeneity. Fine and coarse grains are extremes in a continuum. In many cases, individuals experience environmental variation, but the average environment experienced by all individuals in a deme also varies spatially among demes and temporally across generations. However, it is convenient to organize our exploration of environmental grain through these two extremes. Coupling fine and coarse grain with spatial and temporal variation leads to a total of four combinations of environmental heterogeneity. However, because an individual can only be in one place at one time, the individual always experiences fine-grained spatial heterogeneity as a sequence of temporal changes. For example, we can describe the environmental heterogeneity of a grizzly bear wandering through a series of heterogeneous spatial patches as a temporal sequence within that bear's lifetime. Therefore, we need only consider three types of environmental heterogeneity: coarse-grained spatial, coarse-grained temporal, and fine grained.

There is one other type of environmental heterogeneity that deserves special consideration. An important aspect of the environment of any species is the other species with which it coexists and interacts. Once again, we have already seen this in previous chapters in our discussion of the impact of the malarial parasite upon the selective environment that induces in its human host. There is nothing unusual about one species constituting an important part of the environment of another. When two or more species are interacting with one another, there is the possibility that all species constitute part of each other's environment. Thus, when one species evolves in response to its interspecific interactions, that evolutionary change constitutes a changed environment for the other species. The other species can then evolve in response to its altered biological environment, which in turn constitutes a changed environment for the species of interest. **Coevolution** occurs when two or more species mutually adapt to one another through interspecific interactions. This is a special form of environmental heterogeneity because the "environment" is changing because it is capable of evolving.

COARSE-GRAINED SPATIAL HETEROGENEITY

Consider a landscape subdivided into spatial patches or habitats that induce different fitness responses from specific genotypes. We further assume that these genotypes experience only one of these patches in their lifetime (or at least, the selectively relevant portion of their lifetime for the unit of selection of interest). As an example, consider the northern acorn barnacle (*Semibalanus balanoides*) on the northeastern coast of the United States (Schmidt and Rand 2001). This, as well as many other marine species, has planktonic larval dispersal, resulting in high levels of gene flow and negligible population subdivision at the larval stage. The larvae then settle and enter a completely sessile stage for the remainder of their lives. The intertidal region that they settle in is a mosaic of habitats that greatly differ in environmental parameters that affect barnacle survivorship. A specific individual only experiences the habitat in which it settled, so this is coarse-grained, spatial heterogeneity. Schmidt and Rand (2001) performed genetic surveys on larvae and upon different aged adults found in four habitat types (Table 14.1). They estimated habitat-specific viabilities by measuring how genotype frequencies change with age within a particular habitat type. When genetic surveys were performed on mtDNA and upon an isozyme marker *Gpi* that codes for the enzyme glucose-6-phosphate isomerase, they could detect no heterogeneity either through space or age. These markers indicate the panmictic nature of this population and confirm its high level of gene flow and lack of population subdivision. However, a third genetic marker, the *Mpi* locus that codes for the enzyme mannose-6-phosphate isomerase, showed considerable heterogeneity. In particular, the age-specific surveys indicated a pulse of genotype-specific mortality that occurred over a two-week interval subsequent to metamorphosis from the larval to the adult form, leading to the estimated viabilities given in Table 14.1, although only the viabilities in the high intertidal zones deviated significantly from neutrality. This variation in viability associated with this locus arises in part from the differential ability to process mannose-6-phosphate through the glycolytic pathway. When mannose was supplemented in the diet and barnacles were exposed to temperature

Table 14.1. Habitat-Specific Viability Estimates for *Mpi* Genotypes in Northern Acorn Barnacle

Habitat	SS	SF	FF
Exposed substrate in high intertidal zone	0.696	1	1.424
Exposed substrate in low intertidal zone	0.898	1	1.012
Under algal canopy in high intertidal zone	1.519	1	0.880
Under algal canopy in low intertidal zone	0.913	1	0.976

Source: From Schmidt and Rand (2001).
Note: All values are measured relative to the fitness of the heterozygote.

and desiccation stress, the *FF* genotype grew faster than the *SS* genotype, with *SF* being intermediate. This result is consistent with the pattern of viabilities observed in Table 14.1, as the exposed habitats experience more stressful temperature and desiccation conditions.

Applying equation 11.5 to the viabilities from just one habitat leads to the prediction of fixation for the *F* allele in the exposed, high-intertidal habitat and fixation for the *S* allele in the algal, high-intertidal habitat, the only two habitats with significant evidence for selection. Despite no single habitat having selective forces that would maintain a balanced polymorphism, this polymorphism seems to be stable. How can this be?

We will answer this question with a model known as the Levene model after the person who did the original work (Levene 1953). There are many variants of this model, but Levene's original model starts with a single-locus, two-allele unit of selection with the fitnesses in habitat i being as follows:

Genotype	AA	Aa	aa
Fitness in habitat i	v_i	1	w_i

Levene next let c_i be the proportion of the total population that comes from habitat i. Note that c_i is a function only of the habitat type and not of the genotypic composition within the habitat. This is an example of **soft selection**, in which some factor not related to the genotypes of interest is density limiting in the habitat. In the Levene model, *any* genotypic composition can survive and reproduce in niche i, producing the same absolute number of offspring. The fitness differences shown above are therefore only *relative* differences among the genotypes within a habitat and are in no way measures of an absolute ability to produce offspring.

Levene next assumed a completely panmictic, random-mating global population. The zygotes produced by this single random-mating population are then randomly distributed over the available habitats. Essentially, the population structure of the original Levene model is an island model with $m = 1$ (see Chapter 6). Therefore, the initial genotype frequencies found in all habitats before selection is given by the Hardy–Weinberg law: p^2 for *AA*, $2pq$ for *Aa*, and q^2 for *aa*, where p is the frequency of the *A* allele in the total population. Under these assumptions, we can now predict the change in allele frequency *within* habitat i from equation 11.5 as

$$\Delta p_i = \frac{p}{\bar{W}_i} a_{A,i} \tag{14.1}$$

where

$$\bar{W}_i = p^2 v_i + 2pq + q^2 w_i \quad \text{and} \quad a_{A,i} = p(v_i - \bar{W}_i) + q(1 - \bar{W}_i) = pv_i + 1 - p - \bar{W}_i$$

The quantity $\bar{W}_i$ is the average relative fitness within habitat i, and $a_{A,I}$ is the average excess of the *A* allele for fitness in habitat i. The change in the frequency of the *A* allele in the total population is the average over all habitats of all the allele frequency changes in a specific habitat (Δp_i) as weighted by that habitat's output to the total population (c_i); that is,

$$\Delta p = \sum_i c_i \, \Delta p_i = p \sum_i \frac{c_i}{\bar{W}_i}(pv_i + 1 - p - \bar{W}_i) \tag{14.2}$$

Even this simple model can generate considerable complexity, so Levene only evaluated the conditions for a protected polymorphism (Chapter 13). A polymorphism will be protected if equation 14.2 is positive when p is close to 0 and negative when p is close to 1.

As p approaches 0, most individuals in the population are aa and the average fitness therefore approaches w_i. Taking the limit of the average excess in habitat i as p approaches 0 yields $a_{A,I} = 1 - w_i$. Hence we have

$$\lim_{p \to 0} \Delta p = p \sum_i \frac{c_i}{w_i}(1 - w_i) = p \sum_i c_i \left(\frac{1}{w_i} - 1\right) \quad (14.3)$$

Since p is small but positive, the sign of equation 14.3 is determined only by the summation. Hence, the condition for protecting the A allele when it is rare in the total population is

$$\sum_i c_i \left(\frac{1}{w_i} - 1\right) > 0 \Leftrightarrow \sum_i \frac{c_i}{w_i} > 1 \Leftrightarrow \frac{1}{\sum_i c_i/w_i} < 1 \quad (14.4)$$

The last representation of inequality 14.4 is the harmonic mean across habitats of the fitness of genotype aa. (The harmonic mean of a set of finite numbers is the reciprocal of the sum of the reciprocals of each number in the set.) When the harmonic mean fitness of aa is less than 1, the fitness of the Aa genotype, then the A allele is protected against loss when rare. This makes sense because when A is rare, almost all copies of A-bearing gametes come to be in Aa individuals, with a harmonic mean fitness of 1, and are being selected relative to aa individuals, with a harmonic mean fitness less than 1. Similarly, by taking the limit of equation 14.2 as p goes to 1, the condition for protecting the a allele when it is rare is

$$\sum_i c_i \left(\frac{1}{v_i} - 1\right) > 0 \Leftrightarrow \sum_i \frac{c_i}{v_i} > 1 \Leftrightarrow \frac{1}{\sum_i c_i/v_i} < 1 \quad (14.5)$$

Thus, the polymorphism is protected in this spatially heterogeneous environment when the harmonic mean fitnesses of the homozygotes across all habitats are less than the harmonic mean fitnesses of the heterozygotes.

We can now return to the barnacle example and see if a stable polymorphism is possible with the viabilities given in Table 14.1 under the Levene model. Figure 14.1 shows a plot of equation 14.2 using the viabilities given in Table 14.1. Unfortunately, we do not know what the proportional outputs are from each habitat type, so Figure 14.1 presents the results for two hypothetical cases of the c's. In Figure 14.1a the arrows reveal that there is a stable polymorphism at $p = 0.74$. However, this case does not satisfy Levene's conditions for protection because if the allele frequency is below 0.52, the allele is driven to fixation and is therefore not protected. This case illustrates that, depending upon initial conditions, environmental heterogeneity could maintain a stable polymorphism even when it is not a protected polymorphism. Hence, the conditions for protection are a conservative indicator of the ability of coarse-grained spatial heterogeneity to maintain polymorphisms.

Figure 14.1b shows a protected polymorphism. In this case, whenever the allele frequency is close to 0 or 1, selection pushes it away and toward the interior. There is also a stable polymorphism in this case at $p = 0.86$. A contrast of Figures 14.1a and b reveals the difficulty of trying to predict evolutionary outcomes in heterogeneous environments. Even an

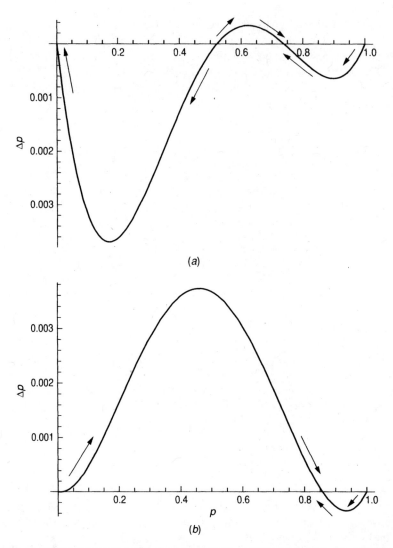

Figure 14.1. Illustration of Levene model of coarse-grained spatial heterogeneity through plot of equation 14.2 against allele frequency. Viability estimates of the northern acorn barnacle are used, as given in Table 14.1. (*a*) The following *c* values are used for the four habitats listed in Table 14.1, going from the top to the bottom of the habitat column: 0.4, 0.05, 0.5, and 0.05. (*b*) The *c* values are 0.15, 0.375. 0.325, and 0.15. Arrows indicate the direction of allele frequency change from possible equilibrium points.

excellent knowledge of fitnesses in all possible habitats is insufficient to make evolutionary predictions in this simple model. Indeed, as the contrast of Figures 14.1*a* and *b* shows, qualitatively different evolutionary outcomes are possible from the same fitnesses, in this case as a function of the ecological parameters *c*, parameters that are difficult to know or estimate.

One general conclusion from Levene's simple model is that coarse-grained spatial heterogeneity tends to broaden the conditions under which polymorphisms are protected. The

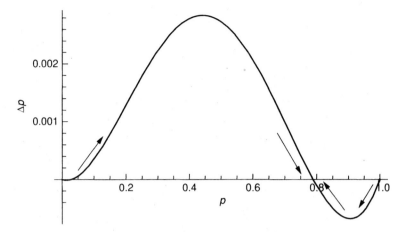

Figure 14.2. Illustration of protected and stable polymorphism with Levene model in which there is no heterozygote superiority in fitness in any habitat. Viability estimates of the northern acorn barnacle are used, as given in Table 14.1, with the following c values: 0.12, 0.58, 0.30, and 0. By assigning $c = 0$ to the last habitat in Table 14.1, the only habitat showing heterozygote superiority is eliminated.

normal criterion for protection in a constant-fitness model with no environmental heterogeneity is that $v < 1$ and $w < 1$, where the relative fitness of the heterozygote is 1. However, inequalities 14.4 and 14.5 can be satisfied even when no single habitat satisfies the normal conditions for protection. Figure 14.2 illustrates this property with the barnacle data by setting the c associated with the algal low-intertidal zone equal to 0. This is the only row in Table 14.1 that has heterozygote superiority. All the other rows favor one homozygote or the other, often very strongly. Yet, Figure 14.2 shows not only a protected polymorphism in this case but a stable one as well. Therefore, coarse-grained spatial heterogeneity can maintain polymorphisms even when there is no environmental condition under which heterozygotes have superior fitness, a situation quite different from the constant-environment models in which heterozygote superiority is required for a balanced polymorphism. However, so far we have only shown this to be true for the simple Levene model. We now examine some extensions of this simple model to see if this conclusion is true under more general conditions.

The Levene model is one of soft selection in which the output of a habitat is not affected by the genotypic composition found in the habitat. We now consider **hard selection**, in which the number of offspring from a population depends upon the genetic composition of the population. In the hard-selection version of the Levene model, let z_i be the proportion of zygotes that enter (and subsequently remain) in habitat i. We next assume that the reproductive output of habitat i is directly proportional to the number going in *and* the average fitness of the population with respect to the unit of selection of interest in habitat i. We can now calculate c_i, the proportion of the total population that comes from habitat i, as

$$c_i = \frac{z_i \bar{W}_i}{\sum_i z_i \bar{W}_i} = \frac{z_i \bar{W}_i}{\bar{\bar{W}}} \qquad (14.6)$$

where $\bar{\bar{W}} = \sum_i z_i \bar{W}_i$ is the average fitness across all habitats. We now substitute equation 14.6 into equation 14.2 to obtain

$$\Delta p = \sum_i c_i \, \Delta p_i$$

$$= p \sum_i \frac{z_i \bar{W}_i / \bar{\bar{W}}}{\bar{W}_i} (pv_i + 1 - p - \bar{W}_i)$$

$$= \frac{p}{\bar{\bar{W}}} \sum_i z_i (pv_i + 1 - p - \bar{W}_i)$$

$$= \frac{p}{\bar{\bar{W}}} \left(p \sum_i z_i v_i + (1 - p) \sum_i z_i - \bar{\bar{W}} \right)$$

$$= \frac{p}{\bar{\bar{W}}} \left(p \bar{v} + q - \bar{\bar{W}} \right)$$

$$= \frac{p}{\bar{\bar{W}}} a_A \qquad (14.7)$$

where $\bar{v} = \sum_i z_i v_i$ is the average fitness of AA across all habitats as weighted by zygotic inputs. Note that equation 14.7 depends only upon the arithmetic means of fitnesses across habitats and not the harmonic means. Indeed, there is really no difference at all between equations 14.7 and 11.5 because all fitnesses assigned to genotypes in Chapter 11 were arithmetic averages across all individuals bearing that genotype; that is, fitness is a genotypic value which explicitly averages over all environmental deviations (Chapter 8). The conditions for protection of the polymorphism are now the standard ones from the constant-fitness model; namely, homozygote fitnesses must be less than the heterozygote fitness: $\bar{v} = \sum_i z_i v_i < 1$ and $\bar{W} = \sum_i z_i w_i < 1$. Thus, coarse-grained spatial heterogeneity does *not* broaden the conditions for polymorphism under hard selection. This is illustrated in Figure 14.3. This figure plots equation 14.7 against p for the same fitness model shown

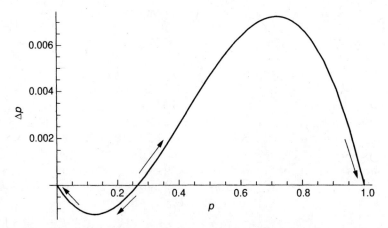

Figure 14.3. Illustration of hard selection. The same model described in Figure 14.2 is used here except that soft selection is replaced by hard selection and the c values in Figure 14.2 are now interpreted as z values. Unlike the soft-selection case that yields a stable, protected polymorphism, now the polymorphism is unstable and unprotected.

in Figure 14.2. The c's in the soft-selection model are now interpreted as the z's in the hard-selection model. A contrast of Figures 14.2 and 14.3 reveals a drastic alteration in going from soft to hard selection: The stable, protected polymorphism in Figure 14.2 is transformed both quantitatively and qualitatively, now being unstable and unprotected.

Now we will introduce restricted gene flow into the Levene model. Given that the Levene model is an island model of population structure with $m = 1$, the simplest extension with restricted gene flow is an island model with $m < 1$. Under this model of population structure with soft selection, Christiansen (1975) has shown that the conditions for protecting the A allele from loss when it is rare (and similar conditions hold for the a allele when it is rare) are now

There exists at least one niche such that $w_i < 1 - m$

OR

$$\frac{1}{\sum_i c_i / [1 - (1 - w_i)/m]} < 1 \qquad (14.8)$$

When p is close to zero, most individuals are aa with only a few Aa. Hence, as before, the conditions for protecting A when it is rare depends only on how the heterozygote class is doing relative to the aa homozygotes. The first condition in inequalities 14.8 states that if there is just one habitat in which the heterozygote is favored over the predominant aa genotype sufficiently to overcome gene flow, then A is protected from loss. Note that this condition becomes more and more likely to be satisfied as m decreases; that is, the more restricted the gene flow, the broader the conditions for protection. In the extreme case in which the population is fragmented into isolates ($m = 0$), the A allele is protected if just one habitat exists in which the heterozygote fitness is greater than the aa homozygote fitness. If no habitat satisfies this first condition, inequalities 14.8 show that the A allele can still be protected if a harmonic mean fitness condition is satisfied. Note here that the harmonic mean condition incorporates both fitness and gene flow. When $m = 1$ this second condition reduces to inequality 14.4, so the original Levene model is just a special case of this more general model. When m decreases in the second inequality in 14.8, the fitness conditions that allow protection of A become broader. Hence, coarse-grained spatial heterogeneity coupled with restricted gene flow creates broad conditions favoring the protection of polymorphisms under soft selection.

We saw with equation 14.7 that the conclusion that coarse-grained spatial heterogeneity favors the protection of polymorphisms vanishes completely when one goes from soft to hard selection. Christiansen (1975) explored this problem as well by doing the hard-selection version of the island model. Under hard selection, the conditions for protecting the A allele now become:

There exists at least one niche such that $\omega_i \leq 1 - m$
where $\omega_i = (1 - m)w_i + m \sum_i z_i w_i$

OR

$$\frac{1}{\sum_i z_i / [1 - (1 - \omega_i)/m]} < 1 \qquad (14.9)$$

Note that as gene flow becomes more and more restricted (m approaches zero), ω_i converges to w_i. This means that conditions 14.8 and 14.9 become nearly identical when gene flow is highly restricted. Thus, *under highly restricted gene flow the ecological distinction between soft and hard selection becomes irrelevant for the conditions under which coarse-grained spatial heterogeneity protects polymorphisms.* This work therefore illustrates the danger of considering one factor in isolation, such as hard versus soft selection. What we see here is strong interaction between ecology (hard and soft selection), population structure (here measured by m), and natural selection. Even a complete knowledge of the spatial heterogeneity of fitnesses and of the ecological conditions determining local densities is insufficient to predict the evolutionary outcome: You also must know about population structure.

One general conclusion to emerge from the contrast of inequalities 14.8 and 14.9 is that as gene flow becomes more and more restricted, populations show greater and greater adaptation to the local rather than the global environment. Indeed, for highly restricted gene flow in both cases, it is local conditions and only local conditions that determine if a polymorphism is protected (the first of the alternative inequalities in 14.8 and 14.9). The role of gene flow as a modulator of local adaptation in the face of spatial heterogeneity is illustrated by studies on industrial melanism.

Industrial melanism is the evolution of dark coloration in various species (mostly insects) in response to a darkening caused by air pollution of the background color upon which the species rests or lives. Industrial melanism is one of the classic examples of natural selection (Kettlewell 1955, 1973; Lees 1981). Although this example has been questioned (Majerus 1998; Hooper 2002), an exhaustive review of the literature and the continuing studies on this system strongly support the role of natural selection, of which predation is one part (Cook 2003). Several species of moths in England have melanic (dark) morphs that are controlled by a single autosomal locus with two alleles. Many of these moths rest upon tree trunks and limbs during the day. There they are subject to predation by a variety of bird species, and such bird predation has been directly observed and recorded under natural conditions (Murray et al. 1980), including on the trunk in an area before any experimentation was performed [Murray, personal communication, and contrary to claims reported in Hooper (2002)]. One type of protection from bird predation during that time is simply being cryptic, that is, matching the background upon which they are resting such that they are difficult to see. The trees normally had a light-colored background due to the growth of epiphytic lichens and bryophytes upon the bark, but air pollution, particularly sulfur oxides and acid rain, can kill these epiphytes, leading to a dark background on the tree trunks. Great Britain has a long history of people collecting butterflies and moths, so collections were available to show that before industrialization the light-colored morphs dominated throughout the island and melanic forms were extremely rare. However, with industrialization and its attendant air pollution, there was a marked increase in the frequency of melanic forms in many species. Of course, not all areas of Great Britain are equally polluted, and different local areas vary from extreme air pollution to areas, mostly in the countryside, that remained unpolluted. As a consequence, as industrialization proceeded, Great Britain became a spatial patchwork of polluted and unpolluted areas, and many moth species responded to this spatial heterogeneity with a corresponding genetic heterogeneity in the frequency of melanic alleles. For example, Figure 14.4 shows the frequency of melanic forms in different local populations of the peppered moth, *Biston betularia*. The melanic morphs are most frequent in areas with much industrial activity and air pollution, whereas the light morphs are most frequent in nonindustrialized areas. Thus, on this coarse

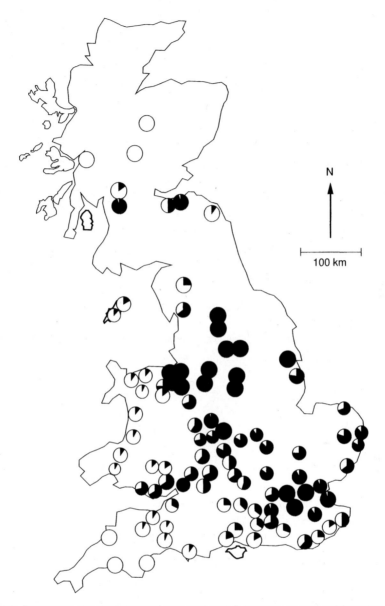

Figure 14.4. Distribution of melanic forms of peppered moth B. betularia on island of Great Britain. Black segments of each pie diagram indicate phenotype frequency of melanic form in sample at particular location. Modified from Fig 5.10 in Lees (1981). Copyright © 1981 by Academic Press, Inc. (London). Reprinted by permission of Elsevier.

geographical scale of the entire island of Great Britain, the moths are adapting to local conditions.

There is considerable spatial variation in the degree of pollution on a finer geographical scale as well. For example, the nearby cities of Manchester and Liverpool have much pollution as compared to the nearby Welch countryside (Bishop and Cook 1975; Bishop

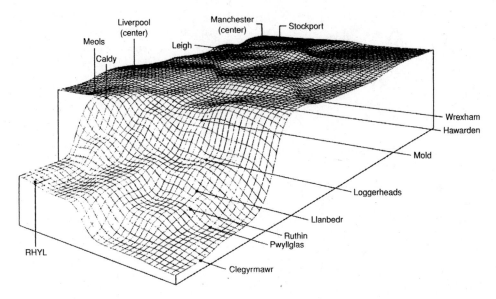

Figure 14.5. Distribution of melanic peppered moths in Manchester/Liverpool area in England and adjacent areas of Wales. The height of the surface gives the proportion of the population in that locale that is melanic. The surface is viewed from the southwest, from rural Wales looking toward Liverpool and Manchester. Modified from Bishop and Cook (1975). Copyright © 1975 by Gabor Kiss.

et al. 1978). Figure 14.5 shows the frequency of the melanic form of *B. betularia* in this area of England and Wales. As can be seen, the frequency surface has two distinct parts: a plateau of more than 90% melanic in Liverpool, Manchester, and the area between them and a descending portion that drops to less than 10% in Northern Wales. However, there is even finer scale heterogeneity in the degree of pollution in the Manchester/Liverpool area. There are some wooded areas to the south and east of this urban area, and even in the urban corridor there is variation in the amount of pollution. However, the peppered moth adapts to the Manchester/Liverpool area as if it were uniformly polluted.

The scalloped hazel moth also lives in this region, and it is also polymorphic for a melanic locus subject to the same type of selective pressures as observed in the peppered moth. As was the case with the peppered moth, this species also shows geographical variation in the frequency of the melanic forms such that melanic forms are most common in polluted areas. Figure 14.6 shows the geographical distribution of the frequency of melanic forms in the scalloped hazel moth in the Manchester/Liverpool urban area, a subset of the area shown in Figure 14.5 that is fully contained with the >90% melanic plateau of the peppered moth. The scalloped hazel moth shows considerable variation in the frequency of melanic forms in this area of England in a manner that corresponds to polluted and relatively unpolluted local areas, but the peppered moth does not show any significant variation in the frequency of melanic forms in this same area. Why do these two moth species, each with a melanic polymorphism subject to similar selection, respond to spatial heterogeneity in such distinct fashions? The answer seems to stem from differences in the amount of gene flow between the two species. The scalloped hazel moth populations are very dense in this area, with as many as 50,000–100,000 moths per square kilometer. In contrast, the peppered moth density is about 10 per square kilometer. This implies that a peppered moth is likely to have to travel

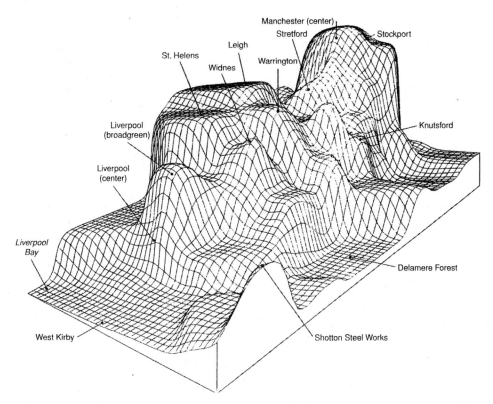

Figure 14.6. Distribution of melanic scalloped hazel moths (*Gonodontis bidentata*) in Manchester/Liverpool area in England. The height of the surface gives the proportion of the population in that locale that is melanic. The surface is viewed from the southwest, looking toward Liverpool center and Manchester. Modified from Bishop and Cook (1975). Copyright © 1975 by Gabor Kiss.

much farther than a scalloped hazel moth to encounter a mate. Mark/recapture experiments confirm this difference, with many peppered moths flying more than a kilometer but with scalloped hazel moths rarely flying more than 150 meters. As shown by conditions 14.8 or 14.9, when the gene flow parameter is very small, the top condition of having just a single local environment favoring an allele is sufficient to protect it. This occurs because with very restricted gene flow the local population is able to adapt to its local environment in a manner not strongly influenced by the influx of genes from other local populations adapted to a different environment. As the amount of gene flow increases, the role of the local environment in protecting polymorphisms is diminished and is replaced increasingly by the lower conditions in 14.8 or 14.9. In this case, the adaptive outcome is determined more by an averaging across several local environments rather than the local conditions per se. In the case of the scalloped hazel moth, the amount of dispersal and gene flow is so low and the densities so high that populations can adaptively respond to local environments on the scale of just a kilometer of two. In contrast, the peppered moth experiences spatial heterogeneity at this small geographical scale as individuals rather than as breeding populations and hence as a population adapts to coarse-grained spatial heterogeneity on a much larger geographical scale. Conditions 14.8 and 14.9 and this example all show that *the adaptive response to spatial heterogeneity arises from an interaction of natural selection and gene flow.* Thus, we see a reinforcement of the theme developed in Chapter 12: *Adaptation arises from the interaction of natural selection with population structure such that natural selection alone cannot explain how a population adapts to its environment.*

The peppered moth (Figure 14.5) and the scalloped hazel moth (Figure 14.6) illustrate how the evolutionary response to spatial heterogeneity emerges from the balance of selection favoring local differentiation versus gene flow which diminishes local differentiation. Another way of characterizing spatial heterogeneity that is sensitive to the balance of selection versus gene flow is ecotones versus gradients. **An ecotone** is a spatially abrupt change from one environment (and hence selective regimen) to another, whereas **a gradient** is a gradual change over space from one environment to another. There is no absolute geographical scale that distinguishes between an ecotone and a gradient because that distinction depends upon how the organism moves and reproduces through space. To see this, consider a single-locus, two-allele codominant model with a spatial transition between two environments that influences the fitnesses such that

$$\left.\begin{array}{l} w_{AA}(x) = 1 - \tfrac{1}{2}b\Delta \\ w_{Aa}(x) = 1 \\ w_{aa}(x) = 1 + \tfrac{1}{2}b\Delta \end{array}\right\} \quad \text{for } x < -x_0$$

$$\left.\begin{array}{l} w_{AA}(x) = 1 + bx \\ w_{Aa}(x) = 1 \\ w_{aa}(x) = 1 - bx \end{array}\right\} \quad \text{for } -x_0 < x < x_0 \quad (14.10)$$

$$\left.\begin{array}{l} w_{AA}(x) = 1 + \tfrac{1}{2}b\Delta \\ w_{Aa}(x) = 1 \\ w_{aa}(x) = 1 - \tfrac{1}{2}b\Delta \end{array}\right\} \quad \text{for } x > x_0$$

where x is the spatial position in a transect between the two environments in which one environment ends at point $-x_0$ and the second environment begins at point x_0, b measures the slope of the fitness effect in the transition space between the two environmental types, and Δ is the transition distance between the two environments (Endler 1977). If fitnesses are changing continuously along some transect, then Δ is the distance between the two most distant points in the transect. Let gene flow be characterized by an isolation-by-distance model in which an individual born at a particular point has a probability m of dispersing from the deme of birth and given dispersal it moves an average distance of d. Then gene flow is measured by the root-mean-square gene flow distance (Endler 1977):

$$\ell = d\sqrt{m} \qquad (14.11)$$

The parameter ℓ is the same parameter as σ in the continuous isolation-by-distance models discussed in Chapter 6 (equations 6.37). Hence, ℓ is a direct measure of gene flow restricted by isolation by distance. This gene flow tends to diminish local adaptation. The measure of the strength of selection in counteracting gene flow is $s = b\Delta$, which is the magnitude of the maximum fitness change in a homozygote in response to this spatial heterogeneity. The balance of gene flow and selection is then given by (Endler 1977)

$$\ell_c = \frac{\ell}{\sqrt{s}} = \frac{\ell}{\sqrt{b\Delta}} = d\sqrt{\frac{m}{b\Delta}} \qquad (14.12)$$

The quantity ℓ_c is known as the characteristic length of variation in allele frequencies and represents the spatial scale over which selection is effectively averaged by gene flow. Note that this characteristic length is directly proportional to our measure of gene flow (equation 14.11) and inversely proportional to the intensity of the spatial change in fitnesses (measured by b) and the spatial scale over which the fitnesses change (measured by Δ). If $\ell_c > \Delta$, the organism experiences the environment as an ecotone because the transition between the two environments occurs on a spatial scale less than individual gene flow relative to selection. If $\ell_c < \Delta$, the organism experiences the environmental transition as a gradient in which adaptation to transitional, intermediate environments is possible. Either situation can result in a **genetic cline**, a gradual shift of gamete frequencies over geographical space. In the case of an ecotone, populations far away from the transitional zone will tend to be fixed for either the A or a alleles depending upon which environment they are in, but populations near the environmental transition zone will have intermediate allele frequencies due to the mixing via gene flow of gametes coming from the two alternative environments. The width of this cline is on the order of ℓ_c. In the case of a gradient, there is more potential for geographic differentiation for a given strength of selection s. Populations in the transition zone can show local adaptation to the transitional environment. This means that the width of the cline is greater than ℓ_c and tends to approach Δ.

Note that equation 14.12 tells us that there is no absolute difference between a gradient and an ecotone. The distinction between the two depends upon the physical scale of the environmental heterogeneity (Δ), the population attribute of degree of isolation by distance (ℓ), and the locus-specific attribute of phenotypic response to the spatial heterogeneity (b). Hence, not only can different species experience the same physical heterogeneity (measured by Δ) in different fashions because they differ in ℓ but also even within a species one locus may respond to the spatial heterogeneity as if it were an ecotone, another respond to the same heterogeneity as if it were a gradient, and yet other loci not respond at all to the spatial heterogeneity because of variation in b across loci.

These points can be illustrated by an examination of a population of the fruit fly *Drosophila mercatorum* in the Kohala Mountains on the Island of Hawaii near the town of Kamauela (also known as Waimea) that was discussed in Chapter 6 (see Figure 6.9). There is an extreme rainfall and humidity gradient on the windward side of Kohala, with a rainforest existing at the top and extending down the slope and then rapidly transitioning into a desert. Site A in Figure 6.9 is close to the rainforest and is extremely humid. However, site B, only 300 meters downhill from site A, is much drier, and sites F and IV are extremely dry. As the distance from site A to F is only about 1 km, this environmental transition from rainforest to desert seems very abrupt and dramatic to most human observers. However, how does a fly experience this 1-km transition? As discussed in Chapter 6, there is sufficient gene flow across this space to result in a nonsignificant f_{st} for nuclear genes, but gene flow is sufficiently restricted to result in f_{st} statistics that are significantly different from zero for mtDNA and Y-DNA, indicating that the variance effective number of migrants across this area is about 8.5 using the island model as our ideal reference population (Chapter 6). The inference of restricted but significant gene flow over this area is confirmed by direct studies on dispersal (Johnston and Templeton 1982). This area of Hawaii is very windy, as noted in a traditional Hawaiian chant about this locale: *Hole waimea ika ihe a ka makani* ("Tousled is Waimea by spear sharp thrusts of wind"). Under normal conditions, the wind blows down the mountainside shown in Figure 6.9 between 15 and 35 km/h, occasionally gusting higher. The flies live in patches of the prickly pear cactus *Opuntia megacantha*, and these patches block the wind. However, both laboratory and field observations reveal that the flies will not disperse between these cactus patches unless the wind speed drops and remains below 10 km/h (Johnston and Templeton 1982). Days with such low winds are relatively rare in this area, occurring on average only about four to five days per month. When these relatively still days occur, 31% of the adult *D. mercatorum* population disperse, and the dispersing adults move an average of 43 meters/day, roughly the average distance between neighboring large cactus patches. This nearest-neighbor pattern of dispersal represents a type of isolation by distance. The amount of gene flow associated with this dispersal is also a function of how long the flies live as adults, for it is only in the adult stage that flies can disperse. Adult survivorship data from the field (to be discussed in Chapter 15) indicates that most flies will have only one or two days on average in their lifetime in which dispersal is possible. Putting all these data together yields $\ell = 28.8$ meters. In terms of the two-dimensional neighborhood model in equation 6.37, this means that the radius of the neighborhood area in which parents can be treated as if drawn at random is $2\sigma = 2\ell = 57.6$ meters. Hence, there is isolation by distance over the 1-km transect, but apparently the densities are sufficiently high (recall that f_{st} emerges from the balance of gene flow to drift and not just gene flow alone) that there is no significant subdivision over the transect for most nuclear loci (DeSalle et al. 1987).

However, what would happen at a locus for which the genotype–fitness relationship is directly affected by the humidity gradient to create fitness differences across this gradient? One such locus is *abnormal abdomen* (*aa*). As will be detailed in Chapter 15, *aa* is not really a single locus, but rather a cluster of genes that are tightly linked on the X chromosome. Recombination is sufficiently rare that we can treat this X-linked region for now as a single Mendelian superlocus (Chapter 13) with two alleles, aa^+ and *aa*. In Chapter 15 we will investigate in detail the phenotypic consequences of genetic variation at this superlocus, but for now it suffices to say that the *aa* allele is favored by natural selection under dry environmental conditions whereas the aa^+ allele is favored under humid conditions. Templeton et al. (1990a) estimated that the magnitude of the selective difference between *aa* homozygotes at the extreme of the transect is $s = 0.0245$. Hence, the characteristic length for the *aa* locus over this transect is $28.8/\sqrt{0.0245} = 184$ meters, which is much less than the length $\Delta = 1000$ meters of the transect. With respect to the *aa* locus, these populations of *D. mercatorum* experience this abrupt environmental transition not as an ecotone, but as a gradient in which local adaptation should result in a genetic cline within the transitional zone. Indeed, there is a significant genetic cline for *aa* across this humid–dry environmental transition (Figure 14.7). However, there was no significant cline for the allozymes, considered either together or individually. One of the isozyme loci is another X-linked locus, *glucose-6-phosphate dehydrogenase* (*G6PD*), which is located at the other end of the X from *aa*. The allele frequency changes for the *G6PD S* allele are also shown in Figure 14.7. In this case, there is no cline and no statistically significant differentiation across this transect, illustrating how gradients are locus specific, as expected from equation 14.12.

Although *G6PD* appears to be a neutral marker in the Kohala populations of *D. mercatorum*, the X-linked *G6PD* locus in humans is involved with malarial resistance (Chapter 11). The island of Sardinia lies off the west coast of Italy. The coastal areas of Sardinia historically have had a high incidence of malaria, but the central mountainous region does not. The estimated relative fitnesses of the genotypes associated with the active (A^+) and deficient (A^-) alleles at this locus in Sardinia are given in Table 14.2 (from Livingstone 1973). Figure 14.8 shows the frequencies of the A^- allele along the 130-km transect bisecting the island going from the east coast through the central mountains to the west coast. The actual transition from the lowland malarial areas to the highland nonmalarial region occurs in just

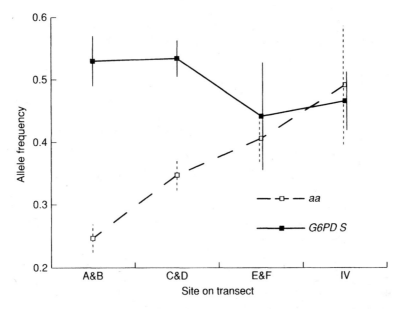

Figure 14.7. Frequencies of the *aa* allele at X-linked *abnormal abdomen* locus and *S* allele at X-linked *G6PD* locus in populations of *D. mercatorum* over transect on leeward side of Kohala in Hawaii. The position of the sites is indicated in Figure 6.9. Vertical solids lines indicate ± one standard deviation above and below the estimated frequency of the *S* allele, and vertical dotted lines indicate ± one standard deviation above and below the estimated frequency of the *aa* allele.

a few kilometers on both the east and west sides of this transect. People have inhabited this area for over 2000 years with relatively constant densities. Livingstone (1973) simulated the evolutionary response to the fitness values shown in Table 14.2 in a manner that tried to mimic the movements of peasant populations in Europe. A good fit to the observed pattern shown in Figure 14.8 was obtained by assuming that the environmental transition was sharp with $\Delta \leq 2.65$ km (the average distance between adjacent villages in the simulations) and that 25% of the people left their village of birth, with those dispersing going primarily to adjacent villages and other nearby villages, yielding an average dispersal distance of $d = 3.34$ km, which yields $\ell = 3.34\sqrt{\frac{1}{4}} = 1.67$ km. The cline in this case is driven by the fitness effects of A^- in hemizygous males and in heterozygous females (Table 14.2). From Table 14.2, $s = 0.97 - 0.90 = 0.07$ in males, and the s for females is $1.07 - 0.98 = 0.09$.

Table 14.2. Relative Fitness of Male and Female Genotypes Created by A^+ and A^- Alleles at Human X-Linked *G6PD* Locus in Malarial (Coastal) and Nonmalarial (Mountain) Regions on Island of Sardinia

Environment	Fitness for Male Genotype		Fitness for Female Genotype		
	A^+	A^-	A^+/A^+	A^+/A^-	A^-/A^-
Coastal	1	0.97	1	1.07	0.97
Mountains	1	0.90	1	0.98	0.90

Source: Livingstone (1973).

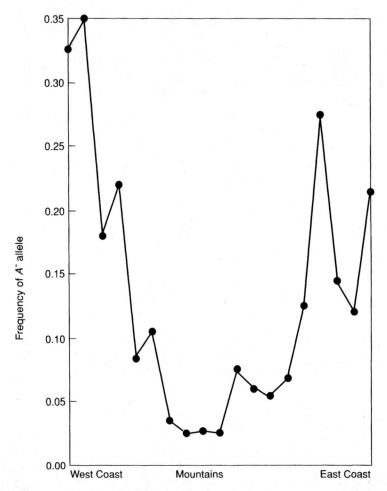

Figure 14.8. Frequencies of the A^- alleles at X-linked *G6PD* locus in human populations along east–west transect of island of Sardinia. Modified from Fig. 8 in Livingstone (1973). Copyright © 1973 by the School of American Research, University of New Mexico Press.

Averaging across the two sexes, $s = 0.08$. Then, from equation 14.12, $\ell_c = 5.91$. Hence, the characteristic length of this cline is more than twice the transitional distance Δ, implying that the ecotone model explains the cline observed in Figure 14.8. Note that the geographic distance of the transition in the environment is larger in this human example than the transition distance in the *Drosophila* example, yet the larger physical distance in the human example defines an ecotone whereas the smaller physical distance in the *Drosophila* example defines a gradient. The distinction between an ecotone and a gradient is not a function of absolute distance but rather depends upon the balance among physical distance, gene flow, and selective strength.

In addition to genetic clines, there are also **phenotypic clines,** gradual shifts of phenotypic frequencies or mean phenotypes over geographical space. In the case of *abnormal abdomen*, the genetic cline is accompanied by a phenotypic cline (the phenotypes associated with *aa* will be discussed in Chapter 15). However, premise 3 (Chapter 1) states that phenotypes arise from the interaction of genotypes and environments. The interactions of genotypes with environments that are changing over ecotones or gradients can result in complex, even counterintuitive relationships between genetic and phenotypic clines. For example, populations of the green frog *Rana clamitans* live on the east coast of the United States and into the Appalachian Mountains (Berven et al. 1979). Populations were sampled in ponds along an elevational transect from 10 to 1250 meters above sea level. The growth rates of the tadpoles were measured, and it was discovered that tadpoles in the lowland ponds had the largest growth rates and tadpoles in the montane ponds had the smallest growth rates. However, growth rates in amphibians are strongly influenced by temperature, and the average temperature varies greatly over this transect. In particular, low temperatures decrease the growth rate whereas warm temperatures increase it. Consequently, the phenotypic cline in growth rates along this elevational transect could be due just to the temperature effects with no genetic component at all. To test this hypothesis, Berven et al. (1979) performed laboratory experiments under controlled temperature conditions on egg masses sampled from both lowland and montane ponds. They discovered that for a given temperature the montane forms had higher growth rates than the lowland forms—exactly the opposite of the observed phenotypic cline!

De Jong (1988) produced a simple model to show how genetic and phenotypic clines can go in opposite directions. Consider first a one-locus model with two alleles, A and a. Let all genotypes have linear norms of reaction (Chapter 10) to temperature such that the genotypic values of growth are

$$G_{AA} = a + cT \qquad G_{Aa} = cT \qquad G_{aa} = -a + cT \qquad (14.13)$$

where T is the temperature, c is the slope of the linear norm of reaction to temperature and is shared in common by all genotypes, and $a > 0$ defines the genotypic specific intercepts of the norm of reaction (note that the heterozygote is always intermediate between the two homozygotes such that no heterozygote superiority for growth rate is possible in this model). For simplicity, we also assume that the environmental variance is zero for any given temperature; that is, all individuals sharing the same genotype have the same phenotype at a specific temperature.

Under equations 14.13, the norms of reaction of the genotypes are all parallel lines when plotted against temperature, but with some genotypes having uniformly higher or lower growth rates for a specific temperature, as shown in Figure 14.9. In this simple model, the AA genotype always has the highest growth rate at any given temperature, the aa genotype the lowest, and the Aa genotype an intermediate growth rate. Now suppose that the fitnesses assigned to individuals on the basis of their actual growth rates are constant throughout the transect with a single optimal growth rate, say β, being associated with the highest fitness, as also shown in Figure 14.9. A fitness function that has a single optimal value β with fitness dropping off symmetrically about β is given by

$$w(P) = 1 - \alpha(\beta - P)^2 \qquad (14.14)$$

where P is the individual's phenotype (in this case growth rate) and α determines how rapidly fitness drops off in individuals with growth rates that deviate from β. Note that this is a fitness of the type commonly used in ecological theory; it is a fitness assigned to an individual's phenotype for some trait besides fitness, and it is *not* a fitness assigned to genotypes. However, in order to predict the selective response at this locus, we need to

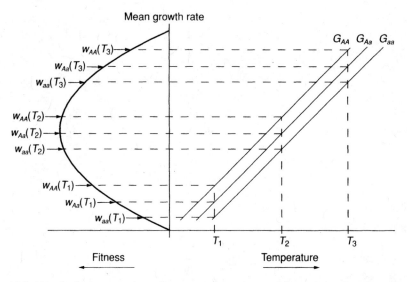

Figure 14.9. Hypothetical norms of reaction over temperature and fitnesses assigned to phenotype of growth rate for one-locus, two-allele model. The fitness curve assigned to the phenotype of growth rate is shown on the left of the figure. The right side plots the growth rate response of each genotype to temperature. Three points on the temperature gradient are indicated, along with their effects on the fitnesses assigned to genotypes.

assign genotypic values of fitness. This is done by using the norm of reaction for frogs reared at any particular temperature. Figure 14.9 shows three possible temperatures. Even though fitness for growth rates is constant regardless of temperature, the genotypic values of fitness vary as temperature changes. The AA genotype has the highest fitness at temperature 1 in Figure 14.9, the Aa genotype at temperature 2, and the aa genotype at temperature 3. Hence, the relative genotypic fitnesses vary dramatically over this temperature gradient. This discrepancy between phenotypic and genotypic fitnesses shows why it is essential to distinguish the cases in which fitness is assigned to another phenotype versus the cases in which fitness is assigned to genotypes.

Fitnesses alone do not determine the evolutionary response to natural selection. Another important factor in determining that response is population structure. We will assume that the species is distributed over this temperature regimen such that there is random mating in any local area, the local densities are large so that drift can be ignored, and there is substantial isolation by distance such that local allele frequencies reflect the local selective conditions. Under these assumptions, selection is expected to drive the local allele frequency to 1 (fixation for A) whenever the fitness of AA is greater than the fitness of Aa and aa. Combining equations 14.13 and 14.14, this occurs whenever

$$1 - \alpha(\beta - a - cT)^2 > 1 - \alpha(\beta - cT)^2 \qquad T < \frac{2\beta - a}{2c} \qquad (14.15)$$

Similarly, selection will cause the local fixation of the a allele whenever

$$1 - \alpha(\beta + a - cT)^2 > 1 - \alpha(\beta - cT)^2 \qquad T > \frac{2\beta + a}{2c} \qquad (14.16)$$

Within the temperature range $(2\beta - a)/2c < T < (2\beta + a)/2c$ the heterozygote has highest fitness (even though the heterozygote is always intermediate between the two homozygotes for the phenotype of growth rate), and from equation 11.13, we predict that selection will stabilize the local allele frequency at

$$p = \frac{1}{2} + \frac{\beta - cT}{a} \qquad (14.17)$$

The average phenotype of growth rate in these locally polymorphic populations is

$$\bar{P} = 2\beta - cT \qquad (14.18)$$

Figure 14.10 shows a plot of how allele frequency and mean phenotype change over a temperature gradient. As the temperature increases, the local populations go from fixation of the A allele through a steadily declining allele frequency and end up with fixation of the a allele. Thus, there is a genetic cline such that those genotypes with the slower growth rates increase in frequency with increasing temperature. However, the mean growth rate

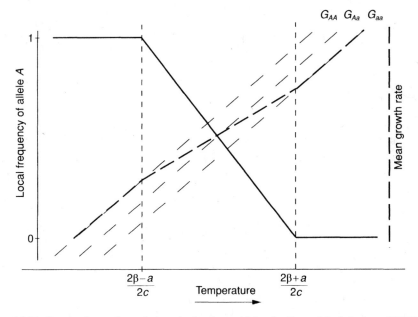

Figure 14.10. Graph of genetic and phenotypic clines obtained with model of de Jong (1988). The temperature axis is subdivided into three sections: $T < (2\beta - a)/(2c)$, in which there is fixation of the A allele ($p = 1$); $(2\beta - a)/(2c) < T < (2\beta + a)/(2c)$, in which there is a balanced polymorphism with $0 < p < 1$; and $T > (2\beta + a)/(2c)$, in which there is fixation of the a allele ($p = 0$). The solid line indicates the allele frequency cline over these temperatures. The thick dashed line indicates the mean phenotype of the local population as a function of the temperature gradient. For $T < (2\beta - a)/(2c)$ the phenotypic response follows the norm of reaction for the AA genotype, the sole genotype in the population under those conditions, and for $T > (2\beta + a)/(2c)$ the phenotypic response follows the norm of reaction for the aa genotype, the sole genotype in the population under those conditions. The norms of reaction for all the genotypes are indicated by the thin dashed lines.

is also increasing with increasing temperature. The phenotypic and genetic clines are going in exactly opposite directions! As the mean growth rate increases in the intermediate temperature range, there is actually an increase in the frequency of genotypes with *slower* growth rates. In the extremes of the temperature range, there is a continuation of the phenotypic cline with temperature but there is no genetic cline whatsoever because there is no genetic variation in the extremes of the temperature range. Figure 14.10 and the example of *R. clamitans* illustrate well the dangers of equating phenotypic clines to genetic clines.

Figure 14.10 represents some of the complexity that emerges from an interaction effect, in this case the interaction of genotypes with environments to produce phenotypes. Epistasis is another type of interaction that can modulate the evolutionary response to selection and gene flow in spatially heterogeneous environments. Suppose the unit of selection is a multilocus architecture with much epistasis for the phenotype of fitness. Hadany (2003) considered the evolutionary fate of such a coadapted gene complex in an environment showing heterogeneity in the intensity of selection rather than direction, with some areas showing intense selection and other areas showing weak selection. Such multilocus, epistatic complexes generate the rugged adaptive landscapes associated with Wright's shifting balance theory (Chapter 12). Hadany modeled shifting balance by defining a two-locus, two-allele genetic architecture (say A and a at locus 1 and B and b at locus 2) with epistasis in a haploid population such that the cis AB and ab combinations had higher fitness than the trans Ab and aB combinations. However, the combination AB had higher fitness than ab, but the populations started out fixed for ab. Mutation and recombination create variation in this model, and a shift to a higher peak occurs when AB arises, spreads, and becomes the most common form. Two neighboring subpopulations were then considered, each with the same fitness pattern, but one in which the fitness differences between cis and trans were large and the other in which the fitness differences between cis and trans were small. The effectiveness of shifting balance was measured by the average waiting time to a peak shift (the ab-to-AB transition). Hadany discovered that shifting balance was most effective when there was some, but limited, gene flow between the two subpopulations. The waiting times were increased if there was extensive gene flow between the subpopulations, and likewise they were increased if there were no gene flow between the subpopulations (that is, two independently evolving subpopulations, one subject to intense selection and one to weak selection). The reason for this is that selective intensity has contrasting effects at different stages of the shifting balance process. When selection is weak, the intermediate Ab and aB forms are more frequent, and the advantageous combination AB appears easily. However, weak selection also means that this advantageous combination is readily broken down by recombination (Chapter 13), making it difficult for this combination to persist and spread. The opposite is true when selection is strong. Now the intermediate Ab and aB forms are extremely rare, making it very difficult to generate the favored AB combination, but once that combination arises, strong selection can cause its rapid spread. When there is spatial heterogeneity in selective intensities and positive but limited gene flow, the AB combination can arise in those areas with weak selection, and when placed in areas of strong selection via gene flow, the favored combination can rapidly increase in frequency and then spread back into the areas of weak selection by subsequent gene flow after the favored combination has become the common form in areas of intense selection. The adaptive response emerges in this model from a three-way interaction between selection, population structure, and genetic architecture. Adaptive evolution cannot be explained in terms of natural selection alone.

COARSE-GRAINED TEMPORAL HETEROGENEITY

Just as a population can move through space via the dispersal of its constituent individuals, a population can also move through time via acts of reproduction to create generation after generation. The environment can change as a population moves through time such that different generations experience different environments, and because fitness is a genotype-by-environment interaction, the fitnesses associated with particular genotypes can also change across the generations.

Gene pools do not change instantaneously in response to a changed environment but rather at a rate proportional to the magnitude of the average excesses of the gametes. As a result, whenever there is a significant environmental change over time, there is usually a time lag before the gamete frequencies can fully adjust. These time lags in turn are strongly influenced by the genetic architecture. We saw in Chapter 11 that there was a rapid increase in S alleles at the β-Hb locus after the introduction of the Malaysian agricultural complex, but there was hardly any initial response at all in the C allele frequency. As discussed in Chapter 11 this difference in the relative time lags to the altered environment was due to the initial allele frequencies, population structure, and details of genetic architecture. In this particular case, the critical feature of the genetic architecture emerges from the fact the S allele behaves as a dominant allele for malarial resistance, whereas the C allele behaves as a recessive allele. These features of genetic architecture, when coupled with initial rare allele frequencies and random mating, lead to orders-of-magnitude differences in the initial adaptive response to the novel environment associated with the Malaysian agricultural complex. Recall also that genetic architecture includes the nature of the genotype-to-phenotype relationship, which itself can be directly altered by an environmental change. Thus, we already saw in Chapter 8 that the S allele is a recessive allele for viability in the nonmalarial environment but an overdominant allele in the malarial environment. Evolution in coarse-grained temporal heterogeneity can become complex and difficult to predict because the adaptive response depends upon a potentially changing genetic architecture, population structure, and historical factors (e.g., the initial composition of the gene pool).

The ever-present time lags in evolutionary response to temporal changes in the environment also mean that the current genetic state of a population is not always well adapted to the current environment. A past environment can continue to affect the genetic composition of a population for long periods of time. We also see this with the S allele. Most African Americans are no longer subject to death via malaria nor have they been for centuries, yet the S allele still persists in high frequency, although its frequency has been reduced (Table 12.1). Thus, the key to understanding the current high frequency of the S alleles in African Americans lies in an understanding of the past environments experienced by this population and not just the current environment.

Time lag effects can be particularly strong for coarse-grained cyclical temporal variation. For example, the twin-spotted ladybug beetle, *Adalia bipunctata*, has at least two generations per year in Germany (Timofeef-Ressovsky 1940). One generation hibernates over winter as adults and comes out in the spring. The second generation lives over the summer and into the autumn. There is also a genetically based color polymorphism in this species, with red and black forms. Populations of these beetles were monitored near Berlin, Germany, to reveal that the black forms survive better in the summer than the red forms, but the red forms survive hibernation much better than the black forms. This seasonal reversal of viabilities results in an annual cycle such that the red forms constitute

63.4% of the population in April (the beetles emerging from hibernation), whereas the black forms predominate by autumn, being some 58.7% of the population in October. Note that the red form is most common in the spring, just as the environmental conditions favoring the black forms are beginning. By autumn, the black forms predominate, yet it is the red form that is better adapted to the hibernation phase that will soon commence. Thus, the time lags inherent in any evolutionary response can yield seemingly maladaptive consequences.

Insight into the evolutionary implications of coarse-grained seasonal selection can be obtained through a simple one-locus, two-allele model (Hoekstra 1975). In most models in population genetics, the basic temporal unit is a point in the life cycle at one generation to the corresponding point in the next generation. However, for a cyclical selection model, Hoekstra chose as his basic unit one complete cycle of the environmental changes, which corresponds to more than one generation in the coarse-grained case. The special case of a cycle of two environments and two generations (such as with the ladybug beetles) is shown in Table 14.3. Notice that the frequencies of the genotypes after one complete cycle are of the same form as the standard single-generation models if we use the cycle fitnesses w_{AA}, w_{Aa}, and w_{aa}. However, these cycle fitnesses are not the standard single-generation fitnesses but rather are nonlinear functions of the fitnesses that occur in both environments in the cycle as weighted by the zygotic genotype frequencies at the beginning of the cycle. Hence, much biological complexity is buried in these seemingly simple equations. Fortunately, we already have the tools to reveal that complexity. Note that the cycle fitnesses are all of the form $w_i = p^2 \omega_{i2} + 2pq\omega_{i1} + q^2 \omega_{i0}$, where we let $i = 2$ correspond to AA, $i = 1$ to Aa, and $i = 0$ to aa. This mathematical form is identical to that of the model of competitive selection of Cockerham et al. (1972) given in Table 13.6. Thus, although these two models

Table 14.3. Hoekstra's (1975) Model of Coarse-Grained Cyclical Selection at One Locus with Two Alleles (*A* and *a*) over Two-Environment Cycle with Each Environment Experienced by Different Generation

Genotype	AA	Aa	aa
Zygotic frequency at beginning of cycle	p^2	$2pq$	q^2
Fitness in environment 1	w_1	1	v_1
Genotype frequency after selection	$\dfrac{p^2 w_1}{\bar{w}_1}$	$\dfrac{2pq}{\bar{w}_1}$	$\dfrac{q^2 v_1}{\bar{w}_1}$
Zygotic frequency at second generation	$\dfrac{p^2 (w_1 p + q)^2}{\bar{w}_1^2}$	$\dfrac{2pq(w_1 p + q)(v_1 q + p)}{\bar{w}_1^2}$	$\dfrac{q^2 (v_1 q + p)^2}{\bar{w}_1^2}$
Fitness in environment 2	w_2	1	v_2
Genotype frequency after one cycle	$\dfrac{p^2 w_{AA}}{\bar{\bar{w}}}$	$\dfrac{2pq w_{Aa}}{\bar{\bar{w}}}$	$\dfrac{q^2 w_{aa}}{\bar{\bar{w}}}$
where	$\bar{w}_1 = p^2 w_1 + 2pq + q^2 v_1$		
	$w_{AA} = w_2 (w_1 p + q)^2 = p^2 w_2 w_1^2 + 2pq w_2 w_1 + q^2 w_2$		
	$w_{Aa} = p^2 w_1 + 2pq \dfrac{w_1 v_1 + 1}{2} + q^2 v_1$		
	$w_{aa} = p^2 v_2 + 2pq v_2 v_1 + q^2 v_2 v_1^2$		
	$\bar{\bar{w}} = p^2 w_{AA} + 2pq w_{Aa} + q^2 w_{aa}$		

deal with different biological situations, they end up having identical mathematical forms. This means that all the results inferred from the frequency-dependent model of competition can be applied to this model of cyclical selection. For example, by simply equating the fitness components in the cyclical model shown at the bottom of Table 14.3 to the corresponding fitness components shown at the bottom of Table 13.6, we have from inequality 13.21 that the conditions for protecting a polymorphism when there is no dominance or recessiveness are

$$v_1 > v_2 v_1^2 \Rightarrow v_1 v_2 < 1 \qquad w_1 > w_2 w_1^2 \Rightarrow w_1 w_2 < 1 \qquad (14.19)$$

Inequalities 14.19 mean that the polymorphism is protected when the geometric mean of the homozygote fitnesses over the environment cycle is less than that of the heterozygote. Moreover, cyclical selection also shares many other evolutionary features typical of frequency-dependent selection (Chapter 13), such as having the potential for multiple equilibria (and hence, the initial state of the gene pool can influence the evolutionary outcome), violating Fisher's fundamental theorem, and displaying chaotic dynamic behavior.

The time lags inherent in coarse-grained temporal models can explain why many seemingly maladaptive traits can be in high frequency in a population. For example, Haldane and Jayakar (1963) showed how a trait that is normally mildly selected against but that is strongly selected for about once in every 20 generations can persist in high frequencies in a population. A possible example of this is the trait type 2 diabetes mellitus. Type 2 diabetes is an adult-onset alteration in insulin secretion and insulin resistance (that is, cells do not respond effectively to insulin, a hormone responsible for mediating the uptake by cells of glucose from the blood). Adult-onset diabetes is one of the more common diseases affecting humanity, with at least 250 million cases worldwide and increasing at an alarming rate (Alper 2000). Both unmeasured and measured genotype approaches have shown that genetic variation at many loci influences the risk for adult-onset diabetes, including several candidate loci such as the human insulin gene region, the calpain-10 locus, and loci coding for transcription factors that influence insulin secretion (Cox 2001; Marx 2002; Meirhaeghe et al. 2001; Pillay et al. 1995; van Tilburg et al. 2001). Both the human insulin gene region and the calpain-10 locus have haploytpe trees and haplotype distributions with the molecular signals indicating recent positive selection, particularly in those populations most susceptible to diabetes (Fullerton et al. 2002; Stead et al. 2003; Vander Molen et al. 2005). The symptoms of this disease show much variation but often lead to many serious complications and even death. As a result of its severity and frequency, adult-onset diabetes alone accounts for 15% of the total health care costs in the United States (Diamond 2003). So why are the genes that contribute to such a deleterious trait so common in human populations and show evidence for positive selection?

Neel (1962) suggested a possible answer to this question: the thrifty-genotype hypothesis. This hypothesis postulates that the same genetic states that predispose one to diabetes also result in a quick insulin trigger even when the phenotype of diabetes is not expressed. Such a quick trigger is advantageous when individuals suffer periodically from famines since it would minimize renal loss of precious glucose and result in more efficient food utilization. When food is more plentiful, selection against these genotypes would be mild because the age of onset of the diabetic phenotype is typically after most reproduction (in the next chapter, we will see how deleterious phenotypes associated with late ages have little impact on fitness) and because the high-sugar, high-calorie diets found in modern societies that help trigger the diabetic phenotype are very recent in human evolutionary history.

The thrifty-genotype hypothesis has much support (Neel et al. 1998; Diamond 2003). One prediction of this hypothesis is that those human populations that regularly suffered from famines in their recent history should be most prone to diabetes today. The Pima Indians in Arizona are one such population. The Pimas were formerly hunter-gatherers and farmers who used irrigation to raise a variety of crops, but principally maize. However, they were living in an arid part of the country, and their maize-based agricultural system was subject to periodic failures during times of drought. This was accentuated in the late nineteenth century when European American immigrants diverted the headwaters of the rivers used by the Pimas for irrigation, resulting in widespread starvation. With the collapse of their agricultural system, the surviving Pimas were dependent on a government-dispensed diet that consisted of high-fat, highly refined foods. Currently among adult Pima Indians, 37% of the men and 54% of the women suffer from type 2 diabetes, one of the highest incidences known in human populations.

Another example is provided by the human population on the Micronesian island of Naura (Diamond 1992, 2003). The Naurans suffered from two extreme bouts of natural selection for thrifty genotypes in their recent history. First, their population was founded by people who undertook interisland canoe voyages lasting several weeks. In numerous attested examples of such lengthy canoe voyages, many voyagers died of starvation. Second, the Naurans were then set apart from most other Pacific Islanders by their extreme starvation and mortality during the World War II. Both of these episodes would have resulted in strong selection for thrifty genotypes. After World War II, an external mining company signed a lucrative deal with the Naurans for the rights to potassium-rich bird guano. With their newfound wealth, food became superabundant. In this new dietary environment, some 28% of the adult population suffers from type 2 diabetes.

Under the thrifty-genotype hypothesis, the extremely high incidence of diabetes in both the Pima Indians and the Naurans is due to their recent evolutionary history of high mortality from starvation. However, most human populations have experienced some famine over a time scale of centuries, so the thrifty-genotype hypothesis also explains why diabetes is so common in human populations in general, although not at the rates seen in the Pima Indians and the Naurans. Moreover, many of the same genes associated with diabetes risk are also associated through pleiotropy with risk for hypertension (high blood pressure) and obesity (Neel et al. 1998), so the thrifty-genotype hypothesis also provides an explanation for these other common human maladies.

A variant of the thrifty-genotype hypothesis also explains why humans are so prone to coronary artery disease (CAD). Coronary artery disease is initiated by injuries to the endothelial lining of the coronary arteries, followed by the deposition of lipids from low-density lipoprotein (LDL) particles. This results in an atherosclerotic plaque. As the plaque grows, it restricts blood flow and changes the mechanical characteristics of the artery wall. These events facilitate plaque rupture, which in turn induces clotting and partial or total blockage of the flow of blood to some heart muscle cells. Depending upon the extent and location of the blockages, symptoms range from mild pain to sudden death. Coronary artery disease accounts for about one-third of total human mortality in western, developed societies, making it the most common cause of death. Both genetic and environmental factors contribute to this disease (e.g., the *ApoE* locus as discussed in Chapter 8). The lateness in life with which CAD typically occurs and the recentness of the environmental situation in which it is common (Western, developed societies) imply that it is unlikely that CAD itself has been the direct target of natural selection during human evolution. Rather, it is more likely that the genes that predispose one to CAD have effects on other traits that were subject to natural selection in past environments. Past human evolution has been characterized by the rapid and dramatic expansion of our brain and cognitive abilities. The development and maintenance of a large brain create a high demand for cholesterol. Because the diet of early humans had much less cholesterol in it than the current diets of people living in developed countries, selection would favor those genotypes that were "thrifty" in their absorption and production of cholesterol (Mann 1998), and specifically the *ApoE ε4* allele (Chapter 8) is hypothesized to be associated with such thrifty lipid genotypes (Corbo and Scacchi 1999). When human life span increased and the diet became high in fat, the thrifty *ε4* genotypes led to increased risk for CAD (Stengård et al. 1996) and other lipid-associated maladies such as Alzheimer's disease (Reiman et al. 2001) and Parkinson's disease (Zareparsi et al. 2002). Overall, we see that some common diseases afflicting modern humans in Western societies all seem to be due to natural selection operating on past environmental conditions.

The incidences of many of the diseases and conditions mentioned in the previous paragraph are increasing at an alarming rate. For example, diabetes in adults increased by 49% (Marx 2002) and obesity by 33% (Friedman 2003) between 1991 and 2000 in the United States. Although the thrifty genotype explains the high risk that humans have for these conditions, it alone cannot explain the rapid increase in these phenotypic frequencies over a time period much shorter than that of a single human generation. Such a large increase in only a decade cannot possibly be due to an evolutionary response in the human gene pool in the United States. The answer is that during this time period the eating and physical activity habits of many people in the United States changed in such a manner that more individuals with the risky genotypes were exposed to the environmental conditions that led to the expression of diabetes and obesity (Friedman 2003). Diabetes and obesity, like other phenotypes, emerge from how genotypes interact with the environment (premise 3 in Chapter 1). When environments change over time, such interactions may lead to direct phenotypic alterations without any evolution in the gene pool. Such interactions are explicitly acknowledged in the concepts of norm of reaction and phenotypic plasticity (Chapter 10). We have already seen how such phenotypic plasticity can influence both the genetic and phenotypic responses to a spatially varying environment with the example of the green frog cline and the model of de Jong (1988). Indeed, the de Jong model shown in Figure 14.10 is equally applicable to coarse-grained temporal heterogeneity. For example, suppose that global warming causes an increase in temperature over time for a frog population at a particular location. Then Figure 14.10 shows the genetic trajectory of the evolution of that population as it adapts to an increasing temperature. Recall how this model warns us to be cautious in interpreting phenotypic clines over space. The same warning is applicable to interpreting phenotypic changes over time. For example, Figure 14.10 shows that it is possible for the most rapid phenotypic evolution to have occurred when there is no genetic evolution at all (when populations are fixed for one allele or the other) and the slowest phenotypic evolution to have occurred when there is rapid genetic evolution (when the populations are polymorphic in Figure 14.10). One should never equate phenotypic change over time to evolutionary change. Unfortunately, such an equation is common in the evolutionary literature. For example, Eldredge and Gould (1972) noted fossil evidence for some organisms showing periods of morphological stasis that were punctuated by short bursts of rapid morphological change. Their theory of **punctuated equilibrium** states that the periods of rapid phenotypic change in otherwise phenotypically static lineages is due to rapid bursts of evolutionary change. However, if something similar to the situation shown in Figure 14.10 were occurring, the periods of rapid phenotypic change would happen when there is little to no evolutionary change, and periods of relative phenotypic stasis would have occurred due to continual evolutionary adjustments. This by no means implies that all cases of rapid phenotypic change are associated with no evolutionary change, but it does serve to warn us that *patterns are insufficient to infer processes*. This occurs because there are often several different processes that can generate the same patterns. In this case, one could have rapid phenotypic change due to either rapid evolutionary change or due to rapid environmental change coupled with phenotypic plasticity. Both are biologically plausible explanations that can yield indistinguishable patterns at the phenotypic level.

As the above warning indicates, phenotypic plasticity can cause a population's average phenotype to change over time (or space) without any genetic evolution. Recall that a phenotype is any measurable trait (Chapter 8). Since we can measure norms of reaction, phenotypic plasticity itself is a phenotype. Like any other phenotype, there can be genetic variation for the degree of plasticity among individuals in a population, and hence the trait of phenotypic plasticity can evolve. One example of such a phenomenon is **genetic assimilation**, in which selection acts upon heritable variation in phenotypic plasticity to turn a phenotype directly stimulated by an altered environment (plasticity) into a fixed phenotypic response no longer sensitive to the ancestral environmental triggers (assimilation) (Waddington 1957). Examples of genetic assimilation are found in ambystomid salamanders, such as *Ambystoma tigrinum*, whose phylogeography was discussed in Chapter 7. Most amphibians have an aquatic larval phase followed by metamorphosis to a terrestrial adult phase. Metamorphosis in many amphibians is under the control of hormones from the hypothalamus, pituitary, and thyroid glands, with the hormone thyroxin (TH) being the primary, but not exclusive, mediator of the tissue responses that lead to metamorphosis through TH receptor proteins on the cells of responsive tissues (Rose 1999). Nutrition, photoperiod, and temperature all affect endocrine activity. In particular, poor nutrition, darkness, and low temperature all tend to reduce the production of TH in ambystomid salamanders, resulting in phenotypic plasticity for the timing of metamorphosis. Indeed, metamorphosis can be completely prevented under appropriate environmental conditions, resulting in aquatic larval forms that became sexually mature and thereby bypass the terrestrial adult phase completely. Such sexually mature aquatic salamanders are called paedomorphs (Gould 1977). Most tiger salamanders are phenotypically plastic for the phenotype of paedomorphs versus metamorphic adults.

The climatic changes associated with the end of the last glacial period not only profoundly influenced the phylogeography of the tiger salamander (Figure 7.9) but also probably influenced this salamander's phenotypes through time. During the glacial period, fossils from the Kansas–Oklahoma area consist of giant paedomorphs of tiger salamanders, but as the temperature increased following the Pleistocene, metamorphic forms become more common. This change in phenotypes may or may not be due to genetic evolution; it certainly could be explained entirely in terms of phenotypic plasticity with no genetic evolution at all. Current populations of the tiger salamander from clade 4-2 (Figure 7.9) are still plastic for the phenotype of paedomorphy. In the Rocky Mountains, where ponds are permanent and have low temperatures, paedomorphs are frequent, and clade 4-2 represents a recent range expansion from this environment that favors the production of paedomorphs. Clade 4-2 produces paedomorphs throughout its current geographic range (Figure 7.9) in large, permanent ponds (Templeton 1994). However, the salamanders from clade 4-1 had a very different recent evolutionary history (Figure 7.9). They were fragmented from the western populations in the Pleistocene and lived in the Ozarks. The Ozarks are not high enough to induce cool temperatures and have few permanent ponds. Under these environmental conditions, paedomorphs are not expected nor are they found. However, the range of this clade has also expanded and now overlaps with that of clade 4-2 (Figure 7.9). When both types of salamanders are taken from permanent ponds, all paedomorphs turn out to be from clade 4-2 (Templeton 1994). Consequently, we see a puzzling pattern. The clade 4-1 salamanders recently lived in environments that would make paedomorphy unlikely, but now even when they are living in large, permanent ponds that induce paedomorphy in clade 4-2 salamanders, they are incapable of becoming paedomorphs.

In contrast, consider the closely related species *Ambystoma mexicanum*. This species lives in permanent lakes in the mountainous region of Mexico. This environment favors paedomorphy, but even when these salamanders are placed in environments that favor metamorphosis in other tiger salamanders, *A. mexicanum* normally fails to undergo metamorphosis. Voss et al. (2003) used a candidate locus approach (Chapter 10) to study the genetic basis of metamorphosis in hybrid populations made from laboratory crosses of *A. mexicanum* with *A. tigrinum tigrinum* (clade 4-1, the population that cannot produce paedomorphs) and discovered that two *TH* receptor loci were associated with these phenotypes, although in a manner that suggested strong epistasis with other, unmeasured loci. Thus, we seem to have a strange, almost Lamarckian phenomenon: Paedomorphy and metamorphosis are phenotypically plastic in some salamanders, but when a salamander population is placed in an environment that favors metamorphosis, it becomes genetically incapable of paedomorphy, whereas a second population found in an environment that favors paedomorphy becomes genetically incapable of metamorphosis. Somehow, prolonged exposure to the environment favoring a particular phenotypic response has become "genetically assimilated" and is now expressed (or not expressed) regardless of the environment.

However, there is no strange evolutionary force working here as long as there is genetic variation for the degree of plasticity in the paedomorphic/metamorphic phenotype, as there clearly is (Voss and Shaffer 2000; Voss et al. 2003). Suppose, for example, that most salamanders can only find temporary ponds, as is the case in the Ozarks. In such an environment, any salamander that failed to undergo metamorphosis would die when the pond dried up. In a population that was genetically variable in its phenotypic responses to environmental cues, some salamanders would respond to a particular environment by developing into paedomorphs, whereas others would respond to the same environment by undergoing metamorphosis. In populations living in temporary ponds, there would be selection against all animals that ever give a paedomorphic response to their immediate environment. Therefore, evolution should shift the responsiveness to the environment in the direction favoring metamorphosis (equation 9.22). As long as there was heritable variation, the population would become less and less likely to produce paedomorphs even under environmental conditions that induced paedomorphs with a high probability in the ancestral population. A similar selective scheme could have been happening in *A. mexicanum*, but in this case living in a permanent lake favors the paedomorphs.

Another possible evolutionary mechanism of genetic assimilation in these salamanders is the accumulation of neutral mutations. *Ambystoma mexicanum* now lives in an environment that always leads to paedomorphy. Hence, any genes that deal exclusively with metamorphosis could, under these environmental conditions, be neutral because they are not expressed in any functional fashion. A neutral locus evolves (Chapter 5), and new, neutral mutations should eventually go to fixation. Since random mutations on average diminish functional capability, the metamorphic genes that have been rendered neutral are expected to eventually become fixed for nonfunctional and diminished-functional mutations. Hence, when the animal is now put into an environment that would favor metamorphosis in the ancestral population, it can no longer do so because it lacks one or more functional genes for metamorphosis. These two hypotheses (selection against plasticity or reduced environmental sensitivity and neutral evolution at loci rendered functionless by long-term nonexpression of specific phenotypes) are not mutually exclusive and both could contribute to the process of genetic assimilation.

Coarse-grained heterogeneity and genetic assimilation can make the shifting balance process (Chapter 12) more likely. Consider a population that experiences a novel environment, either temporal or spatial. Because fitness is a genotype-by-environment interaction, a new fitness landscape will be encountered, just as the novel malarial environment induced a new fitness landscape with respect to the genetic variation at the β-Hb locus with the landscape shown in Figure 12.2 being transformed into that shown in Figure 11.8 by the introduction of the Malaysian agricultural complex into Africa. However, the new environment can also induce novel phenotypic responses from the norms of reaction of the genotypes in the population that initially encounters this novel environment. Pál and Miklós (1999) have shown that even if the environment initially induces random phenotypic variants, the increased phenotypic variance nevertheless makes it more likely that *some* favorable phenotype will exist in the novel environment. This in turn can initiate the process of genetic assimilation and trigger a shift to the new adaptive peak that appeared in this novel environment. These evolutionary tendencies are strengthened when the environmentally induced variation is itself subject to epigenetic inheritance. For example, suppose a novel environment induces new methylation patterns in the DNA that in turn affect DNA expression that in turn affect phenotypic variation. Methylation patterns can sometimes be passed on to the next generation, and this initial nongenetic inheritance actually makes the process of genetic assimilation even more likely (Pál and Miklós 1999).

FINE-GRAINED HETEROGENEITY

An individual often experiences environmental heterogeneity within its own lifetime. Because an individual can only be at one place at any given time, an individual experiences both spatial and temporal heterogeneity within its own lifetime as a temporal sequence. Hence, for purposes of microevolutionary modeling, no distinction is necessary between fine-grained spatial and temporal heterogeneity. In many situations, fine-grained heterogeneity needs no special consideration as it can be folded into the constant-fitness models given in Chapter 11. To see this, consider two extreme situations: the case in which every individual in the population experiences the same temporal sequence of fine-grained heterogeneity and the case in which every individual experiences an independent sample of temporal sequences of environments within its lifetime.

The first case would apply to the situation in which an organism has one generation per year but in which seasonal variation influences the viabilities of the genotypes within the population. Thus, every individual in the population experiences the seasonal variation within its own lifetime, and every individual experiences exactly the same sequence of this seasonal heterogeneity. In the previous section, Hoekstra's (1975) model of coarse-grained seasonal variation was given, and Hoekstra also modeled the case of fine-grained seasonal variation. Table 14.4 gives the two-season fine-grained model in which the fitness effects within a season are identical to those given in Table 14.3, the two-season coarse-grained model. Note that Table 14.4 is just the standard, constant-fitness model (Figure 11.2) with the constant viabilities of $\ell_{AA} = w_{AA} = w_1 w_2$ and $\ell_{aa} = w_{aa} = v_1 v_2$. Hence, all of the equations and conclusions of Chapter 11 apply to this case of fine-grained seasonal variation. In particular, the frequency-dependent dynamics that emerge from the

Table 14.4. Hoekstra's (1975) Model of Fine-Grained Cyclical Selection at One Locus with Two Alleles (A and a) over Two-Environment Cycle with Each Environment Experienced by All Individuals within Single Generation

Genotype	AA	Aa	aa
Zygotic frequency at beginning of cycle	p^2	$2pq$	q^2
Fitness in environment 1	w_1	1	v_1
Genotype frequency after Selection in 1	$\dfrac{p^2 w_1}{\bar{w}_1}$	$\dfrac{2pq}{\bar{w}_1}$	$\dfrac{q^2 v_1}{\bar{w}_1}$
Fitness in environment 2	w_2	1	v_2
Genotype frequency after Selection in 1 and 2	$\dfrac{p^2 w_{AA}}{\bar{\bar{w}}}$	$\dfrac{2pq}{\bar{\bar{w}}}$	$\dfrac{q^2 w_{aa}}{\bar{\bar{w}}}$
where	$\bar{w}_1 = p^2 w_1 + 2pq + q^2 v_1$		
	$w_{AA} = w_1 w_2$		
	$w_{aa} = v_1 v_2$		
	$\bar{\bar{w}} = p^2 w_{AA} + 2pq + q^2 w_{aa}$		

coarse-grained analogue of this model do not appear in the fine-grained case. Even for this simple two-season model, we get qualitatively different evolutionary dynamics for the coarse-grained and fine-grained versions. This shows that environmental grain greatly influences the microevolutionary process.

Now consider the second case in which each individual independently samples its own fine-grained temporal sequence of environments. This will induce fitness differences among individuals sharing the same genotype within a generation. Thus, we need to replace a constant fitness for genotype ij, say w_{ij}, with a random fitness with mean w_{ij} (now the average fitness of all individuals with genotype ij) and a variance of v_{ij}. However, we have already modeled this situation as well in our constant-fitness models of Chapter 11. Recall that the fundamental definition of fitness in population genetics is that of a genotypic value assigned to all individuals sharing a common genotype (Chapter 11). The genotypic value for the phenotype of fitness is simply the average fitness for all individuals who share a common genotype. But there is nothing about the concept of genotypic value that requires that all individuals share exactly the same phenotypic value. Quite the contrary, in Chapter 8 we explicitly assumed that the individuals that share a common genotype do *not* have the same phenotype but rather are characterized by a mean phenotype (the genotypic value) coupled with a random, within-genotype environmental deviation (equation 8.6). In Chapter 8 we assumed that the distribution of environmental deviations was identical for all genotypes and had a variance of σ_e^2, the "environmental variance" (equation 8.16). Because fitness is thought of as a genotypic value in population genetics, there has never been the assumption that every individual with the same genotype has to have exactly the same fitness phenotype; only the averages count in the models developed in Chapter 11. Hence, the environmental variance term can accommodate any fitness fluctuations among individuals with the same genotype that is induced by sampling fine-grained environmental variation. Moreover, since only the genotypic values enter into the equations given in Chapter 11, we do not even require the assumption that each genotype has the same environmental variance as we did in Chapter 8; rather, different genotypes can display different environmental variances to fine-grained heterogeneity, but it is still only their genotypic values of fitness that drive the evolutionary response under most conditions.

However, there are exceptions, and in some situations the within-genotype environmental variance of fitness does matter to the evolutionary outcome. One such exception is for newly arisen mutations. As we saw in Chapter 5, genetic drift has a major impact on the survival of a newly arisen neutral mutation even in an effectively infinite sized population. We then saw in Chapter 12 how genetic drift still has a major impact on the survival of a newly arisen selected mutation even in an effectively infinite sized population. For example, if a new allele, A, mutates from the ancestral allele a such that the relative fitnesses are $1 + s$ ($s > 0$) and 1 for Aa and aa, respectively, we showed in Chapter 12 that the probability of survival of the selectively favored A allele in an ideal population of large size is approximately $2s$, which implies that the majority of selectively favored alleles are lost due to genetic drift. The large impact of genetic drift on newly arisen mutants even in effectively infinite sized populations stems from the fact that a newly arisen mutation is initially found in only one copy, so that finite sampling cannot be ignored. Therefore, the random force of genetic drift is powerful regardless of the total population size because the fate of all new mutations depends upon a small, finite number of copies. Similarly, random forces generated by fine-grained heterogeneity in fitness also play a powerful role in influencing the fate of a new mutation (Templeton 1977b). To model fine-grained heterogeneity, let the mean number of offspring of an *individual* bearing the new mutation (an Aa individual) have a mean of $1 + s$ and variance of $1 + s + \sigma_s^2$ in a large, stable population consisting mostly of aa individuals with a mean and variance of offspring number of 1 (recall from Chapter 5 that the offspring number distribution in our ideal population is Poisson, which implies that the mean and variance in the number of offspring are the same, and both are 1 in this case because we are assuming a stable sized overall population). The quantity σ_s^2 measures the variance in the selection coefficient s among Aa individuals that is induced by fine-grained heterogeneity. If all Aa individuals had exactly the same fitness of $1 + s$, then their mean and variance in offspring numbers would both be $1 + s$ (the Poisson assumption). However, fine-grained heterogeneity (σ_s^2) induces more variation in offspring number than would be expected under the Poisson. When s, the average selection coefficient, is close to zero, the survival probability of the A allele is

$$\Pr(A \text{ survives}) = \frac{2s}{1 + s + \sigma_s^2} \qquad (14.20)$$

Note that when there is no fine-grained heterogeneity in fitness ($\sigma_s^2 = 0$), equation 14.20 reduces to $2s/(1 + s) \approx 2s$ since we assumed s was close to zero. Thus, the constant-fitness case given in Chapter 12 is just a special case of equation 14.20. When fine-grained heterogeneity exists, $\sigma_s^2 > 0$ and the probability of a new mutant surviving always *decreases* relative to the case with no fine-grained heterogeneity. This observation has some important implications for how populations evolve in response to fine-grained heterogeneity. Consider two mutations, say A_1 and A_2, each with identical average fitnesses (both $1 + s$), but with $\sigma_{s1}^2 < \sigma_{s2}^2$; that is, the A_1a genotype is more buffered against the fitness fluctuations caused by fine-grained environmental heterogeneity than is the A_2a genotype. Then equation 14.20 implies that A_1 has a greater chance of surviving and going to ultimate fixation in the population than A_2. Hence, natural selection in this case favors those mutations that are associated with fitness phenotypes that are more buffered against responding to fine-grained heterogeneity. Indeed, selection can even favor such fine-grained buffering over the mean selective value. For example, let s_i now be the average selection coefficient associated with the $A_i a$ genotype. Now let us assume that $s_1 > s_2$. Normally, we would expect A_1 to have a higher probability of survival than A_2 because it is associated with a higher average fitness. But equation 14.20 implies that A_2 is more likely to survive than A_1 whenever

$$\sigma_{s2}^2 < \frac{s_2 - s_1}{s_1} + \frac{s_2}{s_1}\sigma_{s1}^2 \qquad (14.21)$$

Inequality 14.21 tells us that if the $A_2 a$ genotype is sufficiently buffered from fine-grained heterogeneity, then the A_2 allele can have a greater chance of surviving in the population than A_1 even though average fitness is lowered in this case (yet another violation of Fisher's fundamental theorem). For example, let $s_1 = 0.02, s_2 = 0.01, \sigma_{s1}^2 = 4$, and $\sigma_{s2}^2 = 0.5$. With no fine-grained heterogeneity, the equation from Chapter 12 tells us that A_1 is twice as likely to survive as A_2, but with fine-grained heterogeneity, equation 14.20 tells us that the survival probability of A_1 is 0.008 and the survival probability of A_2 is 0.013, so now A_2 is the favored allele.

Equation 14.20 also challenges our notion of selective neutrality. Suppose the two mutants have identical average fitness effects such that $s_1 = s_2$. Then, inequality 14.21 shows that natural selection will preferentially favor the survival of the mutant with the lower within-genotype variance. These two alleles are neutral relative to one another in the sense that they have identical fitnesses (genotypic values), but they are not equivalent at all in their evolutionary dynamics under selection in a fine-grained environment. Once again, selection favors the genotypes best buffered against fitness fluctuations.

The importance of the within-genotype variance of fitness induced by fine-grained environments becomes accentuated when the total population size is finite. In this case, the variance of within-genotype fitness never gets completely averaged out even at the total population level. For example, consider a population fixed for allele a at an autosomal locus in a finite, ideal, random-mating deme of size N. Mutation then creates a single copy of the new allele A. Suppose further that A is a favorable allele under the existing environmental conditions such that the relative fitnesses are 1 for aa, $1 + s$ for Aa, and $1 + 2s$ for AA where the selection coefficient s is greater than zero. When there is no variance of fitness within genotypes, the probability of fixation, u, for the favored A allele is given by equation 12.16 as $u = (1 - e^{-2s})/(1 - e^{-4Ns})$. Now let fine-grained heterogeneity exist such that the selection coefficient is a random variable with mean s and variance v at the level of a single *individual*. Then, the probability of fixation of the favored A allele is (Templeton 1977b)

$$u = \frac{1 - e^{-2(s-v/2N)}}{1 - e^{-4N(s-v/2N)}} \qquad (14.22)$$

Note that equations 14.22 and 12.16 are of the same form except that $s - v/2N$ in the fine-grained model replaces the parameter s in the model with no fine-grained heterogeneity. Notice that the mean fitness (measured by s) in the fine-grained model is discounted by $v/(2N)$ in determining the probability of fixation. Hence, as v goes up (less buffering against fine-grained heterogeneity), the probability of fixation of a favorable allele goes down. Once again, selection tends to favor the genotypes that are buffered against fine-grained heterogeneity.

The third case in which we cannot ignore fitness fluctuations induced by fine-grained heterogeneity occurs when we focus on traits contributing to fitness rather than fitness directly. For example, consider again the fitness model given in equations 14.13 and 14.14

that related temperature to developmental time to fitness in the frog for coarse-grained spatial heterogeneity in temperature. Now we will add on fine-grained heterogeneity by assuming that there is within-genotype variance to the temperature response because not all individuals sharing the same genotype experienced exactly the same fine-grained environmental variation in temperature and by letting T in equations 14.13 now be the average temperature at a particular location. Here, T remains a measure of coarse-grained heterogeneity in overall temperature around which there is fine-grained heterogeneity. Moreover, we will now assume that each genotype exhibits a different variance in responding to the fine-grained heterogeneity; that is, the genotypes differ in their capacity to buffer this trait from fine-grained fluctuations in temperature. Equations 14.13 still reflect the expected genotype values for growth rates, but now we assume a variance of σ_{ij}^2 within genotype ij for the phenotype of growth rate. We retain our assumption that the individual growth rates map onto fitness as given by equation 14.14. However, because there is now phenotypic variance within each genotypic class, we must put equation 14.14 into equation 11.25 to find the average fitness associated with each genotype because of this nonlinear mapping of trait onto fitness (recall the seventh implication of Fisher's fundamental theorem of natural selection given in Chapter 11). The second derivative of fitness function 14.14 with respect to the growth rate phenotype P is -2α, which yields

$$w(G_{AA}) = 1 - \alpha(\beta - a - cT)^2 - \alpha\sigma_{AA}^2$$
$$w(G_a) = 1 - \alpha(\beta - cT)^2 - \alpha\sigma_{Aa}^2 \qquad (14.23)$$
$$w(G_{aa}) = 1 - \alpha(\beta + a - cT)^2 - \alpha\sigma_{aa}^2$$

The squared terms in parentheses in equations 14.23 determine the response to coarse-grained heterogeneity, as seen previously. The terms at the end of each fitness equation reflect the impact of fine-grained heterogeneity. Note yet again that variance in temperature response diminishes fitness as weighted by the parameter α that determines the sensitivity of fitness to temperature fluctuations in equation 14.14. The incorporation of fine-grained fitness fluctuations into this model can greatly change the equilibrium. For example, under the same assumptions of population structure given previously for the frog example, there will now be fixation for the A allele when

$$T < \frac{2\beta - a}{2c} + \frac{\sigma_{Aa}^2 - \sigma_{AA}^2}{2ac} \qquad (14.24)$$

The first term on the right-hand side of inequality 14.24 is simply the previous condition for fixation of A in the coarse-grained model (inequality 14.15). The second term in inequality 14.24 depends upon the fine-grained buffering capacities of the AA and Aa genotypes. This second term is negative if the Aa heterozygote is more buffered (smaller variance) against fine-grained fluctuations than the AA homozygote, which decreases the average temperature threshold at which populations begin to become polymorphic. If AA is more buffered than Aa, then the range of temperatures favoring fixation of A is broadened. Similarly, the conditions that allow the fixation of the a allele are

$$T > \frac{2\beta + a}{2c} - \frac{\sigma_{Aa}^2 - \sigma_{aa}^2}{2ac} \qquad (14.25)$$

For intermediate temperatures, local populations are polymorphic with the frequency of A being

$$p = \frac{a^2 - 2acT + 2a\beta - \sigma_{Aa}^2 + \sigma_{aa}^2}{2a^2 + \sigma_{AA}^2 + \sigma_{aa}^2 - 2\sigma_{Aa}^2} \tag{14.26}$$

Note that when all the within-genotype variances are zero or are identical equation 14.26 reduces to equation 14.17. Consequently, when genotypes differ in their ability to buffer themselves against fine-grained heterogeneity, selection shifts the course of adaptation in a manner that favors those genotypes with better buffering capabilities.

In all the above models we see a consistent theme: Selection in fine-grained environments favors those genotypes most buffered against fitness fluctuations. It is important to note that the trait being buffered in this case is fitness. Often, fitness buffering is accomplished by other traits being highly plastic and sensitive to fine-grained heterogeneity. For example, consider fine-grained temperature fluctuations in humans. Because we are a warm-blooded mammal, it is important that we maintain a nearly constant body temperature despite external temperature fluctuations. If our body temperature increases or decreases too much, our viability plummets. Humans have evolved a number of plastic traits that are sensitive to fine-grained environmental fluctuations in temperature that by their very plasticity help maintain a nearly constant body temperature (Roberts 1978). When the external temperature rises, the surface blood vessels dilate, there is a slight increase in blood and plasma volume, and more blood flows near the skin to carry heat to the periphery to be dissipated. If the external temperature rises more, sweat is produced to lose heat by evaporation. When the ambient temperature rises above the body temperature, sweating is the only means of heat loss. With prolonged exposure to high heat, all these mechanisms can break down or become ineffective, and the person experiences discomfort, physical distress, illness, and eventually death if the exposure to high heat continues too long. When a person is exposed to cold ambient conditions, vasoconstriction occurs to limit heat loss through peripheral tissues, and heat production by the body is increased by a variety of means, including shivering. With prolonged exposure to cold, these buffering mechanisms collapse and the core body temperature starts to decline, a process that can eventually lead to death. Thus, humans achieve buffering to external temperature fluctuations through a variety of highly plastic traits that change rapidly in response to ambient temperature. Although buffering and plasticity seem to be opposites, they are often interlinked.

The intensity for selection for buffering against ambient temperature fluctuations depends upon how extreme and prolonged either high or low temperatures persist. Humans living in different parts of the world show evidence of selective buffering against the type of fine-grained extremes they are most likely to encounter (Roberts 1978). For example, sub-Saharan Africans start sweating at a lower body temperature and produce more sweat per unit of surface area than Europeans. In contrast, the average skin temperature at which shivering commenced in a sample of Europeans was 29.5°C, but in a population of people with sub-Saharan African ancestry, their skin temperatures had to drop to 28°C on average before shivering commenced. The common theme in these studies is that the buffering mechanism against extreme temperatures is initiated sooner (with respect to body temperature) in the environmental extreme that the population from which the subjects were drawn had been historically most likely to encounter.

The above discussion also illustrates another important aspect of buffering against fine-grained perturbations; the fitness consequences of an environment depend not only on the nature of the environment but also on how long the environment persists (at least from the individual's perspective). Shivering or sweating can buffer a person for a while against cold or heat, but if the extreme temperature regimen persists long enough, these short-term buffering responses no longer work. When an environmental condition persists for long periods of time, other buffering strategies are used. For example, prolonged exposure to hot conditions causes an increase in the number of sweat glands (Roberts 1978). This pattern found in humans is a general one; many organisms can buffer themselves well against short exposures of even very harsh environmental states through a variety of physiological and behavioral mechanisms, but a different set of buffering mechanisms is used to deal with more prolonged exposure (Magnum and Towle 1977). Moreover, even "normal" environments can have deleterious fitness consequences upon individuals if they persist long enough. For example, most plants need a certain exposure to both wet and dry days in order to survive and grow. Hence, each environmental state is necessary for the plant. However, a long run of wet days could result in flooding conditions, while a long run of dry days could create a drought. Either of these long environmental runs could easily kill the plant. As a consequence, it is not surprising that many plants have a variety of short-term and long-term buffering mechanisms against both flooding and drought conditions (Kozlowski and Pallardy 2002). For example, flooding prevents much oxygen from reaching plant root systems. In response to these conditions, many plants initially switch to anaerobic metabolism in their roots. This is an effective short-term strategy, but if the flooding conditions persist, this buffering strategy breaks down because anaerobic metabolism is much less efficient energetically than aerobic metabolism and because anaerobic metabolism produces alcohols and other toxins that can kill or decay portions of the root system. Hence, if the flooding persists, alternative buffering mechanisms are needed, such as the production of adventitious roots, absorbing oxygen through the stomata in the leaves and transporting it to the roots, and forming structures known as lenticels that promote exchange of dissolved gases in the floodwater and may also release toxic compounds (e.g., acetaldehyde, ethanol, and ethylene) from the plants. This illustrates that the fitness-buffering response of an individual is a function not only of the environmental state but also of how long exposure to that state persists.

Consider a model with two environmental states, say 0 and 1 (Templeton and Rothman 1978). During any particular time unit, the environment is either in one state or the other. Transitions between time units are governed by the transition matrix

$$\begin{array}{c} 0 1 \\ \begin{array}{c} 0 \\ 1 \end{array} \begin{pmatrix} 1-\alpha & \alpha \\ \beta & 1-\beta \end{pmatrix} \end{array} \quad (14.27)$$

where $1 - \alpha$ is the probability of being in state 0 given that the environment was in state 0 during the previous time period, α is the probability of being in state 1 given 0 previously, β is the probability of being in state 0 given 1 previously, and $1 - \beta$ is the probability of being in state 1 given 1 previously. The average frequencies of states 0 and 1 over a large period of time are

$$f_0 = \frac{\beta}{\alpha + \beta} \qquad f_1 = \frac{\alpha}{\alpha + \beta} \quad (14.28)$$

This environment can be described not only by its average state frequencies but also by its "runniness." A run of length n commences when one state makes a transition into the other state followed by exactly $n - 1$ additional time periods of that state followed by a transition back to the original environmental state. The average run length of state 0 is $1/\alpha$, and the average run length of state 1 is $1/\beta$. Suppose an organism lives L time units, where L is some large number such that an individual normally encounters many runs of both environmental states during its lifetime. Let the fitness of genotype ij be given by

$$w_{ij} = c_{ij} \prod_u \xi_{ijx_u} \prod_v \omega_{ijy_v} \qquad (14.29)$$

where c_{ij} is a component of inherent fitness that is solely a function of the genotype and is not affected by the sequences of environmental states encountered, u indexes all runs of 0's that occurred in L, ξ_{ijx_u} is the fitness response of genotype ij to a run of 0's of length x_u, v indexes all runs of 1's that occurred in L, and ω_{ijy_v} is the fitness response of genotype ij to a run of 1's of length y_v. Consider now a one-locus, two-allele (A and a) model. Assume also that L is much larger than $(1/\alpha + 1/\beta)$, the average cycle length (that is, the length of time it takes to have one run of 0's followed by one run of 1's). This assumption ensures that this is a fine-grained model and that each individual experiences many runs of 0's and 1's in its lifetime. Then, Templeton and Rothman (1978) showed that the polymorphism is protected in this fine-grained environment when the expected values of the natural logarithms of the homozygote's fitnesses are both less than the expected value of the natural logarithm of the heterozygote's fitness. They also showed that the expected values of the logarithms of fitness have the form

$$E(\ln w_{ij}) \approx \ln c_{ij} + \frac{L}{1/\alpha + 1/\beta} \left(\overline{\ln \xi_{ij}} + \overline{\ln \omega_{ij}} \right) \qquad (14.30)$$

where $\overline{\ln \xi_{ij}}$ is the average log fitness effect of a run of 0's and $\overline{\ln \omega_{ij}}$ is the average log fitness effect of a run of 1's. Hence, $\overline{\ln \xi_{ij}} + \overline{\ln \omega_{ij}}$ is the average log fitness effect of one cycle and $L/(1/\alpha + 1/\beta)$ is the average number of cycles an organism experiences in its lifetime. When the c_{ij}'s are of equal value, natural selection favors those genotypes that deal best on the average with the environmental cycles that they encounter.

To capture the distinction between short-term and long-term buffering mechanisms, consider the following fitness response model to environmental runs:

$$\xi_{ijx} = \begin{cases} 1 & x \leq d_{ij0} \\ e^{-\lambda_{ij0}(x - d_{ij0})} & x > d_{ij0} \end{cases}$$

$$\omega_{ijy} = \begin{cases} 1 & y \leq d_{ij1} \\ e^{-\lambda_{ij1}(y - d_{ij1})} & y > d_{ij1} \end{cases} \qquad (14.31)$$

In this model, the d's measure the short-term buffering mechanisms that usually have no or low physiological costs. Thus, each genotype ij can endure a run of 0's of length d_{ij0} and a run of 1's of length d_{ij1} without any deleterious consequences at all. The larger the value of d, the better is the short-term buffering capacity. However, physiological constraints ensure that the d's cannot be too large, and when exposure to these environments lasts longer that the relevant d, the short-term buffering mechanisms break down and fitness begins to decline at an exponential rate measured by the relevant λ. Hence, λ measures the long-term buffering capacity of the organism, with small λ's corresponding to better long-term buffering capabilities. The expected value of the logarithm of fitness under fitness model 14.31 is (Templeton and Rothman 1978)

$$E(\ln w_{ij}) \approx \ln c_{ij} - L\left[f_0(1-\alpha)^{d_{ij0}}\lambda_{ij0} + f_1(1-\beta)^{d_{ij1}}\lambda_{ij1}\right] \qquad (14.32)$$

Hence, fine-grained environmental heterogeneity tends to select those genotypes that minimize the quantity in brackets in equation 14.32.

This bracketed quantity can tell us much about how organisms adapt to fine-grained environmental runs. First, we see that selective impact of the environmental states 0 and 1 depend upon their frequencies, f_0 and f_1. Hence, the more an organism encounters a particular environmental state, the more important it is to have a high fitness response to that state. Second, the impact of an environmental state also depends upon how likely it is to generate long runs. Consider state 0. The quantity $1 - \alpha$ is the probability of remaining in state 0 given the environment is in that state already. If $1 - \alpha$ is small, even modest values of d_{ij0} ensure that environmental state 0 has little overall fitness effect. Hence, a short-term buffering strategy is effective. However, if $1 - \alpha$ is large, long runs of 0's are likely to be encountered, and the quantity $(1 - \alpha)^{d_{ij0}}$ could still be substantial even when d_{ij0} is at its maximum physiological limit. In this case, the long-term buffering parameter λ_{ij0} becomes an important contributor to fitness.

These predictions are consistent with observations on natural populations. Plants that live in upland environments far from rivers have little chance of encountering prolonged flooding conditions, and most such plants only have the short-term physiological buffering mechanisms mentioned above. In contrast, plants that live in riparian habitats are much more likely to encounter prolonged floods and tend to have both the short-term and long-term buffering mechanisms against flooding (Kozlowski and Pallardy 2002). These examples and equation 14.32 tell us that fine-grained environmental heterogeneity cannot be adequately described just by the frequencies of the various environmental states, but rather it is necessary to know something about the temporal sequence of states that individuals encounter during their lifetime. A given environmental state can be benign, mildly deleterious, or lethal to an individual depending upon its temporal context, so evolution in fine-grained environments must also be sensitive to this context.

COEVOLUTION

Fitness arises from how genotypes interact with their environments. The environment for any species often includes other species. We have already seen how interactions among species can define the adaptive environment for a particular species. This was shown in Chapter 11 with the discussion of sickle cell anemia and malarial resistance in Africa. An environmental transition was associated with a radical shift in fitnesses for the genotypes associated with the human β-Hb locus (Table 11.1). This environmental transition was defined by the interactions among three species; humans changed their agricultural practices to provide breeding sites and habitats for the mosquito *Anopholes gambiae* and increases in the densities of these two species provided an increased host resource for a third species, *Plasmodium falciparum*, which parasitizes both humans and mosquitoes. Much of the environment for any organism is defined by its interactions with individuals of other species. But species are not static; they are all capable of evolving. Hence, when the environment of one species is defined by other species, evolution in these other species can create a changing environment for the species of interest. As one species adapts to the "environment" defined by the other species, the other species in turn can adapt to the changing environment created by evolution in the first species. We already saw in Chapter 11 how humans adapted to the malarial parasite (see also Fortin et al. 2002), but the malarial parasite in turn has adapted to humans. For example, the *apical membrane antigen 1 (AMA1)* locus codes for a surface-accessible protein in *P. falciparum* that can serve as a target for the human immune response to malarial infection. The McDonald–Kreitman and Tajima D tests (Chapter 12) indicated strong selection to maintain polymorphism in this malarial gene, presumably driven by selection induced by the human immune response (Polley and Conway 2001). Indeed, many other regions of the *P. falciparum* genome bear the signature of recent and strong selection since the parasite has begun specializing on humans (Conway et al. 2000; Volkman et al. 2001). Humans have also interacted with the malarial parasite by altering the environment, in this case through the development and use of antimalarial drugs. These human-produced drugs have also invoked selective sweeps (Chapter 12) in specific genes in *P. falciparum* for drug resistance (Wootton et al. 2002). Hence, humans are adapting to malaria, and malaria is adapting to humans. Mode (1958) first coined the term **coevolution** to describe the situation when two or more species mutually adapt to one another through interspecific interactions. Coevolution is simply natural selection operating within each of the interacting species, recognizing that each species constitutes part of the environment of the other species.

Mode's original models dealt only with host–parasite interactions (such as humans and malaria), but the concept has been generalized to include any potential interaction among individuals of different species (Table 14.5). Of the terms listed in Table 14.5, only the bottom three are true interactions, so the primary focus of coevolutionary models is upon those last three. The term "species interaction" has many different meanings in ecology and population genetics. In ecology, the traditional meaning of an interspecific interaction is an interaction that influences population dynamics (size, density, growth rates). However, in evolutionary models, the relevant criterion of an interaction is that the relative fitnesses *within* a species are influenced by the interactions of individuals with other species. Such relative fitness interactions can have a strong evolutionary effect even if they have no impact at all on population dynamics. A further complication arises from consideration of units

Table 14.5. Types of Interspecific Interactions

Species 1	Species 2	Type of Interaction
0	0	Neutralism
+	0	Commensalism
−	0	Amensalism
+	−	Predator–prey
		Pathogen host
−	−	Competition
+	+	Mutualism

Note: A plus sign means that the indicated species benefits (either in terms of increased growth of population size and/or individual fitness) from the interaction, and a minus means that the indicated species is harmed (either in terms of a reduction in population size and/or reduced individual fitness) from the interaction. A zero means no effect at all upon the indicated species from the presence of the other species.

and targets of selection. The target of selection induced by an interspecific interaction may well be the individual, but in any model of evolutionary response, the unit of selection is generally much smaller than the individual's intact, multilocus genotype. Indeed, many different units of selection can be influenced by the interactions of individuals of different species, and each unit of selection may respond to that interaction in a qualitatively different fashion. Consequently, the appropriate focus for models of coevolution is at the level of traits that influence fitness through interactions with individuals of another species and their underlying units of selection.

Heliconius butterflies illustrate that different traits can simultaneously exist within the same individuals that have qualitatively different coevolutionary responses (Templeton and Gilbert 1985). *Heliconius* is a genus of New World butterflies found mostly in the tropics. The larvae of these butterflies feed on various species of the plant genus *Passiflora*. Generally, sympatric species (those living in the same area) of *Heliconius* use nonoverlapping sets of host species. Hence, from the point of view of larval feeding traits, there is no interspecific interaction among sympatric *Heliconius* species. Moreover, the population sizes of the various species are probably determined by density-dependent factors acting on the larval stages (Gilbert 1983). Hence, the most likely *population dynamic* interaction between these species is neutralism.

The adult *Heliconius* butterflies use the cucurbit vines from the genera *Gurania* and *Anguria* as a source of nectar and pollen, and usually all sympatric species use the same plants as pollen and nectar sources. Moreover, the nectar and pollen are critical for maintaining adult viability and in producing eggs. Hence, any trait that increases adult foraging efficiency or competitive ability for these resources should be strongly selected. The nature of the coevolutionary interaction for such traits would be one of competition. To see this, consider a two-species model with the following fitness function for such traits within species 1:

$$w(\theta_{12}) = 1 + a_1(K_1 - N_1 - \theta_{12}N_2) \qquad (14.33)$$

where a_1 is a constant for the focal species 1, K_1 is the amount of resource (say pollen) available to species 1 (as well as to the other species), N_1 is the density of conspecific individuals from species 1 that are using this resource, N_2 is the density of individuals from competing species 2, and θ_{12} measures the competitive impact of individuals of species 2 upon individuals of species 1 in obtaining the resource. Notice that equation 14.33 attributes fitness to the phenotype θ_{12}, so this is an unmeasured genotype approach. We therefore assume that there is some heritability to θ_{12} in species 1, and the response to selection is given by equation 11.22, Fisher's fundamental theorem of natural selection. Note also that this is a linear fitness model with respect to θ_{12}, so the complications due to nonlinearity are not applicable in this model. As can be seen from equation 14.33, fitness is a decreasing function of θ_{12}. Therefore, S in equation 11.22 is negative, and given some heritability for this measure of competition, the intraspecific response to selection, R, is also negative. That is, natural selection in this model tends to reduce θ_{12}. Similar selective forces would be operating in species 2. Thus, selection is operating in the same direction within both species, even though the nature of the interaction between the two species is antagonistic. A reduction in θ_{12} can be achieved in many different ways, such as increasing competitive ability against the other species (which in turn increases the θ term in the other species and thereby could induce more intense selection in that species) or specializing on a part of the resource that the other species does not use well or efficiently (which reduces the θ terms for both species). Traits in *Heliconius* that can be explained by such competition selection include their highly developed visual system, learning ability, early morning flight and traplining behavior, and the fact that different sympatric species have significantly differing abilities to utilize small- or large-grained pollen particles.

The adult *Heliconius* butterflies are distasteful to bird predators because of amino acids derived from pollen and from allelochemicals derived from *Passiflora* and stored from the larval stage. *Heliconius* butterflies are brightly colored and have wing patterns that attract attention. Such wing patterns are regarded as an example of **aposematic coloration**, or coloration that warns potential predators that the individual is distasteful and/or poisonous. Although there is much diversity in coloration pattern in this genus, and even between different geographical populations within a species, many sympatric *Heliconius* species tend to look alike with respect to wing coloration and pattern. This is an example of **Müllerian mimicry**, in which two or more aposematic species share a similar warning pattern. Such common warning signals allow potential predators to learn the pattern more efficiently and thereby avoid individuals displaying that pattern. Hence, it is to the mutual benefit of all potential prey to share a common warning pattern. To see how different species could coevolve to converge upon the same pattern, we will develop another fitness model, but this time taking a measured genotype approach based upon recent studies on the genetic architecture underlying Müllerian mimicry in two sympatric species, *Heliconius cydno* and *Heliconius melpomene*. Naisbit et al. (2003) identified 10 autosomal loci, with most of the loci clustered into two genomic regions characterized by tight linkage and extensive epistasis and thereby defining two supergenes (Chapter 13). The unit of selection is therefore not a single locus but rather these multilocus supergenes. These supergenes have "alleles" (particular multilocus states of the genes within a supergene) that display dominance for the mimicry phenotypes, but this dominance has evolved under the influence of identifiable "modifier" loci rather than being a fixed characteristic of each locus. The two Müllerian mimics in this case use different genetic architectures to achieve the same mimetic patterns, although many homologous loci are used in common.

In light of these genetic considerations, consider the following model of coevolution of Müllerian mimicry between two species, say 1 and 2 (Templeton and Gilbert 1985). As before, let N_i be the density of species i, and we will regard these densities as fixed constants (determined by larval food resource availability). In species 1 let there be a recessive wing pattern phenotype associated with the genotype aa at a supergene, and likewise assume that there is a recessive phenotype associated with the genotype bb at a supergene in species 2. It makes no difference whether the supergene in species 2 is homologous or not to that found in species 1, as both supergenes evolve independently in the two species, which are assumed to be reproductively isolated. Within each species, assume there is a dominant allele (A in species 1 and B in species 2). Potential predators are assumed to perceive some resemblance between the dominant phenotypes, a lesser degree of resemblance between the dominant and recessive phenotypes within a species, and the least degree of resemblance between the dominant and recessive phenotypes between species. Predators are assumed to see no resemblance at all between the two recessive phenotypes found in the different species. This means that the dominant phenotypes represent an intermediate trait state between the two recessive phenotypes with respect to predator perception. The fitness effect associated with a given wing phenotype is assumed to be proportional to the number of individuals resembling that phenotype times a coefficient measuring the degree of resemblance. Thus, the fitnesses of the genotypes defined by a single supergene in species 1 are

$$w_{aa} = G_{aa}N_1 + a(1 - G_{aa})N_1 + b(1 - G_{bb})N_2$$
$$w_{A-} = (1 - G_{aa})N_1 + aG_{aa}N_1 + c(1 - G_{bb})N_2 + dG_{bb}N_2 \quad (14.34)$$

where G_{ij} is the genotype frequency of genotype ij, and a, b, c, and d are constants that measure the degree of perceived resemblance of the phenotypes by potential predators such that $1 > a, c > b, d > 0$, where a perfect perceptional match is given a score of 1. Note that the fitnesses within species are both frequency and density dependent and moreover depend upon the genotype frequencies and density of the other species. The average excess of the A allele for fitness is given by

$$a_A = G_{aa}[(1-a)(1-2G_{aa})N_1 + (c-b)(1-G_{bb})N_2 + dG_{bb}N_2] \quad (14.35)$$

Given the assumed patterns of resemblance, $1 - a, c - b$, and d are all positive. Hence, the average excess of A is always positive (and hence increasing in frequency due to natural selection) when $G_{aa} < \frac{1}{2}$. However, under these conditions, the increase in the frequency of the A allele occurs because the dominant phenotype is the most common warning pattern within species 1, and the A allele will be favored even in the absence of any specific interaction (e.g., $N_2 = 0$ in equation 14.35). When $G_{aa} \geq \frac{1}{2}$, the A allele can only increase in species 1 if the interspecific terms weighted by the density of the other species, N_2, are sufficiently large to overcome the intraspecific liability of having the rarer phenotype within species 1. Since it is reasonable to assume that neither species initially resembled one another, it is only the interspecific interactions that allow the evolution of a common warning pattern in this model. When the genotype frequencies of both aa and bb are initially close to 1, the average excess in equation 14.35 will be positive only when

$$dN_2 > (1-a)N_1 \quad (14.36)$$

that is, when the interspecific resemblance of the dominant phenotype in species 1 to the recessive phenotype in species 2 as weighted by the density of species 2 outweighs the lack of resemblance $(1 - a)$ of the dominant phenotype in species 1 to the recessive phenotype in species 1 as weighted by the density of species 1. Hence, the course of evolution within species 1 is determined by a combination of both inter- and intraspecific factors, and coevolution towards a common warning pattern can only occur when the interspecific interactions are strong. Moreover, if one species is much more common than the other, an inequality of the form 14.36 is likely to be satisfied only for the rarer species; that is, the rarer species will be selected to resemble the more common species but not vice versa. True coevolution, where both species are evolving toward a common pattern, is more likely if their densities are comparable.

It is important to keep in mind that the evolutionary processes defined by equations 14.33 and 14.35 can operate simultaneously upon different units of selection. Adult sympatric individuals of different *Heliconius* species are therefore simultaneously competitors for pollen and nectar resources and mutualists for wing color and patterns. These different traits can coevolve in qualitatively and quantitatively different fashions in the same populations at the same time. It is erroneous and misleading to speak of a single interspecific interaction as characterizing the net relationship among these butterflies with respect to coevolution.

Because units of selection are generally less than an individual's intact, total genotype, the very idea of a net or overall interaction among individuals makes no sense when modeling coevolution.

The above model assumes that the mutualistic allele, A, is associated with a dominant wing phenotype. However, what if A were associated with a recessive phenotype and the population were randomly mating? The fitness model now becomes

$$w_{a^-} = (1 - p^2)N_1 + ap^2N_1 + b(1 - G_{bb})N_2$$
$$w_{AA} = p^2N_1 + a(1 - p^2)N_1 + c(1 - G_{bb})N_2 + dG_{bb}N_2 \quad (14.37)$$

where p is the frequency of the A allele. The average excess of the A allele now becomes

$$a_A = p(1 - p)\left[(1 - a)(2p^2 - 1)N_1 + (c - b)(1 - G_{bb})N_2 + dG_{bb}N_2\right] \quad (14.38)$$

The term in the brackets in equation 14.38 is similar to the bracketed term in equation 14.35. However, the condition for the bracketed term always being positive is now $p > 1/\sqrt{2} = 0.7071$. When the mutualistic phenotype was dominant, the corresponding condition was $G_{aa} < \frac{1}{2}$, which under random mating implies $p > 1 - 1/\sqrt{2} = 0.2929$. Hence, the conditions under which natural selection always favors the mutualistic phenotype have become much narrower. Note also that the bracketed term in the average excess equation 14.35 is weighted by G_{aa}, whereas the bracketed term in the average excess equation 14.38 is weighted by $p(1 - p)$. When the A allele is initially rare, G_{aa} is close to 1, and if the interspecific components of the average excess are sufficiently strong to make the bracketed term positive, the magnitude of that positive effect is almost completely translated into increasing p through equation 11.5. In the recessive case, $p(1 - p)$ is close to zero when p is very small, so even if the interspecific components make the bracketed term positive, the average excess of the A allele is still close to zero. Hence, regardless of the advantage of interspecific resemblance, equation 11.5 would predict extremely little selective pressure to increase the frequency of the A allele. Even when positive, the average excesses for the mutualistic allele will differ by several orders of magnitude in the dominant versus recessive case, making it far less likely for mutualism to evolve when A is recessive. Hence, the genetic architecture within a species is an important constraint on the coevolution between species. Indeed, this simple model can explain why the coevolution of Müllerian mimicry in these butterflies has been primarily accomplished through the replacement of recessive alleles by dominant or codominant alleles (Sheppard et al. 1985) and that epistatic modifiers have been favored that strengthen the degree of dominance during this evolutionary process (Naisbit et al. 2003).

If the system of mating is modified to include inbreeding or assortative mating ($f > 0$, Chapter 3), then the average excess for a mutualistic, recessive A allele when mutualistic alleles are rare in both species is (Templeton and Gilbert 1985)

$$f[dN_2 - (1 - a)N_1] \quad (14.39)$$

Hence, the same condition as described in inequality 14.36 must still be satisfied for the evolution of a recessive mutualistic trait, but now we must also have $f > 0$. The intraspecific system of mating also constrains the course of coevolution. This serves to remind us that coevolution is just standard natural selection operating upon the *intraspecific* fitness effects of an *interspecific* interaction. Like all the intraspecific evolutionary processes discussed in this book, coevolution is constrained by intraspecific parameters, such as genetic architecture and population structure. In particular, the selective direction of coevolution is determined by the sign of the average excesses for intraspecific fitness for the intraspecific units of selection. The gamete's perspective still rules in the selective response to environmental heterogeneity.

15

SELECTION IN AGE-STRUCTURED POPULATIONS

Up to now we have assumed discrete generations. Under this assumption, all individuals are born at the same time and then reproduce at the same time followed by complete reproductive senescence or death. Such a model approximates reality for some species. For example, many insects and plants have only one generation per year that is synchronized by the seasons, and the discrete-generation model can approximate their evolution. However, as pointed out in Chapter 2, individuals in many species can reproduce at multiple times throughout their life, can mate with individuals of different ages, can survive beyond their age of reproduction, and can coexist with their offspring and other generations. We do not have to look far to find such a species; our own falls into this category of overlapping generations. In species with overlapping generations, an important component of population structure is **age structure**, the distribution of the ages of the individuals found in the population at a given time. The age distribution can have a large impact on whose gametes are transmitted to the next generation, particularly when an individual's chances of survival, mating, and reproducing are all influenced by age, as they are in humans. In this chapter we will examine the evolutionary impact of age structure and age-dependent fitness components in demes with overlapping generations. Such models also lie at the interface of population genetics and population ecology, a field that deals extensively with age-structured populations. We therefore start this chapter by an examination of some of the fundamental demographic parameters that ecologists have used to characterize populations with overlapping generations. We will then introduce genetic variation that influences the phenotypic variation in these demographic parameters to look at the impact of selection in age-structured populations. Finally, a detailed example of such selection will be given, an example that makes use of many of the concepts developed in the previous chapters.

Population Genetics and Microevolutionary Theory, By Alan R. Templeton
Copyright © 2006 John Wiley & Sons, Inc.

LIFE HISTORY AND FITNESS

Life history is the progress of an individual throughout his or her life. An individual is first born, then grows into an adult or fails to survive to adulthood. If the individual survives to adulthood, then the individual perhaps mates and reproduces at specific ages and finally dies at some age. In Chapter 11, we defined the fitness components of viability, mating success, and fertility/fecundity. To examine life history and its evolutionary implications, we must first make each of these fitness components an explicit function of age. We start with the fitness component of viability. Viability can be measured in an age-specific fashion by the **age-specific survivorship**, ℓ_x, the probability of an individual surviving to age x. Ideally, age should be measured from fertilization in order to cover the diploid individual's entire life history, but in practice a time point well after conception is used in many species. For mammals, and humans in particular, the initial time point is usually birth. This obviously misses any deaths that occur between conception and birth.

Table 15.1 illustrates the concept of age-specific survivorship through the example of females from the United States as determined by the 2000 U.S. census data. As shown in this table, age is often treated not as a continuous variable but rather as a series of consecutive categories or ranges of ages. This represents the practical constraint of how such data are gathered, but in theory one could treat age as a continuous variable. However, in this chapter we bow to the reality of actual data and will treat age as an ordered categorical variable.

Table 15.1. Life History Table for U.S. Females Based on 2000 Census Data

Age Range (years)	Assigned Age, x	ℓ_x	$m_x b_x$	$\ell_x m_x b_x$	$x \ell_x m_x b_x$
<1	0.5	1	0	0	0
1–4	2.5	0.99376	0	0	0
5–9	7	0.99261	0	0	0
10–14	12	0.99189	0	0	0
15–19	17	0.99107	0.14925	0.1479	2.5146
20–24	22	0.98909	0.22950	0.2270	4.9939
25–29	27	0.98671	0.26975	0.2662	7.1865
30–34	32	0.98392	0.21975	0.2162	6.9189
35–39	37	0.98021	0.11275	0.1105	4.0892
40–44	42	0.97460	0.02725	0.0266	1.1154
45–49	47	0.96623	0.00250	0.0024	0.1135
50–54	52	0.95398	0	0.0000	0.0000
55–59	57	0.93561	0	0.0000	0.0000
60–64	62	0.90716	0	0.0000	0.0000
65–69	67	0.86344	0	0.0000	0.0000
70–74	72	0.79983	0	0.0000	0.0000
75–79	77	0.70983	0	0.0000	0.0000
80–84	82	0.58563	0	0.0000	0.0000
85–89	87	0.42145	0	0.0000	0.0000
90–94	92	0.23936	0	0.0000	0.0000
95–99	97	0.09669	0	0.0000	0.0000
≥100	102	0.02479	0	0.0000	0.0000
Total				0.9968	26.9320

Table 15.1 shows the age ranges used for this human example. Although each category is a range of ages (generally five years in Table 15.1), it is often convenient to assign a single age number to characterize numerically the entire age range, and that is done in Table 15.1 in the column labeled "assigned age." Obviously, the finer the age range, the closer the discrete age categories correspond to continuous variation in age. The U.S. census data are based on such large samples that we could have used age ranges far less than five years, but our purpose here is simply to illustrate the meaning and use of life history parameters, and the coarse age ranges used in Table 15.1 are fine for this purpose.

Some conventions must be adopted to define survivorship within an age range. These conventions vary from one study to the next, so it is important for users of life history data to know the conventions used in tables such as Table 15.1. In our case, we use the convention that the age range starts exactly at the age given on the left and continues to include all individuals up to the age indicated on the right plus any fraction of a year above that value. For example, the age range 5–9 includes all females whose age is 5 years or above to 9 years plus any fraction of a year such that the age is less than 10 years. Another convention that we adopt is that any individual who survives into any portion of this age range is counted as surviving in that age range. For example, if an individual died at the age of 5 years plus one day, that individual would still be counted as surviving to the age range 5–9 in Table 15.1. This means the probability of surviving into the initial age range (birth to any age less than 1 year) is always 1 because we only count live births in this study, and by convention all individuals are considered alive in the initial age category. For all subsequent age categories, $\ell_x < 1$, reflecting the fact that some of the females did not survive throughout the entire previous age category and hence did not enter as survivors into the next age category. Note that the ℓ_x's define a monotonically decreasing series of probabilities. This reflects the fact that once an individual is dead, they remain dead. Accordingly, the survivorship up to age x must be a nonincreasing function of age. This is shown more clearly in Figure 15.1, which plots the l_x's versus age, known as a **survivorship curve**.

Given that an individual is alive at age x, let m_x be the probability that an individual successfully mates at age x. Given that the individual is alive and has mated at age x, let

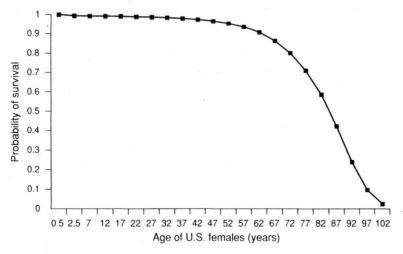

Figure 15.1. Survivorship curve for U.S. females based on the data from the 2000 census given in Table 15.1.

b_x be the number of offspring born to the successfully mated individual at age x. Here, we encounter another convention in counting offspring. In a diploid species with separate sexes, each offspring has two parents, not one. How do we divide the offspring between the two parents for purposes of counting? Here we adopt the convention of splitting the offspring count equally between the two parents. Hence, in Table 15.1, b_x is actually one-half the average number of births to a successfully mated female of age x. Also, in many species births can be counted accurately, but mating cannot. Hence, it is commonplace to report only the product $m_x b_x$, which represents one-half the average number of births to a female of age x regardless of whether or not she had mated. The product $\ell_x m_x b_x$ represents one-half the average number of births to a female born x age units ago regardless of whether or not she survived. This product therefore includes the age-specific impacts of viability, mating success, and fertility—all of the fitness components introduced in Chapter 11 but now in an age-specific form. This triple product of age-specific fitness is also given in Table 15.1. Note in Table 15.1 that the double product $m_x b_x$ defines a prereproductive period of zeros that reflects the inability to reproduce until sexual maturity is achieved and a postreproductive period of zeros that reflects the inability to reproduce at older ages. Prereproductive periods are found in most species, whereas postreproductive periods are not as universal. Table 15.1 does not capture the indirect effects upon reproduction that are possible in a social species such as humans. As Table 15.1 shows, human females have a prolonged postreproductive period. Lahdenperä et al. (2004) used complete multigenerational demographic records to show that women with a prolonged postreproductive life span have more grandchildren, and hence greater fitness, in premodern populations of both Finns and Canadians. This fitness benefit arises because postreproductive mothers enhance the lifetime reproductive success of their offspring by allowing them to breed earlier, more frequently, and more successfully. Such effects show that targets of selection can emerge at the level of interacting individuals across generations in age-structured populations, but these higher targets of selection will be ignored in this chapter and tables such as Table 15.1 are regarded as a complete description of age-specific reproductive success.

A **life history table** gives the age-specific fitness components of ℓ_x, m_x, and b_x (or more commonly the product $m_x b_x$), so Table 15.1 is the life history table of U.S. females as estimated from the 2000 census. Life history tables can be defined for many cohorts (a population followed from birth through death), including groups of individuals that share a common genotype. Hence, the life history table will be our fundamental descriptor of the age-specific fitness effects of genotypes or other cohorts throughout this chapter. However, it is often cumbersome to measure fitness by an entire table of values rather than a single number. Accordingly, there are several measures of fitness that combine the age-specific fitness components over the entire life span into a single number. Two such measures are considered in this book. The first is the **net reproductive rate**, R_0, which is the sum of the triple product $\ell_x m_x b_x$ over all ages and represents the average number of offspring born to members of the cohort over their entire lifetime:

$$R_0 = \sum_{x=0}^{\text{max age}} \ell_x m_x b_x \tag{15.1}$$

As can be seen from Table 15.1, the net reproductive rate of U.S. females from the 2000 census is 0.9968; that is, females had an average of 0.9968 "half offspring" over their entire life span. (Note, a replacement rate is 1 half offspring per female, but the value slightly less

Table 15.2. Life History Table of Two Hypothetical Phenotypes

	Phenotype 1			Phenotype 2		
Age x	ℓ_x	$m_x b_x$	$\ell_x m_x b_x$	ℓ_x	$m_x b_x$	$\ell_x m_x b_x$
0	1	0	0	1	0	0
1	1	2	2	1	0	0
2	0	0	0	1	2	2
3	0	0	0	0	0	0
Sum (R_0)			2			2

than 1 does not mean that the U.S. population is near a stable size, as immigration is not taken into account in Table 15.1).

The net reproductive rate reflects the overall probability of living long enough to become sexually mature, find a mate, and successfully reproduce, perhaps multiple times if one lives long enough. As such, it is a measure of lifetime fitness. However, it has some serious inadequacies as a fitness measure in populations with continuous reproduction rather than a fixed generation length. In continuously reproducing populations, an individual's overall reproductive success depends not only on *how many* offspring they bear (the value measured by R_0) but also on *how fast* they bear their offspring. To see this, consider two phenotypes with the life history tables given in Table 15.2. Note that the net reproductive rate is two offspring per lifetime for both phenotypes, and hence the two phenotypes have the same fitness by this measure. However, do these phenotypes truly have the same fitnesses? To examine this question, suppose that each phenotype is reproducing in a closed manner (either clonal reproduction or 100% assortative mating with each phenotype corresponding to a homozygous genotype). Starting with N individuals of each phenotype in the age 0 category, Figure 15.2 shows how the population sizes change over time for each phenotype

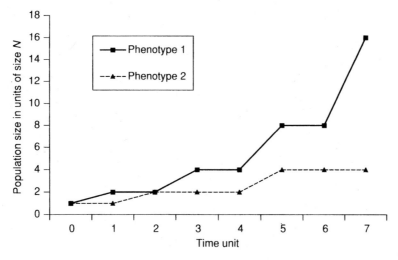

Figure 15.2. Growth of two phenotypes, each starting with N individuals of age 0 at time 0 using life tables given in Table 15.2.

based on the life history data in Table 15.2. Note that phenotype 1 soon becomes much more abundant than phenotype 2 despite the fact that both phenotypes have the same number of offspring per lifetime. The equality of the net reproductive rate in this case fails to detect the reproductive advantage of phenotype 1 that is obvious from Figure 15.2. As shown in Table 15.2, phenotype 1 does all of its reproduction at age 1, whereas phenotype 2 requires an additional unit of age before it can begin to reproduce. Although both phenotypes produce the same number of offspring within their lifetimes, phenotype 1 reproduces faster than phenotype 2. Hence, when measured in units of absolute time rather than per lifetime, phenotype 1 out-reproduces phenotype 2 (Figure 15.2).

We therefore need a measure of fitness that takes into account not only how *many* offspring are produced but also how *fast* offspring are produced when generation time is not fixed. To do so, we first need to define some additional terms that relate to the age structure of a population. Let $n_x(t)$ be the number of individuals of age x who are alive at time t and let $n_0(t)$ be the number of individuals born at time t (that is, those individuals that enter the age class of $x = 0$). The individuals born into time t come from the older individuals in the population who are reproducing such that

$$n_0(t) = \sum_{x=0}^{\text{max age}} n_x(t) m_x b_x \qquad (15.2)$$

Note also that the number of individuals of age x at time t is the number of individuals born at time $t - x$ that survived to live to age x; that is,

$$n_x(t) = n_0(t - x)\ell_x \qquad (15.3)$$

Substituting equation 15.3 into 15.2 yields

$$n_0(t) = \sum_{x=0}^{\text{max age}} n_0(t - x)\ell_x m_x b_x \qquad (15.4)$$

When equation 15.4 is iterated over a large number of time intervals, a remarkable property emerges: The *proportion* of the total population in an age class converges to a constant and the number of individuals in age class x at time $t + 1$ is a constant times the number in age class x at time t, with the constant being the same for every age class. This state is called the **stable age distribution**, in which the numbers in all age classes are growing (or declining or remaining the same) at the same rate, thereby preserving fixed proportions of the total population in all age classes. Note that the stable age distribution does not mean that the population is stable in overall size, only that the proportions in the age classes are stable.

Assuming that a population has reached its stable age distribution, by definition

$$\frac{n_y(t)}{n_y(t-1)} = \Re \text{ for every age class } y \qquad (15.5)$$

where $\Re$ is a constant. Because $\Re$ is a constant for all times, note that

$$\frac{n_y(t)}{n_y(t-2)} = \frac{n_y(t)}{n_y(t-1)} \times \frac{n_y(t-1)}{n_y(t-2)} = \Re \times \Re = \Re^2 \qquad (15.6)$$

and in general

$$\frac{n_y(t)}{n_y(t-x)} = \Re^x \Rightarrow n_y(t-x) = \Re^{-x} n_y(t) \quad (15.7)$$

Using the special case of equation 15.7 where $y = 0$ and substituting it into equation 15.4 yield

$$n_0(t) = \sum_{x=0}^{\text{max age}} \Re^{-x} n_0(t) \ell_x m_x b_x$$

$$1 = \sum_{x=0}^{\text{max age}} \Re^{-x} \ell_x m_x b_x \quad (15.8)$$

Equation 15.8 is known as Euler's equation. Note that once the life history parameters ℓ_x, m_x, and b_x are specified, Euler's equation determines the value of $\Re$ implicitly. Because every age class changes by an amount proportional to $\Re$ over one time unit (equation 15.5), the total population is also changing by an amount proportional to $\Re$ over one time unit. In particular, if $\Re > 1$, the population is growing; if $\Re = 1$, the population size is stable; and if $\Re < 1$, the population is declining. Unlike R_0, the net reproductive rate per lifetime, $\Re$ is measuring the growth of the population in absolute time. For example, using the life history parameters given in Table 15.2, Euler's equation for phenotype 1 becomes $1 = \Re_1^{-1} \times 2$ where the subscript 1 indicates the phenotype. Solving this yields $\Re_1 = 2$. In contrast, using the life parameters given in Table 15.2 for phenotype 2, Euler's equation becomes $1 = \Re_2^{-2} \times 2$, so $\Re_2 = \sqrt{2} = 1.412$. Note that the $\Re$'s correctly predict that phenotype 1 grows faster than phenotype 2, even though both phenotypes have identical net reproductive rates.

In a population of genetically variable individuals in which different genotypes have different life history parameters, the $\Re$ defined by the life history parameters of a particular genotype no longer reflects the growth of that genotype in the population because, under Mendelian genetics, many of the offspring will have a genotype that differs from that of their parents. Rather, the $\Re$ for a particular genotype measures the reproductive output of that genotypic class. Fisher (1930) therefore concluded that $\Re$ is a better measure of fitness in a population with overlapping of generations than is R_0. It is also more convenient mathematically to measure fitness as an exponential rate rather than a multiplicative constant such as $\Re$, so Fisher defined the **Malthusian parameter** r such that $e^{-r} = \Re$ with the cohorts being genotypic classes and not the entire population. With this exponential transformation, a genotype is producing an excess of offspring when $r > 0$, it is producing offspring at a replacement rate when $r = 0$, and it is producing a deficiency of offspring when $r < 0$. Substituting this exponential transformation of $\Re$ into Euler's equation (equation 15.8) yields

$$1 = \sum_{x=0}^{\text{max age}} e^{-rx} \ell_x m_x b_x \quad (15.9)$$

Equation 15.9 provides a method of implicitly calculating r for a genotype or any other cohort as characterized by a set of life history parameters. For example, using the data on U.S.

females given in Table 15.1, the implicit solution for r from equation 15.9 is -0.0001186. An approximate solution to equation 15.9 can be derived by using the linear Taylor's series approximation to $e^{-rx} \approx 1 - rx$. Substituting this approximation into equation 15.9 yields

$$1 = \sum_{x=0}^{\text{max age}} (1-rx)\ell_x m_x b_x = \sum_{x=0}^{\text{max age}} \ell_x m_x b_x - r \sum_{x=0}^{\text{max age}} x\ell_x m_x b_x$$

$$\Rightarrow r = \frac{\sum_{x=0}^{\text{max age}} \ell_x m_x b_x - 1}{\sum_{x=0}^{\text{max age}} x\ell_x m_x b_x} = \frac{1 - \frac{1}{R_0}}{\overline{T}} \quad \text{where } \overline{T} = \frac{\sum_{x=0}^{\text{max age}} x\ell_x m_x b_x}{\sum_{x=0}^{\text{max age}} \ell_x m_x b_x} \quad (15.10)$$

For example, applying equation 15.10 to the U.S. female life history data given in Table 15.1 yields $r = -0.0001193$, which is very close to the exact value of -0.0001186. Equation 15.10 allows us to gain greater biological insight into the meaning of the Malthusian parameter. First, r is an increasing function of R_0; that is, r increases as the number of offspring produced over a lifetime increases. Second, r is a decreasing function of $\overline{T}$. The biological meaning of $\overline{T}$ becomes clear when we note that

$$\frac{\ell_x m_x b_x}{\sum_x \ell_x m_x b_x} \quad (15.11)$$

is the proportion of offspring born to a female of age x relative to her total lifetime reproductive output. Hence, equation 15.11 represents a probability distribution over age of the number of offspring born to a female in this cohort. The expected value of this distribution is $\overline{T}$; that is, $\overline{T}$ is the average age at which a female gives birth and is called the **average generation time**. Therefore, r increases as the total number of births increases and r decreases as the average generation time increases. The Malthusian parameter is a measure of fitness that takes into account both the number of offspring produced as well as how rapidly they are produced.

We now have two fitness measures for dealing with age structure: the net reproductive rate (equation 15.1) when all the genotypes in the population have similar generation times and the Malthusian parameter (equation 15.9) when the genotypes differ in their average generation times. In deriving the Malthusian parameter, we assumed a stable age distribution, but there are many biologically realistic situations in which this assumption can be violated. For example, a survey of 27 insect species living in seasonal environments revealed that none of them ever experience a stable age distribution (Taylor 1979). More complicated measures of fitness can be derived from the life history parameters that take into account deviations from the stable age distribution (Demetrius 1975, 1985; Templeton 1980b), but these lie outside the scope of this book. In the remainder of this chapter, we will use either the net reproductive rate or the Malthusian parameter as our measure of fitness in age-structured populations.

EVOLUTION OF SENESCENCE

Why do we grow old? Should not natural selection favor an *ageless* phenotype in which there is no decline of vigor or reproductive output with age? How could it be an evolutionary advantage for individuals to lose their vigor and reproductive capabilities as they age? In this section, we will see how the fitness measures derived above allow us to address these important questions.

Let us start with a population of ageless individuals who show no senescence over their entire lifetime. Being ageless is not the same as being immortal. Individuals who do not age still can die through accidents, predation, disease, and so on. They are ageless in the sense that the chances of dying in an interval of time do *not* depend upon their age. Let d be the probability of an individual dying in a unit of time. We regard d as being independent of age and a constant throughout the entire lifetime, reflecting the ageless phenotype of the individual. The individual is also regarded as being ageless with respect to reproduction by letting mb, the probability of having mated times the expected number of offspring in a time unit, also be independent of age and a constant throughout the entire lifetime. Given these ageless parameters, the probability of an individual living to age x is

$$\ell_x = \prod_{i=0}^{x}(1-d) = (1-d)^x \tag{15.12}$$

Then, the net reproductive rate of an ageless individual is

$$R_0 = \sum_{x=0}^{\infty}\ell_x mb = mb\sum_{x=0}^{\infty}(1-d)^x = \frac{mb}{d} \tag{15.13}$$

using the well-known formula for the sum of a geometric series $[s_n = a + ag + ag^2 + \cdots + ag^{n-1} = a(1-g^n)/(1-g) \to a/(1-g)$ as $n \to \infty$, where a and g are constants with $-1 < g < 1]$. We can also apply the sum of a geometric series to Euler's equation to obtain the Malthusian parameter for this ageless population as

$$1 = \sum_{x=0}^{\infty} e^{-rx}\ell_x mb = mb\sum[(1-d)e^{-r}]^x = \frac{mb}{1-(1-d)e^{-r}}$$
$$\Rightarrow r = \ln(1-d) - \ln(1-mb) \tag{15.14}$$

When d and mb are small numbers, Taylor's series approximation of r from equation 15.14 is $r \approx mb - d$.

Now suppose a mutation occurs in this ageless population such that the bearers of this mutation senesce and die at age $n-1$. The net reproductive rate of the mutant individuals is

$$R_0' = mb\sum_{x=0}^{n-1}(1-d)^x = \frac{mb}{d}\left[1-(1-d)^n\right] \tag{15.15}$$

For large n and any $d < 1$ (that is, some death occurs from causes unrelated to age), the term $(1-d)^n$ goes to 0, and hence the term in brackets in equation (15.15) goes to 1. Thus, if n is large enough (depending on d), then $R_0 \approx R_0'$ and the mutation is selectively neutral as measured by the net reproductive rate. Similarly, one can show that the mutant's Malthusian parameter obeys the implicit approximation

$$r' \approx mb\{1-[(1-d)e^{-r'}]^n\} - d \approx mb - d \quad \text{for large } n \tag{15.16}$$

Once again, as long as senescence is delayed to an old age, the mutant phenotype is neutral. As we saw in Chapter 5, neutral mutations will inevitably become fixed in a population over long periods of time. This means that if mutations can occur that kill their bearers at a sufficiently advanced age, such mutations are effectively neutral and some will go to fixation, thereby destroying the agelessness of the initial population.

In Chapter 13 we discussed some of the selective pressures on Huntington's chorea, one of several neurodegenerative diseases in humans associated with trinucleotide repeats which have a late age of onset. Langbehn et al. (2004) found that the empirical relationship between age of onset and CAG repeat number is well described by the equation

$$S(\text{Age}, \text{CAG}) = \left(1 + \exp\left\{\frac{\pi}{\sqrt{3}} \frac{[-21.54 - \exp(9.56 - 0.146\text{CAG}) + \text{Age}]}{\sqrt{35.55 + \exp(17.72 - 0.327\text{CAG})}}\right\}\right)^{-1} \tag{15.17}$$

where Age is the age of the individual, CAG is the number of CAG repeats (see Chapter 13 and Figure 13.11), and S (Age, CAG) is the probability of having no neurological symptoms to the given age with the given repeat number. If we make the assumption that all reproduction stops with the onset of the neurological symptoms, the net reproductive rate for a bearer of Huntington's chorea is

$$R'_0(\text{CAG}) = \sum_{x=0}^{\text{max age}} \ell_x m_x b_x S(x, \text{CAG}) \tag{15.18}$$

Using the life history data in Table 15.1 in equation 15.18, we can calculate the net reproductive rate of bearers of a newly formed Huntington's allele (one that just crossed the threshold repeat number and reached a value of 36 repeats, as discussed in Chapter 13) to be 0.9941 versus the normal net reproductive rate from Table 15.1 of 0.9968. The neurological symptoms are relatively mild when they first occur and then get progressively worse, eventually resulting in death. The assumption that all reproduction stops with the onset of symptoms is therefore overly conservative, so the actual difference in net reproductive rates associated with a newly formed Huntington's is even less than that indicated above. Hence, the lethal neurodegeneration of Huntington's disease is essentially neutral with respect to natural selection when an allele reaches the 36 repeat threshold. Of course, as discussed in Chapter 13, there are other targets of selection on Huntington's disease, including the family and the repeats themselves. Focusing just upon the selection on the repeats themselves and ignoring the family-level selection, we saw in Chapter 13 that selection at the genomic level favors an increase in repeat number, which in turn is associated with an earlier age of onset (equation 15.18 and Figure 13.11). As the age of onset is lowered, there is now stronger individual-level selection against Huntington's disease. For example, the net reproductive rate for bearers of a Huntington's allele with a CAG repeat number of 56 using equation 15.18 and the life history data in Table 15.1 is 0.4407 versus the normal net reproductive rate of 0.9968, resulting in substantial selection at the individual level against the Huntington allele. These calculations show how important the age of onset is in determining the fitness impact of an allele that affects life history parameters. Even lethal genetic diseases are effectively neutral when the age of onset is old enough. As a result, individual selection alone cannot prevent the evolution of senescence via genetic drift leading to the fixation of nearly neutral alleles with deleterious effects of late age of onset.

We now turn our attention to another class of mutations with life history effects. Suppose, as before, a mutation occurs that kills its bearers at age $n - 1$. However, we now assume that this same mutation increases earlier reproduction from mb to mb' such that $mb' > mb$. For example, suppose this mutation is associated with transferring the energy used in maintaining viability after age $n - 1$ to reproduction at earlier ages. This mutation is therefore associated with a pattern of antagonistic pleiotropy (Chapter 11) because it is associated with traits that have opposite effects on fitness. The net reproductive rate of the individuals with this antagonistic pleiotropic mutant is

$$R_0'' = mb' \sum_{x=0}^{n-1} (1 - d)^x = \frac{mb'}{d} \left[1 - (1 - d)^n \right] \tag{15.19}$$

As before, the term in brackets in equation 15.19 goes to 1 as n increases, so if the age of onset of the deleterious effects of this mutant is old enough, then its net reproductive rate is approximately mb'/d, which is greater than the net reproductive rate of the nonmutants of mb/d. Similarly, one can show that the Malthusian parameter for this pleiotropic mutant is, for n large, approximately $mb' - d > mb - d$. Once again, by either fitness criterion, bearers of this pleiotropic mutant are actually favored by natural selection as long as the deleterious effects have a late age of onset. In this case, our initial ageless population will evolve senescence due to the positive action of natural selection; that is, it is adaptive to senescence.

Many mutations have been found that are associated with beneficial effects early in life and deleterious effects later in life. For example, in Chapter 11 we discussed several gene loci associated with resistance to falciparum malaria. Since most of the mortality associated with this parasite occurs in childhood, the beneficial effects of these malarial resistance genes are primarily expressed at an early age. However, the deleterious effects of these same genes (often associated with the chronic effects of anemia) are often not clinically significant until later in life. Many life history trade-offs have been documented in the fruit fly *Drosophila melanogaster* and in particular between larval survival and adult size. Bochdanovits and de Jong (2004) examined the trade-off between larval survival and adult size by an analysis of global gene expression. This quantitative genomic approach revealed 34 genes whose expression explained 86.3% of the genetic trade-off between larval survival and adult size. Fourteen of these genes had known functions that suggest that the trade-off is at the level of cellular metabolism and is due to shifts between energy metabolism and protein biosynthesis regulated by the RAS signaling pathway. These and other studies indicate that mutations associated with patterns of antagonistic pleiotropy are common.

Both effectively neutral, late-age-of-onset mutations and antagonistic pleiotropic mutations will lead to the evolution of senescence. An empirical demonstration of the evolution of senescence is provided by an experiment on the flour beetle *Tribolium castaneum* (Sokal 1970). Sokal allowed adults of two strains (wildtype and black bodied) to live three days after emergence from the pupal stage and then killed them. As a control, he allowed adults of the same two strains to live as long as they were able under identical laboratory conditions. He ran these lines for 40 generations and then let adults from all lines, both experimentals and controls, to live as long as they could under the specified laboratory conditions. The results are shown in Table 15.3.

Table 15.3 reveals that the experimental populations had a shorter life span than the controls for both strains and both sexes. Thus, in just 40 generations, these stocks evolved

Table 15.3. Adult Life Spans of Flour Beetles from Two Strains Raised for 40 Generations with All Adults Either Killed at Three Days of Age (Experimental) or Allowed to Live as Long as Possible under Laboratory Conditions (Control)

		Average Adult Life Span in Days	
Strain	Sex	Experimental	Control
Wildtype	Female	5.0	7.5
Wildtype	Male	12.5	14.5
Black bodied	Female	5.6	5.9
Black bodied	Male	5.0	8.8

increased senescence in response to an environmental regimen in which no adult lived past three days of age, thereby rendering any deleterious fitness effects past that age irrelevant as a contributor to fitness. These effects could be due to a combination of mutants of either type discussed above, although the speed of the response indicates selection on at least some genetic variants with antagonistic pleiotropy. Note also that these results are yet another example of genetic assimilation (Chapter 14). In this case, an environmentally imposed early death (by the experimenter) becomes genetically assimilated as an innate decrease in life span. These experimental results indicate that we grow old because the evolutionary forces of genetic drift and natural selection actually favor senescence over agelessness. Once again, if we take the gametic perspective rather than the individual perspective, there is no mystery as to why selection favors those phenotypes that transmit more gametes even at the price of the senescence of the individuals that are the temporary bearers of those gametes.

ABNORMAL ABDOMEN: EXAMPLE OF SELECTION IN AGE-STRUCTURED POPULATION

In Chapter 14 we discussed the *abnormal abdomen* (*aa*) supergene polymorphism that is found in populations of the fruit fly *Drosophila mercatorum* living near the town of Kamuela (also known as Waimea) on the Island of Hawaii (see Figure 6.9). We now return to this system for a more detailed examination because it is an excellent example of natural selection in an age-structured population. Moreover, the *abnormal abdomen* story will illustrate many of the other themes about natural selection that have been developed in the last several chapters.

Genetic Architecture and Units and Targets of Selection below Level of Individual

The *abnormal abdomen* supergene gets its name from the fact that it is often associated with the retention of juvenile cuticle (larval-type cuticle) on the adult abdomen. Another phenotype associated with this genetic syndrome is a slowdown in egg-to-adult developmental time. Both of these phenotypes are also associated with the *bobbed* (*bb*) locus in *D. melanogaster*. *Bobbed* is known to be due to deletions in the X-linked 18S/28S ribosomal DNA (rDNA) (Ritossa et al. 1966), and the *aa* supergene also maps to the *D. mercatorum* X chromosome in a region associated with the nucleolar organizer, which is where the

18S/28S rDNA is located (Templeton et al. 1985). Normally the phenotypic expression of *aa* is limited to females, but there is a Y-linked modifier that allows expression in males (Templeton et al. 1985). One of the few functional regions on the *Drosophila* Y chromosome is another cluster of 18S/28S rDNA. For all of these reasons, the 18S/28S rDNA became a candidate locus (Chapter 10) for this phenotypic syndrome.

The 18S/28S rDNA in *Drosophila* exists as a tandem, multigene family of a repeating unit containing the DNA that codes for the 18S, 5.8S, and 28S RNA subunits of the ribosome, all separated by short, transcribed spacer sequences that are removed during the processing of the primary transcript (Figure 15.3*a*). Each unit is separated from adjacent units by a nontranscribed spacer (NTS). There are normally about 200–300 repeats in the rDNA cluster on the X, and a smaller cluster of rDNA units is found on the Y chromosome (Figure 15.3*f*) (Ritossa 1976). The *bb* mutations of *D. melanogaster* represent deletions in the rDNA on the X that severely reduce the number of 18S/28S units in the cluster. DeSalle et al. (1986) therefore examined the amount of rDNA in *aa* flies displaying the aa phenotype and aa^+ flies without the aa phenotype. They found no deficiency of X-linked rDNA in either group of flies. However, they did find that the Y chromosomes that allowed expression of *aa* in males (Y^{aa}) did indeed have a severe deletion of the Y-linked DNA, thereby indicating that rDNA was involved in the aa syndrome.

Because the quantity of X-linked rDNA was normal in *aa* flies, DeSalle et al. (1986) next turned their attention to the quality of the rDNA. They discovered that all lines showing the aa phenotype had at least a third or more of their 28S genes disrupted by a 5-kb intervening sequence, now known as an R1 insert (Figure 15.3*b*). The R1 inserts are a type of retrotransposon (Chapter 13) found in arthropods that retrotranspose into a particular site in the rDNA 28S subunit (Figure 15.3*b*). Genetic variation at an *Eco*R1 restriction site (Appendix 1) exists among the R1 elements even within a single rDNA cluster of *D. mercatorum* (Figures 15.3*b* and *c*). Another type of retrotransposon, the R2 insert, also has this specificity for 28S rDNA, and the R2 elements insert in a site just 5' of the R1 insertion site in *D. mercatorum* (Figure 15.3*d* and *e*). As shown in Figure 15.3, both types of inserts, both singly and doubly, exist in *D. mercatorum*, even on a single X chromosome (Malik and Eickbush 1999). Such inserts functionally inactivate the 18S/28S unit in which they are imbedded by disrupting normal transcription and processing (DeSalle et al. 1986), so the presence of these inserts functionally inactivates much of the rDNA, thereby explaining the similarities of *aa* to *bb* even though there is no physical deletion of rDNA.

Such inserts have two ways of spreading within the rDNA cluster of the X chromosome. First, they can retrotranspose. The act of transposition often creates some genetic heterogeneity at the 5' end of the R1 elements and the 3' end of the R2, and such heterogeneity is found among different R1 and R2 copies, thereby indicating that these elements have been actively retrotransposing in *D. mercatorum* (Malik and Eickbush 1999). The R1 and R2 elements can also spread within a tandem, multigene family via mechanisms of unequal exchange (Chapter 13). Spread via unequal exchange creates blocks of inserted versus noninserted 28S subunits, as spread via unequal exchange is to adjacent subunits (see Figure 13.5 for an example), whereas retrotransposition should lead to a nonclustered distribution within the rDNA. The inserted and noninserted 28S repeats are highly clustered within the rDNA (DeSalle et al. 1986), indicating that unequal exchange is quantitatively more important than retrotransposition as a mechanism for spread within X chromosomes. Because the R1 and R2 elements can spread within the multigene family by a combination of transposition and unequal exchange, these retrotransposons are both units and targets of selection below the level of the individual (Chapter 13).

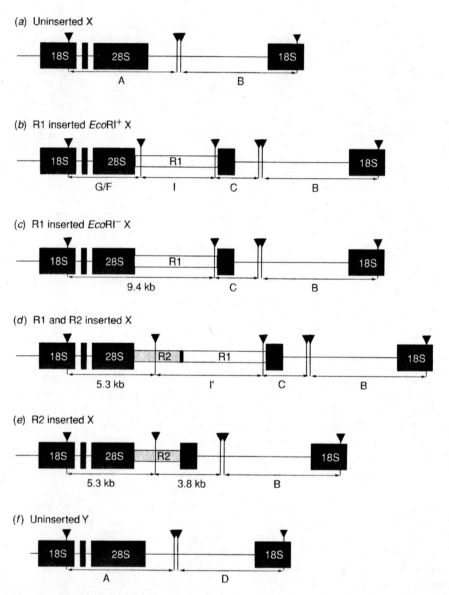

Figure 15.3. 18S/28S rDNA repeat types found in *D. mercatorum*. Blocks labeled R1 indicate R1-type retrotransposon insertions, blocks labeled R2 indicate R2-type retrotransposon insertions. *Eco*R1 restriction sites are shown, and the restriction fragments are designated by letters (DeSalle *et al.* 1986). Those fragments labeled with their length in kilobases are fragments expected to appear on Southern blots using a complete uninserted 18S/28S repeat as a probe. Based on data from DeSalle *et al.* (1986) and Malik and Eickbush (1999).

If the molecular mechanisms of spread are strong, we would expect little variation among the paralogous copies within a chromosome but variation among homologous rDNA families on different X chromosomes at the population level. In contrast, if the molecular mechanisms of spread were weak, we would expect much variation among the paralogous copies within a chromosome as well (Chapter 13; Weir et al. 1985). Of course, flies that had all their 28S genes bearing an insert would be nonviable, so selection at the individual level would prevent inserted repeats being in every paralogous copy. Genetic variation among inserted 28S repeats certainly exists [Figure 15.3 and R1 elements with large 5′ truncations reported in Malik and Eickbush (1999)], so if molecular mechanisms of spread are strong, we would expect only one insert type to dominate the inserted subset of repeats. Hence, the strength of the molecular mechanisms for insert spread can be assessed by examining the patterns of variation of inserted repeats both within and among chromosomes.

All the variants shown in Figure 15.3 were isolated using polymerase chain reaction (PCR) techniques. Such techniques can find even extremely rare variants, but because of biases in the PCR procedure, relative abundance of the alternative forms cannot be estimated reliably. However, for the standard *aa* test strain (derived from the Kamuela population but subjected to intense selection in the laboratory for morphological expression of juvenilized abdominal cuticle), Malik and Eickbush (1999) estimated that 86% of the X-linked 28S repeats had inserts with any of the insert patterns shown in Figure 15.3. Much of the original work on *aa* was performed in the 1980s and utilized Southern blot techniques (Appendix 1). This technique has the advantage that quantitative information exists about the relative abundance of various repeat types if they produce different band lengths that hybridize with the probe being used. Using this technique, the banding pattern associated with Figure 15.3*b* seemed to be dominant among the inserted class. However, band intensity in a Southern blot depends upon many factors in addition to the relative abundance of various DNA fragments, such as the size of the fragment, what proportion of the fragment hybridizes with the probe being used, and how far the fragment migrates in the agarose gel. Taking all these factors into account, Hollocher et al. (1992) and Templeton et al. (1993) developed an unbiased estimator with high replicability for the proportion of X-linked repeats bearing the $EcoR1^+$ R1 insert shown in Figure 15.3*b* from densitometry scans of Southern blots on *D. mercatorum* males (which have only one X chromosome):

$$\text{Prop}(Eco\ R1^+\ R1) = \frac{G}{[A(B-C)/(B-C+D)]A + G} \qquad (15.20)$$

where the italicized letters refer to the densitometry readings on the Southern blot of the various bands indicated in Figure 15.3. Because of length variation in the 5′ end of the R1 sequence, several different bands are often seen close to one another (indicated by the G/F notation in Figure 15.3), so the G in equation 15.20 refers to the sum of the densitometry readings of all these bands. Using equation 15.10 upon replicate scans from the *aa* stock reveals that 76% of the X-linked repeats bear the $EcoR1^+$ R1 insert with a standard deviation of 7%. The value of 76% is not statistically different from 86% (t test = 1.36 with nine degrees of freedom), implying that the vast majority of inserted repeats have just one type of insert, the $EcoR1^+$ R1 element. This means that inserts of type C, D, or E in Figure 15.3 and large 5′ R1 truncations are minor contributors to inserted rDNA elements in the *aa* stock.

The more important question is the contribution of the various insert types to variation in X chromosomes from the natural population. Figure 15.4 shows the distribution of the proportion of 28S rDNA repeats that bear the $EcoR1^+$ R1 insert in 1036 X chromosomes extracted from the Kamuela natural population of *D. mercatorum*. As is readily seen, there is extensive variation among X chromosomes for the proportion of $EcoR1^+$ R1 inserted repeats, a result consistent with the strong degree of within-chromosome dominance of the

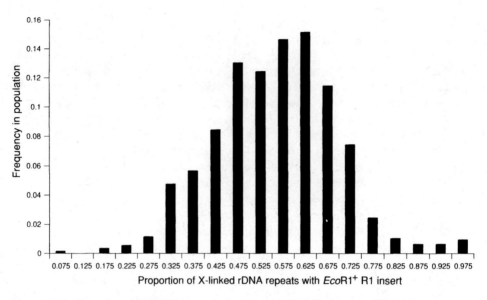

Figure 15.4. Proportion of 28S rDNA repeats that bear EcoR1$^+$ R1 insert in 1036 X chromosomes surveyed from natural population of *D. mercatorum* living in vicinity of Kamuela, Hawaii.

EcoR1$^+$ R1 insert found in the *aa* stock in implying strong molecular mechanisms for the spread of the EcoR1$^+$ R1 insert. As shown by the model of Weir et al. (1985), the same selective forces at the molecular level that would lead to the within-chromosome dominance of the EcoR1$^+$ R1 insert are also expected to lead to much interchromosomal variation (Chapter 13). As Figure 15.4 shows, there is indeed much interchromosomal variation in EcoR1$^+$ R1 insert proportion. However, Figure 15.4 does not directly address the relative abundance of all the insert types shown in Figure 15.3 in the natural population. To test the prediction that the EcoR1$^+$ R1 insert dominates overall on X chromosomes extracted from the natural population, consider another estimator of the proportion of inserted repeats:

$$\text{Prop(inserted repeats)} = \frac{C}{B} \tag{15.21}$$

This estimator uses fragments near the 3′ end of each repeat element and thereby detects the inserted elements of types B, C, and D in Figure 15.3 but fails to detect type F (R2 alone inserted). However, the exclusion of type F is expected to produce little error because Malik and Eickbush (1999) noted that most elements bearing an R2 insert also bore an R1 insert. Moreover, elements bearing just an R2 insert should yield a 3.8-kb band on a Southern blot (see Figure 15.3), but such a band was not visible on almost all autoradiographs and was at most very faint on a few. Hence, most R2 inserts occur in conjunction with an R1 insert in the natural population. Finally, because estimator 15.21 uses a 3′ fragment, it would not be affected by 5′ truncations of the R1 elements. Hence, estimator 15.21 should capture virtually all of the heterogeneity in inserted elements described by Malik and Eickbush (1999).

Estimator 15.21 was not used in the original analysis because the C band has a weak hybridization signal and is more diffuse because of the greater distance it must travel through the agarose gel (Hollocher et al. 1992). These properties cause estimator 15.21 to

Table 15.4. Number of ur^+ and ur^{aa} X Chromosomes with Their Percentage of Inserted 28S Genes as Estimated from Equation 15.10 that Are Above and Below Overall Median Value for All Chromosomes

Allele at ur locus on X	Below Insert Median	Above Insert Median
+	281	251
aa	137	167

Source: From Hollocher et al. (1992).

be severely biased. However, Malik and Eickbush (1999) made a suggestion that allows us to use estimator 15.21 to test for whether or not there is significant variation in nature for the insert types other than $EcoR1^+$ R1. As will be discussed shortly, there is another X-linked element that is necessary for expression of the aa phenotype called the *underreplication locus* (ur), which has two alleles in the natural population, ur^+ and ur^{aa}, that can be scored by test crosses with the *aa* stock. Hollocher et al. (1992) measured the degree of association (a type of linkage disequilibrium) between the $EcoR1^+$ R1 insert proportions of each male shown in Figure 15.4 with the ur allele borne by each male as determined by a test cross with *aa* females. Their significance and strength of the association was measured through a median test (Appendix 2). In this test, all the X chromosomes scored for both insert proportion and ur are ranked by insert proportion and divided into those above and below the median. The X chromosomes are then subdivided into those that are ur^+ and those that are ur^{aa} to yield a 2 × 2 categorical table. The degree of association between insert proportion and the allelic state at the ur locus can then be measured by a standard chi-square homogeneity test of the null hypothesis of no association. Table 15.4 gives the results when insert proportion is estimated using equation 15.20, which yields a chi square of 4.65 with one degree of freedom ($p = 0.03$). Malik and Eickbush (1999) predicted that this statistic of association should be strongly influenced if the inserted types ignored by equation 15.20 also show significant variation in natural populations.

Although estimator 15.21 is biased, it can still be used in statistics that are invariant to shifts in mean, which includes the median test. Therefore, the median test statistic of association was recalculated using equation 15.21 to yield the results shown in Table 15.5. The overall numbers are somewhat different in Tables 15.4 and 15.5 because each table includes only those individuals who were successfully scanned for the bands relevant to a particular estimator. As a consequence, there are few individuals included in one analysis that are not in the other. Despite these minor differences in the samples, the results in both tables are virtually identical. In Table 15.5 the chi-square statistic is 5.58 with one degree of freedom ($p = 0.02$). The near identity of the results using these two extremely different estimators indicates that the inserts other than the $EcoR1^+$ R1 insert make no significant contribution to variation in the X-linked rDNA found in the natural population

Table 15.5. Number of ur^+ and ur^{aa} X Chromosomes with Their Percentage of Inserted 28S Genes as Estimated from Equation 15.11 Above and Below Overall Median Value for All Chromosomes

Allele at ur locus on X	Below Insert Median	Above Insert Median
+	287	254
aa	136	169

near Kamuela. The overall pattern is therefore one of homogeneity among the insert types found within a chromosome (dominated by the $EcoR1^+$ R1 insert) and heterogeneity among X chromosomes in the proportion of $EcoR1^+$ R1 inserts (Figure 15.4). This is the pattern expected when the molecular mechanisms that allow the spread of the $EcoR1^+$ R1 insert are strong (Chapter 13). This conclusion was empirically confirmed by the direct monitoring of strains. For example, a parthenogenetic strain (Chapter 13) was established that had only a few of its 28S genes bearing the $EcoR1^+$ R1 insert. Within three years, a parthenogenetic sublineage derived from this strain had developed the aa phenotype, and upon examination it was found that most of its 28S X-linked repeats now had the $EcoR1^+$ R1 insert, indicating a drastic turnover in its rDNA complement in a relatively short time interval (DeSalle et al. 1986). These direct observations confirm that strong forces for paralogous spread exist for the $EcoR1^+$ R1 insert, and this conclusion is compatible with the population-level observations on intra- and interchromosomal patterns of variation. Hence, the $EcoR1^+$ R1 insert (hereafter called the *aa* insert) is a unit of selection in itself and is also a target of strong selective forces below the level of the individual because it has the ability to replicate and spread within the genome.

Evidence of strong selection directly on the insert below the level of the individual does *not* mean that molecular-level forces overwhelm the evolutionary forces occurring at the level of individuals in a reproducing population, a frequent misinterpretation (Dover 1982; Malik and Eickbush 1999). Rather, as we saw in Chapter 13, the strong molecular-level selection means that this multigene system behaves as a single locus with respect to genetic drift at the population level. Moreover, the strong molecular-level forces create much genetic variation among chromosomes at the population level (Figure 15.4), which *enhances* the response to natural selection for targets of selection at the individual level and above. Thus, the strong molecular-level forces seen in *aa* augment, not diminish, the ability of this system to be subject to population-level evolutionary forces, such as genetic drift and selection targeted at the level of individuals. This is true not just for the R1 retrotransposons but is true for many other transposable elements as well (Kidwell and Lisch 2000).

Genetic Architecture and Units and Targets of Selection at Level of Individual

Surveys of several strains displaying the aa phenotype revealed that they all had one-third or more of their 28S genes with the *aa* insert (DeSalle et al. 1986). Thus, having a third or more of *aa*-inserted 28S genes appears to be necessary for the aa syndrome. However, these same surveys revealed that not all strains with more than a third of *aa*-inserted 28S genes display the aa phenotype. Hence, having a third or more of the 28S genes bearing the *aa* insert is *necessary* but *not sufficient* for the syndrome. As discussed in Chapter 8, when the causes of phenotypic variation are not both necessary and sufficient, interactions among variable, causative factors are frequently implicated. Indeed, DeSalle and Templeton (1986) discovered that a second, interactive criterion must also be satisfied: There must be no preferential underreplication of inserted rDNA repeats in the polytene tissues of the larval fat body. To understand the molecular nature of this second necessary, interactive factor, we must first briefly review what happens to rDNA in somatic tissues in the genus *Drosophila*.

The rDNA codes for components of the ribosomes that in turn are necessary for translating messenger RNA into proteins. The ability to synthesize proteins is such a critical and necessary function of most cells that organisms have evolved several mechanisms to buffer

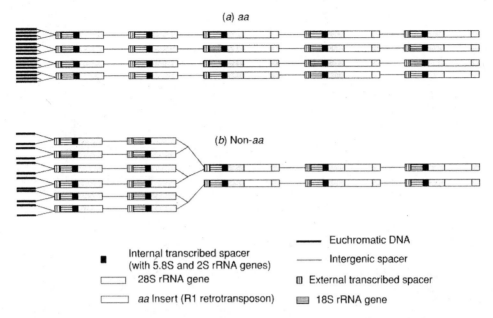

Figure 15.5. Cartoon of X-linked rDNA multigene family showing lack of selective underreplication of 28S repeats bearing *aa* R1 retrotransposon in X chromosomes bearing ur^{aa} (a) and occurrence of selective underreplication in X chromosomes bearing ur^+ allele (b).

themselves against physical or functional deletions of rDNA. In *Drosophila*, diploid cells display a type of somatic amplification of their rDNA in response to physical or functional deletions of rDNA repeats known as compensatory response (Tartof 1971). *Drosophila mercatorum* flies with the aa syndrome also show compensatory response in their diploid tissues, indicating that the *aa* insert is indeed causing a functional deficiency of ribosomal RNA, but one that can be compensated for in diploid somatic cells. *Drosophila* species also have polytene tissues in which there is much endoreplication of DNA, creating the giant polytene chromosomes of *Drosophila* and many other insects (Chapter 5). Euchromatic DNA is greatly amplified in polytene tissues, but the rDNA tends to be underreplicated relative to the euchromatin. For example, in the polytene salivary glands of *D. melanogaster* larvae, most euchromatic DNA is at a level of polytenization of 1024 copies, but the 18S/28S rDNA is present in only about 128 copies (Ritossa 1976). This is also true for *D. mercatorum*, and Figure 15.5 shows a cartoon version of this rDNA underreplication. One way to compensate for a deficiency in rDNA in polytene tissues is to have less underreplication of the rDNA overall, and this occurs in *D. mercatorum* (Malik and Eickbush 1999). However, the degree of underreplication need not be uniform over all rDNA repeats. It has been well documented in *D. melanogaster* that selective replication can favor certain repeat types within the rDNA (Spradling 1987), and the same is true for *D. mercatorum* (DeSalle and Templeton 1986). In flies that do not express *aa* but nevertheless have a large portion of their 28S genes bearing inserts, there is preferential underreplication of the inserted 28S repeats (DeSalle and Templeton 1986). Because of this preferential underreplication coupled with diminished overall underreplication, the uninserted functional 28S repeats are effectively overreplicated relative to the nonfunctional inserted repeats (Figure 15.5*b*), and the resulting tissue seems not to be affected by any functional deficiency of ribosomes. In contrast, *aa* flies have a uniform underreplication across the rDNA cluster in the polytene fat body tissue (Figure 15.5a), although Malik and Eickbush (1999) report preferential underreplication of the inserted 28S repeats in polytene salivary glands of *aa* flies. However, as will soon be apparent, the phenotypic significance of *aa* at the individual level arises from its effects in the fat body, not the salivary gland, so it is the absence of preferential underreplication in this tissue that is critical to the aa phenotype.

The presence or absence of preferential underreplication in the fat body is controlled by an X-linked locus (the *underreplication*, or *ur*, locus) with two alleles, ur^+ (which allows for preferential underreplication) and ur^{aa} (which leads to uniform underreplication). Hence, the expression of *aa* requires two molecular conditions: (1) a third or more of the 28S repeats must bear the *aa* insert and (2) the uniform underreplication must occur in the fat body. Both of these molecular elements show genetic variation in a natural population living near Kamuela, Hawaii (Hollocher et al. 1992). Although these X-linked components are separable by recombination, the linkage is tight with a recombination frequency of 0.004 (Templeton et al. 1985). As indicated above, these two elements also display extensive epistasis at the molecular level, and if the aa syndrome is selected at the individual level, this should translate into strong fitness epistasis as well. That such is the case is indicated by the linkage disequilibrium between these two elements in nature (Tables 15.4 and 15.5). In particular, the nature of the disequilibrium is such that X chromosomes bearing the ur^{aa} allele have *higher* proportions of inserted rDNA. As a consequence, virtually every X chromosome in the natural population with an ur^{aa} allele is also above the one-third threshold level of inserted repeats. Thus, in nature, both of the necessary elements for the aa syndrome cosegregate because of disequilibrium and tight linkage. As a consequence of this disequilibrium and the strong molecular-level forces, this multigene system is effectively a supergene complex that displays inheritance patterns close to that of a single Mendelian locus (Hollocher et al. 1992). Hence, at the level of populations of genetically variable individuals, the combination of tight linkage and strong epistasis makes our unit of selection for the phenotypic consequences of *aa* the entire *ur*/rDNA supergene complex (Chapter 13) on the X chromosome when the individual is the target of selection. For much of the subsequent discussion, we will treat this supergene complex as if it were a single X-linked locus with two alleles, *aa*, which has both the ur^{aa} allele and a third or more of the 28S genes with the *aa* insert, and +, which does not satisfy one or both of these conditions. However, toward the end of the discussion, we will look more carefully at how selection has shaped the different components of this supergene complex.

Phenotypes and Potential Targets of Selection at Level of Individual

The tissue-level consequence of *aa* in the fat body is to produce a deficiency of functional rDNA, which in turn can lead to a deficiency of ribosomes in the fat body. This in turn can lead to reduced rates of protein translation in the fat body. However, not all proteins are sensitive to translational control from a deficiency of ribosomes. The proteins that are expected to be most sensitive to a ribosomal deficiency are those that are synthesized in large quantities over a short time interval. One such protein is juvenile hormone (JH) esterase, which is synthesized in the fat body in large quantities during the late third larval instar (the last instar before pupation in *Drosophila*) and prepupal stages. To understand the significance of this esterase protein, we must first briefly discuss the role of JH in *Drosophila*. Juvenile hormone is typically in high titers during the larval phase of life. The transition from the larval to the pupal stage of life is triggered by an increasing ratio of another insect

ABNORMAL ABDOMEN: EXAMPLE OF SELECTION IN AGE-STRUCTURED POPULATION **517**

hormone ecdysone relative to JH. Part of this changing ratio is due to the production of JH esterase by the fat body starting in the late third instar stage. This enzyme degrades JH, thereby causing the JH titer to decline and the ecdysone–JH ratio to increase. JH esterase is a common physiological mechanism in holometabolous insects (those undergoing full metamorphosis) to reduce JH titers at critical developmental times.

The production of JH esterase should be sensitive to translational control as this is a protein that is produced in the fat body in large quantities during the late third instar stage. Templeton and Rankin (1978) measured the activity of JH esterase in late third instar larvae reared at 25°C by topically applying tritiated *Cecropia* JH (JH is readily absorbed through the cuticle) followed by cytosolic extractions 3 hours later that were then fractionated to separate the amount of labeled JH from its degradation products (Figure 15.6). By summing the total number of counts in the JH fractions and in the JH degradation fractions, the percent

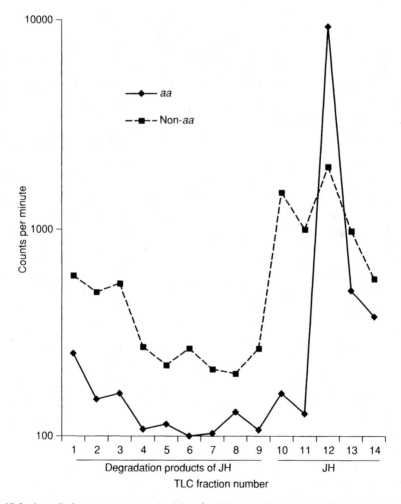

Figure 15.6. Juvenile hormone esterase activity after 3 hours of incubation of late third instar larvae with ^{3}H-*Cecropia* JH in *aa* and non-*aa* stocks of *D. mercatorum*. Thin layer chromatography (TLC) fraction numbers 10–14 correspond predominantly to intact JH, whereas numbers 1–9 correspond to JH degradation products.

of labeled JH that was degraded in this 3-hour period was estimated to be 10.4% in *aa* flies and 33.7% in non-*aa* flies, indicating that non-*aa* larvae degraded JH 3.23-fold greater than *aa* larvae in the late third instar. Hence, the aa syndrome is indeed characterized by a large reduction in the amount of JH esterase in late third instar larvae. This observation explains two of the phenotypic characteristics of the aa syndrome. First, reduced amounts of JH esterase in late third instar larvae would lead to a slower decay of JH, which in turn would cause a prolongation of the larval phase. Such a prolongation of the larval phase is indeed observed in *aa* flies. Second, the abdominal histoblasts undergo their adult differentiation in the late third instar/prepupal phase and hence are extremely sensitive to JH titer at this stage (Riddiford et al. 2003). Juvenilized adult abdominal cuticle can be induced as a phenocopy by topical application of JH at this stage of development. The juvenilized adult abdominal cuticle that gives the aa syndrome its name can therefore also be explained by high titers of JH in the late third instar that result from the reduced translation of JH esterase proteins in the fat body of *aa* flies.

Premise 3 (Chapter 1) states that phenotypes arise from interactions between genotypes and environment. The phenotype of juvenilized adult abdominal cuticle is no exception. When reared in a laboratory environment at 25°C, there is extensive expression of juvenilized abdominal cuticle in *aa* flies. However, rearing larvae at 22°C slows down the rate of larval development and prolongs the larval phase. In this temperature-induced slowdown of larval development, *aa* flies rarely display juvenilized cuticle. Whether or not the phenotype of juvenilized cuticle is expressed in natural populations is critical for understanding the role of natural selection on this syndrome because the phenotype of juvenilized cuticle induces strong selection. At the end of the pupal phase, flies inhale air to expand their abdomens to help them literally pop out of the pupal case. When flies with juvenilized abdominal cuticle do this, the top of the pupal case pops open as normal, but the juvenilized cuticle of the inflated abdomen often sticks to the inside of the pupal case, preventing the fly from emerging from the pupal case and causing it to die. Moreover, of those flies that successfully emerge from the pupal case, flies with juvenilized abdominal cuticle are more prone to death by desiccation. Indeed, the selective pressures are so intense against this phenotype in the laboratory at 25°C that the phenotype of juvenilized cuticle is rapidly eliminated unless actively counteracted by artificial selection every generation favoring flies with juvenilized cuticle. Interestingly, many of the *aa* stocks that revert to a normal cuticle phenotype with the cessation of artificial selection rapidly evolve selective underreplication of inserted 18S/28S rDNA units in the polytene fat body rather than any changes in the rDNA itself (DeSalle and Templeton 1986). As noted earlier, Malik and Eickbush (1999) found selective underreplication of inserted rDNA units in the larval polytene salivary gland in an *aa* strain. One interpretation of this observation is that the failure of selective underreplication in *aa* occurs only in fat body but not in salivary gland polytene tissues. However, Malik and Eickbush (1999) failed to artificially select for juvenilized cuticle after receiving the *aa* strain, and when they examined the *aa* strain after their experiments, its abdominal cuticle was normal (Eickbush, personal communiction). Unfortunately, Malik and Eickbush did not monitor the cuticle as they bred the *aa* flies in their laboratory, so there is no way of knowing when the reversion to wildtype cuticle occurred relative to their experiments. Hence, their experiments could mean that there is selective underreplication in salivary polytene tissues or they could mean that selective underreplication had evolved in the polytene tissues of their uncontrolled stock. The results of Malik and Eickbush (1999) are therefore biologically uninterpretable because they ignored the strength of selection against the phenotype of juvenilized cuticle in a laboratory environment.

ABNORMAL ABDOMEN: EXAMPLE OF SELECTION IN AGE-STRUCTURED POPULATION   519

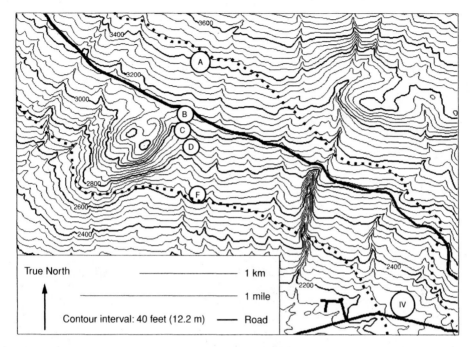

Figure 15.7. Map of distribution of *O. megacantha* near Kamuela, Hawaii. The dotted lines enclose the area in which the cactus was found during much of the 1980s. A transect of collecting sites on the slopes of the Kohalas are indicated by the letters A, B, C, D, and F, and a collecting site in the saddle at the base of the Kohalas is indicated by IV.

Because the phenotype of juvenilized cuticle can induce strong selection under suitable environmental conditions, it is critical to examine wild-caught individuals for this phenotype. As pointed out in Chapter 6, populations of *D. mercatorum* live in the Kohala Mountains on the Island of Hawaii (Figure 6.9). The sole larval food resource in this region is rotting cladodes of the prickly pear cactus *Opuntia megacantha*. Figure 15.7 shows the distribution of the cacti in this area. As pointed out in Chapter 14, the range of the cacti on the mountainside spans a dramatic humidity gradient, and likewise there is a temperature gradient (Figure 15.8). The cladodes of this cactus are large (Figure 15.9), and their large size should dampen considerably the temperature fluctuations experienced throughout the daily cycle. Consequently, it is doubtful if a *Drosophila* larva in a cladode would experience the maximum or minimum temperatures or rather only intermediate temperatures. As can be seen from Figure 15.8, this implies that at all sites the larval rearing temperature is well below 25°C. Based upon the laboratory norm of reaction, this implies that juvenilized abdominal cuticle should rarely be expressed under natural environment conditions. In addition, autosomal loci have been identified in the natural population that have alleles that suppress the juvenilized cuticle trait in *aa/aa* flies (Templeton et al. 1993). Given the low larval temperatures in nature and the existence of epistatic suppressors of juvenilized cuticle in the gene pool, it is not surprising that only a handful of wild-caught flies had small patches of juvenilized abdominal cuticle out of tens of thousands examined even though they were drawn from populations with a high frequency of the *aa* superallele. Thus, the phenotype of juvenilized cuticle and the strong selective forces associated with it appear to play no role in selection on *aa* in natural populations.

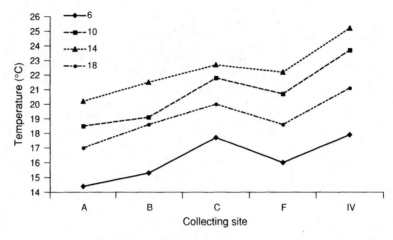

Figure 15.8. Mean temperatures as measured by hygrothermographs placed within shaded cactus patches at collecting sites shown in Figure 15.6 over period 1980–1990. Four different times are plotted for each site: the temperature at 6:00 hours (close to the minimum for the entire day–night cycle), 10:00 hours, 14:00 hours (close to the maximum for the entire day–night cycle), and 8:00 hours. Over the nighttime hours, there is generally a steady decline of the temperature at 8:00 hours to that shown at 6:00 hours.

The other phenotype that is predicted to occur from the low JH esterase activity in third instar larvae (Figure 15.6) is the prolongation of the larval phase, leading to an increase in egg-to-pupal developmental time. This phenotype is also sensitive to temperature and density conditions in the laboratory environment, but the relative slowdown of *aa* flies to non-*aa* flies is robust under a broad range of laboratory conditions, amounting to 1.28 days longer for flies from *aa* lines relative to non-*aa* lines when averaged across both sexes (Templeton et al. 1993). Only the average across sexes was observed in these experiments

Figure 15.9. Picture of small portion of cactus patch at site B. The *Drosophila* collector is Dr. Bonnie Templeton, who is 1.57 meters in height.

because pupae are difficult to sex. However, sex effects were expected. In general males do not display the aa syndrome because there is a cluster of rDNA on the Y chromosome that appears to be immune to insertion by the R1 and R2 elements (Figure 15.3f). This Y-linked rDNA cluster suppresses the effects of *aa* in males, although some Y chromosomes exist in the natural population that have a deletion of this rDNA cluster, thereby allowing expression of *aa* in males (Hollocher et al. 1992). However, the laboratory lines of *aa* used in these experiments all had Y chromosomes with the rDNA cluster. Hence, males should not be affected, which implies that the average developmental delay in reaching the pupal stage in these experiments was actually around 2.56 days in *aa* females.

Although the egg-to-pupa developmental delay appears to be a more robust phenotype of *aa* in the laboratory, the story of the juvenilized cuticle phenotype and premise 3 warn us that it is important to measure the phenotypes of interest under natural environmental conditions and genetic backgrounds. Unfortunately, the phenotype of egg-to-pupa developmental time is difficult to measure in nature, but it is feasible to measure egg-to-adult developmental time. Laboratory experiments indicate that this phenotype also differentiates *aa* from non-*aa* flies in a robust manner, with *aa* females taking an average of 0.52 days longer to reach the adult stage than non-*aa* females, but with no developmental slowdown in males. To see whether or not this developmental delay occurs in nature, patches of the cactus *O. megacantha* at the sites shown in Figure 15.7 were inspected between 1982 and 1990 for rotting cladodes, the sole larval food resource for the Kamuela population of *D. mercatorum*. The rots were bagged and thereafter inspected daily, and all adult flies that emerged over the last 24 hours were aspirated out and genotyped. To ensure that we sampled the entire emergence from a rot, we included in this analysis only those rots for which the first flies emerged after three or more days after bagging, indicating that the rot was bagged early enough to capture the complete emergence profile. If there is a difference in emergence times between *aa* and non-*aa* females, it could be due to a difference either in egg-to-adult developmental time or between *aa* and non-*aa* females in the stage of the rot at which they tend to oviposit. Fortunately, the males serve as a control of this later possibility. Because of the suppression of *aa* caused by most Y chromosomes, male emergence should not be affected by *aa* genotype if both *aa* and non-*aa* females tend to oviposit at the same times. If, however, *aa* and non-*aa* females tend to oviposit at different times, then male emergence should be associated with their *aa* genotype. Finally, because the rots differed greatly in size and in the state of the rot at the time of bagging, all comparisons on emergence time are among genotypes of the same sex within the same rot. These observations on bagged rot emergence revealed that $aa/-$ females take 0.92 days longer to emerge than $+/+$ females (significant at the 0.0001 level), whereas there was no effect of male genotype on emergence time (Templeton et al. 1993). Hence, the developmental slowdown associated with the *aa* superallele is expressed under natural environmental conditions and, indeed, it is expressed more strongly in the field than in the laboratory. Moreover, the autosomal suppressors of juvenilized cuticle that are found in the natural populations have no effect on this life history trait (Templeton et al. 1993). This is an example of **differential epistasis**, in which an epistatic modifier locus alters the phenotypic expression of another locus with respect to some traits but not all traits. This illustrates that the pattern of pleiotropy itself is a genetically variable trait and can evolve as part of an adaptive gene complex. Given the expression of the delay in the time from egg to adult under natural environmental conditions and genetic backgrounds, Equation 15.10 implies that such a delay, by itself, decreases fitness. Hence, if the slowdown in egg-to-adult developmental time were the only phenotype associated with *aa* in this natural population, we would expect *aa* to be selected against.

A slowdown in egg-to-adult developmental time and juvenilized cuticle are not the only phenotypes associated with the aa syndrome under laboratory conditions. The JH/JH esterase system that is so strongly influenced by *aa* in the late third instar (Figure 15.6) is reactivated in the adult stage after successful pupation. JH is a primary coordinator of reproductive processes in insects (Wyatt 1997). All tissues that are directly or indirectly involved in reproduction can be targets for JH action. This reproductive role of JH probably preceded its metamorphic role in the course of insect evolution. *Drosophila mercatorum* is part of the virilis/repleta radiation within the genus and subgenus *Drosophila*, and adult JH titer has been shown to be controlled by JH esterase in *Drosophila virilis* (Khlebodarova et al. 1996). Thomas (1991) measured JH esterase activity in one-day-old adult *D. mercatorum* and found that non-*aa* flies metabolized JH 2.5-fold greater than *aa* flies, thus paralleling the effects seen in the third larval instar.

Juvenile hormone has many functions in adult *Drosophila*, including influencing the onset of oviposition (Khlebodarova et al. 1996), controlling the transcription of specific proteins called vitellogenins that are transported into eggs (Dubrovsky et al. 2002), and influencing the amount of egg production (Wilson et al. 1983). Thomas (1991) therefore measured the titer of JH bound in adult female ovariole tissue. As expected from the deficiency of JH esterase in young adults, the titer of bound JH was significantly higher in adult *aa* females over non-*aa* females within the first week after eclosion, but by two weeks there was no significant difference between the genotypes. These differences in adult JH titer and JH esterase activity should result in increased fecundity in young adult females. This expectation is borne out in the laboratory environment as *aa* flies have increased egg-laying output over non-*aa* females for the first 10 days after eclosion, but with the fecundity differences between the strains diminishing to nonsignificance by 11–14 days after eclosion (Table 15.6).

We next need to see if this life history phenotype of increased early fecundity is also expressed under natural conditions. It is virtually impossible to count the number of eggs a wild female lays in nature on a day-by-day basis as a function of her age. However, it is still possible to test the hypothesis of increased early fecundity in nature (Templeton et al. 1990a). We first need to be able to measure the age of wild-caught females. This can be done by examining apodemes, inner thoracic extensions of cuticle that serve as sites for muscle attachment (Johnston and Ellison 1982). Apodemes accumulate daily growth layers after eclosion (Figure 15.10). Hence, it is possible to count the number of layers and thereby estimate the age of the fly in days since eclosion. Unfortunately, the bands become too small to accurately count after about 12 days, and this aging technique requires that the fly be killed. Nevertheless, it does allow the age structure of a sample of wild-caught flies to be determined. As will be described shortly, there is much spatial and temporal heterogeneity

Table 15.6. Mean Number of Eggs Lain by *aa* and + (non-*aa*) Females in Laboratory Environment

Time Interval (from Eclosion)	+ Females		*aa* Females	
	Eggs/Female	Sample Size	Eggs/Female	Sample Size
Days 2–4	71.05	19	72.90	20
Days 5–7	174.53	19	213.75	20
Days 8–10	174.89	19	210.84	19
Days 11–14	244.00	19	242.88	17

Source: From Templeton et al. (1993).

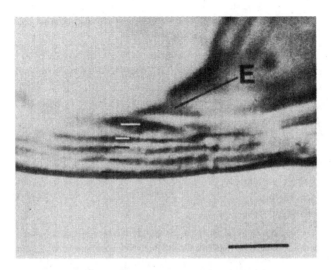

Figure 15.10. Nomarski differential interference contrast picture of apodeme from female *D. mercatorum* six days after eclosion showing eclosion layer (E) and five subsequent growth layers (marked by white or black lines). The bar in the lower right corner is 10 μm.

in the environment in this study area, and much of this heterogeneity can have a profound impact on the age structure of the population at a particular site and time. In particular, we often encounter conditions in which a majority of the wild-caught flies are less than a week of age from eclosion (which we will call "young" populations) versus conditions in which a majority of the adult flies are older than a week of age from eclosion ("old" populations).

Recall now the bagged rot experiments that were used to examine the relative egg-to-adult developmental time in flies bearing *aa* versus those that were +/+. Rots were bagged at an early stage and almost always had many females ovipositing on them at the time of bagging. A collection of these females was taken at the time of bagging and each wild-caught female was then placed in a vial and allowed to lay eggs on an artificial food. The male offspring emerging from these vials bear one X chromosome from their wild-caught mother, so by characterizing the genetic state of these laboratory-reared sons we can infer the genotype of the mother. Similarly, we scored the genetic state of the X chromosomes of the males emerging from the bagged rots. Given that the abnormal abdomen syndrome is generally suppressed in males, *aa* should be a neutral allele in its male carriers. If females with and without *aa* X chromosomes have equal fecundity, the frequency of *aa* in the laboratory-reared sons of wild-caught females should be the same as the frequency of *aa* in the males that emerge from the rots bagged at the same site and time as when the females were captured. However, if fecundity differences exist, equation 11.5 predicts that the *aa* allele frequency should also differ. Moreover, equation 11.5 can be solved for the average excess of fecundity as

$$a_{aa} = \frac{\overline{w}\,\Delta p}{p} \qquad (15.22)$$

where a_{aa} is the average excess of fecundity for *aa*-bearing gametes and Δp is the difference in the frequency of *aa* in mothers (as estimated from their laboratory-reared sons) versus their male offspring in nature (as estimated from the males emerging from bagged rots).

Table 15.7. Estimates of Average Excess of Female Fecundity Associated with *aa* Super allele in Natural Populations with Differing Age Structures

Cactus Site	Year	Age Structure	Frequency of *aa* in Sons	Frequency of *aa* in Males from rots	$\hat{a}_{aa} = \Delta p/p$
B	1982	Old	0.32	0.22	−0.32
IV	1982	Old	0.32	0.31	−0.04
C	1983	Young	0.38	0.67	+0.78
F	1984	Young	0.56	0.70	+0.23
IV	1984	Old	0.36	0.31	−0.15
B	1987	Young	0.30	0.53	+0.77
B	1989	Old	0.31	0.36	+0.15
Pooled for old age structures					−0.09
Pooled for young age structures					+0.59

Note: "Young" refers to an age structure in which a majority of the flies are less than one week of age since eclosion, and "old" is an age structure in which a majority of the flies are older than one week of age since eclosion.

Since we are only interested in relative differences among the genotypes, we can always define the average fitness to be 1. Hence, we can estimate the average excess of relative fecundity associated with *aa* as

$$\hat{a}_{aa} = \frac{\Delta p}{p} = \frac{p(\text{males from rots}) - p(\text{lab-reared sons})}{p(\text{lab-reared sons})} \quad (15.23)$$

Finally, we relate the estimates of the average excess of fecundity to the age structure found in the sample of ovipositing females. Such data are presented in Table 15.7. As can be seen, when the females are primarily a week of adult age or less, there is a strong, positive, and statistically significant average excess of fecundity as estimated by equation 15.23, with an average value of 0.59. In contrast, when the females are primarily older than a week of adult age, the average excess of fecundity is slightly negative but is not significantly different from zero in any old sample (Templeton et al. 1993). Hence, the field data (Table 15.7) are concordant with the laboratory data (Table 15.6) and indicate that flies bearing the *aa* allele have a fecundity advantage during early adulthood that is lost as they age.

The laboratory data indicate yet another life history phenotype: *aa* females have a significant decreased adult survivorship relative to non-*aa* flies (Figure 15.11), but *aa*/Y and +/Y males show no difference in survivorship (Templeton et al. 1993). The molecular and developmental mechanisms for this difference in adult longevity are unknown, but recent work indicates that the vitellogenin gene family plays a central role in regulating life span in a variety of organisms, including both humans and insects (Brandt et al. 2005). As mentioned above, the *aa* system does influence the expression of these genes through its impact on JH titer, and this in turn may influence adult longevity.

As with the other traits associated with the aa syndrome, we would like to measure these longevity effects in the natural population. Unfortunately, it was impossible to simultaneously score the age and genotype of wild-caught flies with the techniques available in the 1980s, so we could not directly test whether or not this life history phenotype was also expressed in nature. However, as we will soon see, in some cases the natural environment is very harsh for the adult *Drosophila*, with few living past a week of adult age. This would imply that the longevity differences shown in Figure 15.11 are not important in nature, at least when the natural conditions are harsh.

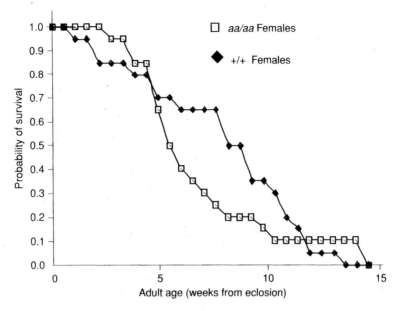

Figure 15.11. Probability of survival of *aa/aa* females versus +/+ females under laboratory conditions (modified from Templeton et al. 1993).

The above considerations about the phenotypes associated with the aa syndrome indicate an abundance of potential targets of selection at the level of the individual (female egg-to-adult developmental time, female fecundity, and female survivorship) in addition to the ones previously discussed below the level of the individual (e.g., the ability of the *aa* R1 insert to spread to other 28S units). We will now turn our attention to whether or not selection is indeed occurring on this genetic syndrome, with our initial focus being upon the target at the individual level by regarding the unit of selection as the supergene, that is, the *aa*-versus-+ superalleles. We will then look at the R1 insert as the unit of selection, with possible targets at both the individual and genomic levels.

Natural Selection on *aa* Supergene in Spatially and Temporally Heterogeneous Environment

Natural selection always occurs in the context of an organism interacting with its environment, so we must first examine the environment experienced by the populations of *D. mercatorum* that live in the Kohala Mountains of Hawaii. The environment in the Kohalas is highly heterogeneous, both spatially and temporally. As already mentioned, there is normally a strong temperature gradient as one goes down the mountainside (Figure 15.8) and a strong humidity/rainfall gradient, with the mean annual precipitation averaging 160 mm near the coast and rising to more than 3000 mm near the summit (Chadwick et al. 2003). The strong winds associated with this area limit dispersal (Chapter 14), so that even the spatial heterogeneity on the kilometer scale of the transect shown in Figure 15.7 is experienced by the flies as coarse grained (Chapter 14). Adaptive responses are determined by how selective forces interact with population structure, among other factors (Chapter 12). Even though dispersal is limited enough to allow adaptive differentiation, the population sizes are sufficiently large such that $N_{ev}m$, the effective amount of gene flow, is between 4 and 8, a

Figure 15.12. Picture of cactus patch at site IV. The *Drosophila* collector is Dr. Rob DeSalle.

value that should yield no significant genetic differentiation for nuclear neutral alleles over the sites shown in Figure 15.7, as is indeed observed for isozyme loci (Chapter 6).

The rainfall/humidity gradient has a large impact on the growth form of the cactus *O. megacantha*, which in turn defines other critical aspects of the fly's environment. The rotting cladodes of this cactus are the sole larval food resource for *D. mercatorum* in this area, and adult flies spend most of their time within cactus patches, which offer protection against desiccation and the wind. As shown in Figure 15.9, the cacti that grow in the upper, humid parts of the gradient are extremely large, with individual rotting cladodes able to support hundreds of larvae. As one goes down the mountainside, the cacti become much smaller (Figure 15.12). As a result, the population sizes per cactus decline as one goes down the mountainside by up to three orders of magnitude, and the flies in the drier parts of the humidity gradient have less protection against desiccation.

The ecological impact of this environmental gradient is also evident from the relative proportions of the species of *Drosophila* collected in cacti along this transect (Figure 15.13). The two species that tend to dominate these collections are *D. mercatorum* and *Drosophila hydei*, and mark/recapture experiments indicate that the relative proportions of these two species to one another in the collections accurately reflect their relative proportions in abundance in the cactus patches. There is a dramatic shift in species composition across this transect, with *D. hydei* dominating at the top but being increasingly replaced by *D. mercatorum* as one goes down the mountainside.

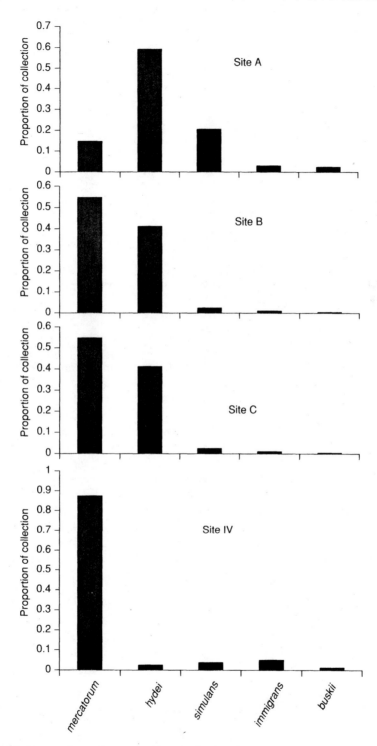

Figure 15.13. *Drosophila* species composition of collections from cactus patches at sites A, B, C, and IV in 1980.

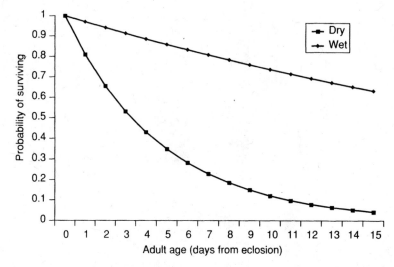

Figure 15.14. Predicted adult survivorship curves based upon observed daily survivorship rates of 0.81 at dry sites (F and IV) and of 0.97 at wet site (B).

The environmental changes on this gradient also strongly affect adult survivorship and age structure within *D. mercatorum*. Mark/recapture experiments on natural populations (Johnston and Templeton 1982) reveal that adult survivorship is strongly affected by humidity, with the per-day survivorship being only 0.81 under the driest conditions where cacti are small (normally, sites F and IV in Figure 15.7) and increasing to 0.97 as one goes up the mountainside to site B. These survivorship estimates result in extremely different predicted age structures across this transect. Using the observed survivorships as estimates of the quantity $1 - d$ in equation 15.12, the predicted age-specific survivorship curves under these two different environments are shown in Figure 15.14. As can be seen, we expect few adult flies to live longer than one week after eclosion under the dry conditions, whereas under the wet conditions found at the top of the transect we expect most adult flies to survive longer than two weeks from eclosion. These expectations based upon field survivorship data are concordant with the observed age structures estimated by apodeme band counting, as shown in Figure 15.15. Although the physical distances are small to a human (Figure 15.7), there is a dramatic shift in the age structure of the population over this humidity gradient, with most flies too old to age by apodeme ridge counting at the humid top of the transect (ages greater than 15 days from eclosion) to almost all flies being less than a week of age at the lower dry elevations.

We now have sufficient information to estimate the fitnesses associated with the aa genotypes. Let x be the age from hatching, e the age at eclosion, $a = x - e$ the adult age from eclosion, n the maximal adult age, and $\ell_a = \ell_{a+e}/\ell_e$ the adult survivorship to age a given eclosion. With these definitions and noting that there is no reproduction until after eclosion, Euler's equation (equation 15.9) becomes

$$1 = \sum_{x=0}^{\text{max age}} e^{-rx}\ell_x m_x b_x = \ell_e \sum_{a=0}^{n} e^{-r(a+e)}\ell_a m_a b_a \qquad (15.24)$$

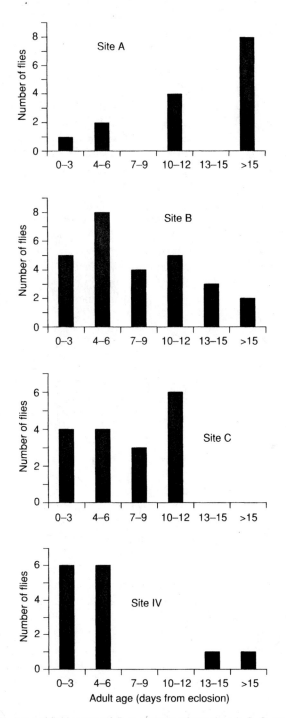

Figure 15.15. Ages of field-captured *D. mercatorum* from sites A, B, C, and IV in 1980.

and the approximation to the Malthusian parameter (equation 15.10) becomes

$$r \approx \frac{\sum_{a=0}^{n} \ell_a m_a b_a - 1/\ell_e}{\sum_{a=0}^{n} \ell_a m_a b_a (a + e)} \qquad (15.25)$$

with approximation 15.25 holding whenever r is close to zero in magnitude (Templeton et al. 1990a). Equation 15.25 shows that the developmental slowdown associated with aa (an increase in the age of eclosion e) tends to reduce fitness, whereas the increased early fecundity associated with aa (Table 15.6) tends to increase fitness. Exactly how these two opposing fitness effects are weighted into overall fitness depends critically upon adult survivorship.

To evaluate equation 15.25, we can use the laboratory data showing that the time to eclosion for female flies bearing the aa superallele, $e(aa/-)$, was 12.2 days (1.74 weeks) whereas $e(+/+) = 10.5$ days or 1.50 weeks for $+/+$ females. These laboratory-based figures are concordant with the developmental slowdown observed under field conditions, as noted above. We have no information on larval viability in the field (ℓ_e), so we will assume that there are no genotypic differences in this trait. Also, to ensure that r is close to zero for both $aa/-$ and $+/+$ genotypes, we also assume that

$$\frac{1}{\ell_e} = \frac{R_a(aa/-) + R_a(+/+)}{2} \qquad (15.26)$$

where

$$R_a(j) = \sum_{a=0}^{n} \ell_a(j) m_a(j) b_a(j) \qquad j = aa/-, +/+$$

Given that virtually all wild-caught females that are not newly eclosed (observable from their cuticle) are inseminated, we set $m_a(j) = 1$ for both females with genotype $j = aa/-$ and $j = +/+$. We use the laboratory-based fecundities (Table 15.6) to estimate the b_a's for the $aa/-$ and $+/+$ genotypes as these values are also concordant with field data, as noted above. The only component remaining that is needed to evaluate equation 15.25 is adult survivorship. Consider first the fitnesses of the genotypes under the "wet" conditions found at the upper sites. These are optimal conditions for adult survivorship, as indicated by the old age distribution (Figure 15.15) and high daily survivorships (Figure 15.14). Accordingly, we estimate the genotypic specific survivorships by the laboratory survivorships shown in Figure 15.11, curves that are concordant with the field data. Inserting all these numbers into equation 15.25 yields $r(aa/-) = -0.012$ and $r(+/+) = +0.010$. Under these wet conditions with an old age structure the $+/+$ genotype has a strong fitness advantage. This is not surprising. The only fitness advantage associated with the $aa/-$ genotype is increased early fecundity, but with an old age structure, early fecundity is given less weight. Moreover, there is a survivorship disadvantage for older $aa/-$ flies, and this disadvantage is given increased weight by an old age structure.

Now consider the dry conditions found at the lower sites that result in a low daily survivorship probability (the "dry" curve in Figure 15.14). Under these conditions, few adult flies live longer than a week, and any difference in survivorship in old adults is probably irrelevant. In this case, desiccation-driven mortality dominates adult survivorship for all genotypes, so we now use the dry survivorship curve in Figure 15.14 in equation 15.25, with all other numbers being the same as before. This survivorship curve yields $r(aa/-) = +0.012$ and $r(+/+) = -0.014$. Hence, there is a complete reversal of the fitnesses in sign in going from the top to the bottom of the transect.

The adaptive response to natural selection is influenced only in part by fitness differences; we must also consider the interactions with other evolutionary forces (Chapter 12). Since spatial heterogeneity is defining the fitness differences in this case, dispersal and gene flow are of primary importance in modulating the adaptive response to the fitness differences. As already shown in Chapters 6 and 14, dispersal and gene flow are such that these fitness differences should induce a cline in the frequency of *aa* over this transect, and indeed, significant differences in the *aa* superallele frequency are found that define such a cline (Figure 14.7).

The environment near Kamuela is characterized not only by extreme spatial heterogeneity but also by temporal heterogeneity that is coarse grained to these short-lived flies. For example, 1980 was a "normal" year in the sense that the overall weather that year deviated little from the long-term averages. The weather was dominated by the trade winds, and the spatial heterogeneity was characterized by the strong humidity/rainfall gradient described above. However, 1981 was the third driest year to that time on the Island of Hawaii, and that year all sites on the transect experienced low humidity and rainfall. This drought had barely ended when, on April 4, 1982, Mexico's El Chichón volcano had an explosive eruption that ejected the largest plume of dust into the atmosphere since 1912. The plume cloud covered more than a quarter of the earth's surface and blocked out as much as 10% of the sun's total radiation between the equator and 30° north, including Hawaii. This induced one of the wettest springs and summers in Hawaii's history. Far from going back to normal, 1982–1983 turned out to be an El Niño year, a recurrent fluctuation of currents and temperatures in the eastern equatorial Pacific that induces major dislocations of the rainfall regimens in the tropics (Cane 1983). This El Niño event induced another dry year in the Kohalas. It was not until 1984 that the weather returned to a normal pattern.

These weather fluctuations occurred on a time scale of many generations from the perspective of a fly, so there was time to adaptively respond to these altered weather regimens (Templeton et al. 1987b; Templeton and Johnston 1988). This is shown in Figure 15.16, which graphs the frequency of the *aa* superallele over the mountainside transect for the normal years (1980, 1984, and 1985), the two drought years (1981 and 1983), and the wet year of 1982. During the drought years, there was no significant change in *aa* allele frequency over the mountainside, in contrast to the normal years. All sites during the drought years had a uniformly high frequency of *aa* that was not significantly different from the frequency of *aa* at the lower dry sites during normal years. There was also no significant cline during the wet year of 1982, but now the frequency of *aa* was uniformly low and not significantly different from the frequency of *aa* at the upper wet sites during normal years. Hence, these populations of *D. mercatorum* are adaptively tracking these coarse-grained weather fluctuations.

Although the weather was normal on the mountainside in 1984 and 1985, it was not at site IV. As can be seen from Figure 15.7, sites A through F are located on the slope of the Kohalas, whereas IV is located on the relatively flat saddle between Kohala and the volcano Mauna Kea. Winds normally blow down the mountainside from A to F, but winds also blow through the saddle. In 1984 and 1985, the wind blowing through the saddle was stronger than normal, causing site IV to be more humid and to have more rainfall than normal. Consistent with the pattern shown in Figure 15.16, site IV had a significantly higher frequency of *aa* during the dry years of 1980, 1981, and 1983 ($p_{aa} = 0.47$) than the wet years of 1982, 1984, and 1985 ($p_{aa} = 0.33$). Hence, there was adaptive tracking of coarse-grained spatial variation induced by coarse-grained temporal variation.

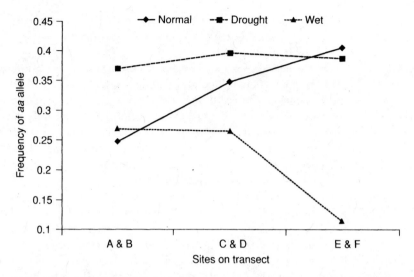

Figure 15.16. Frequency of *aa* on mountainside transect shown in Figure 15.7 for years 1980–1985. The years 1980, 1984, and 1985 are pooled together as they all experienced normal weather, and the years 1981 and 1983 are pooled together because they were both drought years. The year 1982 was an abnormally wet year due the explosive eruption of a volcano in Mexico.

Most of the collections between 1980 and 1985 had been done in the spring or summer months, but in 1986 and 1987 the collections were made in December. The native Hawaiians recognized only two seasons. Kau was the fruitful season, when the weather was warmer and the trade winds were most reliable. The collections up to 1986 were made during Kau. The other season was Hoo-ilo, when the weather was cooler and the trade winds were most often interrupted by other winds. The 1986 and 1987 collections were made in Hoo-ilo and hence had the potential of revealing the impact of coarse-grained (to a fly) seasonal variation. Kona wind is an alternative wind that is more likely in Hoo-ilo. Normally the trade winds blow from the northeast to the southwest, but during Kona weather, the winds tend to blow from the west to the east. Such episodes of Kona weather often last from one to several days, and during that time there is often much rain over the entire leeward slopes of the Kohalas, where all the collection sites are located. There are typically one to two Kona storms per year, sometimes more. Both the 1986 and the 1987 collections were made shortly after major Kona storms.

From an ecological perspective, the wet Kona weather in 1986 and 1987 created an extremely different environment for the *Drosophila* than that found in the wet year of 1982. First, as can be seen from Figure 15.13 for a normal year, the two most abundant species of *Drosophila* overall are *D. mercatorum* and *D. hydei*, the only two *repleta* group *Drosophila* (the *repleta* group as a whole is more adapted to using cacti than other *Drosophila*). Of these two *repleta* group species, *D. hydei* dominates at the top of the transect where conditions are most wet and drops out as one goes down the mountainside, being found only occasionally at the lower sites, whereas *D. mercatorum* is relatively more abundant at the drier sites. During 1982 when conditions were wet over the entire transect for several months, *D. hydei* was able to expand down the mountainside. For example, in 1980, *D. mercatorum* constituted 58% of the *repleta* group flies captured at site B and 99% at site IV. But during the prolonged wet year of 1982, *D. mercatorum* was only 7% of the *repleta* group flies at site B and 27%

at site IV. The only year in which *D. hydei* dominated the collections at all sites was 1982. However, during and immediately after the wet episodes of Kona weather *D. mercatorum* constituted 99% of the *repleta* group collection at site B and 100% at site IV, and the respective figures for the Kona year 1987 were 99.6% at B and 100% at IV. Consequently, from the perspective of the *Drosophila* community, the prolonged wet year of 1982 is not at all comparable to the wet Kona episodes of 1986 or 1987.

The reason for these contrasting ecological responses to the two different types of wet weather lies in the fact that the Kona weather is only a wet episode (about only one generation long for the flies) in an otherwise normal weather pattern. As noted earlier, the sole larval food resource in this area for *D. mercatorum* (and *D. hydei* as well) is rotting cladodes of the prickly pear cactus. The extremely wet conditions associated with a Kona storm create a superabundance of larval food resource, both in the number of cladodes simultaneously rotting and in the size of each individual rot. As a result, there is the potential for a population explosion. For example, the number of adult *D. mercatorum* living in cactus B-1 (the one shown in Figure 15.9) was estimated through mark/recapture experiments to be 971 ± 76 in 1980 and 420 ± 58 in 1984, the only two normal years in which the size was estimated for the flies living in this cactus patch. In the drought year of 1981 the B-1 size was 180 ± 73 and it was 543 ± 121 in the El Niño year of 1983. The B-1 population size was 61 ± 15 in the wet year of 1982, reflecting the dominance of *D. hydei* that year. In the generation after the 1988 Kona storm, the estimated size was 8133 ± 1120 in B-1 (Templeton et al. 1989), by far the largest size ever estimated for B-1 during the entire course of the study. Indeed, the most divergent population sizes in B-1 are between the two wet conditions—the prolonged wet year of 1982 versus the Kona episode in 1988. Hence, these two wet years are quite distinct ecologically.

Much of this distinction between the ecological impact of prolonged wet versus episodic Kona wet conditions appears to be due to interactions between the two *repleta* group species, *D. mercatorum* and *D. hydei*. With its shorter generation time and higher fecundities, *D. mercatorum*, can better exploit the temporary abundance of larval food sources associated with a Kona episode than *D. hydei*. Since the wet conditions disappear shortly after the Kona storm and the larval food resources decline back to normal levels, *D. hydei* never has the chance to expand under these wet conditions as it did under the prolonged wet weather of 1982. This pattern is clearly shown by monitoring the flies that emerge from the bagged rots. Figure 15.17a shows the fly species that emerged from bagged rots from site B cacti during the drier season collecting periods of normal years. As can be seen, *D. mercatorum* is the most common fly to emerge from the rots, with *D. hydei* a close second. However, during the wet year of 1982, *D. hydei* strongly dominates the rots, and few *D. mercatorum* emerged, a fact consistent with the extremely low population size of *D. mercatorum* in B-1 that year (61 adult flies). The wet weather associated with the Kona storm creates exactly the opposite situation: Now *D. mercatorum* dominates and *D. hydei* is almost eliminated from the rots. Not only are there more rots available after the Kona storm, but the rots are almost exclusively used by *D. mercatorum*. This explains the change in the average number of *D. mercatorum* to emerge from a rot at site B from 14.3 flies per rot during the normal years in the Kau season, down to 0.7 fly per rot during the wet year of 1982, and up to 283 flies per rot after the Kona storm of 1987. Hence, the Kona weather episode was characterized by a temporary but very dramatic population explosion of *D. mercatorum*. Such a temporary abundance of larval food resources and its attendant population explosion result in a very young age structure, which in turn would be expected to favor the $aa/+$ genotypes with their higher early fecundity through equation 15.15.

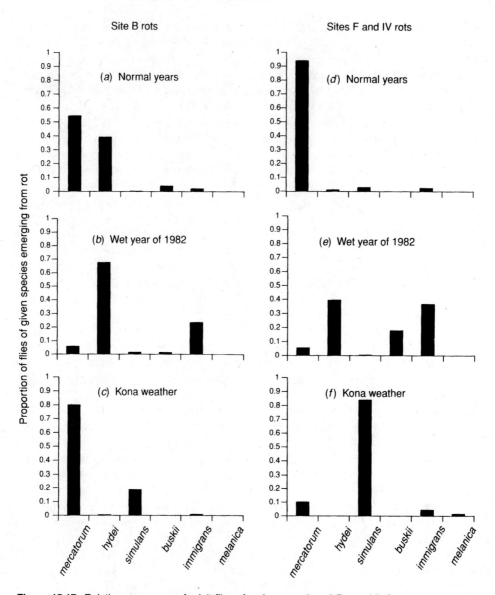

Figure 15.17. Relative emergence of adult flies of various species of *Drosophila* from rots bagged at site B versus F and IV as function of weather conditions at time of bagging.

This temporary pulse of wet, Kona conditions also interacts with the physical structure and location of the cacti in influencing the environment of the *Drosophila*. As noted earlier, the cacti grow to huge sizes in the upper part of the transect (Figure 15.9) but are small in the lower sections (Figure 15.12). The large cacti accumulate many fallen cladodes, and many of these begin to rot simultaneously as a result of the Kona storm. Moreover, each rotting cladode can support hundreds of larvae because of their large size, as noted above. In contrast, the small cacti at the lower sites do not accumulate many fallen pads, and each rot is smaller. Hence, the potential for population growth is not as great at the sites with small cacti (sites F and IV) as compared to the sites with large cacti (sites A to D). A further complicating factor is shown in Figure 15.17d–f. These figures show the emergence of adult flies from rots bagged at sites F and IV. During the normal years, *D. mercatorum* strongly dominates the use of the rots at these normally dry lower sites, so that even though the rots are smaller, the number of *D. mercatorum* that emerge from these small rots is actually larger on the average (34.6 flies per rot) than the average number emerging from site B rots (14.3 flies per rot) during the same time periods. However, during the wet year of 1982, *D. hydei* swept through all sites and became the most common *Drosophila* to emerge even from the rots at sites F and IV, and other, non-*repleta* group *Drosophila* also became common. As a consequence, only an average of 2 *D. mercatorum* per rot emerged at sites F and IV during 1982. During the Kona year, there was an explosive exploitation of the rots by the non-*repleta* species *Drosophila simulans*, and very few *D. mercatorum* emerged from the rots at the lower sites. Accordingly, the average emergence of *D. mercatorum* at sites IV and F was 6.8 flies per rot, well below the normal year average of 34.6 flies per rot. During the normal year of 1984, the population size of cactus IV-1 (shown in Figure 15.12) was estimated to be 51 ± 14.7, but during the Kona weather collections, it was impossible to estimate the size because so few *D. mercatorum* could be collected at this site. Hence, in great contrast to the upper sites with the large cacti that had explosive population growth and a superabundance of larval food resources for *D. mercatorum*, the lower sites experienced a reduction in population size and less larval food resources for *D. mercatorum*. These ecological conditions would lead to an older age structure. So during the Kona years, despite the wet, humid conditions, we expect a reverse of the cline in *aa* frequency observed in the normal years; now, *aa*/+ should have higher fitness at the upper sites and lower fitness at the lower sites due to this reversal of the typical age structure cline. Indeed, this is just what happened, as shown in Figure 15.18.

The reversal of the cline after Kona weather relative to the normal weather cline clearly shows that the *aa* is not an adaptation to low-humidity conditions, but rather *aa* is an adaptation to the age structure itself. The aa syndrome is favored whenever the ecological conditions are such that the adult age structure is young and it is selected against when the adult age structure is old, regardless of the humidity conditions.

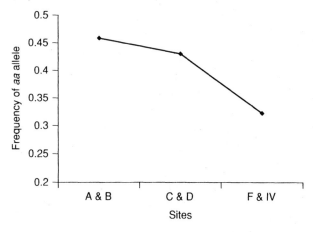

Figure 15.18. Cline in frequency of *aa* during years when collection was made shortly after Kona storms.

Natural Selection on Components of *aa* Supergene

As explained earlier, the *aa* supergene consists of two molecular components: the presence of R1 inserts in a third of more of the 28S ribosomal genes on the X chromosome and the presence of the X-linked ur^{aa} allele that codes for the failure of selective underreplication of inserted rDNA repeats in the fat body polytene tissue. Because of the linkage disequilibrium between the rDNA and the *ur* locus described earlier, virtually all X chromosomes with the ur^{aa} allele also have a third or more of their 28S genes bearing the R1 insert. Hence, the testcross procedure used to score X chromosomes by their phenotypic effect on adult abdominal cuticle effectively only scores for the presence or absence of the ur^{aa} on the X, with variation in the insert proportion only affecting the degree of morphological expression in the testcross progeny (Templeton et al. 1989). To examine the effects of the insert proportion separately from ur^{aa}, Templeton et al. (1989) also scored the proportion of inserted 28S genes using the Southern blot procedure from wild-caught males from the 1986 collection who were also test crossed. In this manner, a sample of X chromosomes was obtained for which information existed both on ur^{aa} and insert proportion. As mentioned in the previous section, 1986 was a Kona year, and the *aa* supergene cline was reversed from the normal pattern (Figure 15.18).

Confining inference just to the 1986 sample, Figure 15.19 shows a statistically significant cline in the frequency of ur^{aa}, with this allele declining in frequency as we went from

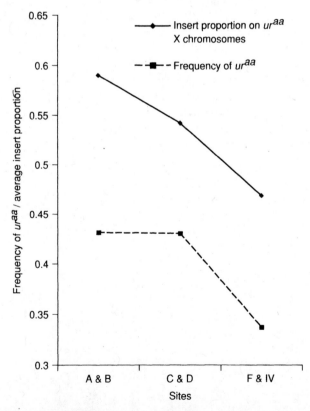

Figure 15.19. Frequency of ur^{aa} allele at *underreplication* locus and mean proportion of inserted 28S in 1986 sample of *D. mercatorum* X chromosomes plotted against collecting site location.

the upper sites with the large cacti to the lower sites with the small cacti. There was also a statistically significant cline in the average proportion of inserted 28S genes that paralleled the ur^{aa} cline. This cline in insert proportions is expected just from the linkage disequilibrium that exists in this supergene. In order to see if there were any patterns that could not be explained by this disequilibrium, Templeton et al. (1989) separated the X chromosomes into those bearing the ur^{aa} allele and those bearing the ur^+ allele and tested each subset of X chromosomes for significant site variation in the insert proportion. A dramatic contrast was revealed in the geographical pattern of insert proportion between ur^{aa} and ur^+ X chromosomes. There was no significant geographical heterogeneity in insert proportion in the ur^+ X chromosomes, whereas there was an even sharper and more significant cline in the insert proportions in the ur^{aa} X chromosomes (Figure 15.19). This cline in insert proportions is *within* ur^{aa} X chromosomes and therefore cannot be explained simply by the disequilibrium between the two major genetic components of the *aa* syndrome. Instead, this pattern suggests that natural selection is operating upon the insert proportion at the individual level as a function of the coarse-grained spatial heterogeneity across the transect, but *only* in the context of an X chromosome that codes for uniform underreplication. When selective underreplication occurs, there is no cline in insert proportion, indicating that the inserts are selectively neutral at the individual level in this genetic background. However, when selective underreplication fails to occur (on ur^{aa} X chromosomes), selection favors increased insert proportions in the same environmental contexts that favors the *aa* syndrome. Note that the selection on the R1 inserts depends both upon the genetic background and upon the external environment.

We have previously discussed evidence that the $EcoR1^+$ R1 elements are favored targets of selection below the level of the individual. They have the ability to spread throughout the rDNA complex via both transposition and tandem duplication and dominate intrachromosomally over alternative transposable elements. The results shown in Figure 15.19 indicate that individual-level, interchromosomal selection also occurs on the $EcoR1^+$ R1 elements, but this selection on the $EcoR1^+$ R1 elements is strictly modulated by fitness epistasis with the *ur* locus and occurs only when there is uniform underreplication. This result is consistent with the molecular biology of the syndrome, since selective underreplication buffers the organism against the impact of having a large proportion of the 28S genes inactivated by the insert. When there is uniform underreplication, the severity of the syndrome in the laboratory environment increases with increasing insert proportion (Templeton et al. 1989), so it is not surprising that individual-level selection could act upon insert proportion in the molecular context of uniform underreplication. Because of this epistasis and its demonstrable impact on joint allele frequency patterns, the *aa* represents an example of a coadapted gene complex (Chapter 12). These results also make it clear that an understanding of the evolutionary significance of these R1 transposable elements requires studies at both the molecular and population levels. These R1 elements are targets of selection both below the level of the individual and at the level of the individual, and studies directed at only one of these biological levels will always yield an incomplete picture.

OVERVIEW

The *aa* story touches upon most of the major points made in this book. The distinction between proximate and ultimate causation was made in Chapter 1, in which proximate causation describes necessary and/or sufficient conditions for a phenomenon as revealed by reductionistic studies on the content of a biological system, and ultimate causation describes how placing a biological system into the context of a higher level, interacting whole yields predictable emergent patterns. These two perspectives are complementary and require integration in population genetic studies. This was exactly what was done in the studies of *aa*. The proximate causation studies were motivated by some of the phenotypic similarities between *aa* and the well-characterized system of *bobbed* in *D. melanogaster* that suggested the X-linked rDNA as a likely candidate locus (in this case, a cluster of tandem repeats in a multigene family). Using the candidate locus approach (Chapter 10), the proximate story quickly focused upon transposable elements that specifically insert in the rDNA. At the molecular level, *aa* begins with the transcriptional inactivation of 18/28S rDNA repeats by the insertion of an R1 transposable element. The thread of proximate causation then tracks to the cellular level through the tissue-specific modulation rDNA underreplication in the fat body as controlled by the *ur* locus. These tissue-specific effects on the amount of functional rDNA transcripts next influence the development of the fly through the physiological role of the fat body as a modulator of JH titer at critical stages of development and ends with the life history consequences of altered JH metabolism as mediated by the normal responses to JH displayed throughout the genus *Drosophila*. These studies on proximate causation revealed that R1 insertion of a third or more of the 28S repeats within the X-linked rDNA and the uniform underreplication of inserted repeats were necessary conditions for the phenotypic expression of *aa*. Thus, the molecular and genetic *content* of the system defines the proximate causation of *aa*.

These proximate findings shaped and directed the studies on ultimate causation, studies seeking to answer why *aa* is present in this natural population and why it displays a dynamic pattern of spatial and temporal shifts in frequency. The proximate studies revealed the spectrum of pleiotropic effects associated with *aa*, which in turn suggested candidate phenotypes that could influence various components of fitness (Chapter 11), particularly those related to viability and fecundity. Moreover, potentially important genotype-by-environment interactions emerged from the proximate studies. Chapter 8 emphasized the importance of genotype-by-environment interactions in shaping the phenotypic variation, and such interactions were important in understanding the *aa* system. Phenotypes expressed in the laboratory were not necessarily expressed in nature, and climatic, weather, and ecological variables modulated the phenotypic expression in nature. Genetic architecture (Chapter 10) was also important in modulating phenotypic variation, as epistatic modifiers were revealed that affect the expression of some but not all pleiotropic traits. These studies on interactions with the environment and with modifier loci would have been impossible without having the ability to measure the appropriate genotypes, an ability that emerged from the proximate studies. The proximate studies also suggested the most appropriate genetic measures for monitoring the population-level responses over time and space.

The proximate studies were also critical in defining a variety of targets and units of selection (Chapter 13): from the R1 elements being both a unit and target of selection with respect to spreading within a chromosome below the level of the individual to the developmental delays and increased early fecundity found at the level of the individual. By defining the genetic elements, their linkage relationships, and strong epistasis, the proximate studies indicate that the appropriate unit of selection for the individual-level phenotypes is not a single locus but rather a supergene consisting of the *ur* locus and the multigene family of X-linked rDNA. Overall, the proximate studies set the groundwork for the investigations on the ultimate causation of why *aa* is polymorphic and shows dynamic temporal and spatial patterns. Reductionism and holism need not be warring philosophies of science;

rather, their integration leads to far richer insights than would be possible with only one approach.

As shown in Chapter 12, even a complete knowledge of the relationship between genotype and phenotype and its fitness consequences is insufficient for ultimate causation of allele frequency patterns in space and time. As shown in Chapters 8 and 11, the evolutionary effects of fitness differences are channeled through gametes through the average excess, which is a function not only of fitness but also of gamete frequencies and system of mating. Moreover, Chapter 12 emphasized that multiple evolutionary forces operate in an interactive fashion upon a genetic system. As a consequence, ultimate causation of temporal and spatial patterns for *aa* could not be understood until the evolutionary context of population structure (Chapters 2–7) was first defined. One of the most important interactions influencing population structure is the balance between genetic drift and gene flow (Chapter 6). The balance between these two evolutionary forces was measured in Chapter 6 for the *D. mercatorum* population living near Kamuela such that the balance across the entire sampling area results in an $N_{ev}m$ [the parameter measuring the ratio of the strength of gene flow, m, to the strength of genetic drift, $1/N_{ev}$, the inverse of the variance effective size (Chapter 4)] equal to 8.5. This value implies that gene flow dominates over drift in the population living in this study site such that there should be no significant spatial variation for a neutral locus; yet, an $N_{ev}m$ value of 8.5 and the measured dispersal are such that strong selection could disturb this neutral expectation for selected loci (Chapter 14). This background knowledge of population structure provides a critical context for interpreting the spatial and temporal heterogeneities observed in the *aa* supergene allele frequencies.

Because natural selection always occurs in an environmental context, the search for ultimate causation was successful in this case because population genetic surveys were coupled with ecological studies on weather, microclimates, dispersal, food resources, population demography, and even community structure. These studies clearly revealed that the fitness consequences of *aa* were strongly interacting with environmental factors that varied over both space and time, creating a highly hetereogeneous selective environment (Chapter 14). The ultimate causation of the spatial and temporal dynamics of *aa* arose from the interactions among the fitness effects of *aa* interacting with spatial heterogeneity in the environment upon the substrate of population structure that modulated the fitness response into spatial clines that tracked temporal changes in the environment. As shown by the *aa* system, population genetic studies are inherently a cross-disciplinary endeavor.

The great population geneticist Theodosius Dobzhansky (1962) stated that "nothing in biology makes sense except in the light of evolution." Dobzhansky's statement emphasizes that evolution is the core of biology. All aspects of biology are informed by evolution because evolutionary processes operating in conjunction with physicochemical properties have shaped all attributes of living organisms. Population genetics is at the core of evolution. All evolutionary change ultimately traces to evolution within populations. Population genetics describes the mechanisms by which evolutionary change occurs within populations as emerging from just three, well-established properties of DNA (or occasionally RNA): *DNA can replicate, DNA can mutate and recombine, and the information in DNA interacts with the environment to produce phenotypes.* This is an amazing accomplishment.

APPENDIX 1

GENETIC SURVEY TECHNIQUES

Population genetics deals with the distribution of genetic variation in space and through time. As a consequence, the specific questions addressed by population genetic studies have always been influenced and constrained by the techniques used to measure genetic variation within a population. This appendix will not present a comprehensive discussion of all the techniques used in population genetics but rather will focus on those techniques currently used or used in the recent past. Moreover, this will not be a cookbook of how to implement these survey techniques, particularly given the rapidity with which many of the techniques are being refined and altered. Instead, the purpose of this appendix is to familiarize the reader with the properties and limitations of the basic genetic survey techniques.

The 1960s marked a major change in population genetic survey techniques with the advent of protein electrophoresis, as discussed in Chapter 5. We begin our survey of techniques for measuring genetic variation with protein electrophoresis and then work our way to the present, when studies of genetic variation at the DNA level are now dominating the field.

PROTEIN ELECTROPHORESIS

Protein electrophoresis was first introduced as a major genetic survey technique in the 1960s (Harris 1966; Johnson et al. 1966; Lewontin and Hubby 1966) and is still used as a tool for screening populations for genetic variation. Protein electrophoresis is based on the fact that nondenatured proteins with different net charges at a specified pH migrate at different rates through various media (paper, cellulose acetate, starch gels, polyacrylamide gels). Because the proteins must be nondenatured, great care is needed in collecting and preparing samples. In many cases, fresh samples are used, such as a blood sample (often separated into the white and red blood cells), whole organisms (such as individual *Drosophila*), or tissues (liver, root tips, etc.). If fresh samples cannot be used, the samples must be quickly frozen after removal

---

Population Genetics and Microevolutionary Theory, By Alan R. Templeton
Copyright © 2006 John Wiley & Sons, Inc.

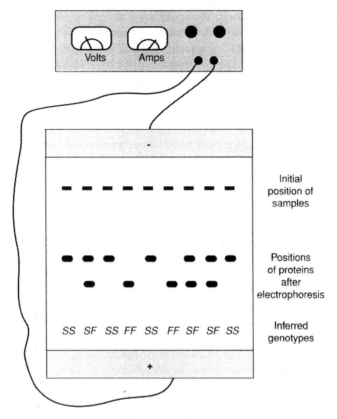

Figure A1.1. Protein electrophoresis. The results shown are for a monomer protein coding for a locus with two alleles, S and F, standing for slow and fast (referring to the relative mobilities of the proteins they code).

from the organism and maintained at low temperatures, typically −70 to −80°C. The samples are homogenized in a buffer solution to release the proteins, and multiple individual samples (often around 20–25 individuals) are generally placed along a line on the buffered medium (Figure A1.1). A power supply is then used to run an electrical current through the medium, and the proteins will migrate into this medium, with the distance of the migration being a function of the running conditions (the medium itself, the buffer used and its pH, and the length of time allowed for the proteins to migrate) and the charge on the protein.

Not all amino acids are charged under the pH ranges mostly commonly used. Most of the charge differences are due to 5 of the 20 amino acids: lysine, arginine and histidine, which tend to have a positive charge, and glutamic acid and aspartic acid, which tend to have a negative charge. Moreover, substitutions involving these amino acids are more effective at changing the charge of the molecule if they are located on or near the external surfaces of the protein. However, other amino acid changes can also alter the rate at which the protein moves through a medium through conformational changes. Overall, only about a third of all amino acid substitutions alter electrophoretic mobility, so protein electrophoresis only detects a subset of the nonsynonymous variants in protein-coding genes.

Once the proteins have migrated to their final positions in the medium, they need to be visualized in order to be scored by a human observer. Sometimes, the protein acts as its own stain. For example, hemoglobin molecules are a bright red, so their position in the medium is indicated by a red band. More generally, the gel or other media are immersed in a histochemical stain that singles out a particular protein from among the thousands of other proteins that have migrated in the same medium. Such stains often take advantage of the fact that many proteins are enzymes that catalyze a specific biochemical reaction. The substrates and cofactors for a specific enzymatic reaction can be added to a staining solution along with a stain that is chemically coupled to the enzymatic reaction such that the stain is deposited only where that reaction is occurring. Some media, such as starch gels, can be sliced into multiple thin sheets and each sheet can be immersed into a different staining solution. In this manner, a single gel can be used to score many different loci. Once stained, bands appear on the medium that indicate the positions to which a specific protein and its variants have migrated (Figure A1.1). The different allelic protein variants revealed by this technique are often called allozymes.

The resulting banding patterns typically define a codominant Mendelian system of variation. Figure A1.1 shows the typical pattern obtained for a monomer protein at a locus polymorphic for two electrophoretic alleles. More complicated patterns can be obtained for proteins that are dimers or tetramers, but in general a set of distinct bands is associated with all the electrophoretic alleles and genotypes. Another advantage of protein electrophoresis is that one chooses the protein-coding loci to be scored on the basis of the stains used, and not necessarily on those loci being electrophoretically polymorphic. Thus, one can survey a population for a specific protein and conclude that there was no electrophoretically detectable genetic variation at that locus, allowing one to estimate the proportion of loci that are polymorphic versus monomorphic. The implications of this feature were discussed in Chapter 5.

One limitation of this method is that the evolutionary relationships among the alleles cannot be reliably inferred from the banding patterns. Hence, protein electrophoretic data cannot generally be used to construct haplotype trees (Chapter 5). However, population genetic distances (Box 6.1) can be calculated from these data, so trees of populations can be estimated using algorithms such as neighbor joining (Box 5.3), although one should check that a tree of populations is a biologically plausible possibility (Chapter 7).

RESTRICTION ENDONUCLEASES

The 1960s also marked the discovery of restriction endonucleases (commonly called restriction enzymes), enzymes that cleave duplex DNA at particular oligonucleotide sequences, usually of 4, 5, or 6 bp in length (Linn and Arber 1968; Meselson and Yuan 1968). For example, the restriction enzyme *Eco*R1 (named from the bacteria, *Escherichia coli*, from which it was isolated) cuts double-stranded DNA where the nonmethylated sequence 5'-GAATTC-3' occurs. Hundreds of other restriction enzymes have been isolated from bacteria, each with a specific but often different recognition sequence. These enzymes have and continue to play an important role in population genetic surveys ever since the 1970s. Many different techniques have been developed that use restriction enzymes for genetic surveys, as will now be outlined.

Restriction Fragment Length Polymorphisms of Purified DNA

The earliest genetic surveys using restriction enzymes began with highly purified DNA, usually mtDNA (Avise et al. 1979). The purified DNA is cut with one or more restriction enzymes, and the resulting fragmented DNA is placed in wells in an agarose or polyacrylimide gel. The gel is then subject to electrophoresis, and the negatively charged

DNA fragments move toward the anode. Such gels are a complex molecular network, so smaller DNA fragments move through it more rapidly than large fragments; hence, the fragments are separated by their molecular weight. The position of the fragments within the gel is visualized by stains such as ethidium bromide or by using radioactively labeled DNA. In the later case, the gel is dried after being run and an X-ray film is overlaid upon it within a light-proof container. After exposing the film to the radioactive gel, the film is developed into an autoradiograph to reveal the position of the fragments.

Variation is observed as different banding patterns, as shown in Figure A1.2, and such variation is known as a restriction fragment length polymorphism (RFLP). Because the recognition sequence is a specific nucleotide sequence, any nucleotide polymorphism or insertion/deletion polymorphisms whose alternative states are associated with the presence

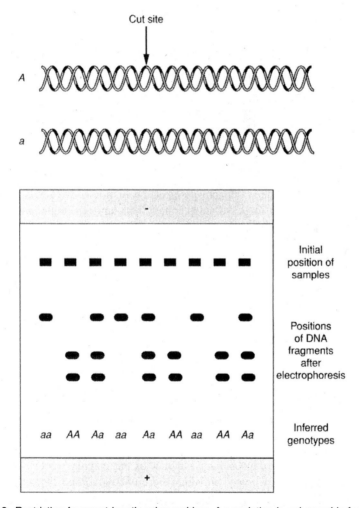

Figure A1.2. Restriction fragment length polymorphism. A population is polymorphic for a single restriction recognition sequence in a specified purified piece of DNA, such that those molecules with the recognition sequence are designated by *A* and those molecules without the recognition sequence are designated by *a*. Upon cutting with the restriction enzyme, molecules of type *A* yield two fragments, whereas molecules of type *a* yield a single large fragment. The resulting banding patterns are shown in the bottom half of the figure.

versus absence of the recognition sequence will define a RFLP. The RFLPs can be located in both protein-coding and noncoding DNA and within coding regions can be due to both synonymous and nonsynonymous mutations. Hence, a broader class of polymorphisms can be detected with this technique. As with allozymes, RFLPs are codominant Mendelian markers, and also like protein electrophoresis, DNA regions can be surveyed to reveal no polymorphism.

Southern Blots

The initial limitation of RFLPs to purified DNA greatly narrowed its applicability because few classes of DNA could be purified at that time. For example, although nuclear DNA could be purified, small homologous regions within the nuclear DNA could not. Consequently, one could digest the entire nuclear genome with one or more restriction enzymes, but this resulted in a highly heterogeneous pool of DNA fragments that would yield a continuous smear in an autoradiograph. Southern (1975) developed a technique that would allow a specific DNA region to be scored even when the initial sample consisted of the entire nuclear genome.

Southern's technique begins with the same procedures described above, except that a sample of purified DNA is no longer required. After the gel electrophoresis, the DNA fragments in the gel are denatured in a basic salt solution and then transferred as single-stranded DNA fragments to a nylon or nitrocellulose membrane through capillary action by blotting the salt solution through the gel and membrane into paper towels placed on top; hence, this technique is known as Southern blotting. The next step requires that the specific region of DNA to be surveyed had previously been cloned or otherwise isolated and purified. This cloned or purified DNA is then radioactively labeled (nonradioactive labels were later developed) and denatured into single-stranded copies. This labeled, single-stranded DNA is known as the "probe," and the membrane with the single-stranded fragments transferred from the gel is incubated with the probe under conditions in which strands that are complementary to those of the probe will hybridize and thereby form labeled duplexes in the membrane. Thus, the probe picks out those sequences that are homologous to it and ideally no others. The fragments that bind to the probe are then visualized by autoradioagraphy or some other appropriate technique for nonradioactive labels.

The development of the Southern blotting technique greatly expanded the utility of RFLPs as a means of surveying for polymorphisms. The entire genome was now theoretically open for investigation.

Restriction Site Mapping

A restriction site map shows the physical position of all the restriction sites relative to one another. Sometimes these relative positions can be inferred just from the digestion profiles and the estimated fragment sizes, but often this is not possible. For example, Figure A1.3 shows a hypothetical region of DNA that is cut once by the enzyme *Eco*R1 and once by the enzyme *Bam*H1. There are two ways in which these two restriction sites can be oriented relative to one another that are both consistent with the single-enzyme fragment patterns. Hence, the restriction site map cannot be inferred from the single-digest fragment data. Additional data are required to form the map in such cases. Often the needed information can be generated by doing a double digest. Figure A1.3 shows the results of digesting the DNA with both *Eco*R1 and *Bam*H1 simultaneously, and as can be seen, the two possible

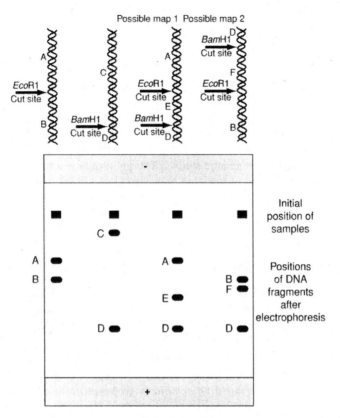

Figure A1.3. Restriction site mapping through double digests. The left two lanes show the fragment length profiles with single digestions of a region of DNA with either *Eco*R1 or *Bam*H1. All the fragments are indicated by capital letters. The single-digestion profiles are compatible with two different restriction site maps, as shown in the right two lanes. The two possible maps are readily distinguished by their double-digestion profiles.

restriction site maps produce different double-digest patterns. Hence, the restriction site map can be generated by producing different kinds of digestion profiles.

All the genetic information available in the RFLP analysis is still present when one goes to the extra effort of producing a map, but in addition the map generates haplotype data, that is, the simultaneous genetic state at two or more polymorphic sites on the same DNA molecule. Haplotype trees and linkage disequilibrium can therefore be estimated from the restriction site haplotypes.

DNA Fingerprinting

Jeffreys et al. (1985) isolated some DNA probes from humans that had a conserved core sequence of 10–100 bp that was widely scattered throughout the human genome, although it was soon discovered that many of these probes worked well on a variety of organisms. At each genomic location, the core sequence tended to exist as tandem repeats, but with the number of tandem repeats varying from location to location and even from individual to individual at the same location. These tandem repeat regions were called minisatellites or VNTRs (variable number of tandem repeats). When nuclear DNA is digested with restriction enzymes that cut outside the core region and then is subject to Southern blots, many fragments of diverse length will hybridize with the probe as a function of the variation of both the number of tandem repeats and the locations of the cut sites outside of the core region. This results in a complex gel profile influenced by many loci scattered throughout the genome, and such multilocus profiles are called DNA fingerprints.

DNA fingerprinting reveals extensive genetic variation in most populations, often to the extent that no two individuals are the same in their gel profiles (except for identical twins). Hence, this is an excellent technique for individual identification, and DNA fingerprinting was used extensively in forensics. However, DNA fingerprinting has some serious disadvantages for many population genetic studies. The gel profile is a multilocus phenotype, and there is generally no way of knowing how many loci are involved and which bands correspond to alleles and which are associated with different loci. Hence, one cannot even determine allele or genotype frequencies from such data, much less more complicated population genetic parameters.

Polymerase Chain Reaction

The original RFLP analyses required purified DNA. At that time, one could purify mtDNA by centrifugation and some other DNA regions by laborious processes. The development of the polymerase chain reaction (PCR) allowed the purification through amplification of small, well-defined regions of the genome that could then be subject to restriction site analysis (Saiki et al. 1985). The PCR technique involves four main steps (Figure A1.4). First, the double-stranded DNA from the sample is denatured (made into single strands) by heating. Second, single-stranded DNA primers are annealed to complementary DNA from the sample that flanks the region to be amplified. Third, the now double-stranded primer regions are extended by use of the thermostable *Taq* DNA polymerase to synthesize complementary strands to the region between the primers. Fourth, steps 1–3 are repeated multiple times to produce a large quantify of the DNA in the region flanked by the primers. Once amplified, the DNA can be subjected to restriction site analysis and mapping, as discussed above.

Saiki et al. (1985) pointed out that the PCR technique was not necessarily limited to restriction enzyme analysis. They were certainly right: PCR revolutionized genetics. It soon became obvious that this technique could be applied in many diverse ways in both molecular and population genetics. Several genetic survey techniques were soon developed that greatly expanded our abilities to survey genetic variation in natural populations. Most of these PCR-based techniques reveal variation at the DNA level without the use of restriction enzymes, and will be discussed in the next section of this appendix, but the last technique to be described in this section combines PCR with restriction enzyme analysis.

Amplified Fragment Length Polymorphisms

This technique begins with a restriction digest of genomic DNA, usually a pair of enzymes with one having a four-base recognition sequence and the other having a six-base recognition sequence. Such a double digestion of genomic DNA would create a complex mix of many different genomic fragments that would yield a smear after gel electrophoresis. To produce a readable gel, the number of visualized bands has to be greatly reduced. The Southern blotting technique solved this problem with the use of a probe that would anneal to only a specific subset of the DNA. A PCR primer can also be thought of as a probe that will

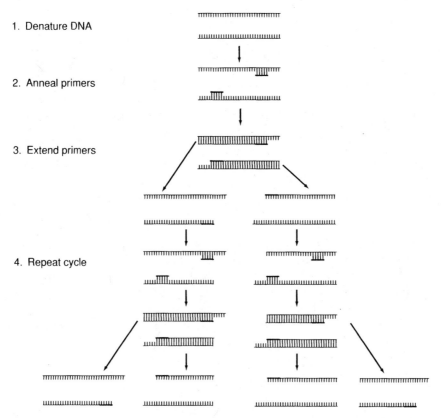

Figure A1.4. Polymerase chain reaction.

anneal to only a subset of the DNA. The task is to choose a primer that will anneal to many fragments in order to detect much polymorphism but not so many fragments as to produce an unreadable gel. This balancing act is achieved by attaching short synthetic DNA sequences known as adapters to the cut ends of the restriction fragments. The primers are then designed to match the known synthetic adapter sequence plus some additional nucleotides. Hence, the only fragments that will amplify under PCR are those that had these additional nucleotide sequences next to the recognition sequences of the restriction enzymes used in the original genomic digestion. Thus, only a small subset of the restriction fragments will amplify, thereby producing a readable gel. As was the case with DNA fingerprinting, this primer strategy generally will amplify DNA fragments from multiple locations within the genome, so the resulting bands on the gels represent a multilocus phenotype. Most of the polymorphism observed with this technique is nucleotide polymorphism in the bases adjacent to the restriction cut sites. A fragment amplifies when its sequence matches the primer, but if an alternative polymorphic state in that set of nucleotides does not match the primer, then no amplification occurs. As a result, the polymorphisms are exhibited as the presence or absence of specific bands, and the presence of a band is a dominant phenotype. Because of the dominant, multilocus nature of the polymorphisms, amplified fragment length polymorphisms (AFLPs) cannot be used to estimate many genetic parameters, such as heterozygosity. As was the case with DNA fingerprinting, AFLPs are most useful for identifying individuals genetically and estimating relatedness between individuals.

DNA POLYMORPHISMS

DNA is the primary genetic material of most organsisms, so scoring genetic variation directly at the level of DNA is ideal. All the methods of surveying genetic variation given above ultimately score genetic variation at the level of DNA, but often with some imprecision as to the nature of the polymorphism at the nucleotide sequence level. For example, when a restriction enzyme cuts a specific region of DNA, we know that the DNA sequence at the cut site must match the recognition sequence of the enzyme. However, given that the site is polymorphic, we do not know the precise nature of the polymorphism. For example, consider a six-base cutter. Wherever it cuts, we know the precise state of six nucleotides, but in those homologous pieces of DNA in which it does not cut, we do not know the nucleotide state. Indeed, a mutation to any other of the three possible nucleotides at any one of the six nucleotides at the recognition site would destroy the ability of the restriction enzyme to cut at that site. Hence, there are 18 different possible single-nucleotide substitutions that could underlie the RFLP. Moreover, the recognition sequence could be either created or destroyed by an appropriate insertion or deletion, so there are many ways at the DNA level to yield an RFLP.

There are now many PCR-based techniques that allow us to survey polymorphisms directly at the level of DNA without having to use the intermediaries of proteins or restriction sites. Many of these techniques reveal the precise molecular nature of the polymorphism, although we begin this section by describing a technique that does not and instead bears many similarities to AFLPs.

Randomly Amplified Polymorphic DNAs

The randomly amplified polymorphic DNA (RAPD) technique uses short PCR primers of about 10–20 arbitrary base pairs in length. Such short sequences will have matching complementary sequences at multiple places within a genome just by chance alone; that is, primers will anneal at places that just happen to match their sequence and not necessarily because of underlying homology (identity by descent). Hence, many DNA fragments are randomly amplified. Polymorphisms are frequently revealed when these randomly amplified products are separated by size with electrophoresis, usually due to underlying polymorphisms in the primer recognition sites, although the precise nature of these polymorphisms at the DNA level is not defined. This technique results in multilocus, dominant gel pattern phenotypes, so it bears many similarities to AFLPs and also shares its constraints.

DNA Resequencing

The most direct and complete method for measuring genetic variation is to completely sequence the DNA section that has been amplified by a pair of PCR primers. Moreover, by designing a set of PCR primers that anneal to conserved regions of the DNA in a manner that creates overlaps in the amplified pieces, it is possible to sequence long stretches of DNA. There are many ways of sequencing amplified DNA, but as the technology in this area changes so rapidly, no attempt will be made to describe the details other than to note that such a sequencing ability exists. Polymorphisms are discovered by resequencing the homologous pieces of DNA from multiple individuals. Resequencing reveals all the polymorphisms in the sampled individuals in the DNA region being sequenced: coding and noncoding, synonymous and nonsynonymous, single-nucleotide substitutions and insertions/deletions. Moreover, the precise molecular nature of the mutation is identified, for example, an A-to-G single-nucleotide substitution or a specific insertion/deletion polymorphism of a specified sequence of nucleotides. The primary limitations of this genetic survey technique are labor (mostly in getting all the primer reactions to work) and cost.

However, the genetic resolution of most DNA sequencing techniques is less than perfect for the simple reason that most techniques do *not* always produce a DNA sequence. DNA sequences are simple to obtain for haploid (or effectively haploid) DNA regions such as mitochondrial DNA (mtDNA), most Y-chromosomal DNA (Y-DNA), or X-chromosomal DNA scored in males. The problem arises when surveying diploid genes located in diploid tissues. For example, suppose an individual is heterozygous at two nucleotide sites in an autosomal region. A standard DNA sequencing gel displays heterozygous sites as double nucleotide scores, say nucleotides A and T at one site and G and C at the second site. However, the sequencing gel does *not* indicate if the A at site 1 is located on the same DNA molecule as the G or the C at site 2. As a result, a double-heterozygote individual such as A/T and G/C could have either the haplotype pair AG and TC (where underlining indicates the pair of polymorphic nucleotides found on a single DNA molecule) or AC and TG. Thus, most sequencing techniques when applied to diploid DNA do *not* yield the sequence of any single DNA molecule but rather reveal the diploid genotype at each heterozygous site but with no phase information between sites. Additional molecular and statistical techniques exist to obtain autosomal haplotypes in cases such as this, but as of this writing, molecular haplotyping techniques for diploid DNA are laborious and/or expensive, which greatly diminishes their utility in large-scale genetic surveys. As a consequence, most inferences of DNA sequences in diploid regions are done statistically by using population genetic information.

Templeton et al. (1988) used an algorithm known as *estimation–maximization* (EM) to statistically phase DNA sequences or haplotypes from diploid data, and the EM algorithm is still widely used for this purpose. To illustrate this algorithm, consider the hypothetical data set presented in Table A1.1. As can be seen from that table, the haplotypes (DNA sequences) of some individuals are known. For example, the homozygous individuals bear only one haplotype, and it can be read directly from the DNA sequencing data. Similarly, the single heterozygotes have two haplotypes, and they also can be inferred directly from the sequencing data. The problem arises with the double and triple heterozygotes. Here the lack of phased data means that these genotypes are compatible with multiple haplotype states, only one pair of which is correct. In general, if an individual is heterozygous at n sites, there are 2^n possible haplotypes compatible with the unphased data, only two of which are true.

Knowing which haplotypes were present and estimating their frequencies would be a trivial problem if only the data contained phased information. The EM algorithm is based

Table A1.1. Hypothetical Data Set from Genetic Survey of 100 Individuals for Autosomal Region

Genotype	Nature of Genotype	Number	Possible Haplotypes
A/A A/A T/T	Homozygote	21	AAT
G/G A/A T/T	Homozygote	19	GAT
A/A C/A T/T	Single heterozygote	9	ACT/GAT
A/G A/A T/T	Single heterozygote	39	AAT/GAT
A/G C/A T/T	Double heterozygote	9	ACT/GAT or AAT/GCT
A/G A/A T/C	Double heterozygote	2	AAT/GAC or AAC/GAT
A/T C/A T/C	Triple heterozygote	1	ACT/GAC or AAT/GCC or ACC/GAT or AAC/GCT

Note: It is assumed that DNA sequencing revealed three polymorphic nucleotide sites in this region, and only the genetic states of these polymorphic sites are indicated.

Table A1.2. Pretend Version of Data Set in Table A1.1

Unphased Genotype	Observed Number	Observed or Pretend Phased Genotype	Observed or Pretend Number
A/A A/A T/T	21	AAT/AAT	21
G/G A/A T/T	19	GAT/GAT	19
A/A C/A T/T	9	ACT/GAT	9
A/G A/A T/T	39	AAT/GAT	39
A/G C/A T/T	9	ACT/GAT*	4.5*
		AAT/GCT*	4.5*
A/G A/A T/C	2	AAT/GAC*	1*
		AAC/GAT*	1*
A/T C/A T/C	1	ACT/GAC*	0.25*
		AAT/GCC*	0.25*
		ACC/GAT*	0.25*
		AAC/GCT*	0.25*

Note: An asterisk by a genotype or number indicates that it was not observed directly in the original data but is only a possibility.

on wishful thinking—we do not have the data that we want, so we will pretend that we do! This exercise in wishful thinking actually works in many situations and is a type of algorithm to obtain maximum-likelihood estimators (Appendix 2). The algorithm is initiated by pretending that we know the haplotypes for all individuals. If an individual has known haplotypes, they are preserved, but if the individual has more than one pair of potential haplotypes, we will choose among the possibilities. It is generally best to be rather wishy-washy in the first round of wishful thinking, so, we will pretend that every unknown-phased genotype consists of equal numbers of all possible phased genotypes consistent with the unphased data. In the case of the data set given in Table A1.1, our initial "pretend" data set is shown in Table A1.2. Note that because we have pretended that the data consist only of known phase genotypes, it is trivial to estimate the haplotype frequencies for this pretend population just by counting up all the numbers of each haplotype and dividing by $2N$, where N is the sample size. These estimated haplotype frequencies from the pretend data are shown in Table A1.3.

The second step in the EM algorithm is to assume a model of how to go from haplotype frequencies to genotype frequencies, such as the random-mating Hardy–Weinberg model

Table A1.3. Estimated Gene Pool from Pretend Version of Data Set in Table A1.2

Haplotype	Haplotype Count	Estimated Haplotype Frequency
ACT	$9 + 4.5^* + 0.25^* = 13.75$	0.069
AAT	$42 + 9 + 39 + 4.5^* + 1^* + 0.25^* = 95.75$	0.479
GAT	$38 + 39 + 4.5^* + 1^* + 0.25^* = 82.75$	0.414
GAC*	$1^* + 0.25^* = 1.25$	0.006
GCT*	$4.5^* + 0.25^* = 4.75$	0.024
AAC*	$1^* + 0.25^* = 1.25$	0.006
GCC*	0.25^*	0.001
ACC*	0.25^*	0.001

Note: An asterisk by a haplotype or number indicates that it was not observed directly in the original data but is only a possibility.

(you should test this first on the individual sites to see if it is a good model—it is in this case). For the unphased genotypes that are compatible with multiple phased genotypes, we now revise our pretending by estimating the conditional probabilities of each possible genotype using our pretend haplotype frequencies and the assumption of random mating. For example, the unphased genotype A/G C/A T/T, which occurred nine times, can either be ACT/GAT or AAT/GCT, and we initially pretended in Table A1.2 that these two possible genotypes were equally frequent. Under Hardy–Weinberg, the frequency of ACT/GAT is proportional to

$$\text{Freq(ACT) Freq(GAT)}$$

and likewise the frequency of AAT/GCT is proportional to

$$\text{Freq(AAT) Freq(GCT)}$$

Let SUM = Freq(ACT) Freq(GAT) + Freq(AAT) Freq(GCT). Then, from Bayes' theorem (Appendix 2)

$$\text{Prob(ACT/GAT given ACT/GAT or AAT/GCT)} = \text{Freq(ACT) Freq(GAT)/SUM}$$

$$\text{Prob(AAT/GCT given ACT/GAT or AAT/GCT)} = \text{Freq(ACT) Freq(GAT)/SUM}$$

We estimate these conditional probabilities using the estimated haplotype frequencies from the initial round of pretend data and then allocate the real number 9 to these two genotypic categories proportional to these conditional probabilities. We do this for all other ambiguous genotypes and thereby generate a second round of pretend data. This second round of pretend data is given in Table A1.4, and this is used to generate new estimates of the haplotype frequencies, also shown in Table A1.4. The entire process is repeated until the frequencies converge to a stable value. This occurs in 26 rounds, with the final results shown in Table A1.5. The EM algorithm is known to converge to the maximum-likelihood estimates (Appendix 2) of the haplotype frequencies.

Table A1.4. Second Round of EM Algorithm Starting with Hypothetical Data Set in Table A1.1

Phased Genotype	Observed or Pretend Number	Haplotype	Estimated Haplotype Frequency
AAT/AAT	21	ACT	0.078
GAT/GAT	19	AAT	0.470
ACT/GAT	9	GAT	0.423
AAT/GAT	39	GAC*	0.007
ACT/GAT*	6.43* } 9	GCT*	0.013
AAT/GCT*	2.57*	AAC*	0.005
AAT/GAC*	1.07* } 2	GCC*	0.002
AAC/GAT*	0.93*	ACC*	0.002
ACT/GAC*	0.25*		
AAT/GCC*	0.35* } 1		
ACC/GAT*	0.31*		
AAC/GCT*	0.09*		

Note: An asterisk by a genotype, haplotype, or number indicates that it was not observed directly in the original data but is only a possibility.

Table A1.5. Round 26 of EM Algorithm Starting with Hypothetical Data Set in Table A1.1

Phased Genotype	Observed or Pretend Number		Haplotype	Estimated Haplotype Frequency
AAT/AAT	21		ACT	0.093
GAT/GAT	19		AAT	0.462
ACT/GAT	9		GAT	0.430
AAT/GAT	39		GAC*	0.013
ACT/GAT*	9.00*	} 9	GCT*	0.000
AAT/GCT*	0.00*		AAC*	0.000
AAT/GAC*	2.00*	} 2	GCC*	0.002
AAC/GAT*	0.00*		ACC*	0.000
ACT/GAC*	0.51*	} 1		
AAT/GCC*	0.43*			
ACC/GAT*	0.06*			
AAC/GCT*	0.00*			

Note: An asterisk by a genotype, haplotype, or number indicates that it was not observed directly in the original data but is only a possibility.

The EM algorithm not only provides maximum-likelihood estimates (Appendix 2) of the haplotype frequencies but also quantifies the degree of ambiguity in inferring the haplotypes borne by specific genotypes. For example, in Table A1.5, both double heterozygotes are completely resolved by the EM algorithm. Consequently, we can be confident of these phased genotypes and confident that the haplotype GAC exists even though it was not unambiguously observed. The only uncertainty is in the single individual who is a triple heterozygote, and even here one of the four possibilities has been eliminated, and only two of the possibilities have a high probability.

Stephens et al. (2001) use a Bayesian (Appendix 2) algorithm to estimate phase rather than a maximum-likelihood algorithm. Here is a simplified description of their algorithm. As with EM, first divide the individuals into those with unambiguous haplotypes and those with ambiguous haplotypes. Then:

1. Let G be the vector of genotypes and $H(0)$ some initial guess of the vector of haplotypes (just like EM).
2. Choose an individual, i, uniformly and at random from the set of ambiguous individuals.
3. Draw a possible haplotype pair for individual i from the probability distribution for the possible haplotypes of individual i given the genotypes and assuming that all the other haplotypes are correctly reconstructed.
4. Substitute the sampled haplotypes for individual i into $H(0)$, keeping the haplotypes assigned to all other individuals the same, to create a new vector, $H(1)$.
5. Go back to step 2 and keep repeating until the haplotype vector converges to a stationary distribution (they show this will always occur).

The difficult part of this algorithm is step 3—the Bayesian part. In general, we do not know the probability of a particular haplotype given an ambiguous genotype and a sample of known (at least pretend-known) haplotypes. Stephens et al. (2001) use neutral coalescent theory to generate this prior probability. Basically, their prior probability distribution gives most weight to resolutions based on unambiguous haplotypes (like the EM algorithm, and

Table A1.6. Comparison of Estimated Haplotypes and Their Frequencies under EM Algorithm and Bayesian Algorithm in Program PHASE Using Hypothetical Data in Table A1.1

Haplotype	EM Estimated Haplotype Frequency	PHASE Estimated Haplotype Frequency
ACT	0.093	0.093
AAT	0.462	0.461
GAT	0.430	0.430
GAC*	0.013	0.014
GCT*	0.000	0.001
AAC*	0.000	0.001
GCC*	0.002	0.000
ACC*	0.000	0.000

illustrated by the first double heterozygote in Table A1.1, which can be resolved in terms of two haplotypes, ACT and GAT, that are known to exist). For ambiguous genotypes that cannot be resolved with unambiguous haplotypes, it gives more weight to those potential haplotypes most like a known haplotype with high frequency (under neutral coalescent theory, most rare haplotypes are one-step or a few-step mutational derivatives of a common haplotype because common haplotypes are more likely to be hit by a mutational event than a rare haplotype simply because there are more copies at risk for mutation).

In our hypothetical example, this Bayesian procedure gives similar results to the EM algorithm (Table A1.6). The largest difference is in the estimated frequency of haplotype GCC. This difference primarily reflects how these two procedures resolved the triple heterozygote, the most ambiguous genotype in the sample. Note from Table A1.5 that the two resolutions ACT/GAC and AAT/GCC of the triple heterozygote are both given high probabilities under EM (0.51 and 0.43, respectively). Note that GAC is only one mutational step from GAT, the second most common haplotype in the known sample. In contrast, GCC is two steps away from any known haplotype. Because of this evolutionary consideration, the Bayesian procedure gives a probability of 0.92 to the ACT/GAC resolution and 0.07 to the AAT/GCC resolution. In general, because the PHASE algorithm uses prior information from evolutionary theory, it does a better job in resolving the most ambiguous genotypes.

Both the EM and PHASE algorithms perform well when the phase of a substantial portion of the data set is known without ambiguity. Similarly, both algorithms perform poorly when the proportion of ambiguous genotypes is high. For example, applying the EM and PHASE algorithms to two portions of human chromosome 19 that were molecularly phased in 20 subjects and that were polymorphic for about 150 nucleotides (thereby ensuring that all individuals had ambiguous genotypes), EM assigned an incorrect haplotype pair to all 20 individuals, and PHASE assigned an incorrect pair to 19 out of the 20 (Clark, personal communication). *No* statistical inference method works well when the initial data set consists mostly of ambiguous genotypes. In such circumstances, a combination of molecular haplotyping (to "seed" the known phase portion of the data) and statistical phasing is generally required (for example, see Clark et al. 1998).

Single-Nucleotide Polymorphisms

DNA resequencing reveals all polymorphisms in the surveyed region, and the most common class of polymorphisms is typically single nucleotides with alternative polymorphic nucleotide states. This class of polymorphisms is particularly valuable in population genetics because the scoring of genotypes at single nucleotides is amenable to rapid and massive automated screening. For example, one method of automated, massive screening is through the single-nucleotide polymorphism (SNP) chip (pronounced "snip chip") that is a type of DNA microarray. A microarray is usually a rectangular array of units. Each unit is a spot less than 200 μm in diameter that is placed on a specific point in the microarray grid. A DNA sequence of known state is affixed to a unit. These known DNA molecules are designed to anneal to the DNA surrounding and including a SNP that is amplified from a subject. A specific unit in the array will only anneal to subject DNA that has an exact nucleotide match, and the spots that anneal are coupled with a labeling technique (usually involving fluorescence) that can be read automatically. The SNP chip is designed to result in a unique labeling pattern for each potential genotype at the polymorphic nucleotide site. A single chip can have up to 100,000 units on it, so tens of thousands of SNP genotypes can be "read" off a single chip. Many other automated methods exist or are being developed (Schneider 2004), and this is an area of rapid technological development. As of this writing, this technology is only practical for humans and a handful of other model organisms.

Gene Expression Microarrays

The DNA sequences affixed to a DNA microarray (DNA chip) can come from the coding regions of a specified set of genes. Messenger RNA (mRNA) is then extracted from a sample, which could range from a whole organism to a specific tissue at a specific time in development or a tissue subjected to a specific environmental manipulation, such as a heat shock. The mRNA or cDNA made from mRNA is then allowed to hybridize with the DNA probes on the microarray such that a fluorescent label marks all successful hybridizations. The DNA chip is then scanned to record the positions and intensities of the fluorescent labels. In this manner, we can estimate which genes are being transcribed in the sample and their relative levels of expression. Such expression arrays can be used to identify candidate loci, as discussed in Chapter 10.

Microsatellites or Short Tandem Repeats

A final PCR-based technique that is extensively used in population genetics is to assay short tandem repeats (STRs) or microsatellites. Scattered throughout the genome of most organisms are tandem repeats of short sequences, often only two to four nucleotides long (Hamada et al. 1984). Many of these sites have polymorphic variation in the number of tandem copies. Primers are developed for invariant regions in the DNA that flank such a microsatellite region, and PCR amplification with such flanking primers will produce DNA products that differ in size as a function of the number of repeats. Fragments of different sizes can be separated by gel electrophoresis, with different sizes corresponding to different alleles. Many such microsatellite regions are highly polymorphic with many alleles. The advantage of this system is that high levels of genetic variation can be detected in such a manner that individual, codominant alleles can be scored at individual loci, in great contrast to other high-variation alternatives such as DNA fingerprinting, AFLPs, or RAPDs. In this regard, microsatellites are more similar to allozymes, although the levels of variation tend to be much higher. Because changes in repeat number are so common and the number of size classes is finite and small (usually less than 20), the same size-class allele can originate from multiple, independent mutations. Hence, the infinite-alleles model is generally inapplicable to microsatellite data.

APPENDIX 2

PROBABILITY AND STATISTICS

Population genetics deals with sampling genetic variation from populations, and hence the fundamental observations gathered in population genetics are subject to sampling error. Moreover, many aspects of the evolutionary process itself are subject to random processes, such as mutation (Chapter 1) and genetic drift (Chapters 4 and 5). Hence, random factors and uncertainty often need to be incorporated even into theoretical population genetics. Randomness in biological processes and sampling error can both be described by probability measures. Therefore, this appendix begins with a brief overview of probability theory. Population geneticists also use their data and models to make inferences about the evolutionary process, and this requires the use of statistics—functions of data that represent realizations of random processes. Therefore, this appendix will also provide a brief outline of some essential statistical concepts and tools used in population genetics.

PROBABILITY

A probability is a measure assigned to an event that is not certain to occur. There are many definitions of exactly what a probability measures, but the most common and straightforward interpretation is that the probability of an event represents the frequency with which the event would occur in a large number of independent trials. Probabilities have some well-defined mathematical properties that can be used to combine simple events into more complex events. Consider two events, called A and B. These events are said to be independent if

$$P(A \text{ and } B) = P(A \cap B) = P(A)P(B) \qquad (A2.1)$$

Population Genetics and Microevolutionary Theory, By Alan R. Templeton
Copyright © 2006 John Wiley & Sons, Inc.

where $P(A$ and $B)$ is the probability that events A and B are simultaneously true. Two events are said to be mutually exclusive if

$$P(A \text{ and } B) = 0 \text{ and } P(A \text{ or } B) = P(A \cup B) = P(A) + P(B) \tag{A2.2}$$

A set of events A_i is said to be mutually exclusive and exhaustive if each pair of events satisfies equations A2.2 and

$$P(A_1 \text{ or } A_2 \text{ or } A_3 \text{ or} \ldots A_n) = P\left(\bigcup_{i=1}^{n} A_i\right) = \sum_{i=1}^{n} P(A_i) = 1 \tag{A2.3}$$

where n is the number of events in the set (either finite or infinite).

Now consider the probability that event A occurs given that event B is known to have occurred. Then

$$P(A \text{ given } B) = P(A|B) = \frac{P(A \cap B)}{P(B)} \tag{A2.4}$$

where $P(A|B)$ is known as the conditional probability of A given B. Equation A2.4 is known as Bayes's theorem.

Consider the special case of Bayes's theorem in which the two events are independent. In that case, equation A2.1 can be substituted into the numerator of equation A2.4 to yield

$$P(A|B) = \frac{P(A \cap B)}{P(B)} = \frac{P(A)P(B)}{P(B)} = P(A) \tag{A2.5}$$

that is, when two events are independent, the probability of one event is not affected at all by the occurrence of the other event.

Random Variables, Probability Distributions, and Expectation

A random variable is simply a number that is assigned to an event. In many cases, a number naturally describes the event (such as an allele frequency in models of genetic drift, as in Chapter 4), but even in the case of discrete, qualitative events, it is always possible to assign a numerical value to describe the different event outcomes, as is the case in Box 3.1 in which a value of 1 was assigned to gametes bearing an A allele and a value of 0 was assigned to gametes bearing an a allele. A probability distribution is the set of the probabilities assigned to a mutually exclusive and exhaustive set of random variables. In many cases, it is possible to describe the entire probability distribution as a closed mathematical function of the random variable (say x) given a set of one or more parameters (say ω). Such a closed form is symbolized by $f(x|\omega)$ in this appendix. If the random variable is discrete, $f(x|\omega)$ gives the probability of the event described by the random variable taking on the value x. If the random variable is continuous, then $f(x|\omega)dx$ is the probability of the event that the random variable takes on a value in the interval between x and $x + dx$ as dx approaches zero. In general for a continuous random variable,

$$\int_{x_1}^{x_2} f(x|\omega)\,dx \tag{A2.6}$$

is the probability of the event that the random variable takes on a value in the interval between X_1 and X_2.

By combining equations A2.2 and A2.3, probability distributions always satisfy the following constraint:

$$\sum_x f(x|\omega) = 1 \quad \text{for a discrete random variable}$$

$$\int_x f(x|\omega)\,dx = 1 \quad \text{for a continuous random variable} \tag{A2.7}$$

where the summation or integration is over all possible values of the random variable x. This means that one of the possible values of the random variable will certainly occur (a probability of 1).

Because random variables are numbers, they can be manipulated by other mathematical functions. Let $g(x)$ be some function of the random variable x. Then the expectation of $g(x)$ is

$$E[g(x)] = \sum_x g(x)f(x|\omega) \quad \text{for a discrete random variable}$$

$$E[g(x)] = \int_x g(x)f(x|\omega)\,dx \quad \text{for a continuous random variable} \tag{A2.8}$$

where the summation or integration is over all possible values of the random variable x. The expectation of $g(x)$ represents the average value of $g(x)$ over all possibilities of x weighted by the probabilities of x. Two special cases of expectations are frequently used to characterize probability distributions. The first is called the mean and it is the expectation of $g(x) = x$ and is symbolized by μ:

$$\mu = \begin{cases} E[x] = \sum_x xf(x|\omega) & \text{for a discrete random variable} \\ E[x] = \int_x xf(x|\omega)\,dx & \text{for a continuous random variable} \end{cases} \tag{A2.9}$$

where the summation or integration is over all possible values of the random variable x. The mean μ is also called the first moment of the probability distribution. The second special case is called the variance and it is the expectation of $g(x) = (x - \mu)^2$ and is symbolized by σ^2:

$$\sigma^2 = \begin{cases} E\left[(x-\mu)^2\right] = \sum_x (x-\mu)^2 f(x|\omega) & \text{for a discrete random variable} \\ E\left[(x-\mu)^2\right] = \int_x (x-\mu)^2 f(x|\omega)\,dx & \text{for a continuous random variable} \end{cases}$$

$$\tag{A2.10}$$

where the summation or integration is over all possible values of the random variable x. The variance σ^2 is also called the second central moment (or second moment around the mean) of the probability distribution. Note that the mean is the average value of the random variable itself whereas the variance is the average value of the squared deviation of the random variable from its own average value. The mean measures the central tendency of the probability distribution, whereas the variance measures the tendency to deviate from the mean of the probability distribution.

There are a large number of probability distributions, but only a few are widely used in population genetics and in commonly used statistics. These basic probability distributions are described below, starting with discrete random variables and then moving on to continuous random variables.

Discrete Probability Distributions

Hypergeometric Distribution. Suppose there are N objects divided into two types, with a of the objects being of type 1 and b of the objects of type 2. Note that $a + b = N$. Now suppose that n objects are drawn from this population of N objects without replacement; that is, if one draws an object of type 1 on the first draw, then there are only $a - 1$ objects of type 1 left on the second draw. Now let the random variable be the number of objects of type 1 in the sample of n objects. The probability distribution that describes this situation is called the hypergeometric distribution and has the form

$$f(x|a,b,n) = \frac{\binom{a}{x}\binom{b}{n-x}}{\binom{a+b}{n}}$$

$$\text{where } \binom{c}{d} = \frac{c!}{d!(c-d)!} \quad y! = y(y-1)(y-2)\ldots(1) \quad \text{(A2.11)}$$

Note that a, b, and n are the parameters of this probability distribution. The random variable can take on any integer value between zero and the minimum of n and a, inclusively. The mean and variance of the hypergeometric are

$$\mu = n\frac{a}{a+b} = np \quad \text{where } p = \frac{a}{a+b}$$

$$\sigma^2 = \left(\frac{a+b-n}{a+b-1}\right)npq \quad \text{where } q = 1-p$$

(A2.12)

Binomial Distribution. If the size of the population being sampled (N) is much larger than the sample size (n) (formally, the limit as $N/n \to \infty$), the hypergeometric distribution converges to the form

$$f(x|p,n) = \binom{n}{x} p^x q^{n-x} \quad \text{(A2.13)}$$

where the random variable can take on any integer value between zero and n inclusively. The parameters of this distribution are n and p, and the mean and variance are given by

$$\mu = np \quad \sigma^2 = npq \quad \text{(A2.14)}$$

Poisson Distribution. A Poisson distribution arises from the binomial when the sample size of the binomial becomes very large (formally, $n \to \infty$) but the probability of drawing an object of type 1 becomes very small ($p \to 0$) with their product constant ($np = \lambda$).

Under these limiting conditions, the binomial takes on the Poisson form

$$f(x|\lambda) = \frac{\lambda^x e^{-\lambda}}{x!} \tag{A2.15}$$

where the random variable x can take on any integer value between zero and infinity. The Poisson has only a single parameter, λ, and its mean and variance are

$$\mu = \lambda \qquad \sigma^2 = \lambda \tag{A2.16}$$

Negative Binomial Distribution. The form of this distribution is

$$f(x|P, n) = \binom{n + x - 1}{n - 1} \left(\frac{P}{Q}\right)^x \left(1 - \frac{P}{Q}\right)^n \quad \text{where } Q = P + 1 \tag{A2.17}$$

where the random variable can take on any integer value between zero and infinity. This distribution also yields the Poisson distribution in the limit as $n \to \infty$, $P \to 0$, and $nP = \lambda$. The mean and variance of the negative binomial are

$$\mu = nP \qquad \sigma^2 = nPQ \tag{A2.18}$$

The four discrete distributions discussed above span increasing amounts of variance for a given mean. To see this, note that all the expressions for the variance of these distributions include the mean. Now substitute μ for the parameters that define the mean within the equations for the variances to yield the relative rankings of the variances for a given mean as

$$\underbrace{\left(\frac{a + b - n}{a + b - 1}\right)(1 - p)\mu}_{\text{hypergeometric}} < \underbrace{(1 - p)\mu}_{\text{binomial}} < \underbrace{\mu}_{\text{Poisson}} < \underbrace{(1 + P)\mu}_{\text{negative binomial}} \tag{A2.19}$$

Inequalities A2.19 show that these four interrelated probability distributions can collectively describe a broad range of random processes involving discrete random variables with little to high levels of variance and uncertainty for a given mean.

Continuous Probability Distributions

Normal Distribution. The normal distribution plays a central role in statistics because it arises under broad conditions. This has been shown by the central limit theorem. Let $f(x)$ be a probability distribution function, either discrete or continuous, with a finite mean μ and variance σ^2. Let $x_1, x_2, x_3, \ldots, x_n$ be a sample of n independent observations from $f(x)$. Let

$$z = \frac{\bar{x} - \mu}{\sigma / \sqrt{n}} \qquad \text{where } \bar{x} = \frac{\sum_{i=1}^{n} x_i}{n} \tag{A2.20}$$

where σ is the square root of the variance of the distribution from which the x's were sampled and is known as the standard deviation of that distribution. Then, as $n \to \infty$ (as the sample size gets large), the probability distribution function of z converges to

$$f(z) = \frac{e^{-\frac{1}{2}z^2}}{\sqrt{2\pi}} \qquad -\infty < z < \infty \qquad (A2.21)$$

Equation A2.21 is known as a normal zero, one distribution because it has a mean of 0 and a variance of 1. The normal can be generalized to an arbitrary mean of μ and variance of σ^2:

$$f(x|\mu, \sigma^2) = \frac{\exp\{-(1/2)[(x-\mu)/\sigma]^2\}}{\sqrt{2\pi}\sigma} \qquad -\infty < x < \infty \qquad (A2.22)$$

Because the central theorem indicates that the normal distribution can arise under broad conditions with the appropriate mathematical transformations, the normal distribution plays a central role in classical statistical theory and is often the first choice to represent the distribution that emerges from the effects of many small random factors. Much of classical statistics begins with the assumption of normality, and additional statistics and their distributions are derived by simple mathematical transformations of the normal distribution, as we will now see.

Chi-Square and Gamma Distributions. One of the simplest mathematical transformations of a normal zero–one random variable is to square it. Let $y = x^2$ where x is a normal zero–one variable, then

$$f(y) = \frac{y^{-1/2}e^{-y/2}}{\sqrt{2\pi}} \qquad 0 < y < \infty \qquad (A2.22)$$

Equation A2.22 is called a chi-square distribution with one degree of freedom, and it has a mean of 1 and a variance of 2. By squaring the original normal variable, y is a measure of deviation from the original normal mean of 0 and its expectation is the original normal variance of 1.

Now suppose we had n independent normal zero, one variables and squared each of them to result in n independent variables $(y_1, y_2, \ldots, y_n)$, each distributed as a chi-square with one degree of freedom. These n variables are transformed by taking their sum, $S = \sum_{i=1}^{n} y_i$, yielding

$$f(S|n) = \frac{S^{(\frac{n}{2}-1)}e^{-S/2}}{\Gamma(n/2)2^{n/2}} \qquad 0 < S < \infty \qquad (A2.23)$$

where Γ is the gamma function, a standard function in mathematics, defined by

$$\Gamma(x) = \int_0^\infty t^{x-1}e^{-t}dt \qquad (A2.24)$$

When x is an integer, say n, $\Gamma(n) = (n-1)!$. The gamma function is therefore a generalization of the factorial operator. Equation A2.23 is called a chi square with n degrees of freedom, and it has a mean of n and a variance of $2n$.

The chi-square distribution can be generalized to yield what is known as the gamma distribution:

$$f(x|\alpha, \beta) = \frac{x^{(\alpha-1)} e^{-x/\beta}}{\Gamma(\alpha) \beta^\alpha} \qquad 0 < x < \infty \qquad (A2.25)$$

which has a mean of $\alpha\beta$ and a variance of $\alpha\beta^2$. The gamma distribution with its two parameters provides a robust descriptor for continuous variables that must be positive.

Beta Distribution. Suppose x_1 is a random variable distributed as a gamma with parameters α_1 and $\beta = 1$, and x_2 is an independent random variable distributed as a gamma with parameters α_2 and $\beta = 1$. Transform these two variables by

$$y = \frac{x_1}{x_1 + x_2} \qquad (A2.26)$$

The distribution of y is

$$f(y|\alpha_1, \alpha_2) = \frac{\Gamma(\alpha_1 + \alpha_2)}{\Gamma(\alpha_1)\Gamma(\alpha_2)} y^{(\alpha_1-1)} (1-y)^{(\alpha_2-1)} \qquad 0 \le y \le 1 \qquad (A2.27)$$

and is called a beta distribution. The mean and variance of the beta distribution are

$$\mu = \frac{\alpha_1}{\alpha_1 + \alpha_2} \qquad \sigma^2 = \frac{\mu(1-\mu)}{\alpha_1 + \alpha_2 + 1} \qquad (A2.28)$$

The beta distribution with its two parameters provides a robust descriptor for continuous variables that are constrained to be between 0 and 1.

The t Distribution. If x is a random variable distributed as a normal zero–one and y is an independent random variable distributed as a chi-square with m degrees of freedom, then the function

$$t = \frac{x}{\sqrt{y/m}} \qquad (A2.29)$$

is distributed as a t distribution with m degrees of freedom:

$$f(t|m) = \frac{\Gamma[(m+1)/2]}{\sqrt{\pi m}\,\Gamma(m/2)} \left(1 + \frac{t}{m}\right)^{-(m+1)/2} \qquad -\infty < t < \infty \qquad (A2.30)$$

The t distribution has a mean of zero and a variance of $m/(m-2)$. As the degrees of freedom, m, go to infinity, the t distribution converges to a normal zero–one.

The F Distribution. Let u be distributed as a chi square with m degrees of freedom and v be distributed as a chi square with n degrees of freedom, with u and v being mutually independent. Then the function

$$F = \frac{u/m}{v/n} \qquad (A2.31)$$

is distributed as an F distribution with m and n degrees of freedom:

$$f(F|m,n) = \frac{\Gamma[(m+n)/2]\,(m/n)^{m/2}\,F^{m/2-1}}{\Gamma(m/2)\Gamma(n/2)(1+mF/n)^{(m+n)/2}} \qquad 0 < F < \infty \qquad (A2.32)$$

with a mean and variance of

$$\mu = \frac{n}{n-2} \qquad \sigma^2 = \frac{2n^2(m+n-2)}{m(n-2)^2(n-4)} \qquad (A2.33)$$

As n goes to infinity, the F distribution with m and n degrees of freedom converges to a chi square with m degrees of freedom.

Multivariate Probability Distributions

Bivariate Normal. Multivariate distributions define the joint probabilities of two or more random variables simultaneously. The random variables in a multivariate distribution need not be independent. Multivariate distributions can be discrete, continuous, or mixed. Many multivariate distributions exist. One example is the bivariate normal distribution:

$$f(x,y|\mu_x,\sigma_x^2,\mu_y,\sigma_y^2,\rho)$$
$$= \frac{\exp\left\{-\frac{1}{2}\left[(x-\mu_x)^2/\sigma_x^2 - 2\rho(x-\mu_x)(y-\mu_y)/(\sigma_x\sigma_y) + (y-\mu_y)^2/\sigma_y^2\right]\right\}}{\sqrt{2\pi}\,\sigma_x\sigma_y\sqrt{1-\rho^2}}$$
$$-\infty < x < \infty, -\infty < y < \infty \qquad (A2.34)$$

Note that in addition to the mean and variance of both x and y, the bivariate normal distribution has an extra parameter, ρ. This parameter determines the degree and direction to which the two random variables tend to covary. To see this, we introduce an expectation of a function of both random variables called the covariance:

$$\text{Cov}(x,y) = E\left[(x-\mu_x)(y-\mu_y)\right] \qquad (A2.35)$$

If y tends to fall above (or below) its mean when x falls above (or below) its mean, then the product in A2.35 will be positive, and there will be a positive covariance between x and y. On the other hand, if y tends to fall below (or above) its mean when x falls above (or below), then the product in A2.35 tends to be a product of a positive number with a negative number, and hence there will be a negative covariance between x and y. When x and y are independent, it is equally likely for either of the above two scenarios to occur, so the covariance is an average of positive and negative numbers that cancel one another out. Hence, the covariance of independent variables is zero.

The covariance of x and y from a bivariate normal (A2.34) turns out to be $\rho\sigma_x\sigma_y$. The correlation between x and y is defined to be $\text{Cov}(x,y)/(\sigma_x\sigma_y)$, so the correlation between x and y from a bivariate normal is ρ. In general, the correlation is used to measure the association between two random variables much more frequently than the covariance. The correlation is scaled to range from -1 to $+1$, whereas there is no universal scale for the covariance.

Multinomial. One commonly used multivariate discrete distribution is the multinomial, a generalization of the binomial distribution (equation A2.13):

$$f(x_1,\ldots,x_r|p_1,\ldots,p_r,n) = \frac{n!}{\prod_{i=1}^{r} x_i!} \prod_{i=1}^{r} p_i^{x_i} \qquad (A2.36)$$

where the range of each x_i is $0, \ldots,$ to n and the sum of all the x's is n. The mean of x_i is np_i and the variance is $np_i(1-p_i)$. The covariance between x_i and x_j is $-np_i p_j$.

STATISTICS

The probability distribution functions discussed above describe the probability measures assigned to the random variable as a function of constant parameters. However, once the data have been gathered, the random variables have been replaced by a set of known observations: the realized values of the random variables. Statistics are functions of the known observations. Statistics are used for inference (estimation of unknown parameters or tests of hypotheses) in a manner that explicitly treats the observed, known numbers as only one possible realization of a random process. Often the objects of statistical inference are the parameters of the probability distribution function, which often are not known.

The essential dilemma of relating probability distribution theory to statistical inference lies in the fact that the observations, once the data have been gathered, are no longer variables but known constants, whereas the parameters of the probability distributions are often not known and hence are more like variables. Several solutions to this dilemma have been proposed, but we will only examine three of them that are commonly used in statistics. The first solution is to simply treat the data points in the statistics as if they were still unobserved; that is, the data are regarded as random variables. The properties of the statistics, including their probability distribution functions, are deduced by applying expectation operators, probability distribution transformations, and so on. Unknown parameters needed for statistical inference are estimated by constructing a loss or error function and then finding the statistic that minimizes the expected loss or error. Another alternative is to base inference on nonparametric statistics, statistics whose properties do not depend upon any unknown parameters.

The second solution is to take the same functions used to describe probability distributions but now regard the random variables as known parameters and the original parameters as variables. Thus, the function appears the same, but its mathematical meaning has been fundamentally altered by this exchange in the status of variables and parameters. This altered function is called the likelihood function to distinguish it from the probability distribution function. Estimation and hypothesis testing are accomplished by finding statistics that maximize the value of the likelihood function, so this subdiscipline of statistics is called maximum-likelihood theory.

Probability theory purists objected to both of the above solutions. The first simply ignores the problem by continuing to treat the data as if they were unobserved random variables, and the second transforms parameters into variables and variables into parameters with no formal mathematical justification. The third solution, the Bayesian approach, keeps statistics completely and formally within the domain of probability theory. This is accomplished through the use of equation A2.4, where A now refers to the set of parameters and B refers to the data. Hence, Bayesians formally calculate the probability that the parameters can take on specific values given the observed data. There are a variety of ways of using these conditional probabilities for estimation and hypothesis testing. However, in order to use Bayes's Theorem (equation A2.4), it is necessary to have the probability distributions of the data *and* the parameters. The standard probability distribution functions give the probability of the data *given* the parameters, so the only piece missing from this inference puzzle is the probability distribution of the parameters. This distribution is known as the prior. The prior either is determined from prior knowledge about the parameter values or, more commonly, is given as a reasonable guess by the investigator.

It is important to keep in mind that these three approaches can all be used in statistical analyses. Together, they constitute a robust toolbox of analytical methods. We now give a brief description of some of the commonly used statistics for estimation and hypothesis testing in population genetics that have been produced by these three approaches.

Treating Data as Random Variables

Minimum-Variance, Unbiased Estimator of Mean. As an initial example of the approach that treats the data as if they were random variables, consider the problem of estimating the mean and variance from a random sample of n observations $(x_1, x_2, \ldots, x_n)$ each independently and identically distributed from some probability distribution with a mean of μ and a variance of σ^2. Consider first the problem of estimating the mean μ. One standard estimator of μ is the sample mean:

$$\hat{\mu} = \bar{x} = \frac{\sum_{i=1}^{n} x_i}{n} \tag{A2.37}$$

If we treat the x's as random variables, we can take the expectation of equation A2.37 as

$$E(\hat{\mu}) = E\left(\frac{\sum_{i=1}^{n} x_i}{n}\right) = \frac{E\left(\sum_{i=1}^{n} x_i\right)}{n} = \frac{\sum_{i=1}^{n} E(x_i)}{n} = \frac{\sum_{i=1}^{n} \mu}{n} = \frac{n\mu}{n} = \mu \tag{A2.38}$$

since n is a known constant not subject to the expectation operator and all the x's are independent. Note that the expectation of the sample mean is the true mean μ, so $\bar{x}$ is said to be an unbiased estimator of the mean μ. Since $\bar{x}$ is regarded as a function of random variables, $\bar{x}$ itself can also be treated as a random variable. The variance of the random variable $\bar{x}$ is (using the fact that all covariances are zero because all x's are independent)

$$\operatorname{Var}(\bar{x}) = E\left(\frac{\sum_{i=1}^{n} x_i}{n} - \mu\right)^2 = \frac{E\left(\sum_{i=1}^{n}(x_i - \mu)\right)^2}{n^2}$$

$$= \frac{\sum_{i=1}^{n} E(x_i - \mu)^2}{n^2} = \frac{n\sigma^2}{n^2} = \frac{\sigma^2}{n} \tag{A2.39}$$

Theorems exist in statistics that show that equation A2.39 represents the minimum variance possible for the class of all unbiased estimators of μ, so equation A2.37 is said to be a minimum-variance, unbiased estimator. Note that in this case the variance of the estimator is the average squared deviation of the estimator from the parameter that it is estimating, so the variance of the estimator can be interpreted as a measure of the average degree of error the estimator will have. Hence, the attribute of having minimum variance can also be thought of as an optimization property for the estimator in minimizing error.

Tests of Hypotheses about Mean. Under broad conditions, the estimator $\bar{x}$ will also converge, when regarded as a random variable, to a normal distribution with a mean of μ and a variance of σ^2/n as n gets large. Hence, using the results from the section on probability distributions,

$$\frac{\bar{x}-\mu}{\sqrt{\sigma^2/n}} = \frac{(\bar{x}-\mu)\sqrt{n}}{\sigma} \tag{A2.40}$$

is distributed as a normal zero–one random variable. Because the normal zero–one distribution has been put into tables and into many standard programs and calculators, equation A2.40 can be used to derive tests of hypotheses about the value of μ. For example, the upper 2.5% tail of a normal zero–one distribution is defined by all values greater than 1.96, and the lower 2.5% tail is defined by all values less than -1.96. Thus, if the mean is hypothesized to be a specific value, say μ_0, we would accept that hypothesis with 95% confidence if statistic A2.40 takes on a value between -1.96 and $+1.96$ when μ_0 is substituted for μ in equation A2.40. Similarly, we would reject the hypothesis that the mean is μ_0 at the 5% level (often called the p level of the test) if statistic A2.40 were less than -1.96 or greater than 1.96 when evaluated at $\mu = \mu_0$. Note that statistical tests do not deal with absolute truths but only probability statements. Thus, a 5% p level means that the test statistic will reject the hypothesis 5% of the time when the hypothesis is true. The rejection of a hypothesis that is true is known as a type I error. Similarly, accepting the hypothesis does not necessarily mean that the hypothesis is true; rather, it means that we have failed to reject the hypothesis at the stated p level. Sometimes the hypothesis is false, but the test statistic still lies within the range of acceptance of the hypothesis. Failing to reject the hypothesis when it is in fact false is known as a type II error.

Confidence Intervals for Mean. Equation A2.40 can also be used to generate confidence intervals for our estimate of the mean. Note that any value of μ that lies between the following values would be accepted with at least 95% confidence when using statistic A2.40 as a test:

$$\frac{(\bar{x}-\mu)\sqrt{n}}{\sigma} = -1.96 \quad \text{to} \quad \frac{(\bar{x}-\mu)\sqrt{n}}{\sigma} = 1.96 \quad \text{or} \quad \bar{x} \pm 1.96\frac{\sigma}{\sqrt{n}} \tag{A2.41}$$

The interval given in A2.41 is called a 95% confidence interval for the unknown mean; that is, one can be 95% sure that the true mean lies within that interval.

Both hypothesis testing and confidence intervals can be generalized by changing the p value, which is the same as the type I error probability. Let the p value be set to α; then one can always find the numbers $\pm d_\alpha$ from a normal zero–one table or function. As already seen, when $\alpha = 0.05$, then $d_\alpha = 1.96$. Setting the type I error to be 0.01, $d_\alpha = 2.58$. If 2.58 were substituted for 1.96 in range A2.41, we would have a 99% confidence interval for the unknown mean, which is broader than the 95% interval. Note also that by reducing the type I error rate we broaden the range of values that leads to the acceptance of the hypothesis through test statistic A2.40; that is, it is more difficult to reject the hypothesis. Generally, reducing the type I error tends to increase the type II error, so some balance is generally sought. By convention, the balance most often used is to set $\alpha = 0.05$.

Estimation and Testing When Variance Is Unknown. In order to evaluate the test A2.40 or the confidence interval A2.41, we need to know the value of the parameter σ^2. Often, however, we do not know either the mean or the variance of the sampled probability distribution, so σ^2 also needs to be estimated. The minimum-variance, unbiased estimator of the parameter σ^2 is

$$\hat{\sigma}^2 = s^2 = \frac{\sum_{i=1}^{n}(x_i - \bar{x})^2}{n-1} \quad \text{(A2.42)}$$

and under the same asymptotic normality assumptions that yield the normal distribution for statistic A2.40, the numerator in estimator A2.42 converges to a chi-square distribution with $n-1$ degrees of freedom. Moreover, the numerator turns out to be independent of $\bar{x}$. Our estimator of the variance of $\bar{x}$ is s^2/n when the variance has to be estimated, so the statistic

$$t = \frac{\bar{x} - \mu}{s/\sqrt{n}} \quad \text{(A2.43)}$$

is distributed as a t distribution (A2.30) with $n-1$ degrees of freedom. The t distribution is also a standard probability distribution, so it is easy to obtain d_α for any specified α. For example, suppose we have a sample of 20 observations ($n=20$); then $d_\alpha = 2.093$ when $\alpha = 0.05$. Note that when the variance was regarded as known, the 5% critical values for test A2.40 were given by ± 1.96, but now they are given by ± 2.093. Hence, the range of acceptance of the hypothesis is broader than in the case where the variance was known and not estimated. Likewise, the 95% confidence interval for the unknown mean is also broader in this case:

$$\bar{x} \pm 2.093 \frac{s}{\sqrt{n}} \quad \text{(A2.44)}$$

These broader test ranges and confidence intervals reflect the added uncertainty caused by having to estimate the variance rather than knowing the variance a priori. However, as n gets large, the critical values converge to the normal results. For example, if the sample size were increased to 100, then the t distribution with 99 degrees of freedom yields $d_\alpha = 1.984$ when $\alpha = 0.05$, and the limit as n goes to infinity is $d_\alpha = 1.96$.

Testing Difference between Means in Two Populations. The t distribution can also be used to test the significance of the difference between observed means from different populations. Let $x_i, i = 1, 2, \ldots, n_x$, be an independent and identically distributed sample of observations from a population characterized by a probability distribution with mean μ_x and variance σ^2 and let $y_i, i = 1, 2, \ldots, n_y$, be an independent and identically distributed sample of observations from a second population (independent from the first populations) characterized by a probability distribution with a potentially different mean μ_y but the same variance σ^2. Then, under the conditions of asymptotic normality, the statistic

$$t = \frac{\bar{x} - \bar{y}}{\sqrt{\frac{(n_x-1)s_x^2+(n_y-1)s_y^2}{n_x+n_y-2}\left(\frac{1}{n_x}+\frac{1}{n_y}\right)}} \tag{A2.45}$$

has a t distribution with $n_x + n_y - 2$ degrees of freedom. Hence, if the value of statistic A2.45 lies outside the range of $\pm d_\alpha$ for a specified α from a t distribution with $n_x + n_y - 2$ degrees of freedom, we reject the null hypothesis that the means of the two populations are the same; that is, we reject $\mu_x = \mu_y$.

Testing Equality of Variances in Two Populations. Statistic A2.45 assumes that both populations share a common variance even though their means may differ. We can test the hypothesis that the variances are indeed the same by taking advantage of the fact that the numerators of s_x^2 and s_y^2 are asymptotically distributed as independent chi squares. From this, it follows (from equation A2.31) that the statistic

$$F = \frac{s_x^2}{s_y^2} \tag{A2.46}$$

has an F distribution (A2.32) with $n_x - 1$ and $n_y - 1$ degrees of freedom under the null hypothesis that both populations have equal variance. The F distribution is also a standard probability distribution that has been tabulated and is a standard function in many statistical programs and calculators. For given α and degrees of freedom it is therefore possible to determine d_α from the F distribution, and we reject the hypothesis of the equality of the variances when $F > d_\alpha$.

One-Way Analysis of Variance. Suppose that instead of just two populations with potentially different means we had independent samples from r different populations, say $y_{ij}, i = 1, 2, \ldots, n_j, j = 1, 2, \ldots, r$. We now want to test the hypothesis that some or all of the means of these different populations (or factors or treatments, as they are commonly called) are different from one another. This can be done through an analysis of variance (ANOVA). The simplest ANOVA model regards the observed value of y in the ith observation for treatment j as the sum of the mean for treatment j, say μ_j, and an error term ε_{ij} such that all the errors are distributed as a normal with a mean of zero and a variance of σ^2:

$$y_{ij} = \mu_j + \varepsilon_{ij} \tag{A2.47}$$

Since $E(\varepsilon_{ij}) = 0$, $E(y_{ij}) = \mu_j$. Thus, all the observations for the jth treatment have the same expected value of μ_j. Since μ_j is a constant, it follows that the variance of y_{ij} is the variance of $\varepsilon_{ij} = \sigma^2$. Thus, all the observations have the same variance, regardless of the treatment. Moreover, since the ε_{ij} are distributed as normal random variables, so are the y_{ij}, and since we assume that all ε_{ij} are independent, it follows that the y_{ij} are distributed as independent normals with mean μ_j and variance σ^2.

The parameters in the ANOVA model are generally estimated by the least-squares procedure; that is, the parameter estimators are those that minimize the sum of the squared deviations of the observations around their expected values. Let this sum of squared deviations be designated by Q. For the simple ANOVA model given in A2.47, we have

$$Q = \sum_{j=1}^{r} \sum_{i=1}^{n_j} (y_{ij} - \mu_j)^2 \qquad (A2.48)$$

To find the minimum, we take the partial derivatives of Q with respect to the μ_j, set the partials equal to zero, and solve:

$$\frac{\partial Q}{\partial \mu_j} = -2 \sum_{i=1}^{n_j} (y_{ij} - \hat{\mu}_j) = 0 \Rightarrow \sum_{i=1}^{n_j} (y_{ij} - \hat{\mu}_j) = 0$$

$$\Rightarrow \sum_{i=1}^{n_j} y_{ij} = n_j \hat{\mu}_j \Rightarrow \hat{\mu}_j = \frac{\sum_{i=1}^{n_j} y_{ij}}{n_j} \equiv \bar{y}_{.j} \qquad (A2.49)$$

Note that the estimator of μ_j is the sample mean (equation A2.37) for the population with treatment j, the least-squares estimators of the means in this case are also minimum-variance, unbiased estimators. The estimator for the error term for a specific observation is

$$\hat{\varepsilon}_{ij} = y_{ij} - \bar{y}_{.j} \qquad (A2.50)$$

Equation A2.50 is called the residual for observation i in population j.

In analogy to equation A2.42, an estimator for the variance in the total population that includes both the effects of the error terms and the potential differences in means among the treatments will be proportional to

$$\text{SSTO} = \sum_{j=1}^{r} \sum_{i=1}^{n_r} (y_{ij} - \bar{y}_{..})^2 \quad \text{where} \quad \bar{y}_{..} = \frac{\sum_{j=1}^{r} \sum_{i=1}^{n_r} y_{ij}}{\sum_{j=1}^{r} n_r} \qquad (A2.51)$$

where SSTO stands for the *total sum of squares*. To disentangle the contributions of differences in means among treatments and the error terms within a treatment population, note that the total deviation can be written as a sum of the deviation of the treatment mean from the overall mean and the deviation around the treatment mean due to the error term:

$$y_{ij} - \bar{y}_{..} = (\bar{y}_{.j} - \bar{y}_{..}) + (y_{ij} - \bar{y}_{.j}) \qquad (A2.52)$$

Substituting the right-half of A2.52 into equation A2.51, we have

$$\text{SSTO} = \sum_{j=1}^{r} \sum_{i=1}^{n_r} [(\bar{y}_{.j} - \bar{y}_{..}) + (y_{ij} - \bar{y}_{.j})]^2$$

$$= \sum_{j=1}^{r} \sum_{i=1}^{n_r} (\bar{y}_{.j} - \bar{y}_{..})^2 + 2 \sum_{j=1}^{r} \sum_{i=1}^{n_r} (\bar{y}_{.j} - \bar{y}_{..})(y_{ij} - \bar{y}_{.j}) + \sum_{j=1}^{r} \sum_{i=1}^{n_r} (y_{ij} - \bar{y}_{.j})^2$$

$$= \sum_{j=1}^{r} n_j (\bar{y}_{.j} - \bar{y}_{..})^2 + 2 \sum_{j=1}^{r} (\bar{y}_{.j} - \bar{y}_{..}) \sum_{i=1}^{n_r} (y_{ij} - \bar{y}_{.j}) + \sum_{j=1}^{r} \sum_{i=1}^{n_r} (y_{ij} - \bar{y}_{.j})^2$$

$$= \sum_{j=1}^{r} n_j (\bar{y}_{.j} - \bar{y}_{..})^2 + \sum_{j=1}^{r} \sum_{i=1}^{n_r} (y_{ij} - \bar{y}_{.j})^2$$

$$= \text{SSTR} + \text{SSE} \qquad (A2.53)$$

where SSTR is called the *treatment sum of squares* and measures the extent of differences between the treatment means and the overall mean and SSE is called the *error sum of squares* that is proportional to the error variance σ^2. From the assumed normality of the random variables, these squares of the normal variables will have distributions proportional to a chi square, with the degrees of freedom being $r - 1$ for the SSTR and $\sum_{j=1}^{r} n_r - r$ for the SSE. The mean squares are the sums of squares divided by their degrees of freedom. Note that the mean-square error is given as

$$\text{MSE} = \frac{\sum_{j=1}^{r} \sum_{i=1}^{n_r} (y_{ij} - \overline{y}_{.j})^2}{\sum_{j=1}^{r} n_r - r} \quad (A2.54)$$

is analogous to equation A2.41, and it is an unbiased estimator of the error variance σ^2. One can show that the treatment mean square has the expected value

$$E(\text{MSTR}) = E\left[\frac{\sum_{j=1}^{r} n_j (\overline{y}_{.j} - \overline{y}_{..})^2}{r - 1}\right] = \sigma^2 + \frac{\sum_{j=1}^{r} n_j (\mu_j - \mu_.)^2}{r - 1}$$

$$\text{where } \mu_. = \frac{\sum_{j=1}^{r} n_j \mu_j}{\sum_{j=1}^{r} n_j} \quad (A2.55)$$

Note that the ratio of the expected mean-square treatment to the expected mean-square error is 1 under the null hypothesis that all treatment means are the same (all $\mu_j = \mu$). Moreover, the ratio of the mean-square treatment to the mean-square error is of the form given by equation A2.31 and therefore has an F distribution with $r - 1$ and $\sum_{j=1}^{r} n_r - r$ degrees of freedom. Hence, by partitioning the total variance into two components, MSTR and MSE, we can test the null hypothesis that all treatments have the same mean by an F test based on the ratio of these two variance components, MSTR/MSE. This is called a one-way ANOVA.

Permutation Testing. The ANOVA can be extended in many ways, but most extensions retain the assumption of normality of the errors. However, in some cases an assumption of normality cannot be justified. An excellent alternative in this situation is a permutation test. This test is based upon the statistical concept of exchangeability. Consider the one-way ANOVA when the null hypothesis that all treatment means are the same is true. Under this null hypothesis, all y_{ij} are sampled from the same probability distribution and the j label assigned to each observation should have no effect. Indeed, one could in theory exchange the labels of different observations and the resulting statistics would still have the same expected values and distributions when the null hypothesis is true. Hence, under the null hypothesis, the observations are said to be exchangeable across the treatment labels.

To test the null hypothesis of exchangeability, consider writing the data in the form of a single long vector, $Y = (y_{11}, y_{21}, \ldots, y_{n_1 1}, y_{12}, \ldots, y_{n_r r})$. Now consider a second vector Y^* that consists of a random permutation of all the elements in the original vector Y; that is, the observations in y are exchanged in position at random in the vector Y^*. Now, assign the first n_1 elements in Y^* to treatment label 1, the next n_2 elements to treatment label 2, and so on. In this manner, we have randomly permuted the treatment labels over the observations but have preserved the sample sizes for each population. Hence, if any associations existed between the observations and the treatments, we have broken them up by this permutation procedure and have simulated exchangeability. If the null hypothesis of exchangeability of the unpermuted data is true, then any statistic that is potentially sensitive to differences across the treatments should not be altered significantly by this permutation, but if the treatments do indeed influence the underlying probability distributions, then such statistics could be altered in a systematic fashion. For example, in the standard one-way ANOVA, we measured the potential differences in the means across treatments by the statistic MSTR/MSE. We can recalculate this statistic with the randomly permutated data Y^*, say MSTR*/MSE*. If the null hypothesis of exchangeability is true, MSTR*/MSE* would still be expected to be different from MSTR/MSE due to the individual observation error terms, but the differences should be slight. However, if the means did indeed differ across the treatments, then we would expect MSTR/MSE to be larger systematically than MSTR*/MSE*. In order to quantify the significance of the difference between MSTR/MSE and MSTR*/MSE*, the random permutation procedure is repeated a large number of times, say 10,000, with MSTR*/MSE* recalculated after each permutation. In this manner, we can obtain the empirical probability distribution MSTR*/MSE* under the null hypothesis of no effect of the treatments (which is the same as exchangeability of the treatment labels). We then place the observed statistic based on the original data, MSTR/MSE, into this distribution. For example, suppose K of the 10,000 MSTR*/MSE* values are as large or larger than those of the MSTR/MSE. Then the observed statistic is said to deviate from the null hypothesis at the $K/10,000\, p$ level.

The permutation test has some advantages over the traditional ANOVA. First, the permutation procedure is flexible. In the above example, we used the standard ANOVA statistic MSTR/MSE to focus on the potential effects of the factors on differences in means, but we could have just as easily used a different statistic to measure some other potential way in which the treatments could differ. Second, the permutation test is robust. Unlike the ANOVA in which we had to make assumptions about the underlying probability distributions and had to invoke a variety of parameters to describe these distributions (the μ_j's and σ^2), here we did not assume a particular form for the underlying probability distributions and did not have to invoke any parameters whatsoever, thereby making this test robust to deviations from underlying normality. Because no parameters are invoked and no specific probability distribution is assumed, the permutation test is an example of a *nonparametric test*. Nonparametric tests are in general more robust than their parametric analogues, but when the probability model used in the parametric analogue is a good approximation to reality, the nonparametric version tends to have lower power (greater type II error) than the parametric alternative. Hence, the trade-off between parametric versus nonparametric statistics is generally one of power versus robustness. The judgment about this trade-off depends on the confidence (or lack thereof) of the investigator in being able to choose a realistic probability model.

The permutation test is also an example of a *sample reuse statistic*. Note that although we have only one sample, namely Y, we reused this sample to generate an empirical probability distribution of the statistic under the null hypothesis of no treatment effects. Sample reuse statistics now have great popularity because the accessibility of high-speed computers makes them easy to execute.

Regression and Correlation. Regression is similar to ANOVA. As with ANOVA, we start with independent samples from r different treatments, say $y_{ij}, i = 1, 2, \ldots, n_j, j = 1, 2, \ldots, r$. Also, like ANOVA, we are interested in the case in which each treatment potentially has a different mean. However, unlike ANOVA, we also assign a number, say x_j, to each treatment. Note that this x is not a random variable. The x's are commonly called the independent variables, but most often the x's are treated as known parameters. This x number is not just an arbitrary label but is a quantitative measure of the effect that that treatment should have on the mean through some mathematical function of x. Thus, the treatments are now quantitatively ordered by the x's. The simplest function relating the x's to the treatment means is a linear one: $\mu_j = \alpha + \beta x_j$. Substituting this expected mean for treatment j into equation A2.47, we have as the basic linear regression model

$$y_{ij} = \alpha + \beta x_j + \varepsilon_{ij} \quad (A2.56)$$

with all other assumptions being the same as in the one-way ANOVA model. The least-squares estimators of α and β are

$$\hat{\beta} = \frac{\sum_{j=1}^{r} n_j(x_j - \bar{x})(\bar{y}_{.j} - \bar{y})}{\sum_{j=1}^{r} n_j(x_j - \bar{x})^2} \quad \text{where} \quad \bar{x} = \frac{\sum_{j=1}^{r} n_j x_j}{\sum_{j=1}^{r} n_j}$$

$$\hat{\alpha} = \bar{y}_{..} - \hat{\beta}\bar{x} \quad (A2.57)$$

and the estimator of the variance is still given by equation A2.54.

In some cases the x's are regarded as random variables rather than known parameters. Consider a model in which n pairs of observations (x_i, y_i) are sampled from a bivariate normal distribution (equation A2.34). The estimators for the means and variances of x and y are the same minimum-variance, unbiased estimators given earlier. In addition, we now need to estimate the correlation coefficient ρ, and this estimator is given by

$$\hat{\rho} = \frac{\sum_{i=1}^{n}(x_i - \bar{x})(y_i - \bar{y})}{\sqrt{\sum_{i=1}^{n}(x_i - \bar{x})^2 \sum_{i=1}^{n}(y_i - \bar{y})^2}} \quad (A2.58)$$

We can also express the relationship between x and y in terms of a linear regression model, $y_i = \alpha + \beta x_i$. In this case,

$$\hat{\beta} = \hat{\rho}\frac{s_y}{s_x} \quad \hat{\alpha} = \bar{y}_{..} - \hat{\beta}\bar{x} \quad (A2.59)$$

Regression equations are useful when we want to predict the value of y we would expect to obtain from a value of x that was not part of the original sample. Once the regression parameters have been estimated, the expected value of y for a new value of x, say x_{new}, is

$$E(y) = \hat{\alpha} + \hat{\beta}x_{\text{new}} \quad (A2.60)$$

The concept of regression can be extended in many ways. One common extension is to have a single random variable, y, but to have many independent variables, say $x_1, \ldots, x_k$. This is known as multiple regression. It can also be extended by using nonlinear mathematical equations to describe the relationship between the random variable and the independent variable(s).

Contingency Tests. Consider again the situation in which we have independent samples from r different populations and want to test the null hypothesis that they all are samples from the same underlying probability distribution. Unlike the one-way ANOVA, we now assume that we are observing discrete categorical variables and not continuous random variables. Suppose that there are c different types of categories. If a sample of $n_{.j}$ observations is made from population j, then the random variables are the number of these observations that fall into each of the c different categories, say n_{ij}, where i indicates category i. The observations are often summarized in a two-way table with c columns and r rows called an $r \times c$ contingency table. The null hypothesis specifies that the probability of an observation falling into category i is the same for all r populations. Assuming that each population is sampled from a multinomial distribution (equation A2.36), it can be shown that the statistic

$$\sum_{i=1}^{c}\sum_{j=1}^{r} \frac{(n_{ij} - n_{i.}n_{.j}/n_{..})^2}{n_{i.}n_{.j}/n_{..}} \quad \text{where } n_{i.} = \sum_{j=1}^{r} n_{ij}, \; n_{.j} = \sum_{i=1}^{c} n_{ij}$$

$$\text{and } n_{..} = \sum_{i=1}^{c}\sum_{j=1}^{r} n_{ij} \quad (A2.61)$$

converges with increasing sample size to a chi-square distribution with $(c-1)(r-1)$ degrees of freedom under the null hypothesis that all r multinomial distributions are identical. Accordingly, the null hypothesis of homogeneity of all r populations is rejected when test statistic A2.61 exceeds the critical value d_α for a specified α from a standard chi-square distribution with $(c-1)(r-1)$ degrees of freedom.

The convergence of statistic A2.61 to a chi-square distribution under the null hypothesis depends upon having a large sample size. The approximation can also break down when the total data set is large but unbalanced; that is, one or more rows or columns have very few observations. One frequently used rule of thumb to both judge the adequacy of the sample size and ensure balance is to check that the terms $n_{i.}n_{.j}/n_{..}$ are all ≥ 5 for each cell in the contingency table. The approximation can also break down when the data set is sparse; that is, it contains many cells with few or no observations. Alternative procedures exist when dealing with small data sets, unbalanced data sets, or sparse data sets. For example, Fisher came up with an algorithm for calculating the exact probability of a deviation as larger or larger than that observed in a 2×2 contingency table under the null hypothesis of homogeneity, and this is known as a Fisher's exact test. Before the advent of computers, Fisher's algorithm was practical only when the sample size was small (say, a total sample size of less than 25 observations), but now exact probabilities can be calculated for all 2×2 tables and many larger tables as well. When the exact algorithms are not computationally feasible to execute, a random permutation test that preserves the row and column marginal totals can be used instead.

Because of the many computational options available for contingency tables, it is sometimes desirable to transform an ANOVA-like problem with continuous random variables into a discrete contingency table format. This is particularly desirable when there are serious questions about the underlying normality assumptions of the ANOVA and the investigator is willing to sacrifice potential power (if the normality assumptions are true) for robustness (in case the normality assumptions are violated). One such nonparametric test is the median test. Starting with the same sampling situation as the one-way ANOVA, independent samples from r different populations ($y_{ij}, i = 1, 2, \ldots, n_j, j = 1, 2, \ldots, r$), we want to test the null hypothesis that all populations are sampled from the same underlying probability distribution but we do not wish to assume any particular parametric form for that distribution.

We pool all populations together and determine the median value for the pooled observations. The median is the number such that half of the y_{ij}'s are below this number and half are above. Designate this median number as M. If the null hypothesis is true, then M is an estimator of the median value for all populations; that is, we expect half of the observations to fall below M and half above *within every* population under the null hypothesis. Thus, for each j, we calculate the number of observations for that population that are below M (say n_{1j}) and the number above M (say n_{2j}) and then repeat this calculation for $j = 1, \ldots, r$. The n_{ij}, $i = 1, 2$, $j = 1, \ldots, r$, define a $2 \times r$ contingency table, and the null hypothesis of homogeneity of all r populations can be tested with either a contingency chi square, exact test, or permutation test, depending upon the nature of the sample. This is another nonparametric alternative to the one-way ANOVA. The median test can be extended by determining the tertile, quartile, ... values of the pooled observations instead of just the median, which are then used to split up the observations within each population into more than just two categories, say c categories. Now one has a $c \times r$ contingency table. In splitting the pooled observations into more than two equal-sized categories, one should keep in mind that if the resulting contingency table becomes sparse, there tends to be low power to reject the null hypothesis.

Goodness-of-Fit Tests. Suppose a sample of n independent observations is taken from a discrete probability distribution whose random variable can be one of c discrete categories. We want to test the hypothesis that the underlying probability distribution is a multinomial (equation A2.36) with parameters n and p_i, $i = 1, \ldots, c$. Let n_i be the number of observations in the sample of n that fall into category i. Under the null hypothesis of the specified multinomial, the statistic

$$\sum_{i=1}^{c} \frac{(n_i - np_i)^2}{np_i} = \sum_{i=1}^{c} \frac{\left(\text{observed}_i - \text{expected}_i\right)^2}{\text{expected}_i} \qquad (A2.62)$$

converges to a chi-square distribution with $c - 1$ degrees of freedom as n gets large. Equation A2.62 is called the goodness-of-fit chi square of the observations to the specified multinomial distribution.

In the above test, the null hypothesis specified the multinomial distribution, including all of its parameters. Sometimes, we do not know all of the parameters. Consider the extreme case in which we have to estimate all the p's in the multinomial. The minimum-variance, unbiased estimator of p is n_i/n. Moreover, it is necessary to only estimate $c - 1$ of the p's because the p's must sum to 1. A degree of freedom is lost every time a parameter has to be estimated from the observations in order to generate the expected values. In this case, $c - 1$ degrees of freedom are used to estimate the p's, but since test A2.62 starts with $c - 1$ degrees of freedom, we end up with no degrees of freedom. This means that the null hypothesis is untestable. To see why, simply substitute the estimated p's into equation A2.62. The expected values then turn out to be n_i, so every component of the goodness-of-fit statistic is zero; that is, the fit is perfect. Such a situation is said to be overdetermined; that is, so many parameters are estimated that the fit will always be excellent (in this case perfect), thereby preventing any meaningful test. Fortunately, in many cases, the p's are themselves derived from some underlying model with fewer than $c - 1$ parameters, as was the case in testing the fit of the Hardy–Weinberg model given in Box 2.1. Let k be the number of parameters that have to be estimated to generate the p's. Then the goodness-of-fit chi square has $c - 1 - k$ degrees of freedom.

Test A2.62 can be used to test the goodness of fit of many other distributions by simple transformations of the original data. For example, suppose our null hypothesis is that the observations come from a Poisson distribution with mean λ (equation A2.15). One way of testing the goodness of fit is to make a few discrete categories and calculate their probabilities from the Poisson. A good rule of thumb in making these categories is to make sure that each category has five or more observations to ensure convergence to a chi-square distribution. For example, suppose the categories of the Poisson variable being 0, being 1, being 2–5, and being >5 all produce categories with more than five observations. From equation A2.15, we have that $p_0 = e^{-\lambda}$, $p_1 = \lambda e^{-\lambda}$, $p_{2-5} = \sum_{i=2}^{5} \lambda^i e^{-\lambda}/i!$, and $p_{>5} = 1 - p_0 - p_1 - p_{2-5}$. These p's now define a multinomial distribution, so the goodness of fit to a Poisson can now be tested with a chi square with $4 - 1 = 3$ degrees of freedom. If we did not specify the value of the parameter λ, we would have to estimate it by $\bar{x}$, the minimum-variance, unbiased estimate of the mean. Using $\bar{x}$ instead of λ in the equations for the p's, we could now test the goodness of fit to the Poisson with $3 - 1 = 2$ degrees of freedom.

We can use test A2.62 to test the goodness of fit of even continuous probability distributions. For example, suppose we transform an original set of random variables, say x_i, $i = 1, \ldots, n$, with the function

$$z_i = \frac{x_i - \bar{x}}{s} \tag{A2.63}$$

where s is the square root of equation A2.42. Such a transformation is often called a normalizing transformation because in many, but not all, cases the z's will tend to have a normal distribution with mean 0 and variance 1. Suppose we want to see if this transformation did indeed normalize our data. We can use the standard normal function to subdivide our continuous variables into c equally probable discrete categories, once again with the provision that we want at least five observations in each cell. Suppose this is achieved by dividing the normal into quartiles. From the standard normal function, we expect 25% of the observations to be less than -0.674, 25% to be between -0.674 and 0, 25% to be between 0 and 0.674, and 25% above 0.674. Hence, let n_1 be the number of times $z_i \leq -0.675$, n_2 the number of times $-0.675 < z_i \leq 0$, n_3 the number of times $0 < z_i \leq 0.675$, and n_4 the number of times $z_i > 0.675$. The goodness of fit of these n's to a normal zero–one is then tested with A2.62 using $p_i = \frac{1}{4}$ for $i = 1, 2, 3, 4$. The test in this case has three degrees of freedom. As these examples show, test A2.62 is a highly versatile test for goodness of fit.

Maximum Likelihood

In the previous section, the data were treated as if they were random variables. We took expectations of the data and of statistics derived from the data, we assigned probability distributions to the data and to statistics, and so on. However, once the data have been gathered, they are no longer random variables but rather are known constants. What is generally unknown and uncertain are the values of one or more parameters. Likelihood acknowledges the certain status of the data versus the uncertain status of some or all of the parameters. This acknowledgment is accomplished by a mathematical fiat: The parameters of the original probability distribution are now regarded as variables, and the random variables are now regarded as known parameters. However, the form of the distribution appears identical.

These look-alike functions of parameters that are now variables and variables that are now parameters are called likelihoods.

For example, suppose we were going to obtain a sample from a binomial probability distribution (equation A2.13). The binomial depends upon two parameters, n (the sample size) and p (the probability that on a single observation the random variable takes on the value of 1). Generally, the n parameter is regarded as known, but the p parameter is not. After the sample was obtained and the original random variable x (the total number of observations of category 1) takes on a specific, known value, say X, the likelihood of the parameter p given the data X and the sample size n is

$$L(p|X, n) = \binom{n}{X} p^X q^{n-X} \tag{A2.64}$$

Note that the right-hand side of equation A2.64 appears to be identical to that of A2.13, except that x (a variable) has been replaced by X (a known number). However, the left-hand sides of these two equations are extremely different from a mathematical point of view: In equation A2.13, x is the variable and p and n are the parameters; in equation A2.64, p is the variable and X and n are the parameters. Because the likelihood looks like a probability distribution, many people mistakenly equate likelihoods to probabilities. However, there is no random variable in equation A2.64, so a likelihood is definitely not a probability. Indeed, the mathematical meaning of a likelihood is not well defined, but it is often stated that in some sense likelihood measures the relative probabilities of obtaining the observed data as a function of possible parameter values. Regardless, the justification for the use of likelihoods stems not from their mathematical meaning but rather from the fact that likelihoods can be used to obtain estimators and tests that have a variety of optimal statistical properties.

Maximum-Likelihood Estimation. Estimators are obtained through the principle of maximum likelihood; that is, we estimate parameters by finding the value of the parameters in the probability distribution function (which are the variables in the likelihood) that maximizes the likelihood function. In practice, it is often easier to find this maximum by using the natural logarithm of the likelihood, which is symbolized by $\ell n\, L$. To illustrate the principle of maximum likelihood, consider estimating the unknown parameter/variable p using equation A2.64. First, transform A2.64 into the log likelihood:

$$\ell n\, L(p|n, X) = \ell n \binom{n}{X} + X \ell n\, p + (n - X) \ell n (1 - p) \tag{A2.65}$$

Next, we find the value of the variable p that maximizes the log likelihood. There are many ways of doing so, but one simple method from calculus is to take the derivative of A2.65 relative to p, set it equal to zero, and solve for p (technically, you should also evaluate the second derivative to ensure that this is a maximum and not a minimum or saddle point):

$$\frac{d\,\ell n\, L(p|n, X)}{dp} = -X\frac{1}{p} + (n - X)\frac{1}{1 - p} = 0 \Rightarrow (n - X)p = X(1 - p)$$

$$\Rightarrow np = X \quad \Rightarrow \hat{p} = \frac{X}{n} \tag{A2.66}$$

The estimator $\hat{p}$ is the maximum-likelihood estimate of the parameter p (from the original probability distribution, A2.13). Theorems exist in statistics that show that the maximum-likelihood estimators have the following properties:

- Asymptotically Efficient: As the sample size gets large, the error associated with maximum-likelihood estimators becomes as small as one can get with any other estimator.
- Consistent: As more and more data are gathered, the maximum-likelihood estimator converges with probability 1 to the true state.
- Sufficient: All the information in the data about the parameter(s) being estimated is used by the maximum-likelihood estimator.

It must be emphasized that these optimal properties are known to hold *only* when the original probability distribution chosen is indeed an accurate model of the sampling distribution. An incorrect probability model can yield inefficient, inconsistent, and/or insufficient estimators.

One nonoptimal property of maximum-likelihood estimators is that they are sometimes biased. For example, consider a sample of size n from a normal distribution with mean μ and variance σ^2. The log likelihood after the data have been gathered is

$$\ln L(\mu, \sigma^2 | n, X) = -\frac{n}{2}\ln 2\pi - \frac{n}{2}\ln \sigma^2 - \frac{\sum_{i=1}^{n}(X_i - \mu)^2}{2\sigma^2} \qquad (A2.67)$$

To compute the maximum-likelihood estimators, we take the first derivatives of A2.67, set them equal to zero, and solve:

$$\frac{\partial \ln L}{\partial \mu} = \frac{\sum_{i=1}^{n}(X_i - \mu)}{\sigma^2} = 0 \Rightarrow \hat{\mu} = \frac{\sum_{i=1}^{n} X_i}{n} = \overline{X}$$

$$\frac{\partial \ln L}{\partial \sigma^2} = -\frac{n}{2}\frac{1}{\sigma^2} + \frac{\sum_{i=1}^{n}(X_i - \mu)^2}{2\sigma^4} = 0 \Rightarrow \hat{\sigma}^2 = \frac{\sum_{i=1}^{n}(X_i - \overline{X})^2}{n} \qquad (A2.68)$$

Note that the maximum-likelihood estimator of the mean is the same as for the minimum-variance, unbiased estimator of the mean (equation A2.37), but the maximum-likelihood estimator of the variance is not the same as the minimum-variance, unbiased estimator (equation A2.42). In particular, the expected value of the maximum-likelihood estimator of the variance is $(n-1)\sigma^2/n$, so this estimator is biased. As the sample size gets larger and larger, this bias becomes smaller and smaller, so this estimator has the property of consistency. However, the minimum-variance, unbiased estimator is also efficient, consistent, and sufficient in this case, and in addition it is also unbiased. Hence, although maximum likelihood produces statistics with many optimal properties, this technique does not always yield the most optimal estimator even when the assumed probability model is true.

In general, one cannot always solve for the maximum-likelihood estimator analytically as done in equations A2.66 or A2.68. In such cases, some other way must be found to discover the maximum-likelihood estimator. For example, the EM iterative algorithm discussed in Appendix 1 for estimating haplotypes and haplotype frequencies is a method for finding the maximum-likelihood estimator. Other techniques include the Newton–Raphson iterative algorithm, grid searches, and the Gibbs sampler (a way of resampling the data to approximate the likelihood surface and discover its high points). Because there may be multiple local maxima, these techniques may not always find the true maximum-likelihood estimator but only a local maxima.

Hypothesis Testing. Another great benefit of the maximum-likelihood framework is that it also provides a mechanism for hypothesis testing called log likelihood ratio tests. Suppose we have two models of reality, one called Ω and the other called ω, where ω is a proper subset of Ω (that is, if Ω has k parameters, then ω has j parameters with $j < k$, with the j parameters all being part of Ω). Then, the log likelihood ratio test statistic of these two models is

$$-2\ln\frac{L(\hat{\omega})}{L(\hat{\Omega})} = -2\left[\ln L(\hat{\omega}) - \ln L(\hat{\Omega})\right] \quad (A2.69)$$

where $L(\hat{\Omega})$ is the likelihood function evaluated with the maximum-likelihood estimators of the k parameters in Ω and $L(\hat{\omega})$ is the likelihood function evaluated with the maximum-likelihood estimators of the j parameters in ω. Statistic A2.69 is asymptotically distributed as a chi-square distribution with $k - j$ degrees of freedom under the null hypothesis that ω is true. Hence, the likelihood ratios provide a general method of testing nested hypotheses against one another.

For example, suppose we want to test the hypothesis that the binomial parameter p has the specific value of 0.25. Let Ω be the model where p can be any number between 0 and 1 inclusively (which depends therefore upon one unknown parameter, p, so $k = 1$), and let ω be the model $p = 0.25$ (which has no unknown parameters, so $j = 0$). Note that ω is a special case of Ω so the log likelihood ratio test can be used. Now suppose that 100 observations were sampled, and 30 times the binomial random variable took on the value of 1 ($X = 30$). Then, the maximum-likelihood estimator of p is $30/100 = 0.30$ (equation A2.66). Using the likelihood function given in equation A2.65, the log likelihood ratio test of the null hypothesis that $p = 0.25$ is

$$\text{Test} = -2[\ln L(p = 0.25|n, X) - \ln L(\hat{p}|n, X)]$$
$$= -2[X\ln(0.25) + (n-X)\ln(0.75) - X\ln\hat{p} - (n-X)\ln(1-\hat{p})]$$
$$= 1.28 \quad (A2.70)$$

The value of 1.28 is not significant at the 5% level from a standard chi-square distribution with $k - j = 1 - 0 = 1$ degree of freedom, so we cannot reject the null hypothesis of $p = 0.25$.

As a second example, suppose we sample two populations (say 1 and 2), each assumed to be a binomial sample, with the results $X_1 = 40$, $n_1 = 100$ and $X_2 = 60$, $n_2 = 100$. Let Ω be the model in which the two populations can have different p's, say p_1 and p_2, and ω the model in which both samples share a common p; that is $p_1 = p_2 = p$. Given that the two samples are independent, the likelihoods for each sample can be multiplied (or the log likelihoods added) to obtain the maximum-likelihood estimators under Ω as $\hat{p}_1 = 40/100 = 0.4$ and $\hat{p}_2 = 60/100 = 0.6$. Under ω, the two samples can be combined into a single binomial sample, so $\hat{p} = (40 + 60)/(100 + 100) = 0.5$. Hence, the test of the null hypothesis that two binomials have the same p is

$$\text{Test} = -2[\ln L(\hat{p}_1 = 0.4|100, 40) + \ln L(\hat{p}_2 = 0.6|100, 60) - \ln L(\hat{p} = 0.5|200, 100)]$$
$$= 8.06 \quad (A2.71)$$

which has $k - j = 2 - 1 = 1$ degree of freedom. Test A2.71 is significant at the 1% level using the chi-square distribution with one degree of freedom, so we reject the null hypothesis in this case that the two populations share a common p.

As the above two examples illustrate, likelihood ratio tests are versatile and relatively simple to implement and evaluate. They are constrained to testing nested hypotheses, but many hypotheses naturally fall into this domain, so likelihood ratio tests have many applications in population genetics.

Bayesian Inference

Bayesian inference is accomplished completely within the domain of probability theory. To accomplish this, several probability distributions are used in Bayesian statistics. First is the standard probability distribution function, say $f(x|\omega)$. This describes the probabilities with which the random variable, x, takes on particular values as a function of one or more parameters, indicated by ω. Because the parameters are often not known, Bayesian statistics explicitly acknowledge this uncertainty about the parameter values by invoking another probability distribution, $f(\omega)$, that treats the parameters as random variables. This distribution is called the *prior* probability distribution of the parameters. The prior reflects our knowledge about the parameter values before making observations. These two probability distributions can be combined to yield the *joint* probability distribution of the random variable and the parameters, now treated as random variables, as $f(x, \omega) = f(x|\omega) f(\omega)$. From this joint probability distribution, we can also obtain the *marginal* distribution of x by integrating (or summing for discrete probability distributions on the parameters) the joint probability distribution over all possible parameter values as weighted by the prior on the parameter values: $f(x) = \int_\omega f(x, \omega) d\omega$. Once the observations have been made, we have the known observations X, not the random variable x. Given the observations, we now use Bayes's theorem (equation A2.4) to generate the *posterior* probability distribution, that is, the probability of the parameters (treated as random variables) given the observed data:

$$f(\omega|X) = \frac{f(x, \omega)}{f(x = X)} \quad \text{(A2.72)}$$

The posterior probability distribution quantifies the uncertainty of the parameters given the data. Note that the posterior distribution is similar to the likelihood function in that the original, probability distribution parameters are now the variables and the observed data are known constants. However, unlike the likelihood function, the parameters are true random variables in the posterior distribution and equation A2.72 defines true probabilities. Once the posterior distribution has been obtained, it can be used in a variety of ways for estimating and testing the parameters.

To illustrate this approach, consider again the problem of obtaining a sample from a binomial probability distribution. The probability distribution of the random variable x is given by equation A2.13. We next need a prior distribution for the parameter p (we still regard n as always known). The specification of the prior has been the most controversial aspect of Bayesian statistics. In some cases, we truly have prior knowledge about the system. For example, suppose earlier experiments have indicated estimated values of p of 0.22, 0.40, and 0.31. We could incorporate this prior information into a probability distribution on p that puts more weight in the range of 0.22–0.4. Sometimes we have prior information, but it is less direct. For example, suppose we were going to perform an isozyme survey and wanted to estimate the parameter of expected heterozygosity. As shown in Figure 5.6, many previous isozyme surveys all yield values of this parameter in the range of 0–0.25. Hence, even if the proposed isozyme survey is being done on a new species, this prior experience would indicate that a reasonable prior would place most of its weight in the range of 0–0.25. Sometimes we have prior knowledge from theory. For example, in Appendix 1, we described a Bayesian algorithm for haplotype phasing. Although we do not know the specific phases of the haplotypes being sampled, we can use coalescent theory to place a prior distribution on the haplotypes consistent with the evolutionary process. As these examples show, many cases of inferences do indeed have some degree of prior information, and the Bayesian procedure allows this information to be used rather than ignored. However, in some cases, there is no prior information from previous experiments, from similar observations, or from theory. In cases such as these, a prior is chosen to reflect this lack of knowledge (for example, a uniform probability distribution over the entire range of the possible parameter values) and/or is chosen such that the prior is quickly overwhelmed by the data in shaping the posterior distribution and hence has minimal impact on inference after the data have been gathered.

A natural prior for the problem of inferring p from a binomial distribution is the beta distribution (equation A2.27). In particular, we will assume the prior for p is

$$f(p|\alpha_1, \alpha_2) = \frac{\Gamma(\alpha_1 + \alpha_2)}{\Gamma(\alpha_1)\Gamma(\alpha_2)} p^{(\alpha_1 - 1)} (1 - p)^{(\alpha_2 - 1)} \quad (A2.73)$$

The beta is a natural prior for p because, as we will see, the beta and binomial distributions go well together mathematically, the random variable of a beta distribution ranges from 0 to 1 (the same potential range as p), and the two parameters of the beta distribution allow great flexibility in incorporating prior information. Figure A2.1 shows the shapes of some prior distributions on p that can be obtained by using different values of the beta parameters. The prior associated with $\alpha_1 = 1$ and $\alpha_2 = 1$ is known as a flat prior and represents the case where there is no true prior knowledge. The prior associated with $\alpha_1 = 1$ and $\alpha_2 = 2$ reflects weak prior knowledge that p is more likely to be small than to be large. Finally, the prior associated with $\alpha_1 = 9$ and $\alpha_2 = 2$ indicates strong prior knowledge that the p parameter should be around 0.9.

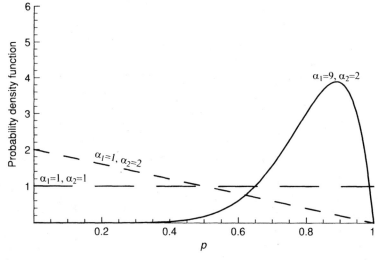

Figure A2.1. Some prior distributions on binomial parameter p. All priors are beta distributions with parameter values as given.

The next step is to obtain the marginal distribution of x:

$$f(x) = \int_0^1 f(p|\alpha_1, \alpha_2) f(x|n, p)\, dp = \binom{n}{x} \frac{\Gamma(\alpha_1 + \alpha_2)\Gamma(\alpha_1 + x)\Gamma(n + \alpha_2 - x)}{\Gamma(\alpha_1)\Gamma(\alpha_2)\Gamma(n + \alpha_1 + \alpha_2)}$$

(A2.74)

Using Bayes's theorem, the posterior distribution of p given X is

$$f(p|X, n, \alpha_1, \alpha_2) = \frac{f(p|\alpha_1, \alpha_2) f(x|n, p)}{f(x = X)}$$

$$= \frac{\Gamma(n + \alpha_1 + \alpha_2)}{\Gamma(X + \alpha_1)\Gamma(n - X + \alpha_2)} p^{X+\alpha_1-1}(1-p)^{n-X+\alpha_2-1} \quad \text{(A2.75)}$$

Note that the posterior distribution is also a beta distribution. Figure A2.2 shows the posterior distributions for the three prior distributions shown in Figure A2.1 for an observed data set with $n = 100$ and $X = 30$. Note that the flat prior and the prior with weak knowledge that p was more likely to be small than large give similar posterior distributions. Even the posterior distribution associated with the prior that assumed much knowledge that $p = 0.9$ has been shifted downward and overlaps extensively with the other two posterior distributions once the data have been obtained.

Once the posterior distribution has been determined, a variety of estimation and testing procedures can be implemented. For example, the Pitman estimate of the parameter p is the expected value of p with respect to its posterior distribution:

$$\hat{p} = \int_0^1 p \frac{\Gamma(n + \alpha_1 + \alpha_2)}{\Gamma(X + \alpha_1)\Gamma(n - X + \alpha_2)} p^{X+\alpha_1-1}(1-p)^{n-X+\alpha_2-1}\, dp = \frac{\alpha_1 + X}{\alpha_1 + \alpha_2 + n}$$

(A2.76)

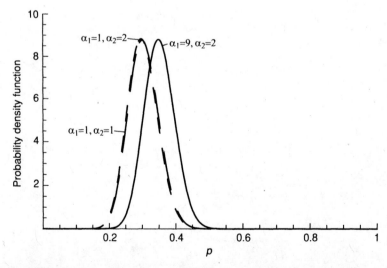

Figure A2.2. Posterior distributions of binomial parameter p given observed data set with $X = 30$ and $n = 100$ with three priors shown in Figure A2.1.

For example, the Pitman estimates of p for the three priors shown in Figure A2.1 are 0.3039, 0.3010, and 0.3514, in contrast to the maximum-likelihood estimate of 0.3000. Other estimation procedures are also possible. For example, akin to least squares, we can estimate p by the estimator that minimizes the expected squared-error loss:

$$E(\hat{p} - p)^2 = \int_0^1 (\hat{p} - p)^2 \frac{\Gamma(n + \alpha_1 + \alpha_2)}{\Gamma(X + \alpha_1)\Gamma(n - X + \alpha_2)} p^{X+\alpha_1-1}(1 - p)^{n-X+\alpha_2-1} \, dp \quad \text{(A2.77)}$$

It turns out that the estimator that minimizes the expected squared-error loss is the Pitman estimator in this case.

We can also test hypotheses with the posterior distribution. For example, to test the hypothesis that p is <0.25, we can find the probability tail from 0–0.25 of the posterior distribution. For the three posterior distributions shown in Figure A2.2, these tail probabilities are 0.115, 0.127, and 0.009, so only in the last case (the one with the strong prior that $p = 0.9$) would we reject this hypothesis at the 5% level or less. This example does show that priors can influence inference, particularly when the prior strongly favors some outcomes over others. Hence, priors should always be constructed with care and thought.

It is often not mathematically tractable to solve for the posterior distribution analytically, as was done in equation A2.75. Because of this common mathematical difficulty, the Bayesian approach was not used extensively in statistics until the advent of high-speed computers and the development of iterative and simulation algorithms that will produce posterior distributions. An example of the iterative simulation approach was given in Appendix 1 for the problem of haplotype phasing. Bayesian approaches are now being applied to many problems in population genetics.

REFERENCES

Abzhanov, A., M. Protas, B. R. Grant, P. R. Grant, and C. J. Tabin. 2004 Bmp4 and morphological variation of beaks in Darwin's finches. *Science* 305:1462–1465.

Adams, J., and R. H. Ward. 1973. Admixture studies and the detection of selection. *Science* 180:1137–1143.

Agrawal, A. 2000. Transposition and evolution of antigen-specific immunity. *Science* 290:1715–1716.

Akey, J. M., M. A. Eberle, M. J. Rieder, C. S. Carlson, M. D. Shriver, D. A. Nickerson, and L. Kruglyak. 2004. Population history and natural selection shape patterns of genetic variation in 132 genes. *PLOS Biology* 2:1591–1599.

Alper, J. 2000. Biomedicine—New insights into type 2 diabetes. *Science* 289:7.

Amante, A., F. Gloria-Bottini, and E. Bottini. 1990. Intrauterine growth: Association with acid phosphatase genetic polymorphism. *Journal of Perinatal Medicine* 18:275–282.

Ammerman, A. J., and L. L. Cavalli-Sforza. 1984. *The Neolithic Transition and the Genetics of Populations in Europe*. Princeton University Press, Princeton, NJ.

Anderson, W. W. 1969. Polymorphism resulting from the mating advantage of rare male genotypes. *Proceedings of the National Academy of Sciences* USA 64:190–197.

Anxolabehere, D., M. Kidwell, and G. Periquet. 1988. Molecular characteristics of diverse populations are consistent with the hypothesis of a recent invasion of *Drosophila melanogaster* by mobile P elements. *Molecular Biology and Evolution* 5:252–269.

Aoki, K., and M. W. Feldman. 1994. Cultural transmission of a sign language when deafness is caused by recessive alleles at two independent loci. *Theoretical Population Biology* 45:101–120.

Aquadro, C. F., S. F. Desse, M. M. Bland, C. H. Langley, and C. C. Laurie-Ahlberg. 1986. Molecular population genetics of the alcohol dehydrogenase gene region of *Drosophila melanogaster*. *Genetics* 114:1165–1190.

Averhoff, W. W., and R. H. Richardson. 1974. Pheromonal control of mating patterns in *Drosophila melanogaster*. *Behavior Genetics* 4:207–225.

Averhoff, W. W., and R. H. Richardson. 1975. Multiple pheromone system controlling mating in *Drosophila melanogaster*. *Proceedings of the National Academy of Sciences USA* 73:591–593.

Avise, J. C. 1994. *Molecular Markers, Natural History and Evolution*. Chapman & Hall, New York.

Avise, J. C., R. A. Lansman, and R. O. Shade. 1979. The use of restriction endonucleases to measure mitochondrial DNA sequence relatedness in natural populations. I. Population structure and evolution in the genus *Peromyscus*. *Genetics* 92:279–295.

Balciuniene, J., L. Emilsson1, L. Oreland, U. Pettersson, and E. E. Jazin. 2002. Investigation of the functional effect of monoamine oxidase polymorphisms in human brain. *Human Genetics* 110:1–7.

Bamshad, M. J., and S. P. Wooding. 2003. Signatures of natural selection in the human genome. *Nature Reviews Genetics* 4:99–111.

Barbujani, G., A. Magagni, E. Minch, and L. L. Cavalli-Sforza. 1997. An apportionment of human DNA diversity. *Proceedings of the National Academy of Sciences, USA* 94:4516–4519.

Barton, N. H., and S. Rouhani. 1993. Adaptation and the "shifting balance." *Genetical Research* 61:57–74.

Baudry, E., and F. Depaulis. 2003. Effect of misoriented sites on neutrality tests with outgroup. *Genetics* 165:1619–1622.

Bennett, S. T., and J. A. Todd. 1996. Human type 1 diabetes and the insulin gene—Principles of mapping polygenes. *Annual Review of Genetics* 30:343–370.

Bergman, A., D. B. Goldstein, K. E. Holsinger, and M. W. Feldman. 1995. Population structure, fitness surfaces, and linkage in the shifting balance process. *Genetical Research* 66:85–92.

Berry, M., F. Grosveld, and N. Dillon. 1992. A single point mutation is the cause of the Greek form of hereditary persistence of fetal haemoglobin. *Nature* 358:499–502.

Berven, K. A., D. E. Gill, and S. J. Smith-Gill. 1979. Countergradient selection in the green frog, *Rana clamitans*. *Evolution* 33:609–623.

Bishop, J. A., and L. M. Cook. 1975. Moths, melanism and clean air. *Scientific American* 232: 90–99.

Bishop, J. A., L. M. Cook, and J. Muggleton. 1978. The response of two species of moths to industrialization in northwest England. *Philosophical Transactions of the Royal Society of London B* 281:489–542.

Boag, P. T. 1983. The heritability of external morphology in Darwin's ground finches (*Geospiza*) on Isla Daphne Major, Galápagos. *Evolution* 37:877–894.

Boag, P. T., and P. R. Grant. 1981. Intense natural selection in a population of Darwin's finches (Geospizinae) in the Galápagos. *Science* 214:82–85.

Bochdanovits, Z., and G. de Jong. 2004. Antagonistic pleiotropy for life-history traits at the gene expression level. *Proceedings of the Royal Society of London Series B-Biological Sciences* 271:S75–S78.

Bodmer, J. G., and W. F. Bodmer. 1999. HLA polymorphism and evolution. In S. P. Wasser, Ed. *Evolutionary Theory and Processes: Modern Perspectives, Papers in Honour of Eviatar Nevo.* Kluwer Academic Publishers, Dordrecht, pp. 3–21.

Boerwinkle, E. and G. Utermann. 1988. Simultaneous effects of the apolipoprotein E polymorphism or apolipoprotein E, apolipoprotein B, and cholesterol metabolism. *American Journal of Human Genetics* 42:104–112.

Boerwinkle, E., S. Visvikis, D. Welsh, J. Steinmetz, S. M. Hanash, and C. F. Sing. 1987. The use of measured genotype information in the analysis of quantitative phenotypes in man. II. The role of the apolipoprotein E polymorphism in determining levels, variability, and covariability of cholesterol, betalipoprotein, and triglycerides in a sample of unrelated individuals. *American Journal of Medical Genetics* 27:567–582.

Bottini, E., G. Gerlini, N. Lucarini, A. Amante, and F. Gloria-Bottini. 1991. Evidence of selective interaction between adenosine deaminase and acid phosphatase polymorphisms in fetuses carried by diabetic women. *Human Genetics* 87:199–200.

Bottini, E., F. Gloria-Bottini, and P. Borgiani. 1995. ACP1 and human adaptability. 1. Association with common diseases: A case-control study. *Human Genetics* 96:629–637.

Bottini, E., P. Lucarelli, R. Agostino, R. Palmarino, L. Businco, and G. Antognoni. 1971. Favism: Association with erythrocyte acid phosphatase phenotype. *Science* 171:409–411.

Bottini, E., R. Palmarino, P. Lucarelli, F. Lista, and N. Bottini. 2001a. ACP1 and human adaptability: Association with past malarial morbidity in the Sardinian population. *American Journal of Human Biology* 13:753–760.

Bottini, N., G. F. Meloni, A. Finocchi, G. Ruggiu, A. Amante, T. Meloni, and E. Bottini. 2001b. Maternal-fetal interaction in the ABO system: A comparative analysis of healthy mothers and couples with recurrent spontaneous abortion suggests a protective effect of B incompatibility. *Human Biology* 73:167–174.

Bowcock, A. M., J. R. Kidd, J. L. Mountain, J. M. Hebert, L. Carotenuto, K. K. Kidd, and L. L. Cavalli-Sforza. 1991. Drift, admixture, and selection in human evolution: A study with DNA polymorphisms. *Proceedings of the National Academy of Sciences USA* 88:839–843.

Boyd, W. C. 1950. *Genetics and the Races of Man*. Little, Brown, Boston.

Brandt, B. W., B. J. Zwaan, M. Beekman, R. G. J. Westendorp, and P. E. Slagboom. 2005. Shuttling between species for pathways of lifespan regulation: A central role for the vitellogenin gene family? *BioEssays* 27:339–346.

Brisson, D., K. Ledoux, Y. Bosse, J. St-Pierre, P. Julien, P. Perron, T. J. Hudson, M. C. Vohl, and D. Gaudet. 2002. Effect of apolipoprotein E, peroxisome proliferator-activated receptor alpha and lipoprotein lipase gene mutations on the ability of fenofibrate to improve lipid profiles and reach clinical guideline targets among hypertriglyceridemic patients. *Pharmacogenetics* 12:313–320.

Brodie, E. D., III. 2000. Why evolutionary genetics does not always add up. In J. B. Wolf, I. E. D. Brodie, and M. J. Wade, Eds. *Epistasis and the Evolutionary Process*. Oxford University Press, Oxford, pp. 3–19.

Buri, P. 1956. Gene frequency in small populations of mutant *Drosophila*. *Evolution* 10:367–402.

Cai, L. Q., Y. S. Zhu, M. D. Katz, C. Herrera, J. Baez, M. DeFillo-Ricart, C. H. Shackleton, and J. Imperato-McGinley. 1996. 5-Alpha-reductase-2 gene mutations in the Dominican Republic. *Journal of Clinical Endocrinology and Metabolism* 81:1730–1735.

Cane, M. A. 1983. Oceanographic events during El Nino. *Science* 222:1189–1195.

Cann, R. L., M. Stoneking, and A. C. Wilson. 1987. Mitochondrial DNA and human evolution. *Nature* 325:31–36.

Carey, N., K. Johnson, P. Nokelainen, L. Peltonen, M. L. Savontaus, V. Juvonen, M. Anvret, U. Grandell, K. Chotai, E. Robertson, H. Middletonprice, and S. Malcolm. 1995. Meiotic drive and myotonic dystrophy—Reply. *Nature Genetics* 10:133.

Carson, H. L. 1967. Selection for parthenogenesis in *Drosophila mercatorum*. *Genetics* 55:157–171.

Carson, H. L. 1968. The population flush and its genetic consequences. In R. C. Lewontin, Ed. *Population Biology and Evolution*. Syracuse University Press, Syracuse, NY, pp. 123–127.

Carson, H. L., and A. R. Templeton. 1984. Genetic revolutions in relation to speciation phenomena: The founding of new populations. *Annual Review of Ecology and Systematics* 15:97–131.

Carvalho-Silva, B. R., F. R. Santos, J. Rocha, and S. D. J. Pena. 2001. The phylogeography of Brazilian Y-chromosome lineages, *American Journal of Human Genetics* 68:281–286.

Castelloe, J., and A. R. Templeton. 1994. Root probabilities for intraspecific gene trees under neutral coalescent theory. *Molecular Phylogenetics and Evolution* 3:102–113.

Cavalieri, D., J. P. Townsend, and D. L. Hartl. 2000. Manifold anomalies in gene expression in a vineyard isolate of *Saccharomyces cerevisiae* revealed by DNA microarray analysis. *Proceedings of the National Academy of Sciences USA* 97:12369–12374.

Cavalli-Sforza, L., P. Menozzi, and A. Piazza. 1994. *The History and Geography of Human Genes*. Princeton University Press, Princeton, NJ.

Cavalli-Sforza, L. L., and W. F. Bodmer. 1971. *The Genetics of Human Populations*. W. H. Freeman, San Francisco.

Chadwick, O. A., R. T. Gavenda, E. F. Kelly, K. Ziegler, C. G. Olson, W. C. Elliott, and D. M. Hendricks. 2003. The impact of climate on the biogeochemical functioning of volcanic soils. *Chemical Geology* 202:195–223.

Chagnon, N. A., J. V. Neel, L. Weitkamp, H. Gershowitz, and M. Ayres. 1970. The influence of cultural factors on the demography and pattern of gene flow from the Makiritare to the Yanomama Indians. *American Journal of Physical Anthropology* 32:339–350.

Chartier-Harlin, M., M. Parfitt, S. Legrain, J. Pérez-Tur, T. Brousseau, A. Evans, C. Berr, O. Vidal, P. Roques, V. Gourlet, J. Fruchart, A. Delacourte, M. Rossor, and P. Amouyel. 1994. Apolipoprotein E, $\varepsilon 4$ allele as a major risk factor for sporadic early and late-onset forms of Alzheimer's disease: Analysis of the 19q13.2 chromosomal region. *Human Molecular Genetics* 3:569–574.

Chasman, D., and R. M. Adams. 2001. Predicting the functional consequences of non-synonymous single nucleotide polymorphisms: Structure-based assessment of amino acid variation. *Journal of Molecular Biology* 307:683–706.

Cheng, D., R. Huang, I. S. Lanham, H. M. Cathcart, M. Howard, E. H. Corder, and S. E. Poduslo. 2005. Functional interaction between APOE4 and LDL receptor isoforms in Alzheimer's disease. *Journal of Medical Genetics* 42:129–131.

Chesser, R. K., M. H. Smith, and J. I. L. Brisbin. 1980. Management and maintenance of genetic variability in endangered species. *International Zoo Yearbook* 20:146–154.

Chetverikov, S. S. 1926. On certain aspects of the evolutionary process from the standpoint of modern genetics. *Zhurnal Eksperimental' noi Biologii* A2:3–54.

Cheverud, J. M. 2001. A simple correction for multiple comparisons in interval mapping genome scans. *Heredity* 87:52–58.

Cheverud, J. M., and E. J. Routman. 1995. Epistasis and its contribution to genetic variance components. *Genetics* 139:1455–1461.

Cheverud, J. M., and E. J. Routman. 1996. Epistasis as a source of increased additive genetic variance at population bottlenecks. *Evolution* 50:1042–1051.

Cheverud, J. M., E. J. Routman, F. A. M. Duarte, B. Vanswinderen, K. Cothran, and C. Perel. 1996. Quantitative trait loci for murine growth. *Genetics* 142:1305–1319.

Cheverud, J. M., T. T. Vaughn, L. S. Pletscher, K. King-Ellison, J. Bailiff, E. Adams, C. Erickson, and A. Bonislawski. 1999. Epistasis and the evolution of additive genetic variance in populations that pass through a bottleneck. *Evolution* 53:1009–1018.

Chong, S. S., E. Almqvist, H. Telenius, L. Latray, K. Nichol, B. Bourdelatparks, Y. P. Goldberg, B. R. Haddad, F. Richards, D. Sillence, C. R. Greenberg, E. Ives, G. Vandenengh, M. R. Hughes, and M. R. Hayden. 1997. Contribution of DNA sequence and CAG size to mutation frequencies of intermediate alleles for Huntington disease: Evidence from single sperm analyses. *Human Molecular Genetics* 6:301–309.

Christiansen, R. B. 1975. Hard and soft selection in a subdivided population. *American Naturalist* 109:11–16.

Clark, A. G., S. Glanowski, R. Nielsen, P. D. Thomas, A. Kejariwal, M. A. Todd, D. M. Tanenbaum, D. Civello, F. Lu, B. Murphy, S. Ferriera, G. Wang, X. G. Zheng, T. J. White, J. J. Sninsky, M. D. Adams, and M. Cargill. 2003. Inferring nonneutral evolution from human-chimp-mouse orthologous gene trios. *Science* 302:1960–1963.

Clark, A. G., K. M. Weiss, D. A. Nickerson, S. L. Taylor, A. Buchanan, J. Stengard, V. Salomaa, E. Vartianinen, M. Perola, E. Boerwinkle, and C. F. Sing. 1998. Haplotype structure and population genetic inferences from nucleotide sequence variation in human lipoprotein lipase. *American Journal of Human Genetics* 63:595–612.

Clausen, J., D. D. Keck, and W. W. Hiesey. 1958. *Experimental Studies on the Nature of Species*, Vol. 3: *Environmental Responses of Climatic Races of Achillea*. Publ. No. 581, Carnegie Institute of Washington, Washington, D.C., 1–129.

Clayton, G. A., J. A. Morris, and A. Robertson. 1957. An experimental check on quantitative genetical theory. I. Short-term responses to selection. *Journal of Genetics* 55:131–151.

Cook, L. M. 2003. The rise and fall of the Carbonaria form of the peppered moth. *Quarterly Review of Biology* 78:399–417.

Cockerham, C. C., P. M. Burrows, S. S. Young, and T. Prout. 1972. Frequency-dependent selection in randomly mating populations. *American Naturalist* 106:493–515.

Cockerham, C. C., and B. S. Weir. 1987. Correlations, descent measures: Drift with migration and mutation. *Proceedings of the National Academy of Sciences USA* 84:8512–8514.

Comeron, J. M., and M. Kreitman. 2000. The correlation between intron length and recombination in *Drosophila*: Dynamic equilibrium between mutational and selective forces. *Genetics* 156:1175–1190.

Conway, D. J., D. R. Cavanagh, K. Tanabe, C. Roper, Z. S. Mikes1, N. Sakihama, K. A. Bojang, A. M. J. Oduola, P. G. Kremsner, D. E. Arnot, B. M. Greenwood, and J. S. McBride. 2000. A principal target of human immunity to malaria identified by molecular population genetic and immunological analyses. *Nature Medicine* 6:689–692.

Cooper, D. N. 1999. *Human Gene Evolution*. BIOS Scientific Publishers, Oxford.

Copin, B., A. P. Brézin, F. Valtot, J.-C. Dascotte, A. Béchetoille, and H.-J. Garchon. 2002. Apolipoprotein E promoter single-nucleotide polymorphisms affect the phenotype of primary open-angle glaucoma and demonstrate interaction with the myocilin gene. *American Journal of Human Genetics* 70:1575–1581.

Corbo, R. M., and R. Scacchi. 1999. Apolipoprotein E (APOE) allele distribution in the world. Is APOE*4 a "thrifty" allele? *Annals of Human Genetics* 63:301–310.

Corder, E. H., K. Robertson, L. Lannfelt, N. Bogdanovic, G. Eggertsen, J. Wilkins, and C. Hall. 1998. HIV-infected subjects with the $\epsilon 4$ allele for ApoE have excess dementia and peripheral neuropathy. *Nature Medicine* 4:1182–1184.

Cornelis, F., S. Faure, M. Martinez, J. F. Prudhomme, P. Fritz, C. Dib, H. Alves, P. Barrera, N. Devries, A. Balsa, D. Pascualsalcedo, K. Maenaut, R. Westhovens, P. Migliorini, T. H. Tran, A. Delaye, N. Prince, C. Lefevre, G. Thomas, M. Poirier, S. Soubigou, O. Alibert, S. Lasbleiz, S. Fouix, C. Bouchier, et al. 1998. New susceptibility locus for rheumatoid arthritis suggested by a genome-wide linkage study. *Proceedings of the National Academy of Sciences USA* 95:10746–10750.

Cornuet, J. M., and G. Luikart. 1996. Description and power analysis of two tests for detecting recent population bottlenecks from allele frequency data. *Genetics* 144:2001–2014.

Coveney, P., and R. Highfield. 1995. *Frontiers of Complexity*. Fawcett Columbine, New York.

Cox, N. J. 2001. Challenges in identifying genetic variation affecting susceptibility to type 2 diabetes: Examples from studies of the calpain-10 gene [Review]. *Human Molecular Genetics* 10:2301–2305.

Coyne, J. A., N. H. Barton, and M. Turelli. 2000. Is Wright's shifting balance process important in evolution? *Evolution* 54:306–317.

Crandall, K. A., and A. R. Templeton. 1993. Empirical tests of some predictions from coalescent theory with applications to intraspecific phylogeny reconstruction. *Genetics* 134:959–969.

Crandall, K. A., and A. R. Templeton. 1999. Statistical approaches to detecting recombination. In K. A. Crandall, Ed. *The Evolution of HIV*. Johns Hopkins University Press, Baltimore, MD, pp. 153–176.

Crawford, D. C., T. Bhangale, N. Li, G. Hellenthal, M. J. Rieder, D. A. Nickerson, and M. Stephens. 2004. Evidence for substantial fine-scale variation in recombination rates across the human genome. *Nature Genetics* 36:700–706.

Crow, J. F. 1957. Genetics of DDT resistance in *Drosophila*. *Proceedings of the International Genetics Symposia*, 1956:408–409.

Crow, J. F., and M. Kimura. 1970. *An Introduction to Population Genetic Theory*. Harper & Row, New York.

Crow, J. F., and T. Maruyama. 1971. The number of neutral alleles maintained in a finite, geographically structured population. *Theoretical Population Biology* 2:437–453.

Crow, J. F., and T. Nagylaki. 1976. The rate of change of a character correlated with fitness. *American Naturalist* 110:207–213.

Crow, T. J. 1997. Current status of linkage for schizophrenia—Polygenes of vanishingly small effect or multiple false positives. *American Journal of Medical Genetics* 74:99–103.

Curtsinger, J. W. 1984. Evolutionary landscapes for complex selection. *Evolution* 38:359–367.

Czeizel, A. 1989. Application of DNA analysis in diagnosis and control of human diseases. *Biologisches Zentralblatt* 108:295–301.

Davis, S. K., J. E. Strassmann, C. Hughes, L. S. Pletscher, and A. R. Templeton. 1990. Population structure and kinship in *Polistes* (Hymenoptera, Vespidae): An analysis using ribosomal DNA and protein electrophoresis. *Evolution* 44:1242–1253.

Dayhoff, M. O. 1969. *Atlas of Protein Sequence and Stucture*. National Biomedical Research Foundation, Silver Springs, MD.

de Jong, G. 1988. Consequences of a model of counter-gradient selection. In G. d. Jong, Ed. *Population Genetics and Evolution*. Springer-Verlag, Berlin, pp. 264–277.

de la Fuente, A., P. Brazhnik, and P. Mendes. 2002. Linking the genes: Inferring quantitative gene networks from microarray data. *Trends in Genetics* 18:395–398.

Demetrius, L. 1975. Natural selection and age-structured populations. *Genetics* 79:535–544.

Demetrius, L. 1985. The units of selection and measures of fitness. *Proceedings of the Royal Society of London B* 225:147–159.

DeSalle, R., J. Slightom, and E. Zimmer. 1986. The molecular through ecological genetics of abnormal abdomen. II. Ribosomal DNA polymorphism is associated with the abnormal abdomen syndrome in *Drosophila mercatorum*. *Genetics* 112:861–875.

DeSalle, R., and A. R. Templeton. 1986. The molecular through ecological genetics of abnormal abdomen in *Drosophila mercatorum*. III. Tissue-specific differential replication of ribosomal genes modulates the abnormal abdomen phenotype in *Drosophila mercatorum*. *Genetics* 112:877–886.

DeSalle, R., A. Templeton, I. Mori, S. Pletscher, and J. S. Johnston. 1987. Temporal and spatial heterogeneity of mtDNA polymorphisms in natural populations of *Drosophila mercatorum*. *Genetics* 116:215–223.

Devlin, B., M. Daniels, and K. Roeder. 1997. The heritability of IQ. *Nature* 388:468–471.

Devlin, B., K. Roeder, and L. Wasserman. 2001. Genomic control, a new approach to genetic-based association studies. *Theoretical Population Biology* 60:155–166.

Diamond, J. 2003. The double puzzle of diabetes. *Nature* 423:599–602.

Diamond, J. M. 1992. Diabetes running wild. *Nature* 357:362–363.

Dickerson, R. E. 1971. The structure of cytochrome c and the rates of molecular evolution. *Journal of Molecular Evolution* 1:26–45.

Dobzhansky, T. 1962. *Mankind Evolving*. Yale University Press, New Haven, CT.

Dover, G. 1982. Molecular drive: A cohesive mode of species evolution. *Nature* 299:111–116.

Drasinover, V., S. Ehrlich, N. Magal, E. Taub, V. Libman, T. Shohat, G. J. Halpern, and M. Shohat. 2000. Increased transmission of intermediate alleles of the FMR1 gene compared with normal alleles among female heterozygotes. *American Journal of Medical Genetics* 93:155–157.

Dressler, W. W. 1996. Hypertension in the African American community – social, cultural, and psychological factors. *Seminars in Nephrology* 16:71–82.

Drysdale, C. M., D. W. McGraw, C. B. Stack, J. C. Stephens, R. S. Judson, K. Nandabalan, K. Arnold, G. Ruano, and S. B. Liggett. 2000. Complex promoter and coding region beta 2-adrenergic receptor haplotypes alter receptor expression and predict in vivo responsiveness. *Proceedings of the National Academy of Sciences USA* 97:10483–10488.

Dubrovsky, E. B., V. A. Dubrovskaya, and E. M. Berger. 2002. Juvenile hormone signaling during oogenesis in *Drosophila melanogaster*. *Insect Biochemistry and Molecular Biology* 32:1555–1565.

Ehrman, L. 1966. Mating success and genotype frequency in *Drosophila*. *Animal Behavior* 14:332–339.

Ehrman, L. 1967. Further studies on genotype frequency and mating success in *Drosophila*. *American Naturalist* 101:415–424.

Eisenbarth, I., G. Vogel, W. Krone, W. Vogel, and G. Assum. 2000. An isochore transition in the *nf1* gene region coincides with a switch in the extent of linkage disequilibrium. *American Journal of Human Genetics* 67:873–880.

Eldredge, N., and S. J. Gould. 1972. Punctuated equilibria: An alternative to phyletic gradualism. In T. J. M. Schopf, Ed. *Models in Paleobiology*. Freeman, Cooper & Co., San Francisco, pp. 82–115.

el-Hazmi, M. A., H. M. Bahakim, and A. S. Warsy. 1992. DNA polymorphism in the beta-globin gene cluster in Saudi Arabs: Relation to severity of sickle cell anaemia. *Acta Haematologica* 88:61–66.

el-Hazmi, M. A., and A. S. Warsy. 1996. Frequency of sickle cell gene in Arabia. *Gene Geography* 10:87–91.

el-Hazmi, M. A., A. S. Warsy, M. H. Addar, and Z. Babae. 1994a. Fetal haemoglobin level—effect of gender, age and haemoglobin disorders. *Molecular and Cellular Biochemistry* 135:181–186.

el-Hazmi, M. A., A. S. Warsy, A. R. al-Swailem, F. Z. al-Faleh, and F. A. al-Jabbar. 1994b. Genetic compounds—Hb S, thalassaemias and enzymopathies: Spectrum of interactions. *Journal of Tropical Pediatrics* 40:149–156.

Eller, E. 1999. Population substructure and isolation by distance in three continental regions. *American Journal of Physical Anthropology* 108:147–159.

Eller, E. 2001. Estimating relative population sizes from simulated data sets and the question of greater African effective size. *American Journal of Physical Anthropology* 116:1–12.

El-Maarri, O., A. Olek, B. Balaban, M. Montag, H. v. d. Ven, B. Urman, K. Olek, S. H. Caglayan, J. Walter, and J. Oldenburg. 1998. Methylation levels at selected CpG sites in the factor VIII and FGFR3 genes, in mature female and male germ cells: Implications for male-driven evolution. *American Journal of Human Genetics* 63:1001–1009.

Endler, J. A. 1977. *Geographic Variation, Speciation, and Clines*. Princeton University Press, Princeton, NJ.

Excoffier, L., P. E. Smouse, and J. M. Quattro. 1992. Analysis of molecular variance inferred from metric distances among DNA haplotypes: Application to Human Mitochondrial-DNA Restriction Data. *Genetics* 131:479–491.

Falconer, D. S., and T. F. C. Mackay. 1996. *Introduction to Quantitative Genetics*. Addison-Wesley Longman Limited, Essex, England.

Falush, D., E. W. Almqvist, R. R. Brinkmann, Y. Iwasa, and M. R. Hayden. 2001. Measurement of mutational flow implies both a high new-mutation rate for Huntington Disease and substantial underascertainment of late-onset cases. *American Journal of Human Genetics* 68:373–385.

Fay, J. C., and C. I. Wu. 2000. Hitchhiking under positive Darwinian selection. *Genetics* 155:1405–1413.

Fay, J. C., G. J. Wyckoff, and C. I. Wu. 2001. Positive and negative selection on the human genome. *Genetics* 158:1227–1234.

Ferea, T. L., D. Botstein, P. O. Brown, and R. F. Rosenzweig. 1999. Systematic changes in gene expression patterns following adaptive evolution in yeast. *Proceedings of the National Academy of Sciences USA* 96:9721–6.

Filippi, G., A. Rinaldi, R. Palmarino, E. Seravalli, and M. Siniscalco. 1977. Linkage disequilibrium for two X-linked genes in *Sardinia* and its bearing on the statistical mapping of the human X chromosome. *Genetics* 86:199–212.

Filosa, S., V. Calabro, G. Lania, T. J. Vulliamy, C. Brancati, A. Tagarelli, L. Luzzatto, and G. Martini. 1993. G6PD haplotypes spanning Xq28 from F8C to red/green color vision. *Genomics* 17:6–14.

Fisher, R. A. 1918. The correlation between relatives on the supposition of Mendelian inheritance. *Transactions of the Royal Society of Edinburgh* 52:399–433.

Fisher, R. A. 1930. *The Genetical Theory of Natural Selection.* Clarendon, Oxford.

Flory, J. D., S. B. Manuck, R. E. Ferrell, C. M. Ryan, and M. F. Muldoon. 2000. Memory performance and the apolipoprotein E polymorphism in a community sample of middle-aged adults. *American Journal of Medical Genetics* 96:707–711.

Fortin, A., M. M. Stevenson, and P. Gros. 2002. Susceptibility to malaria as a complex trait: Big pressure from a tiny creature. *Human Molecular Genetics* 11:2469–2478.

Franco, M. H. L. P., T. A. Weimar, and F. M. Salzano. 1982. Blood polymorphisms and racial admixture in two Brazilian populations. *American Journal of Physical Anthropology* 58:127–132.

Frankel, W. N., and N. J. Schork. 1996. Whos afraid of epistasis. *Nature Genetics* 14:371–373.

Friedman, J. M. 2003. A war on obesity, not the obese. *Science* 299:856–858.

Friedman, M. J., and W. Trager. 1981. The biochemistry of resistance to malaria. *Scientific American* 244:154–155, 158–164.

Frigerio, R., G. Sole, M. Lovicu, and G. Passiu. 1994. Molecular and biochemical data on some glucose-6-phosphate dehydrogenase variants from southern Sardinia. *Haematologica* 79:319–321.

Fullerton, S. M., A. Bartoszewicz, G. Ybazeta, Y. Horikawa, G. I. Bell, K. K. Kidd, N. J. Cox, R. R. Hudson, and A. D. Rienzo. 2002. Geographic and haplotype structure of candidate type 2 diabetes—susceptibility variants at the Calpain-10 locus. *American Journal of Human Genetics* 70:1096–1106.

Fullerton, S. M., A. Bernardo Carvalho, and A. G. Clark. 2001. Local rates of recombination are positively correlated with GC content in the human genome. *Molecular Biology Evolution* 18:1139–1142.

Fullerton, S. M., A. G. Clark, K. M. Weiss, D. A. Nickerson, S. L. Taylor, J. H. Stengård, V. Salomaa, E. Vartiainen, M. Perola, E. Boerwinkle, and C. F. Sing. 2000. Apolipoprotein E variation at the sequence haplotype level: Implications for the origin and maintenance of a major human polymorphism. *American Journal of Human Genetics* 67:881–900.

Futuyma, D. J. 1986. *Evolutionary Biology.* Sinauer Associates, Sunderland, MA.

Gabunia, L., A. Vekua, D. Lordkipanidze, C. C. Swisher, R. Ferring, A. Justus, M. Nioradze, M. Tvalchrelidze, S. C. Anton, G. Bosinski, O. Joris, M. A. de Lumley, G. Majsuradze, and A. Mouskhelishvili. 2000. Earliest Pleistocene hominid cranial remains from Dmanisi, Republic of Georgia: Taxonomy, geological setting, and age. *Science* 288:1019–1025.

Gavrilets, S., and A. Hastings. 1995. Intermittency and transient chaos from simple frequency-dependent selection. *Proceedings of the Royal Society of London B* 261:233–238.

Georgiadis, N., L. Bischof, A. Templeton, J. Patton, W. Karesh, and D. Western. 1994. Structure and history of African elephant populations: I. Eastern and Southern Africa. *Journal of Heredity* 85:100–104.

Gerdes, L. U., C. Gerdes, P. S. Hansen, I. C. Klausen, and O. Faergeman. 1996. Are men carrying the apolipoprotein epsilon-4- or epsilon-2 allele less fertile than epsilon-3-epsilon-3 genotypes. *Human Genetics* 98:239–242.

Ghai, G. L. 1973. Limiting distribution under assortative mating. *Genetics* 75:727–732.

Giblett, E. R. 1969. *Genetic Markers in Human Blood.* F. A. Davis, Philadelphia, PA.

Gibson, G. 1996. Epistasis and pleiotropy as natural properties of transcriptional regulation. *Theoretical Population Biology* 49:58–89.

Gilbert, L. E. 1983. Coevolution and mimicry. In D. Futuyma and M. Slatkin, Eds. *Coevolution.* Sinauer, Sunderland, MA, pp. 263–281.

Gilbert, S. F. 2000. *Developmental Biology.* Sinauer Associates, Sunderland, MA.

Gloria-Bottini, F., G. Gerlini, N. Lucarini, P. Borgiani, M. C. Gori, A. Amante, and E. Bottini. 1989a. Enzyme polymorphism and clinical variability of diseases: A study of diabetes mellitus. *Human Biology* 61:571–89.

Gloria-Bottini, F., P. Lucarelli, A. Amante, N. Lucarini, G. Finocchi, and E. Bottini. 1989b. Interaction at clinical level between erythrocyte acid phosphatase and adenosine deaminase genetic polymorphisms. *Human Genetics* 82:213–215.

Goodnight, C. J. 1995. Epistasis and the increase in additive genetic variance: Implications for phase 1 of Wright's shifting-balance process. *Evolution* 49:502–511.

Goodnight, C. J. 2000. Quantitative trait loci and gene interaction: The quantitative genetics of metapopulations. *Heredity* 84:587–598.

Goodnight, C. J., and M. J. Wade. 2000. The ongoing synthesis: A reply to Coyne, Barton, and Turelli. *Evolution* 54:317–324.

Gould, S. J. 1977. *Ontogeny and Phylogeny*. Belknap, Cambridge, MA.

Grant, B. R., and P. R. Grant. 1993. Evolution of Darwin's finches caused by a rare climatic event. *Proceedings of the Royal Society of London B* 251:111–117.

Grant, P. R. 1986. *Ecology and Evolution of Darwin's Finches*. Princeton University Press, Princeton, NJ.

Grant, P. R., and B. R. Grant. 2002. Unpredictable evolution in a 30-year study of Darwin's finches. *Science* 296:707–711.

Gray, D. A., and W. H. Cade. 1999. Sex, death and genetic variation: Natural and sexual selection on cricket song. *Proceedings of the Royal Society of London B* 266:707–709.

Green, G. E., D. A. Scott, J. M. McDonald, G. G. Woodworth, V. C. Sheffield, and R. J. Smith. 1999. Carrier rates in the midwestern United States for GJB2 mutations causing inherited deafness. *Journal of the American Medical Association* 281:2211–2216.

Greene, L. S., N. Bottini, P. Borgiani, and F. Gloria-Bottini. 2000. Acid phosphatase locus 1 (ACP1): Possible relationship of allelic variation to body size and human population adaptation to thermal stress—A theoretical perspective. *American Journal of Human Biology* 12:688–701.

Gusella, J. F., N. S. Wexler, P. M. Conneally, S. L. Naylor, M. A. Anderson, R. E. Tanzi, P. C. Watkins, K. Ottina, M. R. Wallace, A. Y. Sakaguchi, A. B. Young, I. Shoulson, E. Bonilla, and J. B. Martin. 1983. A polymorphic DNA marker genetically linked to Huntington's disease. *Nature* 306:234–238.

Guss, K. A., C. E. Nelson, A. Hudson, M. E. Kraus, and S. B. Carroll. 2001. Control of a genetic regulatory network by a selector gene. *Science* 292:1164–1167.

Hadany, L. 2003. Adaptive peak shifts in a heterogenous environment. *Theoretical Population Biology* 63:41–51.

Haile-Selassie, Y. 2001. Late Miocene hominids from the Middle Awash, Ethiopia. *Nature* 412:178–181.

Haldane, J. B. S. 1949. Disease and evolution. *Ricerca Scientifica* 19:3–10.

Haldane, J. B. S., and S. D. Jayakar. 1963. Polymorphism due to selection of varying direction. *Journal of Genetics*. 58:237–242.

Hallman, D. M., E. Boerwinkle, N. Saha, S. Sandholzer, H. J. Menzel, A. Csazur, and G. Utermann. 1991. The apolipoprotein E polymorphism: A comparison of allele frequencies and effects in nine populations. *American Journal of Human Genetics* 49:338–349.

Hallman, D. M., S. Visvikis, J. Steinmetz, and E. Boerwinkle. 1994. The effect of variation in the apolipoprotein B gene on plasmid lipid and apolipoprotein B levels. I. A likelihood-based approach to cladistic analysis. *Annals of Human Genetics* 58:35–64.

Hamada, H., M. G. Petrino, T. Kakunaga, M. Seidman, and B. D. Stollar. 1984. Characterization of genomic poly (dT-dG) poly (dC-dA) sequences: Structure, organization, and conformation. *Molecular and Cell Biology* 4:2610–2621.

Hamilton, W. D. 1964. The genetical evolution of social behavior, I and II. *Journal of Theoretical Biology* 7:1–52.

Hammer, M. F., T. Karafet, A. Rasanayagam, E. T. Wood, T. K. Altheide, T. Jenkins, R. C. Griffiths, A. R. Templeton, and S. L. Zegura. 1998. Out of Africa and back again: Nested cladistic analysis of human Y chromosome variation. *Molecular Biology and Evolution* 15:427–441.

Hardy, G. H. 1908. Mendelian proportions in a mixed population. *Science* 28:49–50.

Harris, H. 1966. Enzyme polymorphisms in man. *Proceedings of the Royal Society of London B* 164:298–310.

Harrison, R. G., and A. T. Vawter. 1977. Allozyme differentiation between pheromone strains of the European corn borer, *Ostrinia nubilasis*. *Annals of the Entomological Society of America* 70:717–727.

Haviland, M. B., A. M. Kessling, J. Davignon, and C. F. Sing. 1991. Estimation of Hardy-Weinberg and pairwise disequilibrium in the apolipoprotein AI-CIII-AIV gene cluster [published erratum appears in *American Journal of Human Genetics* 51(6):1457 (1992)]. *American Journal of Human Genetics* 49:350–365.

Haviland, M. B., A. M. Kessling, J. Davignon, and C. F. Sing. 1995. Cladistic analysis of the apolipoprotein AI-CIII-AIV gene cluster using a healthy French Canadian sample. I. Haploid analysis. *Annals of Human Genetics* 59:211–231.

Hegele, R. A., J. Wang, S. B. Harris, J. H. Brunt, T. K. Young, A. J. G. Hanley, B. Zinman, P. W. Connelly, and C. M. Anderson. 2001. Variable association between genetic variation in the CYP7 gene promoter and plasma lipoproteins in three Canadian populations. *Atherosclerosis* 154:579–587.

Herrnstein, R. J., and C. A. Murray. 1994. *The Bell Curve: Intelligence and Class Structure in American Life*. Free Press, New York.

Hillis, D. M. 1984. Misuse and modification of Nei's genetic distance. *Systematic Zoology* 33:238–240.

Hoekstra, R. F. 1975. A deterministic model of cyclical selection. *Genetical Research* 25:1–15.

Hollocher, H., and A. R. Templeton. 1994. The molecular through ecological genetics of abnormal abdomen in *Drosophila mercatorum* VI. The nonneutrality of the Y-chromosome rDNA polymorphism. *Genetics* 136:1373–1384.

Hollocher, H., A. R. Templeton, R. DeSalle, and J. S. Johnston. 1992. The molecular through ecological genetics of *abnormal abdomen*. IV. Components of genetic-variation in a natural-population of *Drosophila mercatorum*. *Genetics* 130:355–366.

Holsinger, K. E. 1991. Inbreeding depression and the evolution of plant mating systems. *Trends in Evolution and Ecology* 6:307–308.

Hooper, J. 2002. *Of Moths and Men: An Evolutionary Tale*. Norton, New York, London.

Huang, W., Y.-P. Sher, K. Peck, and Y. C. B. Fung. 2002. Matching gene activity with physiological functions. *Proceedings of the National Academy of Sciences USA* 99:2603–2608.

Hudson, R. R. 1993. Levels of DNA polymorphism and divergence yield important insights into evolutionary processes. *Proceedings of the National Academy of Sciences USA* 90:7425–7426.

Hudson, R. R., D. D. Boos, and N. L. Kaplan. 1992. A statistical test for detecting geographical subdivision. *Molecular Biology and Evolution* 9:138–151.

Hudson, R. R., M. Kreitman, and M. Aguade. 1987. A test of neutral molecular evolution based on nucleotide data. *Genetics* 116:153–159.

Husband, B. C., and S. C. Barrett. 1992. Effective population-size and genetic drift in tristylous eichhornia-paniculata (pontederiaceae). *Evolution* 46:1875–1890.

Husband, B. C., and S. C. H. Barrett. 1995. Estimating efective population size—A reply to Nunney. *Evolution* 49:392–394.

Hutchison, D. W. 2003. Testing the central/peripheral model: Analyses of microsatellite variability in the eastern collared lizard (*Crotaphytus collaris collaris*). *American Midland Naturalist* 149:148–162.

Hutchison, D. W., and A. R. Templeton. 1999. Correlation of pairwise genetic and geographic distance measures: Inferring the relative influences of gene flow and drift on the distribution of genetic variability. *Evolution* 53:1898–1914.

Imperato-McGinley, J., L. Guerrero, T. Gautier, and R. Peterson. 1974. Steroid 5 α-reductase deficiency in man: An inherited form of male pseudohermaphroditism. *Science* 186:1213–1215.

Innan, H., K. Zhang, P. Marjoram, S. Tavare, and N. A. Rosenberg. 2005. Statistical tests of the coalescent model based on the haplotype frequency distribution and the number of segregating sites. *Genetics* 169:1763–1777.

Ioerger, T. R., A. G. Clark, and T.-H. Kao. 1991. Polymorphism at the self-incompatibility locus in Solonaceae predates speciation. *Proceedings of the National Academy of Sciences USA* 87:9732–9735.

Jacquard, A. 1970. *Structures Génétique des Populations*. Masson & Cie, Paris.

Jacquard, A. 1974. *The Genetic Structure of Populations*. Springer-Verlag, New York.

Jacquard, A. 1975. Inbreeding: One word, several meanings. *Theoretical Population Biology* 7:338–363.

Jain, S. K., and R. W. Allard. 1966. The effects of linkage, epistasis, and inbreeding on population changes under selection. *Genetics* 53:633–659.

Jannink, J. L., and R. Jansen. 2001. Mapping epistatic quantitative trait loci with one-dimensional genome searches. *Genetics* 157:445–454.

Jeffreys, A. J., V. Wilson, and S. L. Thein. 1985. Individual-specific "fingerprints" of human DNA. *Nature* 316:76–79.

Jehle, R., J. W. Arntzen, T. Burke, A. P. Krupa, and W. Hödl. 2001. The annual number of breeding adults and the effective population size of syntopic newts (*Triturus cristatus, T. marmoratus*). *Molecular Ecology* 10:839–850.

Johnson, F. M., C. G. Kanapi, R. H. Richardson, M. R. Wheeler, and W. S. Stone. 1966. An operational classification of *Drosophila* esterases for species comparisons. In M. R. Wheeler, Ed. *Studies in Genetics* Vol. III. Morgan Centennial Issue. University of Texas, Austin, pp. 517–532.

Johnston, J. S., and J. R. Ellison. 1982. Exact age determination in laboratory and field-caught *Drosophila*. *Journal of Insect Physiology* 28:773–780.

Johnston, J. S., and A. R. Templeton. 1982. Dispersal and clines in *Opuntia* breeding *Drosophila mercatorum* and *D. hydei* at Kamuela, Hawaii. In J. S. F. Barker and W. T. Starmer, Eds. *Ecological Genetics and Evolution: The Cactus-Yeast-Drosophila Model System*. Academic, Sydney, pp. 241–256.

Jones, P. A., W. M. d. Rideout, J. C. Shen, C. H. Spruck, and Y. C. Tsai. 1992. Methylation, mutation and cancer. *Bioessays* 14:33–36.

Jorde, P. E., and N. Ryman. 1995. Temporal allele frequency change and estimation of effective size in populations with overlapping generations. *Genetics* 139:1077–1090.

Jukes, T. H., and C. R. Cantor. 1969. Evolution in protein molecules. In H. N. Munro, Ed. *Mammalian Protein Metabolism*. Academic, New York, pp. 21–123.

Kamboh, M. I., D. K. Sanghera, R. E. Ferrell, and S. T. Dekosky. 1995. APOEε-4 associated Alzheimer's disease risk is modified by alpha-1-antichymotrypsin polymorphism. *Nature Genetics* 10:486–488.

Karlin, S. 1969. *Equilibrium Behavior of Population Genetic Models with Non-Random Mating*. Gordon and Breach Science Publishers, New York.

Kehoe, P., M. Krawczak, P. S. Harper, M. J. Owen, and A. L. Jones. 1999. Age of onset in Huntington disease: Sex specific influence of apolipoprotein E genotype and normal CAG repeat length. *Journal of Medical Genetics* 36:108–111.

Kelly, J. K. 2000. Epistasis, linkage, and balancing selection. In J. B. Wolf, I. E. D. Brodie, and M. J. Wade, Eds. *Epistasis and the Evolutionary Process*. Oxford University Press, Oxford, pp. 146–157.

Kettlewell, H. B. D. 1955. Selection experiments on industrial melanism in the Lepidoptera. *Heredity* 9:323–342.

Kettlewell, H. B. D. 1973. *The Evolution of Melanism*. Clarendon, Oxford.

Khlebodarova, T. M., N. E. Gruntenko, L. G. Grenback, M. Z. Sukhanova, M. M. Mazurov, I. Y. Rauschenbach, B. A. Tomas, and B. D. Hammock. 1996. A comparative analysis of juvenile hormone metabolizing enzymes in two species of *Drosophila* during development. *Insect Biochemistry and Molecular Biology* 26:829–835.

Kidwell, M. G., and D. R. Lisch. 2000. Transposable elements and host genome evolution. *Trends in Evolution and Ecology* 15:95–99.

Kimura, M. 1968a. Evolutionary rate at the molecular level. *Nature* 217:624–626.

Kimura, M. 1968b. Genetic variability maintained in a finite population due to mutational production of neutral and nearly neutral isoalleles. *Genetical Research* 11:247–269.

Kimura, M. 1970. The length of time required for a selectively neutral mutant to reach fixation through random frequency drift in a finite population. *Genetical Research* 15:131–133.

Kimura, M. 1979. Model of effectively neutral mutations in which selective constraint is incorporated. *Proceedings of the National Academy of Sciences USA* 76:3440–3444.

King, J. L., and T. H. Jukes. 1969. Non-Darwinian evolution: Random fixation of selectively neutral mutations. *Science* 164:788–798.

Kingman, J. F. C. 1982a. The coalescent. *Stochastic Processes and their Applications* 13:235–248.

Kingman, J. F. C. 1982b. On the genealogy of large populations. *Journal of Applied Probability* 19A:27–43.

Kitada, S., T. Hayashi, and H. Kishino. 2000. Empirical bayes procedure for estimating genetic distance between populations and effective population size. *Genetics* 156:2063–2079.

Klaver, C. C. W., M. Kliffen, C. M. Vanduijn, A. Hofman, M. Cruts, D. E. Grobbee, C. Vanbroeckhoven, and P. Dejong. 1998. Genetic association of apolipoprotein E with age-related macular degeneration. *American Journal of Human Genetics* 63:200–206.

Knoblauch, H., A. Bauerfeind, C. Krahenbuhl, A. Daury, K. Rohde, S. Bejanin, L. Essioux, H. Schuster, F. C. Luft, and J. G. Reich. 2002. Common haplotypes in five genes influence genetic variance of LDL and HDL cholesterol in the general population. *Human Molecular Genetics* 11:1477–1485.

Knudsen, B., and M. M. Miyamoto. 2001. A likelihood ratio test for evolutionary rate shifts and functional divergence among proteins. *Proceedings of the National Academy of Sciences USA* 98:14512–14517.

Koehn, D., K. Morgan, and F. C. Fraser. 1990. Recurrence risks for near relatives of children with sensori-neural deafness. *Genetic Counseling* 1:127–132.

Kozlowski, T. T., and S. G. Pallardy. 2002. Acclimation and adaptive responses of woody plants to environmental stresses. *Botanical Review* 68:270–334.

Krawczak, M., and D. N. Cooper. 1991. Gene deletions causing human genetic disease: Mechanisms of mutagenesis and the role of the local DNA sequence environment. *Human Genetics* 86:425–441.

Kremer, B., E. Almquist, J. Theilmann, N. Spence, H. Telenius, Y. P. Goldberg, and M. R. Hayden. 1995. Sex-dependent mechanisms for expansions and contractions of the CAG repeat on affected Huntington Disease chromosomes. *American Journal of Human Genetics* 57:343–350.

Krimbas, C. B., and S. Tsakas. 1971. The genetics of *Dacus oleae*. V. Changes of esterase polymorphism following insecticide control—Selection or drift? *Evolution* 25:454–460.

Lahdenperä, M., V. Lummaa, S. Helle, M. Tremblay, and A. F. Russell. 2004. Fitness benefits of prolonged post-reproductive lifespan in women. *Nature* 428:178–181.

Lander, E. S., and L. Kruglyak. 1995. Genetic dissection of complex traits: Guidelines for interpreting and reporting linkage results. *Nature Genetics* 11:241–247.

Langbehn, D. R., R. R. Brinkman, D. Falush, J. S. Paulsen, and M. R. Hayden. 2004. A new model for prediction of the age of onset and penetrance for Huntington's disease based on CAG length. *Clinical Genetics* 65:267–277.

Lapoumeroulie, C., O. Dunda, R. Ducrocq, G. Trabuchet, M. Monylobe, J. M. Bodo, P. Carnevale, D. L. J. Labie, and R. Krishnamoorthy. 1992. A novel sickle-cell mutation of yet another origin in Africa—The Cameroon type. *Human Genetics* 89:333–337.

Larson, A. 1984. Neontological inferences of evolutionary pattern and process in the salamander family Plethodontidae. *Evolutionary Biology* 17:119–217.

Lasker, G. W., and D. E. Crews. 1996. Behavioral influences on the evolution of human genetic diversity. *Molecular Phylogenetics and Evolution* 5:232–240.

Lautenberger, J. A., J. C. Stephens, S. J. O'Brien, and M. W. Smith. 2000. Significant admixture linkage disequilibrium across 30 cM around the FY locus in African Americans. *American Journal of Human Genetics* 66:969–978.

Lederberg, J., and E. M. Lederberg. 1952. Replica plating and indirect selection of bacterial mutants. *Journal of Bacteriology* 63:399–406.

Leeflang, E. P., S. Tavare, P. Marjoram, C. O. S. Neal, J. Srinidhi, M. E. MacDonald, M. de Young, N. S. Wexler, J. F. Gusella, and N. Arnheim. 1999. Analysis of germline mutation spectra at the Huntington's disease locus supports a mitotic mutation mechanism. *Human Molecular Genetics* 8:173–183.

Lees, D. R. 1981. Industrial melanism: Genetic adaptation of animals to air pollution. In J. A. Bishop and L. M. Cook, Eds. *Genetic Consequences of Man Made Change*. Academic, London, pp. 129–176.

Lerman, D. N., P. Michalak, A. B. Helin, B. R. Bettencourt, and M. E. Feder. 2003. Modification of heat-shock gene expression in *Drosophila melanogaster* populations via transposable elements. *Molecular Biology and Evolution* 20:135–144.

Levene, H. 1953. Genetic equilibrium when more than one ecological niche is available. *American Naturalist* 87:331–333.

Levin, D. A., and H. Kerster. 1969. Density-dependent gene dispersal in *Liatris*. *American Naturalist* 103:61–74.

Levy, H. L., and S. Albers. 2000. Genetic screening of newborns. *Annual Review of Genomics and Human Genetetics* 1:139–73.

Levy-Lahad, E., A. Lahad, S. Eisenberg, E. Dagan, T. Paperna, L. Kasinetz, R. Catane, B. Kaufman, U. Beller, P. Renbaum, and R. Gershoni-Baruch. 2001. A single nucleotide polymorphism in the *RAD51* gene modifies cancer risk in *BRCA2* but not *BRCA1* carriers. *Proceedings of the National Academy of Sciences USA* 98:3232–3236.

Lewontin, R. C., and J. L. Hubby. 1966. A molecular genetic approach to the study of genic heterozygosity in natural populations. II. Amount of variation and degree of heterozygosity in natural populations of *Drosophila pseudoobscura*. *Genetics* 54:595–609.

Li, C. C. 1955. *Population Genetics*. University of Chicago Press, Chicago.

Liberles, D. A., D. R. Schreiber, S. Govindarajan, S. G. Chamberlin, and S. A. Benner. 2001. The Adaptive Evolution Database (TAED). *Genome Biology* 2(8): research 0028.1–0028.6.

Linn, S., and W. Arber. 1968. Host specificity of DNA produced by *Escherichia coli*. X. In vitro restriction of phage fd replicative form. *Proceedings of the National Academy of Sciences USA* 59:1300–1306.

Liu, N. J., S. L. Sawyer, N. Mukherjee, A. J. Pakstis, J. R. Kidd, K. K. Kidd, A. J. Brookes, and H. Y. Zhao. 2004. Haplotype block structures show significant variation among populations. *Genetic Epidemiology* 27:385–400.

Livingstone, F. B. 1973. Gene frequency differences in human populations: Some problems of analysis and interpretation. In M. H. Crawford and P. L. Workman, Eds. *Methods and Theories of Anthropological Genetics*. University of New Mexico Press, Albuquerque, pp. 39–66.

Llopart, A., and M. Aguade. 1999. Synonymous rates at the *RpII215* gene of *Drosophila*: Variation among species and across the coding region. *Genetics* 152:269–280.

Llopart, A., and M. Aguade. 2000. Nucleotide polymorphism at the *RpII215* gene in *Drosophila subobscura*: Weak selection on synonymous mutations. *Genetics* 155:1245–1252.

Long, J. R., P. Y. Liu, Y. J. Liu, Y. Lu, D. H. Xiong, L. Elze, R. R. Recker, and H. W. Deng. 2003. APOE and TGF-beta 1 genes are associated with obesity phenotypes. *Journal of Medical Genetics* 40:918–924.

Lucarini, N., M. Nicotra, F. Gloria-Bottini, P. Borgiani, A. Amante, C. Muttinelli, F. Signoretti, M. La Torre, and E. Bottini. 1995. Interaction between ABO blood groups and ADA genetic polymorphism during intrauterine life. A comparative analysis of couples with habitual abortion and normal puerperae delivering a live-born infant. *Human Genetics* 96:527–531.

Luikart, G., J. M. Cornuet, and F. W. Allendorf. 1999. Temporal changes in allele frequencies provide estimates of population bottleneck size. *Conservation Biology* 13:523–530.

Luikart, G., W. B. Sherwin, B. M. Steele, and F. W. Allendorf. 1998. Usefulness of molecular markers for detecting population bottlenecks via monitoring genetic change. *Molecular Ecology* 7:963–974.

Lynch, M., and T. J. Crease. 1990. The analysis of population survey data on DNA sequence variation. *Molecular Biology and Evolution* 7:377–394.

Lynch, M., and W. G. Hill. 1986. Phenotypic evolution by neutral mutation. *Evolution* 40:915–935.

Lynch, M., and B. Walsh. 1998. *Genetics and Analysis of Quantitative Traits*. Sinauer Associates, Sunderland, MA.

MacCleur, J. W., J. L. Vandeberg, B. Read, and O. A. Ryder. 1986. Pedigree analysis of computer simulation. *Zoo Biology* 5:147–160.

Mackay, T. F. C. 2001. The genetic architecture of quantitative traits. *Annual Review of Genetics* 35:303–339.

Magnum, C., and D. Towle. 1977. Physiological adaptation to unstable environments. *American Scientist* 65:67–75.

Mahley, R. W., and S. C. Rall, Jr. 2000. Apolipoprotein E: Far more than a lipid transport protein. *Annual Review Genomics and Human Genetics* 1:507–573.

Maitland-van der Zee, A. H., B. H. C. Stricker, O. H. Klungel, A. K. Mantel-Teeuwisse, J. J. P. Kastelein, A. Hofman, H. G. M. Leufkens, C. M. van Duijn, and A. de Boer. 2003. Adherence to and dosing of beta-hydroxy-beta-methylglutaryl coenzyme A reductase inhibitors in the general population differs according to apolipoprotein E-genotypes. *Pharmacogenetics* 13:219–223.

Majerus, M. E. N. 1998. *Melanism: Evolution in Action*. Oxford University Press, Oxford, New York.

Malausa, T., M. T. Bethenod, A. Bontemps, D. Bourguet, J. M. Cornuet, and S. Ponsard. 2005. Assortative mating in sympatric host races of the European corn borer. *Science* 308:258–260.

Malécot, G. 1950. Quelques schémas probabilistes sur la variabilitié des populations naturelles. *Annales de L'universite de Lyon, Sciences A* 13:37–60.

Malik, H. S., and T. H. Eickbush. 1999. Retrotransposable elements R1 and R2 in the rDNA units of *Drosophila mercatorum*: Abnormal abdomen revisited. *Genetics* 151:653–665.

Mann, F. D. 1998. Animal fat and cholesterol may have helped primitive man evolve a large brain. *Perspectives in Biology and Medicine* 41:417–425.

Marazita, M. L., L. M. Ploughman, B. Rawlings, E. Remington, K. S. Arnos, and W. E. Nance. 1993. Genetic epidemiological studies of early-onset deafness in the U.S. school-age population. *American Journal of Medical Genetics* 46:486–491.

Marchini, J., P. Donnelly, and L. R. Cardon. 2005. Genome-wide strategies for detecting multiple loci that influence complex diseases. *Nature Genetics* 37:413.

Markham, R. B., W. Wang, A. E. Weisstein, Z. Wang, A. Munoz, A. R. Templeton, J. Margolick, D. Vlahov, T. Quinn, H. Farzadegan, and X. Yu. 1998. Patterns of HIV-1 evolution in individuals with differing rates of CD4 T cell decline. *Proceedings of the National Academy of Sciences USA* 95:12568–12573.

Marshall, A. 1997. Laying the foundations for personalized medicines. *Nature Biotechnology* 15:954–957.

Martin, E. R., E. H. Lai, J. R. Gilbert, A. R. Rogala, A. J. Afshari, J. Riley, K. L. Finch, J. F. Stevens, K. J. Livak, B. D. Slotterbeck, S. H. Slifer, L. L. Warren, P. M. Conneally, D. E. Schmechel, I. Purvis, M. A. Pericak-Vance, A. D. Roses, and J. M. Vance. 2000. SNPing away at complex diseases: Analysis of single-nucleotide polymorphisms around APOE in Alzheimer disease. *American Journal of Human Genetics* 67:383–394.

Martorelli, L., C. Virgos, J. Valero, G. Coll, L. Figuera, J. Joven, M. Pocovi, A. Labad, and E. Vilella. 2001. Schizophrenic women with the APOE epsilon 4 allele have a worse prognosis than those without it. *Molecular Psychiatry* 6:307–310.

Maruyama, T. 1972. Rate of decrease of genetic variability in a two-dimensional continuous population of finite size. *Genetics* 70:639–651.

Maruyama, T. 1977. *Stochastic Problems in Population Genetics*. Springer-Verlag, Berlin.

Marx, J. 2002. Unraveling the causes of diabetes. *Science* 296:686–689.

Mauch, D. H., K. Nagler, S. Schumacher, C. Goritz, E. C. Muller, A. Otto, and F. W. Pfrieger. 2001. CNS synaptogenesis promoted by glia-derived cholesterol. *Science* 294:1354–1357.

Maynard Smith, J. 1970. Population size, polymorphism, and the rate of non-Darwinian evolution. *American Naturalist* 104:231–236.

Mayr, E. 1959. Where are we? Cold Spring Harbor *Symposia on Quantitative Biology* 24:1–14.

Mayr, E. 1970. *Populations, Species, and Evolution*. The Belknap Press of Harvard University Press, Cambridge, MA.

McCauley, D. E. 1993. Evolution in metapopulations with frequent local extinction and recolonization. *Oxford Surveys in Evolutionary Biology* 9:109–134.

McDonald, J. H., and M. Kreitman. 1991. Adaptive protein evolution at the Adh locus in *Drosophila*. *Nature* 351:652–654.

McKeigue, P. M. 1998. Mapping genes that underlie ethnic differences in disease risk—methods for detecting linkage in admixed populations, by conditioning on parental admixture. *American Journal of Human Genetics* 63:241–251.

McVean, G. A. T., S. R. Myers, S. Hunt, P. Deloukas, D. R. Bentley, and P. Donnelly. 2004. The fine-scale structure of recombination rate variation in the human genome. *Science* 304:581–584.

Meirhaeghe, A., N. Helbecque, D. Cottel, D. Arveiler, J.-B. Ruidavets, B. Haas, J. Ferrières, J.-P. Tauber, A. Bingham, and P. Amouye. 2001. Impact of sulfonylurea receptor 1 genetic variability on non-insulin-dependent diabetes mellitus prevalence and treatment: A population study. *American Journal of Medical Genetics* 101:4–8.

Merilä, J., and B. C. Sheldon. 1999. Genetic architecture of fitness and nonfitness traits: Empirical patterns and development of ideas. *Heredity* 83:103–109.

Meselson, M., and R. Yuan. 1968. DNA restriction enzyme from *E. coli*. *Nature* 217:1110–1114.

Miall, W. E., and P. D. Oldham. 1963. The hereditary factor in arterial blood pressure. *British Medical Journal* 19:75–80.

Michalakis, Y., and M. Slatkin. 1996. Interaction of selection and recombination in the fixation of negative-epistatic genes. *Genetical Research* 67:257–269.

Miller, M. P., and S. Kumar. 2001. Understanding human disease mutations through the use of interspecific genetic variation. *Human Molecular Genetics* 10:2319–2328.

Mode, C. J. 1958. A mathematical model for the co-evolution of obligate parasites and their hosts. *Evolution* 12:158–165.

Modiano, D., G. Luoni, B. S. Sirima, J. Simporé, F. Verra, A. Konaté, E. Rastrelli, A. Olivieri, C. Calissano, G. M. Paganotti, L. D'urbano, I. Sanou, A. Sawadogo, G. Modiano, and M. Coluzzi.

2001. Haemoglobin C protects against clinical *Plasmodium falciparum* malaria. *Nature* 414: 305–308.

Mokdad, A. H., J. S. Marks, D. F. Stroup, and J. L. Gerberding. 2004. Actual causes of death in the United States, 2000. *Journal of the American Medical Association* 291:1238–1245.

Mondy, M. D. 1970. A Distributional Study of the Collared Lizard, *Crotaphytus collaris collaris*. Southern Illinois University, Edwadsville.

Murray, N. D., J. A. Bishop, and M. R. MacNair. 1980. Melanism and predation by birds in the moths *Biston betularia* and *Phigalia pilosaria*. *Proceedings of the Royal Society of London B* 210:277–283.

Naisbit, R. E., C. D. Jiggins, and J. Mallet. 2003. Mimicry: Developmental genes that contribute to speciation. *Evolution and Development* 5:269–280.

Nakagawa, H., K. Koyama, Y. Miyoshi, H. Ando, S. Baba, M. Watatani, M. Yasutomi, N. Matsuura, M. Monden, and Y. Nakamura. 1998. Nine novel germline mutations of *Stk11* in ten families with Peutz-Jeghers-Syndrome. *Human Genetics* 103:168–172.

Neel, J. V. 1962. Diabetes mellitus: a "thrifty genotype" rendered detrimental by "progress." *American Journal of Human Genetics* 14:353–362.

Neel, J. V., M. Kodani, R. Brewer, and R. C. Anderson. 1949. The incidence of consanguineous matings in Japan. *American Journal of Human Genetics* 1:156–178.

Neel, J. V., and W. J. Schull. 1954. *Human Heredity*. University of Chicago Press, Chicago.

Neel, J. V., A. B. Weder, and S. Julius. 1998. Type II diabetes, essential hypertension, and obesity as "syndromes of impaired genetic homeostasis": the "thrifty" genotype enters the 21st century. *Perspectives in Biology and Medicine* 42:44–74.

Nei, M. 1972. Genetic distance between populations. American Naturalist 89:583–590.

Nei, M., and D. Graur. 1984. Extent of protein polymorphism and the neutral mutation theory. *Evolutionary Biology* 17:73–118.

Nei, M., and F. Tajima. 1981. Genetic drift and estimation of effective population size. *Genetics* 98:625–640.

Nei, M., and N. Takezaki. 1996. The root of the phylogenetic tree of human populations. *Molecular Biology and Evolution* 13:170–177.

Nekrutenko, A., K. D. Makova, and W.-H. Li. 2001. The K_A/K_S ratio test for assessing the protein-coding potential of genomic regions: An empirical and simulation study. *Genome Research* 12:198–202.

Nelson, M. R., S. L. R. Kardia, R. E. Ferrell, and C. F. Sing. 2001. A combinatorial partitioning method to identify multilocus genotypic partitions that predict quantitative trait variation. *Genome Research* 11:458–470.

Nickerson, D. A., S. L. Taylor, K. M. Weiss, A. G. Clark, R. G. Hutchinson, J. Stengard, V. Salomaa, E. Vartiainen, E. Boerwinkle, and C. F. Sing. 1998. DNA sequence diversity in a 9.7-Kb region of the human *Lipoprotein lipase* gene. *Nature Genetics* 19:233–240.

Nielsen, R. 2001a. Mutations as missing data: Inferences on the ages and distributions of nonsynonymous and synonymous mutations. *Genetics* 159:401–411.

Nielsen, R. 2001b. Statistical tests of selective neutrality in the age of genomics. *Heredity* 86: 641–647.

Nunney, L. 1995. Measuring the ratio of effective population size to adult numbers using genetic and ecological data. *Evolution* 49:389–392.

Ober, C., M. Abney, and M. S. McPeek. 2001. The genetic dissection of complex traits in a founder population. *American Journal of Human Genetics* 69:1068–1079.

Ober, C., A. Tsalenko, R. Parry, and N. J. Cox. 2000. A second-generation genomewide screen for asthma-susceptibility alleles in a founder population. *American Journal of Human Genetics* 67:1154–1162.

Ober, C., L. R. Weitkamp, N. Cox, H. Dytch, D. Kostyu, and S. Elias. 1997. HLA and mate choice in humans. *American Journal of Human Genetics* 61:497–504.

Odenheimer, D. J., C. F. Whitten, D. L. Rucknagel, S. A. Sarnaik, and C. F. Sing. 1983. Heterogeneity of sickle-cell anemia based on a profile of hematological variables. *American Journal of Human Genetics* 35:1224–1240.

Ohta, T. 1976. Role of very slightly deleterious mutations in molecular evolution and polymorphism. *Theoretical Population Biology* 10:254–275.

Oner, C., A. J. Dimovski, N. F. Olivieri, G. Schiliro, J. F. Codrington, S. Fattoum, A. D. Adekile, R. Oner, G. T. Yuregir, C. Altay, A. Gurgey, R. B. Gupta, V. B. Jogessar, M. N. Kitundu, D. Loukopoulos, G. P. Tamagnini, M. Ribeiro, F. Kutlar, L. H. Gu, K. D. Lanclos, and T. Huisman. 1992. Beta-S haplotypes in various world populations. *Human Genetics* 89:99–104.

Orr, H. A. 2000. Adaptation and the cost of complexity. *Evolution* 54:13–20.

Pál, C., and I. Miklós. 1999. Epigenetic inheritance, genetic assimilation and speciation. *Journal of Theoretical Biology* 200:19–37.

Palmarino, R., R. Agostino, F. Gloria, P. Lucarelli, L. Businco, G. Antognoni, G. Maggioni, P. L. Workman, and E. Bottini. 1975. Red cell acid phosphatase: Another polymorphism correlated with Malaria? *American Journal of Physical Anthropology* 43:177–186.

Palsbøll, P. J., M. Berube, A. Aguilar, G. Notarbartolo-Di-Sciara, and R. Nielsen. 2004. Discerning between recurrent gene flow and recent divergence under a finite-site mutation model applied to North Atlantic and Mediterranean Sea fin whale (*Balaenoptera physalus*) populations. *Evolution* 58:670–675.

Parra, E. J., R. A. Kittles, G. Argyropoulos, C. L. Pfaff, K. Hiester, C. Bonilla, N. Sylvester, D. Parrish-Gause, W. T. Garvey, L. Jin, P. M. McKeigue, M. I. Kamboh, R. E. Ferrell, W. S. Pollitzer, and M. D. Shriver. 2001. Ancestral proportions and admixture dynamics in geographically defined African Americans living in South Carolina. *American Journal of Physical Anthropology* 114:18–29.

Paterson, H. 1985. The recognition concept of species. In E. Vrba, Ed. *Species and Speciation*. Transvaal Museum, Pretoria, South Africa, pp. 21–29.

Patrinos, G. P., A. Loutradianagnostou, and M. N. Papadakis. 1996. A new base substitution in the 5' regulatory region of the human (A) gamma globin gene is linked with the beta(S) gene. *Human Genetics* 97:357–358.

Paxinos, E. E., H. F. James, S. L. Olson, J. D. Ballou, J. A. Leonard, and R. C. Fleischer. 2002. Prehistoric decline of genetic diversity in the nene. *Science* 296:7.

Payne, F. 1918. The effect of artificial selection on bristle number in *Drosophila ampelophila* and its interpretation. *Proceedings of the National Academy of Sciences USA* 4:55–58.

Payseur, B. A., and M. W. Nachman. 2000. Microsatellite variation and recombination rate in the human genome. *Genetics* 156:1285–1298.

Peck, S. L., S. P. Ellner, and F. Gould. 1998. A spatially explicit stochastic model demonstrates the feasibility of Wright's Shifting Balance Theory. *Evolution* 52:1834–1839.

Peck, S. L., S. P. Ellner, and F. Gould. 2000. Varying migration and deme size and the feasibility of the shifting balance. *Evolution* 54:324–327.

Pedersen, J. C., and K. Berg. 1989. Interaction between low-density lipoprotein receptor (LDLR) and apolipoprotein-E (ApoE) alleles contributes to normal variation in lipid level. *Clinical Genetics* 35:331–337.

Pedersen, J. C., and K. Berg. 1990. Gene-gene interaction between the low-density-lipoprotein receptor and apolipoprotein-E loci affects lipid levels. *Clinical Genetics* 38:287–294.

Peltonen, L. 2000. Positional cloning of disease genes: Advantages of genetic isolates. *Human Heredity* 50:66–75.

Peltonen, L., A. Palotie, and K. Lange. 2000. Use of population isolates for mapping complex traits. *Nature Reviews Genetics* 1:182–190.

Peripato, A. C., R. A. De Brito, S. R. Matioli, L. S. Pletscher, T. T. Vaughn, and J. M. Cheverud. 2004. Epistasis affecting litter size in mice. *Journal of Evolutionary Biology* 17:593–602.

Perondini, A. L. P., P. A. Otto, A. R. Templeton, and A. Rogatko. 1983. Evidence for assortative mating systems related to the polytene chromosome-band polymorphism in *Sciara ocellaris*. *Journal of Heredity* 74:283–288.

Petren, K., B. R. Grant, and P. R. Grant. 1999. A phylogeny of Darwin's finches based on microsatellite DNA length variation. *Proceedings of the Royal Society of London Series B—Biological Sciences* 266:321–329.

Pickford, M., and B. Senut. 2001. "Millennium Ancestor," a 6-million-year-old bipedal hominid from Kenya—Recent discoveries push back human origins by 1.5 million years. *South African Journal of Science* 97:22.

Pillay, T. S., W. J. Langlois, and J. M. Olefsky. 1995. The genetics of non-insulin-dependent diabetes mellitus. *Advances in Genetics* 32:51–98.

Polley, S. D., and D. J. Conway. 2001. Strong diversifying selection on domains of the *Plasmodium falciparum* apical membrane antigen 1 gene. *Genetics* 158:1505–1512.

Posada, D. 2002. Evaluation of methods for detecting recombination from DNA sequences: Empirical Data. *Molecular Biology and Evolution* 19:708–717.

Posada, D., and K. A. Crandall. 1998. ModelTest: Testing the model of DNA substitution. *Bioinformatics* 14:817–818.

Posada, D., and K. A. Crandall. 2002. The effect of recombination on the accuracy of phylogeny estimation. *Journal of Molecular Evolution* 54:396–402.

Posada, D., K. A. Crandall, and E. C. Holmes. 2002. Recombination in evolutionary genomics. *Annual Review of Genetics* 36:75–97.

Posada, D., K. A. Crandall, and A. R. Templeton. 2000. GeoDis: A program for the cladistic nested analysis of the geographical distribution of genetic haplotypes. *Molecular Ecology* 9:487–488.

Potts, W. K., and E. K. Wakeland. 1993. Evolution of MHC genetic diversity: A tale of incest, pestilence and sexual preference. *Trends in Genetics* 9:408–412.

Powell, J. R. 1997. *Progress and Prospects in Evolutionary Biology: The Drosophila Model*. Oxford University Press, Oxford.

Powell, J. R., and L. Morton. 1979. Inbreeding and mating patterns in *Drosophila pseudoobscura*. *Behavior Genetics* 9:425–429.

Prak, E. T. L., and H. H. Kazazian. 2000. Mobile elements and the human genome. *Nature Reviews Genetics* 1:134–144.

Pritchard, J. K., M. Stephens, N. A. Rosenberg, and P. Donnelly. 2000. Association mapping in structured populations. *American Journal of Human Genetics* 67:170–181.

Ralls, K., J. D. Ballou, and A. Templeton. 1988. Estimates of lethal equivalents and the cost of inbreeding in mammals. *Conservation Biology* 2:185–193.

Ramana, G. V., G. R. Chandak, and L. Singh. 2000. Sickle cell gene haplotypes in Relli and Thurpu Kapu populations of Andhra Pradesh. *Human Biology* 72:535–540.

Rannala, B., and G. Bertorelle. 2001. Using linked markers to infer the age of a mutation. *Human Mutation* 18:87–100.

Rao, P. S. S., and S. G. Inbaraj. 1980. Inbreeding effects on fetal growth and development. *Journal of Medical Genetics* 17:27–33.

Rask-Nissila, L., E. Jokinen, J. Viikari, A. Tammi, T. Ronnemaa, J. Marniemi, P. Salo, T. Routi, H. Helenius, I. Valimaki, and O. Simell. 2002. Impact of dietary intervention, sex, and apolipoprotein E phenotype on tracking of serum lipids and apolipoproteins in 1-to 5–year-old children—The Special Turku Coronary Risk Factor Intervention Project (STRIP). *Arteriosclerosis Thrombosis and Vascular Biology* 22:492–498.

Redkar, A. A., Y. Si1, S. N. Twine, S. H. Pilder, and P. Olds-Clarke. 2000. Genes in the first and fourth inversions of the mouse *t* complex synergistically mediate sperm capacitation and interactions with the oocyte. *Developmental Biology* 226:267–280.

Reed, T. E. 1969. Caucasian genes in American Negroes. *Science* 165:762–768.

Reed, T. E., and J. V. Neel. 1959. Huntington's chorea in Michigan. 2. Selection and mutation. *American Journal of Human Genetics* 11:107–136.

Reich, D. E., S. F. Schaffner, M. J. Daly, G. McVean, J. C. Mullikin, J. M. Higgins, D. J. Richter, E. S. Lander, and D. Altshuler. 2002. Human genome sequence variation and the influence of gene history, mutation and recombination. *Nature Genetics* 32:135–142.

Reiman, E. M., R. J. Caselli, K. Chen, G. E. Alexander, D. Bandy, and J. Frost. 2001. Declining brain activity in cognitively normal apolipoprotein E varepsilon 4 heterozygotes: A foundation for using positron emission tomography to efficiently test treatments to prevent Alzheimer's disease. *Proceedings of the National Academy of Sciences USA* 98:3334–3339.

Rendine, S., A. Piazza, P. Menozzi, and L. L. Cavalli-Sforza. 1999. A problem with synthetic maps: Reply to Sokal et al. *Human Biology* 71:15–25.

Resnick, H. E., B. Rodriguez, R. Havlik, L. Ferrucci, D. Foley, J. D. Curb, and T. B. Harris. 2000. Apo E genotype, diabetes, and peripheral arterial disease in older men: The Honolulu Asia-Aging study. *Genetic Epidemiology* 19:52–63.

Richards, M., V. Macaulay, E. Hickey, E. Vega, B. Sykes, V. Guida, C. Rengo, D. Sellitto, F. Cruciani, T. Kivisild, R. Villems, M. Thomas, S. Rychkov, O. Rychkov, Y. Rychkov, M. Gölge, D. Dimitrov, E. Hill, D. Bradley, V. Romano, F. Calì, G. Vona, A. Demaine, S. Papiha, C. Triantaphyllidis, G. Stefanescu, J. Hatina, M. Belledi, A. D. Rienzo, A. Novelletto, A. Oppenheim, S. Nørby, N. Al-Zaheri, S. Santachiara-Benerecetti, R. Scozzari, A. Torroni, and H.-J. Bandelt. 2000. Tracing European founder lineages in the near eastern mtDNA pool. *American Journal of Human Genetics* 67:1251–1276.

Riddiford, L. M., K. Hiruma, X. F. Zhou, and C. A. Nelson. 2003. Insights into the molecular basis of the hormonal control of molting and metamorphosis from *Manduca sexta* and *Drosophila melanogaster*. *Insect Biochemistry and Molecular Biology* 33:1327–1338.

Rinaldo, A., S. A. Bacanu, B. Devlin, V. Sonpar, L. Wasserman, and K. Roeder. 2005. Characterization of multilocus linkage disequilibrium. *Genetic Epidemiology* 28:193–206.

Risch, N., E. Burchard, E. Ziv, and H. Tang. 2002. Categorization of humans in biomedical research: Genes, race and disease. *Genome Biology* 3(7): comment2007.1–2007.12.

Ritossa, F. 1976. The *bobbed* locus. In M. Ashburner and E. Novitski, Eds. *The Genetics and Biology of Drosophila*. Academic, London, pp. 801–846.

Ritossa, F. M., K. C. Atwood, and S. Spiegelman. 1966. A molecular explanation of the bobbed mutants of *Drosophila* as partial deficiencies of "ribosomal" DNA. *Genetics* 54:819–834.

Roberts, D. F. 1967. The development of inbreeding in an island population. *Ciência e Cultura* 19:78–84.

Roberts, D. F. 1968. Genetic effects of population size reduction. *Nature* 220:1084–1088.

Roberts, D. F. 1978. *Climate and Human Variability*. Cummings Publishing, Menlo Park, CA.

Rohde, D. L. T., S. Olson, and J. T. Chang. 2004. Modelling the recent common ancestry of all living humans. *Nature* 431:562–566.

Rohlf, F. J. 1993. *NTSYS-pc: Numerical Taxonomy and Multivariate Analysis System*, Version 1.80. Exeter Software, Setauket, NY.

Rose, C. S. 1999. Hormonal control in larval development and evolution—Amphibians. In B. K. Hall and M. H. Wake, Eds. *The Origin and Evolution of Larval Forms*. Academic, San Diego, pp. 167–216.

Rosenberg, N. A., J. K. Pritchard, J. L. Weber, H. M. Cann, K. K. Kidd, L. A. Zhivotovsky, and M. W. Feldman. 2002. Genetic structure of human populations. *Science* 298:2381–2385.

Rubinsztein, D. C. 1999. Trinucleotide expansion mutations cause diseases which do not conform to classical Mendelian expectations. In D. B. Goldstein and C. Schötterer, Eds. *Microsatellites: Evolution and Applications*. Oxford University Press, Oxford, pp. 80–97.

Rubinsztein, D. C., W. Amos, J. Leggo, S. Goodburn, R. S. Ramesar, J. Old, R. Bontrop, R. Mcmahon, D. E. Barton, and M. A. Fergusonsmith. 1994. Mutational bias provides a model for the evolution of Huntington's disease and predicts a general increase in disease prevalence. *Nature Genetics* 7:525–530.

Rubinsztein, D. C., J. Leggo, M. Chiano, A. Dodge, G. Norbury, E. Rosser, and D. Craufurd. 1997. Genotypes at the GluR6 kainate receptor locus are associated with variation in the age of onset of Huntington disease. *Proceedings of the National Academy of Sciences USA* 94:3872–3876.

Ruiz, A., J. M. Ranz, M. Caceres, C. Segarra, A. Navarro, and A. Barbadilla. 1997. Chromosomal evolution and comparative gene mapping in the *Drosophila repleta* species group. *Brazilian Journal of Genetics* 20:553–565.

Ruwende, C., and A. Hill. 1998. Glucose-6-phosphate dehydrogenase deficiency and malaria. *Journal of Molecular Medicine* 76:581–588.

Saiki, R. K., S. Scharf, F. Faloona, K. B. Mullis, G. T. Horn, H. A. Erlich, and N. Arnheim. 1985. Enzymatic amplification of β-globin genomic sequences and restriction site analysis for diagnosis of sickle cell anemia. *Science* 230:1350–1354.

Saitou, N., and M. Nei. 1987. The neighbor-joining method: A new method for reconstructing phylogenetic trees. *Molecular Biology and Evolution* 4:406–425.

Santos, E. J. M., J. T. Epplen, and C. Epplen. 1997. Extensive gene flow in human populations as revealed by protein and microsatellite DNA markers. *Human Heredity* 47:165–172.

Scarr, S., A. J. Pakstis, S. H. Katz, and W. B. Barker. 1977. Absence of a relationship between degree of white ancestry and intellectual skills within a black population. *Human Genetics* 39:69–86.

Schaner, P., N. Richards, A. Wadhwa, I. Aksentijevich, D. Kastner, P. Tucker, and D. Gumucio. 2001. Episodic evolution of pyrin in primates: Human mutations recapitulate ancestral amino acid states. *Nature Genetics* 27:318–321.

Schimenti, J. 2000. Segregation distortion of mouse *t* haplotypes. *Trends in Genetics* 16:240–243.

Schmidt, P. S., and D. M. Rand. 2001. Adaptive maintenance of genetic polymorphism in an intertidal barnacle: Habitat- and life-stage-specific survivorship of *MPI* genotypes. *Evolution* 55:1336–1344.

Schmidt, S., L. F. Barcellos, K. DeSombre, J. B. Rimmler, R. R. Lincoln, P. Bucher, A. M. Saunders, E. Lai, E. R. Martin, J. M. Vance, J. R. Oksenberg, S. L. Hauser, M. A. Pericak-Vance, and J. L. Haines. 2002. Association of polymorphisms in the apolipoprotein E region with susceptibility to and progression of multiple sclerosis. *American Journal of Human Genetics* 70:708–717.

Schneider, I. 2004. DNA variation on chips, beads, and arrays. *Genetic Engineering News* 24:32–33, 36.

Scriver, C. R., and P. J. Waters. 1999. Monogenic traits are not simple: Lessons from phenylketonuria. *Trends in Genetics* 15:267–272.

Sekiguchi, H., and K. Sekiguchi. 1951. On consanguineous marriage in the upper Ina Valley. *Minzoku Eisei* 17:117–127.

Selander, R. K., and R. O. Hudson. 1976. Animal population structure under close inbreeding: The land snail *Rumina* in southern France. *American Naturalist* 110:695–718.

Selander, R. K., S. Y. Yang, and W. G. Hunt. 1969. Polymorphisms in esterases and hemoglobins in wild populations of the house mouse. In M. R. Wheeler, Ed. *Studies in Genetics*, Vol. V. University of Texas Press, Austin, pp. 271–328.

Seltman, H., K. Roeder, and B. Devlin. 2001. Transmission/disequilibrium test meets measured haplotype analysis: Family-based association analysis guided by evolution of haplotypes. *American Journal of Human Genetics* 68:1250–1263.

Serre, D., and S. Pääbo. 2004. Evidence for gradients of human genetic diversity within and among continents. *Genome Research* 14:1679–1685.

Sexton, O. J., R. M. Andrews, and J. E. Bramble. 1992. Size and growth rate characteristics of a peripheral population of *Crotaphytus collaris* (Sauria; Iquanidae). *Copeia* 1992:968–980.

Sheppard, P. M., J. R. G. Turner, K. S. Brown, W. W. Benson, and M. C. Singer. 1985. Genetics and the evolution of Muellarian mimicry in *Heliconius* butterflies. *Philosophical Transactions of the Royal Society of London B* 308:433–610.

Shokier, M. H. K. 1975. Investigation of Huntington's disease in the Canadian prairies. II. Fecundity and fitness. *Clinical Genetics* 7:349–353.

Siegismund, H. R., and P. Arctander. 1995. Structure of African elephant populations. *Journal of Heredity* 86:467–468.

Siest, G., E. Jeannesson, H. Berrahmoune, S. Maunus, J. B. Marteau, S. Mohr, and S. Visvikis. 2004. Pharmacogenomics and drug response in cardiovascular disorders. *Pharmacogenomics* 5:779–802.

Silva, J. C., and M. G. Kidwell. 2000. Horizontal transfer and selection in the evolution of P elements. *Molecular Biology and Evolution* 17:1542–1557.

Sing, C. F., E. Boerwinkle, and P. P. Moll. 1985. Strategies for elucidating the phenotypic and genetic heterogeneity of a chronic disease with a complex etiology. In R. Chakraborty and J. E. Szathmary, Eds. *Disease of Complex Etiology in Small Populations: Ethnic Differences and Research Approaches.* Alan R. Liss, New York, pp. 39–66.

Sing, C. F., and J. Davignon. 1985. Role of the apolipoprotein E polymorphism in determining normal plasma lipid and lipoprotein variation. *American Journal of Human Genetics* 37:268–285.

Sing, C. F., M. B. Haviland, and S. L. Reilly. 1996. Genetic architecture of common multifactorial diseases. In G. Cardew, Ed. *Variation in the Human Genome.* Wiley, Chichester, pp. 211–232.

Sing, C. F., M. B. Haviland, A. R. Templeton, and S. L. Reilly. 1995. Alternative genetic strategies for predicting risk of atherosclerosis. In F. P. Woodford, J. Davignon, and A. D. Sniderman, Eds. *Atherosclerosis X. Excerpta Medica International Congress Series.* Elsevier Science, Amsterdam, pp. 638–644.

Singer, G. A. C., A. T. Lloyd, L. B. Huminiecki, and K. H. Wolfe. 2005. Clusters of co-expressed genes in mammalian genomes are conserved by natural selection. *Molecular Biology and Evolution* 22:767–775.

Singh, R. S., and L. R. Rhomberg. 1987. A comprehensive study of genic variation in natural populations of *Drosophila melanogaster*. II. Estimates of heterozygosity and patterns of geographic differentiation. *Genetics* 117:255–271.

Siniscalco, M., L. Bernini, G. Filippi, B. Latte, P. M. Khan, S. Piomelli, and M. Rattazzi. 1966. Population genetics of haemoglobin variants, thalassaemia and glucose-6-phosphate dehydrogenase deficiency, with particular reference to the malaria hypothesis. *Bulletin of the World Health Organization* 34:379–393.

Skodak, M., and H. M. Skeels. 1949. A final follow-up study of one hundred adopted children. *Journal of Genetic Psychology* 75:85–125.

Slatkin, M. 1981. Populational heritability. *Evolution* 35:859–871.

Slatkin, M. 1991. Inbreeding coefficients and coalescence times. *Genetical Research* 58:167–175.

Slatkin, M. 1996. In defense of founder-flush theories of speciation. *American Naturalist* 147:493–505.

Slatkin, M. 2000. Allele age and a test for selection on rare alleles. *Philosophical Transactions of the Royal Society of London B* 355:1663–1668.

Slatkin, M., and B. Rannala. 2000. Estimating allele age. *Annual Review of Genomics and Human Genetics* 1:225–273.

Smith, N. G. C., and A. Eyre-Walker. 2002. Adaptive protein evolution in *Drosophila*. *Nature* 415:1022–1024.

Sokal, R. R. 1970. Senescence and genetic load. *Science* 167:1733–1734.

Sokal, R. R., N. L. Oden, and B. A. Thomson. 1999a. A problem with synthetic maps. *Human Biology* 71:1–13.

Sokal, R. R., N. L. Oden, and B. A. Thomson. 1999b. Problems with synthetic maps remain: Reply to Rendine et al. *Human Biology* 71:447–453.

Soodyall, H. 1993. *Mitochondrial DNA Polymorphisms in Southern African Populations*. University of the Witwatersrand, Johannesburg.

Southern, E. M. 1975. Detection of specific sequences among DNA fragments separated by gel electrophoresis. *Journal of Molecular Biology* 98:503–517.

Spielman, R. S., R. E. McGinnis, and W. J. Ewens. 1993. Transmission test for linkage disequilibrium: The insulin gene region and insulin-dependent diabetes mellitus (IDDM). *American Journal of Human Genetics.* 52:506–516.

Spradling, A. 1987. Gene amplification in Dipteran chromosomes. In W. Hennig, ed. *Structure and Function of Eukaryotic Chromosomes*. Springer-Verlag, New York, pp. 199–212.

Stead, J. D. H., M. E. Hurles, and A. J. Jeffreys. 2003. Global haplotype diversity in the human insulin gene region. *Genome Research* 13:2101–2111.

Steinberg, A. G., H. K. Bleibtreu, T. W. Kurczynski, A. O. Martin, and E. M. Kurczynski. 1966. Genetic studies on an inbred human isolate. In *Proceedings of the Third International Congress of Human Genetics*. University of Chicago, Chicago, IL, pp. 267–289.

Steinmetz, I. M., H. Sinha, D. R. Richards, J. I. Spiegelman, P. J. Oefner, J. H. Mccusker, and R. W. Davis. 2002. Dissecting the architecture of a quantitative trait locus in yeast. *Nature* 416:326–330.

Stengård, J. H., J. Pekkanen, C. Ehnholm, A. Nissinen, and C. F. Sing. 1996. Genotypes with the apolipoprotein epsilon-4 allele are predictors of coronary heart disease mortality in a longitudinal study of elderly Finnish men. *Human Genetics* 97:677–684.

Stephens, M., N. J. Smith, and P. Donnelly. 2001. A new statistical method for haplotype reconstruction from population data. *American Journal of Human Genetics* 68:978–989.

Stine, G. J. 1977. *Biosocial Genetics*. Macmillan, New York.

Stoneking, M., K. Bhatia, and A. C. Wilson. 1986. Rate of sequence divergence estimated from restricted maps of mitochondrial DNAs from Papua New Guinea. *Cold Spring Harbor Symposium on Quantitative Biology* 51:433–439.

Storm, K., S. Willocx, K. Flothmann, and G. Van Camp. 1999. Determination of the carrier frequency of the common GJB2 (connexin-26) 35delG mutation in the Belgian population using an easy and reliable screening method. *Human Mutation* 14:263–266.

Storz, J. F., B. A. Payseur, and M. W. Nachman. 2004. Genome scans of DNA variability in humans reveal evidence for selective sweeps outside of Africa. *Molecular Biology and Evolution* 21:1800–1811.

Stringer, C. 2002. Modern human origins: Progress and prospects. *Philosophical Transactions of the Royal Society of London B* 357:563–579.

Sturtevant, A. H., and T. Dobzhansky. 1936. Inversions in the third chromosome of wild races of *Drosophila pseudoobscura*, and their use in the study of the history of the species. *Proceedings of the National Academy of Sciences USA* 22:448–450.

Sunyaev, S., F. A. Kondrashov, P. Bork, and V. Ramensky. 2003. Impact of selection, mutation rate and genetic drift on human genetic variation. *Human Molecular Genetics* 12:3325–3330.

Sunyaev, S., V. Ramensky, I. Koch, W. Lathe, A. S. Kondrashov, and P. Bork. 2001. Prediction of deleterious human alleles. *Human Molecular Genetics* 10:591–597.

Swofford, D. 1997. *PAUP*: Phylogenetic Analysis Using Parsimony (*and Other Methods)*. Sinauer, Sunderland, MA.

Tajima, F. 1983. Evolutionary relationship of DNA sequences in finite populations. *Genetics* 105:437–460.

Tajima, F. 1989a. Statistical-method for testing the neutral mutation hypothesis by DNA polymorphism. *Genetics* 123:585.

Tajima, F. 1989b. The effect of change in population size on DNA polymorphism. *Genetics* 123:597–601.

Takahata, N., S.-H. Lee, and Y. Satta. 2001. Testing multiregionality of modern human origins. *Molecular Biology and Evolution* 18:172–183.

Takano, K., and J. R. Miller. 1972. ABO incompatibility as a cause of spontaneous abortion: Evidence from abortuses. *Journal of Medical Genetics* 9:144–150.

Tang, G., H. Xie, L. Xu, Y. Hao, D. Lin, and D. Ren. 2002. Genetic study of apolipoprotein E gene, alpha-1 antichymotrypsin gene in sporadic Parkinson disease. *American Journal of Medical Genetics* 114:446–449.

Tartof, K. D. 1971. Increasing the multiplicity of ribosomal RNA genes in *Drosophila melanogaster*. *Science* 171:294–297.

Taylor, F. 1979. Convergence to the stable age distribution in populations of insects. *American Naturalist* 113:511–530.

Templeton, A. R. 1977a. Analysis of head shape differences between two interfertile species of Hawaiian *Drosophila*. *Evolution* 31:630–642.

Templeton, A. R. 1977b. Survival probabilities of mutant alleles in fine-grained environments. *American Naturalist* 111:951–966.

Templeton, A. R. 1979a. The unit of selection in *Drosophila mercatorum*. II. Genetic revolution and the origin of coadapted genomes in parthenogenetic strains. *Genetics* 92:1265–1282.

Templeton, A. R. 1979b. A frequency dependent model of brood selection. *American Naturalist* 114:515–524.

Templeton, A. R. 1980a. The theory of speciation via the founder principle. *Genetics* 94:1011–1038.

Templeton, A. R. 1980b. The evolution of life histories under pleiotropic constraints and r-selection. *Theoretical Population Biology* 18:279–289.

Templeton, A. R. 1981. Mechanisms of speciation—A population genetic approach. *Annual Review of Ecological Systems* 12:23–48.

Templeton, A. R. 1982. Adaptation and the integration of evolutionary forces. In R. Milkman, Ed. *Perspectives on Evolution*. Sinauer, Sunderland, MA, pp. 15–31.

Templeton, A. R. 1983a. Phylogenetic inference from restriction endonuclease cleavage site maps with particular reference to the evolution of humans and the apes. *Evolution* 37:221–244.

Templeton, A. R. 1983b. Natural and experimental parthenogenesis. In M. Ashburner, H. L. Carson, and J. N. Thompson, Eds. *The Genetics and Biology of Drosophila*. Academic, London, pp. 343–398.

Templeton, A. R. 1987a. Nonparametric phylogenetic inference from restriction cleavage sites. *Molecular Biology and Evolution* 4:315–319.

Templeton, A. R. 1987b. Genetic systems and evolutionary rates. In K. S. W. Campbell and M. F. Day, Eds. *Rates of Evolution*. Allen & Unwin, London, pp. 218–234.

Templeton, A. R. 1987c. The general relationship between average effect and average excess. *Genetical Research* 49:69–70.

Templeton, A. R. 1989. The meaning of species and speciation: A genetic perspective. In D. Otte and J. A. Endler, Eds. *Speciation and Its Consequences*. Sinauer, Sunderland, MA, pp. 3–27.

Templeton, A. R. 1993. The "Eve" hypothesis: A genetic critique and reanalysis. *American Anthropologist* 95:51–72.

Templeton, A. R. 1994. The role of molecular genetics in speciation studies. In B. Schierwater, B. Streit, G. P. Wagner, and R. DeSalle, Eds. *Molecular Ecology and Evolution: Approaches and Applications*. Birkhäuser-Verlag, Basel, pp. 455–477.

Templeton, A. R. 1995. A cladistic analysis of phenotypic associations with haplotypes inferred from restriction endonuclease mapping or DNA sequencing. V. Analysis of case/control sampling designs: Alzheimer's disease and the apoprotein E locus. *Genetics* 140:403–409.

Templeton, A. R. 1996a. Experimental evidence for the genetic-transilience model of speciation. *Evolution* 50:909–915.

Templeton, A. R. 1996b. Contingency tests of neutrality using intra/interspecific gene trees: The rejection of neutrality for the evolution of the mitochondrial cytochrome oxidase II gene in the hominoid primates. *Genetics* 144:1263–1270.

Templeton, A. R. 1997a. Testing the out of Africa replacement hypothesis with mitochondrial DNA data. In G. A. Clark and C. M. Willermet, Eds. *Conceptual Issues in Modern Human Origins Research*. Aldine de Gruyter, New York, pp. 329–360.

Templeton, A. R. 1997b. Out of Africa? What do genes tell us? *Current Opinion in Genetics and Development* 7:841–847.

Templeton, A. R. 1998a. Human races: A genetic and evolutionary perspective. *American Anthropologist* 100:632–650.

Templeton, A. R. 1998b. Species and speciation: Geography, population structure, ecology, and gene trees. In D. J. Howard and S. H. Berlocher, Eds. *Endless Forms: Species and Speciation*. Oxford University Press, Oxford, pp. 32–43.

Templeton, A. R. 1998c. Nested clade analyses of phylogeographic data: Testing hypotheses about gene flow and population history. *Molecular Ecology* 7:381–397.

Templeton, A. R. 1999a. Uses of evolutionary theory in the human genome project. *Annual Review of Ecology and Systematics* 30:23–49.

Templeton, A. R. 1999b. Experimental tests of genetic transilience. *Evolution* 53:1628–1632.

Templeton, A. R. 1999c. Using gene trees to infer species from testable null hypothesis: Cohesion species in the *Spalax ehrenbergi* complex. In S. P. Wasser, Ed. *Evolutionary Theory and Processes: Modern Perspectives, Papers in Honour of Eviatar Nevo*. Kluwer Academic, Dordrecht, pp. 171–192.

Templeton, A. R. 2000. Epistasis and complex traits. In J. B. Wolf, I. E. D. Brodie, and M. J. Wade, Eds. *Epistasis and the Evolutionary Process*. Oxford University Press, Oxford, pp. 41–57.

Templeton, A. R. 2001. Using phylogeographic analyses of gene trees to test species status and processes. *Molecular Ecology* 10:779–791.

Templeton, A. R. 2002a. The Speke's gazelle breeding program as an illustration of the importance of multilocus genetic diversity in conservation biology: Response to Kalinowski *et al. Conservation Biology* 16:1151–1155.

Templeton, A. R. 2002b. Out of Africa again and again. *Nature* 416:45–51.

Templeton, A. R. 2003. Human races in the context of recent human evolution: A molecular genetic perspective. In A. H. Goodman, D. Heath, and M. S. Lindee, Eds. *Genetic Nature/Culture*. University of California Press, Berkeley, pp. 234–257.

Templeton, A. R. 2004a. Statistical phylogeography: Methods of evaluating and minimizing inference errors. *Molecular Ecology* 13:789–809.

Templeton, A. R. 2004b. A maximum likelihood framework for cross validation of phylogeographic hypotheses. In S. P. Wasser, Ed. *Evolutionary Theory and Processes: Modern Horizons*. Kluwer Academic, Dordrecht, The Netherlands, pp. 209–230.

Templeton, A. R. 2005. Haplotype trees and modern human origins. *Yearbook of Physical Anthropology* 48:33–59.

Templeton, A. R., E. Boerwinkle, and C. F. Sing. 1987a. A cladistic analysis of phenotypic associations with haplotypes inferred from restriction endonuclease mapping. I. Basic theory and an analysis of alcohol dehydrogenase activity in *Drosophila*. *Genetics* 117:343–351.

Templeton, A. R., A. G. Clark, K. M. Weiss, D. A. Nickerson, J. Stengård, E. Boerwinkle, and C. F. Sing. 2000a. Recombinational and mutational hotspots within the human *lipoprotein lipase* gene. *American Journal of Human Genetics* 66:69–83.

Templeton, A. R., K. A. Crandall, and C. F. Sing. 1992. A cladistic analysis of phenotypic associations with haplotypes inferred from restriction endonuclease mapping and DNA sequence data. III. Cladogram estimation. *Genetics* 132:619–633.

Templeton, A. R., T. J. Crease, and F. Shah. 1985. The molecular through ecological genetics of *abnormal abdomen* in *Drosophila mercatorum*. I. Basic genetics. *Genetics* 111:805–818.

Templeton, A. R., and N. J. Georgiadis. 1996. A landscape approach to conservation genetics: Conserving evolutionary processes in the African Bovidae. In J. C. Avise and J. L. Hamrick, Eds. *Conservation Genetics: Case Histories from Nature*. Chapman & Hall, New York, pp. 398–430.

Templeton, A. R., and L. E. Gilbert. 1985. Population genetics and the coevolution of mutualisms. In D. H. Boucher, Ed. *The Biology of Mutualism: Ecology and Evolution*. Crom Helm, London, pp. 128–144.

Templeton, A. R., H. Hollocher, and J. S. Johnston. 1993. The molecular through ecological genetics of abnormal abdomen in *Drosophila mercatorum*. V. Female phenotypic expression on natural genetic backgrounds and in natural environments. *Genetics* 134:475–485.

Templeton, A. R., H. Hollocher, S. Lawler, and J. S. Johnston. 1989. Natural selection and ribosomal DNA in *Drosophila*. *Genome* 31:296–303.

Templeton, A. R., H. Hollocher, S. Lawler, and J. S. Johnston. 1990a. The ecological genetics of abnormal abdomen in *Drosophila mercatorum*. In J. S. F. Barker, Ed. *Ecological and Evolutionary Genetics of Drosophila*. Plenum, New York, pp. 17–35.

Templeton, A. R., and J. S. Johnston. 1988. The measured genotype approach to ecological genetics. In G. d. Jong, Ed. *Population Genetics and Evolution*. Springer-Verlag, Berlin, pp. 138–146.

Templeton, A. R., J. S. Johnston, and C. F. Sing. 1987b. The proximate and ultimate control of aging in *Drosophila* and humans. In A. D. Woodhead and K. H. Thompson, Eds. *Evolution of Longevity in Animals*. Plenum, New York, pp. 123–133.

Templeton, A. R., S. D. Maskas, and M. B. Cruzan. 2000c. Gene trees: A powerful tool for exploring the evolutionary biology of species and speciation. *Plant Species Biology* 15:211–222.

Templeton, A. R., T. Maxwell, D. Posada, J. H. Stengard, E. Boerwinkle, and C. F. Sing. 2005. Tree scanning: A method for using haplotype trees in genotype/phenotype association studies. *Genetics* 169:441–453.

Templeton, A. R., and M. A. Rankin. 1978. Genetic revolutions and control of insect populations. In R. H. Richardson, Ed. *The Screwworm Problem*. University of Texas, Austin, pp. 83–112.

Templeton, A. R., and B. Read. 1983. The elimination of inbreeding depression in a captive herd of Speke's gazelle. In C. M. Schonewald–Cox, S. M. Chambers, B. MacBryde, and L. Thomas, Eds. *Genetics and Conservation: A Reference for Managing Wild Animal and Plant Populations*. Addison–Wesley, Reading, Massachusetts, pp. 241–261.

Templeton, A. R., and B. Read. 1984. Factors eliminating inbreeding depression in a captive herd of Speke's gazelle (*Gazella spekei*). *Zoo Biology* 3:177–199.

Templeton, A. R., and B. Read. 1994. Inbreeding: One word, several meanings, much confusion. In V. Loeschcke, J. Tomiuk, and S. K. Jain, Eds. *Conservation Genetics*. Birkhäuser-Verlag, Basel, pp. 91–106.

Templeton, A. R., and B. Read. 1998. Elimination of inbreeding depression from a captive population of Speke's gazelle: Validity of the original statistical analysis and confirmation by permutation testing. *Zoo Biology* 17:77–94.

Templeton, A. R., R. A. Reichert, A. E. Weisstein, X. F. Yu, and R. B. Markham. 2004. Selection in context: Patterns of natural selection in the glycoprotein 120 region of human immunodeficiency virus 1 within infected individuals. *Genetics* 167:1547–1561.

Templeton, A. R., R. J. Robertson, J. Brisson, and J. Strasburg. 2001. Disrupting evolutionary processes: The effect of habitat fragmentation on collared lizards in the Missouri Ozarks. *Proceedings of the National Academy of Sciences USA* 98:5426–5432.

Templeton, A. R., and E. D. Rothman. 1978. Evolution in fine-grained environments. I. Environmental runs and the evolution of homeostasis. *Theoretical Population Biology* 13:340–355.

Templeton, A. R., E. Routman, and C. Phillips. 1995. Separating population structure from population history: A cladistic analysis of the geographical distribution of mitochondrial DNA haplotypes in the Tiger Salamander, *Ambystoma tigrinum*. *Genetics* 140:767–782.

Templeton, A. R., K. Shaw, E. Routman, and S. K. Davis. 1990b. The genetic consequences of habitat fragmentation. *Annals of the Missouri Botanical Garden* 77:13–27.

Templeton, A. R., and C. F. Sing. 1993. A cladistic analysis of phenotypic associations with haplotypes inferred from restriction endonuclease mapping. IV. Nested analyses with cladogram uncertainty and recombination. *Genetics* 134:659–669.

Templeton, A. R., C. F. Sing, and B. Brokaw. 1976. The unit of selection in *Drosophila mercatorum*. I. The interaction of selection and meiosis in parthenogenetic strains. *Genetics* 82:349–376.

Templeton, A. R., C. F. Sing, A. Kessling, and S. Humphries. 1988. A cladistic analysis of phenotypic associations with haplotypes inferred from restriction endonuclease mapping. II. The analysis of natural populations. *Genetics* 120:1145–1154.

Templeton, A. R., K. M. Weiss, D. A. Nickerson, E. Boerwinkle, and C. F. Sing. 2000b. Cladistic structure within the human lipoprotein lipase gene and its implications for phenotypic association studies. *Genetics* 156:1259–1275.

Thomas, M. G., T. Parfitt, D. A. Weiss, K. Skorecki, J. F. Wilson, M. I. Roux, N. Bradman, and D. B. Goldstein. 2000. Y chromosomes traveling south: The Cohen modal haplotype and the origins of the Lemba—the "Black Jews of Southern Africa." *American Journal of Human Genetics* 66:674–686.

Thomas, R. R. 1991. Ecological aspects of longevity, fecundity, and desiccation and the role of juvenile hormone in the abnormal abdomen syndrome of *Drosophila mercatorum* (Diptera: Drosophilidae). Texas A&M University, Ph.D. thesis.

Timofeef-Ressovsky, N. W. 1940. Zur analyse des polymorphismus bei *Adalia bipunctata* L. *Biologishe Zentralblatt* 60:130–137.

Tishkoff, S. A., R. Varkonyi, N. Cahinhinan, S. Abbes, G. Argyropoulos, G. Destro-Bisol, A. Drousiotou, B. Dangerfield, G. Lefranc, J. Loiselet, A. Piro, M. Stoneking, A. Tagarelli, G. Tagarelli, E. H. Touma, S. M. Williams, and A. G. Clark. 2001. Haplotype diversity and linkage disequilibrium at human G6PD: Recent origin of alleles that confer malarial resistance. *Science* 293:455–462.

Todorova, A., and G. A. Danieli. 1997. Large majority of single-nucleotide mutations along the dystrophin gene can be explained by more than one mechanism of mutagenesis. *Human Mutation* 9:537–547.

Tong, A. H. Y., M. Evangelista, A. B. Parsons, H. Xu, G. D. Bader, N. Page, M. Robinson, S. Raghibizadeh, C. W. V. Hogue, H. Bussey, B. Andrews, M. Tyers, and C. Boone. 2001. Systematic genetic analysis with ordered arrays of yeast deletion mutants. *Science* 294:2364–2368.

Tong, A. H. Y., G. Lesage, G. D. Bader, H. M. Ding, H. Xu, X. F. Xin, J. Young, G. F. Berriz, R. L. Brost, M. Chang, Y. Q. Chen, X. Cheng, G. Chua, H. Friesen, D. S. Goldberg, J. Haynes, C. Humphries, G. He, S. Hussein, L. Z. Ke, N. Krogan, Z. J. Li, J. N. Levinson, H. Lu, P. Menard, C. Munyana, A. B. Parsons, O. Ryan, R. Tonikian, T. Roberts, A. M. Sdicu, J. Shapiro, B. Sheikh, B. Suter, S. L. Wong, L. V. Zhang, H. W. Zhu, C. G. Burd, S. Munro, C. Sander, J. Rine,

J. Greenblatt, M. Peter, A. Bretscher, G. Bell, F. P. Roth, G. W. Brown, B. Andrews, H. Bussey, and C. Boone. 2004. Global mapping of the yeast genetic interaction network. *Science* 303:808–813.

Torroni, A., T. G. Schurr, C. C. Gan, E. Szathmary, R. C. Williams, M. S. Schanfield, G. A. Troup, W. C. Knowler, D. N. Lawrence, K. M. Weiss, and D. C. Wallace. 1992. Native American mitochondrial DNA analysis indicates that the Amerind and the Nadene populations were founded by two independent migrations. *Genetics* 130:153–162.

Tvrdik, T., S. Marcus, S. M. Hou, S. Falt, P. Noori, N. Podlutskaja, F. Hanefeld, F. Stromme, and B. Lambert. 1998. Molecular characterization of two deletion events involving alu-sequences, one novel base substitution and two tentative hotspot mutations in the hypoxanthine phosphoribosyltransferase (*HPRT*) gene in five patients with Lesch-Nyhan-Syndrome. *Human Genetics* 103:311–318.

Twells, R. C. J., C. A. Mein, M. S. Phillips, J. F. Hess, R. Veijola, M. Gilbey, M. Bright, M. Metzker, B. A. Lie, A. Kingsnorth, E. Gregory, Y. Nakagawa, H. Snook, W. Y. S. Wang, J. Masters, G. Johnson, I. Eaves, J. M. M. Howson, D. Clayton, H. J. Cordell, S. Nutland, H. Rance, P. Carr, and J. A. Todd. 2003. Haplotype structure, LD blocks, and uneven recombination within the LRP5 gene. *Genome Research* 13:845–855.

Val, F. C. 1977. Genetic analysis of the morphological differences between two interfertile species of Hawaiian *Drosophila*. *Evolution* 31:611–629.

Vallender, E. J., and B. T. Lahn. 2004. Positive selection on the human genome. *Human Molecular Genetics* 13:R245–R254.

van der Knaap, E., Z. B. Lippman, and S. D. Tanksley. 2002. Extremely elongated tomato fruit controlled by four quantitative trait loci with epistatic interactions. *Theoretical and Applied Genetics* 104:241–247.

Vander Molen, J., L. M. Frisse, S. M. Fullerton, Y. Qian, L. Del Bosque-Plata, R. R. Hudson, and A. Di Rienzo. 2005. Population genetics of CAPN10 and GPR35: Implications for the evolution of type 2 diabetes variants. *American Journal of Human Genetics* 76:548–560.

Van Eerdewegh, P., R. D. Little, J. Dupuis, R. G. D. Mastro, K. Falls, J. Simon, D. Torrey, S. Pandit, J. Mckenny, K. Braunschweiger, A. Walsh, Z. Liu, B. Hayward, C. Folz, S. P. Manning, A. Bawa, L. Saracino, M. Thackston, Y. Benchekroun, N. Capparell, M. Wang, R. Adair, Y. Feng, J. Dubois, M. G. Fitzgerald, H. Huang, R. Gibson, K. M. Allen, A. Pedan, M. R. Danzig, S. P. Umland, R. W. Egan, F. M. Cuss, S. Rorke, J. B. Clough, J. W. Holloway, S. T. Holgate, and T. P. Keith. 2002. Association of the ADAM33 gene with asthma and bronchial hyperresponsiveness. *Nature* 418:426–430.

van Tilburg, J., T. W. van Haeften, P. Pearson, and C. Wijmenga. 2001. Defining the genetic contribution of type 2 diabetes mellitus. *Journal of Medical Genetics* 38:569–578.

Vekemans, X., and M. Slatkin. 1994. Gene and allelic genealogies at a gametophytic self-incompatibility locus. *Genetics* 137:1157–1165.

Vigilant, L., M. Stoneking, H. Harpending, K. Hawkes, and A. C. Wilson. 1991. African populations and the evolution of human mitochondrial DNA. *Science* 253:1503–1507.

Vitart, V., A. D. Carothers, C. Hayward, P. Teague, N. D. Hastie, H. Campbell, and A. F. Wright. 2005. Increased level of linkage disequilibrium in rural compared with urban communities: A factor to consider in association-study design. *American Journal of Human Genetics* 76:763–772.

Vohl, M. C., F. Szots, M. Lelievre, P. J. Lupien, J. Bergeron, C. Gagne, and P. Couture. 2002. Influence of LDL receptor gene mutation and apo E polymorphism on lipoprotein response to simvastatin treatment among adolescents with heterozygous familial hypercholesterolemia. *Atherosclerosis* 160:361–368.

Volkman, S. K., A. E. Barry, E. J. Lyons, K. M. Nielsen, S. M. Thomas, M. Choi, S. S. Thakore, K. P. Day, D. F. Wirth, and D. L. Hartl. 2001. Recent origin of *Plasmodium falciparum* from a single progenitor. *Science* 293:482–484.

Voss, S. R., K. L. Prudic, J. C. Oliver, and H. B. Shaffer. 2003. Candidate gene analysis of metamorphic timing in ambystomatid salamanders. *Molecular Ecology* 12:1217–1223.

Voss, S. R., and H. B. Shaffer. 2000. Evolutionary genetics of metamorphic failure using wild-caught vs. laboratory axolotls (*Ambystoma mexicanum*). *Molecular Ecology* 9:1401–1407.

Waddington, C. H. 1957. *The Strategy of the Genes*. George Allen & Unwin, London.

Wade, M. J., and C. J. Goodnight. 1991. Wright's shifting balance theory: An experimental study. *Science* 253:1015–1018.

Wade, M. J., and C. J. Goodnight. 1998. Perspective: The theories of Fisher and Wright in the context of metapopulations: When nature does many small experiments [Review]. *Evolution* 52:1537–1553.

Wagner, G. P. 1988. The influence of variation and of developmental constraints on the rate of multivariate phenotypic evolution. *Journal of Evolutionary Biology* 1:45–66.

Wallace, D. C. 1976. The social effect of Huntington's chorea on reproductive effectiveness. *Annals of Human Genetics* 39:375–379.

Walsh, J. B. 1983. Role of biased gene conversion in one-locus neutral theory and genome evolution. *Genetics* 105:461–468.

Wang, H. Y., F. C. Zhang, J. J. Gao, J. B. Fan, P. Liu, Z. J. Zheng, H. Xi, Y. Sun, X. C. Gao, T. Z. Huang, Z. J. Ke, G. R. Guo, G. Y. Feng, G. Breen, D. St Clair, and L. He. 2000. Apolipoprotein E is a genetic risk factor for fetal iodine deficiency disorder in China. *Molecular Psychiatry* 5:363–368.

Waples, R. S. 1989. A generalized approach for estimating effective population size from temporal changes in allele frequency. *Genetics* 121:379–391.

Ward, R. H., and J. V. Neel. 1976. The genetic structure of a tribal population, the Yanomama Indians. XIV. Clines and their interpretation. *Genetics* 82:103–121.

Waterhouse, J. A. H., and L. Hogben. 1947. Incompatibility of mother and foetus with respect to the iso-agglutinogen A and its antibody. *British Journal of Preventative and Social Medicine* 1:1–17.

Weatherall, D. J. 2001. Phenotype–genotype relationships in monogenic disease: Lessons from the thalassaemias. *Nature Reviews Genetics* 2:245–255.

Wedekind, C., and S. Füri. 1997. Body odour preferences in men and women: Do they aim for specific MHC combinations or simple heterozygosity? *Proceedings of the Royal Society of London B* 264:1471–1479.

Wedekind, C., T. Seebeck, F. Bettens, and A. J. Paepke. 1995. MHC-dependent mate preferences in humans. *Proceedings of the Royal Society of London B* 260:245–249.

Wedell, N. 2001. Female remating in butterflies: Interaction between female genotype and nonfertile sperm. *Journal of Evolutionary Biology* 14:746–754.

Weinberg, S. 1977. *The First Three Minutes: A Modern View of the Origin of the Universe*. Basic Books, New York.

Weinberg, W. 1908. Über den Nachweis der Vererbung beim Menschen. *Naturkunde Württemberg* 64:368–382.

Weir, B. S., R. W. Allard, and A. L. Kahler. 1974. Further analysis of complex allozyme polymorphisms in a barley population. *Genetics* 78:911–919.

Weir, B. S., T. Ohta, and H. Tachida. 1985. Gene conversion models. *Journal of Theoretical Biology*. 116:1–8.

Weiss, G. H., and M. Kimura. 1965. A mathematical analysis of the stepping stone model of genetic correlation. *Journal of Applied Probability* 2:129–149.

Whitlock, M. C. 1999. Neutral additive genetic variance in a metapopulation. *Genetical Research* 74:215–221.

Whitlock, M. C., and P. C. Phillips. 2000. The exquisite corpse: A shifting view of the shifting balance. *Trends in Evolution and Ecology* 15:347–348.

Wiesenfeld, S. L. 1967. Sickle-cell trait in human biological and cultural evolution. *Science* 157:1134–1140.

Williamson, E. G., and M. Slatkin. 1999. Using maximum likelihood to estimate population size from temporal changes in allele frequencies. *Genetics* 152:755–761.

Wilson, J. F., and D. B. Goldstein. 2000. Consistent long-range linkage disequilibrium generated by admixture in a Bantu-Semitic hybrid population. *American Journal of Human Genetics* 67:926–935.

Wilson, T. G., M. H. Landers, and G. M. Happ. 1983. Precocene I and II inhibition of vitellegenic oocyte deveopment in *Drosophila melanogaster*. *Journal of Insect Physiology* 29:249–254.

Wolpoff, M. H., J. Hawks, and R. Caspari. 2000. Multiregional, not multiple origins. *American Journal of Physical Anthropology* 112:129–136.

Wood, T. K., and M. C. Keese. 1990. Host-plant-induced assortative mating in *Enchenopa* treehoppers. *Evolution* 44:619–628.

Wootton, J. C., X. Feng, M. T. Ferdig, R. A. Cooper, J. Mu, D. I. Baruch, A. J. Magill, and X.-Z. Su. 2002. Genetic diversity and chloroquine selective sweeps in *Plasmodium falciparum*. *Nature* 418:320–323.

Workman, P. L. 1973. Genetic analysis of hybrid populations. In M. H. Crawford and P. L. Workman, Eds. *Methods and Theories of Anthropological Genetics*. University of New Mexico Press, Albuquerque, pp. 117–150.

Worobey, M. 2001. A novel approach to detecting and measuring recombination: New insights into evolution in viruses, bacteria, and mitochondria. *Molecular Biology and Evolution* 18:1425–1434.

Wozniak, M. A., E. B. Faragher, J. A. Todd, K. A. Koram, E. M. Riley, and R. F. Itzhaki. 2003. Does apolipoprotein E polymorphism influence susceptibility to malaria? *Journal of Medical Genetics* 40:348–351.

Wright, S. 1931. Evolution in Mendelian populations. *Genetics* 16:97–159.

Wright, S. 1932. The roles of mutation, inbreeding, crossbreeding, and selection in evolution. *Proceedings of the Sixth International Congress on Genetics* 1:356–366.

Wright, S. 1943. Isolation by distance. *Genetics* 28:114–138.

Wright, S. 1946. Isolation by distance under diverse systems of mating. *Genetics* 31:39–59.

Wright, S. 1969. *Evolution and the Genetics of Populations*, Vol. 2: *The Theory of Gene Frequencies*. University of Chicago Press, Chicago.

Wu, R. L. 2000. Partitioning of population genetic variance under multiplicative-epistatic gene action. *Theoretical and Applied Genetics* 100:743–749.

Wu, X. Z. 2004. On the origin of modern humans in China. *Quaternary International* 117:131–140.

Wyatt, G. R. 1997. Juvenile hormone in insect reproduction—A paradox. *European Journal of Entomology* 94:323–333.

Yang, W. L., B. Whitea, E. K. Spicer, B. L. Weinstein, and J. D. Hildebrandt. 2004. Complex haplotype structure of the human GNAS gene identifies a recombination hotspot centred on a single nucleotide polymorphism widely used in association studies. *Pharmacogenetics* 14:741–747.

Yokoyama, S., and A. R. Templeton. 1980. The effect of social selection on the population dynamics of Huntington's disease. *Annals of Human Genetics* 43:413–417.

Zareparsi, S., R. Camicioli, G. Sexton, T. Bird, P. Swanson, J. Kaye, J. Nutt, and H. Payami. 2002. Age at onset of Parkinson disease and apolipoprotein E genotypes. *American Journal of Medical Genetics* 107:156–161.

Zaykin, D. V., P. H. Westfall, S. S. Young, M. A. Karnoub, M. J. Wagner, and M. G. Ehm. 2002. Testing association of statistically inferred haplotypes with discrete and continuous traits in samples of unrelated individuals. *Human Heredity* 53:79–91.

Zeyl, C., and J. DeVisser. 2001. Estimates of the rate and distribution of fitness effects of spontaneous mutation in *Saccharomyces cerevisiae*. *Genetics* 157:53–61.

Zhu, X., A. Luke, R. S. Cooper, T. Quertermous, C. Hanis, T. Mosley, C. Charles Gu, H. Tang, D. C. Rao, N. Risch, and A. Weder. 2005. Admixture mapping for hypertension loci with genome-scan markers. *Nature Genetics* 37:177.

Zhu, Z., V. Vincek, F. Figueroa, C. Schonbach, and J. Klein. 1991. MHC-DRB genes of the pigtail macaque (*Macaca nemestrina*): Implications for the evolution of human DRB genes. *Molecular Biology and Evolution* 8:563–578.

Zuckerkandl, E., and L. Pauling. 1962. Molecular disease, evolution, and genetic heterogeneity. In M. Kasha and B. Pullman, Eds. *Horizons in Biochemistry*. Academic, New York, pp. 189–225.

PROBLEMS AND ANSWERS

CHAPTER 2

Problems

1. A population of infants from Musoma, Tanzania, was scored for the β-hemoglobin locus genotype (determined by a single autosomal locus with two alleles, A and S) as follows:

Genotype	AA	AS	SS	Total
Number	189	89	9	287

    Characterize this population by its genotypic frequencies. Characterize the gene pool by the allele frequencies for A and S. Using the Hardy–Weinberg law, predict the genotypic frequencies.

2. Among people of southern Italian and Sicilian ancestry living in Rochester, New York ($N = 10,000$), about one birth in 2500 has thalassemia major (a type of anemia) and about one birth in 25 has a milder anemia known as thalassemia minor. Are these data compatible with a single-locus hypothesis as a basis for the heredity of these anemic conditions? Why or why not?

3. Calculate the genotype frequencies expected under Hardy-Weinberg to determine which of the following populations are at Hardy-Weinberg genotypic frequencies. For any population not at Hardy-Weinberg, state whether there is an excess or a deficiency of heterozygotes.

Genotype	AA	Aa	aa	Total
a.	50	20	30	100
b.	25	10	1	36
c.	20	20	5	45
d.	9	10	81	100
e.	5625	3750	625	10,000

4. Two populations are examined for the same gene locus with the following results:

    Population 1: AA 162; Aa 36; aa 2

    Population 2: AA 18; Aa 84; aa 98

Population Genetics and Microevolutionary Theory, By Alan R. Templeton
Copyright © 2006 John Wiley & Sons, Inc.

Now suppose these two populations are combined to form a new population:

Population 3: *AA* 180; *Aa* 120; *aa* 100

a. What system(s) of mating do populations 1 and 2 have for this locus? Are their different genotype frequencies explained by different systems of mating? If not, why do their genotype frequencies differ?

b. Is population 3 at Hardy–Weinberg equilibrium? If not, indicate whether heterozygotes are in excess or in deficiency.

c. For how many generations will the effect of this single episode of admixture of populations be detectable in a population established from population 3 with respect to genotype frequencies if mating is at random?

5. A population is polymorphic for an *Eco*R1 restriction site in a defined chromosomal region. Of the chromosomes 75% have the cut site and 25% do not. A closely linked mutation then occurs that causes a *Hin*dIII cut site to appear. This mutation occurs on a chromosome without the *Eco*R1 cut site. With subsequent evolution, the frequency of the *Hin*dIII cut site evolves to a current value of 0.1, although the frequency of the *Eco*R1 cut site remains at 0.75. During this time, no recombination events occur between the *Eco*R1 and *Hin*dIII sites.

   a. Calculate the gametic-phase imbalance (linkage disequilibrium) between the two restriction sites in the current population.

   b. Suppose that the initial and current frequencies of the *Eco*R1 cut site had both been 0.2 instead of 0.75 but all else is as described above. Calculate the gametic-phase imbalance between the two restriction sites in the current population.

6. One thousand individuals were typed for two traits determined by two different autosomal loci, each with two alleles (*A* and *a*; *B* and *b*) with the following table giving the two-locus genotype numbers:

	BB	Bb	Bb	Sum
AA	36	48	16	100
Aa	72	96	32	200
Aa	252	336	112	700
Sum	360	480	160	1000

   a. Determine if *each locus* is at a single-locus Hardy–Weinberg equilibrium. If not, what would the genotype frequencies be at each locus considered separately?

   b. Are these two loci in linkage equilibrium? If not, what would each gametic combination frequency be if they were in linkage equilibrium?

   c. Is this array of nine genotypes consistent with random mating and neutrality? If not, what would be the expected frequencies of these nine biloci genotypes?

7. The frequency of phenylketonuria (PKU, caused by an autosomal recessive allele) is 0.00004 at birth. Assuming Hardy–Weinberg, what is the frequency of the PKU allele?

What is the expected Hardy–Weinberg ratio of PKU carriers (heterozygotes) to affecteds (PKU homozygotes)?

8. Given random mating and the absence of mutation migration and selection, what will be the genotype frequencies in the zygotes of the next generation of a very large population if the initial population had the following genotype frequencies?

Genotype	AA	Aa	aa
a.	0.16	0.48	0.36
b.	0.30	0.00	0.70
c.	0.20	0.40	0.40

9. A person has type O blood only if they are homozygous for the O allele at the ABO locus. The frequency of the O allele is 0.67 in the United States. A person's blood is Rh+ if he or she is homozygous or heterozygous for the D allele at the Rh locus, which has an allele frequency of 0.6. Assuming no linkage disequilibrium between these two unlinked loci, what is the frequency of people with blood type O+, assuming a random-mating population?

10. Two populations are examined for the same pair of loci with the following results in terms of two-locus gamete numbers:

   Population 1: AB 720; Ab 180; aB 80; ab 20
   Population 2: AB 300; Ab 2700; aB 700; ab 6300

   These two populations are now combined to form a third:

   Population 3: AB 1020; Ab 2880; aB 780; ab 6320

   a. Calculate the two-locus gamete frequencies for each population and the gametic-phase imbalance (linkage disequilibrium). Is any population out of linkage equilibrium?
   b. What is the effect of pooling of the two populations on linkage equilibrium?
   c. Suppose that the recombination rate between loci A/a and B/b is 0.1. What is the expected disequibrium in the offspring of population 3 (the parental gene pool), assuming random mating? In the second generation?
   d. Can this single episode of admixture be detected in the population established from population 3 after two generations of random mating? Can it be detected in the genotype frequencies at the A/a locus after two generations of random mating? Can it be detected in the genotype frequencies at the B/b locus after two generations of random mating?

11. Consider two unlinked loci each with two alleles: A and a; B and b. The frequencies of the nine observable genotypes in a population with respect to these loci are:

	AA	Aa	aa
BB	0.20	0.40	0.04
Bb	0.05	0.07	0.20
bb	0.00	0.03	0.01

a. Is this population at Hardy–Weinberg equilibrium for each locus considered separately?

b. Is this population at Hardy–Weinberg equilibrium with no gametic-phase imbalance with respect to the nine, two-locus genotypic frequencies?

c. If it is not in two-locus equilibrium, what are the equilibrium genotype frequencies and how long will it take to dissipate the gametic-phase imbalance to less than 5% of its initial value?

d. If the loci are linked with a recombination frequency of 0.001, how will this change the equilibrium genotype frequencies?

12. A population of Greenland Eskimos, when tested for the blood antigens M and N (determined by a single locus with two codominant alleles), was categorized as follows:

Blood type	M	MN	N	Total
Number	61	64	27	152

a. Characterize this population by its genotypic frequencies.

b. Characterize the gene pool by the allele frequencies for M and N.

c. Using the Hardy–Weinberg law, predict the genotypic frequencies.

d. Test the goodness of fit of this population to the Hardy–Weinberg expectations.

13. Suppose that in the above population it was not possible to test for the M antigen, but instead it was only possible to test for the presence or absence of antigen N.

a. Assuming Hardy–Weinberg, what is the frequency of the allele coding for the M antigen?

b. What is the expected Hardy–Weinberg ratio of *MN* heterozygotes to *MM* homozygotes?

14. A person has type M blood only if he or she is homozygous for the *M* allele at the *MN* locus. The frequency of the *M* allele is 0.24 in European Americans and 0.09 in African Americans. A person's blood is Rh+ if homozygous or heterozygous for the *D* allele at the *Rh* locus, which has an allele frequency of 0.6 in European Americans and 0.84 in African Americans.

a. Assuming no linkage disequilibrium between these two unlinked loci within European Americans and within African Americans, what are the frequencies of gametes bearing the *M* and *D* alleles in European Americans and African Americans?

b. Suppose all the people in a town are surveyed for these alleles, and the population consists of 70% European Americans and 30% African Americas. Do the *M* and *D* alleles show linkage disequilibrium in the total town population, and if so, what is its value?

c. Assuming random mating within European Americans and African Americans, what is the frequency of Rh+ people in the total town population? What is the frequency of the *D* allele in the total town population? Is the frequency of Rh+ people in the total town population predicted well by applying the Hardy–Weinberg law to the total town gene pool (don't test statistically, just look at exact numbers)?

d. Suppose the town population becomes a single random-mating population. What is the frequency of Rh+ people in the town after a single generation of random mating? What is the linkage disequilibrium between the *M* and *D* alleles after a single generation of random mating given that the *MN* and *Rh* loci are on separate autosomes?

15. The disease erythroblastosis fetalis occurs when an Rh− woman has an Rh+ child. If there is some leakage across the placenta, the Rh− mother can produce antibodies against the Rh+ fetus. This usually does not affect the first Rh+ child, but subsequent pregnancies with Rh+ fetuses can be severely affected by this immune reaction. Given the data in problem 14, what are the frequencies of families at risk for this disease in European Americans and in African Americans, assuming random mating within these two populations?

16. Consider a population with the following gamete frequencies at a pair of loci: *AB* 0.4; *Ab* 0.3; *aB* 0.2; *ab* 0.1.
    a. Calculate the linkage disequilibrium as both D and D' (equation 2.15).
    b. Suppose that the recombination rate between loci A/a and B/b is 0.2. What is the expected disequibrium (D) in the next generation, assuming random mating? In the second generation?
    c. Now suppose that the recombination rate is 0.01. Repeat the calculations in part b under this assumption.

Answers

1. First, you must calculate the genotypic frequencies, since your calculations from this point on rely on frequencies, not absolute numbers:

Genotype	AA	AS	SS	Total
Number	189	89	9	287
Frequency	189/287 = 0.66	0.31	0.03	1.0

Next, calculate allele frequencies:

$$p(A) = G_{AA} + \frac{1}{2}G_{AS} = 0.66 + 0.155 = 0.815$$

$$q(S) = G_{SS} + \frac{1}{2}G_{AS} = 0.03 + 0.155 = 0.184$$

Using the values you obtain for p and q above, determine the Hardy–Weinberg genotype frequencies:

AA	AS	SS
p^2	$2pq$	p^2
0.662	0.303	0.035

2. To answer this question, construct a model with all the elements of the basic Hardy–Weinberg model and see if the data you have been given fit. Assume two alleles and one locus and arbitrarily name the alleles A and a. You have the genotypic frequencies of two genotypes from the question: $Aa = 1/25 = 0.04$; $AA = 1/2500 = 0.0004$.

Therefore:

AA (thalassemia major)	Aa (thalassemia minor)	aa (no anemia)
0.0004	0.04	0.9596 $(1-G_{AA}-G_{Aa})$

Now, calculate p and q:

$$p = 0.0004 + 0.02 = 0.0204$$
$$q = 0.9596 + 0.02 = 0.9796$$

Given these values for p and q, the expected genotype frequencies under Hardy–Weinberg are $p^2 = 0.0004$; $2pq = 0.04$; $q^2 = 0.9596$. These match the given values in the population. Therefore, this trait fits the Hardy–Weinberg two-allele, one-locus model.

3. These calculations should be done as shown in problem 1:

Genotype	AA	Aa	Aa	Heterozygote
a. Observed number	50	20	30	
Observed frequency	0.5	0.2	0.3	
Hardy–Weinberg expected	0.36	0.48	0.16	Deficiency
b. Observed number	25	10	1	
Observed frequency	0.694	0.278	0.028	
Hardy–Weinberg expected	0.694	0.278	0.028	In Hardy–Weinberg
c. Observed number	20	20	5	
Observed frequency	0.444	0.444	0.111	
Hardy–Weinberg expected	0.444	0.444	0.111	In Hardy–Weinberg
d. Observed number	9	10	81	
Observed frequency	0.09	0.10	0.81	
Hardy–Weinberg expected	0.02	0.24	0.78	Deficiency
e. Observed number	5625	3750	625	
Observed frequency	0.563	0.375	0.063	
Hardy–Weinberg expected	0.563	0.375	0.063	In Hardy–Weinberg

4. a. To determine the system of mating that these populations are experiencing, first calculate their Hardy–Weinberg parameters:

Population 1:	AA	Aa	aa	
	162	36	2	
G_{ij} observed	$162/200 = 0.81$	0.18	0.01	$p = 0.81 + 0.09$
G_{ij} expected	$p^2 = 0.81$	$2pq = 0.18$	$q^2 = 0.01$	$= 0.9; q = 0.1$

Population 2:   AA   Aa   aa

	AA	Aa	aa	
	18	84	98	
G_{ij} observed	$18/200 = 0.09$	0.42	0.49	$p = 0.09 + 0.21$
G_{ij} expected	$p^2 = 0.09$	$2pq = 0.42$	$q^2 = 0.49$	$= 0.3; q = 0.7$

Both of these populations meet Hardy–Weinberg expectations: Random mating cannot be rejected in either of these populations. The genotype frequencies between populations 1 and 2 differ because of different allele frequencies, not because of different systems of mating.

b. Again, first calculate the Hardy–Weinberg parameters:

	AA	Aa	aa	
Population 3:				
	180	120	100	
G_{ij} observed	$180/400 = 0.45$	0.30	0.25	$p = 0.45 + 0.15$
G_{ij} expected	$p^2 = 0.36$	$2pq = 0.48$	$q^2 = 0.16$	$= 0.6; q = 0.4$

This population does not meet Hardy–Weinberg expectations. The observed frequency of heterozygotes is 0.30, but the Hardy–Weinberg expected frequency is 0.48. Therefore, there is a deficiency of heterozygotes.

c. With random mating after the admixture event, the admixture will be undetectable after the first generation.

5. a. It may be helpful to draw a picture in order to conceptualize this problem correctly:

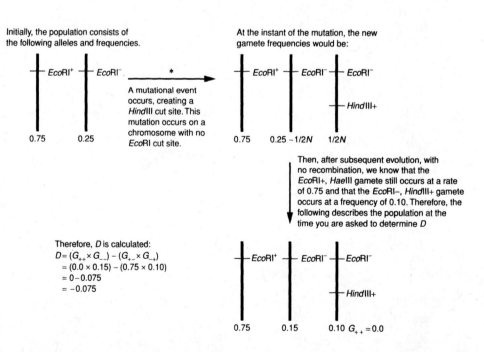

b. This part is done exactly as part a except that the initial frequencies of the alleles in the population are changed. The final result:

The gamete frequencies are now:

	EcoR1$^+$	EcoR1$^-$
HindIII$^+$	0.00	0.10
HindIII$^-$	0.20	0.70

So, $D = (0.00)(0.70) - (0.10)(0.20) = -0.020$.

6. a. Approach this part the same way you did problem 1. For the A locus (row sums), $p = 0.2$ and the Hardy–Weinberg (HW) frequencies are 0.04, 0.32, and 0.64, respectively. For the B locus (column sums), $p = 0.6$ and it is exactly at HW.

b. b. At linkage equilibrium the two-locus gamete frequencies are given by the products of the marginal allele frequencies. Therefore, to test if there really is no linkage disequilibrium, we must determine if the gamete frequencies are equal to the products of the marginal frequencies. We can test this by determining the actual gamete frequencies from the data by summing over gamete types:

	B	b	Total
A	$36 + 1/248 + 1/272 + 1/496 = 120$	80	200
a	$252 + 1/272 + 1/2336 + 1/496 = 480$	320	800
Total	600	400	1000

At linkage equilibrium, the two-locus gamete frequencies are given by the products of the allele (marginal) frequencies, which they are. Therefore, this system has no linkage disequilibrium.

c. Take the product of the marginal HW frequency to get the total HW expected frequencies:

A Locus Genotype	Two-Locus Expected Genotype Frequency			A Locus HW
AA	0.0144	0.0192	0.0064	0.04
Aa	0.1152	0.1536	0.0512	0.32
Aa	0.2304	0.3072	0.1024	0.64
B locus genotype	BB	Bb	Bb	
B locus HW	0.36	0.48	0.16	

These do NOT correspond to the observed frequencies because locus A is not in HW.

7. $q^2 = 0.00004$, so $q = 0.0063$, so $2pq = 0.0126$. Ratio $= 2pq/q^2 = 315 : 1$.

8.

		AA	Aa	aa
a. $p = (.16) + \frac{1}{2}(0.48) = 0.4$	Next generation frequencies:	$(0.4)^2 = 0.16$	$2(0.4)(0.6) = 0.48$	$(0.6)^2 = 0.36$
b. $p = (.30) + \frac{1}{2}(0.00) = 0.3$	Next generation frequencies:	$(0.3)^2 = 0.09$	$2(0.3)(0.7) = 0.42$	$(0.7)^2 = 0.49$
c. $p = (0.20) + \frac{1}{2}(0.40) = 0.4$	Next generation frequencies:	$(0.4)^2 = 0.16$	$2(0.4)(0.6) = 0.48$	$(0.6)^2 = 0.36$

9. The genotype frequency under random mating of OO is $(0.67)^2 = 0.45$. The frequency of Rh+ individuals is $(0.6)^2 + 2(0.6)(0.4) = 0.84$. With no linkage disequilibrium, the frequency of O+ individuals is $(0.45)(0.84) = 0.38$.

10. a. $D(1) = (0.72)(0.02) - (0.18)(0.08) = 0$
    $D(2) = 0$
    $D(3) = (0.09273)(0.57455) - (26,182)(0.07091) = 0.03471$
    Population 3 is not in linkage equilibrium.
    b. Creates linkage disequilibrium.
    c. $D_1 = (1-r)D_0 = 0.9(0.03471) = 0.03124$; $D_2 = 0.9(0.03124) = 0.02812$
    d. $D_t \neq 0$ at $t = 2$, so admixture is still detectable as a nonequilibrium state. Admixture is not detectable with the single-locus genotype frequencies (see problem 4).

11. Let p be the frequency of A, q of a, r of B, and s of b.
    a. First determine the genotype frequencies for just the A/a locus by adding up the frequencies in the columns and the B/b locus genotype frequencies by adding up the rows. Then determine allele frequencies by adding up appropriate genotype frequencies (dividing by 2 for heterozygotes) to obtain $p = 0.5$, $q = 0.5$, $r = 0.8$, and $s = 0.2$. Both single loci are in Hardy–Weinberg.
    b. No. This is immediately apparent from the complete absence of the $AAbb$ genotype.
    c. Multiply the appropriate single-locus Hardy–Weinberg genotype frequencies:

	AA	Aa	aa
BB	$(0.25)(0.64) = 0.16$	$(0.5)(0.64) = 0.32$	$(0.25)(0.64) = 0.16$
Bb	$(0.25)(0.32) = 0.08$	$(0.5)(0.32) = 0.16$	$(0.25)(0.32) = 0.08$
bb	$(0.25)(0.04) = 0.01$	$(0.5)(0.04) = 0.02$	$(0.25)(0.04) = 0.01$

    d. Linkage affects only the time to equilibrium, not the equilibrium itself. So there is no effect of $r = 0.001$ upon the equilibrium genotype frequencies.

12.
Genotype	MM	MN	NN
a. Frequency	$61/152 = 0.4013$	0.4211	0.1776
c. Hardy–Weinberg	$(0.612)^2 = 0.3744$	$2(0.612)(0.388) = 0.475$	$(0.388)^2 = 0.1507$
d. Expectation	$0.3744(152) = 560.901$	72.197	22.901
$(o-e)^2/e$	0.2952	0.9307	0.7335
Sum = 1.9595,	df = 1,	$p = 0.16$	

Gametes	M	N
b. $4013 + 0.5(0.4211) =$	0.612	0.388

Fail to reject Hardy–Weinberg for this population at the 5% level.

13. a. $p^2 = 0.4013$ (the non-N phenotype = MM), so $p = 0.6335$
    b. $2pq = 0.4644$. Ratio = $2pq/p^2 = 1.1572 : 1$

14. a. Frequency of MD in European Americans = $(0.24)(0.6) = 0.144$
    Frequency of MD in African Americans = $(0.09)(0.84) = 0.0756$
    b. Frequency of MD in total population = $0.7(0.144) + (0.3)(0.0756) = 0.1235$.
    Frequency of M in total population = $0.7(0.24) + (0.3)(0.09) = 0.195$
    Frequency of D in total population = $0.7(0.6) + (0.3)(0.84) = 0.672$

From equation 2.9, $D = 0.1235 - (0.195)(0.672) = -0.0075$. The town population has linkage disequilibrium.

c. Frequency of $Rh+$ within European Americans $= 1 - (1 - p)^2 = 1 - 0.4^2 = 0.84$
Frequency of $Rh+$ within African Americans $= 1-(1-p)^2 = 1-0.16^2 = 0.9744$
Frequency of $Rh+$ in town $= 0.7(0.84) + (0.3)(0.9744) = 0.8803$
Frequency of D in total population $= 0.672$, so frequency of $Rh+$ in town with Hardy–Weinberg is $1 - (1 - p)^2 = 1 - 0.328^2 = 0.8924$
There are fewer $Rh+$'s than expected under Hardy–Weinberg.

d. Expected under random mating $= 0.8924$ (see above)
$D_1 = (1 - r)D_0 = 0.5(-0.0075) = -0.0038$

15. The matings at risk are (mother × father) $dd \times DD$ and $dd \times Dd$. Under random mating, the probability of an at-risk mating is (see Table 2.3), in European Americans,

$$(0.4)^2 \times (0.6)^2 + (0.4)^2 \times [2(0.4)(0.6)] = 0.1344$$

Under random mating, the probability of an at-risk mating is, in African Americans,

$$(0.16)^2 \times (0.84)^2 + (0.16)^2 \times [2(0.16)(0.84)] = 0.0249$$

16. a. $D = (0.4)(0.1) - (0.3)(0.2) = -0.02$
$p_A = 0.4 + 0.3 = 0.7$; $p_a = 0.2 + 0.1 = 0.3$; $p_B = 0.4 + 0.2 = 0.6$;
$p_b = 0.3 + 0.1 = 0.4$
$p_A p_B = (0.7)(0.6) = 0.42$; $p_a p_b = (0.3)(0.4) = 0.12$
$D' = -0.02/(0.12) = -0.1667$

b. $D_1 = (1 - r)D_0 = 0.8(-0.02) = -0.016$; $D_2 = 0.8(-0.016) = -0.0128$

c. $D_1 = (1 - r)D_0 = 0.99(-0.02) = -0.0198$; $D_2 = 0.99(-0.0198) = -0.0196$

CHAPTER 3

Problems

1. Consider the following two pedigrees:

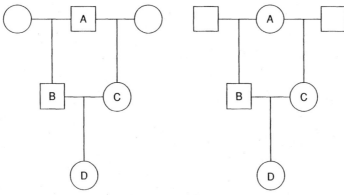

a. b.

What is the probability of identity by descent in individual D for a randomly chosen X-linked locus in pedigrees a and b?

2. Consider the pedigree below:

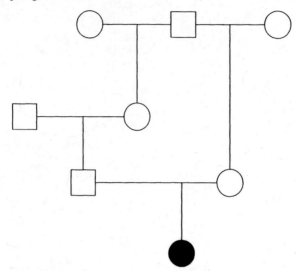

What is the pedigree inbreeding coefficient F of the female shown by the solid circle under the assumption that all relevant genetic relationships are shown in the pedigree.

3. Given the following allele frequencies for an autosomal locus with two alleles (A and a, with p being the frequency of A) and inbreeding coefficients (measured as a deviation from Hardy–Weinberg proportions in all problems in this set), calculate the genotype frequencies.
   a. $p = 0.1, f = 0.5$
   b. $p = 0.8, f = 0.1$
   c. $p = 0.3, f = 0.05$
   d. $p = 0.3, f = -0.05$

4. Estimate the value of f for each population given the following genotype frequencies.

	GENOTYPE		
	AA	Aa	aa
a.	0.30	0.40	0.30
b.	0.20	0.60	0.20
c.	0.056	0.288	0.656
d.	0.024	0.352	0.624

5. Estimate the value of f for each population given the following genotype numbers.

	GENOTYPE		
	AA	Aa	aa
a.	60	0	100
b.	20	60	20
c.	50	50	50

Pool all of these populations together and calculate the value of f for the combined population.

6. Consider the model of assortative mating shown in Figure 3.6. Calculate the genotype and allele frequencies for the first three generations of assortative mating starting with a population in Hardy–Weinberg equilibrium and $p = 0.6$.

7. Consider the model of disassortative mating shown in Figure 3.8. Calculate the genotype and allele frequencies for the first three generations of disassortative mating starting with a population in Hardy–Weinberg equilibrium and $p = 0.6$.

8. The following data were collected on species X at two locations. At this autosomal locus, there are two alleles, A and B.

Location	AA	AB	BB
1	144	471	384
2	64	32	4

(a) For each population, calculate f, the system of mating "inbreeding" coefficient that measures deviations from random mating.
(b) Pool the two populations together and estimate f for the pooled population. Is the deviation from random-mating expectations reduced or intensified by this pooling?

9. Suppose a deme is scored for three randomly chosen loci. The genotypic frequency results are as follows:

Locus i	$A_i A_i$	$A_i a_i$	$a_i a_i$
1	0.09	0.42	0.49
2	0.36	0.48	0.16
3	0.111	0.378	0.511

What do you conclude about this population's system of mating (assume no selection or drift) and why?

10. Same as problem 9, but now assume the results were:

1	0.09	0.42	0.49
2	0.16	0.48	0.36
3	0.024	0.352	0.624

11. Same as problem 9, but now assume the results were:

1	0.1005	0.3990	0.5005
2	0.1720	0.4560	0.3720
3	0.0480	0.3040	0.6480

12. Consider two loci each with two alleles (A and a; B and b) with recombination frequency of 0.3 between them. Population 1 has the frequency of A as 0.5 and the frequency of

B as 0.1, with no linkage disequilibrium, whereas population 2 has the frequency of A as 0.9 and the frequency of B as 0.9, once again with no linkage disequilibrium. These two populations are brought together to form a new admixed population consisting of 60% population 1 and 40% population 2.

a. What is the linkage disequilibrium in the initial admixed population?

b. Suppose population 1 speaks Frigglepix and population 2 speaks Tharkian and there is 100% assortative mating by language in the admixed population. What is the linkage disequilibrium after one generation of assortative mating?

c. Suppose instead that there is random mating in the admixed population. What is the linkage disequilibrium after one generation of random mating?

Answers

1. a. Individual B receives his only X chromosome from his mother, who is not a common ancestor. Therefore, the probability of identity by descent in D is zero.

   b. Individual A passes on X chromosomes to both B and C, and the probability of a match is $\frac{1}{2}$. Individual B passes on his X with probability 1 to D, and C passes on her X obtained from A with probability $\frac{1}{2}$. Hence, the total probability of identity by descent is $\frac{1}{4}$.

2.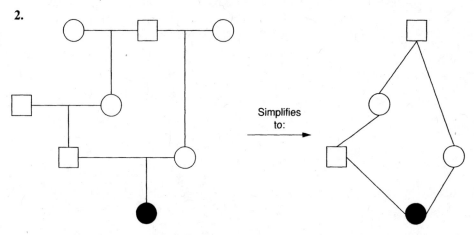

   Calculate F as $\left(\frac{1}{2}\right)^4 = 0.0625$.

3.
	AA	Aa	aa
a. $p = 0.1$, $f = 0.5$	0.055	0.09	0.855
b. $p = 0.8$, $f = 0.1$	0.656	0.288	0.056
c. $p = 0.3$, $f = 0.05$	0.1005	0.399	0.5005
d. $p = 0.3$, $f = -0.05$	0.0795	0.441	0.4795

4.
$p = G_{AA} + \left(\frac{1}{2}\right)G_{Aa}$	$f = 1 - G_{Aa}/(2pq)$
a. 0.5	0.2
b. 0.5	-0.2
c. 0.2	0.1
d. 0.2	-0.1

5.

	GENOTYPE			
	AA	Aa	aa	
a.	60	0	100	$f = 1 = 1 -$ observed heterozygote/(expected $= 2pq$)
b.	20	60	20	$f = -0.2$
c.	50	50	50	$f = \frac{1}{3} = 0.3333$

Pool all of these populations together and calculate the value of f for the combined population.

	AA	Aa	aa	
Pooled population	130	110	170	$f = 0.4583$

6.

Generation	AA	Aa	aa	A	a
0	0.36	0.48	0.16	0.6	0.4
1	0.48	0.24	0.28	0.6	0.4
2	0.54	0.12	0.34	0.6	0.4
3	0.57	0.06	0.37	0.6	0.4

7.

Generation	AA	Aa	aa	A	a
0	0.36	0.48	0.16	0.6	0.4
1	0.28125	0.59375	0.125	0.578125	0.421875
2	0.30212	0.56360	0.13428	0.583922	0.416078
3	0.29714	0.57079	0.13206	0.582540	0.417460

8. a. First, calculate the observed genotype frequencies. Then, using the allele frequencies calculated from the genotype frequencies, determine the expected Hardy–Weinberg heterozygote frequencie. For example, for locality 1:

AA	AB	BB
144/999	471/999	384/999
0.144	0.471	0.384

Frequency of $A = p = 0.144 + 0.5(0.471) = 0.38$
Frequency of $B = q = 0.62$
Expected heterozygotes $= 2pq = 0.471$

$$f = 1 - \frac{\text{observed heterozygotes}}{\text{expected heterozygotes}} = 1 - \frac{0.471}{0.471} = 0.0$$

If you follow these steps for the other population, you will find $p = 0.8$ and $f = 0.0$.

b. $p = 0.4181$ and $f = 0.0594$. The pooling of the populations has created a heterozygote deficiency from Hardy–Weinberg.

9. $f = 0$ for $i = 1, 2$; $f = 0.1$ for $i = 3$. Implies random mating at loci 1 and 2; assortative mating at locus 3. It cannot be inbreeding because inbreeding affects all loci simultaneously.

10. $f = 0$ for $i = 1, 2$; $f = -0.1$ for $i = 3$. Implies random mating at loci 1 and 2; disassortative mating at locus 3.

11. $f = 0.05$ for all i. Implies inbreeding, not assortative mating given three "randomly chosen loci."

12. a. What is the linkage disequilibrium in the initial admixed population? Use $D_{admixture} = m(1 - m)(p_1 - p_2)(k_1 - k_2)$, $m = 0.6$, $p_1 = 0.5$, $p_2 = 0.9$, $k_1 = 0.1$, $k_2 = 0.9$, so $D = 0.0768$.
b. Same as previous generation, so $D = 0.0768$.
c. $D_1 = (0.0768)(1 - 0.3) = 0.0538$

CHAPTER 4

Problems

1. An allele has a frequency of 0.01 in a population. If genetic drift is the only force operating, what is the probability under random mating that the allele will ultimately be lost if the population size is 50? Repeat the calculation for a population size of 5,000,000. Repeat both calculations under the assumption that the population is now inbreeding with $f = 0.1$.

2. One hundred populations of size 10 are all started with two alleles (A and a) at a locus, with the A allele having a frequency of 0.2. Eventually, all populations are fixed for one allele or the other, with 32 being fixed for A. Test the hypothesis (using a chi-square statistic, stating degrees of freedom) that this result is consistent with genetic drift being the sole evolutionary force operating on this locus.

3. Eight thousand isolated populations of size 4, each with the same initial frequency of allele A (p), are simulated with the following results:

NUMBER OF ISOLATES WITH y A ALLELES

Generation	$y = 0$	1	2	3	4	5	6	7	8
5	5516	537	564	446	335	238	175	105	84
10	6403	179	221	196	174	158	149	119	401
20	6887	39	47	51	33	55	42	42	804
40	7029	3	2	4	4	3	4	4	947
60	7038	0	0	0	0	0	0	0	962

For the questions below, you will need to calculate means and variances of allele frequencies. Let p_y be the frequency of A in an isolate with y copies of A, n_y be the number of simulated populations with allele frequency p_y, and $n = \sum n_y = 8000$ the total number of simulations. Then the mean and variance of allele frequencies are given by

$$\bar{p} = \sum_{y=0}^{8} \frac{n_y p_y}{n} \quad \text{Var}(p) = \sum \frac{n_y(p_y - \bar{p})^2}{n - 1}$$

a. What is the average frequency of A in each generation? Does the average allele frequency over all isolates change much with increasing generation time?

b. What is the most likely original allele frequency?

c. What is the variance in allele frequency among isolates at generations 5, 20, and 60?

d. What proportion of the demes have fixation or loss of A at generations 5, 20, and 60?

e. What evolutionary force is causing these simulated populations to evolve? Justify your answer in terms of the results obtained in parts a, c, and d.

4. Ten replicates of a population evolving at a single locus with two alleles (A and a) are simulated in a computer with the following results for the A allele frequencies:

Trial	Generation 2	Generation 25	Generation 50
1	0.53	0.56	0
2	0.55	0.15	1
3	0.4	0.52	0
4	0.45	0.5	0.67
5	0.48	0.43	0.3
6	0.5	0.61	0.83
7	0.47	0.42	0.67
8	0.48	0.32	0.59
9	0.53	0.36	1
10	0.55	0.7	0.35

What is the variance effective size of the populations at generations 2, 25, and 50 relative to generation 0 (the initial generation at which all replicates had identical gene pools)? Is there any evidence for a bottleneck, and if so, between what generations did it occur?

5. a. Calculate $\bar{F}(1)$ through $\bar{F}(5)$ for populations of sizes $N = 2, 10, 50$ from equation 4.3. Assume $\bar{F}(0) = 0$.

b. Calculate the average inbreeding coefficients for a population of size 50 at generations 0, 1, and 2 that is decreased to size 2 at generations 3, 4, and 5. What is the inbreeding effective size of this population at generation 5 relative to generation 0?

6. Five isolated demes are drawn at random from zygotes produced by 10,000 unrelated adults mating at random for a locus with an allele having an initial frequency of 0.5. Subsequently, the allele frequencies in the individuals drawn to form the five demes are measured to be 0.9, 0.7, 0.6, 0.2, and 0.1. Assume all demes are of equal size (both census and effective in all senses) and have equivalent population structures and genetic drift is the only evolutionary force operating. Use the population of 10,000 unrelated adults as the reference generation.

a. What is the inbreeding effective size of each deme?

b. What is the variance in observed allele frequencies? What is the variance effective size of each deme?

7. a. A captive population of an endangered species is started at a zoo with eight founding individuals. After five generations in captivity, the average F is found to be 0.1. Using the founders as the reference generation, what is the inbreeding effective size over this time period?

b. Assume that a second, isolated population was started with 12 founders at another zoo. After five generations in captivity, this population also had an average F of 0.1.

Using its 12 founders as the reference generation, what is the inbreeding effective size over this time period?

8. Ten unrelated, hermaphroditic individuals from an idealized population produce 100 progeny under random mating.
   a. What are the inbreeding effective and variance effect sizes of the progeny generation relative to the parental generation?
   b. Suppose the parental population consisted of 10 full sibs from a mated pair of two unrelated individuals (the grandparents of the progeny generation). What is the inbreeding effective size of the 100 progeny relative to their grandparental generation?

9. What are the relative sizes of the inbreeding and variance effective sizes given the following demographic events?
   a. A founder event, using the large population the generation before the founder event as the reference generation and the founder generation itself for calculating effective sizes
   b. A founder event, using the founder generation as the reference generation and calculating effective sizes 10 generations later, during which time the population grew at a rapid rate
   c. A large idealized population of 10,000 with the sole exception that $f = 0.1$.
   d. A steadily declining population size

10. Five isolated demes are drawn at random from zygotes produced by 5000 unrelated adults mating at random (the reference population) for a locus with an allele having a frequency of 0.5 in the reference population. Subsequently, the allele frequencies in the individuals drawn to form the five demes are measured to be 0.7, 0.6, 0.5, 0.4, and 0.3. Assume all demes are of equal size (both census and effective in all senses) and have equivalent population structures and genetic drift is the only evolutionary force operating.
    a. What is the inbreeding effective size of each deme?
    b. What is the variance effective size of each deme?

Answers

1. The probability of ultimate loss $= 1-$ Prob(ultimate fixation)$= 1 - p_0 = 0.99$. This is true for all population sizes and values of f.

2. The probability of fixation of A under drift $= p_0 = 0.2$. So the expected number of populations fixed for A under drift $= 100(0.2) = 20$ and the expected number fixed for a is $100(0.8) = 80$. The observed numbers are 32 fixed for A and 68 fixed for a. The chi-square statistic is therefore $(32 - 20)^2/20 + (68 - 80)^2/80 = 9$. There is one degree of freedom, so we reject the null hypothesis of drift with a p value of 0.0027.

3.

Generation	$\bar{p}$	Var(p)	Proportion fixed
5	0.125	0.0527	$(5516 + 84)/8000 = 0.70$
10	0.119		
20	0.120	0.0982	$(6887 + 804)/8000 = 0.96$
40	0.120		
60	0.120	0.1058	1

a. See the table above; $\bar{p}$ does not change much.

b. Since $2N = 8$, the most likely $p_o = 0.125 (= \frac{1}{8})$ because $\bar{p} = 0.125$ at generation 5, the generation closest to the initial conditions and therefore the most likely to be closest to the initial conditions.

c. See the table above.

d. See the table above.

e. Genetic drift because $\bar{p} \approx$ constant, var(p) increases with time, and the number of fixed demes increases with time.

4.

ALLELE FREQUENCY

Trial/Generation	2	25	50
1	0.53	0.56	0
2	0.55	0.15	1
3	0.4	0.52	0
4	0.45	0.5	0.67
5	0.48	0.43	0.3
6	0.5	0.61	0.83
7	0.47	0.42	0.67
8	0.48	0.32	0.59
9	0.53	0.36	1
10	0.55	0.7	0.35
Average p	0.494	0.457	0.541
Variance p	0.002293333	0.024823333	0.13561
N_{ev}	54.49796512	114.1050038	31.26706232

Use

$$N_{ev} = \frac{1}{2\left\{1 - \left[1 - \sigma_t^2/(pq)\right]^{1/t}\right\}}$$

The results are given in the table above. As can be seen, the effective size drops between generations 25 and 50, indicating a bottleneck some time in that period.

5. a. Below are the values for $\bar{F}(i)$:

POPULATION SIZE

Generation	2	10	50
0	0	0	0
1	0.25	0.05	0.01
2	0.4375	0.0975	0.0199
3	0.578125	0.142625	0.029701
4	0.68359375	0.18549375	0.03940399
5	0.762695313	0.226219063	0.04900995

b. $\bar{F}(i)$ depends on the number of parents at $i - 1$. For the individuals produced at generation 3, there are still 50 parents from the previous generation. Hence,

Generation	$\bar{F}(i)$	Number of Parents
0	0	50
1	0.01	50
2	0.0199	50
3	0.029701	50
4	0.27227575	2
5	0.454206813	2

Use

$$N_{ef} = \frac{1}{2\left\{1 - \left[1 - \bar{F}(t)\right]^{1/t}\right\}}$$

with $\bar{F}(5) = 0.45420681$, $t = 5$, to yield $N_{ef} = 4.3838$.

6. a. $N_{ef} = 10{,}000$ (the number of parents)
  b. $\sigma_p^2 = 0.2 \sum (p_i - 0.5)^2 = 0.092$
  $p_o q_o / (2N_{ev}) = 0.25/(2N_{ev}) = 0.092$
  $N_{ev} = 0.125/0.092 = 1.36$

7. a. Use

$$N_{ef} = \frac{1}{2\left\{1 - \left[1 - \bar{F}(t)\right]^{1/t}\right\}}$$

with $t = 5$ and $\bar{F} = 0.1$ to yield $N_{ef} = 24$.
  b. Nothing has changed: $t = 5$ and $\bar{F} = 0.1$ to yield $N_{ef} = 24$.

8. a. Since the population is ideal and $\bar{F}(0) = 0$, $N_{ef} =$ number of parents $= 10$ and $N_{ev} =$ number of offspring $= 100$.
  b. The $\bar{F}$ (offspring generation) $= \frac{1}{20} + (1 - \frac{1}{20})(0.25)$ [the probability of selfing times $\frac{1}{2}$ (the inbreeding coefficient of offspring from a self mating) + the probability of not inbred through selfing times $\frac{1}{4}$ (the inbreeding coefficient for offspring of full sibs)] $= 0.2875$. Use

$$N_{ef} = \frac{1}{2\left\{1 - \left[1 - \bar{F}(t)\right]^{1/t}\right\}}$$

with $t = 2$ (two generations from grandparents) and $\bar{F} = 0.2875$ to yield $N_{ef} = 3.21$.

9. a. For the founder generation N_{ef} is minimally affected because the number of parents in the reference generation is large; N_{ev} is strongly affected and will be very small because the number of offspring (the founder generation) is small.

b. Here N_{ef} is very small because of the founder event 10 generations ago (the number of ancestors is small); N_{ev} is larger than N_{ef} because the population size is growing, reducing the variance in allele frequency due to less drift with increasing sample size.

c. Here N_{ef} is small because identity by descent accumulates rapidly due to the system of mating; N_{ev} is large because it is not as sensitive to f as N_{ef}.

d. Here N_{ef} will be larger than N_{ev} because the number of parents is larger than the number of offspring.

10. a. The inbreeding effective size in this case depends only upon the number of unrelated parents, which is 5000.

b. First, calculate the variance of allele frequencies across the replicates: $\sigma_p^2 = \sum 1/5(p_i - 0.5)^2 = 0.02$.

Second, from the fundamental definition of variance effective size, we can solve for N_{ev} : $p_o q_o/(2N_{ev}) = 0.25/(2N_{ev}) = 0.02$. So $N_{ev} = 0.125/0.02 = 6.25$.

CHAPTER 5

Problems

1. Let $\mu = 10^{-6}$ be the neutral mutation rate at a locus. Assuming all alleles are neutral and mating is at random, what is the expected average equilibrium probability of identity by descent at this locus for inbreeding effective sizes of 1000, 10,000 and 100,000?

2. a. A population of size 1000 is ideal except that the number of offspring per individual has a mean of 2 and a variance of 4. Assume that the neutral mutation rate is 10^{-6} for a particular locus. What is the rate of neutral molecular evolution at this locus? What is the expected average inbreeding coefficient in this population at equilibrium between mutation and drift? What is the expected equilibrium heterozygosity for neutral alleles at this locus in this population?

b. Assume the population size is changed to 100,000. Redo all the calculations made in part a.

c. Assume that the population size is still 100,00 but that the neutral mutation rate is now 10^{-5}. Redo all the calculations made in part a.

3. Two homologous, autosomal genes are sampled from a population with an inbreeding effective size of 100. What is the probability that these two genes coalesced in the previous generation? What is the probability that they coalesced 100 generations ago?

4. Two homologous, autosomal genes are sampled from a population with an inbreeding effective size of 1000 with the mutation rate at the locus being 10^{-5} per generation.

a. What is the probability that these two genes coalesced before mutation 500 generations ago?

b. What is the probability that one of these gene lineages experienced mutation before coalescence 500 generations ago?

c. What is the probability of mutation before coalescence given that either mutation or coalescence occurred 500 generations ago?

d. What is the expected heterozygosity for this system under the neutral theory?

5. **a.** A population of size 100 is ideal except for a deviation from random mating of $f = 0.1$. Assume that variance effective size and eigenvalue effective size are identical in this population. Assume that the neutral mutation rate is 10^{-7} for a particular locus. What is the rate of neutral molecular evolution at this locus?

   b. Assume the population size is changed to 1000. What is the rate of neutral molecular evolution at this locus?

6. A DNA region of nine nucleotides is sequenced, with variable sites being found that define five haplotypes as follows:

   1. A C C G T T G C A
   2. T C G G T C G C A
   3. A C G G T T G C A
   4. A C C G T T G C C
   5. A C G G T C G T A

   Construct the maximum-parsimony haplotype network for these sequences assuming no recombination and no multiple mutation events. Can you infer any haplotypes that must have existed but are not present in this sample? What is/are the sequence(s) of such haplotype(s)?

7. **a.** Two autosomal genes are sampled at random from a deme of inbreeding effective size 500 and variance effective size 1000. What are the mean and variance to the time of coalescence of these two genes?

   b. Assume that 10 genes are sampled. What are the mean and variance to the first coalescent event involving these 10 genes?

   c. Assume again that 10 genes are sampled. What are the mean and variance to the time of coalescence for all 10 genes?

8. Some VNTR loci (Appendix 1) have a mutation rate of 10^{-2}. In a population of inbreeding effective size of 25, calculate the probability of coalescence for two randomly drawn alleles for one, two, and three generations in the past. Calculate the probability of coalescence before mutation and the probability of mutation before coalescence for two randomly drawn alleles. What is the expected heterozygosity at this VNTR locus in this population if it were randomly mating for this locus?

Answers

1. $F_{eq} = 1/[4N\mu + 1]$, $\mu = 10^{-6}$, so:

N	F_{eq}
10^3	0.996
10^4	0.9615
10^5	0.7143

2. **a.** The rate of neutral molecular evolution = the neutral mutation rate = 10^{-6}. Use equation 4.31 with $k = 2$ and $v = 4$ to get $N_{ef} = 666.78$. From equation 5.6, $F_{eq} = 0.9973$ and heterozygosity = 0.0027.

b. The rate of neutral molecular evolution = the neutral mutation rate = 10^{-6}:

N	100,000
N_{ef}	66,666.77778
F	0.789473407
H	0.210526593

c. The rate of neutral molecular evolution = the neutral mutation rate = 10^{-5}:

N	100,000
N_{ef}	66,666.77778
F	0.272726942
H	0.727273058

3. Use Prob(coalesce at t) = $[1 - 1/(xN_{ef})]^{t-1}[1/(xN_{ef})]$ with $N_{ef} = 100$, $x = 2$, and $t = 1$ to yield 0.005 and with $t = 100$ to yield 0.00304.

4. **a.** Use equation 5.11 to get the probability of coalescence before mutation to be 0.000385695.

   b. Use equation 5.12 to get the probability of mutation before coalescence to be 1.54×10^{-5}.

   c. Probability of mutation before coalescence given one has occurred is $1.54 \times 10^{-5}/(1.54 \times 10^{-5} + 0.000385695) = 0.038443417$.

   d. Use equation 5.14 or 5.7 to get expected heterozygosity. First, $\theta = 4 \times 1000 \times 10^{-5} = 0.04$. Expected heterozygosity is $\theta/(1 + \theta) = 0.038452$.

5. **a.** Rate of neutral evolution = neutral mutation rate = 10^{-7}.

   b. Rate of neutral evolution = neutral mutation rate = 10^{-7}.

6. The tree is $4 \leftrightarrow 1 \leftrightarrow 3 \leftrightarrow 0 \leftrightarrow 2$ where arrows indicate single nucleotide changes
$$\downarrow$$
$$5$$
and zero an intermediate haplotype not present in the sample.
   Yes, and the sequence must be A C G G T C G C A.

7. **a.** Expected coalescence time = $2N_{ef} = 1000$ generations. Var = $2N_{ef}(2N_{ef} - 1) = 1000(999) = 999{,}000$.

   b. For $n = 10$, the expected time for first coalescence is $4N_{ef}/(10 \times 9) = 44.44$ generations.

   $$\text{Var} = \frac{4N_{ef}}{10 \times 9}\left(\frac{4N_{ef}}{10 \times 9} - 1\right) = 471.60$$

   c. For $n = 10$, the expected coalescence time = $4N_{ef}(1 - 1/n) = 2000(1 - 0.1) = 1800$ generations.

   $$\text{Var} \approx 16 N_{ef}^2 \sum_{i=2}^{10} \frac{1}{(i)^2(i-1)^2} = 2316.28$$

8. Prob(coalescence at generation 1) = $1/(2N_{ef}) = 1/50 = 0.02$
   Prob(coalescence at generation 2) = $[1 - 1/(2N_{ef})][1/(2N_{ef})] = 0.01960$
   Prob(coalescence at generation 3) = $[1 - 1/(2N_{ef})]^2[1/(2N_{ef})] = 0.01921$

   $\theta = 4N_{ef}\mu = 1$
   Prob(mutation before coalescence) = $\theta/(1+\theta) = 0.5$
   Prob(coalescence before mutation) = 0.5
   Expected heterozygosity = $\theta/(1+\theta) = 0.5$

CHAPTER 6

Problems

1. Suppose that the neutral allele A has a frequency of 0.4 in population 1 and 0.7 in population 2. After a single generation of gene flow between these populations, the frequency of A in population 1 is 0.46. What proportion of gametes (m) entered population 1 from population 2 to explain this result (assume that the two populations are effectively infinite in size).

2. Two populations exchange gametes at a rate of 0.01 every generation. Given that the two populations were initially fixed for different alleles at an autosomal locus, what is the expected difference (assuming neutrality and infinite population size in each deme) in the allele frequency between the demes at generation 10 and at generation 100? Now assume that the rate of exchange is 0.001. Redo the calculations for generations 10 and 100. What are the expected differences in allele frequency at equilibrium for $m = 0.01$ and $m = 0.001$?

3. Two autosomal genes picked at random from within a randomly chosen local deme coalesce on the average at 150 generations ago. Two autosomal genes picked at random from the entire species coalesce on the average at 1000 generations ago. What is the F_{st} for this species?

4. Three populations are screened for genetic variability at an autosomal locus with two alleles, A and a, with the following results:

	AA	Aa	aa
Population 1	12	176	312
Population 2	90	220	90
Population 3	469	462	69

Calculate f_{is}, f_{st}, and f_{it}.

5. A species fits the one-dimensional stepping-stone model with each deme having a variance effective size of 50 and with $m_1 = 0.1$ and $m_\infty = 0.001$.
   a. What is the equilibrium f_{st} for this species?
   b. Double the amount of local gene flow and recalculate f_{st}.

c. Go back to the original parameters, but now double the long-distance gene flow parameter and recalculate f_{st}. Which parameter doubling had the greatest effect on f_{st}? Which doubling has the greatest effect on the number of gametes being exchanged between demes?

6. Suppose that the allele A is fixed in population 1 and absent in population 2. Now assume that the two populations exchange 10% of their genes every generation and each population is effectively infinite in size.
   a. What are the allele frequencies in each population after one and after two generations of gene flow?
   b. What will the allele frequency of A be in populations 1 and 2 after a large number of generations of gene flow?

7. Four populations are screened for genetic variability at an autosomal locus with two alleles, A and a, with the following results:

Population	Population 1			Population 2			Population 3			Population 4		
Genotype	AA	Aa	aa	AA	Aa	aa	AA	Aa	aa	AA	Aa	aa
Frequency	0.019	0.162	0.819	0.184	0.432	0.384	0.275	0.45	0.275	0.511	0.378	0.111
Population size	2000			3000			3000			2000		

   a. Calculate f_{is}, f_{st}, and f_{it}.
   b. Calculate the effective number of migrating individuals among these populations under an island model. Is this a variance or inbreeding effective number of migrants and why?

8. A population has a 50–50 sex ratio, and males disperse twice as much as females (for females $m = 0.001$), all according to an island model and in which dispersal equals gene flow. All local populations are "ideal" in an effective size sense and each has a size of 1000.
   a. What is the expected F_{st} for an autosomal locus?
   b. What is the expected F_{st} for mtDNA (assume standard maternal, haploid inheritance)?
   c. What is the expected F_{st} for Y-DNA (assume a standard XY sytem of sex determination)?

9. Several replicates are taken from a reference population such that each replicate has a variance effective size of 10. An autosomal locus with two alleles has an allele frequency of 0.8 in the reference population.
   a. What is the variance in allele frequency among the replicates at this locus?
   b. What value of F_{st} do you expect among the replicates?
   Imagine that all replicates are carried on for two more generations (three generations from the ancestral reference population) and that the variance effective size is maintained

to be a constant 10 each generation. At this third generation:
c. What is the variance in allele frequency among the replicates at this locus?
d. What value of F_{st} do you expect among the replicates?

10. The following allele frequencies for the M–N blood group locus were estimated for these populations:

Population	West Africa	African Americans, Georgia	European Americans, Georgia
Frequency of M allele	0.476	0.484	0.507

Calculate M, the net amount of gene flow from the Georgia European American population into the Georgia African American population (assume gene flow has been essentially a one-way phenomenon from European Americans to African Americans).

11. A species is subdivided into four subpopulations with the following genotype frequencies at a locus:

	AA	Aa	aa
Population 1	0.1005	0.3990	0.5005
Population 2	0.1720	0.4560	0.3720
Population 3	0.0480	0.3040	0.6480
Population 4	0.8145	0.1710	0.0145

Assume all populations are of equal size. Calculate f_{is}, f_{st}, and f_{it}.

12. A population is continuously distributed over a one-dimensional habitat with a density of two individuals per meter. The standard deviation (i.e., the square root of the variance) between place of birth and place of reproduction is 4 meters. If 0.1% of the individuals engage in long-distance dispersal at random over the entire population, what is the expected equilibrium value of f_{st}? Redo the calculations assuming 1% of the individuals engage in long-distance dispersal.

Now suppose that all is the same as above, except that the habitat is two dimensional and the density is $2^2 = 4$ individuals per square meter. Calculate the equilibrium f_{st} values of 0.1% and 1% long-distance dispersal.

Answers

1. Use equation 6.1 and solve for m: $p'_1 = (1-m)p_1 + mp_2$ or $0.46 = (1-m)0.4 + m0.7$, $m = 0.2$.

2. Use equation 6.8 with $d_0 = 1$, $d_t = (1-2m)^t$. It is easiest to use logarithms; $\ln d_t = t\ln(1-2m)$.
   For $m = 0.01$, $\ln d_{10} = 10(\ln 0.98) = -0.202$, so $d_{10} = 0.8171$. At $t = 100$, $d_{10} = 0.1326$.
   For $m = 0.001$, $\ln d_{10} = 10(\ln 0.998)$, so $d_{10} = 0.9802$. At $t = 100$, $d_{10} = 0.8186$.
   For both m's, the equilibrium difference is zero.

3. Use equation 6.14: $F_{st} = (1000 - 150)/1000 = 0.85$.

4. Convert from numbers to frequencies by dividing by the population size, and also divide each population size by the sum of all population sizes to get the w_i's. Then calculate the average allele frequency (0.5263) and the variance in allele frequency (0.0440 from equation 6.22) to yield $f_{st} = 0.1767$. Within each population, calculate f (equation 3.3) to get -0.1 in each case, so $f_{is} = -0.1$. Apply equation 3.3 to the genotype numbers in the total population (the sum of all three local populations) or use the definition of f_{it} below equation 6.30 to get $f_{it} = 0.0943$.

5. a. Use equation 6.34 to get $f_{st} = 0.2612$.
   b. When $m_1 = 0.2$, using equation 6.34 now yields $f_{st} = 0.2$.
   c. With $m_\infty = 0.002$, using equation 6.34 now yields $f_{st} = 0.2$. Hence, the two doublings have identical effects on f_{st}. However, the doubling of m_1 increases the number of gametes being exchanged much more than the doubling of m_∞.

6. a. $p_1(t=1) = (0.9)(1) + (0.1)(0) = 0.9,\quad p_2(t=1) = (0.9)(0) + (0.1)(1) = 0.1$
   $p_1(t=2) = (0.9)(.9) + (0.1)(1) = 0.82,\quad p_2(t=2) = (0.9)(0.1) + (0.1)(0.9) = 0.18$

   b. $p_1(t=\infty) = p_2(t=\infty) = 0.5$

7. a.
Population	1	2	3	4
f_{is}	$1 - 0.162/.18 = 0.1$	0.1	0.1	0.1

   f_{st}: Average $p = \bar{p} = (2000/10{,}000)(0.1) + (0.3)(0.4) + (0.3)(0.5) + (0.2)(0.7) = 0.43$

   $\text{Var}(p) = (0.2)(0.1 - 0.43)^2 + (0.3)(0.4 - 0.43)^2 + (0.3)(0.5 - 0.43)^2 + (0.2)(0.7 - 0.43)^2 = 0.0381$

   $\bar{p}\bar{q} = 0.2451$

   So $f_{st} = 0.0381/0.2451 = 0.155$.

   $f_{it} = f_{st} + f_{is}(1 - f_{st}) = 0.24$

   Alternative. average frequency of Aa in total population $= 0.373$, so $f_{it} = 1 - 0.373/[2(0.43)(0.57)] = 0.24$.

   b. Under an island model, $f_{st} = 1/(1 + 4Nm)$, so $Nm = (1/f_{st} - 1)/4 = 1.36$. This is a variance effective number of migrants because the genetic parameter we are using is the variance of allele frequency.

8. a. For an autosomal locus, $m = (0.001 + 0.002)/2$.
   $F_{st} = 1/(1 + 4Nm) = 1/[1 + 4(1000)(0.0015)] = 0.143$

   b. For mtDNA, $m = 0.001$ (female only) and $N = 500$ (given the 50–50 sex ratio). Because mtDNA is haploid:

   $$F_{st} = 1/(1 + 2Nm) = 1/[1 + 2(500)(0.001)] = 0.5$$

   c. For Y-DNA, $m = 0.002$ (male only) and $N = 500$. Because it is haploid:

   $$F_{st} = 1/(1 + 2Nm) = 1/[1 + 2(500)(0.002)] = 0.333$$

9. a. The expected variance in allele frequency is $pq/(2N_{ev}) = (0.8)(0.2)/(20) = 0.008$.
   b. $f_{st} = \sigma_p^2/(pq) = 0.008/(0.8)(0.2) = 0.05$
   c. Var(allele freq. after t generations) $= pq\{1 - [1 - 1/(2N_{ev})]^t\}$, so

   $$\text{Var}(t) = (0.8)(0.2)\{1 - [1 - 1/20]^3\} = 0.023$$

   d. $f_{st} = \sigma_p^2/(pq) = 0.023/(0.8)(0.2) = 0.143$

10. $M = (p_B - p_{Af})/(p_W - p_{Af}) = 0.258$

11.
Population	$f = f_{is} = 1 - \text{freq}_i(Aa)/(2p_i q_i)$
1	0.05
2	0.05
3	0.05
4	0.05

$f_{is} = 0.05$ for total population

Var$(p) = 0.0725$; $f_{st} = \text{Var}(p)/(\bar{p}\bar{q}) = 0.293$
$f_{it} = f_{st} + f_{is}(1 - f_{st}) = 0.328$

Alternative: calculate f_{st} as above, but calculate $f_{it} = 1 - \text{freq}(Aa)/(2\bar{p}\bar{q})$ [note freq(Aa), $\bar{p}$, and $\bar{q}$ refer to the *total* population]. Then calculate f_{is} from $f_{it} = f_{st} + f_{is}(1 - f_{st})$.

12.
Dimensionality	m_∞	f_{st}
1	0.001	0.411
1	0.01	0.181
2	0.001	0.0038
2	0.01	0.0024

CHAPTER 7

Problems

1. A DNA region of nine nucleotides is sequenced, with variable sites being found that define five haplotypes as follows:

   1. A C C G T T G C A
   2. A C G G T C G C A
   3. A C G G T T G C A
   4. A C C G T T A C A
   5. A C G G T C G T A

   a. Construct a cladogram of the haplotypes assuming the least number of mutational events and no recombination.
   Suppose that sequence 5 came from another species and can be regarded as a good "outgroup" for the remaining haplotypes.

b. Under an isolation-by-distance model, which haplotype would be expected to have the most restricted geographical distribution and which the broadest?

2. Several Baltic Sea populations of the fish *Zoarces viviparus* were scored for the frequency of the $EstIII^1$ allele at one locus, and the frequency of the HbI^2 allele at another locus with the results shown below:

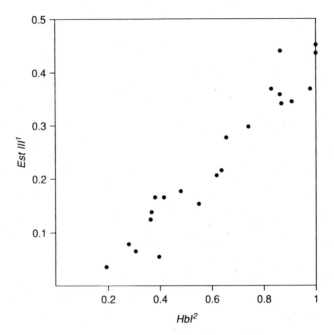

Draw a rotated set of axes on this diagram that should approximate the results obtained from a principal-component analysis. Identify which of your rotated axes corresponds to the first principal component.

3. A DNA region of 1000 nucleotides is screened with a battery of restriction enzymes, each with a recognition sequence of length 4. A total of 100 chomosomes were screened with the following haplotypes being discovered:
Haplotype (+ = site cut, − = site not cut)

1. + + + − + + − − + + + − − + + − +
2. + + − − + + − − + + + + + + + − +
3. + + + − + + − − + + + − − + + + +
4. + + + − + + − + + + + − − + + − +
5. + + + − + + − − + + + + + + + − +
6. + + + + + + − − + + + − − + + + +
7. + + + − + + − − + + + + − + + − +
8. + + − − + + + − + + + + + + + − +
9. − + + − + + − − + + + − − + + + +

a. Construct a haplotype tree assuming the least number of mutational events and no recombination.

b. Assume that the samples were taken over the entire geographical range of the species. Assume further that all subpopulations within the species are genetically interconnected but that gene flow is restricted by isolation by distance. Assume further that haplotype 1 is closest to the haplotypes found in an outgroup species. Under these assumptions, which haplotypes would be expected to have the more restricted geographical distribution relative to the other and why?

i. 3 vs. 6;   ii. 1 vs. 4;   iii. 2 vs. 8;   iv. 6 vs. 9;   v. 4 vs. 8

Which of the following groups (clades) of haplotypes collectively would be expected to have the most widespread geographical distribution and why?

vi. {3,6,9}, {1,4,7}, (5,2,8)

c. Assume that factors other than isolation by distance could be operating and that haplotype 4 has the most widespread geographical distribution and is found in some areas where none of the other haplotypes are found. What does this pattern suggest about the evolutionary history of this species and why?

4. Five loci showing no internal recombination are surveyed in populations spanning the geographical range of species X. Nested-clade analyses of these five loci reveal that each infers a range expansion event into the northern part of its current range from the southern part of its range. A molecular clock is used to time these inferences (t_i) from k_i, the average nucleotide divergence at locus i, with the following results:

Locus	t_i (years before present)	k_i
1	12,000	20
2	20,000	5
3	10,000	10
4	13,000	8
5	21,000	7

a. Assuming all the loci are detecting the same range expansion event, what is the estimate of the time of this range expansion based upon all five loci and what are the variance and standard deviation (square root of the variance) of this estimated time?

b. Test the assumption that all five loci are detecting the same range expansion event using the 5% level of significance.

Answers

1. a. cladogram: 4–1–3–2–5

   b. Since 5 is the outgroup, haplotype 2 is the oldest and 4 the youngest in the ingroup. Under isolation by distance, we expect the geographical range of a haplotype to be positively correlated with its age, so haplotype 4 should have the most restricted geographical range and haplotype 2 the broadest.

2.

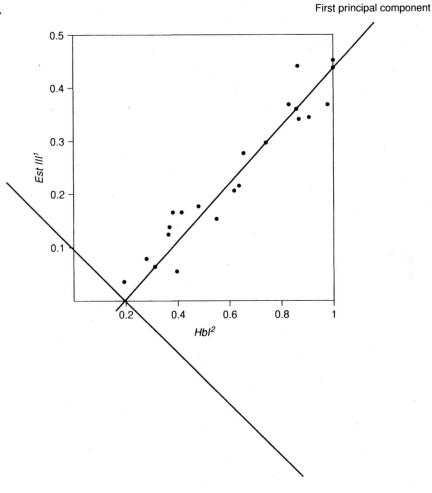

3. a. 6—3—1—7—5—2—8
     | |
     9   4

b. Expect tips to be more restricted than their interiors under isolation by distance. Therefore:
  i. 6 (tip) should be more restricted than 3 (its interior);
  ii. 4 (tip) should be more restricted than 1 (its interior);
  iii. 8 (tip) should be more restricted than 2 (its interior);
  iv. no predictions because 6 and 9 are both tips;
  v. no predictions because 4 and 8 are both tips.

Let clade $1 - 1 = \{3, 6, 9\}$, clade $1 - 2 = \{1, 4, 7\}$, and clade $1 - 3 = \{5, 2, 8\}$, the tree of these clades is $1 - 1 \leftrightarrow 1 - 2 \leftrightarrow 1 - 3$, so $1 - 2$ is the interior clade to the tip clades $1 - 1$ and $1 - 3$. Moreover $1 - 2$ is the oldest clade because it contains the connection to the outgroup. Therefore, expect $1 - 2$ to have the widest distribution.

c. This pattern suggests a range expansion because 4 is the tip to 1, yet 4 is more widespread and found in areas in which the other haploytpes, including 1, are absent. This indicates 4 was in the founding population of the new areas and went to fixation.

4.
a.

Pooled T (equation 7.9)	13945.45455
Variance of pooled T (equation 7.11)	3825785.124
Standard deviation of pooled T	1955.961432

b. G test statistic (equation. 7.8) = 12.52 with four degrees of freedom, which yields a p level of 0.0139. Therefore, we reject the null hypothesis of a single range expansion event.

CHAPTER 8

Problems

1. The *pygmy* locus in mice has two alleles, A and a, that affect body size, with the following genotypic values:

Genotype	Average Body Weight (grams)
AA	14
Aa	12
Aa	6

   Suppose a mouse population is randomly mating with $p = 0.6$ (the frequency of A).
   a. Calculate the population mean and all genotypic deviations, average excesses, average effects, additive genotypic deviations, nonadditive genotypic deviations, genetic variance, additive genetic variance, and nonadditive genetic variance. Given that the environmental variance is 10 g^2, calculate the broad-sense heritability and the heritability.
   b. Redo part a under the assumption that inbreeding, rather than random mating, is occurring with $f = 0.2$ (as a deviation from Hardy–Weinberg genotype frequencies).

2. The following data set is a slightly numerically simplified version of a real data set. A human population was scored for their LDL cholesterol levels and for their genotype at the *ApoE* locus. The *ApoE* locus codes for apoprotein E, which is not found on LDL cholesterol but does competitively bind the LDL receptor protein. There are three common alleles at this locus: $\varepsilon 2$, $\varepsilon 3$, and $\varepsilon 4$. The population was found to be in Hardy–Weinberg equilibrium frequencies at this locus with the following allele frequencies: 0.1 for $\varepsilon 2$, 0.7 for $\varepsilon 3$, and 0.2 for $\varepsilon 4$. The mean phenotypes (measured as mg LDL cholesterol/dl blood plasma) for the various genotypes were as follows:

Genotype	$\varepsilon 2/\varepsilon 2$	$\varepsilon 2/\varepsilon 3$	$\varepsilon 3/\varepsilon 3$	$\varepsilon 2/\varepsilon 4$	$\varepsilon 3/\varepsilon 4$	$\varepsilon 4/\varepsilon 4$
LDL cholesteral	76	90	100	115	110	106

   The total phenotypic variance of LDL cholesterol is 554.2 mg^2/dl^2.
   a. Calculate the genotypic deviations for all genotypes.
   b. Calculate the average excesses and average effects for all alleles.
   c. Calculate the breeding values and dominance deviations for all genotypes.
   d. Calculate the environmental, genetic, additive genetic, and dominance variances. Give the broad-sense and narrow-sense heritabilities.

e. Increased levels of LDL cholesterol are associated with increased risks of coronary heart disease. Which genotype is at greatest risk? Which genotype has his or her children at greatest risk?

f. Redo parts a through e under the assumption that inbreeding, rather than random mating, is occurring with $f = 0.1$ (as a deviation from Hardy–Weinberg genotype frequencies).

3. Thalassemia is an inherited, autosomal hemoglobin abnormality associated with both anemia and malarial resistance. It is frequent in many human populations in Asia and the Mediterranean where malaria is common. Suppose in Sardinia that the various genotypes have the following average life spans in years:

Genotype	AA	ATh	ThTh
Life span	60	65	20

Assume that the standard deviation (square root of the variance) around each of these means is 5 years for each genotypic class (that is, the environmental variance is 25 years^2).

a. What are the genotypic values for the three genotypes?

b. If the allele frequency of Th in Sardinia is 0.1 and the population is in Hardy–Weinberg equilibrium (retain these assumptions in all subsequent questions unless stated otherwise), what is the average life span for an individual from the population?

c. What are the genotypic deviations of life span for the three genotypes?

d. Given a $ThTh$ individual that died at age 27, what is that individual's environmental deviation?

e. Calculate the average excesses for each gamete type and the breeding value for each genotype.

f. Repeat the calculations in parts b, c and e for allele frequencies of 0.2 and 0.05 for Th. At which of these three allele frequencies (0.05, 0.1, 0.2) does Th have the smallest average effect in absolute value? At which allele frequency does the population have the highest average life span?

g. Calculate the narrow-sense heritability of life span in this population at the allele frequencies of 0.05, 0.1, and 0.2. What does the heritability value at the allele frequency of 0.1 tell you about the inheritance of life span in this population?

Answers

1. a.

	GENOTYPE			
	AA	Aa	aa	
Genotype frequency	0.36	0.48	0.16	$f = 0$
Genotypic mean	14	12	6	$p = 0.6$
Genotypic deviation	2.24	0.24	−5.76	Overall mean = 11.76
Average excess	for A:	1.44	for a:	−2.16
Average effect	for A:	1.44	for a:	−2.16
Additive deviation	2.88	−0.72	−4.32	

Dominance deviation   −0.64   0.96   −1.44
Genetic variance   7.1424
Additive variance   6.2208
Dominance variance   0.9216
Heritability   Environmental variance = 10
  Broad   0.4167
  Narrow   0.3629

b.

	GENOTYPE			
	AA	Aa	aa	
Genotype frequency	0.408	0.384	0.208	$f = 0.2$
Genotypic mean	14	12	6	$p = 0.6$
Genotypic deviation	2.432	0.432	−5.568	Overall mean = 11.568
Average excess	for A: 1.792		for a: −2.688	
Average effect	for A: 1.4933		for a: −2.24	
Additive deviation	2.9867	−0.7467	−4.48	
Dominance deviation	−0.5547	1.1787	−1.088	
Genetic variance	8.9334			
Additive variance	8.0282			
Dominance variance	0.9052			
Heritability				Environmental variance = 10
Broad	0.4718			
Narrow	0.4240			

2.

	GENOTYPE						
	$\varepsilon 2/\varepsilon 2$	$\varepsilon 2/\varepsilon 3$	$\varepsilon 3/\varepsilon 3$	$\varepsilon 2/\varepsilon 4$	$\varepsilon 3/\varepsilon 4$	$\varepsilon 4/\varepsilon 4$	Variance, Average, or Sum
Hardy–Weinberg genotype	0.010	0.140	0.490	0.040	0.280	0.040	1 (sum)
Genotypic mean	76	90	100	115	110	106	102.0 (ave.)
a. Genotypic deviation	−26.0	−12.0	−2.0	13.0	8.0	4.0	54.2 (var.)
b. Average excess = Average effect	$\varepsilon 2$: −8.4		$\varepsilon 3$: −1.0		$\varepsilon 4$: 7.7		
c. Breeding value	−16.8	−9.4	−2.0	−0.7	6.7	15.4	39.2 (var.)
Dominance deviation	−9.2	−2.6	0.0	13.7	1.3	−11.4	15.0 (var.)

d. From above, $V_g = 54.2$, $V_e = 554.2 − 54.2 = 500$, $V_a = 39.2$, $V_d = 15.0$.
Heritability: broad: 54.2/554.2 = 0.098; narrow = 39.2/554.2 = 0.071.

e. $\varepsilon 2/\varepsilon 4$ has the highest genotypic value (115), and therefore has the greatest risk. $\varepsilon 4/\varepsilon 4$ has the highest breeding value (15.4), and therefore its offspring are at greatest risk.

f. for inbreeding with $f = 0.1$:

GENOTYPE

	$\varepsilon 2/\varepsilon 2$	$\varepsilon 2/\varepsilon 3$	$\varepsilon 3/\varepsilon 3$	$\varepsilon 2/\varepsilon 4$	$\varepsilon 3/\varepsilon 4$	$\varepsilon 4/\varepsilon 4$	Variance, Average, or Sum
Genotype frequency	0.019	0.126	0.511	0.036	0.252	0.056	1 (sum)
Genotypic mean	76	90	100	115	110	106	101.7 (ave.)
Genotypic deviation	−25.68	−11.68	−1.68	13.32	8.32	4.32	56.0 (var.)
Average excess	$\varepsilon 2$: −9.840		$\varepsilon 3$: −0.780		$\varepsilon 4$: 7.650		
Average effects	−8.945		−0.709		6.955		
Breeding value	−17.891	−9.655	−1.418	−1.991	6.245	13.909	39.7 (var.)
Dominance deviation	−7.789	−2.025	−0.262	15.311	2.075	−9.589	16.4 (var.)

From above, $V_g = 56.0$, $V_e = 554.2 − 56.0 = 498.2$, $V_a = 39.7$, $V_d = 16.4$.
Heritability: broad $= 56.0/554.2 = 0.101$; narrow $= 39.7/554.2 = 0.072$.
$\varepsilon 2/\varepsilon 4$ is still at greatest risk, and the offspring of $\varepsilon 4/\varepsilon 4$ are still at greatest risk.

3. a. AA: 60 years; ATh: 65 years; $ThTh$: 20 years
b. $\mu = (0.9)2(60) + 2(0.9)(0.1)(65) + (0.1)2(20) = 60.5$ years
c. $g_i = G_i − \mu$: AA: $60 − 60.5 = −0.5$: ATh: 4.5: $ThTh$: −40.5
d. $P_{ij} = \mu + g_i + e_j$; $27 = 60.5 + (−40.5) + e_j$, so $e_j = 7$ years
e. $a_A = (0.9)(−0.5) + (0.1)(4.5) = 0$
$a_{Th} = (0.9)(4.5) + (0.1)(−40.5) = 0$
Because of random mating, average effects = average excesses, so
Breeding value $(AA) = a_A + a_A = 0 + 0 = 0$
Breeding value $(ATh) = a_A + a_{Th} = 0 + 0 = 0$
Breeding value $(ThTh) = a_{Th} + a_{Th} = 0 + 0 = 0$

f. For $p = 0.2$, $\mu = 60$ years, and for $p = 0.05$, $\mu = 60.375$.

	$p = 0.2$			$p = 0.05$		
	AA	ATh	ThTh	AA	ATh	ThTh
g_i	0	5	−40	−0.375	4.625	−40.375
g_{ai}	2	−3	−8	−0.25	2.25	4.75

Average excess:
$a_A = (0.8)(0) + (0.2)(5) = 1$ $a_A = (0.95)(−0.375) + (0.05)(4.625) = −0.125$
$a_{Th} = (0.8)(5) + (0.2)(−40) = −4$ $a_{Th} = (0.95)(4.625) + (0.05)(−40.375) = 2.375$

The average excess of *Th* is minimum in absolute value at $p = 0.1$.
The average life span is maximum At $p = 0.1$.

g. At $p = 0.1$, $h^2 = 0$ because all breeding values are zero. At $p = 0.2$:

$$V_a = (0.8)^2(2)^2 + 2(0.8)(0.2)(-3)^2 + (0.2)^2(-8)^2 = 8$$
$$V_g = (0.8)^2(0)^2 + 2(0.8)(0.2)(5)^2 + (0.2)^2(-40)^2 = 72$$
$$V_e = 25, \text{ so } V_p = 72 + 25 = 97$$

Hence, $h^2 = V_a/V_p = 8/97 = 0.0825$.
Similarly, at $p = 0.05$, $h^2 = 0.0190$.
What does $h^2 = 0$ tell you about the inheritance of life span—NOTHING!

CHAPTER 9

Problems

1. Estimate the heritability of male height (in centimeters) from the following data:

Father's Height	Son's Height
180	175
170	172
167	165
177	168
182	180
165	172
172	180
177	172
170	175
179	176

Use the following equation to estimate the correlation coefficient where appropriate:

$$\text{corr}(X, Y) = \frac{\sum (x_i - \bar{x})(y_i - \bar{y})}{\sqrt{\sum (x_i - \bar{x})^2 \sum (y_i - \bar{y})^2}} \quad \bar{x} = \frac{\sum x_i}{n} \quad \bar{y} = \frac{\sum y_i}{n}$$

2. A population was subjected to selection on a quantitative trait with the following results:

Generation	Total Population Mean	Mean of Selected Parents
0	2.16	2.32
1	2.26	2.34
2	2.30	2.37
3	2.33	2.41
4	2.36	2.47
5	2.40	

a. Estimate the heritability for every generation for which it is possible.
b. Are there any trends in heritability over the generations?

3. Two inbred lines of beans have mean weights of 35 and 65 grams. When crossed, the phenotypic variance of the F_1 was 8, and the phenotypic variance of the F_2 was 41. What is the genetic variance in the F_2 and the broad-sense heritability in the F_2?

4. What is the expected phenotypic covariance between half sibs (offspring who share only one parent in common) in a random-mating population?

5. The highly inbred Canton-S strain of *Drosophila melanogaster* with a high sternopleural bristle number was crossed to the highly inbred Lausanne-S strain with a low bristle number. The F_1's were then used to generate an F_2 and both backcrosses. The mean and variance in sternopleural bristle number in 100 males from each strain or cross are given as follows:

Strain	Mean	Variance
Laussane-S	8.34	0.841
Canton-S	12.12	1.102
F_1	9.48	1.000
F_2	8.89	1.283
BC_1	8.76	1.134
BC_2	10.59	1.180

For the trait of male sternopleural bristle number, use the above data:
a. Estimate the environmental variance.
b. Estimate the additive variance.
c. Estimate the dominance variance.
d. What do the above estimates tell you about the heritability of sternopleural bristle number within the Canton-S and Lausanne-S strains?

6. Following are data on the means and variances of skin color (antilog of reflectance at 685 nanometers) for people of various self-reported backgrounds:

Population	Sample Size	Mean	Variance ($\times 1000$)
European	105	0.421	1.09
African	106	0.225	1.05
F_1 African European hybrid	94	0.334	1.59
African backcross	26	0.304	1.71
European backcross	30	0.382	2.00
F_2 hybrid	14	0.346	1.99

For the trait of skin color, use the above data to estimate:
a. The overall dominance deviation, d
b. The environmental variance
c. The additive variance
d. The dominance variance

7. Individuals from a population of lizards have a mean adult weight of 40 grams. However, of the lizards that actually bred during a particular year, the mean weight is 50 grams. The mean adult weight of the resulting offspring is 42 grams. Assuming that the parents and offspring experienced identical environmental conditions, what is the heritability of adult weight?

8. The following data were collected from a study on diastolic blood pressure in 612 families. The correlation between a parent and an offspring was 0.183, and the correlation between full sibs was 0.265.
   a. Estimate the proportions of the total phenotypic variance that are due to additive effects, dominance effects, and environmental effects.
   b. What are the narrow-sense and broad-sense heritabilities of this trait?
   c. Repeat parts a and b, but now assume that 0.183 is the correlation between midparent and offspring.

Answers

1. Correlation   0.447607044
   heritability = 2 × corelation   0.895214087

2. a.

Generation	Total Population Mean	Mean of Selected Parents	Selection Coefficient	Response	Heritability
0	2.16	2.32	0.16		
1	2.26	2.34	0.08	0.1	0.625
2	2.3	2.37	0.07	0.04	0.5
3	2.33	2.41	0.08	0.03	0.428571429
4	2.36	2.47	0.11	0.03	0.375
5	2.4			0.04	0.363636364

   b. The heritability is getting smaller each generation.

3. From the F_1, the environmental variance is 8. Therefore the genetic variance in the F_2 is $41 - 8 = 33$, and the broad-sense heritability is $33/41 = 0.805$.

4. Half sibs share only a fourth of their alleles. They can share only the half of their alleles they get from their common parent, and Mendelian segregation ensures they share only half of that half. They do not share any genotypic states because that requires that they share both parents. Hence, their expected covariance is $\frac{1}{4}\sigma_a^2$.

5. a. Use equation 9.23 to get $V_e = 0.981$.
   b. Use equation 9.24 to get $V_a = 2(1.283) - (1.134 + 1.18) = 0.252$.
   c. Use equation 9.25 to get $V_d = 0.05$.
   d. Nothing, this experiment only looks at between-strain differences, not within.

6. a. $d = \mu_{F_1} - \frac{1}{2}(\mu_{Euro} + \mu_{African}) = 0.334 - 0.323 = 0.011$
   b. Use equation 9.23 to get $V_e = 1.23 \times 10^{-3}$.
   c. Use equation 9.24 to get $V_a = 0.27 \times 10^{-3}$.
   d. Use equation 9.25 to get $V_d = 0.49 \times 10^{-3}$.

7. $S = 50 - 40 = 10$; $R = 42 - 40 = 2$; $h^2 = R/S = \frac{2}{10} = 0.2$

8. For parts a and b:

$$\text{Correlation of parent and offspring} = (0.5)h^2 = 0.183, \text{ so } h^2 = 0.366.$$
$$\text{Correlation of sibs} = (V_a/2 + V_d/4)/V_p = 0.265$$
$$= (0.183 + V_d/4)/V_p, \text{ so } V_d/V_p = 0.328.$$
$$\text{Broad sense } h^2 = (V_a + V_d)/V_p = 0.366 + 0.328 = 0.694.$$
$$(V_a + V_d + V_e)/V_p = 0.694 + V_e/V_p = 1, \text{ so } V_e/V_p = 0.306.$$

9. Correlation of midparent and offspring $= \sqrt{\frac{1}{2}h^2}$, so $h^2 = 0.259$; $V_d/V_p = 0.542$; $h_B^2 = 0.801$; $V_e/V_p = 0.199$.

CHAPTER 10

Problems

1. Two inbred lines have mean body masses of 50 and 20 kg, respectively. An F_1 is bred and then backcrossed to the larger inbred line. Two marker loci (A and a; B and b, with lowercase alleles coming from the small line), located at the same chromosome with $r = 0.1$, are scored in the backcross progeny with the following results:

Genotype	AB/AB	AB/Ab	AB/aB	AB/ab
Average size	47	44	40	37

   a. What are the expected average sizes for the two recombinant genotypes AB/Ab and AB/aB if a QTL for body size exists within this interval and with a recombination rate of 0.03 from the A locus?
   b. What are the expected average sizes for the two recombinant genotypes AB/Ab and AB/aB if a QTL for body size exists within this interval and with a recombination rate of 0.05 from the A locus?
   c. Which is the more likely position for the QTL relative to the A locus: $r = 0.03$ or $r = 0.05$? Justify your answer.

2. An X-linked locus is surveyed in a sample of 160 males and five haplotypes are discovered. A phenotype is scored for all 160 males, and the mean phenotypes for males bearing haplotype i (along with the estimated variance, s_i^2, and sample size, n_i, for haplotype i) are as follows:

Haplotype	Mean	s^2	n
1	15	95	20
2	20	100	100
3	13	105	20
4	25	90	10
5	18	110	10

Use the two-sample t test (Appendix 2, equation A2.45) to test the difference between the means of all contrasts between two haplotype categories. Because multiple tests may be used, statistical significance will be corrected by the Bonferroni criteria. Starting with an overall 5% test ($\alpha = 0.05$), no single test will be regarded as significant unless its p value is below 0.05/(number of tests). (Note, ideally the tests corrected in this manner should be independent, which they are not in this case, but this does give a conservative correction and hence is often used.) Are any significant contrasts obtained?

3. Assume the same situation described in problem 2, but now also assume that no recombination has been detected in this X-linked region and that the estimated haplotype tree is

$$\begin{array}{c} 4 \\ | \\ 1-2-3 \\ | \\ 5 \end{array}$$

Use the two-sample t test (Appendix 2, equation A2.45) to test the difference between the means of all evolutionarily relevant contrasts of two haplotypes. Because multiple tests may be used, statistical significance will be corrected by the Bonferroni criteria. Starting with an overall 5% test ($\alpha = 0.05$), no single test will be regarded as significant unless its p value is below 0.05/(number of tests). (Note, the evolutionarily relevant contrasts are all asymptotically independent.) Are any significant contrasts obtained?

4. Suppose further sampling of males in the situation described in problem 2 revealed a sixth haplotype with a mean phenotype of 10, estimated variance of 100, and sample size of 5. Suppose further that the haplotype tree is now

$$\begin{array}{c} 4 \\ | \\ 1-2-3-6 \\ | \\ 5 \end{array}$$

The nested-clade design groups haplotypes 1, 2, 4, and 5 together into one-step clade 1-1 and groups haplotypes 3 and 6 together into one-step clade 1-2. The nested analysis first performs all the evolutionarily relevant contrasts among haplotypes *within* each one-step clade and next contrasts the two one-step clades against one another. Use t tests to perform all the necessary contrasts required by the nested design and do a Bonferroni correction (all contrasts in a nested-clade analysis are asymptotically independent, so a simple Bonferroni correction is appropriate). To perform the contrast of the one-step clades, you will need to pool data from two or more haplotype categories. Let J be a set of haplotypes found in one-step clade $1 - j$. Then the pooled mean phenotype for clade $1 - j$ is

$$\text{Mean}(j) = \frac{\sum_{i \text{ in } J} n_i \bar{x}_i}{\sum_{i \text{ in } J} n_i}$$

The pooled estimated variance is

$$s^2(j) = \frac{\sum_{i \text{ in } J} n_i s_i^2}{\sum_{i \text{ in } J} n_i}$$

and the pooled sample size is $\sum_{i \text{ in } J} n_i$. If any significant contrasts are detected, identify the branch(es) in the haplotype tree with the significant phenotypic association(s).

5. Perform one round of tree scanning on the problem described in problem 4. In this case, the Bonferroni correction will be overly conservative because the contrasts are not independent, but use this correction anyway. If any significant contrasts are detected, identify the branch(es) in the haplotype tree with the significant phenotypic association(s).

Answers

1. For parts a and b, use Table 10.1 (note, you have to reverse the genotypes because this is the opposite backcross to the one shown in that table):

Genotype	AB/AB	AB/Ab	AB/aB	AB/ab	r from A to QTL
Average size	47	44	40	37	
Expected		44	40		0.03
value given r		42	42		0.05

c. $r = 0.03$ is the more likely location as it explains the recombinant means better.

2. There are 10 possible contrasts:

Contrast	t	df	p
1 vs. 2	2.0495	118	0.042629141
1 vs. 3	0.1432	38	0.886870876
1 vs. 4	−0.8449	28	0.405336846
1 vs. 5	−0.2452	28	0.808114383
2 vs. 3	2.8463	118	0.005217409
2 vs. 4	−1.5139	108	0.132977928
2 vs. 5	0.6005	108	0.54941408
3 vs. 4	−0.9789	28	0.336001289
3 vs. 5	−0.3954	28	0.695547805
4 vs. 5	0.5078	18	0.617736579
Bonferroni for 10 tests:			0.005

None of the tests have p values below the Bonferroni p level.

3. Here are all the evolutionarily relevant contrasts and test results:

Contrast	t	df	p
1 vs. 2	2.0495	118	0.042629141
3 vs. 2	2.8463	118	0.005217409
4 vs. 2	−1.5139	108	0.132977928
5 vs. 2	0.6005	108	0.54941408
Bonferroni for 4 tests:		0.0125	

The only test that satisfies the Bonferroni cutoff is the contrast between haplotypes 3 and 2.

4. First are the evolutionarily relevant haplotype contrasts within clades 1–1 and 1–2:

		t	df	p
Clade 1–1	1 vs. 2	2.0495	118	0.042629141
	4 vs. 2	−1.5139	108	0.132977928
	5 vs. 2	0.6005	108	0.54941408
Clade 1–2	3 vs. 6	0.5880	23	0.562275267

To contrast the one-step clades, we must first pool:

	Mean	s^2	N
Clade 1–1	19.5	99.28571429	140
Clade 1–2	12.4	104	25

Finally, we do the final t test to contrast 1–1 vs. 1–2:

	t	df	p
1–1 vs. 1–2	3.270349555	163	0.001310713

There are a total of five t tests in the nested-clade analysis, so the Bonferroni 5% p value is 0.01. The only contrast that is significant is the contrast between clades 1–1 and 1–2. Hence, the only branch in the tree associated with a significant phenotypic change is the branch interconnecting haplotypes 2 and 3.

5. To perform a tree scan, we first must pool some haplotypes:

Pooled Haplotypes	Mean	s^2	n
2, 3, 4, 5, 6	18.89655172	100.6896552	145
1, 2, 3, 5, 6	18	100.6451613	155
1, 2, 3, 4, 6	18.4516129	99.35483871	155
1, 2, 4, 5	19.5	99.28571429	140
3, 6	12.4	104	25
1, 2, 3, 4, 5	18.6875	100	160

Next, do all the contrasts made by cutting one branch and test:

Contrast	t	df	p
1 vs. 2, 3, 4, 5, 6	−1.633353013	163	0.104324891
4 vs. 1, 2, 3, 5, 6	2.144851991	163	0.033445955
5 vs. 1, 2, 3, 4, 6	−0.138456988	163	0.890050165
1, 2, 4, 5 vs. 3, 6	3.270349555	163	0.001310713
1, 2, 3, 4, 5 vs. 6	−1.912924542	163	0.057510967
	Bonferroni for 5 tests:	0.01	

Only one test is significant, and it localizes the association to the branch interconnecting haplotypes 2 and 3.

CHAPTER 11

Problems

1. A population is polymorphic at an autosomal locus with two alleles, *A* and *a*. The probability is 0.1 for an *AA* zygote surviving to adulthood, 0.08 for an *Aa* zygote, and 0.12 for an *aa* zygote. The probability is 0.8 for an *AA* adult to successfully mate, 0.9 for an *Aa* adult, and 0.7 for an *aa* adult. A mated *AA* individual produces an average of 10 offspring, a mated *Aa* individual produces 9 offspring, and a mated *aa* individual produces 8 offspring.
   a. What are the expected numbers of offspring for *AA*, *Aa*, and *aa* zygotes?
   b. Assign a fitness of 1 to *aa*. What are the relative fitnesses of the *AA* and *Aa* genotypes?

2. The zygote-to-adult relative viabilities are 1, 0.5, and 0.25 for the *AA*, *Aa*, and *aa* genotypes, respectively. Assume random mating.
   a. Starting with the zygotic frequency of *A* being 0.2, what are the genotype frequencies in the adult population?
   b. Suppose you did not know the fitnesses but did know the population was randomly mating. Suppose further that you can *only* observe the adult population. Some workers estimate relative viabilities as deviations from Hardy–Weinberg genotype frequencies. Given only what you can observe about this population, what are your estimates of relative viabilities using the Hardy–Weinberg deviation technique?

3. *Drosophila pseudoobscura* is polymorphic for an autosomal inversion, with the two states designated *ST* and *CH*. These two rearrangements behave like alleles in a single-locus system with respect to Mendelian inheritance. A large number of *ST/CH* flies were put in a cage, and the next generation consisted of 69 *CH/CH*, 234 *ST/CH*, and 97 *ST/ST* flies.
   a. What are the relative fitnesses of the three karyotypes when the fitness of *ST/CH* is set to 1?
   b. What is the evolutionary fate of this experimental population under random mating in subsequent generations assuming that selection is the only evolutionary force?

4. The peppered moth is polymorphic at an autosomal locus determining a melanic form. Let the three genotypes be *MM*, *Mm*, and *mm* with estimated viabilities of 0.9, 0.95, and 1, respectively. Assume the frequency of the *m* allele was 0.1 in Oxford in 1940.
   a. What do you expect the frequency of *m* to be in 1941 (one generation per year) under the assumption of random mating?
   b. What do you expect the frequency of *m* to be in 1941 under the assumption of assortative mating yielding an f for this locus of 0.2?

5. The following fitnesses were estimated for humans living in a malarial environment in West Africa as a function of their genotype at the hemoglobin β-chain locus:

Genotype	AA	AS	SS	AC	SC	CC
Fitness	0.9	1	0.2	0.9	0.71	1.31

   Given these fitnesses and the assumption of random mating in an infinitely large population, calculate each population's average fitness and the average excess and rate of change of allele frequency for all three alleles for populations with the following initial gene pools:

	ALLELE FREQUENCY		
	A	S	C
a.	0.9	0.05	0.05
b.	0.85	0.05	0.10
c.	0.75	0.05	0.20

d. Do the initial allele frequencies affect the course of adaptive evolution in the above cases (that is, are the signs in the changes of allele frequencies the same in parts a, b, and c)?

e. Redo parts a, b, c, and d keeping all the same fitnesses and gene pools, but now assuming an inbreeding system of mating for which $f = 0.05$ (measured as a deviation from Hardy–Weinberg genotype frequencies).

f. Does this change in population structure alter the course of adaptive evolution in any of the cases (once again, in terms of the sign of allele frequency change)?

6. Two populations were studied in two environments with the following results (note, treat the two populations as independent evolutionary units):

	Environment 1			Environment 2		
Genotype	AA	AB	BB	AA	AB	BB
Zygote number	160	480	360	160	480	360
Adult number	90	450	300	150	400	340

a. Assuming that viability is the only component of fitness affected by genetic variability at this locus and that generations are discrete, estimate the relative fitnesses of the three genotypes in each environment using the convention that the fitness of the AB heterozygote is 1.

b. Assuming the population structure remains constant and that drift can be ignored, what is the population structure and what is the initial response to selection under the two environments (i.e., which alleles are favored and what is Δp)?

c. Does a polymorphic equilibrium point exist for each of the environments (that is, $\Delta p = 0$ for $0 < p < 1$)? If so, do the initial changes in allele frequency calculated above move the population away or toward any polymorphic equilibria that may exist?

7. A trait x is related to fitness by the function $w(x) = 3x - x^2$ for $0 \leq x \leq 3$.
   a. What value of x results in maximum fitness for bearers of that phenotypic value (hint, remember your calculus and use first and second derivatives)?
   b. Consider a population at selective equilibrium with a positive phenotypic variance. Do you expect this population at selective equilibrium to have the average trait value of x that optimizes $w(x)$ that you calculated above? Why or why not?

8. A population has an average fitness of 0.9, a variance of fitness of 3, and a heritability of fitness of 0.1.
   a. What is the average fitness in the next generation?
   b. What will be the heritability of fitness at equilibrium?

9. Consider the following genotypes and fitnesses:

AA	Aa	aa
1	0.4	1.2

   a. Assuming a random-mating population, does a nontrivial (i.e., $p \neq 0$ or $p \neq 1$) equilibrium exist? (Hint: The average excesses of A and a are equal at equilibrium.) If such a nontrivial equilibrium does exist, show whether or not it is stable (a numerical demonstration is fine, using $p_{eq} \pm 0.1$).
   b. Redo part a, but now assume a system of mating with $f = 0.4$.

10. A trait x is related to fitness by the function $w(x) = 10x - x^2$ for $0 < x < 10$. What value of x results in maximum fitness for bearers of that phenotypic value? Consider a population at selective equilibrium. At this selective equilibrium, the phenotypic variance of x is 0. What is the mean of x at selective equilibrium, and what is the average fitness of the population?

11. The following fitnesses were estimated for a locus with three alleles (A,a,α) in a population:

Genotype	AA	Aa	aa	Aα	aα	$\alpha\alpha$
Fitness	0.2	1.7	0.3	0.1	1.4	0.6

    Given these fitness and the assumption of random mating in an infinitely large population, calculate the rate of change of average fitness using Fisher's fundamental theorem of natural selection for the following initial gene pools:

	ALLELE FREQUENCY		
	A	a	α
a.	0.55	0.40	0.05
b.	0.35	0.60	0.05
c.	0.05	0.20	0.75

    Below is a contour plot of average fitness versus the allele frequencies of A and a ($p + q \leq 1$), with darker shading corresponding to lower average fitness. How many local fitness peaks are in this adaptive landscape, and do they represent stable polymorphic equilibria in a local sense? Give the approximate gene pool states at these equilibria as extrapolated from the plot below. Toward which fitness peak does natural selection take the population in parts a, b, and c. What do the average excesses in parts a, b, and c say about the direction of the initial response to selection?

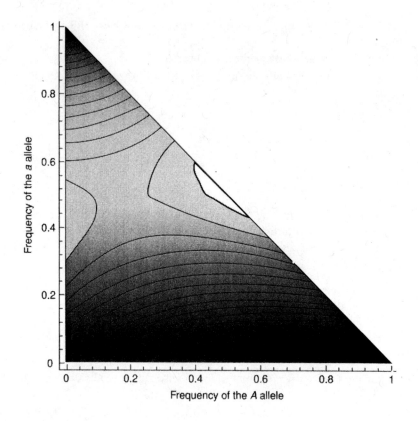

12. For the following fitnesses and genotype frequency arrays, calculate the change in the frequency of the A allele in the following generation:

	Genotype Frequency			Fitness		
	AA	Aa	Aa	AA	Aa	aa
a.	0.4	0.3	0.3	100	135	140
b.	0.8	0.05	0.15	0.005	0.003	0.003
c.	0.35	0.35	0.3	π	2π	6π

13. Consider the following pleiotropic traits influenced by a single autosomal locus with two alleles that is found in a random–mating population:

Genotype	AA	Aa	aa
Viability	0.9	0.5	0.2
Mating success	1	1	1
Fecundity	1	3	4

Calculate the equilibrium allele frequency. Calculate the additive variance in fitness (do not use relative fitness), viability, and fecundity at equilibrium. Are these additive variances consistent with Fisher's fundamental theorem?

Answers

1.

	AA	Aa	aa
l_x	0.1	0.08	0.12
c	0.8	0.9	0.7
b	18	20	10
a. w	1.44	1.44	0.84
b. Relative fitness	1.714285714	1.714285714	1.00

2.

	AA	Aa	aa	p
Zygotic Hardy–Weinberg genotype frequency	0.04	0.32	0.64	0.2
Fitnesses (average fitness = 0.36)	1	0.5	0.25	

a.

Adult genotype frequency:	0.1111	0.4444	0.4444	0.3333
Expected Hardy–Weinberg given adult p	0.1111	0.4444	0.4444	

b. Since the adult genotype frequencies are equal to the expected Hardy–Weinberg frequencies given the adult allele frequency, the estimated relative viabilities would all be equal to 1.

3.

	CH/CH	ST/CH	ST/ST	Sum
Observed	69	234	97	400
Expected	100	200	100	
Fitness = observed/expected	0.69	1.17	0.97	

a.

Relative fitness	0.58974359	1	0.829059829
Selection coefficient	0.41025641		0.170940171

b. Using equation 11.13, a balanced polymorphism, with p at equilibrium = 0.294117647.

4.

	MM	Mm	mm
Hardy–Weinberg genotype frequency	0.81	0.18	0.01
Assortative mating genotype frequency	0.828	0.144	0.028
Fitness	0.9	0.95	1

Average fitness ($f = 0$): 0.91
Average fitness ($f = 0.2$): 0.91

		$f = 0$	$f = 0.2$
Average excess:	A	−0.005	−0.006
	a	0.045	0.054

Δq for $f = 0$	0.004945055	So $q' = 0.104945055$
Δq for $f = 0.2$	0.005934066	So $q' = 0.105934066$

5. Under Random Mating

a.

Genotype	AA	AS	SS	AC	SC	CC	
Fitness	0.9	1	0.2	0.9	0.71	1.31	$\bar{W} = 0.907325$
Frequency	0.81	0.09	0.0025	0.09	0.005	0.0025	

Allele	A	S	C
Average excess	−0.002325	0.038175	0.003675
$\Delta p = pa_x/\bar{W}$	−0.0023062	0.00210371	0.00020252

b.

Genotype	AA	AS	SS	AC	SC	CC	
Fitness	0.9	1	0.2	0.9	0.71	1.31	$\bar{W} = 0.90895$
Frequency	0.7225	0.085	0.0025	0.17	0.01	0.01	

Allele	A	S	C
Average excess:	−0.00395	0.02205	0.02255
$\Delta p = pa_x/\bar{W}$	−0.0036938	0.00121294	0.00248088

c.

Genotype	AA	AS	SS	AC	SC	CC	
Fitness	0.9	1	0.2	0.9	0.71	1.31	$\bar{W} = 0.91835$
Frequency	0.5625	0.075	0.0025	0.3	0.02	0.04	

Allele	A	S	C
Average excess:	−0.01335	−0.01635	0.05415
$\Delta p = pa_x/\bar{W}$	−0.0109027	−0.0008902	0.01179289

d. The initial array of allele frequencies does affect the course of adaptive evolution, with the sign in the change of the S allele frequency being reversed in part c from Parts a and b.

e. Under Inbreeding with $f = 0.05$

Part a:

Genotype	AA	AS	SS	AC	SC	CC	
Fitness	0.9	1	0.2	0.9	0.71	1.31	$\bar{W} = 0.906234$
Frequency	0.8145	0.0855	0.004875	0.0855	0.00475	0.004875	

Allele	A	S	C
Average excess	−0.0014837	0.00199125	0.02471625
$\Delta p = pa_x/\bar{W}$	−0.0014735	0.00010986	0.00136368

Part b:

Genotype	AA	AS	SS	AC	SC	CC	
Fitness	0.9	1	0.2	0.9	0.71	1.31	$\bar{W} = 0.9088025$
Frequency	0.729	0.081	0.0049	0.1615	0.0095	0.0145	

Allele	A	S	C
Average excess	−0.0040524	−0.0143525	0.0416225
$\Delta p = pa_x/\bar{W}$	−0.0037903	−0.0007896	0.00457993

Part c:

Genotype	AA	AS	SS	AC	SC	CC	
Fitness	0.9	1	0.2	0.9	0.71	1.31	$\bar{W} = 0.9197825$
Frequency	0.5147	0.07125	0.001	0.2565	0.0135	0.06288	

Allele	A	S	C
Average excess	-0.0150325	-0.0528825	0.0695925
$\Delta p = pa_x/\bar{W}$	-0.0122577	-0.0028747	0.01513238

Part d: The initial array of allele frequencies does affect the course of adaptive evolution, with the sign in the change of the S allele frequencies being reversed in part a from parts b and c.

Part f: The course of evolution is altered by the system of mating, with the sign of the change in the S allele frequency changing in part b from + (random mating) to − (inbreeding).

6.

	Environment 1			Environment 2		
Genotype	AA	AB	BB	AA	AB	BB
Zygote no. (Z#)	160	480	360	160	480	360
Adult no. (A#)	90	450	300	150	400	340
Viability = A#/Z#	0.5625	0.9375	0.8333	0.9375	0.8333	0.9444
a. Relative fitness	0.6	1	0.8889	1.125	1	1.133

b. Both populations are in Hardy–Weinberg and have $p_A = 0.4$, with genotype frequencies of 0.16, 0.48, and 0.36, respectively.

	Environment 1	Environment 2
$\bar{W}$	0.896	1.0678

Allele	A	B	A	B
Average excess	-0.056	0.0373	-0.0178	0.0112
$\Delta p = pa_x/\bar{W}$	-0.025	0.025	-0.00637	0.00637

c. Polymorphic equilibrium is defined by $a_A = a_B$ or $pw_{AA} + (1-p)w_{AB} = pw_{AB} + (1-p)w_{BB}$. Solving the above equation for p given the fitnesses,

For environment 1, $p_{eq} = 0.2174$ (or for the B allele, 0.7826).
For environment 2, $p_{eq} = 0.5161$ (or the for B allele, 0.4839).

In environment 1, $p_A = 0.4$ is higher than the equilibrium, and the negative average excess causes selection to get closer to the equilibrium.

In environment 2, $p_A = 0.4$ is lower than the equilibrium, and the negative average excess causes selection to get farther away from the equilibrium.

7. a. To find the maximum of $w(x)$, take the first derivative and set it to 0: $w'(x) = 3 - 2x = 0$, so $x_{max} = \frac{3}{2} = 1.5$.

To confirm this is a maximum, take the second derivative, $w''(x) = -2$. This is negative, so it is a maximum.

b. For a population at selective equilibrium, you do not expect $\bar{x} = x_{max}$ because $w(x)$ is nonlinear.

8. a. From FFTNS, $\Delta \bar{W}$ = (additive variance of fitness)/$\bar{W} = h^2$ (variance of fitness)/$\bar{W} = 0.1(3)/0.9 = 0.333$. Hence,

$$\bar{W}' = \bar{W} + \Delta \bar{W} = 0.9 + 0.333 = 1.233$$

b. At equilibrium, $h^2 = 0$ (from FFTNS).

9. a. $a_A = p(1) + (1-p)(0.4) - \bar{W} = p(0.4) + (1-p)(1.2) - \bar{W} = a_a$ at equilibrium, so

$$p(1) + (1-p)(0.4) = p(0.4) + (1-p)(1.2) \Rightarrow p - 0.4p - 0.4p + 1.2p$$
$$= -0.4 + 1.2, \ 1.4p = 0.8 \Rightarrow p_{eq} = 0.8/(1.4) = 0.571$$

at $p = 0.671$, $a_A = 0.054$; at $p = 0.471$ $a_A = -0.061 \Rightarrow$ unstable.

b. With $f = 0.4$:

$$a_A = [f + p(1-f)](1) + (1-p)(1-f)(0.4) - \bar{W}$$
$$= p(1-f)(0.4) + [f + (1-p)(1-f)](1.2) - \bar{W} = a_a$$

so

$$[0.4 + p(0.6)](1) + (1-p)(0.6)(0.4) = p(0.6)(0.4)$$
$$+ [0.4 + (1-p)(0.6)](1.2) \Rightarrow 0.84p = 0.560 \Rightarrow p_{eq} = 0.667$$

at $p = 0.767$, $a_A = 0.026$; at $p = 0.567$, $a_A = -0.036 \Rightarrow$ unstable.

10. $w(x)' = 10 - 2x$. To find max/min, set the derivative to 0: $10 - 2x = 0 \Rightarrow x = 5$. To see if max or min (or saddle point), take second derivative to get $-2 \Rightarrow$ max.

Although $w(x)$ is nonlinear, there is no phenotypic variance, so from equation 11.26, $x_{eq} = 5 = x_{opt}$ and $\bar{W} = w(5) = 25$.

11. a.

							Sum or Average
Genotype	AA	Aa	aa	Aα	aα	$\alpha\alpha$	
Fitness	0.2	1.7	0.3	0.1	1.4	0.6	0.9195
Genotype frequency	0.3025	0.44	0.16	0.055	0.04	0.0025	1
Allele	A	a	α				
Allele frequency	0.55	0.4	0.05				1
	A	a	α				
Average excess	-0.1245	0.2055	-0.2745				

	AA	Aa	aa	Aα	aα	αα	Additive variance:
Breeding value (of fitness)	−0.249	0.081	0.411	−0.399	−0.069	−0.549	0.0583695

FFTNS rate of change of average fitness: 0.063479608

b.

Genotype	AA	Aa	aa	Aα	aα	αα	
Fitness	0.2	1.7	0.3	0.1	1.4	0.6	0.9355
Genotype frequency	0.1225	0.42	0.36	0.035	0.06	0.0025	1

Allele	A	a	α	
Allele frequency	0.35	0.6	0.05	1

	A	a	α
Average excess	0.1595	−0.0905	−0.0305

	AA	Aa	aa	Aα	aα	αα	Additive variance:
Breeding value	0.319	0.069	−0.181	0.129	−0.121	−0.061	0.0277295

FFTNS rate of change of average fitness: 0.029641368

c.

Genotype	AA	Aa	aa	Aα	aα	αα	
Fitness	0.2	1.7	0.3	0.1	1.4	0.6	0.8115
Genotype frequency	0.0025	0.02	0.04	0.075	0.3	0.5625	1

Allele	A	a	α	
Allele frequency	0.05	0.2	0.75	1

	A	a	α
Average excess	−0.3865	0.3835	−0.0765

	AA	Aa	aa	Aα	aα	αα	Additive variance:
Breeding value	−0.773	−0.003	0.767	−0.463	0.307	−0.153	0.0825455

FFTNS rate of change of average fitness: 0.101719655

In the contour plot, there are two peaks separated by a saddle. They are locally stable because they are peaks (all approaches to them represent increasing $\bar{W}$). They are approximately at p_A of about 0.5 and p_a of about 0.5 and at p_A of 0 and p_a of about 0.45.

In Part a you go to the taller peak (about 0.5, 0.5). Note, you start out with the frequency of A too high for this peak, and its average excess is negative (which takes you closer), and a is too low but its average excess is positive (which takes you closer).

In Part b you also go to the taller peak. In this case you start out with the frequency of A too low for this peak, and the average excess is positive; the frequency of a is too high but its average excess is negative.

In Part c you go to the lower peak (0, 0.45). You start out with the frequency of A too high for this peak, and its average excess is negative (which takes you closer), and the frequency of a is too low but its average excess is positive (which takes you closer).

12. **a.** $p = 0.55$, $\bar{W} = 122.5$, $a_A = -12.95$, $\Delta p = pa_A/\bar{W} = -0.058$
    b. $p = 0.825$, $\bar{W} = 0.0535$, $a_A = -0.0041$, $\Delta p = -0.063$
    c. $p = 0.525$, $\bar{W} = 2.85\pi$, $a_A = -1.517\pi$, $\Delta p = -0.279$

13. First, calculate the fitnesses by multiplying the fitness components:

Genotype	AA	Aa	aa
Viability	0.9	0.5	0.2
Mating success	1	1	1
Fecundity	1	3	4
Fitness	0.9	1.5	0.8

    In a random-mating population at equilibrium, $a_A = a_a$, so

    $$p(0.9) + (1-p)(1.5) - \bar{W} = p(1.5) + (1-p)(0.8) - \bar{W}$$

    $$-1.3p = -0.7$$

    $$p_{eq} = 0.5385$$

    At equilibrium, $\bar{W} = 1.1769$, $a_A = a_a = 0$, so all breeding values (additive genotypic deviations) are zero, so the additive variance in fitness is zero, consistent with the fundamental theorem.

    At equilibrium, average viability = 0.5521, $a_A = 0.1633$, $a_a = -0.1905$, so

Genotype	AA	Aa	aa
Breeding value	0.3266	−0.0272	−0.3811

    So the additive variance (variance of breeding values) is 0.0622. This result is also consistent with the fundamental theorem because viability is not the same as fitness and the FFTNS only ensures that the additive variance of fitness is zero at equilibrium.

    At equilibrium, the average fecundity is 2.6331, $a_A = -0.7101$, $a_a = 0.8284$, so

Genotype	AA	Aa	aa
Breeding value	−1.4201	0.1183	1.6568

    So the additive variance (variance of breeding values) is 1.1764. This is also consistent with FFTNS because fecundity is not the same as fitness.

CHAPTER 12

Problems

1. Assume $h = 0.03$ for a mutation that is lethal when homozygous and $h = 0.4$ for a mildly deleterious mutation when homozygous ($s = 0.05$). In both cases, the mutation rate to the deleterious form is 10^{-5}.

a. Assuming random mating, what are the equilibrium frequencies for the mutant allele in these two cases?

b. Assuming $f = 0.1$ and that both mutants are now completely recessive, what are the equilibrium frequencies for the mutant allele in these two cases?

2. Cystic fibrosis (CF) is a fatal disease caused by a recessive mutation. Individuals homozygous for the *CF* allele die before reproducing. The *CF* allele has a frequency of 0.02 in a randomly mating human population. Assume that this is an equilibrium frequency.

   a. Assume in addition that the *CF* heterozygotes have the same fitness as the normal-allele homozygotes. Estimate the mutation rate from the normal to *CF* alleles.

   b. Assume instead that the *CF* heterozygotes have greater viability than either homozygote. Assume further that selection is sufficiently strong that the effects of mutation on the equilibrium frequency can be ignored. Estimate the selection coefficient of the normal-allele homozygote.

   c. There has been controversy over whether the *CF* frequency reflects a mutation/selection balance or an overdominance (heterozygote superiority) balance. In light of your answers to part a and b and your knowledge of mutation rates in general, which seems more reasonable under the assumption of equilibrium?

3. A population of snakes lives on an island and has the following fitnesses at a locus with two alleles, *A* and *a*:

Genotype	AA	Aa	aa
Fitness	1	0.8	0.2

   This island population also receives some gene flow from the mainland population, with 10% of the gametes coming in from the mainland every generation. The frequency of the *a* allele is maintained at 0.7 by overdominant selection in the mainland environment.

   a. Assuming random mating and a frequency of *a* of 0.2 on the island, what is the frequency of the *a* allele on the island in the next generation?

   b. Same as part a, but now assume the frequency of *a* on the island is 0.3.

   c. Assuming avoidance of inbreeding with $f = -0.2$ and a frequency of *a* of 0.2 on the island, what is the frequency of the *a* allele on the island in the next generation?

   d. Same as part c, but now assume the frequency of *a* on the island is 0.3.

   e. Under which system of mating does the equilibrium between selection and gene flow yield an allele frequency between 0.2 and 0.3? Justify your answer.

4. A codominant advantageous mutation, *a*, occurs in a random-mating population with the following fitnesses: *AA*, 1; *Aa*, 1.05, and *aa*, 1.1. What is the probability of fixation of this mutation when:

   a. The population is ideal (all effective sizes = census size) and of size 50?

   b. The population is ideal (all effective sizes = census size) and of size 500?

   c. The variance effective size is 50 and the census size is 200?

5. A molecular genetic survey is conducted at a homologous protein-coding locus in two closely related species, and the distribution of all mutations is mapped onto the branches

of the resulting haplotype tree with the following results:

	Silent Substitutions	Replacement Substitutions
Tip	20	7
Interior	15	16
Fixed	18	24

Use a contingency chi square (Appendix 2) to test the hypothesis of neutrality. If you reject that hypothesis at or below the 5% level of significance, describe the type of selection that is occurring at this locus.

Now test the null hypothesis of neutrality with these data using the contingency test of McDonald and Kreitman. Contrast the results to the previous contingency test.

6. Consider a diploid population with two alleles (A and a) at an autosomal locus and an initial frequency of the A allele of 0.8. Now consider the following genotypes and fitnesses:

AA	Aa	aa
1	0.5	1.5

There are two stable equilibria under selection in this population, at $p = 0$ and $p = 1$.

a. Suppose the population is effectively infinite in size and randomly mating but receives a portion m of its genes from another population fixed for allele a. What is the minimum value of m that will cause the population in the next generation to be in the domain of the $p = 0$ equilibrium?

b. Same as part a, but now assume the system of mating has $f = -0.1$.

Answers

1. a. Use equation 12.8 to get $q_{eq} = 10^{-5}/[(0.03)(1)] = 0.00033$ for the first case and $q_{eq} = 10^{-5}/[(0.4)(0.05)] = 0.0005$ for the second case.
   b. Use equation 12.10 to get $q_{eq} = 10^{-5}/[(0.1)(1)] = 0.0001$ for the first case and $q_{eq} = 10^{-5}/[(0.1)(0.05)] = 0.002$ for the second case.

2. a. Using equation 12.7 and noting that $s = 1$, the estimate of the mutation rate at equilibrium is $\mu = q^2 = 0.0004$.
   b. Using equation 11.13 and noting that $s = 1$, the estimate of t is $0.02/0.98 = 0.0204$.
   c. 0.0004 would represent an extremely high mutation rate and therefore is unlikely. A modest selection coefficient of only 2% could explain the results under the overdominance hypothesis, so this explanation seems more reasonable.

3. a. Use equation 12.13 with $p_2 = 0.7$ and $m = 0.1$. Under random mating with $p_1 = 0.2$, the average fitness is 0.904, the average excess of the a allele is -0.224, so from equation 12.13, the change in p is 0.00044248 and the frequency in the next generation is 0.20044248.
   b. Use equation 12.13 with $p_2 = 0.7$ and $m = 0.1$. Under random mating with $p_1 = 0.3$, the average fitness is 0.844, the average excess of the a allele is -0.224, so from

equation 12.13, the change in p is -0.0396209 and the frequency in the next generation is 0.26037915.

c. Use equation 12.13 with $p_2 = 0.7$ and $m = 0.1$. With $f = -0.2$ and $p_1 = 0.2$, the average fitness is 0.9168, the average excess of the a allele is -0.1408, so from equation 12.13, the change in p is 0.01928447 and the frequency in the next generation is 0.21928447.

d. Use equation 12.13 with $p_2 = 0.7$ and $m = 0.1$. With $f = -0.2$ and $p_1 = 0.3$, the average fitness is 0.8608, the average excess of the a allele is -0.1568, so from equation 12.13, the change in p is -0.0145468 and the frequency in the next generation is 0.2853516.

e. Under both systems of mating, because at $p = 0.2$ the allele frequency increases but at $p = 0.3$ the allele frequency decreases.

4. a. Note that $s = 0.05$; then use equation 12.16 to obtain $u = 0.0951669$.
   b. $u = 0.09516258$
   c. Use equation 12.15 to obtain $u = 0.02469$.

5. Get the marginals:

	Silent Substitutions	Replacement Substitutions	Sum
Tip	20	7	27
Interior	15	16	31
Fixed	18	24	42
Sum	53	47	100

Then generate the expected values:

	Silent Substitutions	Replacement Substitutions
Tip	14.31	12.69
Interior	16.43	14.57
Fixed	22.26	19.74

Calculate the chi square from the observed and expected values as 6.81 with two degrees of freedom, which is significant at the 5% level ($p = 0.033$). Therefore, we reject the null hypothesis of neutrality. The silent substitutions are disproportionately on the tips and the replacements on the fixed branches, so this implies positive selection favoring amino acid change in this protein-coding locus.

The McDonald–Kreitman test pools the categories of tip and interior into polymorphic, so this results in a 2 × 2 contingency table:

	Silent	Replacement	Sum
Polymorphic	35	23	58
fixed	18	24	42
Sum	53	47	100

The expecteds are

	Silent	Replacement
Polymorphic	30.74	27.26
Fixed	22.26	19.74

yielding a chi square of 2.99 with one degree of freedom, which is not significant at the 5% level ($p = 0.0837$). In contrast to the previous test, here the null hypothesis of neutrality cannot be rejected.

6. a. $a_A = p(1) + (1 - p)(0.5) - \bar{W} = p(0.5) + (1 - p)(1.5) - \bar{W} = a_a$ at the unstable equilibrium, so $p + (1 - p)(0.5) = p(0.5) + (1 - p)(1.5) \Rightarrow p - 0.5p - 0.5p + 1.5p = -0.5 + 1.5 \Rightarrow 1.5p = 1 \Rightarrow p_{eq} = 1/1.5 = \frac{2}{3}$.

At $p = 0.8$, $\bar{W} = 0.86$ and $a_A = 0.056$.

$\Delta p = pa_A/\bar{W} + \Delta p$ (gene flow) $= pa_A/\bar{W} - m(p_1 - p_2) = 0.037 - m(0.8 - 0) = 0.037 - 0.8m$.

To put p into the domain of $p_{eq} = 0$, $p' < \frac{2}{3}$, or $\Delta p < 0.667 - 0.8 = -0.133$; so, $0.037 - 0.8m < -0.133 \Rightarrow -0.8m < -0.170 \Rightarrow m > 0.213$.

b. $a_A = [f + p(1 - f)](1) + (1 - p)(1 - f)(0.5) - \bar{W} = p(1 - f)(0.5) + [f + (1 - p)(1 - f)](1.5) - \bar{W} = a_a$, so $[-0.1 + p(1.1)] + (1 - p)(1.1)(0.5) = p(1.1)(0.5) + [-0.1 + (1 - p)(1.1)](1.5) \Rightarrow 1.65p = 1.05 \Rightarrow p_{eq} = 0.636$.

At $p = 0.8$, $\bar{W} = 0.836$ and $a_A = 0.076$.

$\Delta p = pa_A/\bar{W} + \Delta p$ (gene flow) $= pa_A/\bar{W} - m(p_1 - p_2) = 0.052 - m(0.8 - 0) = 0.052 - 0.8m$.

To put p into the domain of $p_{eq} = 0$, $p' < 0.636$, or $\Delta p < 0.636 - 0.8 = -0.164$, so $0.052 - 0.8m < -0.164 \Rightarrow -0.8m < -0.216 \Rightarrow m > 0.27$.

CHAPTER 13

Problems

1. An allele A shows meiotic drive over allele a such that in heterozygotes there is a 4 : 1 segregation in favor of A. Assume for now that this is the only phenotype associated with A.
  a. Starting with an allele frequency of 0.05 for A in a random-mating population, what is the frequency of A in the next generation?
  b. Same as part a, but now assume nonrandom mating with $f = 0.5$.
  c. Assume random mating again but also assume that AA homozygotes are lethal but Aa and aa genotypes are equally fit. What is the frequency of A in the next generation?

2. A tandem, multigene family with 100 copies has a per-locus neutral mutation rate of 10^{-5} per generation in an ideal population of size 10,000. The probability of one copy converting a paralogous copy to its state is 10^{-4} per generation.
  a. What is the rate of neutral evolution for the entire multigene family?
  b. What is the expected time to fixation (in all individuals in the population and at all paralogous sites) of a new neutral mutation?

c. Which is the limiting step in the total fixation time given above: orthologous fixation or paralogous fixation?

3. Given the following fitnesses, show whether or not the polymorphism is protected. If the polymorphism is not protected, identify the alleles that are not protected. In all cases, assume random mating.

   a. Competitive interactions:

Genotype	AA	Aa	aa
AA	0.95	0.9	1.2
Aa	0.9	1	1.7
aa	1.2	1.7	1.6

   b. Fertility:

Mating type	AA×AA	AA×Aa	AA×aa	Aa×Aa	Aa×aa	aa×aa
Fertility	7.5	8	5	6	4	3.9

   c. Family selection:

Parental Mating Type	Offspring Fitnesses		
	AA	Aa	aa
AA × AA	1.05		
AA × Aa	0.8	1.04	
AA × aa	—	1.9	
Aa × Aa	1.5	2.0	1.6
Aa × aa	—	1.5	1.4
aa × aa	—	—	1.6

4. Suppose a human population in a malarial region is at equilibrium for the sickle cell polymorphism with $p = 0.89$, $q = 0.11$ (the frequencies of the A and S alleles, respectively), the population is randomly mating, and the fitnesses are: AA, 0.9; AS, 1; and SS, 0.2. Now suppose the system of mating changes such that females choose males as mates that maximize the fitness of their offspring. For each of the three female genotypes, calculate the average fitness of their progeny when they mate with each of the three types of males. What male maximizes the average fitness of the female's progeny in each case? Should females ever choose as mates the males with the highest fitness?

5. An allele A shows biased gene conversion over allele a such that in heterozygotes there is a 2 : 1 segregation in favor of A in heterozygotes. Assume for now that this is the only phenotype associated with A.

   a. Starting with an allele frequency of 0.1 for A in a random-mating population, what is the frequency of A in the next generation?

   b. Same as part a, but now assume nonrandom mating with $f = 0.5$.

   c. Assume random mating again but also assume that AA homozygotes are lethal but Aa and aa genotypes are viable and equally fit. Starting with the frequency of Aa's at 0.18 in the adult population, what is the frequency of A in the next generation of adults?

6. Consider the following frequency-dependent fitness model:

Genotype	AA	Aa	aa
Fitness	$1 - tp^2$	$1 - tpq$	$1 - tq^2$

$t > 0$

where q is the frequency of the a allele. Assume random mating.

a. Write down the equation for average fitness as a function of q and find the value of q that maximizes average fitness. (Hint: Remember your calculus and show explicitly that it is a maximum.)

b. Write down the equation for the average excesses of the A and a alleles and find the polymorphic equilibrium q value. (Hint: The average excesses for both alleles are the same, namely 0, only at equilibrium. Also, remember about factoring polynomials and solving for roots.)

c. Does the equilibrium q value maximize average fitness and is it stable (look at Δq at ± 0.1 from q_{eq})?

d. Calculate the fitnesses of all genotypes at equilibrium. What would you infer about the presence of natural selection at this locus given only the genotypic fitnesses at equilibrium?

7. Given the following fitnesses, show whether or not the polymorphism is protected. If the polymorphism is not protected, identify the alleles that are not protected. In all cases, assume random mating.

a. Competitive interactions:

Genotype	AA	Aa	aa
AA	0.9	0.9	1.2
Aa	0.95	1	1.7
aa	1.2	1.7	1.6

b. Fertility:

Mating type	AA × AA	AA × Aa	AA × aa	Aa × Aa	Aa × aa	aa × aa
Fertility	7	6	5	6	4	5

c. Family selection:

Parental Mating Type	Offspring Fitnesses		
	AA	Aa	aa
AA × AA	1	—	—
AA × Aa	1	1.05	—
AA × aa	—	1.9	—
Aa × Aa	1.5	2.0	1.6
Aa × aa	—	1.6	1.6
aa × aa	—	—	1.5

8. Consider the following two-locus fitness model:

Genotype at locus 1\2	AA	Aa	aa
BB	1.2	0.7	1.9
Bb	0.3	1	0.8
bb	0.9	0.4	1.1

Given that the initial gene pool consists only of the *AB* and *ab* gametes each at equal frequency, what is the change under random mating in an infinitely large population of the *AB* gamete for a recombination frequency of 0, for a recombination frequency of 0.025, and for a recombination frequency of 0.1? Does the genetic architecture (in this case as measured by r) affect the initial evolutionary direction in this case?

9. For some systems, there are interactions between diploid and haploid phases (as for some self-sterility alleles in plants). As a result, frequency-dependent selection can be induced that depends upon gamete rather than genotype frequencies. Consider the following gamete frequency-dependent fitness model:

Genotype	AA	Aa	aa	
Fitness	$1-s+tq$	1	$1+s-tq$	$s, t > 0$

where q is the frequency of the *a* allele. Assuming random mating, write down the equation for average fitness and find the q value that maximizes average fitness. (Hint: Remember your calculus.) Write down the equation for the average excesses of the *A* and *a* alleles and find the polymorphic equilibrium q value. (Hint: The average excesses for both alleles are the same, namely 0, only at equilibrium.) Is the equilibrium q value stable and does it maximize average fitness? Calculate the fitnesses of all genotypes at equilibrium. What would you infer about the presence of natural selection at this locus given only the genotypic fitnesses at equilibrium? Suppose that $s = 0.25$, $t = 1$, and the initial q starts at a frequency of 0.3. Calculate the change in q and average fitness from this point. Does this example obey Fisher's fundamental theorem of natural selection?

Answers

1. a. $k = \frac{4}{5} = 0.8$; $\Delta p = G_{Aa}(k - \frac{1}{2}) = 2(0.05)(0.95)(0.8 - 0.5) = 0.029$, $p' = p + \Delta p = 0.079$
   b. $k = \frac{4}{5} = 0.8$; $\Delta p = G_{Aa}(k - \frac{1}{2}) = 2(0.05)(0.95)(0.5)(0.8 - 0.5) = 0.014$, $p' = p + \Delta p = 0.064$
   c. $p' = p + G_{Aa}(k - \frac{1}{2}) = 0.05 + 0.029 = 0.079$, $\bar{W} = 0.9938$, $a_A = -0.010$, so $\Delta p' = -0.006$. From part a, $\Delta p = G_{Aa}(k - \frac{1}{2}) = 0.029$, so Δp(total) $= -0.006 + 0.029 = 0.023$, $p'' = p + \Delta p = 0.073$.

2. a. From equation 13.12, it is 10^{-5}.
   b. Using equation 13.13, first note that $\alpha = 10^{-4} > 1/(2N) = 5 \times 10^{-5}$, so the time to fixation is $4N = 40,000$ generations.
   c. Orthologous fixation (whose coalescence time is also $4N$).

3. a. When A is rare, the population consists mostly of aa homozygotes interacting with one another, so $\bar{W} \approx 1.6$. The A allele is found almost exclusively in Aa individuals that interact primarily with aa's. So, when A is rare, the average excess of A is approximately $1.7 - 1.6 = 0.1 > 0 \Rightarrow A$ will increase in frequency. Therefore, A is protected.

When a is rare, the population consists mostly of AA homozygotes interacting with one another, so $\bar{W} \approx 0.95$. The a allele is found almost exclusively in Aa individuals that interact primarily with AA's. So, when a is rare, the average excess of a is approximately $0.9 - 0.95 = -0.05 < 0 \Rightarrow a$ will decrease in frequency. Therefore, a is not protected. *The polymorphism is not protected.*

(Interesting aside, note that fixation of A is stable, resulting in a population with average fitness of 0.95, whereas fixation of a is not stable, preventing the evolution of a population with average fitness of 1.6.)

b. When A is rare, the population consists mostly of $aa \times aa$ mating pairs with a fertility of 3.9, so $\bar{W} \approx 3.9$. The A allele is found almost exclusively in $Aa \times aa$ mating pairs with fertility 4. So, when A is rare, the average excess of A is approximately $4 - 3.9 = 0.1 > 0 \Rightarrow A$ will increase in frequency. Therefore, A is protected.

When a is rare, the population consists mostly of $AA \times AA$ mating pairs, so $\bar{W} \approx 7.5$. The a allele is found almost exclusively in $AA \times Aa$ mating pairs with fertility 8. So, when a is rare, the average excess of a is approximatley $8 - 7.5 = 0.5 > 0 \Rightarrow a$ will increase in frequency. Therefore, a is protected. *The polymorphism is protected.*

c. When A is rare, the population consists mostly of $aa \times aa$ mating pairs with aa offspring of fitness 1.6, so $\bar{W} \approx 1.6$. The A allele is found almost exclusively in Aa offspring of $Aa \times aa$ mating pairs with fitness of 1.5. So, when A is rare, the average excess of A is approximately $1.5 - 1.6 = -0.1 < 0 \Rightarrow A$ will decrease in frequency. Therefore, A is not protected.

When a is rare, the population consists mostly of $AA \times AA$ mating pairs with AA offspring of fitness 1.05, so $\bar{W} \approx 1.05$. The a allele is found almost exclusively in Aa offspring of $AA \times Aa$ mating pairs with fitness of 1.04. So, when a is rare, the average excess of a is approximately $1.04 - 1.05 = -0.01 < 0 \Rightarrow a$ will decrease in frequency. Therefore, a is not protected. *The polymorphism is not protected.*

4.

PROBABILITY OF OFFSPRING GENOTYPES

	AA	AS	SS		
Fitness	0.9	1	0.2		
Mating				Average Fitness	
AA × AA	1	0	0	0.9	
AA × AS	0.5	0.5	0	0.95	
AA × SS	0	1	0	1	maximum
AS × AA	0.5	0.5	0	0.95	maximum
AS × AS	0.25	0.5	0.25	0.775	
AS × SS	0	0.5	0.5	0.6	
SS × AA	0	1	0	1	maximum
SS × AS	0	0.5	0.5	0.6	
SS × SS	0	0	1	0.2	

No female should ever choose the AS males, who have the highest fitness.

5. a.

Genotype	AA	Aa	aa
Starting population	0.01	0.18	0.81

$$p' = p + \text{(frequency of } Aa)(k - \tfrac{1}{2}) = 0.13$$

b.

Genotype	AA	Aa	aa
Starting population	0.055	0.09	0.855

$$p' = p + \text{(frequency of } Aa)(k - \tfrac{1}{2}) = 0.115$$

c.

Genotype	AA	Aa	aa	
Starting population	0.0	0.18	0.82	$p = 0.09$

After conversion: $p' = p + \text{(frequency of } Aa)(k - \tfrac{1}{2}) = 0.09 + (0.18)(0.1667) = 0.12$

Zygotic frequency	0.0144	0.2122	0.7744	
Fitness	0	1	1	$\bar{w} = 1 - (0.12)^2 = 0.9856$
After lethal selection	0.0	0.2143	0.7857	

$$p'(\text{in adults}) = (0.5)(\text{adult frequency of } Aa) = 0.1071$$

6. a.
$$\bar{w} = p^2(1 - tp^2) + 2pq(1 - tpq) + q^2(1 - tq^2) = 1 - t(p^4 + 2p^2q^2 + q^4)$$
$$= 1 - t(p^2 + q^2)^2 = 1 - t(1 - 2pq)^2 = 1 - t(1 - 2q + 2q^2)^2$$

To find the maximum, take the first derivative as $-2(-2 + 4q)(1 - 2q + 2q^2)t$ using the chain rule from introductory calculus (this can also be found by taking the derivative of the first form of $\bar{w}$ given above and using the chain rule to find the derivative; there is no need to expand the function out entirely). Next, set the first derivative to zero and solve. The only real solution is $\tfrac{1}{2}$.

To confirm that this is a maximum and not a minimum, take the second derivative as $-2(-2 + 4q)^2 t - 8(1 - 2q + 2q^2)t$ and evaluate at $q = \tfrac{1}{2}$ to be $-4t < 0$, which implies it is a maximum.

b. Average excesses of the A and a alleles:

$$a_A = p(1 - tp^2) + q(1 - tpq) - \bar{w} \qquad a_a = p(1 - tpq) + q(1 - tq^2) - \bar{w}$$

At equilibrium, $a_A = a_a$, which implies $p(1 - tp^2) + q(1 - tpq) = p(1 - tpq) + q(1 - tq^2)$, or $(-1 + 4q - 6q^2 + 4q^3) = 0$.

$(-1 + 4q - 6q^2 + 4q^3) = (-1 + 2q)(1 - 2q + 2q^2)$, so the roots are $\tfrac{1}{2}$ and $\tfrac{1}{2} \pm \tfrac{i}{2}$. The only real root is $q_{eq} = \tfrac{1}{2}$.

c. $q_{eq} = \tfrac{1}{2}$ does maximize $\bar{w}$ in this case. For stability, look at Δq at ± 0.1 from q_{eq}, that is, $q = 0.4$ and $q = 0.6$.

At $q = 0.4$, $\bar{w} = 1 - (0.2704)t$ and $a_a = (0.0624)t > 0$, which implies q is increasing under selection.

At $q = 0.6$, $\bar{w} = 1 - (0.2704)t$ and $a_a = -(0.0416)t > 0$, which implies q is decreasing. Therefore, q_{eq} is *stable*.

d. Fitnesses of all genotypes at equilibrium:

Genotype	AA	Aa	aa
Fitness	$1 - tp^2 = 1 - \frac{1}{4}t$	$1 - tpq = 1 - \frac{1}{4}t$	$1 - tq^2 = 1 - \frac{1}{4}t$

All genotypes have the same fitness at equilibrium, which creates the false impression that natural selection is not operating at this locus.

7. a. When a is rare, $w_2 \approx 0.9 + 2q(0.95) \approx \bar{w}$ and $w_1 \approx w_{12} + 2qw_{11} = 0.9 + 2q(1)$, so $a_a \approx 0.9 + 2q - 0.9 - 1.9q = 0.1q > 0$, which implies that a is protected.

When A is rare, $w_1 \approx w_{10} = 1.7$ and $w_0 \approx w_{00} = 1.6 \approx \bar{w}$, so $a_A \approx 0.1 > 0$, which implies A is protected.

Both alleles are protected; therefore this is a protected polymorphism.

b. When a is rare, $W_{AA} \approx w_1 = 7 \approx \bar{w}$ and $W_{Aa} \approx w_2 = 6$; so $a_a \approx -1 < 0$ and a is not protected.

When A is rare, $W_{aa} \approx w_6 = 5 \approx \bar{w}$ and $W_{Aa} \approx w_5 = 4$, so $a_A \approx -1 < 0$ and A is not protected.

Therefore, there is no protected polymorphism.

c. When a is rare, $W_{AA} \approx w_{21} = 1 \approx \bar{w}$ and $W_{Aa} \approx w_{12} = 1.05$, so $a_a \approx 0.05 > 0$ and a is protected.

When A is rare, $W_{aa} \approx w_{06} = 1.5 \approx \bar{w}$ and $W_{Aa} \approx w_{15} = 1.6$, so $a_A \approx 0.1 > 0$ and A is protected. Both alleles are protected; therefore this is a protected polymorphism.

8. The initial population is always as follows:

Genotype	AABB	AaBb	aabb	
Fitness	1.2	1	1.1	$\bar{W} = 1.075$
Genotype frequency (Hardy–Weinberg)	0.25	0.5	0.25	

$D = 0.25$

Average excess	AB	ab
	0.025	−0.025

The equation for change in the frequency of AB is always (from equation 13.1)

$$\Delta g_{AB} = \frac{g_{AB}}{\bar{w}} a_{AB} - rD \frac{w_{AB/ab}}{\bar{w}}$$

$$\Delta g_{AB} = \frac{0.5}{1.075} - r(0.25)\frac{1}{1.075}$$

so for $r = 0$ 0.025   0.1
$\Delta g_{AB} = 0.0116$ 0.0058   −0.012

Note that r does affect the initial evolutionary direction (+ for $r = 0$, 0.025; − for $r = 0.1$).

9. $\bar{W} = (1-q)^2(1 - s + tq) + 2(1-q)q + q^2(1 + s - tq) = 1 - s + q(t + 2s) - 2tq^2$. To find the maximum (or minimum), $d\bar{W}/dq = t + 2s - 4tq = 0 \Rightarrow q = (t + 2s)/(4t)$. To show it is the maximum, $d^2\bar{W}/dq^2 = -4t < 0 \Rightarrow$ a maximum.

$a_A = p(1 - s+tq) + q - \bar{W} = -qs + tq^2; a_a = p + q(1 + s-tq) - \bar{W} = -q(s + t) + tq^2 + s$

At equilibrium, $a_A = a_a = 0 \Rightarrow p(1 - s + tq) + q = p + q(1 + s - tq) \Rightarrow q_{eq} = s/t$.

To check for stability, note that $a_A = -qs + tq^2 = q(-s + tq)$. Hence, the sign of a_A (and thus automatically a_a) is determined solely by $-s + tq$. If $q > s/t$, then $a_A > 0 \Rightarrow q$ will decrease. If $q < s/t$, $a_A < 0 \Rightarrow q$ will increase. Hence, the equilibrium is stable.

At equilibrium, $w_{AA} = 1 - s + t(s/t) = 1; w_{Aa} = 1; w_{aa} = 1 + s - t(s/t) = 1$ ⇒ *appears* neutral at equilibrium.

At $q = 0.3$, $s = 0.25$ and $t = 1$; $\bar{W} = 1.02$; $a_a = (0.3)(1 + 0.25 - 0.3) + 0.7(1) - 1.02 = -0.0350$.

Hence $\Delta q = qa_a/\bar{W} = (0.3)(-0.035)/1.02 = -0.0103 \Rightarrow q' = q + \Delta q = 0.2897 \Rightarrow \bar{W}' = 1.0167$.

Hence, $\Delta \bar{W} = -0.0033 < 0 \Rightarrow$ Fisher's Fundamental Theorem of Natural Selection (FFTNS) is violated.

CHAPTER 14

Problems

1. Consider a two-niche Levene model with a single locus and two alleles, with random mating at the total population level. Let $w_1 = v_2 = 1.05$ and $w_2 = v_1 = 0.95$. Assume $c_1 = c_2 = 0.5$.
   a. Will the polymorphism be protected?
   b. Let $w_1 = 1.06$, with all other fitnesses unchanged. Will the polymorphism be protected?
   c. Assume the original fitnesses, but let $c_1 = 0.53$ and $c_2 = 0.47$. Will the polymorphism be protected?
   d. Assume a hard-selection model by regarding all the c's as z's in the sense of equation 14.6. Redo parts a through c under this assumption.
   e. Assume that the population is no longer randomly mating at the total population level but rather that the niches define a subdivided population with $m = 0.1$. Redo parts a through d under this assumption.

2. A population fits the assumptions of a two-niche, soft-selection Levene model with a single locus and two alleles, with random mating at the total population level. Let the fitnesses be:

| Niche | Fitness | | | c_i |
	AA	Aa	aa	
1	1.5	1	0.5	0.25
2	0.85	1	1	0.75

a. Will the polymorphism be protected?

b. Assume that selection is hard and that the c's in the above table now correspond to z's (proportion of zygotes entering the niche). Will the polymorphism be protected?

c. Assume that $c_1 = c_2 = 0.5$. Redo part a with this new assumption.

d. Assume that $z_1 = z_2 = 0.5$. Redo part b with this new assumption.

3. Assume an organism is continuously distributed across a space in which the environment changes over a distance of Δ. Suppose the population is polymorphic for a single locus with two alleles, A and a, with a fitness model given by equation 14.10 and a gene flow model given by equation 14.11. Calculate the characteristic length of variation in allele frequencies for the following parameters and indicate whether this environmental shift is experienced as an ecotone or gradient with respect to this locus.

d	m	b	Δ
50	0.3	0.0001	500
50	0.3	0.001	500
200	0.8	0.0001	500
200	0.8	0.001	500
500	0.5	0.0001	500
500	0.5	0.001	500

4. Assume an organism lives in a seasonal environment and is polymorphic for an autosomal locus with two alleles, A and a, such that the genotypes have the following fitnesses:

Genotype	AA	Aa	aa
Fitness season 1	0.7	1	1.2
Fitness season 2	1.3	1	0.7

a. Is this polymorphism protected when the organism has two generations per year, with one generation experiencing season 1 and the next generation experiencing season 2? Assume random mating at this locus.

b. Is this polymorphism protected when the organism has one generation per year, with all living individuals experiencing both seasons? Assume random mating at this locus.

5. Suppose a new allele, A, mutates from the ancestral allele a such that the relative fitnesses are 1.02 and 1 for Aa and aa, respectively. Assume an idealized, random-mating population of large size. Further assume that fine-grained environmental variation induces a variance in fitness across Aa individuals, with the variance in the number of offspring of Aa individuals being $1.02 + \sigma_s^2$.

a. What is the probability of survival of the A mutant for the following values of σ_s^2: 0, 0.5, 1, 2?

b. Assume that the fitness of the Aa type is 1.04 with a variance of $1.04 + \sigma_s^2$. Redo part a with these new assumptions.

c. Contrast the results of parts a and b to identify situations in which a new mutant of lower average fitness has a greater probability of survival.

d. Assume that the total population has an ideal size of 30. Redo parts a, b, and c.

Answers

1. Soft selection:

	w Harmonic Mean	v Harmonic Mean	Protected?
a.	0.9975	0.9975	Yes
b.	1.0020	0.9975	No
c.	1.0005	0.9945	No

d. Hard selection:

	w Arithmetic Mean	v Arithmetic Mean	Protected?
a.	1.0000	1.0000	No
b.	1.0050	1.0000	No
c.	1.0030	0.9970	No

e. Soft selection with $m = 0.1$:

	w, m Harmonic Mean	v, m Harmonic Mean	Protected?
a.	0.7500	0.7500	Yes
b.	0.7619	0.7500	Yes
c.	0.7732	0.7282	Yes

Hard selection with $m = 0.1$:

	w Arithmetic Mean	v Arithmetic Mean	ω_1	ω_2	Protected?
a.	0.7975	0.7975	1.045	0.955	Yes
b.	0.8166	0.7975	1.0545	0.955	Yes
c.	0.8233	0.7730	1.0453	0.9547	Yes

2. a.

w Harmonic Mean	v Harmonic Mean	Protected?
0.8000	0.9533	Yes

b.

w Arithmetic Mean	v Arithmetic Mean	Protected?
0.8750	1.0125	No

c.

w Harmonic Mean	v Harmonic Mean	Protected?
0.6667	1.0851	No

d.

w Arithmetic Mean	v Arithmetic Mean	Protected?
0.7500	1.1750	No

3.

d	m	b	Δ	ℓ_c	Ecotone/Gradient
50	0.3	0.0001	500	122.4744871	Gradient
50	0.3	0.001	500	38.72983346	Gradient
200	0.8	0.0001	500	800	Ecotone
200	0.8	0.001	500	252.9822128	Gradient
500	0.5	0.0001	500	1581.13883	Ecotone
500	0.5	0.001	500	500	Ecotone

4. Because there is no dominance or recessiveness, the conditions for both the coarse-grained case (a) and the fine-grained case (b) depend upon the following fitness products (equation 14.19 and Table 14.4): $v_1 v_2 = 0.84$ and $w_1 w_2 = 0.91$. As both of these products are less than 1, the polymorphism is protected in both cases.

5. **a.** From equation 14.20:

σ_s^2	Prob(survival)
0	0.039215686
0.5	0.026315789
1	0.01980198
2	0.013245033

b.

σ_s^2	Prob(survival)
0	0.076923077
0.5	0.051948052
1	0.039215686
2	0.026315789

c. The only case in which the mutant with the lower average fitness has a higher survival probability is the case when $\sigma_s^2 = 0$ in part a contrasted to the case of $\sigma_s^2 = 2$ in part b.

d. Use equation 14.22:

Redoing part a:

σ_s^2	Prob(survival)
0	0.04312255
0.5	0.030612056
1	0.020154376
2	0.006836626

Redoing part b:

σ_s^2	Prob(survival)
0	0.077521637
0.5	0.062773753
1	0.048546645
2	0.024052179

Redoing part c: The probability of survival of a mutation with a fitness of 1.04 with $\sigma_s^2 = 2$ is less than the probability of survival of a mutation with a fitness of 1.02 with $\sigma_s^2 = 0$ and $\sigma_s^2 = 0.5$.

CHAPTER 15

Problems

1. The following life history data were gathered by following three genotypic cohorts (no individual lives to age 4):

	COHORT					
	AA		Aa		aa	
x = age	l_x	$m_x b_x$	l_x	$m_x b_x$	l_x	$m_x b_x$
0	1	0	1	0	1	0
1	0.25	2	0.5	2	0.5	0
2	0.25	2	0.15	2	0.25	4
3	0	0	0	0	0.1	4

   Calculate the net reproductive rate and Mathulsian parameter (use the approximation for the Mathulsian parameter rather than solving it implicity) for each genotype.

2. For the situation described in problem 1, assume that the frequency of the A allele is 0.2. Assume that the species is an annual plant with one generation per year, with the ages above corresponding to the ages in months since germination in the spring. All seeds must overwinter before germinating the following spring. Calculate the average excesses of the A and a gametes for the appropriate fitness measures in this situation under random mating and under an inbreeding system with $f = 0.1$. Identify which allele is increasing in frequency due to natural selection or, if the system is at equilibrium, for each system of mating.

3. For the situation described in problem 2, assume that the species is a perennial shrub, with the ages above corresponding to the ages in years since germination. All seeds produced in a given year must overwinter before germinating the following spring. Calculate the average excesses of the A and a gametes for the appropriate fitness measures in this situation under random mating and under an inbreeding system with $f = 0.1$. Identify which allele is increasing in frequency due to natural selection or, if the system is at equilibrium for each system of mating.

4. A population of ageless individuals has a probability of dying of 0.1 in a time interval. The probability of having mated times the expected number of offspring is 0.11 in a time interval.
   a. What is the net reproductive rate of these ageless individuals?
   b. Suppose a mutation occurs such that the bearers of this mutant allele die at age n but the probability of having mated times the expected number of offspring is 0.12 in a

time interval for ages less than n. What value of n results if the mutant and ageless individuals have identical net reproductive rates?

c. When would the mutant described above be favored by natural selection?

Answers

1.

	COHORT		
	AA	Aa	aa
R_0	1	1.3	1.4
m	0	0.1875	0.125

2. In this situation of discrete generations of fixed length, R_0 is the appropriate measure of fitness. Hence:

Genotype	AA	Aa	aa	p	f
Frequency	0.04	0.32	0.64	0.2	0
Fitness	1	1.3	1.4	Average $W = 1.352$	
Genotypic deviation	−0.352	−0.052	0.048		

	A	a	
Average excess	−0.112	0.028	a is increasing in frequency

Genotype	AA	Aa	aa	p	f
Frequency	0.056	0.288	0.656	0.2	0.1
Fitness	1	1.3	1.4	Average $W = 1.3488$	
Genotypic deviation	−0.3488	−0.0488	0.0512		

	A	a	
Average excess	−0.1328	0.0332	a is increasing in frequency

3. In this situation of overlapping generations and no fixed generation length, the Malthusian parameter is the appropriate measure of fitness. Hence:

Genotype	AA	Aa	aa	p	f
Frequency	0.04	0.32	0.64	0.2	0
Fitness	0	0.1875	0.125	Average $W = 0.14$	
Genotypic deviation	−0.14	0.0475	−0.015		

	A	a	
Average excess	0.01	−0.0025	A is increasing in frequency

Genotype	AA	Aa	aa	p	f
Frequency	0.056	0.288	0.656	0.2	0.1
Fitness	0	0.1875	0.125	Average $W = 0.136$	
Genotypic Deviation	−0.136	0.0515	−0.011		

	A	a	
Average excess	−0.001	0.00025	a is increasing in frequency

4. a. From equation 15.13, $R_0 = 1.1$.

b. Use equation 15.19 and set it equal to 1.1:

$$R_0' = mb' \sum_{x=0}^{n-1} (1-d)^x = \frac{mb'}{d}\left[1 - (1-d)^n\right] = 1.2[1 - 0.9^n] = 1.1$$

Solving for n yields $n = 23.585$.

c. It would be favored for any $n > 23.585$.

INDEX

abnormal abdomen (aa) locus. *See also* Supergenes
  age-structured populations, natural selection, 508–539
    genetic architecture and units/targets of selection, 508–516
    phenotypes and potential targets of selection, 516–525
    supergene components, 536–537
    supergene spatiotemporal heterogeneous environment, 525–536
  coarse-grained spatial heterogeneity, 468–474
ABO blood group:
  fertility and, 439–443
  natural selection, genetic architecture, 396–397
  population history and gene flow, linkage disequilibrium, 208–211
Acheulean replacement hypothesis, gene flow and population history, haplotype tree loci/DNA region inferences, 242–243
Achillea plant, quantitative genetics, between-population differences, mean phenotype, 288–290
ACP-1 locus, natural selection, genetic architecture, 396–397
Adalia bipunctata, coarse-grained temporal heterogeneity, 475–482
Adaptation:
  age-structured population selection, *aa* supergene, spatio-temporal heterogeneous environment, 531–535
  coarse-grained spatial heterogeneity, industrial melanism, 465–474
  DNA-environmental interaction and, 8–9
  natural selection, 344–345
    average fitness and, 365
    fitness variation and population structure, 363
    genetic architecture, 394–397
    genetic drift-gene flow interaction, 389–391
    genetic drift interaction, 381–387
    nonadaptive trait evolution, 368

  polygenic process, 358–361
  sickle cell anemia and malarial resistance, 353–358
  natural selection and adaptation, 369–370
Adaptative traits, natural selection, 345
Adaptive Evolution Database, development of, 328
Adaptive surface/landscape:
  genetic drift, natural selection, 383–387
  natural selection, 364–370
  genetic architecture, 392–397
Additive genetic variance:
  natural selection, 364–365
  quantitative genetics, 272–273
    between-population differences, controlled crosses, 292–295, 294–295
    candidate loci, epistasis, 335–337
    correlation between relatives, 280–283
    mutation, drift and gene flow, 295–296
Additive genotypic deviation, quantitative genetics, 269–270
  correlation between relatives, 278–283
Adenosine deaminase locus (ADA), natural selection, genetic architecture, 396–397
Admixture:
  assortative mating and linkage disequilibrium and, 73–77
  gene flow and population history, 215–218
  haplotype tree loci/DNA region inferences, 241–243
  haplotype trees and, 218–221
  linkage disequilibrium, 207–211
  natural selection, gene flow interaction, 379–380
  quantitative genetics:
    between-population differences, controlled crosses, 291–295
    interpopulation disequilibrium marker loci, 299–303
Adoptive children, quantitative genetics, between-population differences, mean phenotype, 287–290

Adult-onset diabetes, coarse-grained temporal heterogeneity, 477–482
Ageless individuals, senescence evolution, 505–508
Age of onset, senescence evolution and, 506–508
Age of variants, natural selection, genetic drift and mutation, 405–406
Age-specific survivorship, age-structured populations, selection in, life history and fitness, 498–504
Age-structured populations, natural selection:
  abnormal abdomen case study, 508–539
    genetic architecture and units/targets of selection, 508–516
    phenotypes and potential targets of selection, 516–525
    supergene components, 536–537
    supergene spatiotemporal heterogeneous environment, 525–536
  life history and fitness, 498–504
  overview, 497
  senescence evolution, 504–508
Agriculture, quantitative genetics, response to selection, 283–284
Alcohol dehydrogenase (Adh) locus:
  activity histogram, 322–328
  evolutionary history, 319–328
Allele frequencies:
  age-structured population selection, ultimate causation patterns, 539
  assortative mating, admixture and linkage disequilibrium, 74–77
  balance of gene flow and drift, one-dimensional stepping-stone model, 186–193
  coarse-grained spatial heterogeneity, 457–474
  coarse-grained temporal heterogeneity, 475–482
  disassortative mating and, 79–81
  fertility, 441–443
  gene flow:
    vs. admixture, 216–218
    genetic impact of, 173–174
    local populations, 169–172
    population history and linkage disequilibrium, 208–211
    Wahlund effect, 180–183
  genetic drift and, 82–83
    founder and bottleneck effects, 89–94
    linkage disequilibrium, 94–97
  Hardy-Weinberg law, 26–30
  inbreeding deviation from random-mating, 56–63
  linkage disequilibrium and, 45–47
  meiotic drive, 421–424
  natural selection:
    evolutionary forces and, 372–373
    gene flow interaction, 378–380
    genetic drift-gene flow interaction, 389–397
    genetic drift interaction, 380–387
    mutation interaction with, 373–375
  quantitative genetics:
    average effect of gamete type i, 268–270
    interpopulation disequilibrium markers, 302–303
    single loci, unit of selection and, 411–419
    targets of sexual selection, 437–439
Allopatric fragmentation, range expansion, 231–233
Altruism, kin/family selection, 448–452
Alu gene family, coalescent theory and, haplotype trees and mutation in, 155–156
Alzheimer's disease, linkage disequilibrium and, 45–47
Ambystoma tigrinum, coarse-grained temporal heterogeneity, 480–482
Amino acid sequencing:
  candidate loci analysis, 328
  limits of neutral theory and, 129–132
  neutral alleles, 120–128
  protein electrophoresis, 541–542
Amplified fragment length polymorphisms (AFLPs), genetic survey techniques, 546–548
Analysis of molecular variance (AMOVA), population subdivision measurements, sequencing and restriction data, 199
Analysis of variance (ANOVA):
  one-way analysis, 567–569
  permutation testing, 669–670
  quantitative genetics, candidate loci, 317
    evolutionary history, 318–328
  regression and correlation, 570–571
Ancestral forms:
  coalescent theory, inversion network, 148–150
  inbreeding and, 48–49
Antagonistic pleiotropy, natural selection, 368
Apical membrane antigen 1 (AMA1) locus, coevolution mechanisms, 491–496
ApoA1-CIII-AIV gene cluster, candidate loci analysis:
  epistasis, 336–337
  evolutionary history, 318–328
Apodemes, age-structured population selection:
  aa supergene, spatio-temporal heterogeneous environment, 528–535
  abnormal abdomen (aa) locus targets of selection, 522–525
Apoprotein E (ApoE):
  germline selection, 449–452
  haplotype trees:
    mutation and, 151–156
    recombination, 158–161
  quantitative genetics:
    between-population differences, mean phenotype, 285–290
    candidate loci, 312–317
      combined pleiotropy and epistasis, 337
      epistasis, 330–337

gene-environment interactions, 337–338
  pleiotropy, 329–330
  mean-related measures, 262–270
Aposematic coloration, coevolution mechanisms, 493–496
Assortative mating, 63–77
  admixture, 73–77
  coevolution mechanisms, 495–496
  gene flow, local populations, 170–172
  inbreeding vs., 71–73
  linkage disequilibrium, 66–71
    genetic drift and, 95–97
  simple model, 63–66
Asymmetrical gene flow, local populations, 171–172
Asymptotic normality:
  maximum-likelihood estimation, 576–577
  means difference testing in two populations, 567
Average effect of gamete type i:
  natural selection, 364
  quantitative genetics:
    correlation between relatives, 278–283
    mean-related measures, 268–270
Average excess of fitness:
  age-structured population selection, fecundity estimates with aa syndrome, 524–525
  fertility, 441–443
  meiotic drive, 422–424
  natural selection:
    evolutionary forces and, 372–373
    gene flow interaction, 379–380
    measured genotypes, 349
    mutation-mating interaction, 376–378
    selective equilibria, 364
    sickle cell anemia, 351–358
Average excess of gamete type:
  coarse-grained temporal heterogeneity, 475–482
  quantitative genetics, 265–270
    between-population differences, mean phenotype, 285–290
Average generation time, age-structured populations, life history and fitness, 504
Average population fitness:
  natural selection, 364–365
  unmeasured genotypes, 361–363

Backcross measurements, quantitative genetics, linkage markers, 307–312
Balanced polymorphism:
  competitive ability, 444–446
  natural selection, sickle cell anemia, 353–358
"Balanced" population genetics theory, genetic drift, neutral alleles, 123–128
Balance of gene flow and drift:
  multiple inheritance modes, 200–202
  population subdivision and, 174–180
  Wahlund effect, 181–183
Bayesian inference:
  DNA resequencing, estimation-maximization algorithm, 552–554
  statistical data, 578–581
Beak depth:
  comparative analysis of, 12–14
  natural population monitoring, 15–18
Beanbag genetics, history of, 407–408
Beta distribution, 561
  Bayesian inference and, 579–581
Between-population differences, quantitative genetics:
  controlled crosses, 290–295
  mean phenotype, 284–290
Biased estimation, maximum-likelihood estimators, 576–577
Biased gene conversion, targets of selection, 424–425
Binomial distribution, 558
  Bayesian inference and, 578–581
  maximum likelihood techniques, 575–577
  negative distribution, 559
Biston betularia, coarse-grained spatial heterogeneity, industrial melanism, 4623–474
Bivariate normal, multivariate probability distribution, 562–563
Blending inheritance, Hardy-Weinberg law and, 34–35
Blood group loci, Hardy-Weinberg modeling of, 30–32
Bmp4 gene expression, comparative analysis of, 12–14
bobbed (bb) locus, age-structured population selection, 508–516
  proximate causation, 538–539
Bonellia, phenotypes, 7–8
Bottleneck effect:
  genetic drift and, 89–94
    mating systems and, 97–98
    newly arisen mutations, 119
  linkage disequilibrium and, 94–97
  population history and gene flow, direct space-time studies, 206–207
  quantitative genetics, intrapopulation disequilibrium markers, 303–306
Breast cancer:
  candidate loci, combined pleiotropy and epistasis, 337
  transposons, 426–429
Breeding value:
  quantitative genetics, 269–270
  unit of selection and, 414–419
Broad-sense heritability, quantitative genetics:
  correlation between relatives, 280–283
  variance and, 272–273
Buffering, fine-grained heterogeneity, 487–490

CAG repeats:
  age-structured populations, senescence evolution, 506–508
  candidate loci, epistasis, 331–337
  germline selection, 450–452
  kin/family selection, 449–452
  meiotic drive, 421–424
Calpain-10 locus, coarse-grained temporal heterogeneity, 477–482
Candidate loci, measured genotypes, 299, 312–340
  alleles, SNPs, and haplotypes, 312–317
  epistasis, 330–339
  evolutionary history, 318–328
  genetic architecture, 329–340
  pleiotropy, 329–339
Central limit theorem, continuous probability distribution, 559–560
CG dinucleotides, coalescent theory and, haplotype trees and mutation in, 155–156
Chi-square statistic:
  age-structured population selection, genetic architecture and targets and units of selection, 513–516
  contingency analysis, 571–573
  continuous probability distribution, 560–561
  goodness-of-fit testing, 573–574
  variance equality testing in two populations, 567
Cholesterol phenotypes, candidate loci, epistasis, 330–337
Clade distance:
  coarse-grained temporal heterogeneity, 480–482
  fragmentation patterns, 229–231
  halotype trees, nested-clade phylogeographic analysis, 225–228
  isolation by distance models, 228
  range expansion, 231–233
"Classical" population genetics, neutral alleles and, 122–128
Climate change:
  age-structured population selection, aa supergene, spatio-temporal heterogeneous environment, 531–535
  coarse-grained temporal heterogeneity, 480–482
  fine-grained heterogeneity, buffering mechanisms and, 488–490
Clinal patterns:
  age-structured population selection, aa supergene: components, 536–537
    spatio-temporal heterogeneous environment, 531–535
  population history and gene flow, principal component analysis, 211
Cluster analysis of genes:
  coarse-grained spatial heterogeneity, 468–474
  unit of selection and, 419
Coadapted gene complex, natural selection, 395–397

Coalescent theory:
  balance of gene flow and drift in, 175–180
  gene flow and population history:
    haplotype tree loci/DNA region inferences, 236–243
    haplotype trees and, 219–220
  genetic drift:
    basic process, 132–142
    gene trees $vs.$ allele/haplotype trees, 143–145
    inversion trees and networks, 145–150
    mutation, 142–143
    N coalescent, 137–138
    restriction site and DNA sequence haplotype trees, 150–167
      Jukes-Cantor molecule genetic distance, 164–165
      molecule genetic distances, 161–167
      mutation effects, 150–156
      neighbor-joining estimation method, 166
      recombination effects, 156–161
  natural selection, genetic drift and mutation, 397–406
  tandem multigene families, unequal exchange, 431–434
Coarse-grained environment:
  defined, 454
  spatial heterogeneity, 455–474
  temporal heterogeneity, 475–482
Codominance:
  electrophoretic mobility phenotype, 251
  Mendelian inheritance, 250
Codon bias, natural selection, genetic drift and mutation, 401–406
Coevolution:
  defined, 455
  heterogeneous environments, 490–496
Coin box simulations, genetic drift, 83–89
Collared lizard example, total population genetic variation, population subdivision and, 195–198
Comparative analysis, in population genetics, 11–14
Competition, natural selection, 443–446
Complete selfing, two-loci architecture, 71–73
Concerted evolution, tandem multigene families, 432–434
Conditional probability:
  DNA resequencing, estimation-maximization algorithm, 550–554
  multilocus evolution, unit of selection and, 411–419
Confidence intervals of mean, 565–566
Consistency, maximum-likelihood estimation, 576–577
Constant-fitness models, fine-grained heterogeneity, 483–490
Context dependence, candidate loci, coronary artery disease, 339–340

Contingency analysis:
  natural selection, genetic drift and mutations, 399–406
  statistics, 571–573
Continuous probability distribution:
  balance of gene flow and drift, one-dimensional stepping-stone model, 188–193
  beta distribution, 561
  chi-square and gamma distribution, 560–561
  F distribution, 561–562
  Fisherian model of quantitative genetics, 258–260
  normal distribution, 559–560
  t distribution, 561
Controlled crosses, between-population differences, quantitative genetics, 290–295
Cophenetic correlation, gene flow and population history, 214–215
Coronary artery disease (CAD):
  coarse-grained temporal heterogeneity, thrifty-genotype hypothesis and, 478–482
  linkage disequilibrium and, 45–47
  quantitative genetics, candidate loci, 313–317
    pleiotropy, 329–330
    reductionist models, 339–340
Correlation between relatives, quantitative genetics, 275–283
Correlation coefficient:
  analysis of variance and, 570–571
  inbreeding deviation from random-mating, 57–63
  quantitative genetics, correlation between relatives, 278–283
Covariance:
  bivariate normal, 562–563
  minimum-variance, unbiased mean estimator, 564–565
  quantitative genetics, correlation between relatives, 275–283
Cross validation, haplotype tree inferences, 234–243
Cyclical fitnesses:
  coarse-grained temporal heterogeneity, 476–482
  fine-grained heterogeneity, 482–490

Darwin's finches:
  comparative analysis of, 12–14
  natural population monitoring of, 15–18
Darwin-Wallace natural selection theory, Hardy-Weinberg law and, 34
Deafness, assortative mating, 65–66
Decomposition of genetic variance, quantitative genetics, candidate loci, epistasis, 333–337
Deletrious alleles, natural selection:
  genetic drift interaction, 382–387
  mutation interaction:
    evolutionary dynamics, 373–375
    mating systems, 376–378

Demes:
  additive genetic variance, 296
  disassortative mating and, 80–81
  gene flow:
    balance of genetic drift and, 174–180
      one-dimensional stepping-stone model, 186–193
    Wahlund effect, 180–183
    natural selection interaction, 378–380, 387–397
  genetic drift:
    isolation of, 86–89
    natural selection interaction, 387–397
    natural selection, metapopulation, 392
  pedigree inbreeding in, 52–55
  population genetics in, 2–3
Density-dependent gene flow, two-dimensional habitats, 188–193
Deoxyribonucleic acid (DNA):
  basic properties, 1
  environmental interaction, 7–9
  mutation and recombination, 3–6
  replication, 2–3
Developmental constraints, natural selection, 368–369
Differential epistasis, age-structured population selection, emergence time characteristics, 521–525
Dimensionality parameters, balance of gene flow and drift, 188–193
Disassortative mating:
  basic principles, 77–81
  gene flow, local populations, 170–172
  genetic drift and, 97–98
  linkage disequilibrium, genetic drift and, 95–97
Discrete probability distribution, 558–559
Disequilibrium:
  genetic drift and, 94–97
  mating system and, 97–98
  interpopulation disequilibrium, marker loci studies, 298–303
  intrapopulation disequilibrium, marker loci studies, 299, 303–306
Dispersal variance:
  age-structured population selection, aa supergene, spatio-temporal heterogeneous environment, 531–535
  balance of gene flow and drift, neighborhood size, 189–193
DNA fingerprinting, genetic survey applications, 545–546
DNA forensics, assortative mating, admixture and linkage disequilibrium, 77
DNA regions, haplotype tree inferences, 234–243
DNA replication, gene flow and population history, haplotype trees, 218–221
DNA resequencing, genetic survey applications, 5r48–554

DNA sequencing:
  haplotype trees, 150–167
    Jukes-Cantor molecule genetic distance, 164–165
    molecule genetic distances, 161–167
    mutation effects, 150–156
    neighbor-joining estimation method, 166
    recombination effects, 156–161
  neutral theory and, 130–132
  population subdivision measurements and, 198–199
DNA transposons, targets of selection, 425–429
Dominance:
  malarial resistance phenotype, 253
  Mendelian inheritance, 250
  quantitative genetics:
    between-population differences, controlled crosses, 294–295
    correlation between relatives, 277–283
    deviation of genotype, 270
    variance and, 273
  sickling phenotype, 252
Double-digest patterns, restriction site mapping, 545
Double heterozygotes, unit of selection and, 410–419
Drosophila hydei, age-structured population selection, *aa* supergene, spatio-temporal heterogeneous environment, 526–535
Drosophila melanogaster:
  age-structured population selection, genetic architecture and targets and units of selection, 515–516, 519–516
  candidate loci, evolutionary history, 319–328
  coalescent theory and, inversion trees and networks, 145–150
  DNA mutation and recombination in, 6
  genetic drift and, 87–89
    coalescent time estimates, 140–141
    disassortative mating systems, 97–98
    founder and bottleneck effects, 89–94
    neutral alleles, 122–128
    polytene chromosomes, 123
  natural selection, genetic drift and mutation, 397–406
  population structure and multiple inheritance modes, 200–202
  quantitative genetics:
    between-population differences, mean phenotype, 289–290
    response to selection, 283–284
  senescence evolution, 507–508
  transposons, 426–429
Drosophila mercatorum:
  age-structured population selection:
    aa supergene, spatio-temporal heterogeneous environment, 525–535
    genetic architecture and targets and units of selection, 508–516, 515–516
  coarse-grained spatial heterogeneity, 467–474
  unit of selection experiment, 413–419

Drosophila miranda, coalescent theory, inversion trees, 146–150
Drosophila pseudoobscura:
  coalescent theory, inversion trees, 146–150
  targets of sexual selection, 436–439
D statistic:
  coevolution mechanisms, 491–496
  natural selection, genetic drift and mutation, 403–406

Ecotone, coarse-grained spatial heterogeneity, industrial melanism and, 466–474
Effective number of migrants, balance of gene flow and drift and, 176–180
Egg-to-pupal developmental time, age-structured population selection, 520–525
Eigenvalue effective size, quantitative genetics, mutation, drift and gene flow, 295–296
Electrophoretic alleles, quantitative genetics, candidate loci:
  genetic architecture, 339–340
  pleiotropy, 329–330
Electrophoretic mobility, phenotype, 250–251
Emergence profiles, age-structured population selection:
  aa supergene, spatio-temporal heterogeneous environment, 532–535
  targets of selection, 521–525
Endpoint analysis, candidate loci, 339–340
Environment:
  age-structured population selection, juvenilized abdominal cuticle phenotype, 519–525
  DNA interaction with, 7–9
  gene-environment interactions, candidate loci studies, 337–338
  heterogeneous environments, selection in:
    aa supergene, spatio-temporal heterogeneous environment, 525–535
    coarse-grained spatial heterogeneity, 455–474
    coarse-grained temporal heterogeneity, 475–482
    coevolution, 490–496
    fine-grained heterogeneity, 482–490
    overview, 453–455
  nature *vs.* nurture theory and, 254–257
  quantitative genetics, between-population differences:
    controlled crosses, 290–295
    mean phenotype, 285–290
Environmental deviation, quantitative genetics:
  correlation between relatives, 276–283
  mean-related measures, 263–270
Environmental grain, natural selection and, 454
Environmental variance:
  fine-grained heterogeneity, 483–490
  quantitative genetics, 271–273

Epistasis:
   candidate loci, genetic architecture, 330–337
   coarse-grained spatial heterogeneity, 474
   Mendelian inheritance, 250
   natural selection, genetic architecture, 394–397
   quantative trait region and, 310–312
   unit of selection and, 416–419
   unity of genotype and, 408
Equilibrium:
   assortative mating, 68–77
   disassortative mating and, 78–781
   genetic drift and, 82–83
      infinite-alleles model, 127–128
   in Hardy-Weinberg law, 30
   natural selection:
      average excess and average effects, 364
      mutation interaction, 374–375
      sickle cell anemia, 354–358
      trait value, 367–368
   punctuated equilibrium, coarse-grained temporal heterogeneity, 479–482
   targets of sexual selection, 439
Error sum of squares (SSE), one-way analysis of variance, 569
Estimation-maximization (EM) algorithm:
   maximum-likelihood estimators, 576–577
   resequencing techniques, 549–554
Euchromatic DNA, age-structured population selection, genetic architecture and targets and units of selection, 515–516
Euler's equation, age-structured populations, life history and fitness, 503–504
Evolution:
   defined, 3
   disequilibrium and, 43–47
   DNA replication and, 2–3
   genetic drift and, 83–89
   Hardy-Weinberg modeling of, 21–47
      basic properties, 23–30
      disequilibrium implications, 43–47
      examples of, 30–32
      linkage disequilibrium, 41–43
      population structure and, 22–23
      single-loci applications, 32–35
      two-loci applications, 35–41
      two-loci frequencies, 38–41
   natural selection interaction with:
      gene flow interaction, 378–387
      genetic architecture, 392–397
      genetic drift:
         gene flow and, 387–397
         mutation and, 397–406
      mating system/mutation interaction, 376–378
      mutation, 373–375, 397–406
      population subdivision, 391–392
      research overview, 372–373
   tandem multigene families, 431–434

Evolutionary history, candidate locus analysis, 318–328
Evolutionary stochasticity, coalescent theory and, 141–142
Evolutionary tree:
   neutral alleles, 121–28
   transmission disequilibrium test (ET-TDT), candidate loci evolutionary history, 322–328
Expectation, probability distribution, 556–558

Family environmental effects, quantitative genetics, between-population differences, mean phenotype, 285–290
Fecundity:
   age-structured population selection:
      aa supergene, spatio-temporal heterogeneous environment, 530–535
      abnormal abdomen (aa) locus targets of selection, 522–525
   natural selection, 343
   phenotypes, 8
Fertility:
   natural selection, 343
   phenotypes, 8
   targets of selection, 439–443
Fibrinopeptide, neutral mutation in, 129–132
Fine-grained environment, defined, 454
Fine-grained heterogeneity, natural selection and, 482–490
Fine-scale sampling, gene flow *vs.* admixture, 215–218
Finite island model, total population genetic variation, population subdivision and, 194–198
Finite population size:
   genetic drift and, 83–89
   linkage disequilibrium and, 41–43
Finite-sites model, coalescent theory, haplotype trees, mutation and, 151–156
Fisherian model of quantitative genetics, 257–260
   mean-related measures, 262–270
   natural selection, genetic architecture, 394–397
   neutral mutation and, 130
   unmeasured genotypes:
      basic principles, 274–275
      between-population differences:
         controlled crosses, 290–295
         mean phenotypes, 284–290
      correlation between relatives, 275–283
      phenotypic variance, mutation, drift, and gene flow, 295–296
      selection response, 283–284
      variance and, 270–273
Fisher's exact test, natural selection, genetic drift and mutation, 400–406

Fisher's fundamental theorem of natural selection:
  coevolution mechanisms, 492–496
  fine-grained heterogeneity, 485–490
  genetic architecture, 392–397
  genetic drift interaction, 381–387
  kin/family selection, 446–452
  targets of sexual selection, 437–439
  trait heritability, 364
  unmeasured genotypes, 362–363
Fitness parameters:
  age-structured populations, selection in, 498–504
    aa supergene, spatio-temporal heterogeneous environment, 528–535
  coarse-grained spatial heterogeneity, 457–474
  coevolution mechanisms, 492–496
  competitive ability, 443–446
  fertility, 439–443
  fine-grained heterogeneity, 485–490
  kin/family selection, 447–452
  meiotic drive, 422–424
  natural selection, 343–344
    gamete transmission, 363
    genetic drift-gene flow interaction, 387–398
    measured genotypes, 348–349
    mutation-mating interaction, 377–378
    sickle cell anemia, 350–358
    trait optimization, 367–378
  neutral theory and, 130–132
  unit of selection:
    epistasis and, 416–419
    multilocus genetic continuity, 412–419
Fixation:
  coarse-grained spatial heterogeneity, 456–474
  fine-grained heterogeneity, 486–490
  genetic drift and, 85–89
  natural selection interaction, 382–387
Fossil traits, gene flow and population history, haplotype tree loci/DNA region inferences, 242–243
Founder effects:
  genetic drift and, 89–94
    mating systems and, 97–98
    newly arisen mutations, 119
  linkage disequilibrium and, 94–97
  quantitative genetics:
    intrapopulation disequilibrium markers, 303–306
    linkage markers, 307–312
Founder-flush model, genetic drift and, newly arisen mutations, 119
Fragile X syndrome, meiotic drive, 421–424
Fragmentation:
  gene flow and population history, 229–231
    haplotype tree loci/DNA region inferences, 238–243
    range expansion, 232–233

gene flow *vs.* admixture, 215–218
genetic drift, 87–89
population history, 205
total population genetic variation, population subdivision and, 195–198
Frequency-dependent selection, competitive ability, 443–446
Frequency of disease:
  candidate loci, epistasis, 334–337
  nature *vs.* nurture theory and, 257
F statistic:
  continuous distribution, 561–562
  gene flow and population history, past population structure, 221–234
  gene flow and population subdivision, 180–183
  variance equality testing, two populations, 567
Fundamental theorem of natural selection:
  basic attributes of, 363–371
  measured genotypes, 347–349
  mutation-mating interaction, 377–378
  unmeasured genotypes, 361–363

Gametic-phase imbalance, Hardy-Weinberg law, two-loci architecture, 38–41
Gamma distribution:
  continuous probability, 560–561
  gene flow and population history, haplotype tree loci/DNA region inferences, 237–243
Gene conversion, targets of selection, 424–425
Gene-environment interactions, candidate loci studies, 337–338
Gene flow:
  age-structured population selection, *aa* supergene, spatio-temporal heterogeneous environment, 531–535
  coarse-grained spatial heterogeneity, 455–474
    industrial melanism and, 466–474
    Levene model, 461–474
  linkage disequilibrium and, 41–43
  natural selection, 378–380
    genetic drift, 387–397
  population history and:
    admixture *vs.*, 215–218
    basic principles, 204–205
    fragmentation patterns, 229–231
    haplotype trees, 218–221
      architecture, 221–228
      loci/DNA region integration, 234–243
    historical events, 243–245
    isolation by distance patterns, 228
    linkage disequilibrium and multilocus associations, 207–211
    nested clade multiple patterns, 233–234
    population trees, 212–215
    range expansion patterns, 231–233
    space and time studies, 205–207

population subdivision:
  drift/flow equations, 185–193
  genetic drift and, 174–180
  genetic impact, 172–174
  limitations of, 203
  mating systems, 183–185
  multiple inheritance models, 200–202
  Nei's population genetic distance, 191–192
  population structure, 173–174
  sequence or restriction site data, 198–199
  total population genetic variation, 193–198
  two populations, 168–172
  Wahlund effect and F statistics, 180–183
  quantitative genetics, phenotypic variance and, 295–296
Gene pool:
  assortative mating, admixture and linkage disequilibrium, 75–77
  coarse-grained temporal heterogeneity, 475–482
  DNA resequencing, estimation-maximization algorithm, 550–554
  founder and bottleneck effects, 91–94
  genetic drift in, 88–89
    linkage disequilibrium and, 96–97
  linkage disequilibrium and, 43
  natural selection:
    measured genotypes, 347–349
    sickle cell anemia, 351–358
  population genetics and, 2–3
General protein (GP) locus, 59–63
Genetic architecture:
  abnormal abdomen (aa) locus, unts and targets of selection, 508–516
  coarse-grained temporal heterogeneity, 475–482
  microevolution modeling and, 21–22
  natural selection:
    adaptation, 369–370
    interaction, 392–397
  quantitative genetics:
    candidate loci, 329–340
      combined pleiotropy and epistasis, 337
      epistasis, 330–337
      gene-environment interactions, 337–338
      pleiotropy, 329–330
      potential interaction effects, 338–340
    intrapopulation disequilibrium markers, 305–306
Genetic assimilation, coarse-grained temporal heterogeneity, 480–482
Genetic cline, coarse-grained spatial heterogeneity, industrial melanism and, 467–474
Genetic drift:
  basic principles, 82–83
  coalescence:
    basic process, 132–142
    gene trees *vs.* allele/haplotype trees, 143–145
    inversion trees and networks, 145–150
    mutation, 142–143

N coalescent, 137–138
    restriction site and DNA sequence haplotype trees, 150–167
      Jukes-Cantor molecule genetic distance, 164–165
      molecule genetic distances, 161–167
      mutation effects, 150–156
      neighbor-joining estimation method, 166
      recombination effects, 156–161
  disequilibrium and, 94–97
    mating systems and, 97–98
  evolutionary properties, 83–89
  fine-grained heterogeneity, 484–490
  found and bottleneck effects, 89–94
  gene flow balance with, 174–180
  large populations, new mutations, 118–119
  natural selection:
    gene flow interaction, 387–397
    interaction with, 380–387
    mutation and, 397–406
    polygenic adaptation, 360–361
    sickle cell anemia, 355–358
  neutral alleles, 120–128
  neutral theory, 128–132
  quantitative genetics, phenotypic variance and, 295–296
  tandem multigene families, 431–434
  total population genetic variation, population subdivision and, 194–198
  unit of selection and, 418–419
Genetic homology, tandem multigene families, 429–434
Genetic survey techniques:
  DNA polymorphisms, 548–554
  DNA resequencing, 548–554
    microsatellites or short tandem repeats, 554
    randomly amplified polymorphic DNAs, 548
    single-nucleotide polymorphisms, 554
  protein electrophoresis, 540–542
  restriction endonucleases, 542–548
    amplified fragment length polymorphisms, 546–548
    DNA fingerprinting, 545–546
    polymerase chain reaction, 546
    restriction fragment length polymorphisms, purified DNA, 542–544
    restriction site mapping, 544–545
    Southern blots, 544
Genetic variance, quantitative genetics, 271–273
  candidate loci, epistasis, 333–337
Genetic variation:
  gene flow and, 172–174
  natural selection:
    genetic drift-gene flow interaction, 388–397
    phenotypic variation, 363
    polygenic adaptation, 360–361

Genetic variation (*Continued*)
  neutral theory and, 130–132
  quantitative genetics, candidate loci, 313–317
  total population, population subdivision and, 193–198
Gene trees, coalescent events and, 143–145
  species tree *vs.*, 149–150
Genome scan, quantitative trait loci, 309–312
  candidate loci, 325–328
Genotype deviation:
  candidate loci, evolutionary history, 324–328
  natural selection, measured genotypes, 349
  quantitative genetics:
    correlation between relatives, 280–283
    mean-related measures, 273–270
Genotypic value of genotype:
  fine-grained heterogeneity, 483–490
  quantitative genetics:
    between-population differences, controlled crosses, 291–295
    candidate loci, epistasis, 331–337
    mean-related measures, 262–270
  unit of selection and, 410–419
Geographical distance:
  coarse-grained spatial heterogeneity, industrial melanism, 463–474
  gene flow and population history:
    fragmentation patterns, 231
    population genetic distance, 212–215
  gene flow *vs.* admixture, 216–218
  halotype trees, nested-clade phylogeographic analysis, 225–228
  total population genetic variation, population subdivision and, 195–198
Germline selection, individual phenotypes, 449–452
Gibbs sampler, 576–577
GJB2 locus, assortative mating, 65–71
Glade populations, total population genetic variation, population subdivision and, 196–198
Glucose-6-phosphate dehydrogenase (G6PD) deficiency:
  coarse-grained spatial heterogeneity, 468–474
  linkage disequilibrium and genetic drift and, 94–97
  natural selection:
    genetic architecture, 396–397
    genetic drift and mutation, 405–406
    mutation interaction, 373–375
  polygenic adaptation, 359–361
  population history and gene flow, linkage disequilibrium, 208–211
Goodness-of-fit testing, 573–574
Gradient, coarse-grained spatial heterogeneity, industrial melanism and, 466–474
Grid searches, 576–577
Growth rate phenotype, fine-grained heterogeneity, 486–490
Gryllus integer, targets of sexual selection, 434–435
Guevedoces founder effect, genetic drift and, 90–91

Habitat dimensionality, balance of gene flow and drift, 188–193
Habitat-specific viability estimates, coarse-grained spatial heterogeneity, 455–474
Haplotype network:
  coalescent theory and, 145
  DNA resequencing, estimation-maximization algorithm, 550–554
Haplotype trees:
  coalescent theory and:
    gene trees *vs.*, 143–145
    monophyletic group, 149–150
    restriction site and DNA sequence data, 150–167
      Jukes-Cantor molecule genetic distance, 164–165
      molecule genetic distances, 161–167
      mutation effects, 150–156
      neighbor-joining estimation method, 166
      recombination effects, 156–161
  DNA mutation and recombination and, 3–5
  gene flow and population history, 218–221
    fragmentation patterns, 229–231
    loci and DNA region integration, 234–243
    nested-clade phylogeographic analysis, 222–228
    multiple patterns, 233–234
    past population structure and, 221–234
    population genetic distances and, 212–215
    range expansion, 231–233
  neutral alleles, genetic drift and mutation, 397–406
  quantitative genetics, candidate loci, 317
Haplotype variation:
  natural selection, genetic architecture, 395–397
  population history and gene flow, direct space-time studies, 205–207
Hard selection, coarse-grained spatial heterogeneity, 459–474
Hardy-Weinberg law:
  assortative mating, 64–77
  admixture and linkage disequilibrium, 75–77
  coarse-grained spatial heterogeneity, 456–474
  disassortative mating and, 78–81
  evolution modeling, 21–47
    basic properties, 23–30
    disequilibrium implications, 43–47
    examples of, 30–32
    linkage disequilibrium, 41–43
    population structure and, 22–23
    single-loci applications, 32–35
    two-loci applications, 35–41
  gene flow:
    mating systems and population subdivision, 184–185
    Wahlund effect, 181–183
  genetic drift and, 82–83
  inbreeding deviation from random-mating, 55–63

natural selection-mutation interaction, mating systems, 376–378
pedigree inbreeding and, 52–55
quantitative genetics:
average effect of gamete type i, 268–270
average excess of gamete type, 266–270
candidate loci, epistasis, 332–337
variance and, 272–273
research background, 23–30
single loci applicability, 407–408
targets of sexual selection, 437–439
Harmonic mean, coarse-grained spatial heterogeneity, 457–474
Health phenotype, quantitative genetics, 254
Heliconius butterflies, coevolution mechanisms, 492–496
Hemoglobin chains:
coevolution mechanisms, 490–496
electrophoretic mobility phenotype, 250–251
natural selection:
genetic architecture, 395–397
genetic drift, 383–387
polygenic adaptation, 358–361
sickle cell anemia, 349–358
sickle cell anemia phenotype, 252–253
tandem multigene families, unequal exchange, 430–434
Heritability. *See* Narrow-sense heritability
HERV transposons, targets of selection, 426–429
Heterozygosity:
assortative mating, 70–77
inbreeding *vs.*, 71–73
Bayesian inference and, 578–581
coarse-grained spatial heterogeneity, 459–474
competitive ability, 444–446
disassortative mating and, 79–81
electrophoretic mobility phenotype, 251
gene flow:
mating subdivisions and population subdivision, 183–185
Wahlund effect, 181–183
genetic drift, infinite-alleles model, 127–128
natural selection:
genetic drift and mutation, 403–406
mutation interaction, 375
polygenic adaptation, 360–361
quantitative genetics, between-population differences, controlled crosses, 291–295
targets of selection, gene conversion, 425
Hierarchical models:
balance of gene flow and drift and, 179–180
population subdivision measurements, sequencing and restriction data, 199
Historical events:
defined, 204–205
gene flow studies, past population structure and, 243–245

Hitchhiking, natural selection, genetic drift and mutation, 404–406
HKA test, natural selection, genetic drift and mutation, 403–406
Hoekstra model:
coarse-grained temporal heterogeneity, 476–482
fine-grained heterogeneity, 482–490
Holism, in population genetics, 10–11
Homogeneity of loci, natural selection, genetic drift and mutation, 402–406
Homologous genes:
age-structured population selection, genetic architecture and targets and units of selection, 510–516
coalescent theory and, DNA replication, 133–142
Homology, neutral alleles, 120
Homoplasy:
candidate loci analysis, 327–328
coalescent theory, 146–150
haplotype trees:
mutation and, 150–156
recombination, 156–161
Homozygosity:
assortative mating, 65–66
inbreeding *vs.*, 71–73
coarse-grained spatial heterogeneity, Levene model, 461–474
electrophoretic mobility phenotype, 251
gene flow, mating subdivisions and population subdivision, 183–185
genetic drift, infinite-alleles model, 127–128
pedigree inbreeding and, 53–55
quantitative genetics, between-population differences, controlled crosses, 291–295
Horizontal transmission, transposons, 427–429
H statistic, natural selection, genetic drift and mutation, 405–406
Human evolution, gene flow and population history, 243–245
Human immunodeficiency (HIV) clones, natural selection, genetic drift and mutation, 402–406
Human population distances:
balance of gene flow and drift, isolation-by-distance models, 192–193
gene flow and population history, pairwise models, 212–215
population history and gene flow, linkage disequilibrium, 209–211
Humidity gradient, age-structured population selection:
aa supergene, spatio-temporal heterogeneous environment, 526–535
juvenilized abdominal cuticle phenotype, 519–525

Huntington's disease:
  age-structured populations, senescence evolution, 506–508
  candidate loci, epistasis, 331–337
  kin/family selection, 449–452
Hutterites:
  candidate loci, epistasis, 335–337
  quantitative trait loci, 307–312
Hybrid populations, quantitative genetics, between-population differences, controlled crosses, 291–295
Hypergeometric distribution, 558
Hypermutability, coalescent theory and, haplotype trees mutation, 155–156
Hypertension, genetic drift and, 96–97
Hypothesis testing of mean, 565
  Bayesian inference and, 581
  maximum-likelihood techniques, 577–578
Hypothetical disease, nature vs. nurture theory and, 256–257

Identity by descent ancestry, 48–49
  balance of gene flow and drift and, 178–180
  gene flow, Wahlund effect, 182–183
  genetic drift and, 87–89
    infinite-alleles model, 127
Identity by state ancestry, 49
  balance of gene flow and drift and, 179–180
Identity concepts, gene flow balance with genetic drift, 175–180
Inbred offspring, pedigree inbreeding, 49–55
Inbreeding:
  assortative mating, 71–73
    admixture and, 74–77
    coefficient, 65–77
  balance of gene flow and drift and coefficient of, 175–180
  basic principles of, 48–63
  coevolution mechanisms, 495–496
  founder and bottleneck effects and, 91–94
  gene flow balance with genetic drift, 174–180
  natural selection-mutation interaction, 376–378
  total population genetic variation, population subdivision and, 193–198
Inbreeding depression, defined, 53–55
Inbreeding effective size, limits of neutral theory concerning, 128–132
Incest taboos, gene flow and population subdivision, mating systems, 184–185
Individuals:
  fertility, 440–443
  kin/family selection, 450–452
  pedigree inbreeding in, 52–55
Industrial melanism, coarse-grained spatial heterogeneity, hard selection, 462–474

Inference key:
  haplotype tree loci/DNA region inferences, 234–243
  nested-clade analysis, 233–234
Infinite-alleles model:
  coalescent theory and, 146–150
    haplotype tree recombination, 158–161
  genetic drift vs. mutation, 126–128
Infinite-sites model:
  candidate loci analysis, 327–328
  coalescent theory, haplotype trees:
    mutation and, 150–156
    recombination, 160–161
  natural selection, genetic drift and mutation, 405–406
Inheritance, quantitative genetics:
  correlation between relatives, 282–283
  marker linkage mapping and, 311–312
Insulin gene region, coarse-grained temporal heterogeneity, 477–482
Intelligence quotient score, quantitative genetics, between-population differences, mean phenotype, 285–290
Intensity of selection:
  multilocus genetic continuity, unit of selection, 412–419
  quantitative genetics, 283–284
Interaction variance of the phenotype:
  candidate loci, epistasis, 333–337, 335–337
  coarse-grained spatial heterogeneity, 474
Interchromosomal variation, age-structured population selection, genetic architecture and targets and units of selection, 512–516
Interpopulation disequilibrium, marker loci studies, 298–303
Intersexual selection, targets of, 434–439
Interspecific genetic variation:
  coevolution mechanisms, 491–496
  limits of neutral theory concerning, 128–132
  neutral alleles, 122–128
    genetic drift and mutation, 397–406
Interval mapping, quantitative trait loci, 307–312
Intrapopulation disequilibrium, marker loci studies, 299, 303–306
Intrasexual selection, targets of, 434–439
Intraspecific genetic variation:
  coalescent theory, inversion network, 147–150
  limits of neutral theory concerning, 128–132
  neutral alleles, 122–128
    genetic drift and mutation, 397–406
Inversion trees:
  coalescent theory and, 145–150
  neutral alleles, genetic drift and mutation, 397–406
Iolation by distance, balance of gene flow and drift, 185–193

Island models:
    balance of gene flow and drift and, 177–180
    coarse-grained spatial heterogeneity, hard selection, 461–474
    gene flow:
        isolation by distance, 185–193
        Wahlund effect, 182–183
Isolated founder populations, intrapopulation disequilibrium markers, 304–306
Isolation-by-distance models:
    balance of gene flow and drift:
        human models, 192–193
        pairwise distribution, 190–193
    gene flow and population history, 228
        range expansion, 232–233
    gene flow vs. admixture, 215–218
Isozyme surveys, Bayesian inference and, 579–581

Jawed vertebrate immune system, transposons, 426–429
Joint probability distribution, Bayesian inference and, 578–581
Jukes-Cantor molecule genetic distance, haplotype tree estimation, 163–166
Juvenile hormone (JH) esterase, age-structured population selection, *abnormal abdomen (aa)* locus targets of selection, 516–525
Juvenilized abdominal cuticle phenotype, age-structured population selection, 518–525

Kin/family selection, fitness parameters, 446–452
kk genotype, phenylketonuria, nature vs. nurture theory and, 255–257
Knockout mutations, pedigree inbreeding and, 54–55

Larval phase prolongation, age-structured population selection, 520–525
Least-squares regression coefficient, quantitative genetics, correlation between relatives, 279–283
Lemba tribe, population history and gene flow:
    haplotypes, 220–221
    linkage disequilibrium, 207–211
Lethal equivalents, pedigree inbreeding, 54–55
Lethality, pedigree inbreeding and, 53–55
Levene model, coarse-grained spatial heterogeneity, 456–474
Life history:
    age-structured populations, selection in, 498–504
Likelihood ratio test. *See also* Maximum likelihood techniques
    quantitative genetics, linkage markers, 307–312
Linkage disequilibrium:
    assortative mating, 66–71
    admixture and, 73–77
    evolutionary history and, 43–47
    genetic disease and, 45–47
    genetic drift and, 94–97
        mating systems and, 97–98
    haplotype tree recombination, 160–161
    Hardy-Weinberg law, two-loci architecture, 38–41
    inbreeding and, 71–73
    marker loci studies, 298, 306–312
    population history and gene flow, multilocus associations, 207–211
    quantitative genetics:
        candidate loci, 313–317
        intrapopulation disequilibrium markers, 305–306
    sources of, 41–43
Lipid phenotypes, quantitative genetics, candidate loci:
    epistasis, 336–337
    pleiotropy, 329–330
Lipoprotein lipase locus *(LPL)*:
    DNA mutation and recombination and, 5–6
    haplotype trees:
        mutation, 150–156
        recombination, 156–161
    Hardy-Weinberg law, two-loci architecture, 39–41
    linkage disequilibrium and, 44–47
    population subdivision measurements, sequencing and restriction data, 199
    quantitative genetics:
        candidate loci, 313–317
        epistasis, 330–337
        intrapopulation disequilibrium markers, 305–306
Local adaptive solutions, natural selection, 365
Local fixation, total population genetic variation, population subdivision and, 195–198
Local populations:
    gene flow in, 168–172
    genetic impact of, 173–174
    population genetics in, 2–3
LOD score, quantitative genetics, linkage markers, 308–312
Logarithms of fitness, fine-grained heterogeneity, 489–490
Log likelihood ratio test:
    haplotype tree loci/DNA region inferences, 239–243
    hypothesis testing, 577–578
    maximum-likelihood estimation, 575–577
Long-distance dispersal parameter:
    balance of gene flow and drift, one-dimensional stepping-stone model, 187–193
    limitations of, 203
Longevity differences, age-structured population selection, fecundity estimates with aa syndrome, 524–525
Long-term effective size, total population genetic variation and, 193–198

Loss:
  of alleles, total population genetic variation,
      population subdivision and, 194–198
    genetic drift and, 85–89
Low density lipoprotein receptor (LDLR), candidate loci:
  combined pleiotropy and epistasis, 337
  epistasis, 331–337

Magnitude of selection, multilocus genetic continuity, unit of selection, 412–419
Major histocompatibility complex (MHC):
  coalescent theory, inversion network, 148–150
  disassortative mating and, 77–81
  DNA mutation and recombination and, 4–6
Major loci, candidate loci studies, epistasis, 331–337
Makiritare tribes, disassortative mating and, 80–81
Maladaptive traits, coarse-grained temporal heterogeneity, 477–482
Malarial resistance phenotype:
  age-structured populations, life history effects, 507–508
  coarse-grained temporal heterogeneity, 475–482
  coevolution mechanisms, 490–496
  heterogeneous environments, 453–455
  natural selection:
    gene flow interaction, 380–387
    genetic architecture, 396–397
    genetic drift-gene flow interaction, 387–397
    polygenic adaptation, 359–361
    sickle cell anemia, 349–358
  quantitative genetics, 253
  viability phenotypes, 254
Malthusian parameter, age-structured populations:
  aa supergene, spatio-temporal heterogeneous environment, 530–535
  life history and fitness, 503–504
  senescence evolution, 505–508
Marginal quantiative genetics, candidate loci, epistasis, 332–337
Mark/capture experiments:
  aa supergene, spatio-temporal heterogeneous environment, 526–535
  coarse-grained spatial heterogeneity, industrial melanism, 465–474
Marker loci studies, measured genotypes:
  basic principles, 298–299
  quantitative genetics, 299–312
    interpopulation disequilibrium, 298–303
    intrapopulation disequilibrium, 299, 303–306
    linkage markers, 299, 306–312
Markovian property, N coalscent, 138–139
Mating pairs:
  fertility, 439–443
  quantitative genetics, correlation between relatives, 277–283

Mating success:
  natural selection, 343
  phenotypes, 8
Mating systems. *See also* Random mating
  assortative mating, 63–77
    admixture, 73–77
    inbreeding *vs.*, 71–73
    linkage disequilibrium, 66–71
    simple model, 63–66
  disassortative mating, 77–81
  gene flow, population structure and, 174
  genetic drift and, 97–98
  Hardy-Weinberg law and, 28–30
  inbreeding, 48–63
    assortative mating *vs.*, 71–73
    random-mating deviations, 55–63
  natural selection interaction with, 376–378
  pedigree inbreeding, 40–55
  population genetics and, 2–3
  population subdivision and, 183–185
  targets of selection, gene conversion, 425
Maximum-likelihood estimator:
  haplotype tree loci/DNA region inferences, 239–243
  resequencing techniques, 550–554
  statistical data, 575–577
Maximum likelihood techniques, statistical analysis, 574–578
  hypothesis testing, 577–578
  maximum-likelihood estimation, 575–577
Maximum parsimony:
  coalescent theory, 146–150
  haplotype trees, mutation and, 151–156
Mean:
  age estimations, gene flow and population history, 237–243
  coarse-grained spatial heterogeneity, fitness estimates, 460–474
  confidence intervals, 565–566
  difference testing in two populations, 566–567
  growth rate, coarse-grained spatial heterogeneity, 473–474
  hypothesis testing, 565
  minimum-variance, unbiased mean estimator, 564–565
  one-way analysis of variance, 567–569
  population fitness, unmeasured genotypes, 361–363
  probability distribution and, 557–558
  quantitative genetic measures and, 260–270
    between-population differences, 284–290
Mean-square error, one-way analysis of variance, 569
Measured genotypes:
  natural selection, 345–349
  quantitative genetics:
    basic principles, 297–299
    candidate loci, 299, 312–340
      alleles, SNPs, and haplotypes, 312–317
      epistasis, 330–339

evolutionary history, 318–328
   genetic architecture, 329–340
   pleiotropy, 329–339
   limitations of, 311–312
   marker loci, 299–312
      interpopulation disequilibrium, 298–303
      intrapopulationi disequilibrium, 299, 303–306
      linkage markers, 299, 306–312
Median test, age-structured population selection, genetic architecture and targets and units of selection, 513–516
Meiosis, unit of selection, multilocus genetic continuity, 411–419
Meiotic drive, targets of selection and, 419–424
Mendelian laws:
   assortative mating, 65–77
   Fisherian model of quantitative genetics and, 257–260
   Hardy-Weinberg assumptions and, 26–30
      two-loci frequencies, 37–41
   meiotic drive and, 424
   microevolution modeling and, 22–23
   natural selection, genetic architecture, 394–397
   pedigree inbreeding, 50–55
   Punnet square and, 32–35
   quantitative genetics:
      correlation between relatives, 276–283
      epistasis, 335–337
   simple Mendelian phenotypes, 250–254
      electrophoretic mobility, 250–251
      health and viability, 254
      malarial resistance, 253
      sickle cell anemia, 252–253
      sickling, 251–252
Mental retardation trait, nature vs. nurture theory, phenylketonuria and, 255–257
Metapopulation, natural selection, 392
Microevolution, models of, 21–23
Microsatellite loci:
   genetic survey applications, 554
   quantitative genetics, interpopulation disequilibrium markers, 302–303
Microvicariance, gene flow and population history, fragmentation patterns, 229–231
Midparent value, quantitative genetics, correlation between relatives, 279–283
Migrants, balance of gene flow and drift and, 176–180
Minimum-variance, unbiased mean estimator, 564–566
   goodness-of-fit testing, 573–574
Minisatellite tandem repeats, genetic survey applications, 546
Minor loci, candidate loci studies, epistasis, 331–337
Mitochondrial DNA (mtDNA):
   coalescent theory and, 135–142
      ploidy level and, 139–140
   fragmentation patterns, 229–231
   gene flow, Wahlund effect, 183
   haplotype tree loci/DNA region inferences, 238–243
   isolation by distance models, gene flow and population history, 228
   multiple inheritance modes and population structure, 202
   nested-clade phylogeographic analysis, 223–228
   population history and gene flow, direct space-time studies, 206–207
   resequencing techniques, 549–554
Mode model, coevolution mechanisms, 491–496
Molecular clock:
   gene flow and population history:
      haplotype tree loci/DNA region inferences, 236–243
      nested-clade phylogeographic analysis, 222–234
      population genetic distances and, 212–215
   genetic drift, neutral alleles, 122–128
   neutral theory and, 129–132
Molecular genetics:
   distances:
      gene flow and population history, 212–215
      haplotype tree estimation, 161–167
         Jukes-Cantor model, 163–166
         neighbor-joining method, 166–167
   measured genotypes and, 297
   natural selection, genetic drift and mutation, 405–406
Monophyletic group, coalescent events and, 149–150
Mpi locus, coarse-grained spatial heterogeneity, 455–474
Müllerian mimicry, coevolution mechanisms, 493–496
Multi-deme coalescence, balance of gene flow and drift and, 177–180
Multilocus evolution:
   assortative mating vs. inbreeding, 71–73
   disassortative mating and, 80–81
   genetic drift and disequilibrium, 94–97
   haplotype tree inferences, 234–243
   linkage disequilibrium and, 43–47
   natural selection, genetic architecture, 394–397
   Nei population genetic distance, 191–193
   population history and gene flow, linkage disequilibrium, 207–211
   unit of selection and, 410–419
Multinomial, multivariate distribution, 563
Multiple equilibria:
   assortative mating, 70–77
   natural selection and genetic architecture, 392–397
Multiple inheritance modes, population structure and, 200–202
Multiple mutations, haplotype tree estimation, Jukes-Cantor molecule genetic distance, 164–166
Multivariate probability distribution, 562–563

Mutation:
    balance of gene flow and drift in, 175–180
    coalescence with, 142–143
        haplotype trees, impact on, 150–156
    of DNA, population genetics and, 3–6
    fine-grained heterogeneity, 483–490
    genetic drift:
        infinite-alleles model, 126–128
        neutral mutations, 118–119, 126–128
    life history effects, 507–508
    linkage disequilibrium and, 41–43
    natural selection:
        genetic architecture, 393–397
        genetic drift and, 397–406
        interaction, 373–375
        mating system interaction, 376–378
        polygenic adaptation, 360–361
    quantitative genetics:
        candidate loci analysis, 324–328
        intrapopulation disequilibrium markers, 305–306
        phenotypic variance and, 295–296
    unit of selection and, 418–419
Mutational substitution rate, natural selection, genetic drift and, 397–406
MX1 autosomal locus, coalescent theory, inversion network, 148–150
Myotonic dystrophy, meiotic drive, 421–424

Narrow-sense heritability:
    natural selection:
        measured genotypes, 349
        unmeasured genotypes, 362–363
    quantitative genetics, 272–273
        between-population differences, mean phenotype, 285–290
        correlation between relatives, 280–283
        interpopulation disequilibrium marker loci, 301–303
Natural population monitoring:
    genetic drift, neutral alleles, 123–128
    population genetics and, 15–18
Natural selection:
    age-structured populations:
        abnormal abdomen case study, 508–539
            genetic architecture and units/targets of selection, 508–516
            phenotypes and potential targets of selection, 516–525
            supergene components, 536–537
            supergene spatiotemporal heterogeneous environment, 525–536
        life history and fitness, 498–504
        overview, 497
        senescence evolution, 504–508
    basic principles, 343–345
    evolutionary forces interaction with:
        gene flow interaction, 378–387
        genetic architecture, 392–397

    genetic drift and gene flow, 387–397
    genetic drift and mutation, 397–406
    mating system/mutation interaction, 376–378
    mutation, 373–375
    population subdivision, 391–392
    research overview, 372–373
fundamental equations, 363–371
Hardy-Weinberg law and, 34–35
heterogeneous environments:
    coarse-grained spatial heterogeneity, 455–474
    coarse-grained temporal heterogeneity, 475–482
    coevolution, 490–496
    fine-grained heterogeneity, 482–490
    overview, 453–455
linkage disequilibrium and, 41–43
measured genotypes, 345–349
polygenic adaptation, 358–361
sickle cell anemia, 349–358
unmeasured genotypes, 361–363
Nature *vs.* nurture theory, basic principles, 254–257
Naura population, coarse-grained temporal heterogeneity, 478–482
N coalescent, genetic drift and, 137–138
Negative binomial distribution, 559
Negative covariance:
    natural selection, pleotropy, 368
    quantitative genetics, correlation between relatives, 276–283
Neighborhood size, balance of gene flow and drift, dimensionality parameters, 189–193
Neighboring demes, balance of gene flow and drift:
    dimensionality parameters, 189–193
    one-dimensional stepping-stone model, 187–193
Neighbor-joining technique:
    gene flow and population history, population genetic distances and, 212–215
    haplotype tree estimation, 166–167
Nei population genetic distance, gene flow and drift, 191–193
Nene mtDNA, population history and gene flow, direct space-time studies, 205–207
Nested analysis of variance (NANOVA), candidate loci, 326–328
Nested-clade distance:
    fragmentation patterns, 229–231
    halotype trees, 226–228
    multiple patterns, 233–234
    range expansion, 231–233
Nested-clade phylogeographic analysis:
    candidate loci, evolutionary history, 319–328
    gene flow and population history:
        fragmentation, 229–231
        historical events and past population structure, 243–245
        isolation by distance models, 228
        multiple patterns, 233–234
        past population structure, 222–234

haplotype tree loci/DNA region inferences, 234–243
Net reproductive rate, age-structured populations:
  life history and fitness, 500–504
  senescence evolution, 505–508
Networks, coalescent theory and, 145–150
Neutral alleles:
  genetic drift and, 120–128
    natural selection interaction, 381–387
  natural selection, genetic drift and mutation, 397–406
Neutrality theory:
  critiques of, 128–132
  fine-grained heterogeneity, 485–490
  genetic drift:
    mutations and, 118
    neutral alleles, 123–128
  mutations and limits of, 128–132
  natural selection, genetic drift and mutation, 399–406
  tandem multigene families, unequal exchange, 431–434
Neutral mutation:
  coarse-grained temporal heterogeneity, 481–482
  natural selection, genetic drift and, 397–406
Newton-Raphson iterative algorithm, 576–577
Nonadaptive traits, natural selection, 368
Nonrandom mating:
  gene flow, population subdivision, 184–185
  linkage disequilibrium and, 41–43
  targets of selection, gene conversion, 425
Nontranscribed spacers (NTS), age-structured population selection, genetic architecture and targets and units of selection, 516–519
Normal distribution:
  continuous probability, 559–560
  Fisherian model of quantitative genetics, 258–260
Normalized linkage disequilibrium, defined, 41–42
Norm of reaction:
  candidate loci, gene-environment interactions, 338
  coarse-grained spatial heterogeneity, 471–474
  coarse-grained temporal heterogeneity, 479–482
  quantitative genetics, between-population differences, mean phenotype, 288–290
Nucleotides:
  haplotype tree estimation:
    Jukes-Cantor molecule genetic distance, 164–166
    neighbor-joining technique, 166–167
  haplotype trees, mutation and, 150–156
  natural selection, genetic drift and mutation, 397–406
Null hypothesis:
  age-structured population selection, genetic architecture and targets and units of selection, 513–516
  goodness-of-fit testing, 573–574

haplotype tree loci/DNA region inferences, 238–243
log likelihood ratio test, 577–578
natural selection, genetic drift and mutation, 403–406
nested-clade phylogeographic analysis, 227–228
one-way analysis of variance, 569
permutation testing, 669–670
quantitative genetics, linkage markers, 308–312

Offspring phenotype, quantitative genetics, correlation between relatives, 280–283
One-dimensional stepping-stone model, balance of gene flow and drift, 186–193
  isolation-by-distance models, 190–193
Optimal trait value, natural selection, 367–368
Optimization, natural selection and, 367–368
Ormia ochracea, targets of sexual selection, 435
Orthology, tandem multigene families, 429–434
Outgroup rooting, coalescent theory, inversion network, 147–150
Out-of-Africa range expansions, haplotype tree loci/DNA region inferences, 239–243
Overdominance, health phenotypes, 254
Oviposition onset, age-structured population selection, *abnormal abdomen (aa)* locus targets of selection, 522–525

Paedomorphic/metamorphic phenotype, coarse-grained temporal heterogeneity, 481–482
Panmictic characteristics, coarse-grained spatial heterogeneity, 455–474
Paracentric inversions, coalescent theory and, 145–150
Paralogy, tandem multigene families, 429–434
Parsimony techniques, haplotype tree estimation, neighbor-joining technique, 166–167
Parthenogenesis, unit of selection and, 413–419
Peak shift, natural selection-genetic drift interaction, 385–387
  gene flow and, 387–397
Pedigree inbreeding, 49–55
  coalescent theory and, 136–142
    gene trees *vs.* allele and haplotype trees, 144–145
  founder and bottleneck effects, 91–94
  quantitative trait loci, linkage markers, 306–312
  total population genetic variation, population subdivision and, 195–198
P elements, transposons, 427–429
Peppered moth, coarse-grained spatial heterogeneity, industrial melanism, 462–474
Permutation testing, analysis of variance and, 669–670
Persistence of fetal hemoglobin, natural selection, supergenes, 396–397

PHASE algorithm, DNA resequencing, estimation-maximization algorithm, 552–554
Phenotypes:
  age-structured population selection, targets of selection, 516–525
  competitive ability, 443–446
  developmental mechanisms for, 21–23
  disassortative mating and, 79–81
  DNA-environmental interaction and, 7–9
  gene flow, local populations, 171–172
  life history table, 500–504
  natural selection, measured genotypes, 345–349
  nature *vs.* nurture theory and, 254–257
  plasticity:
    candidate loci, gene-environment interactions, 338
    coarse-grained temporal heterogeneity, 479–482
  quantitative genetics, 250–254
    candidate loci evolution, 322–328
    correlation between relatives, 279–283
    electrophoretic mobility, 250–251
    health and viability, 254
    interpopulation disequilibrium marker loci, 300–303
    malarial resistance, 253
    mean-related measures, 260–270
    measured marker approach, 311–312
    sickle cell anemia, 252–253
    sickling, 251–252
    variance and, 271–273
  target of selection and, 409
Phenotypic clines, coarse-grained spatial heterogeneity, 470–474
Phenotypic covariance, correlation between relatives, 275–283
Phenotypic variation:
  natural genetics, genetic variation and, 363
  targets of sexual selection, 435–439
Phenylketonuria (PKU):
  nature *vs.* nurture theory and, 254–257
  quantitative genetics, correlation between relatives, 282–283
Pheromone races, gene flow, local populations, 170–172
Phylogenetic Analysis Using Parsimony (PAUP) techniques, haplotype tree estimation, neighbor-joining technique *vs.*, 166–167
Pieris napi, targets of sexual selection, 435
Pima Indian studies, coarse-grained temporal heterogeneity, 478–482
Pitman estimator, Bayesian inference and, 580–581
Pleiotropy:
  age-structured population selection, emergence time characteristics, 521–525
  candidate loci, genetic architecture, 329–330
  life history effects, 507–508
  Mendelian inheritance, 250
  natural selection:
    antagonistic pleiotropy, 368
    genetic architecture, 397
  sickle cell anemia phenotype, 252–253
Poisson distribution:
  discrete probability, 558–559
  fine-grained heterogeneity, 484–490
  genetic drift:
    coalescent time estimates, 141–142
    newly arisen mutations, 118–119
  goodness-of-fit testing, 573–574
  pedigree inbreeding, 53–55
Pollution, coarse-grained spatial heterogeneity, industrial melanism, 462–474
Polygenic adaptation, natural selection, 358–361
Polymerase chain reaction (PCR):
  age-structured population selection, genetic architecture and targets and units of selection, 511–516
  genetic survey applications, 546
Polymorphism:
  coalescent theory, inversion network, 148–150
  coarse-grained spatial heterogeneity, 457–474
  coarse-grained temporal heterogeneity, 475–482
  fine-grained heterogeneity, 488–490
  gene flow and population history, population genetic distances and, 212–215
  genetic drift, neutral alleles, 126–128
  genetic survey applications, 548–554
    DNA resequencing, 548–554
    microsatellites or short tandem repeats, 554
    randomly amplified polymorphic DNAs, 548
    single-nucleotide polymorphisms, 554
  natural selection:
    genetic drift and mutation, 397–406
    supergene architecture, 396–397
  quantitative genetics, candidate loci, 313–317
Polytene chromosomes:
  assortative mating *vs.* inbreeding, 73
  genetic drift research, neutral alleles, 123–128
Pooled haplotype trees:
  candidate loci, evolutionary history, 319–328
  natural selection, genetic drift and mutation, 401–406
Population genetic distance:
  balance of gene flow and drift, 190–193
  Nei distance, 191–193
  gene flow and population history, tree architecture, 212–215
Population genetics:
  basic premises, 1–9
  DNA mutation and recombination, 3–6
  DNA replication, 2–3

methodologies, 9–18
　comparative analysis, 11–14
　holism, 10–11
　natural population monitoring, 15–18
　reductionism, 9–10
　phenotype emergence, 6–9
Population history, gene flow and:
　admixture vs., 215–218
　basic principles, 204–205
　fragmentation patterns, 229–231
　haplotype trees, 218–221
　　architecture, 221–228
　　loci/DNA region integration, 234–243
　historical events, 243–245
　isolation by distance patterns, 228
　linkage disequilibrium and multilocus associations, 207–211
　nested clade multiple patterns, 233–234
　population trees, 212–215
　range expansion patterns, 231–233
　space and time studies, 205–207
Population mean, quantitative genetics, between-population differences, 288–290
Population ratios, Hardy-Weinberg law and, 32–35
Population size:
　founder and bottleneck effects and, 91–94
　gene flow and, Wahlund effect, 183
　genetic drift:
　　coalescence and, 118
　　natural selection, 382–387
　　neutral alleles, 120–128
　　newly arisen mutations, 119
　heterozygosity and, 131–132
Population structure:
　assortative mating, admixture and linkage disequilibrium, 76–77
　coarse-grained spatial heterogeneity:
　　evolutionary response, 472–474
　　Levene model, 461–474
　gene flow, impact of, 174
　gene flow and population history, 221–234
　gene flow vs. admixture, 216–218
　Hardy-Weinberg law and, 24–30
　microevolution modeling and, 22–23
　multiple inheritance modes, 200–202
　natural selection, adaptation and fitness variation, 363
Population subdivision:
　coarse-grained spatial heterogeneity, 455–474
　gene flow:
　　drift/flow equations, 185–193
　　genetic drift and, 174–180
　　genetic impact, 172–174
　　limitations of, 203
　　mating systems, 183–185
　　multiple inheritance models, 200–202
　　Nei's population genetic distance, 191–192

　　population structure, 173–174
　　sequence or restriction site data, 198–199
　　total population genetic variation, 193–198
　　two populations, 168–172
　　Wahlund effect and F statistics, 180–183
　natural selection:
　　genetic drift-gene flow interaction, 389–397
　　interaction, 391–392
　quantitative genetics:
　　average effect of gamete type i, 269–270
　　mutation, drift and gene flow, 296
Population trees, gene flow and population history, population genetic distance, 212–215
Posterior distribution, Bayesian inference and, 579–581
Potential equilibria of evolution, natural selection, 373
　balance of evolutionary forces, 373
Predation, coarse-grained spatial heterogeneity, industrial melanism, 462–474
Preferred substitutions, natural selection, genetic drift and mutation, 401–406
Principal-component analysis, population history and gene flow, linkage disequilibrium, 208–211
Prior probability distribution, Bayesian inference and, 578–581
Probability:
　basic principles, 555–563
　Bayesian inference and, 578–581
　coalescence with mutation, 142–143
　continuous probability distributions, 559–562
　discrete probability distributions, 558–559
　haplotype tree loci/DNA region inferences, 241–243
　of identity:
　　balance of gene flow and drift and, 179–180
　　Nei population genetic distance, 191–193
　multivariate probatility distributions, 562–563
　nature vs. nurture theory and, 256–257
　random variables, distribution, and expectation, 556–558
Proportion of inserted repeats estimator, age-structured population selection, genetic architecture and targets and units of selection, 512–516
Protected polymorphism, competitive ability, 444–446
Protein electrophoresis:
　electrophoretic mobility phenotype, 250–251
　genetic drift, neutral alleles, 123–128
　genetic survey applications, 540–542
Protein synthesis, age-structured population selection, genetic architecture and targets and units of selection, 514–516
Proximate causation, age-structured population selection, *aa* supergene components, 537–539

Punctuated equilibrium, coarse-grained temporal heterogeneity, 479–482
Punnet square, Hardy-Weinberg law and, 32–35

Quantitative genetics:
  definitions, 261
  Fisherian model, 257–260
  mean-related measures, 260–270
  measured genotypes:
    basic principles, 297–299
    candidate loci, 312–340
      alleles, SNPs, and haplotypes, 312–317
      epistasis, 330–339
      evolutionary history, 318–328
      genetic architecture, 329–340
      pleiotropy, 329–339
    marker loci, 299–312
      interpopulation disequilibrium, 298–303
      intrapopulationi disequilibrium, 299, 303–306
      linkage markers, 299, 306–312
  simple Mendelian phenotypes, 250–254
    electrophoretic mobility, 250–251
    health and viability, 254
    malarial resistance, 253
    sickle cell anemia, 252–253
    sickling, 251–252
  unmeasured genotypes:
    basic principles, 274–275
    between-population differences:
      controlled crosses, 290–295
      mean phenotypes, 284–290
    correlation between relatives, 275–283
    phenotypic variance, mutation, drift, and gene flow, 295–296
    selection response, 283–284
    variance-related measures, 270–273
Quantitative trait locus (QTL):
  interval mapping, 307–312
  intrapopulation disequilibrium markers, 304–306
  linkage disequilibrium, 306–312
  natural selection, genetic architecture, 394–397
  unit of selection and, 413–419
Quantitative trait region (QTR):
  candidate loci, 312
  defined, 310–312

Racial characteristics, gene flow, local populations, 172
RAG transposon, targets of selection, 427–429
Rana clamitans, coarse-grained spatial heterogeneity, 471–474
Randomly amplified polymorphic DNA (RAPD), genetic survey applications, 548
Random mating:
  coalescent theory and, 136–142
  coarse-grained spatial heterogeneity, 456–474
  coevolution mechanisms, 495–496
  fertility and, 439–443
  Hardy-Weinberg law and, 23–24
    allele frequency multiplications, 28–30
    blood group loci, 32
    two-loci frequencies, 35–41
  inbreeding deviation from, 55–63
  kin/family selection, 446–452
  natural selection, sickle cell anemia, 352–358
  quantitative genetics:
    average effect of gamete type i, 268–270
    correlation between relatives, 277–283
    mean-related measures, 267–270
Random variables:
  binomial distribution, 558
  contingency analysis, 571–573
  hypergeometric distribution, 558
  negative binomial distribution, 559
  probability distribution, 556–558
  statistical data, 564–574
    contingency tests, 571–573
    goodness-of-fit tests, 573–574
    mean confidence intervals, 565–566
    mean hypothesis testing, 565
    minimum-variance, unbiased mean estimation, 564–565
    one-way analysis of variance, 567–569
    permutation testing, 569–570
    regression and correlation, 570–571
    two-population means testing, 566–567
    unknown variance estimation and testing, 566
    variance equality in two populations, 567
Range expansion:
  gene flow and population history, 231–233
    haplotype tree loci/DNA region inferences, 239–243
  population history, 205
18S/28S rDNA locus, age-structured population selection, genetic architecture and targets and units of selection, 509–516
Recessive alleles:
  health phenotypes, 254
  sickle cell anemia phenotype, 252–253
Recombination:
  assortative mating *vs.* inbreeding, 71–73
  coalescent theory:
    haplotype trees and, 156–161
    inversion trees and networks, 146–150
  Hardy-Weinberg law and, 35–41
  hotspot, haplotype trees, 156–161
  linkage disequilibrium and, 44–47
  natural selection, genetic drift and mutation, 404–406
  population genetics and, 3–6
  quantitative trait loci, linkage disequilibrium, 306–312
  unit of selection and, 418–419

Recurrent evolution:
    haplotype tree loci/DNA region inferences, 236–243
    natural selection, mutation and, 373–375
Recurrent flow models, isolation by distance and, 215–218
Reductionism:
    holism and, 10–11
    in population genetics, 9–10
Regression coefficient:
    analysis of variance and, 570–571
    quantitative genetics, correlation between relatives, 280–283
Relative fitness:
    coarse-grained spatial heterogeneity, 469–474
    natural selection, sickle cell anemia, 350–358
Relative mortality risk, coronary artery disease, candidate loci, 314–317
Replacement events:
    gene flow and population history, haplotype tree loci/DNA region inferences, 242–243
    natural selection, genetic drift and mutation, 400–406
Replica plating, techniques for, 3–4
Replication:
    coalescent theory and, 132–142
    genetic drift and, 86–89
    natural selection, 343–345
    population genetics and, 2–3
Reproductive fitness:
    age-structured populations, 502–504
    natural selection, 344
Residual variance, quantitative genetics, 271–273
    correlation between relatives, 276–283
    mean-related measures, 264–270
Response to selection, quantitative genetics, 283–284
Restriction endonucleases, genetic survey techniques, 542–548
    amplified fragment length polymorphisms, 546–548
    DNA fingerprinting, 545–546
    polymerase chain reaction, 546
    restriction fragment length polymorphisms, purified DNA, 542–544
    restriction site mapping, 544–545
    Southern blots, 544
Restriction fragment length polymorphism (RFLP):
    genetic survey techniques, 542–544
    intrapopulation disequilibrium markers, 304–306
    Southern Blot techniques, 544
Restriction site data:
    genetic survey applications, 544–545
    haplotype trees, 150–167
        Jukes-Cantor molecule genetic distance, 164–165
        loci/DNA region inferences, 241–243
        molecule genetic distances, 161–167
        mutation effects, 150–156

neighbor-joining estimation method, 166
recombination effects, 156–161
linkage disequilibrium and, 44–47
population subdivision measurements and, 198–199
Retrotransposons:
    age-structured population selection, genetic architecture and targets and units of selection, 509–516
    targets of selection, 425–429
Rh locus:
    gene flow, local populations, 172
    genetic drift and, 96–97
Ribosomal DNA (rDNA), age-structured population selection, targets of selection, 516–525
Rooted mitochondrial DNA, gene flow and population history, nested-clade phylogeographic analysis, 223–228
Root-mean-square gene flow distance, coarse-grained spatial heterogeneity, industrial melanism and, 466–474

Saccharomyces cerevisae:
    genetic drift, neutral alleles, 125–128
    natural selection and adaptation, 369
S allele:
    coarse-grained temporal heterogeneity, 475–482
    Mendelian phenotypes, 250–254
        electrophoretic mobility, 250–251
        health and viability, 254
        malarial resistance, 253
        sickle cell anemia, 252–253
        sickling, 251–252
    natural selection:
        average fitness, 365–367
        genetic drift, 383–387
        sickle cell anemia, 350–358
Sample reuse statistics, permutation testing, 570
Sample size:
    contingency analysis, 571–573
    haplotype tree loci/DNA region inferences, 235–243
Scalloped hazel moth, coarse-grained spatial heterogeneity, industrial melanism, 464–474
Scurvy, nature *vs.* nurture theory and, 255–257
Seasonal variation:
    age-structured population selection, *aa* supergene, spatio-temporal heterogeneous environment, 531–535
    fine-grained heterogeneity, 482–490
Seed availability, natural population monitoring and, 17–18
Segregation distortion, targets of selection and, 419–424
Selection:
    quantitative genetics and response to, 283–284
    units and targets, overview, 407–409

Selection coefficients, natural selection, sickle cell anemia, 354–358
Selective equilibrium, mutation-natural selection interaction, 374–375
Selective neutrality, fine-grained heterogeneity, 485–490
Self-sterility loci, disassortative mating and, 79–81
Semibalanus balanoides, coarse-grained spatial heterogeneity, 455–474
Senescence evolution, age-structured populations, 504–508
Sexual selection:
  age-structured populations, life history and fitness, 498–504
  targets of, 434–439
Shifting balance theory:
  coarse-grained temporal heterogeneity and, 481–482
  natural selection, genetic drift-gene flow interaction, 390–391
Short-distance dispersal:
  balance of gene flow and drift, one-dimensional stepping-stone model, 188–193
  haplotype tree loci/DNA region inferences, 240–243
Short tandem repeats (STRs), genetic survey applications, 554
Siblings:
  germline selection and, 450–452
  pedigree inbreeding, 49–55
  quantitative genetics, correlation between relatives, 280–283
Sickle cell anemia:
  coevolution mechanisms, 490–496
  health phenotypes, 254
  as natural selection, 349–358
  pathophysiology, 250
  phenotype, 252–253
Sickling phenotype:
  natural selection, 350–358
    genetic architecture, 395–397
  quantitative genetics, 251–252
Silent/replacement mutations, natural selection, genetic drift and, 400–406
Single loci:
  assortative mating, 65–77
    admixture and linkage disequilibrium, 75–77
  coarse-grained spatial heterogeneity, genetic/phenotypic clines, 471–474
  coarse-grained temporal heterogeneity, two-allele model, 476–482
  disassortative mating and, 77–81
  Hardy-Weinberg law:
    allelic derivation, 25–30
    population ratios, 33–35
    two-loci architecture and, 38–41
  kin/family selection, 446–452
  linkage disequilibrium, 95–97
  mathematical tractability, 407–408
  natural selection, 346–349
    genetic architecture, 394–397
    genetic drift interaction, 385–387
    sickle cell anemia, 353–358
  quantitative genetics:
    between-population differences, controlled crosses, 292–295
    candidate loci, epistasis, 332–337
  space-time continuity in, 408
  tandem multigene families, 433–434
  unit of selection and, 410–419
Single-nucleotide polymorphisms (SNPs):
  genetic survey applications, 554
  quantitative genetics, candidate loci, 313–317
  evolutionary history, 318–328
Socioeconomic status, measured genotypes, interpopulation disequilibrium marker loci, 300–303
Soft selection, coarse-grained spatial heterogeneity, 456–474
Southern Blot techniques:
  age-structured population selection, genetic architecture and targets and units of selection, 511–516
  genetic survey applications, 544
Space/time investigations:
  age-structured population selection, *aa* supergene selection, 525–535
  coarse-grained spatial heterogeneity, 455–474
  coarse-grained temporal heterogeneity, 475–482
  coevolution, 490–496
  continuity in single loci, 408–409
  fine-grained heterogeneity, 482–490
  gene flow and population history, past population structure, 221–234
  haplotype tree loci/DNA region inferences, 234–243
  nested-clade phylgeographic analysis, 225–228
  overview, 453–455
  population history and gene flow:
    basic principles, 205
    direct studies, 205–207
Spatial heterogeneity:
  age-structured population selection, *aa* supergene, spatio-temporal heterogeneous environment, 526–535
  coarse-grained environment, 455–474
  natural selection and, basic principles, 453–455
Spatial sample distribution, haplotype tree loci/DNA region inferences, 235–243
Species trees, coalescent events and, 149–150
Speke's gazelle inbreeding coefficient, 51–63
  disassortative mating and, 80–81
Stable age distribution, age-structured populations, 502–504

Standard deviation, quantitative genetics,
    interpopulation disequilibrium markers,
    303
Standard inversion, coalescent theory, inversion trees,
    146–150
Statistical analysis:
  basic principles, 563–581
  Bayesian inference, 578–581
  maximum likelihood techniques, 574–578
    hypothesis testing, 577–578
    maximum-likelihood estimation, 575–577
  nested-clade phylgeographic analysis, 228
  random variables as data, 564–574
    contingency tests, 571–573
    goodness-of-fit tests, 573–574
    mean confidence intervals, 565–566
    mean hypothesis testing, 565
    minimum-variance, unbiased mean estimation,
        564–565
    one-way analysis of variance, 567–569
    permutation testing, 569–570
    regression and correlation, 570–571
    two-population means testing, 566–567
    unknown variance estimation and testing, 566
    variance equality in two populations,
        567
Statistical parsimony:
  coalescent theory, haplotype trees:
    ApoE recombination, 159–161
    mutation and, 151–156
    recombination and, 156–161
  quantitative genetics, candidate loci, 315–317
Statistic of association, age-structured population
    selection, genetic architecture and targets
    and units of selection, 513–516
Subpopulations:
  gene flow in, 168
  natural selection, genetic drift-gene flow
    interaction, 388–397
  quantitative genetics:
    between-population differences, mean
        phenotype, 287–290
    mutation, drift, and gene flow, 296
Sufficiency, maximum-likelihood estimation, 576–577
Supergenes. *See also abnormal abdomen (aa)* locus
  age-structured population selection:
    aa supergene, spatio-temporal heterogeneous
        environment, 525–535
    aa supergene components, 536–537
    natural selection, 395–397
    unit of selection, 413–419
Survivorship curve:
  aa supergene, spatio-temporal heterogeneous
      environment, 526–535
  age-structured populations, 498–504, 499–504
  fecundity estimates with aa syndrome, 524–525
  natural selection, sickle cell anemia, 353–358

Symmetrical gene flow, local populations, 169–172
Synonymous mutations, natural selection, genetic drift
    and, 399–406
System-of-mating inbreeding coefficient, 58–63
  disassortative mating and, 79–81
  gene flow, local populations, 170–172
  quantitative genetics, average effect of gamete type
      i, 268–270

Tandem multigene family, unequal exchange,
    429–434
Target of selection:
  abnormal abdomen (aa) locus, age-structured
      population selection:
    phenotypes at individual level, 516–525
    sub-individual level, 508–516
  above individual level, 434
    competition, 443–446
    fertility, 439–443
    kin/family selection, 446–452
    sexual selection, 434–439
  coevolution mechanisms, 491–496
  defined, 409
  meiotic drive, 422–424
  sub-individual level, 419–434
    biased gene conversion, 424–425
    meiotic drive, 419–424
    tandem multigene family exchange, 429–434
    transposons, 425–429
Taylor's series expansion:
  balance of gene flow and drift and, 178–180
  genetic drift:
    infinite-alleles model, 127–128
    natural selection, 381–387
Tay-Sachs disease, pedigree inbreeding, 54–55
t distribution, 561
  means difference testing in two populations,
      566–567
  meiotic drive, 422–424
  unknown variance estimation and testing, 566
Temperature gradient:
  age-structured population selection, juvenilized
      abdominal cuticle phenotype, 519–525
  coarse-grained spatial heterogeneity, 473–474
  fine-grained heterogeneity, 487–490
Temporal heterogeneity:
  age-structured population selection, *aa* supergene,
      spatio-temporal heterogeneous
      environment, 531–536
  coarse-grained environment, 474–482
  natural selection and, basic principles, 453–455
Temporal persistence, natural selection, genetic drift
    and mutation, 401–406
Tertile analysis, candidate loci, coronary artery
    disease, 339–340
Thalassemia, natural selection, polygenic adaptation,
    359–361

Thrifty-genotype hypothesis, coarse-grained temporal heterogeneity, 477–482
Thyroxin receptors, coarse-grained temporal heterogeneity, 480–482
Time lag effects, coarse-grained temporal heterogeneity, 475–482
Topological positions, natural selection, genetic drift and mutation, 400–406
Total population genetic variation, population subdivision and, 193–198
Total sum of squares (SSTO), one-way analysis of variance, 567–569
Trait heritability:
  Fisher's fundamental theorem of natural selection, 364
  quantitative genetics, response to selection, 283–284
Transient polymorphism:
  assortative mating, 69–77
  natural selection, sickle cell anemia, 356–358
Transition matrix:
  assortative mating, 68–71
  fine-grained heterogeneity, 488–490
Translational control, age-structured population selection, *abnormal abdomen (aa)* locus targets of selection, 517–525
Transmission disequilibrium test (TDT), candidate loci evolutionary history, 322–328
Transmission mechanisms, targets of selection, transposons, 427–429
Transmission of fitness phenotypes through gametes, natural selection, measured genotypes, 349
Transpecific polymorphisms, coalescent theory, inversion network, 148–150
Transposons:
  age-structured population selection, genetic architecture and targets and units of selection, 509–516
  targets of selection, 425–429
Treatment sum of squares (SSTR), one-way analysis of variance, 569
Tree-estimating algorithms, haplotype trees, molecule genetic distances, 162–167
Treeness property:
  gene flow and population history, 214–215
  gene flow *vs.* admixture, 217–218
Tree scanning, candidate loci, evolutionary history, 323–328
Triallelic frequencies, gene flow *vs.* admixture, 216–218
Tribolium castaneum, senescence evolution, 507–508
Trinucleotide repeats:
  germline selection, 450–452
  meiotic drive, 421–424

Tristan da Cunha population:
  genetic drift in, 91–94
  pedigree inbreeding coefficient for, 60–63
Two-deme model, gene flow and population subdivision, 185–193
Two-loci architecture:
  assortative mating *vs.* inbreeding, 71–73
  Hardy-Weinberg law and, 35–41
  natural selection, 346–349
  population history and gene flow, linkage disequilibrium, 208–211
  quantitative genetics, candidate loci, epistasis, 332–337
  unit of selection and, 409–419
Type 2 diabetes mellitus, coarse-grained temporal heterogeneity, 477–482

Ultimate causation, age-structured population selection, *aa* supergene components, 538–539
underreplication (ur) locus, age-structured population selection:
  aa supergene components, 536–537
  genetic architecture and targets and units of selection, 516
Unequal exchange mechanism, tandem multigene families, 429–434
Unequal gene conversion, targets of selection, 424–425
Unit of selection:
  abnormal abdomen (aa) locus, age-structured population selection, 508–516
  basic properties, 409–419
  coarse-grained spatial heterogeneity, 474
  coevolution mechanisms, 491–496
  defined, 408–409
  meiotic drive, 422–424
  transposons as, 426–429
Unity of the genotype, single loci and, 407–408
Unmeasured genotypes:
  natural selection, 361–363
  quantitative genetics:
    basic principles, 274–275
    between-population differences:
      controlled crosses, 290–295
      mean phenotypes, 284–290
    correlation between relatives, 275–283
    marker linkage mapping and, 311–312
    phenotypic variance, mutation, drift, and gene flow, 295–296
    selection response, 283–284
Unpreferred substitutions, natural selection, genetic drift and mutation, 402–406
Unrooted network, allele/haplotype trees:
  candidate loci, evolutionary history, 318–328
  coalescent theory and, 145
  gene flow and population history, past population structure, 221–222

Variability:
    candidate loci, epistasis, 336–337
    covariance of, quantitative genetics, 276–283
    genetic drift and, 85–89
    nature *vs.* nurture theory and, 254–257
Variable number of tandem repeats (VNTR), genetic survey applications, 546
Variance. *See also* Analysis of variance (ANOVA)
    equality testing in two populations, 567
    minimum-variance, unbiased mean estimator, 564–565
    negative binomial distribution, 559
    probability distribution and, 557–558
    quantitative genetics, 270–273
        between-population differences, controlled crosses, 292–295
        mutation, drift and gene flow, 295–296
        of variable, 276–283
        ratios, gene flow, Wahlund effect, 181–183
    unknown variance estimation and testing, 566
V(D)J recombination, transposons, 427–429
Vertical transmission, transposons, 427–429
Viability:
    coarse-grained spatial heterogeneity, habitat-specific estimates, 455–474
    natural selection, 343
        measured genotypes, 346–349
    phenotypes, 8
    quantitative genetics, 254
    unit of selection and, 415–419
Virilis/repleta radiation, age-structured population selection, *abnormal abdomen (aa)* locus targets of selection, 522–525
Vitellogenins, age-structured population selection, *abnormal abdomen (aa)* locus targets of selection, 522–525

Wahlund effect, gene flow and population subdivision, 180–183
    mating systems and, 184–185
Weinberg's derivation, Hardy-Weinberg genotype frequencies, 29–30

Within-family selection, Huntington's disease, 450–452
Wright adaptive surface metaphor:
    coarse-grained spatial heterogeneity, 474
    natural selection:
        genetic architecture, 392–397
        genetic drift, 383–387
        genetic drift-gene flow interaction, 387–397
Wright's peak climbing metaphor, kin/family selection, 452

X-linked DNA:
    age-structured population selection, genetic architecture and targets and units of selection, 509–516
    coarse-grained spatial heterogeneity, 468–474
    genetic drift, coalescent theory and, 139–140
    resequencing techniques, 549–554
X-linked microsatellite loci, genetic drift and, linkage disequilibrium, 94–97

Yanomana tribes:
    disassortative mating and, 80–81
    gene flow and population subdivision, mating systems, 185–186
Y-chromosome haplotypes:
    DNA resequencing, 549–554
    gene flow and population history, 220–221
Y-DNA:
    genetic drift, coalescent theory and, 139–140
    multiple inheritance modes, 200–202

Zero covariance, quantitative genetics, correlation between relatives, 276–283
Z scores, quantitative genetics, interpopulation disequilibrium markers, 303
Zygotes:
    coarse-grained spatial heterogeneity, 456–474
    Hardy-Weinberg law, allele frequency multiplications, 28–30
    microevolution modeling and, 23
    natural selection, measured genotypes, 346–349